Organic Reactions

Organic Reactions

VOLUME 37

JOHN WILEY & SONS, INC.

New York • Chichester • Brisbane • Toronto • Singapore

Published by John Wiley & Sons, Inc.

Library of Congress Catalog Card Number: 42-20265

ISBN 0-471-50169-7

Printed in the United States of America

10 9 8 7 6 5 4 3 2 1

PREFACE TO THE SERIES

In the course of nearly every program of research in organic chemistry the investigator finds it necessary to use several of the better-known synthetic reactions. To discover the optimum conditions for the application of even the most familiar one to a compound not previously subjected to the reaction often requires an extensive search of the literature; even then a series of experiments may be necessary. When the results of the investigation are published, the synthesis, which may have required months of work, is usually described without comment. The background of knowledge and experience gained in the literature search and experimentation is thus lost to those who subsequently have occasion to apply the general method. The student of preparative organic chemistry faces similar difficulties. The textbooks and laboratory manuals furnish numerous examples of the application of various syntheses, but only rarely do they convey an accurate conception of the scope and usefulness of the processes.

For many years American organic chemists have discussed these problems. The plan of compiling critical discussions of the more important reactions thus was evolved. The volumes of *Organic Reactions* are collections of chapters each devoted to a single reaction, or a definite phase of a reaction, of wide applicability. The authors have had experience with the processes surveyed. The subjects are presented from the preparative viewpoint, and particular attention is given to limitations, interfering influences, effects of structure, and the selection of experimental techniques. Each chapter includes several detailed procedures illustrating the significant modifications of the method. Most of these procedures have been found satisfactory by the author or one of the editors, but unlike those in *Organic Syntheses* they have not been subjected to careful testing in two or more laboratories.

Each chapter contains tables that include all the examples of the reaction under consideration that the author has been able to find. It is inevitable, however, that in the search of the literature some examples will be missed, especially when the reaction is used as one step in an extended synthesis. Nevertheless, the investigator will be able to use the tables and their accompanying bibliographies in place of most or all of the literature search so often required.

Because of the systematic arrangement of the material in the chapters and the entries in the table, users of the books will be able to find information desired by reference to the table of contents of the appropriate chapter. In the interest of economy the entries in the indices have been kept to a minimum, and, in particular, the compounds listed in the tables are not repeated in the indices.

The success of this publication, which will appear periodically, depends upon the cooperation of organic chemists and their willingness to devote time and effort to the preparation of the chapters. They have manifested their interest already by the almost unanimous acceptance of invitations to contribute to the work. The editors will welcome their continued interest and their suggestions for improvements in *Organic Reactions*.

Chemists who are considering the preparation of a manuscript for submission to *Organic Reactions* are urged to write either secretary before they begin work.

CUMULATIVE CHAPTER TITLES BY VOLUME

Volume 1 (1942)

1. **The Reformatsky Reaction**: Ralph L. Shriner

2. **The Arndt-Eistert Reaction**: W. E. Bachmann and W. S. Struve

3. **Chloromethylation of Aromatic Compounds**: R. C. Fuson and C. H. McKeever

4. **The Amination of Heterocyclic Bases by Alkali Amides**: Martin T. Leffler

5. **The Bucherer Reaction**: Nathan L. Drake

6. **The Elbs Reaction**: Louis F. Fieser

7. **The Clemmensen Reduction**: Elmore L. Martin

8. **The Perkin Reaction and Related Reactions**: John R. Johnson

9. **The Acetoacetic Ester Condensation and Certain Related Reactions**: Charles R. Hauser and Boyd E. Hudson, Jr.

10. **The Mannich Reaction**: F. F. Blicke

11. **The Fries Reaction**: A. H. Blatt

12. **The Jacobsen Reaction**: Lee Irvin Smith

Volume 2 (1944)

1. **The Claisen Rearrangement**: D. Stanley Tarbell

2. **The Preparation of Aliphatic Fluorine Compounds**: Albert L. Henne

3. **The Cannizzaro Reaction**: T. A. Geissman

4. **The Formation of Cyclic Ketones by Intramolecular Acylation**: William S. Johnson

5. **Reduction with Aluminum Alkoxides (The Meerwein-Ponndorf-Verley Reduction)**: A. L. Wilds

6. **The Preparation of Unsymmetrical Biaryls by the Diazo Reaction and the Nitrosoacetylamine Reaction**: Werner E. Bachmann and Roger A. Hoffman

7. **Replacement of the Aromatic Primary Amino Group by Hydrogen**: Nathan Kornblum

8. **Periodic Acid Oxidation**: Ernest L. Jackson

9. **The Resolution of Alcohols**: A. W. Ingersoll

10. **The Preparation of Aromatic Arsonic and Arsinic Acids by the Bart, Béchamp, and Rosenmund Reactions**: Cliff S. Hamilton and Jack F. Morgan

Volume 3 (1946)

1. **The Alkylation of Aromatic Compounds by the Friedel-Crafts Method**: Charles C. Price

2. **The Willgerodt Reaction**: Marvin Carmack and M. A. Spielman

3. **Preparation of Ketenes and Ketene Dimers**: W. E. Hanford and John C. Sauer

4. **Direct Sulfonation of Aromatic Hydrocarbons and Their Halogen Derivatives**: C. M. Suter and Arthur W. Weston

5. **Azlactones**: H. E. Carter

6. **Substitution and Addition Reactions of Thiocyanogen**: John L. Wood

7. **The Hofmann Reaction**: Everett S. Wallis and John F. Lane

8. **The Schmidt Reaction**: Hans Solff

9. **The Curtius Reaction**: Peter A. S. Smith

Volume 4 (1948)

1. **The Diels-Alder Reaction with Maleic Anhydride**: Milton C. Kloetzel

2. **The Diels-Alder Reaction: Ethylenic and Acetylenic Dienophiles**: H. L. Holmes

3. **The Preparation of Amines by Reductive Alkylation**: William S. Emerson

4. **The Acyloins**: S. M. McElvain

5. **The Synthesis of Benzoins**: Walter S. Ide and Johannes S. Buck

6. **Synthesis of Benzoquinones by Oxidation**: James Cason

7. **The Rosenmund Reduction of Acid Chlorides to Aldehydes**: Erich Mosettig and Ralph Mozingo

8. **The Wolff-Kishner Reduction**: David Todd

Volume 5 (1949)

Volume 6 (1951)

Volume 7 (1953)

1. **The Pechmann Reaction**: Suresh Sethna and Ragini Phadke
2. **The Skraup Synthesis of Quinolines**: R. H. F. Manske and Marshall Kalka
3. **Carbon-Carbon Alkylations with Amines and Ammonium Salts**: James H. Brewster and Ernest L. Eliel
4. **The von Braun Cyanogen Bromide Reaction**: Howard A. Hageman
5. **Hydrogenolysis of Benzyl Groups Attached to Oxygen, Nitrogen, or Sulfur**: Walter H. Hartung and Robert Simonoff
6. **The Nitrosation of Aliphatic Carbon Atoms**: Oscar Touster
7. **Epoxidation and Hydroxylation of Ethylenic Compounds with Organic Peracids**: Daniel Swern

Volume 8 (1954)

1. **Catalytic Hydrogenation of Esters to Alcohols**: Homer Adkins
2. **The Synthesis of Ketones from Acid Halides and Organometallic Compounds of Magnesium, Zinc, and Cadmium**: David A. Shirley
3. **The Acylation of Ketones to Form β-Diketones or β-Keto Aldehydes**: C. R. Hauser, Frederic W. Swamer, and Joe T. Adams
4. **The Sommelet Reaction**: S. J. Angyal
5. **The Synthesis of Aldehydes from Carboxylic Acids**: Erich Mosettig
6. **The Metalation Reaction with Organolithium Compounds**: Henry Gilman
7. **β-Lactones**: Harold E. Zaugg
8. **The Reaction of Diazomethane and Its Derivatives with Aldehydes and Ketones**: C. David Gutsche

Volume 9 (1957)

1. **The Cleavage of Non-enolizable Ketones with Sodium Amide**: K. E. Hamlin and Arthur W. Weston
2. **The Gattermann Synthesis of Aldehydes**: William E. Truce
3. **The Baeyer-Villiger Oxidation of Aldehydes and Ketones**: C. H. Hassall
4. **The Alkylation of Esters and Nitriles**: A. C. Cope, H. L. Holmes, and H. O. House
5. **The Reaction of Halogens with Silver Salts of Carboxylic Acids**: C. V. Wilson

6. **The Synthesis of β-Lactams**: John C. Sheehan and Elias J. Corey

7. **The Pschorr Synthesis and Related Diazonium Ring Closure Reactions**: DeLos F. DeTar

Volume 10 (*1959*)

1. **The Coupling of Diazonium Salts with Aliphatic Carbon Atoms**: Stanley M. Parmerter

2. **The Japp-Klingemann Reaction**: Robert R. Philips

3. **The Michael Reaction**: Ernst D. Bergmann, David Ginsburg, and Raphael Pappo

Volume 11 (*1960*)

1. **The Beckmann Rearrangement**: L. Guy Donaruma and Walter Z. Heldt

2. **The Demjanov and Tiffeneau-Demjanov Ring Expansions**: Peter A. S. Smith and Donald R. Baer

3. **Arylation of Unsaturated Compounds by Diazonium Salts**: Christian S. Rondestvedt, Jr.

4. **The Favorskii Rearrangement of Haloketones**: Andrew S. Kende

5. **Olefins from Amines: The Hofmann Elimination Reaction and Amine Oxide Pyrolysis**: A. C. Cope and Elmer R. Trumbull

Volume 12 (*1962*)

1. **Cyclobutane Derivatives from Thermal Cycloaddition Reactions**: J. D. Roberts and Clay M. Sharts

2. **The Preparation of Olefins by the Pyrolysis of Xanthates. The Chugaev Reaction**: Harold R. Nace

3. **The Synthesis of Aliphatic and Alicyclic Nitro Compounds**: Nathan Kornblum

4. **Synthesis of Peptides with Mixed Anhydrides**: Noel F. Albertson

5. **Desulfurization with Raney Nickel**: George R. Pettit and Eugene E. van Tamelen

Volume 13 (*1963*)

1. **Hydration of Olefins, Dienes, and Acetylenes via Hydroboration**: George Zweifel and Herbert C. Brown

2. **Halocyclopropanes from Halocarbenes**: Edward E. Schweizer

3. **Free Radical Additions to Olefins to Form Carbon-Carbon Bonds**: Cheves Walling and Earl S. Huyser

4. **Formation of Carbon-Heteroatom Bonds by Free Radical Chain Additions to Carbon-Carbon Multiple Bonds**: F. W. Stacey and J. F. Harris, Jr.

Volume 14 (1965)

1. **The Chapman Rearrangement**: J. W. Schulenberg and S. Archer

2. **α-Amidoalkylations at Carbon**: Harold E. Zaugg and William B. Martin

3. **The Wittig Reaction**: Adalbert Maercker

Volume 15 (1967)

1. **The Dieckmann Condensation**: John P. Schaefer and Jordan J. Bloomfield

2. **The Knoevenagel Condensation**: G. Jones

Volume 16 (1968)

1. **The Aldol Condensation**: Arnold T. Nielsen and William J. Houlihan

Volume 17 (1969)

1. **The Synthesis of Substituted Ferrocenes and Other π-Cyclopentadienyl-Transition Metal Compounds**: Donald E. Bublitz and Kenneth L. Rinehart, Jr.

2. **The γ-Alkylation and γ-Arylation of Dianions of β-Dicarbonyl Compounds**: Thomas M. Harris and Constance M. Harris

3. **The Ritter Reaction**: L. I. Krimen and Donald J. Cota

Volume 18 (1970)

1. **Preparation of Ketones from the Reaction of Organolithium Reagents with Carboxylic Acids**: Margaret J. Jorgenson

2. **The Smiles and Related Rearrangements of Aromatic Systems**: W. E. Truce, Eunice M. Kreider, and William W. Brand

3. **The Reactions of Diazoacetic Esters with Alkenes, Alkynes, Heterocyclic, and Aromatic Compounds**: Vinod David and E. W. Warnhoff

4. **The Base-Promoted Rearrangements of Quaternary Ammonium Salts**: Stanley H. Pine

Volume 19 (1972)

1. **Conjugate Addition Reactions of Organocopper Reagents**: Gary H. Posner

2. **Formation of Carbon-Carbon Bonds via π-Allylnickel Compounds**: Martin F. Semmelhack

3. **The Thiele-Winter Acetoxylation of Quinones**: J. F. W. McOmie and J. M. Blatchly

4. **Oxidative Decarboxylation of Acids by Lead Tetraacetate**: Roger A. Sheldon and Jay K. Kochi

Volume 20 (1973)

1. **Cyclopropanes from Unsaturated Compounds, Methylene Iodide, and Zinc-Copper Couple**: H. E. Simmons, T. L. Cairns, Susan A. Vladuchick, and Connie M. Hoiness

2. **Sensitized Photooxygenation of Olefins**: R. W. Denny and A. Nickon

3. **The Synthesis of 5-Hydroxyindoles by the Nenitzescu Reaction**: George R. Allen, Jr.

4. **The Zinin Reduction of Nitroarenes**: H. K. Porter

Volume 21 (1974)

1. **Fluorination with Sulfur Tetrafluoride**: G. A. Boswell, Jr., W. C. Ripka, R. M. Scribner, and C. W. Tullock

2. **Modern Methods to Prepare Monofluoroaliphatic Compounds**: William A. Sheppard

Volume 22 (1975)

1. **The Claisen and Cope Rearrangements**: Sara Jane Rhoads and N. Rebecca Raulins

2. **Substitution Reactions Using Organocopper Reagents**: Gary H. Posner

3. **Clemmensen Reduction of Ketones in Anhydrous Organic Solvents**: E. Vedejs

4. **The Reformatsky Reaction**: Michael W. Rathke

Volume 23 (1976)

1. **Reduction and Related Reactions of α,β-Unsaturated Compounds with Metals in Liquid Ammonia**: Drury Caine

2. **The Acyloin Condensation**: Jordan J. Bloomfield, Dennis C. Owsley, and Janice M. Nelke

3. **Alkenes from Tosylhydrazones**: Robert H. Shapiro

Volume 24 (*1976*)

1. **Homogeneous Hydrogenation Catalysts in Organic Synthesis**: Arthur J. Birch and David H. Williamson

2. **Ester Cleavages via S_N2-Type Dealkylation**: John E. McMurry

3. **Arylation of Unsaturated Compounds by Diazonium Salts (The Meerwein Arylation Reaction)**: Christian S. Rondestvedt, Jr.

4. **Selenium Dioxide Oxidation**: Norman Rabjohn

Volume 25 (*1977*)

1. **The Ramberg-Bäcklund Rearrangement**: Leo A. Paquette

2. **Synthetic Applications of Phosphoryl-Stabilized Anions**: William S. Wadsworth, Jr.

3. **Hydrocyanation of Conjugated Carbonyl Compounds**: Wataru Nagata and Mitsuru Yoshioka

Volume 26 (*1979*)

1. **Heteroatom-Facilitated Lithiations**: Heinz W. Gschwend and Herman R. Rodriguez

2. **Intramolecular Reactions of Diazocarbonyl Compounds**: Steven D. Burke and Paul A. Grieco

Volume 27 (*1982*)

1. **Allylic and Benzylic Carbanions Substituted by Heteroatoms**: Jean-François Biellmann and Jean-Bernard Ducep

2. **Palladium-Catalyzed Vinylation of Organic Halides**: Richard F. Heck

Volume 28 (*1982*)

1. **The Reimer-Tiemann Reaction**: Hans Wynberg and Egbert W. Meijer

2. **The Friedländer Synthesis of Quinolines**: Chia-Chung Cheng and Shou-Jen Yan

3. **The Directed Aldol Reaction**: Teruaki Mukaiyama

2. **The Persulfate Oxidation of Phenols and Arylamines (The Elbs and The Boyland-Sims Oxidations)**: E. J. Behrman

3. **Fluorination with Diethylaminosulfur Trifluoride and Related Aminofluorosulfuranes**: Miloš Hudlický

Volume 36 (*1988*)

1. **The [3 + 2] Nitrone-Olefin Cycloaddition Reaction:** Pat N. Confalone and Edward M. Huie

2. **Phosphorus Addition at sp^2 Carbon:** Robert Engel

3. **Reduction by Metal Alkoxyaluminum Hydrides. Part II. Carboxylic Acids and Derivatives, Nitrogen Compounds, and Sulfur Compounds**: Jaroslav Málek

CONTENTS

Organic Reactions

CHAPTER 1

CHIRAL SYNTHONS BY ESTER HYDROLYSIS CATALYZED BY PIG LIVER ESTERASE

MASAJI OHNO AND MASAMI OTSUKA

Faculty of Pharmaceutical Sciences, University of Tokyo, Bunkyo-ku, Tokyo, Japan

CONTENTS

ACKNOWLEDGMENT

The authors thank Professor Charles J. Sih for calculating the values of the enantiomeric ratio (E) listed in TABLES VIII–XI.

INTRODUCTION

Development of efficient methodology to produce an optically pure enantiomer is of fundamental importance, particularly for the synthesis of biologically active natural products. Optically active compounds can be obtained by three different approaches, that is, resolution of racemates, use of a chiral pool or "chiron" (which are enantiomerically pure synthons),[1] or asymmetric synthesis. Of these, one of the more challenging tasks involves asymmetric synthesis, which may be carried out either enzymatically or nonenzymatically. Whereas the nonenzymatic method enables us to introduce a chiral center with either a stoichiometric or catalytic amount of a chiral compound, the enzymatic method uses biological systems, such as microorganisms or isolated enzymes, to create the center of asymmetry.[2–10] The purpose of this chapter is to survey asymmetric synthesis via the production of enantiomerically pure or enriched organic molecules by the enzymatic method focusing on the use of a hydrolytic enzyme, pig liver esterase (PLE; Enzyme Commission classification number, E.C. 3.1.1.1).[11]

Enzymes are classified into the following six groups based on the reactions they catalyze:

1. Oxidoreductase (oxidation–reduction reactions).
2. Transferase (transfer of functional groups).
3. Hydrolase (hydrolysis reactions).

4. Lyase (addition to double bonds or the reverse).
5. Isomerase (isomerization reaction).
6. Ligase (formation of bonds coupled with pyrophosphate bond cleavage of ATP).

Virtually all biochemical transformations are catalyzed in vivo by these six groups, and the biochemical aspects of these enzymes have been studied in detail. However, the practical utility of these enzymes in organic synthesis remains to be further exploited and refined. For example, some enzymes require cofactors, which are often expensive and must be regenerated in situ to achieve a catalytic process. The ability of a substrate to associate with an enzyme is also one of the most significant problems. In some cases, these problems have been overcome and several enzyme reactions have been successfully used on a large scale.

An enzyme reaction generally takes place when an intimate interaction between the reactant and the chiral catalyst (enzyme protein) is realized. This enzyme–substrate complex is designated as the Michaelis complex. Certain amino acid residues of the enzyme form a three-dimensional structure as the active site. This site often contains reactive groups of the amino acids such as amino, mercapto, hydroxyl, carbonyl, carboxyl, guanidino, or imidazolyl. When the substrate is bound to the active site in a specific orientation, enantiotopic groups or faces of the substrate molecule are discriminated by the chiral enzyme. This discrimination is sufficiently sensitive to differentiate between the two hydrogen atoms in a methylene group. Thus the reaction proceeds stereospecifically.[12] The enzyme can also distinguish among several substrates competing for an active site. Such substrate specificity is sometimes strict and sometimes broad. Synthetically useful enzymes should accept a wide range of substrates and exhibit high stereospecificity. Many enzymes meet these criteria, and PLE is one of them.

SCOPE AND LIMITATIONS

Enantioselective Hydrolysis of Prochiral Diesters

In 1961 diethyl 3-acetamidoglutarate (**1**) was first hydrolyzed with an enzyme, α-chymotrypsin (CHT), to afford the (*R*)-monoester **2** (Eq. 1).[13]

$$C_2H_5O_2C\text{–}CH_2\text{–}CH(NHCOCH_3)\text{–}CH_2\text{–}CO_2C_2H_5\ (\mathbf{1}) \xrightarrow{\text{CHT}} HO_2C\text{–}CH_2\text{–}CH(NHCOCH_3)\text{–}CH_2\text{–}CO_2C_2H_5\ (\mathbf{2}) \qquad \text{(Eq. 1)}$$

There was no further report on such an enzymatic reaction until 1975, when successful application of PLE to the asymmetric hydrolysis of dimethyl 3-hydroxy-3-methylglutarate (**3**) to monoester **4** with excellent enantiomeric excess (ee) was reported (Eq. 2).[14] The use of CHT is not satisfactory for this substrate because of the low hydrolysis rate and the requirement for a stoichiometric amount of the enzyme. These results show that the constraint of substrate specificity can be overcome by choosing an appropriate enzyme. Although natural substrates for PLE have not been identified in biological systems, it is now known that PLE cleaves a broad range of esters.

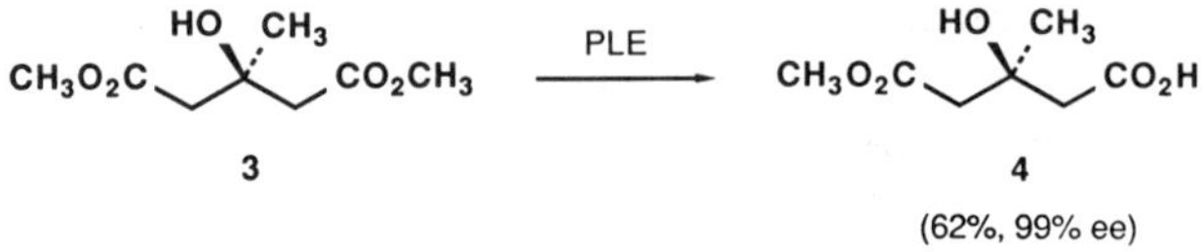

(Eq. 2)

PLE-catalyzed hydrolyses of many substituted glutarate esters are reported (Table I). 3-(Protected amino)glutarates are efficiently hydrolyzed with moderate to excellent enantioselectivity (Eq. 3).[15,16] Although an unprotected 3-aminoglutarate is easily hydrolyzed, the optical yield of the reaction is not high. The optical purity of the monoester decreases considerably as the chain length of the acylamino substituent increases (**5**, R^1 = CH_3CONH, C_2H_5CONH, and *n*-C_3H_7CONH), and the absolute configuration reverses with the pentanoylamino group (albeit ee is very low). Hydrolysis of glutarates with bulky acylamino groups (**5**, R^1 = *i*-C_3H_7CONH, *t*-C_4H_9CONH, and $C_6H_{11}CONH$, R^2 = H) affords the monoesters of *S* configuration **6**. Substrates with an unsaturated small group, such as acrylylamino, are transformed to the (*R*) products **7**, but with the crotonylamino group the monoester of *S* configuration is slowly produced in almost 100% ee. 3-Aminoglutarates with urethane groups afford products with the *S* configuration. 3-Alkylglutarates are also hydrolyzed with PLE in good yields and moderate to good stereoselectivities.[17] The hydrolysis is pro-*S* selective for the diesters with small C-3 alkyl substituents, and the hydrolyzed ester reverses to pro-*R* group when the C-3 group is large.

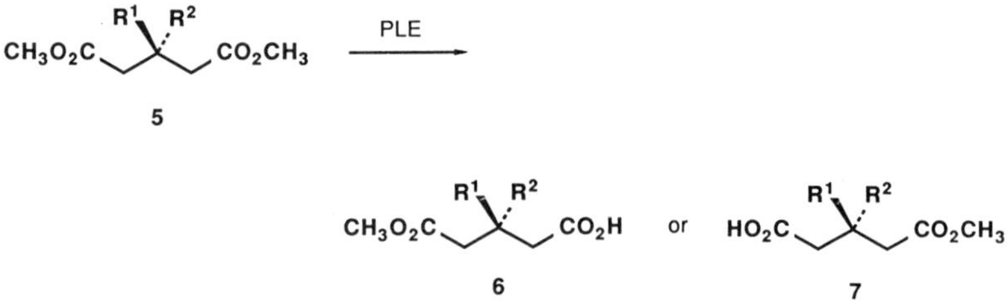

(Eq. 3)

A remarkable reversal of the enantioselectivity is observed when 2,2-disubstituted malonates are hydrolyzed with PLE (Table II).[18] Substrates with a methyl and a short alkyl group give the *S* enantiomer (Eq. 4), whereas the one with a longer alkyl chain gives the *R* enantiomer (Eq. 5). Dimethyl ester substrates give much better enantioselectivity than the corresponding diethyl esters.

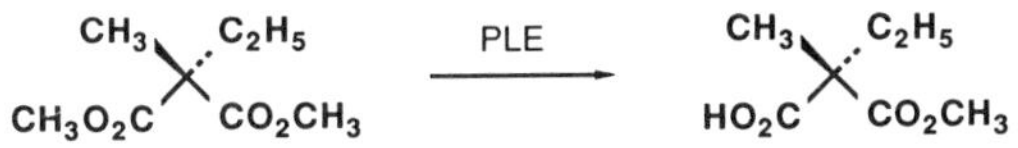

(Eq. 4)

(73% ee)

(Eq. 5)

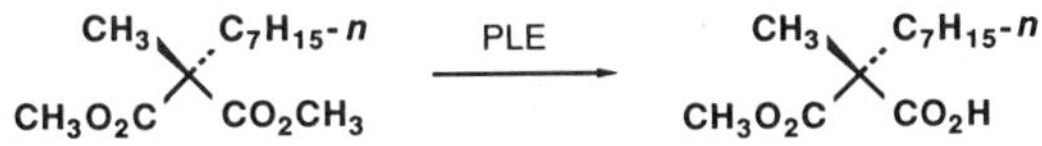

(88% ee)

Enantioselective hydrolysis of diacylated glycerol derivative **8** can be carried out with PLE, pig pancreatic lipase (PPL), lipase from *Candida cylindracea,* or Baker's yeast.[19–21] PLE gives monoacetate **9** with approximately 40% ee (Table III). The antipode of **9** (88% ee) is obtained by use of PPL.[21]

Sulfur-containing diester **10** can be hydrolyzed with PLE. Although the absolute configuration of the hydrolysate **11** remains to be determined, the enantiomeric purity is shown to be 82%, based on NMR measurement.[22]

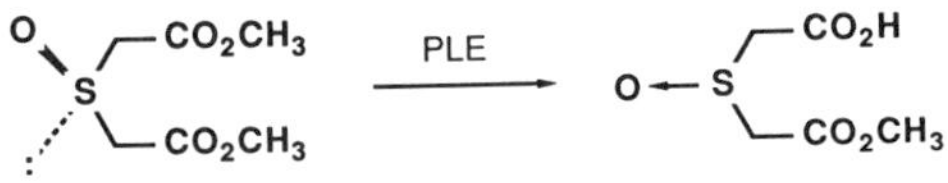

(82% ee)

Enantioselective Hydrolysis of *meso*-Diesters

Linear and cyclic *meso*-diesters are hydrolyzed with PLE to afford enantiomerically pure or enriched monoesters. Various open chain *meso* substrates, such as 2,4-disubstituted glutarates **12** and 2,3-disubstituted succinates **14** and **15**, are hydrolyzed with PLE in good chemical yields and moderate optical yields (Table IV).[23,24]

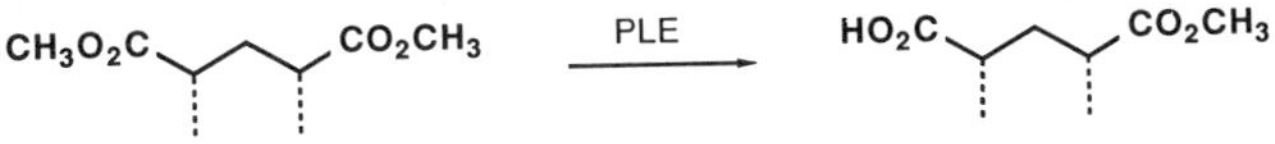

(85%, 64% ee)

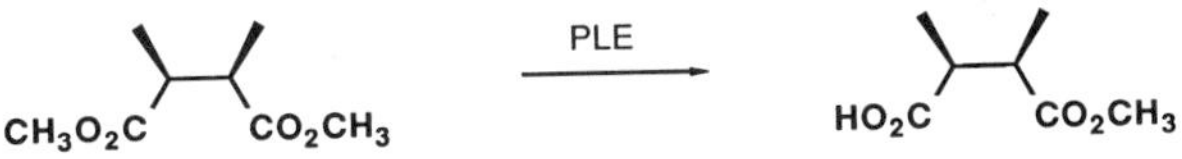

(94%, 18% ee)

HO OH / CH_3O_2C CO_2CH_3 (15) —PLE→ HO OH / CH_3O_2C CO_2H (92%, 48% ee)

Many cyclic *meso*-diesters can be hydrolyzed (Table V). In the hydrolysis of dimethyl *cis*-cycloalkane-1,2-dicarboxylates, enantioselectivity is generally very good except for the cyclopentane derivative **16** (n = 3) (Eq. 6).[24–26] Change of the asymmetric recognition is observed depending on the ring size of the cycloalkane moiety; whereas (1*R*,2*S*)-monoesters **17** are obtained from cyclopropane- and cyclobutanedicarboxylates **16** (n = 1 and 2), (1*S*,2*R*)-monoesters **18** are obtained from **16** (n = 3 and 4).

$(CH_2)_n$ / CH_3O_2C CO_2CH_3 (16) —PLE→ $(CH_2)_n$ / HO_2C CO_2CH_3 (17), n = 1, 2; $(CH_2)_n$ / CH_3O_2C CO_2H (18), n = 3, 4 (Eq. 6)

On the other hand, PLE hydrolysis of the corresponding cyclic acetates, *cis*-1,2-bis(acetoxymethyl)cycloalkanes (**19**), gives monoacetates **20** or **21** in moderate chemical yields and low to moderate optical yields (Eq. 7, Table VII).[27]

$(CH_2)_n$ / CH_3CO_2 O_2CCH_3 (19) —PLE→ $(CH_2)_n$ / HO O_2CCH_3 (20) or $(CH_2)_n$ / CH_3CO_2 OH (21) (Eq. 7)

Dimethyl *cis*-cyclohex-4-ene-1,2-dicarboxylate (**22**) is efficiently hydrolyzed with PLE to the synthetically useful monoester **23** in excellent chemical and optical yields.[24,28–30] However, the efficiency of PLE hydrolysis of the corresponding *cis*-diacetate **24** to monoacetate **25** seems to depend on the condition and the preparation of the enzyme.[27,31]

CO_2CH_3 / CO_2CH_3 (22) —PLE→ CO_2H / CO_2CH_3 (23) (94 - 99%, 85 - 99% ee)

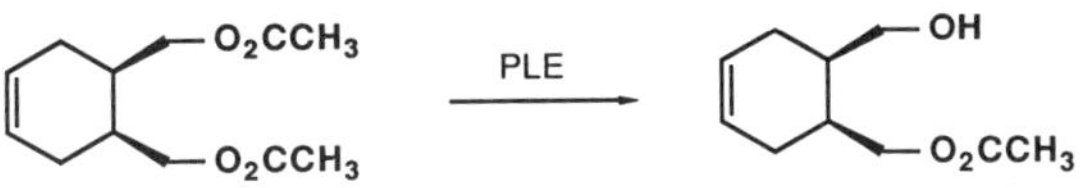

24

PLE

25

(78%, 96% ee)

In the 1,3-dibenzylimidazolidin-2-one system, the optical yield of PLE-catalyzed hydrolysis of the *cis*-diacetate **26**[32] is much better than that of the *cis*-diester **27**[33,34] (Eqs. 8 and 9). Thus PLE hydrolyses of a diester and a diacetate are often complementary to each other in terms of optical yield.

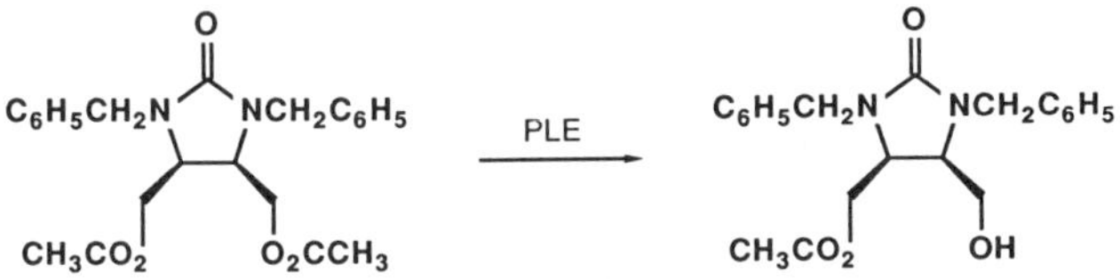

26 —PLE→ (90% ee) (Eq. 8)

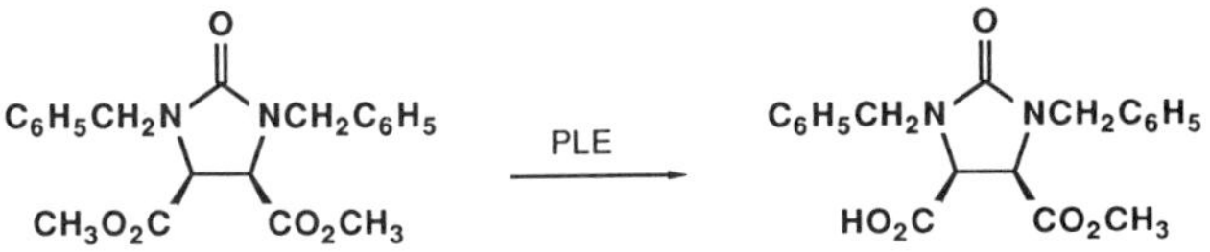

27 —PLE→ (71%, 38% ee) (Eq. 9)

Dimethyl cycloalkanediacetates **28**, **30**, and **32** are hydrolyzed with PLE, and the corresponding monoesters are accessible in good chemical and optical yields.[22] Cyclohexene derivative **29** and cyclopentanone derivatives **31** and **33** thus obtained serve as useful chiral synthons for natural product synthesis.

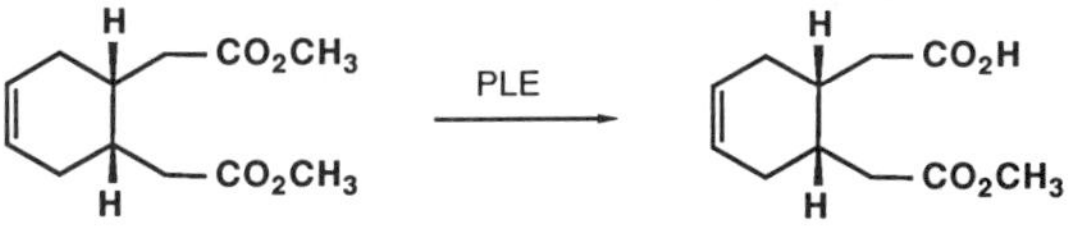

28 —PLE→ 29 (91%, 76% ee)

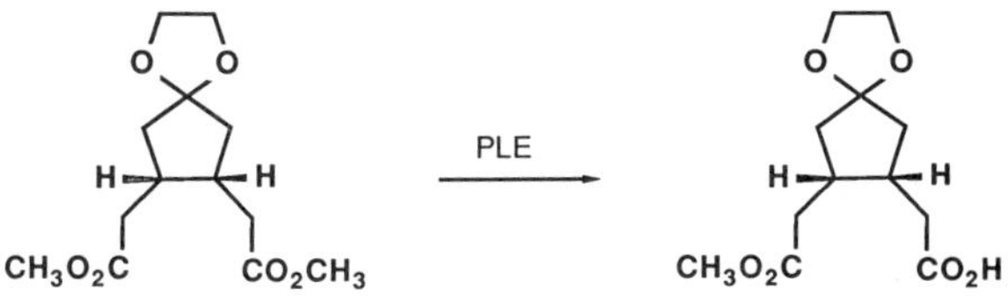

30 —PLE→ 31 (99%, 90% ee)

PLE

32 → 33 (85%, 88% ee)

Cyclic *cis*-1,3-diesters are accepted as substrates by PLE. While dimethyl *cis*-cyclopentane-1,3-dicarboxylate (**34**) gives (1*S*,3*R*)-monoester **35**,[35] its heterocyclic analogs **36** give the antipodal monoesters **37** upon PLE hydrolysis.[35–37]

CH_3O_2C … CO_2CH_3 —PLE→ CH_3O_2C … CO_2H

34 → 35 (82%, 34% ee)

CH_3O_2C … CO_2CH_3 —PLE→ HO_2C … CO_2CH_3

36 → 37

X = O (98%, 42% ee)
X = S (83%, 46% ee)
X = $NCH_2C_6H_5$ (85%, 80% ee)[36]
(39%, 100% ee)[37]

The optical yield in the hydrolysis of *cis*-1,3-diacyloxycyclopent-4-enes depends on the chain length of the acyl group (Eq. 10) (Table VII).[38,39] While diacetate **38** is hydrolyzed to give the monoacetate **41** in about 80% ee, propionate **39** and butyrate **40** are converted into the monoesters **42** and **43** with decreased optical yields.

RCO_2 … O_2CR —PLE→ RCO_2 … OH (Eq. 10)

38 R = CH_3
39 R = C_2H_5
40 R = n-C_3H_7

41 R = CH_3 (80.3% ee)
42 R = C_2H_5 (52%, 66% ee)
43 R = n-C_3H_7 (trace, 30% ee)

Nitro-containing *cis*-1,3-diacetate **44** is hydrolyzed with PLE to afford the monoacetate **45** in a good chemical yield and an excellent optical yield.[40]

CH_3CO_2 … NO_2 … O_2CCH_3 —PLE→ HO … NO_2 … O_2CCH_3

44 → 45 (89%, 98% ee)

7-Oxabicyclo[2.2.1]heptane-2,3-dicarboxylates are hydrolyzed with PLE to give the corresponding monoesters (Table VI). Whereas the hydrolysis of *exo*-diesters **46** and **48** gives the monoesters **47** and **49** with good to excellent ee,[31,41] *endo*-diester **50** is hydrolyzed less selectively to afford monoester **51** with 64% ee.[41]

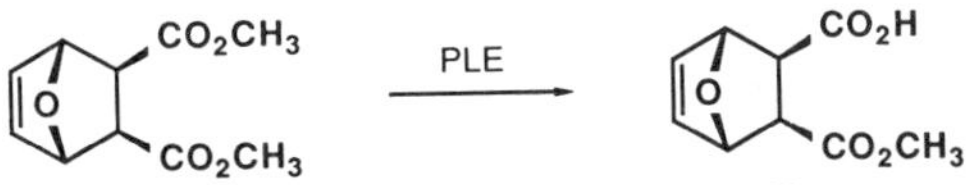

CO_2CH_3 / CO_2CH_3 — PLE → CO_2H / CO_2CH_3

46 → **47** (75 - 82% ee)

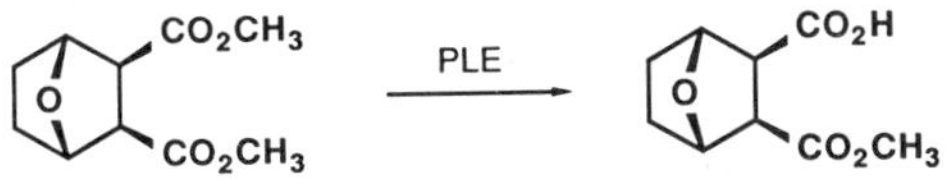

48 → **49** (82%, 98% ee)

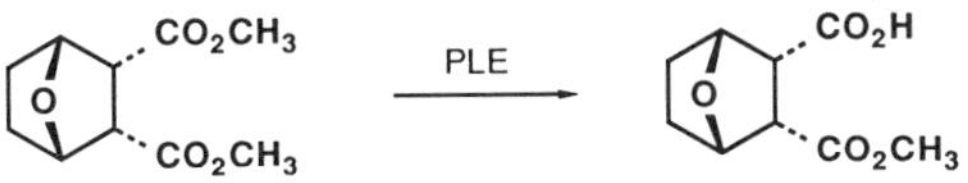

50 → **51** (87%, 64% ee)

Bicyclic and tricyclic diesters containing an unsaturated maleate partial structure can be hydrolyzed with PLE. Bicyclo diester **52** is hydrolyzed with PLE to give monoester **53** in a moderate optical yield.[42] Tricyclic diesters **54**, **56** (R = CH_3), and **58** give the corresponding monoesters **55**, **57** (R = CH_3), and **59** with better optical yields of around 80% ee and quantitative chemical yields.[43,44] In the hydrolysis of ethyl ester **56**, the optical yield is improved at the expense of chemical yield.[16]

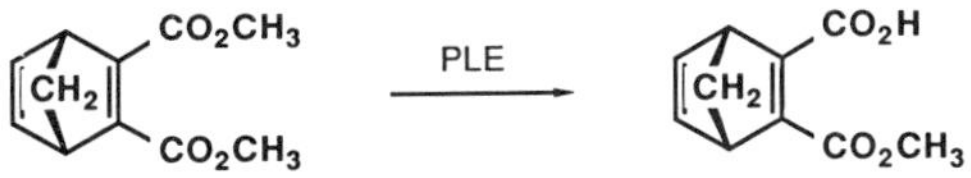

52 → **53** (41% ee)

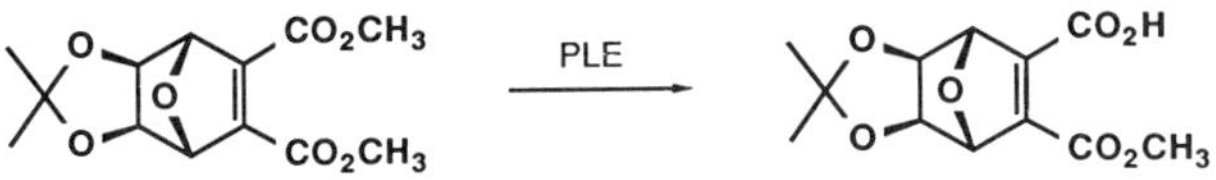

54 → **55** (96%, 77% ee)

56 $R = CH_3, C_2H_5$ —PLE→ 57 $R = CH_3$ (quant, 80% ee) $R = C_2H_5$ (30%, 100% ee)

58 —PLE→ 59 (quant, 77% ee)

Kinetic Resolution of Racemates

Diastereomeric transition states are generated during the reaction of enantiomeric substrates with a chiral enzyme. These transition states have different energy values and the reaction proceeds faster through the favored pathway of lower energy. Kinetic resolution is based on such differences in the rate of enzymatic reaction of each enantiomer via the diastereomeric transition states. Since the enzyme reaction of one enantiomer goes faster than that of the other isomer, the products will be enriched with one enantiomer and most of the other isomeric substrate will remain unreacted.

Several examples of kinetic resolution are reported. PLE preferentially hydrolyzes the *R* enantiomers of several 3-hydroxy-3-methylalkanoic esters **60** to produce (*R*)-acid **61** and (*S*)-ester **60** (Table VIII).[45]

$R^1-C(CH_3)(OH)-CH_2CO_2R^2$ (±)-60 —PLE→ $R^1-C(CH_3)(OH)-CH_2CO_2H$ (*R*)-61 + $R^1-C(CH_3)(OH)-CH_2CO_2R^2$ (*S*)-60

PLE-catalyzed hydrolysis of *trans*-cyclopropanecarboxylates (±)-**62** and (±)-**64** at 50% conversions produces the optically active acids (1*R*,2*R*)-**63** and (1*R*,2*R*)-**65**, and the unreacted esters (1*S*,2*S*)-**62** and (1*S*,2*S*)-**64** (Table IX).[46] The 1*R* enantiomers are preferentially hydrolyzed. The *cis*-isomers of the two methyl esters **62** and **64** are not substrates for PLE and they can be recovered unchanged.

(±)-62 —PLE, 50% Conversion→ (1*R*,2*R*)-63 + (1*S*,2*S*)-62

(±)-64 —PLE, 50% Conversion→ (1*R*,2*R*)-65 + (1*S*,2*S*)-64 *trans*-Cyclohexane-1,2-dicarboxylates (±)-**66** are hydrolyzed with PLE to give (*R*,*R*)-monoester **67** and (*S*,*S*)-diester **66**.[47] The best resolution of the two enantiomers is achieved when the diethyl ester (±)-**66** (R = C_2H_5) is used as substrate (Table XI).[47] Dimethyl *trans*-4-oxocyclopentane-1,2-dicarboxylate [(±)-**68**] is similarly hydrolyzed with PLE to give (+)-monoester **69** and (−)-diester **68** in excellent optical yields.[48] Resolution of methyl (±)-*cis*-4-acetamidocyclopent-2-ene-1-carboxylate **70** with PLE affords (+)-ester **70** and (−)-acid **71** in good chemical and optical yields.[49]

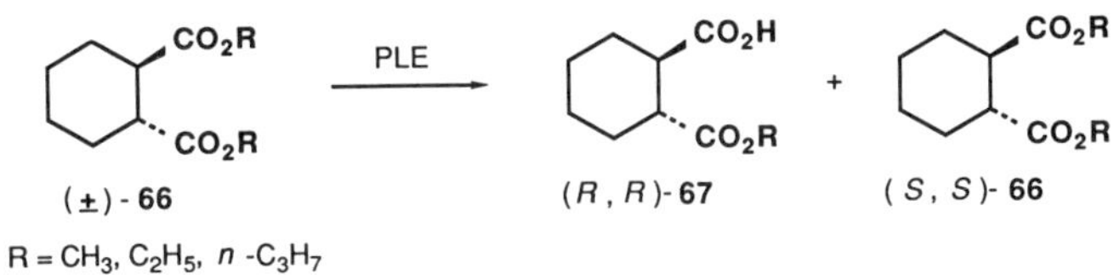

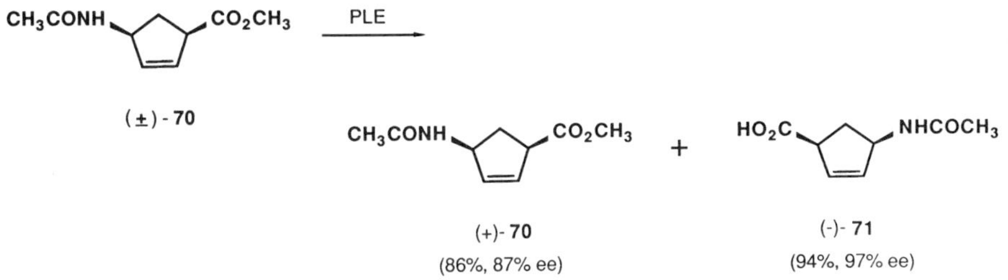

(±)-**66** →(PLE) (*R*, *R*)-**67** + (*S*, *S*)-**66**

R = CH_3, C_2H_5, *n*-C_3H_7

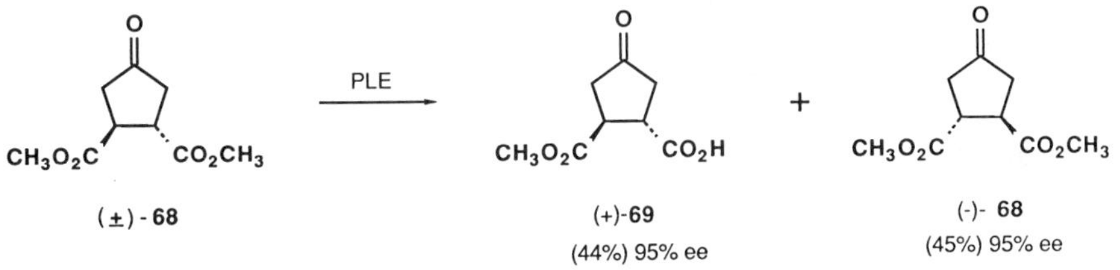

(±)-**68** →(PLE) (+)-**69** (44%) 95% ee + (-)-**68** (45%) 95% ee

(±)-**70** →(PLE) (+)-**70** (86%, 87% ee) + (-)-**71** (94%, 97% ee)

PLE-catalyzed kinetic resolution is successfully applied to epoxy esters. When dimethyl (±)-3,4-epoxyhexanedioate (**72**) is incubated with PLE, (+)-diester **72** and (−)-monoester **73** are obtained in excellent optical yields.[50] Methyl (±)-3,4-epoxybutanoate (**74**) is hydrolyzed with PLE at a somewhat slower rate than (±)-**72** to give (+)-ester **74** (74% ee) and hydrolyzed product **75** (97% ee).

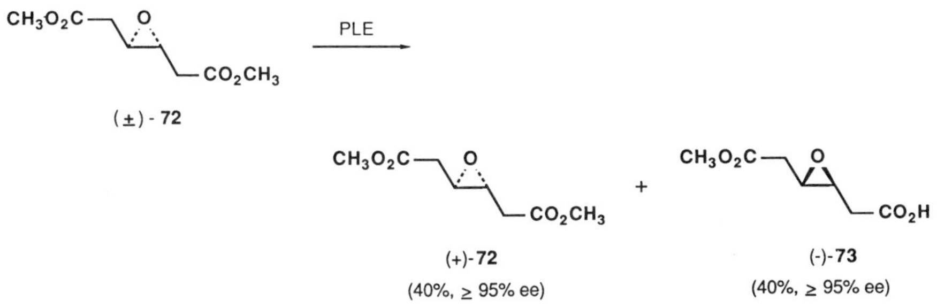

(±)-**72** →(PLE) (+)-**72** (40%, ≥ 95% ee) + (-)-**73** (40%, ≥ 95% ee)

(±)-74 —PLE→ (+)-74 (40%, 74% ee) + 75 (30%, 97% ee)

The broad substrate specificity of PLE is shown by the hydrolysis of ester-containing monomers and polymers.[51] Racemic *N*-methacryloylalanine ethyl ester [(±)-**76**] is hydrolyzed with PLE to afford (*S*)-acid **77** with recovery of (*R*)-**76**. Racemic copolymers (±)-**78** and (±)-**79** undergo ester hydrolysis with PLE to produce optically active species.[51] The degree of optical rotation depends on the weight fraction of amino acid containing monomer in the copolymer.

(±)-76 (*RS*) —PLE→ (*S*)-77 (*S*) + (*R*)-76 (*R*)

(±)-78 R = CH_3, C_2H_5 x : y = 5 : 1

(±)-79 x : y = 10 : 1

Hydrolysis of Ester Groups in Labile Molecules

Many biologically active molecules contain labile groups that undergo facile degradation. In fact, potent biological activities are often attributed to the intrinsic high reactivity of the molecules. For the synthesis of such natural products, it is frequently necessary to selectively hydrolyze ester groups in the presence of other sensitive functional groups. This can often be best achieved by an enzyme.

The instability of prostaglandin E_1 is due to its β-hydroxy ketone structure.

Whereas the usual chemical hydrolysis of ester **80** causes degradation, PLE cleaves the ester group of **80** without destroying the rest of the molecule to afford prostaglandin E_1 in good yield (Eq. 11).[52]

80 $\xrightarrow{\text{PLE}}$ Prostaglandin E_1 (85%) (Eq. 11)

2-Azetidinone-containing carboxylic esters are also cleaved with PLE. Thus methyl ester **81** is treated with PLE to afford a chiral synthon **82** for carbapenem antibiotics in 70% yield (Eq. 12).[53] Attempted alkaline hydrolysis of **81** results in a low yield of the acid, presumably because of the anticipated ring opening. Methyl ester **83** is hydrolyzed with PLE to give selenapenam **84** (Eq. 13).[54]

81 $\xrightarrow{\text{PLE}}$ 82 (71%) (Eq. 12)

83 $\xrightarrow{\text{PLE}}$ 84 (Eq. 13)

PLE-catalyzed cleavage of cyclopropyl acetate **85a** at neutral pH results in a mixture of cyclopropanol **85b** and acetate, from which cyclopropanol can be isolated by simple extraction.[55] Base-catalyzed hydrolysis or reductive cleavage of cyclopropyl acetate afford considerable amounts of ring-opened products.

85a $\xrightarrow{\text{PLE}}$ 85b

Enhancement of Optical Purity

Monoesters of high optical purity can be obtained even by using esterases of low to moderate stereospecificity according to the following concept.[38,56] Suppose that prochiral diester **S** is converted into diacid or diol **R** via two enantiomeric monoesters, **P** (fast forming) and **Q** (slow forming) (Eq. 14). If the same stereochemical preference is maintained, the relative rate constants of hydrolysis would be expected to follow the order $k_1 > k_2$ and $k_4 > k_3$. It follows that **P** is produced faster than **Q** and is transformed into **R** slower than **Q**. Consequently, as the result of enantioselective hydrolyses and the inherent consecutive kinetic resolution step, the concentration of the monoester **P** increases as the enzymatic reaction from **S** to **R** proceeds.

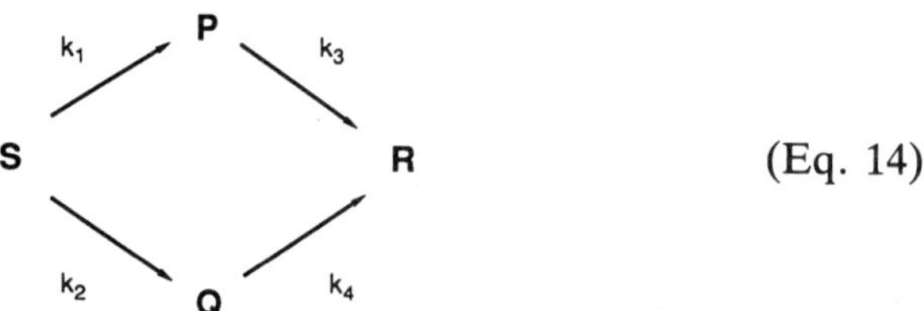

(Eq. 14)

Definition of the following kinetic parameters allows the prediction of the ee of the monoester fraction and the optimization of optical and chemical yields.[38,56]

$$\alpha = k_1/k_2$$
$$E_1 = k_3/(k_1 + k_2)$$
$$E_2 = k_4/(k_1 + k_2)$$

The concentrations of **P**, **Q**, and **R** and the ee are expressed as follows:

$$\mathrm{P} = \frac{\alpha \mathrm{S_0}}{(\alpha + 1)(1 - E_1)}\left[\left(\frac{\mathrm{S}}{\mathrm{S_0}}\right)^{E_1} - \left(\frac{\mathrm{S}}{\mathrm{S_0}}\right)\right]$$
$$\mathrm{Q} = \frac{\mathrm{S_0}}{(\alpha + 1)(1 - E_2)}\left[\left(\frac{\mathrm{S}}{\mathrm{S_0}}\right)^{E_2} - \left(\frac{\mathrm{S}}{\mathrm{S_0}}\right)\right]$$
$$\mathrm{R} = \mathrm{S_0} - \mathrm{S} - \mathrm{P} - \mathrm{Q}$$
$$\mathrm{ee} = (\mathrm{P} - \mathrm{Q})/(\mathrm{P} + \mathrm{Q})$$

The principle is illustrated by the following example.[38] *cis*-3,5-Diacetoxycyclopent-1-ene (**38**) is hydrolyzed with PLE in 0.1 M phosphate buffer (pH 7.0) to give the monoacetate **41** in 80.3% ee. The kinetic constants for the hydrolysis of **38** are $\alpha = 8.44 \pm 0.56$, $E_1 = 0.06 \pm 0.01$, and $E_2 = 0.12 \pm$

0.02. A computer-generated graph shows that the maximal obtainable recovery of the monoacetate fraction is 83% with an ee of 81%.

Reaction Media

Generally enzyme reactions are carried out in aqueous solution, occasionally with an organic cosolvent to dissolve the substrate. Organic cosolvents and other additives can influence the rate and the enantioselectivity of PLE-catalyzed hydrolyses of *meso*-diesters and diacetates.[31,57] In the hydrolysis of diacetate **24** (p. 7), addition of organic solvents to water results in a decrease in the rate of hydrolysis and an increase of chemical and optical yields.[31] The best result is obtained by using 10% of *tert*-butyl alcohol as a cosolvent for this substrate. However, for the PLE-catalyzed hydrolysis of other substrates, the effect on the enantioselectivity does not appear to follow a general trend.[31] In the hydrolysis of dialkylated malonic diesters, increasing the concentration of dimethyl sulfoxide results in the enrichment of the *R* enantiomers of the monoesters.[57] Dimethyl *cis*-*N*-benzylpyrrolidine-2,5-dicarboxylate (**36**, $X = NCH_2C_6H_5$) is hydrolyzed with PLE in an aqueous dimethyl sulfoxide medium to give optically pure monoester **37** ($X = NCH_2C_6H_5$) in 39% yield.[37] Optically active esters and alcohols are produced from racemates by the esterase-catalyzed transesterifications in biphasic aqueous–organic mixtures.[58]

Immobilized Enzyme

The efficiency of the enzyme reaction process is greatly facilitated by an immobilization technique because the immobilized enzyme can be easily recovered from the reaction mixture by simple filtration, making it reusable many times. Covalent immobilization of PLE has been reported.[59] Commercially available PLE is dialyzed against phosphate buffer and then mixed with oxirane-activated acrylic beads to afford immobilized PLE (Eq. 15).

NH_2 - PLE + ⟶ OH, NH - PLE (Eq. 15)

The immobilized PLE retains excellent specific activity (68% of the soluble enzyme) for several months when stored at 7°. Diacetate **38** is hydrolyzed with immobilized PLE to afford the monoester **41** in 87% chemical yield and about 75% ee on a 500-mmol scale in one day.[59]

CH_3CO_2, O_2CCH_3 **38** —Immobilized PLE→ HO, O_2CCH_3 **41** (87%, ~75% ee)

Optically active alkyl hydrogen (*R* or *S*)-3-aminoglutarates, such as **6** or **7** wherein $R^1 = NH_2$, $R^2 = H$ (p. 4), are produced by hydrolysis of dialkyl 3-aminoglutarates with esterase immobilized onto BrCN–Sepharose.[60]

Active Site Models

Whereas the topography of the active site of CHT has been intensively investigated, little is known about PLE. Based on the results of PLE-catalyzed asymmetric hydrolysis of substrates having ester groups, the following structural and stereochemical requirements for substrates are proposed.[24]

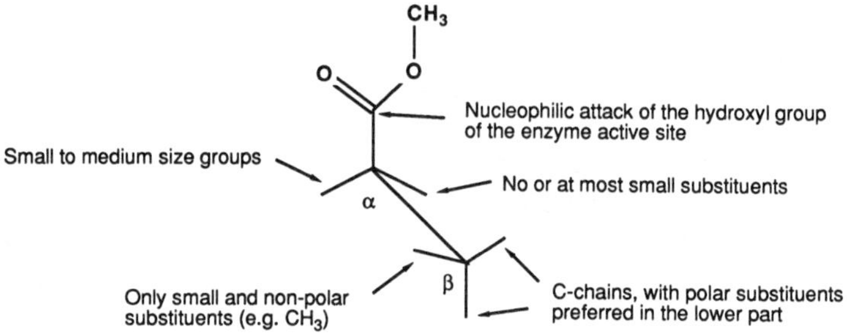

The topography of the active site of PLE can be best inferred from its interaction with rigid systems, such as tricyclic diester substrates. The active site seems to be hydrophobic, and it is likely that a flat region which can easily accommodate a six-membered ring is located near the catalytic serine residue of PLE. The following working model of the active site of PLE is proposed to reasonably explain the observed results.[61]

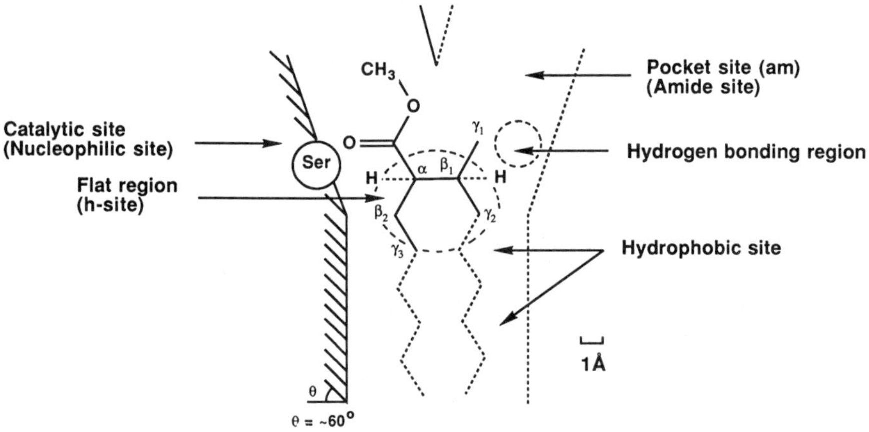

PLE-catalyzed hydrolysis of cyclopentanedicarboxylate **86a** gives monoester **86b** with >95% ee in 40–80% yield.[61a] This finding led to the proposal of a new active site model for PLE having an extra hydrophilic site in addition to the hydrophobic site, ester binding site, and catalytic site.

PLE

86a → **86b**

(40 - 80%, >95% ee)

SYNTHETIC APPLICATIONS

Symmetrical units are frequently present in the molecules of natural products. The efficient construction of new chiral synthons with the desired asymmetric centers and functional groups can be carried out catalytically by choosing an appropriate enzyme. Whereas relatively simple natural products are quasi-symmetrical and have obvious symmetry factors, some complex molecules contain hidden symmetry. The strategy for synthesizing complicated natural products and other compounds can be designed from the following principles.

1. Symmetrization: retrosynthetic analysis to find a latent symmetry in the complex target molecule leading to a simple starting diester having a σ symmetry in the prochiral or *meso* form.

2. Asymmetrization: enzymatic transformation of σ symmetry to C_1 symmetry; creation of chirality in the symmetric diester by means of PLE-catalyzed asymmetric hydrolysis.

3. Nonenzymatic procedures: conversion of the chiral monoester into the target molecule by means of nonenzymatic operations.

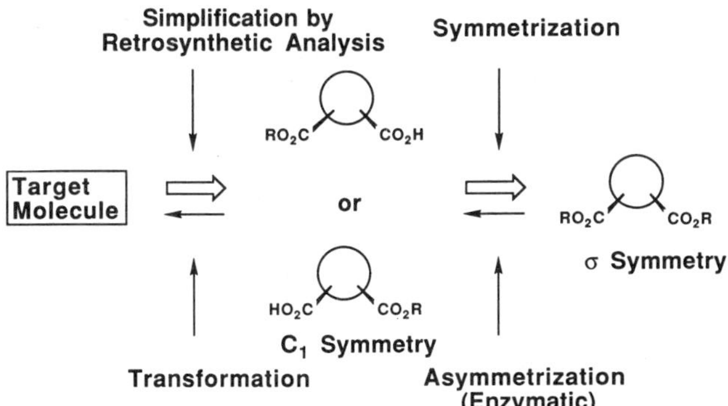

This asymmetrization–symmetrization concept is based on the ability of PLE to preferentially cleave one specific ester group of the two enantiotopic esters of a substrate to afford an optically active monoester. The formal enantiomer conversion of the monoester is easily achieved by the chemical step of formal intramolecular transesterification. For instance, a chiral methyl monoester **87** is transformed into the formally antipodal *tert*-butyl monoester **89** via a mixed diester **88** (Eq. 16). Thus both of the antipodes are in practice interchangeable by a simple esterification–hydrolysis process.

CH_3O_2C / CO_2H **87** —(Isobutene, H^+)→ CH_3O_2C / $CO_2C_4H_9$-*t* **88** —(OH^-)→ HO_2C / $CO_2C_4H_9$-*t* **89** (Eq. 16)

This characteristic of PLE is particularly advantageous for the flexible generation of various chiral synthons. Enantioselective total synthesis of some biologically active molecules is achieved by a combination of enzymatic and nonenzymatic procedures based on the asymmetrization–symmetrization concept.

Glutarate Derivatives

Chiral monoesters of glutarates serve as synthons for the highly functionalized linear sequences of carbon atoms often found in various natural products. (*R*)-Mevalonolactone (**90**) comprises a 3-hydroxy-3-methylpentane unit in which the oxidation level at each carbon terminus is different. This synthon can be made by a combination of enzymatic asymmetrization and selective chemical transformation. A short synthesis of (*R*)-mevalonolactone (**90**) is achieved by PLE-catalyzed asymmetric hydrolysis of dimethyl 3-hydroxy-3-methylglutarate (**3**) followed by selective reduction of the ester group of **4**.[14]

HO CH_3 / CH_3O_2C CO_2CH_3 **3** —(PLE)→ HO CH_3 / CH_3O_2C CO_2H **4** (62%, 99% ee) —($LiBH_4$ or Na - NH_3)→ HO CH_3 / HO CO_2H → HO CH_3 / O O **90** (*R*)- Mevalonolactone (81%) ($LiBH_4$) (73%) (Na - NH_3)

β-Amino acid derivatives are available in excellent optical yields by treatment of 3-aminoglutarate derivatives with PLE. These are plausible precursors for β-lactams. Three representative carbapenems, all having the *R* configuration at the bridgehead carbon, are synthesized from the PLE hydrolysate **91**.[61,62] Cyclization of the β-amino acid is achieved by the condensing system triphenylphosphine–2,2′-dipyridyl disulfide–acetonitrile.[63] (+)-Thienamycin, a *trans*-carbapenem, is synthesized by acetylation of 2-azetidinone **92** followed by selective reduction of the ketone.[61,64] The *cis* side chain of (−)-carpetimycin A is constructed stereoselectively via the key lactone intermediate **93**.[65,66] The olefinic side chain of (−)-asparenomycin C is introduced into the 2-azetidinone **92** by a chelation-controlled Peterson olefination.[66,67]

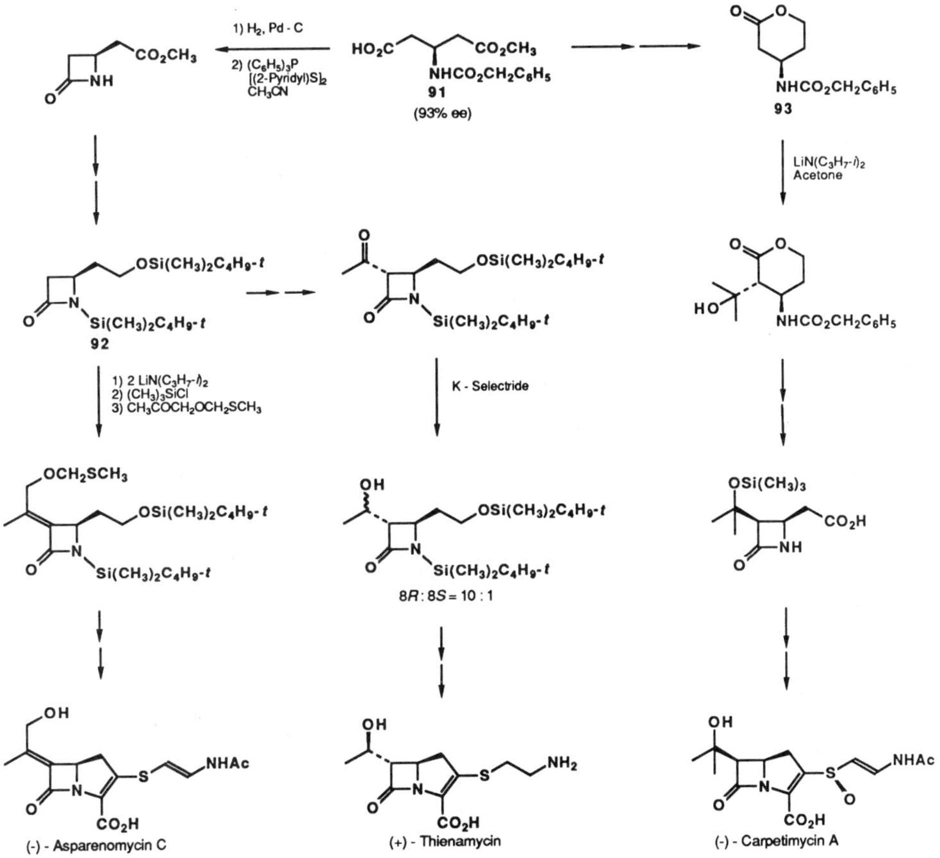

(+)-Negamycin is synthesized from the same monoester **91**. The key feature of the synthesis is the 1,3-asymmetric induction of chiral homoallylamine **94** to afford the key cyclic intermediate **95** in an *R* : *S* ratio of 50 : 1 (Eq. 17).[68,69]

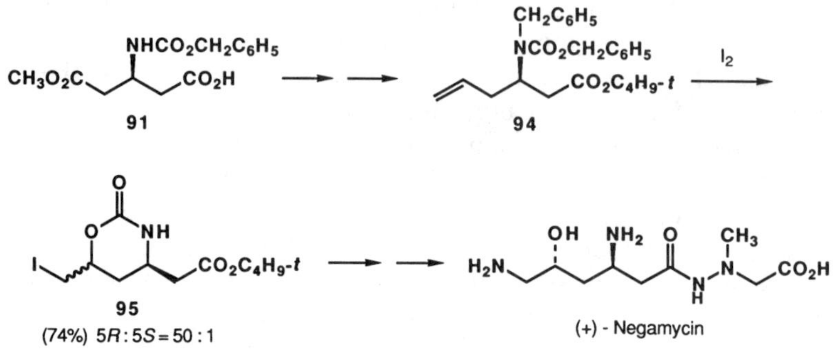

(Eq. 17)

PLE-catalyzed hydrolysis of dimethyl 3-methylglutarate (**96**) followed by borohydride reduction gives the chiral lactone **97**, which is further converted into a chiral synthon **98** for isoprenoids (Eq. 18).[70]

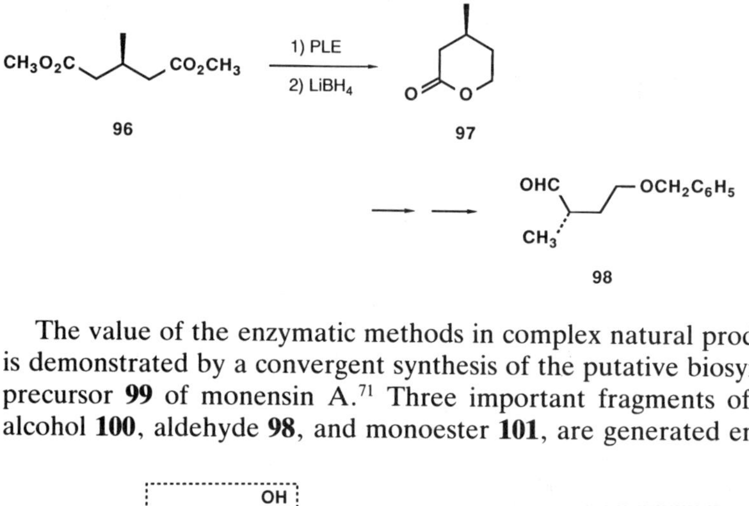

(Eq. 18)

The value of the enzymatic methods in complex natural product synthesis is demonstrated by a convergent synthesis of the putative biosynthetic triene precursor **99** of monensin A.[71] Three important fragments of **99**, namely, alcohol **100**, aldehyde **98**, and monoester **101**, are generated enzymatically.

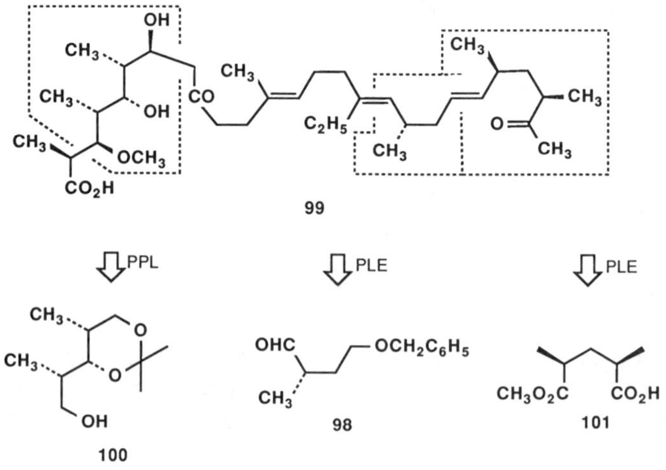

Chiral nonracemic glutarates are useful synthons for macrocyclic natural products. Verrucarin A and 3α-hydroxyverrucarin A are synthesized from a component obtained by the hydroxylation of chiral monoester **102**.[72,73]

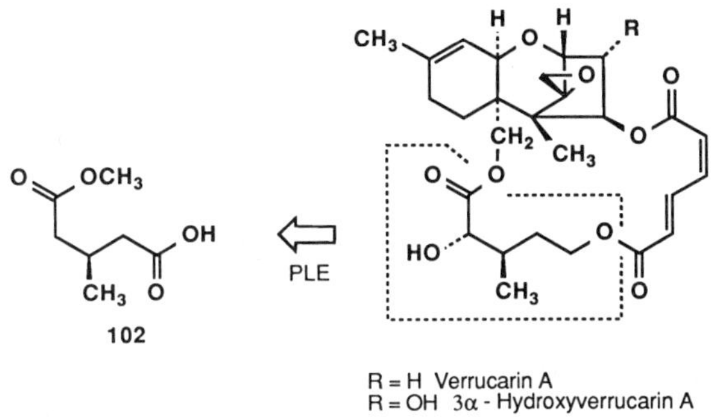

102

PLE

R = H Verrucarin A
R = OH 3α - Hydroxyverrucarin A

Monoester **103** is converted into a rifamycin S fragment epimeric at C-23.[74] Pimaricin contains two molecules of antipodal 3-hydroxyglutarate units **104** and **105**, which are generated by using PLE and CHT, respectively.[75]

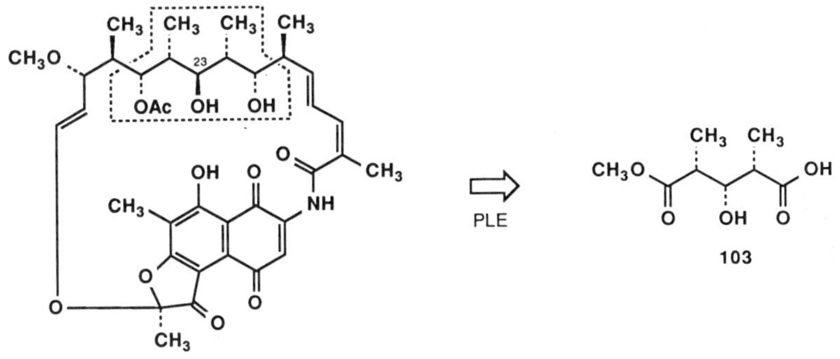

Rifamycin S

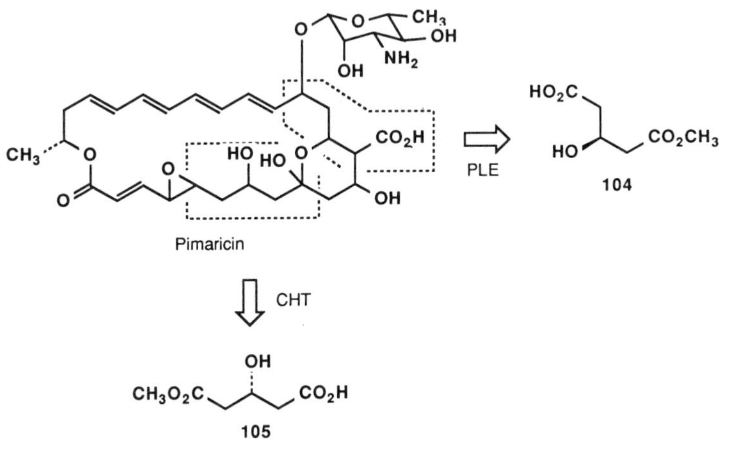

PLE

103

Pimaricin

PLE

104

CHT

105

Bicyclo[2.2.1]hept-2-ene Derivatives

Synthesis of the optically active sugar moiety of nucleosides is carried out by PLE-catalyzed hydrolysis of Diels–Alder adduct **106**. This approach is feasible because of the unusual decarboxylative ozonolysis of monoester **107**, furnishing the ribose skeleton of **109** via intermediate **108** (Eq. 19).[61]

X, CO_2CH_3, CO_2CH_3 **106** —PLE→ X, CO_2H, CO_2CH_3 **107** —O_3→

108 (X, O, O–H, OCH_3, HO, O, O, CH_3O_2C) → **109** (CO_2CH_3, O=C, X, CO_2CH_3, O O)

(Eq. 19)

The usefulness of the chiral synthon **107** is extended by the enantiomer interconversion of **55** into **111**. Thus chiral nonracemic monoester **55** is converted into either (−)-6-azapseudouridine or an unnatural L-ribose derivative.[61,66] On the other hand, the antipodal *tert*-butyl monoester **111**, obtained from **55** by the enantiomer interconversion through **110**, affords a D-ribose derivative or (+)-showdomycin.[43,76]

O, CO_2H, CO_2CH_3 **55** → O, $CO_2C_4H_9$-*t*, CO_2CH_3 **110** → O, $CO_2C_4H_9$-*t*, CO_2H **111**

(-) - 6 - Azapseudouridine ←← **55** →→ L - Ribose derivative

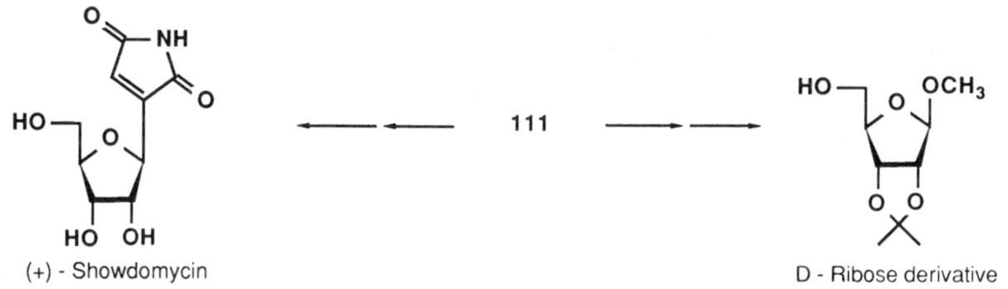

(+) - Showdomycin ← ← 111 → → D - Ribose derivative

(−)-Cordycepin (3′-deoxyadenosine) is synthesized from epoxy monoester **59** (Eq. 20).[43,66,76,77] Monoester **57** with a methano bridge is a valuable precursor for the unusual carbohydrate moiety of optically active carbocyclic nucleosides, because neither the cyclopentylamine moiety of aristeromycin nor the cyclopentenylamine part of neplanocin A is accessible by the degradation of natural products (Eq. 21).[42,44,66,78] This approach allows the systematic synthesis of various neplanocin analogs including cytidine derivative **112**, which is highly active against mouse lymphoma L5178 cell in vitro.[42,69,79,80]

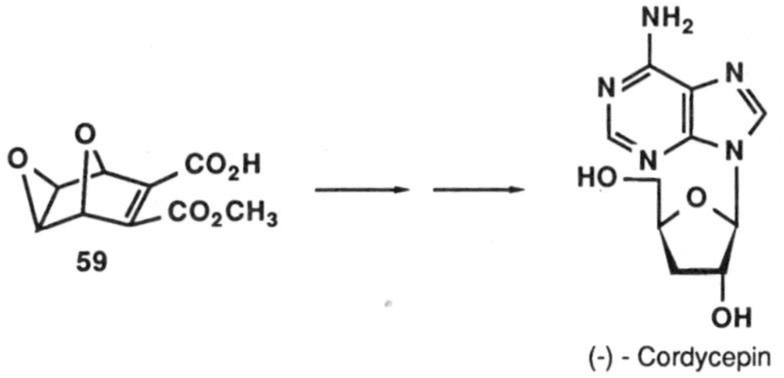

59 → → (-) - Cordycepin (Eq. 20)

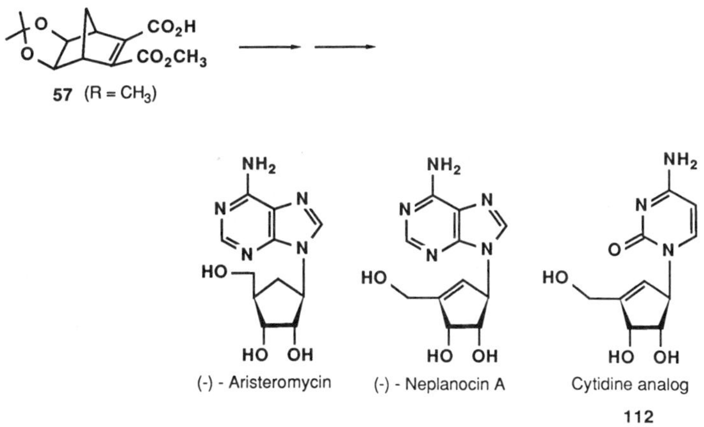

57 (R = CH_3) → →

(-) - Aristeromycin (-) - Neplanocin A Cytidine analog 112 (Eq. 21)

cis-Cyclohex-4-ene-1,2-dicarboxylates

The bicyclic β-lactam **114** is generated as an intermediate for *cis*-carbapenems (Eq. 22).[36] The Curtius rearrangement of the optically active monoester **23** affords β-amino acid **113**, which is cyclized to **114** with triphenylphosphine–2,2′-dipyridyl disulfide–acetonitrile. Since racemic **114** is known to be convertible into racemic *cis*-carbapenem,[81,82] the successful preparation of optically pure **114** constitutes a formal enantioselective synthesis of 6-*epi*-PS-5, a *cis*-carbapenem antibiotic.

CO_2H CO_2CH_3 **23** 96% ee → → NH_2 CO_2CH_3 **113** →

H H NH O **114** —known for racemate→ cis - Carbapenems (Eq. 22)

A synthesis of (−)-fortamine, a unique 1,4-diaminocyclitol skeleton of an antibiotic fortimicin A, is achieved starting with the same optically active monoester **23**. The introduction of each functional group on the cyclohexane ring is accomplished with full stereocontrol via amino acid derivative **115**, bicyclic lactone **116**, and epoxide **117**.[83]

CO_2H CO_2CH_3 **23** → $CO_2C_4H_9$-*t* $NHCO_2CH_3$ **115** → O O NCO_2CH_3 CH_3 **116**

→ O CO_2CH_3 NCH_3 O O **117** → OH H_2N OCH_3 HO $NHCH_3$ OH (-) - Fortamine

Monoester **23** is also a synthon for substituted cyclopentanes.[84] Ring contraction of the cyclohexene moieties of bicyclic lactones **118** and **119** is performed by permanganate oxidation, esterification, and Dieckmann condensation. The regioselective formation of the key intermediates **120** and its enantiomer **121** allows facile access to cyclopentanoid natural products such as prostaglandins,[85,86] carbacyclin,[29,30] brefeldin A,[29,87] and pentalenolactone.[29,30]

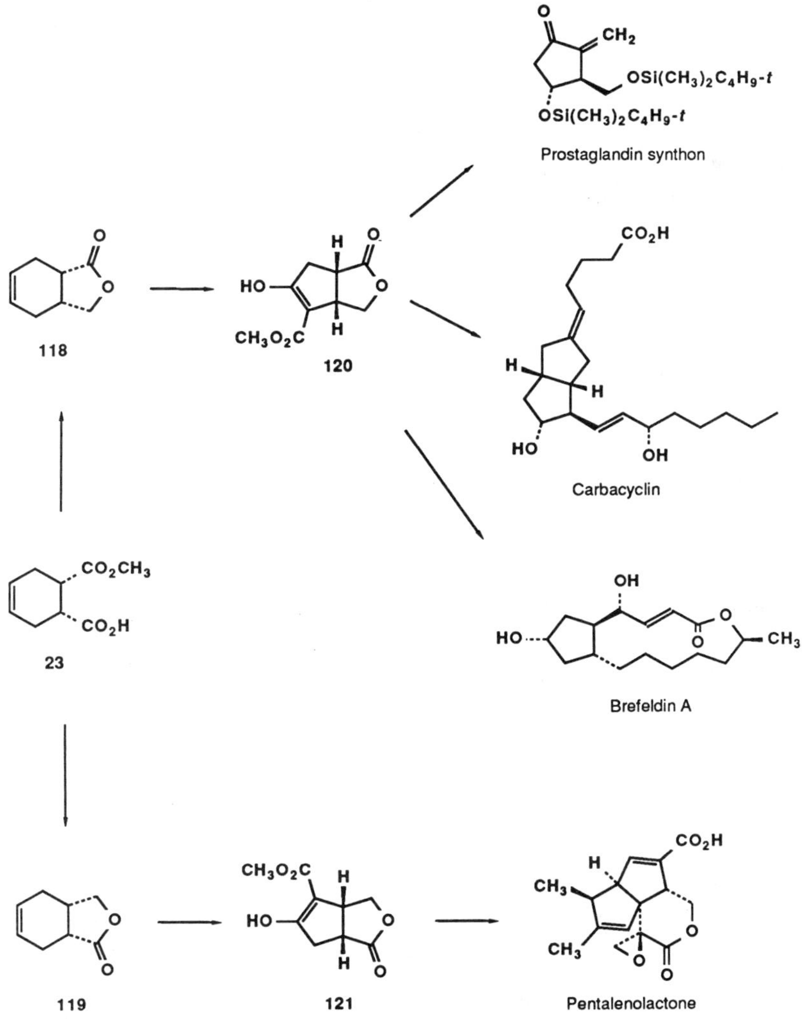

118 120 Prostaglandin synthon

Carbacyclin

23

Brefeldin A

119 121 Pentalenolactone

Miscellaneous

The chiral cyclopentenyl monoacetate **41** is a useful synthon for prostaglandin syntheses. Claisen rearrangement of **41** affords the chiral bicyclic lactone **122**, which is known to be convertible to the "Corey lactone," a key intermediate for prostaglandins.[39,88] On the other hand, the cyclopentenone **123** is a versatile intermediate for cyclopentanoid natural products.[89] Development of a three-component coupling procedure to introduce α- and ωappendages into **123** or its equivalent allows facile access to the prostaglandins.[90,91]

CH_3CO_2 OH
41
CH_3CO_2 O OC_2H_5
Claisen rearrangement
$CO_2C_2H_5$ O_2CCH_3
122
Prostaglandins
RO O
123
R^1Li - CuI - $(n\text{-}C_4H_9)_3P$
RO M R^1
1) $(C_6H_5)_3SnCl$
2) R^2I
RO O R^1 R^2
R^1Li = Li OR^3
R^2I = I CO_2CH_3

The chiral cyclopentanecarboxylic acid derivative **69**, obtained by kinetic resolution of the corresponding *trans*-diester **68** (page 11), is converted into the synthon for prostaglandin A_2.[48]

O
CH_3O_2C CO_2H
69
(95% ee)
CH_3CO_2 O_2CCH_3
CH_3CO_2
O O
Prostaglandin A_2 synthon

Bicyclic lactone **126**, a known intermediate for the synthesis of (+)-biotin, is synthesized by two different chemicoenzymatic approaches, either by the reduction of monoester **124**[33] or, alternatively, by the oxidation of monoacetate **125**.[32] The optical yield of the monoacetate **125** (90%) is better than that of the monoester **124** (75%).

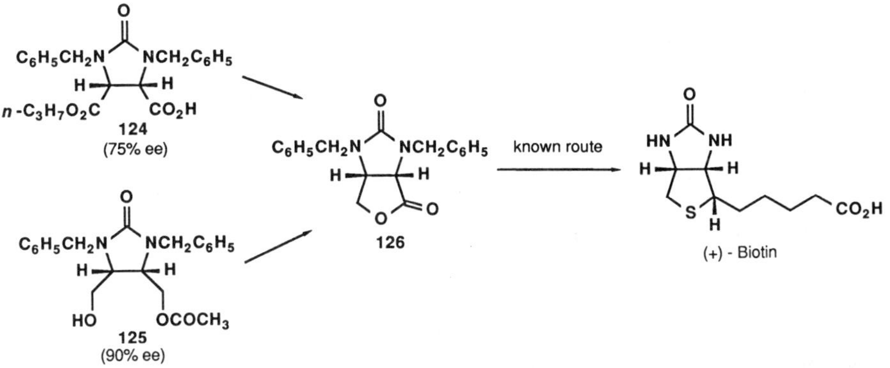

124
(75% ee)

125
(90% ee)

126

known route

(+) - Biotin

(*R*)-Monoester **128**, a known precursor for L-α-methyldopa, is obtained by the PLE-catalyzed hydrolysis of dialkyl malonate **127**. Whereas the optical yield of the monoethyl ester **128** (R = C_2H_5) is 59%,[33] the corresponding monomethyl ester **128** (R = CH_3) is obtained in 93% ee.[92] Carboxylic acid **128** is converted into L-α-methyldopa via acyl azide formation followed by the Curtius rearrangement.

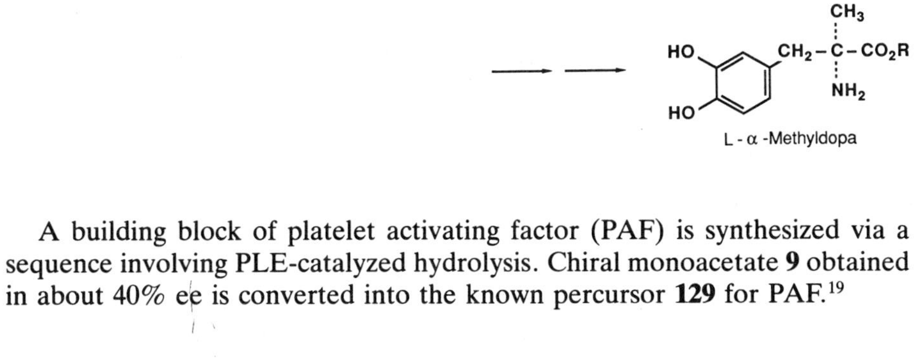

127

PLE

128

L-α-Methyldopa

A building block of platelet activating factor (PAF) is synthesized via a sequence involving PLE-catalyzed hydrolysis. Chiral monoacetate **9** obtained in about 40% ee is converted into the known percursor **129** for PAF.[19]

9 → 129 → Platelet Activating Factor (n = 15 or 17)

Chiral reagents of auxiliaries for organic synthesis are enzymatically prepared. Multiple coupling reagents **130** and **131**[93,94] are prepared from enzymatically generated chiral nonracemic nitroacetate **45**.[40] Several synthetically useful chiral amino alcohols are also synthesized from **45**.[40]

45 (98% ee) → **130**, **131**

Chiral auxiliaries for asymmetric synthesis, **133** and **135**, are practically prepared by enzymatic resolution.[95] Treatment of racemic *trans*-acetate **132** with powdered pig liver (pig liver acetone powder, PLAP) gives a nearly 1:1 mixture of (−)-alcohol **133** and (+)-acetate **132**. Alkaline hydrolysis of (+)-**132** affords (+)-alcohol **133**. Recrystallization of (+)-**133** and (−)-**133** gives optically pure materials. Resolution of racemic *trans*-acetate **134** proceeds similarly. Although the resulting purified enantiomers are not crystalline at room temperature, material of high ee is obtained by carrying the PLAP hydrolysis past the point where the ratio of acetate to alcohol is 1:1. Alkaline hydrolysis of unreacted acetate (−)-**134** affords alcohol (+)-**135** of high ee. The enzymatically produced alcohol **135** enriched in the (−)-enantiomer is reacetylated and carefully rehydrolyzed with PLAP to give alcohol (−)-**135**. The optical rotations of the resulting alcohols (+)-**135** and (−)-**135** are equal in magnitude and opposite in sign.

(±)-132 —PLAP→ (-)-133 (Optically pure after recrystallization (82% yield)) + (+)-132

(+)-132 —OH^-→ (+)-133 (Optically pure after recrystallization (55% yield))

(±)-**134** —PLAP→ (-)-**135** + (-)-**134**

(-)-**135** → acetate —PLAP→ (-)-**135** [α]D -26.3°

(-)-**134** —OH⁻→ (+)-**135** [α]D +26.3°

COMPARISON WITH OTHER METHODS

Other Hydrolytic Enzymes and Microorganisms

Formation of optically enriched molecules with enzymes other than PLE has also been widely investigated, and there are several examples of results superior to those of PLE-catalyzed hydrolysis. For instance, CHT-catalyzed hydrolysis of dialkyl 3-hydroxyglutarates and diethyl 3-acetamidoglutarate (Eqs. 1 and 2) proceeds stereospecifically to give (*R*)-monoester in higher ee than that obtained with PLE.[13,96] However, diethyl acetamidomalonate is hydrolyzed with CHT with virtually no stereospecificity.[97] PPL-catalyzed hydrolysis of *cis*-1,2-bis(acetoxymethyl)cycloalkanes results in better optical yield than that obtained with PLE (Eq. 7).[98,99] The *meso*-dibutyrate **136** is also cleaved with PPL to afford alcohol **137** of 55% ee in 77% isolated yield.[100]

136 —PPL→ **137**

PPL-catalyzed hydrolysis of dimethyl *cis*-cyclohex-4-ene-1,2-diacetate **138** affords monoester **139**, the enantiomer of that obtained with PLE, in excellent chemical and optical yields.[22] Ester **139** is converted into chiral bicyclic diol **140**, a potential synthon for carbacyclin and isocarbacyclin.[101,102]

138 → (PPL) → 139 (97%, 97% ee) → → 140 → → Carbacyclin, Isocarbacyclin

Electric eel acetylcholinesterase is also a valuable enzyme for the hydrolysis of certain diacetates. Treatment of *cis*-1,3-diacetoxycyclopent-4-ene (**38**) with acetylcholinesterase affords monoacetate **141** in excellent chemical and optical yields.[103] PPL is also successfully applied to the hydrolysis of **38**.[104] Bicyclic *meso*-diacetate **142** is hydrolyzed with electric eel acetylcholinesterase to provide monoacetate **143** (80%) which is oxidized with Jones reagent to enone (+)-**144** (95%, 98% ee).[105]

38 → 141

(94%, 96% ee) (Acetylcholinesterase)
(87%, 92% ee) (PPL)

142 → (Acetylcholinesterase) → 143 (80%)

→ (CrO_3, H_2SO_4, acetone) → (+)-144 (95%, 98% ee)

Hydrolysis with appropriate microorganisms is also possible. The diacetate **38** is also hydrolyzed with *Bacillus subtillis var. Niger* to afford chiral monoester **41**.[88] About 500 species of microorganisms have been screened for the large-scale hydrolysis of dimethyl 3-[(benzyloxycarbonyl)amino]glutarate, and *Flavobacterium lutescens* IFO3084 and IFO3085 hydrolyze the pro-*R* ester group most specifically and effectively (better than 97% ee).[106] Whereas the optical yield in the PLE-catalyzed hydrolysis of dimethyl *cis*-2,4-dimethylglutarate (**12**) is not high, an alternative microbial hydrolysis using *Gliocladium roseum* gives monoester **13** with an ee greater than 98%.[23] Incubation of (±)-diacetate **145** with *Trichoderma viride* affords (+)-diacetate **145** (85% ee), (−)-monoacetate **146** (15% ee), and (−)-diol **147** (100% ee) in a ratio of 49:10:41.[107]

C_6H_5 ... C_6H_5 / CH_3CO_2 ... O_2CCH_3 — *Trichoderma viride* →

(±)- **145**

(+)- **145** (85% ee) + (-)- **146** (15% ee) + (-)- **147** (100% ee)

Kinetic resolution studies of racemic acetates and esters using Baker's yeast,[108,109] lipase,[110] *Saccharomyces sp.*,[111] *Bacillus subtilis var. Niger*,[112] *Brevibacterium ammoniagenes*,[113] *Trichoderma konigi*,[114] *Rhizopus nigricans*,[115] and *Klebsiella pneumoniae*[116] are reported. Resolution of racemic acids and alcohols by lipase-catalyzed hydrolysis, esterification, and transesterification are reported.[58,117] Kinetic resolution studies of (±)-menthol and (±)-2-(*p*-chlorophenoxy)propionic acid using a commercial *Candida cylindracea* lipase in biphasic media are reported.[118]

The maximum yield of one enantiomer is 50% in all of the conventional resolution processes. However, it should be possible to transform a racemic ester into one enantiomer product in up to 100% yield if a reaction could be conducted under conditions wherein the substrate may be racemized in situ without racemization of the product. This concept is successfully applied to the *Streptomyces griseus* protease-catalyzed hydrolysis of ester **148** to give (−)-ketrolac in 92% isolated yield.[119]

148 — Protease → (-)-Ketrolac (92%, 85% ee)

Oxidoreductases

An alternative to PLE-catalyzed hydrolysis is the use of oxidoreductases for desymmetrization. Whereas esterase hydrolyzes one of the enantiotopic ester groups (Eq. 23), oxidase discriminates between enantiotopic hydroxyl groups to afford a chiral keto alcohol (Eq. 24).

RO_2C CO_2R —Esterase→ RO_2C CO_2H (Eq. 23)

$HOCH_2$ CH_2OH —Oxidase→ $HOCH_2$ CO_2H (CHO) (Eq. 24)

Whereas (*R*)-mevalonolactone is synthesized via the PLE-catalyzed hydrolysis, (*S*)-mevalonolactone is accessible using *Flavobacterium oxydans* which oxidizes *gem*-hydroxymethyl compounds to the corresponding hydroxymethyl carboxylic acids.[14] When prochiral triol **149** is incubated with lyophilized cells of *F. oxydans,* (*S*)-mevalonolactone is obtained (Eq. 25). On the other hand, treatment of (*RS*)-mevalonolactone with *F. oxydans* affords (*S*)-mevalonolactone. These results suggest the following two possibilities: (1) the pro-*S* hydroxymethyl group of **149** is preferentially oxidized with the microorganism to afford (*S*)-mevalonolactone; or (2) nonspecific oxidation of **149** to (*RS*)-mevalonolactone takes place, followed by selective utilization of the *R* isomer by the microorganism, leaving behind (*S*)-mevalonolactone. 3-Methyl-1,5-pentanediol is similarly converted into the corresponding chiral lactone with the enzyme systems of *Gluconobacter roseus*.[120]

149 —*Flavobacterium oxydans*→ (*S*) - Mevalonolactone (Eq. 25)

Horse liver alcohol dehydrogenase (HLADH) has been most intensively studied for application to organic synthesis. L-Glyceraldehyde is obtained via the HLADH-catalyzed oxidation of pro-*S* hydroxyl groups of glycerol.[121,122] The efficiency of the oxidation is improved using flavin mononucleotide (FMN) recycling of catalytic amounts of expensive NAD^+, the cofactor required for HLADH oxidation. Thus 3-methyl-1,5-pentanediol (**150**) is oxidized to the chiral lactone **151** in 70% chemical and 90% optical yields (Eq.

26).[123,124] This HLADH procedure is especially effective for the oxidation of *meso*-cycloalkanedimethanols **152** to afford the corresponding bicyclic or tricyclic lactones **153** in optical yields better than 97% (Eq. 27).[125–128]

CH_3
HO OH
150
HLADH
NAD$^+$ recycling
CH_3 HO CHO
CH_3 O OH
HLADH
CH_3 O O
151
(70%, 90% ee)

(Eq. 26)

OH OH
152
HLADH
OH CHO
O OH
HLADH
O O
153

(Eq. 27)

Fermenting yeast reduces carbonyl compounds to optically active alcohols,[129] and is applied to synthetic studies on prostaglandins[130,131] and leucotrienes.[116]

Nonenzymatic Methods

The usual optical resolution of racemates inevitably produces the useless enantioisomer theoretically in up to 50% yield. The "mesotrick method"[132] is an alternative approach for asymmetrization starting with a symmetrically functionalized *meso* compound to utilize the total amount of the material. The functionalization of *meso* compound **A** with an equimolar amount of chiral compound **B*** gives a mixture of the two diastereoisomers **C** and **D**, which then can be separated from each other by chromatography or crystallization. As it is possible to transform **C** or **D** into the antipodal compounds **E** or **F** by conventional chemical manipulations, the total amount of **A** can be utilized to produce **E** or **F**.

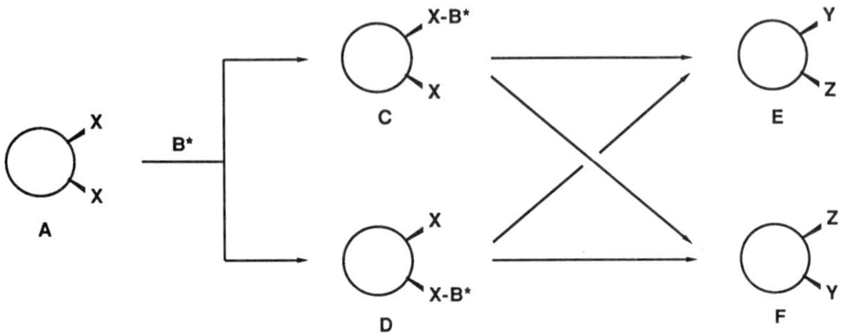

This concept is successfully applied to the preparation of optically pure prostaglandin intermediates starting with *cis*-2-cyclopentene-1,4-diol (**154**).[132] Treatment of *meso*-diol **154** with *N*-methanesulfonyl-L-phenylalanyl chloride gives diester **155** (24%) and a mixture of monoesters **156** and **157** (51%). Fractional crystallization of the oily mixture of **156** and **157** affords optically pure **156** and the mother liquid gives **157** contaminated with a small amount

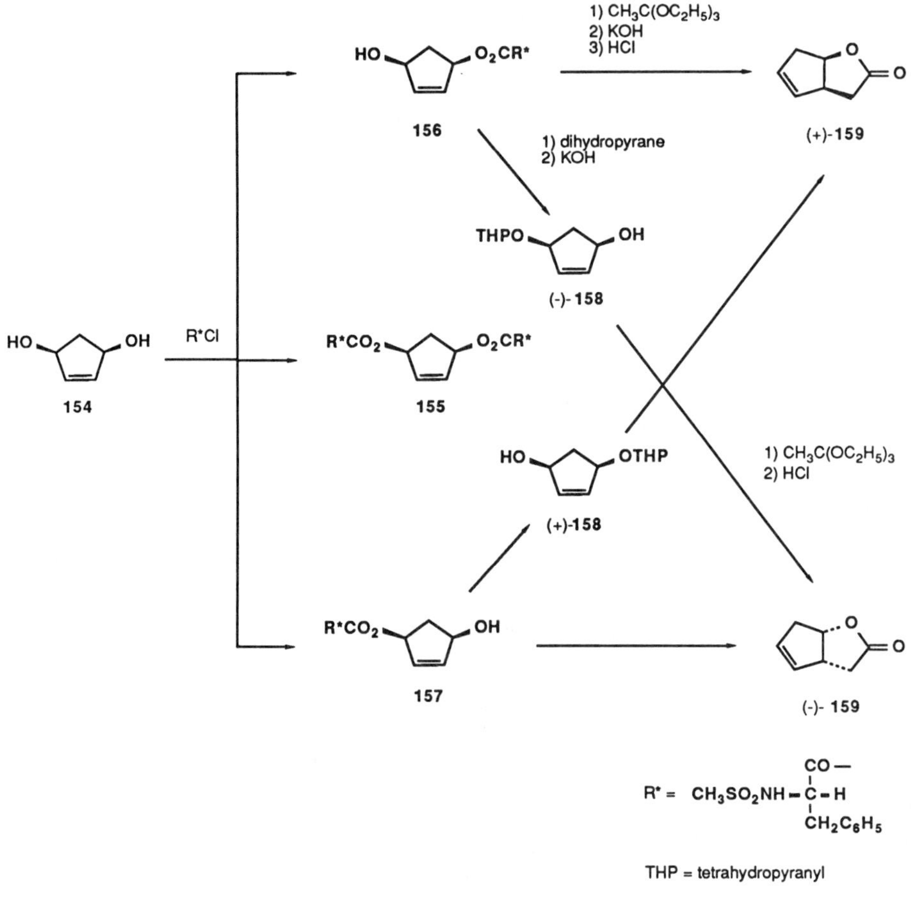

$R^* = CH_3SO_2NH-CH(CO-)(CH_2C_6H_5)$

THP = tetrahydropyranyl

of **156**. The monoester **156** is converted into (+)-**159** by the Claisen rearrangement, hydrolysis, and lactonization. On the other hand, **156** is transformed into the formally antipodal alcohol (−)-**158**, which affords lactone (−)-**159**. Likewise, the other monoester **157** is transformed into either (−)-**159** or (+)-**159**.[132]

This "mesotrick method" is further applied to *cis*-2-cyclohexene-1,4-diol **160**, and either (+)-**161** or (−)-**161** is similarly obtained.[133]

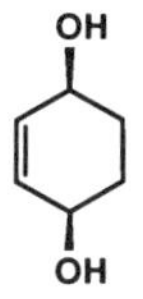

OH OH O O O O

160 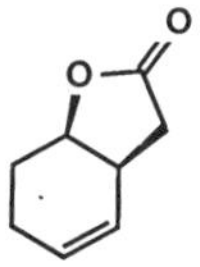 (+)-161 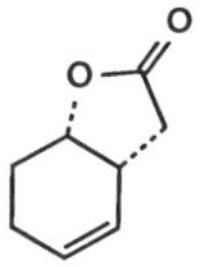 (-)-161

Enantiomers of high optical purity can be obtained by virtue of enantiotopic carboxyl group differentiation in the ring opening of cyclic anhydrides. Symmetrically substituted glutaric anhydrides are treated with (*S*)-phenethylamine,[134] *l*-menthol,[135] binaphthylamine derivatives,[136] or (*R*)-phenethyl alcohol[137] to afford monoamide or monoester derivatives of glutaric acid in a diastereoisomeric ratio of 54:46 to 4:96. *meso*-Cycloalkanedicarboxylic anhydrides (such as **162**) are esterified by the diphenylboric ester of (*R*)-2-methoxy-1-phenylethanol (**163**) in the presence of a catalytic amount of diphenylboryl triflate in a highly stereoselective manner (Eq. 28).[138] Enantiotopic differentiation is observed in the ring opening of prochiral cyclic anhydrides with methanol in the presence of a catalytic amount of cinchona alkaloids with the formation of monoesters in optical yield of up to 70%.[139,140]

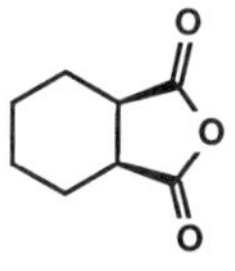

O O O + $OB(C_6H_5)_2$ C_6H_5 OCH_3

162 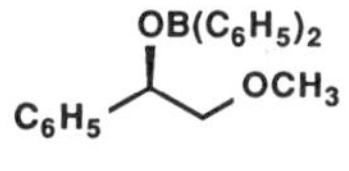 163

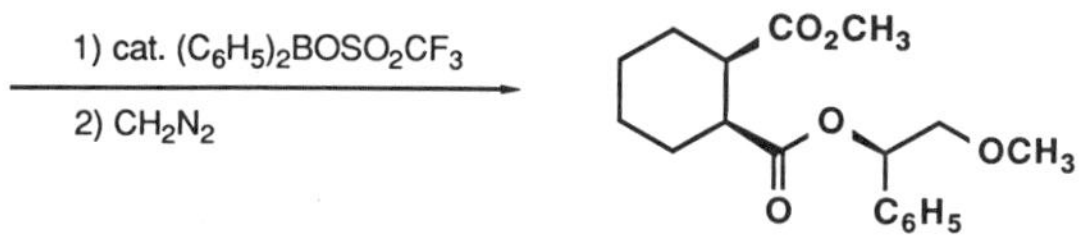

1) cat. $(C_6H_5)_2BOSO_2CF_3$ 2) CH_2N_2 → CO_2CH_3 O OCH_3 O C_6H_5 (Eq. 28)

(90%, 99% de)

The heterocycle (*R*)-4-methoxycarbonyl-1,3-thiazoline-2-thione (MCTT) is an efficient chiral leaving group for the differentiation of the two enantiotopic carboxyl groups in *meso* compounds.[141] 3-Methylglutaric acid (**164**) is coupled with two equivalents of (4*R*)-MCTT to afford the optically active diamide (**165**) in which the stereochemistry of the pro-*S* and the pro-*R* groups is different. Excellent regioselectivity is displayed in the treatment of diamide **165** with a nucleophile Nu_1. Second attack by another nucleophile Nu_2 produces enantioisomeric products **168** and **169** from compounds **166** and **167**.[141] This highly regioselective differentiation can also be applied to *meso*-2,4-dimethylglutaric acid,[142] *cis*-5-norbornene-*endo*-2,3-diacetic acid,[143] and *cis*-cyclohex-4-ene-1,2-diacetic acid.[144]

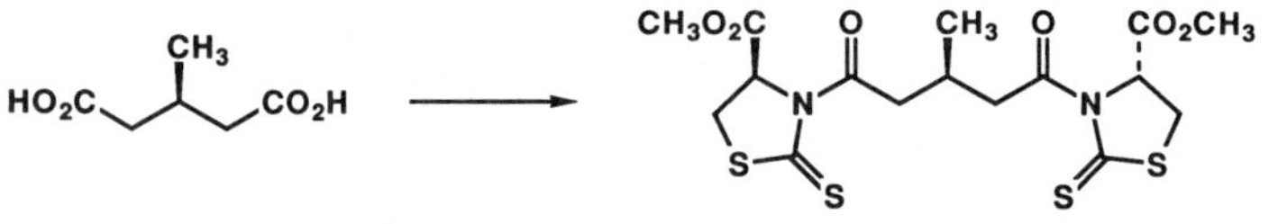

164 → 165

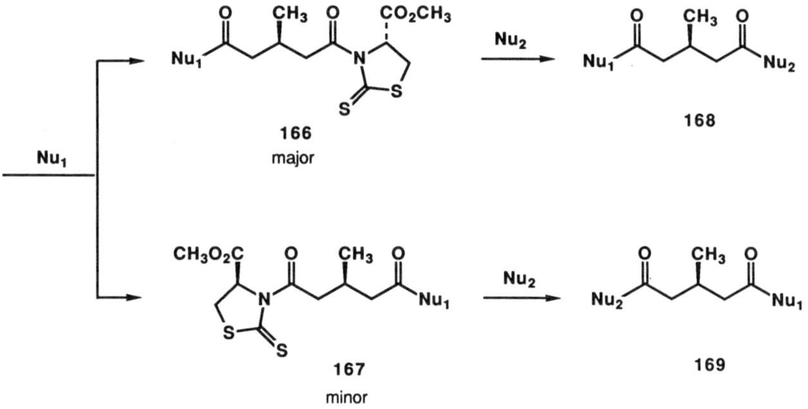

Nu1 → 166 (major), 167 (minor); Nu2 → 168, 169

Optically active γ- and δ-lactones **172** and **173** are obtained either by the catalytic asymmetric hydrogenation of prochiral cyclic anhydrides **170** and **171**[145] or by the homogeneous catalytic dehydrogenation of prochiral diols **174** and **175**[146] using a Ru(II) complex of 2,3-*O*-isopropylidene-2,3-dihydroxy-1,4-bis(diphenylphosphino)butane (DIOP) (Eq. 29).

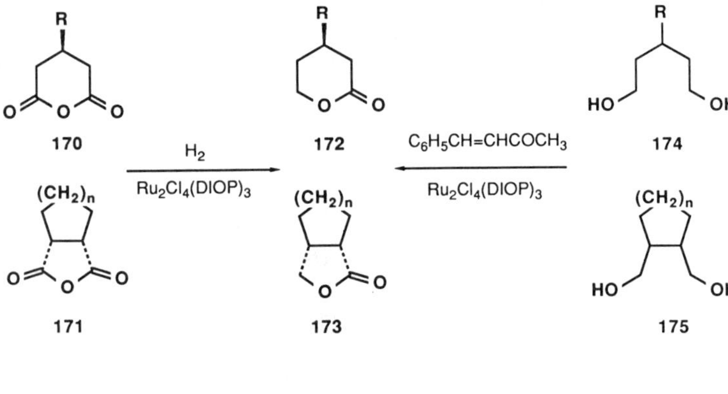

170, 171 —(H_2, $Ru_2Cl_4(DIOP)_3$)→ 172, 173 ←($C_6H_5CH{=}CHCOCH_3$, $Ru_2Cl_4(DIOP)_3$)— 174, 175 (Eq. 29)

Selective acylation of prochiral diols and triols is accomplished by application of tin chemistry. Optically active glycerol derivatives are obtained in up to 80% ee by the selective acylation of Sn(II) alkoxides by the use of a chiral diamine as a ligand (Eq. 30).[147] Cyclic Sn(IV) ethers of glycerol and *meso*-1,2-diols are acylated with optically active acid chlorides in up to 90% de (Eq. 31).[148,149]

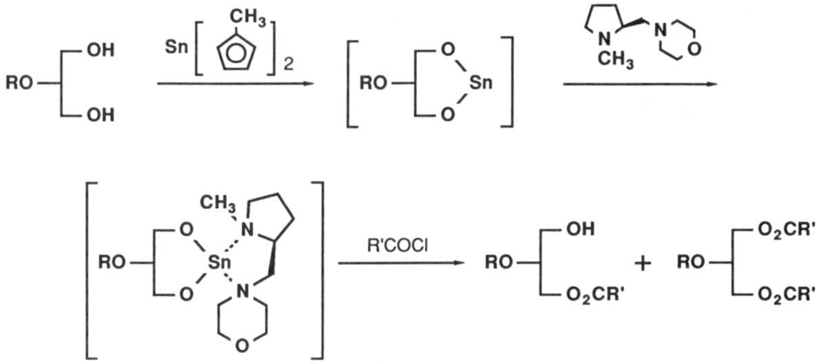

(Eq. 30)

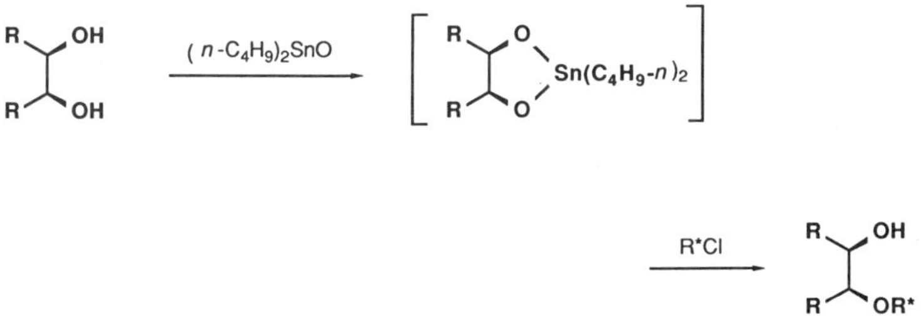

(Eq. 31)

Sharpless asymmetric epoxidation of prochiral allyl alcohol **176** produces the optically active epoxy alcohol **177** (Eq. 32).[150–152]

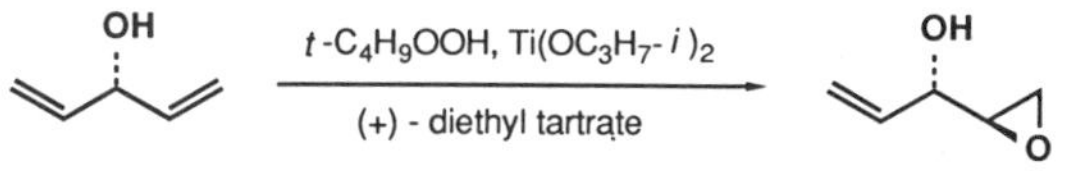

(Eq. 32)

These examples are indeed ingeniously designed and are thought to be chemical counterparts of enzymatic asymmetrization. However, it should be noted that most of the chemical asymmetrizations described above could in principle be achieved enzymatically.

EXPERIMENTAL CONDITIONS

Enzyme reactions are not used extensively by most organic chemists because of the instability of enzymes under acidic or basic conditions, at high temperature, or in organic solvents. Whereas organic reactions are often carried out in water-free organic solvents at various temperatures from $-100°$ to above 100°, enzyme reactions are performed in aqueous solution at physiological pH and temperature. PLE hydrolysis is carried out in phosphate buffer (pH 7–8). Good results are often obtained by maintaining the pH of the solution around 7 or 8 using a pH stat and stopping the reaction when the calculated amount of base is consumed. Solubility of the substrate diesters in the aqueous medium is crucial. Good results are frequently obtained by the addition of a small amount of an appropriate cosolvent, such as methanol, ethanol, *tert*-butyl alcohol, acetone, dimethyl sulfoxide, *N,N*-dimethylformamide, or tetrahydrofuran.[31,57] The reaction is carried out in multigram scales using a round-bottomed flask equipped with an efficient stirrer. Vigorous stirring with a mechanical stirrer increases the efficiency of the reaction.

Commercially available PLE [Sigma E3128, suspension in 3.2 M $(NH_4)_2SO_4$ solution, pH 8] is employed in most of the enzymatic hydrolyses. In some reactions satisfactory results are obtained by using crude preparations of the enzyme.[95] A large-scale preparation of highly purified PLE from minced pig liver has been reported.[153]

EXPERIMENTAL PROCEDURES

Preparation of Crude PLE.[153] Fresh pig liver mince (approximately 1 kg) was homogenized with 5 L of redistilled acetone (-30 to $-10°$) in an explosion-proof Waring Blendor®. The acetone was filtered through a large Büchner funnel (24 cm, Whatman No. 542 paper). The filter cake was washed with cold acetone (-30 to $-10°$) until the filtrate was colorless. Excess acetone was removed by compressing the cake with a rubber dam fitted over the Büchner funnel. The filter cake was then finely divided and dried in vacuo (10°), the residual acetone being condensed in two traps at $-80°$. Final traces of acetone were removed by evacuating the powder over concentrated sulfuric acid (10°). The thoroughly dried preparation was powdered by grinding in a Waring Blendor® and stored in air-tight jars at 4°.

Enzyme Assay.[153] Esterase activity was measured using an automatic titrator. The reaction was initiated by the addition of a 100-μL aliquot of the enzyme solution to 10 mL of 0.0125 M ethyl butyrate at pH 7.5. This pH was maintained by the automatic addition of 0.0100 N sodium hydroxide contained in a micrometer syringe buret. The reaction mixture, in a stoppered tube (2.5 × 5.7 cm), was equilibrated at 38 ± 0.1° for 15 minutes before the addition of enzyme. Under the conditions of the assay, no correction for spontaneous hydrolysis or for the absorption of carbon dioxide was necessary. Further, butyric acid ($pK_a = 4.8$) is completely ionized. One unit of PLE activity is defined as the amount of enzyme required to catalyze the hydrolysis of 1 μmole of ethyl butyrate per minute, under the assay conditions described above. Units of the commercial PLE are shown on the label of the bottle.

Methyl Hydrogen (3*S*)-3-[(Benzyloxycarbonyl)amino]glutarate (Enantioselective Hydrolysis of a Prochiral Glutarate with PLE).[69] To a solution of dimethyl 3-[(benzyloxycarbonyl)amino]glutarate (465 mg, 1.5 mmol) in 1.5 mL of acetone and 45 mL of 0.05 *M* phosphate buffer (pH 8) was added 0.22 mL (300 units) of PLE (Sigma, E3128 Type I). The mixture was incubated for 7 hours at 25° and then acidified to pH 3.0 with hydrochloric acid and extracted with dichloromethane. The organic layer was dried with sodium sulfate, concentrated in vacuo, and purified by chromatography on silica gel (eluted with diethyl ether) to afford 410 mg (93%) of monoester; $[\alpha]_D^{25}$ +0.69° (*c* 7.50, $CHCl_3$). Optically pure monoester was obtained by recrystallization (dichloromethane–*n*-hexane); mp 97.0–97.5°; $[\alpha]_D^{25}$ + 0.72° (*c* 7.5, $CHCl_3$); IR (KBr): 3330, 1740, 1725, 1705 cm^{-1}; ^1H NMR ($CDCl_3$) δ: 2.70 (d, J = 5.8 Hz, 4*H*), 3.64 (s, 3*H*), 4.37 (m, 1*H*), 5.08 (s, 2*H*), 5.66 (br d, 1*H*), 7.35 (s, 5*H*), 8.30 (br s, 1*H*); mass spectrum, m/e 295 (M^+).

Methyl Hydrogen (1*S*,2*R*)-Cyclohex-4-ene-1,2-dicarboxylate (Enantioselective Hydrolysis of a *meso*-Diester with PLE).[24] To 1.98 g (10 mmol) of dimethyl *cis*-cyclohex-4-ene-1,2-dicarboxylate suspended in 50 mL of 0.1 M phosphate buffer (pH 8) was added 500 units of PLE with vigorous stirring. The pH was kept within the range 7.5–8.0 by addition of 1 N sodium hydroxide. After consumption of 1 mol-equivalent of base the mixture was homogeneous. The pH was adjusted to 9, and the aqueous phase was extracted with diethyl ether. The organic layer was washed with water and the combined aqueous solutions were acidified to pH 2.5. The solution was again extracted with diethyl ether, dried, and evaporated in vacuo to yield 95% of the monoester; $[\alpha]_D^{20}$ +14.6° (*c* 0.2, C_2H_5OH); ^1H NMR ($CDCl_3$) δ: 2.0–2.8 (m, 4*H*), 2.8–3.2 (m, 2*H*), 3.7 (s, 3*H*), 5.65 (br s, 2*H*), 10.1 (br s, 1*H*).

(1*S*,4*R*,5*R*,6*S*)-5,6-Dimethylmethylenedioxy-3-methoxycarbonyl-7-oxabicyclo[2.2.1]hept-2-ene-2-carboxylic Acid (Enantioselective Hydrolysis of a *meso*-Diester with PLE).[77] To a solution of dimethyl *exo*-5,6-dimethylmethylenedioxy-7-oxabicyclo[2.2.1]hept-2-ene-2,3-dicarboxylate (3 g, 10.6 mmol) in 30 mL of acetone and 300 mL of 0.1 M phosphate buffer (pH 8) was added 3 mL (4140 units) of PLE. The mixture was incubated for 4 hours at 32°, and then acidified to pH 4 with 2 M hydrochloric acid and extracted with ethyl acetate. The organic layer was washed with water, dried over sodium sulfate, and concentrated in vacuo to afford 2.73 g (96%) of monoester as white solid; $[\alpha]_D^{20}$ −37.1° (*c* 1.0, $CHCl_3$). It was dissolved in hot carbon tetrachloride and the solution was allowed to stand at room temperature. A small amount of solid material of very low optical purity separated and was removed by filtration. The filtrate was concentrated and the residue was recrystallized two times from carbon tetrachloride–*n*-hexane to give optically pure monoester; mp 115.5–117.5°; $[\alpha]_D^{20}$ −49° (*c* 1.0, $CHCl_3$); IR (KBr) 1725, 1650, 1622 cm^{-1}; ^1H NMR ($CDCl_3$) δ: 1.37 (s, 3*H*), 1.52 (s, 3*H*), 4.00 (s, 3*H*), 4.54 (s, 2*H*), 5.20 (s, 1*H*), 5.24 (s, 1*H*).

(1*S*,4*R*,5*R*,6*S*)-5,6-Epoxy-3-methoxycarbonyl-7-oxabicyclo[2.2.1]hept-2-ene-2-carboxylic Acid (Enantioselective Hydrolysis of a *meso*-Diester with PLE).[77] To an emulsion of 51.1 mg (0.23 mmol) of dimethyl 5,6-*exo*-epoxy-7-oxabicyclo[2.2.1]hept-2-ene-2,3-dicarboxylate in 10 mL of 0.1 M phosphate buffer (pH 8) was added 50 μL (65 units) of PLE. The mixture was incubated for 7 hours at 20°, and then acidified with 2 M hydrochloric acid and extracted with ethyl acetate several times. Usual workup afforded 48.1 mg (100%) of the monoester as a white solid; mp 120–122°; $[\alpha]_D^{20}$ −23° (*c* 0.52, $CHCl_3$); IR (KBr) 3350, 1725, 1670, 1615 cm^{-1}; ^1H NMR ($CDCl_3$) δ: 3.80 (s, 2*H*), 4.03 (s, 3*H*), 5.23 (s, 1*H*), 5.42 (s, 1*H*).

(1*S*,4*R*,5*R*,6*S*) - 5,6 - Dimethylmethylenedioxy - 3 - methoxycarbonylbicyclo[2.2.1]hept-2-ene-2-carboxylic Acid (Enantioselective Hydrolysis of a *meso*-Diester with PLE).[44] To a solution of dimethyl *exo*-5,6-dimethylmethylenedioxybicyclo[2.2.1]hept-2-ene-2,3-dicarboxylate (3 g, 10.6 mmol) in 30 mL of acetone and 300 mL of 0.1 M phosphate buffer (pH 8) was added 3 mL (4140 units) of PLE. The mixture was incubated for 5 hours at 30–32° and then acidified to pH 2 with 2 M hydrochloric acid and extracted with dichloromethane. The organic layer was dried over sodium sulfate and concentrated in vacuo to afford 2.84 g (99.6%) of monoester as a white solid; mp 115–118°; $[\alpha]_D^{25}$ −23.8° (*c* 1.17, $CHCl_3$); IR (KBr) 3425, 2990, 2925, 2670, 1725, 1640, 1440, 1385, 1380 cm^{-1}; ^1H NMR ($CDCl_3$) δ: 1.35 and 1.49 (2s, 6*H*), 1.97 (m, 2*H*), 3.40 (m, 2*H*), 3.94 (s, 3*H*), 4.38 (d, *J* = 1 Hz, 2*H*).

(1*S*,4*R*)-4-Hydroxy-2-cyclopentenyl Acetate (Enantioselective Hydrolysis of a *meso*-Diacetate with PLE).[39] A suspension of *cis*-3,5-diacetoxycyclopentene (12.9 g, 70 mmol) in 0.1 M phosphate buffer (140 mL, pH 7) was treated with 10 mg (1000 units) of PLE (Boehringer) at 32°. The pH was kept constant by continuous addition of 1 N sodium hydroxide. After consumption of 74 mL (1.05 eq) of sodium hydroxide the mixture was extracted with diethyl ether. Workup and fractional distillation yielded 8.6 g (86%) of the monoester; bp 82° (0.2 mm); $[\alpha]_D^{20}$ −49.7° (*c* 0.86, $CHCl_3$). One crystallization from diethyl ether–petroleum ether (2 : 1) produced crystalline monoacetate; mp 40–40.5°; $[\alpha]_D^{20}$ −60.4° (*c* 0.27, $CHCl_3$).

Diethyl (3*R*,4*S*)-3,4-Epoxyadipate and (3*R*,4*R*)-3,4-Epoxy-5-methoxycarbonylpentanoic Acid (Kinetic Resolution of a Racemic Diester with PLE).[50] To 1.88 g (10 mmol) of dimethyl (±)-3,4-epoxyadipate suspended in 50 mL of 0.1 M phosphate buffer of pH 7.0 was added 500 units of PLE with vigorous stirring. The pH was kept constant at 7.0 by adding 1 N sodium hydroxide. After consumption of 0.5 equivalent of base, the rate decreased dramatically, and the aqueous solution was extracted with ethyl acetate. Drying and evaporation of the solvent yielded ca. 40% (80% of the theoretical amount) of (+)-diester; (97%, GLC), $[\alpha]_D^{rt}$ +28.4° (*c* 1.27, C_2H_5OH). The aqueous layer was acidified with hydrochloric acid to pH 2.5, and the product was extracted with a large amount of ethyl acetate to afford 40% (80% of the theoretical amount) of (−)-monoester as a colorless oil; IR (film): 3500–2800, 3010, 2960, 1740, 1715 (sh), 1440, 1175, 960 cm^{-1}; ^1H NMR ($CDCl_3$) δ: 2.55 (2d, $2H \times 2$), 3.15 (br t, J = 5 Hz, $2H$), 3.7 (s, $3H$), 10.6 (br s, $1H$). Esterification of the monoester with diazomethane in diethyl ether yielded (−)-diester (97%, GLC), $[\alpha]_D^{rt}$ −27.0° (c 1.48, C_2H_5OH).

TABULAR SURVEY

The following tabular survey is an attempt to cover all the literature to the end of 1986 and some to the middle of 1987. The arrangement of the tables is based on the structural characteristics of the substrates.

TABLE I. Enantioselective Hydrolysis of Prochiral Glutarates

CH_3O_2C–CH_2–$C(R^1)(R^2)$–CH_2–CO_2CH_3 $\xrightarrow{\text{PLE}}$ R^3O_2C–CH_2–$C(R^1)(R^2)$–CH_2–CO_2R^4

Substrate		Product				
R^1	R^2	R^3	R^4	Yield (%)	ee (%)	Refs.
OH	CH_3	CH_3	H	62	99	14
CH_3	H	H	CH_3	86	90	73
OH	H	CH_3	H	95	12	24
CH_3	H	H	CH_3	95	79	17
C_2H_5	H	H	CH_3	67	50	17
n-C_3H_7	H	H	CH_3	78	25	17
i-C_3H_7	H	CH_3	H	98	38	17
C_6H_{11}	H	H	CH_3	95	17	17
C_6H_5	H	CH_3	H	98	42	17
$C_6H_5CH_2$	H	CH_3	H	95	54	17
NH_2	H	H	CH_3	94	41	15, 69
CH_3CONH	H	H	CH_3	81	93	16
C_2H_5CONH	H	H	CH_3	50	6	16
n-C_3H_7CONH	H	H	CH_3	52	15	16
n-C_4H_9CONH	H	CH_3	H	48	2	16
i-C_3H_7CONH	H	CH_3	H	55	54	16
t-C_4H_9CONH	H	CH_3	H	50	93	16
$C_6H_{11}CONH$	H	CH_3	H	52	79	16
CH_2=CHCONH	H	H	CH_3	50	8	16
CH_3CH=CHCONH	H	CH_3	H	60	100	16
C_6H_5CONH	H	CH_3	H	59	72	16
CH_3CO_2NH	H	CH_3	H	60	20	16
$C_2H_5CO_2NH$	H	CH_3	H	70	40	16
t-$C_4H_9CO_2NH$	H	CH_3	H	93	53	16
$C_6H_5CH_2CO_2NH$	H	CH_3	H	93	93	15, 69
$C_6H_5CH_2NH$	H	CH_3	H	58	33	16
(E)-$HOCH_2CH$=$CHCH_2$	H	CH_3	H	99	19	16
(E)-$THPOCH_2CH$=$CHCH_2$[a]	H	CH_3	H	95	74	16
(E)-$C_6H_5CH_2CH$=$CHCH_2$	H	CH_3	H	97	88	16

[a] THP = tetrahydropyranyl.

TABLE II. ENANTIOSELECTIVE HYDROLYSIS OF PROCHIRAL MALONATES

$R^1R^2C(CO_2R^3)_2 \xrightarrow{\text{PLE}} R^1R^2C(CO_2R^4)(CO_2R^5)$

Substrate			Product				
R^1	R^2	R^3	R^4	R^5	Yield (%)	ee (%)	Refs.
OH	CH_3	CH_3	CH_3	H	82.2	46	24
CH_3	C_2H_5	CH_3	H	CH_3	—	73	18
CH_3	n-C_3H_7	CH_3	H	CH_3	—	52	18
CH_3	n-C_4H_9	CH_3	H	CH_3	—	58	18
					—	50	154
CH_3	n-C_5H_{11}	CH_3	CH_3	H	—	46	18
CH_3	n-C_6H_{13}	CH_3	CH_3	H	—	87	18
CH_3	n-C_7H_{15}	CH_3	CH_3	H	—	88	18
CH_3	C_2H_5	C_2H_5	H	C_2H_5	—	15	18
					—	20	154
CH_3	n-C_3H_7	C_2H_5	H	C_2H_5	—	10	18
					—	10	154
CH_3	n-C_4H_9	C_2H_5	H	C_2H_5	—	25	18
					—	38	154
CH_3	n-C_5H_{11}	C_2H_5	C_2H_5	H	—	10	18
CH_3	n-C_8H_{17}	C_2H_5	H	C_2H_5	—	5	18
CH_3	C_6H_5	C_2H_5	H	C_2H_5	—	86	154
C_2H_5	C_6H_5	CH_3	H	CH_3	—	84	18
CH_3	$C_6H_5CH_2$	CH_3	CH_3	H	85–100	45	92
CH_3	4-$CH_3OC_6H_4CH_2$	CH_3	CH_3	H	85–100	82	92
CH_3	$3,4$-$(CH_3O)_2C_6H_3CH_2$	CH_3	CH_3	H	85–100	93	92
C_2H_5	$3,4$-$(CH_3O)_2C_6H_3CH_2$	C_2H_5	C_2H_5	H	86	59	33

TABLE III. ENANTIOSELECTIVE HYDROLYSIS OF DIACYLATED PROCHIRAL DIOLS

Substrate	Product	Yield (%)	ee (%)	Refs.
$CH_3CO_2CH_2CH(OCH_2C_6H_5)CH_2O_2CCH_3$	$CH_3CO_2CH_2CH(OCH_2C_6H_5)CH_2OH$	43	29	19
		54	39	20

TABLE IV. ENANTIOSELECTIVE HYDROLYSIS OF ACYCLIC *meso*-DIESTERS

Substrate	Product	Yield (%)	ee (%)	Refs.
CH_3O_2C–CH(CH₃)–CH₂–CH(CH₃)–CO_2CH_3	HO_2C–CH(CH₃)–CH₂–CH(CH₃)–CO_2CH_3	85	64	23
		98	60	24
		—	60	31
CH_3O_2C–CH(CH₃)–CH(OH)–CH(CH₃)–CO_2CH_3	CH_3O_2C–CH(CH₃)–CH(OH)–CH(CH₃)–CO_2H	95	98	24
CH_3O_2C–CH(CH₃)–CH(CH₃)–CO_2CH_3	HO_2C–CH(CH₃)–CH(CH₃)–CO_2CH_3	94	18	24
CH_3O_2C–CH(OH)–CH(OH)–CO_2CH_3	CH_3O_2C–CH(OH)–CH(OH)–CO_2H	92	48	24

TABLE V. ENANTIOSELECTIVE HYDROLYSIS OF MONOCYCLIC *meso*-DIESTERS

Substrate	Product	Yield (%)	ee (%)	Refs.
cyclopropane-1,2-di(CO_2CH_3)	cyclopropane HO_2C, CO_2CH_3	92	100	24
		—	97	25, 157
		90	94	26, 157[a]
oxirane-2,3-di(CO_2CH_3)	oxirane CH_3O_2C, CO_2H	53	20	155
		69	31	157
3,3-dimethylcyclopropane-1,2-di(CO_2CH_3)	3,3-dimethylcyclopropane HO_2C, CO_2CH_3	23	43	24
		42	80	26
cyclobutane-1,2-di(CO_2CH_3)	cyclobutane HO_2C, CO_2CH_3	99	90	24
		—	97	25, 157
		98	94	26
cyclopentane-1,2-di(CO_2CH_3)	cyclopentane CH_3O_2C, CO_2H	80	9	24
		—	17	25, 157
1,3-bis($C_6H_5CH_2$)-2-oxoimidazolidine-4,5-di(CO_2CH_3)	1,3-bis($C_6H_5CH_2$)-2-oxoimidazolidine HO_2C, CO_2CH_3	71	38	33

TABLE V. ENANTIOSELECTIVE HYDROLYSIS OF MONOCYCLIC *meso*-DIESTERS (*Continued*)

Substrate	Product	Yield (%)	ee (%)	Refs.
1,3-Dibenzyl-2-oxoimidazolidine-4,5-dicarboxylate, di-n-propyl ester ($C_6H_5CH_2N$, $NCH_2C_6H_5$; n-$C_3H_7O_2C$, $CO_2C_3H_7$-n)	Monoacid (HO_2C, $CO_2C_3H_7$-n)	85	75	33
1,3-Dibenzyl-2-oxoimidazolidine-4,5-dicarboxylate, diisopropyl ester (i-$C_3H_7O_2C$, $CO_2C_3H_7$-i)	No hydrolysis			33
cis-Cyclohexane-1,2-di(CO_2CH_3)	CO_2H / CO_2CH_3	98	78	24
		—	97	25, 157
		75	80	26
		—	97	47
		97	80	30
cis-Cyclohexane-1,2-di($CO_2C_2H_5$)	CO_2H / $CO_2C_2H_5$	—	0	47
cis-Cyclohex-4-ene-1,2-di(CO_2CH_3)	CO_2H / CO_2CH_3	95	85	24
		94	94	26
		98	96	28
		99	99	29, 30
cis-Cyclohex-4-ene-1,2-di($CO_2C_2H_5$)	CO_2H / $CO_2C_2H_5$	67	27	16
cis-Cyclohex-4-ene-1,2-di($CO_2C_3H_7$-n)	CO_2H / $CO_2C_3H_7$-n	68	25	16
cis-Cyclohex-4-ene-1,2-di($CO_2C_3H_7$-i)	CO_2H / $CO_2C_3H_7$-i	5	2	16
cis-Cyclohex-4-ene-1,2-di($CO_2C_4H_9$-n)	CO_2H / $CO_2C_4H_9$-n	18	13	16
cis-Cyclopentane-1,3-di(CO_2CH_3) (CH_3O_2C, CO_2CH_3)	CH_3O_2C / CO_2H	82	34	35
cis-Tetrahydrofuran-2,5-di(CO_2CH_3) (CH_3O_2C, O, CO_2CH_3)	HO_2C / CO_2CH_3	98	42	35
cis-Tetrahydrothiophene-2,5-di(CO_2CH_3) (CH_3O_2C, S, CO_2CH_3)	HO_2C / CO_2CH_3	83	46	35
cis-N-Benzylpyrrolidine-2,5-di(CO_2CH_3) ($CH_2C_6H_5$ on N)	HO_2C / CO_2CH_3	85	80	36
		39	100	37
cis-Cyclohex-4-ene-1,2-di($CH_2CO_2CH_3$) (H, H)	CH_2CO_2H / $CH_2CO_2CH_3$	91	76	22

TABLE V. ENANTIOSELECTIVE HYDROLYSIS OF MONOCYCLIC *meso*-DIESTERS (*Continued*)

Substrate	Product	Yield (%)	ee (%)	Refs.
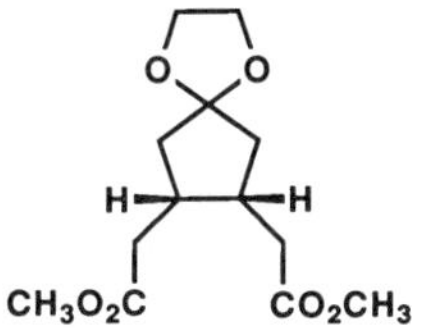 O O H H CH_3O_2C CO_2CH_3	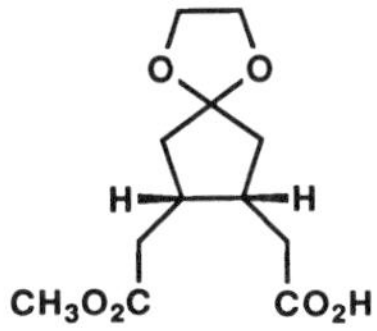 O O H H CH_3O_2C CO_2H	99	90	22
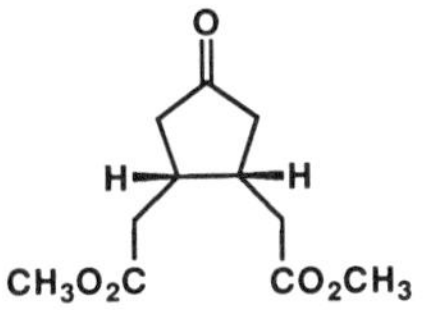 O H H CH_3O_2C CO_2CH_3	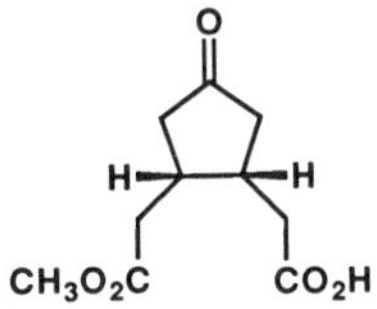 O H H CH_3O_2C CO_2H	85	88	22

[a] Although the opposite enantiomeric configuration is assigned for the monoester in ref. 26, the correctness of the stereochemical assignment depicted herein is reconfirmed in ref. 157.

TABLE VI. ENANTIOSELECTIVE HYDROLYSIS OF BICYCLIC *meso*-DIESTERS

Substrate	Product	Yield (%)	ee (%)	Refs.
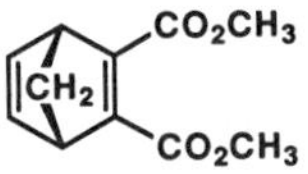 CH_2 CO_2CH_3 CO_2CH_3	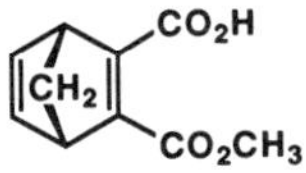 CH_2 CO_2H CO_2CH_3	—	41	42
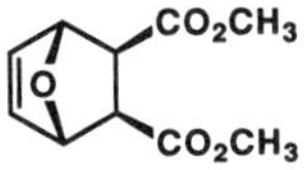 O CO_2CH_3 CO_2CH_3	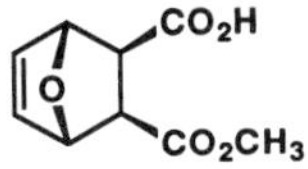 O CO_2H CO_2CH_3	— 86	82 75	31 41
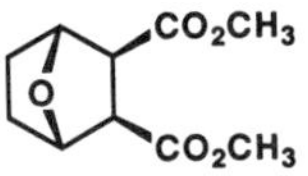 O CO_2CH_3 CO_2CH_3	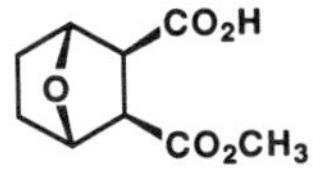 O CO_2H CO_2CH_3	82	98	41
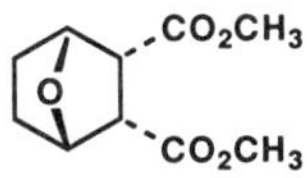 O CO_2CH_3 CO_2CH_3	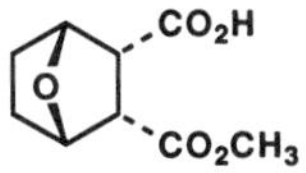 O CO_2H CO_2CH_3	87	64	41
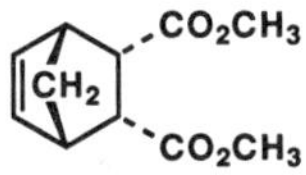 CH_2 CO_2CH_3 CO_2CH_3		No hydrolysis		41
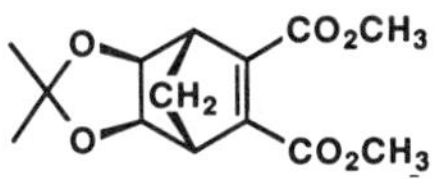 O O CH_2 CO_2CH_3 CO_2CH_3	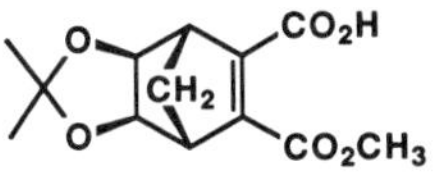 O O CH_2 CO_2H CO_2CH_3	Quant	80	44
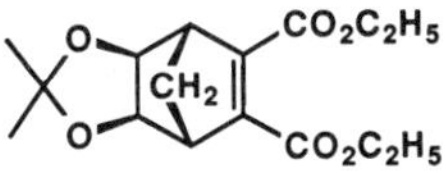 O O CH_2 $CO_2C_2H_5$ $CO_2C_2H_5$	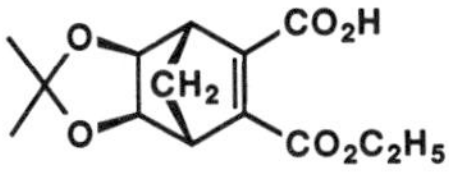 O O CH_2 CO_2H $CO_2C_2H_5$	37	100	16
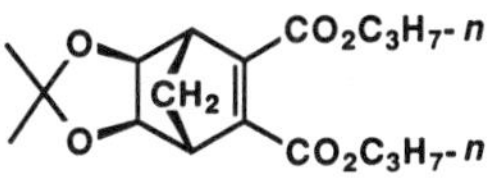 O O CH_2 $CO_2C_3H_7$-*n* $CO_2C_3H_7$-*n*	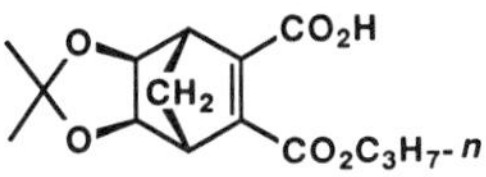 O O CH_2 CO_2H $CO_2C_3H_7$-*n*	1.5	45	16
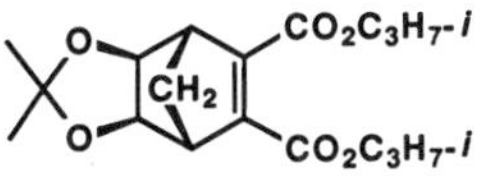 O O CH_2 $CO_2C_3H_7$-*i* $CO_2C_3H_7$-*i*	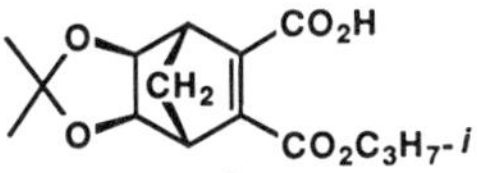 O O CH_2 CO_2H $CO_2C_3H_7$-*i*	22	39	16

TABLE VI. ENANTIOSELECTIVE HYDROLYSIS OF BICYCLIC *meso*-DIESTERS (*Continued*)

Substrate	Product	Yield (%)	ee (%)	Refs.
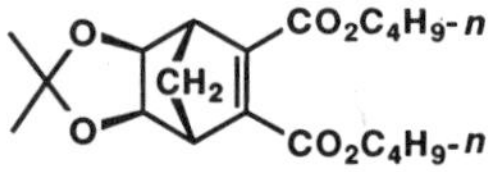 $CO_2C_4H_9$-*n*, CH_2, $CO_2C_4H_9$-*n*	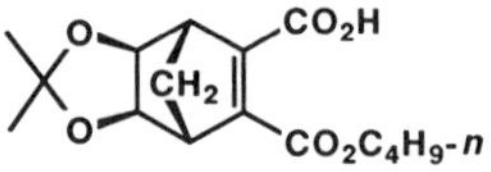 CO_2H, CH_2, $CO_2C_4H_9$-*n*	4.4	73	16
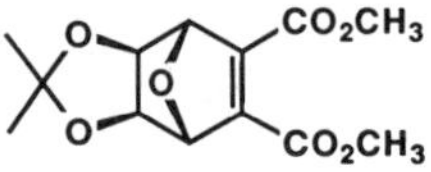 CO_2CH_3, O, CO_2CH_3	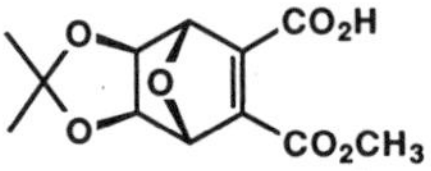 CO_2H, O, CO_2CH_3	96	77	43
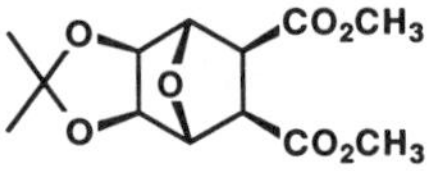 CO_2CH_3, O, CO_2CH_3	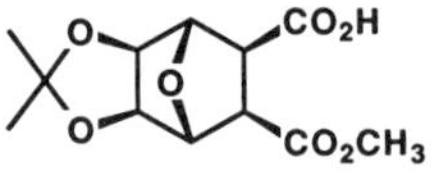 CO_2H, O, CO_2CH_3	10	—	42
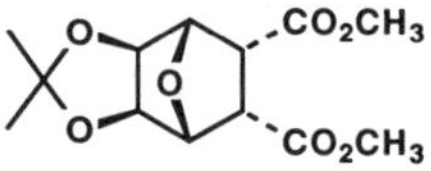 CO_2CH_3, O, CO_2CH_3		No reaction		42
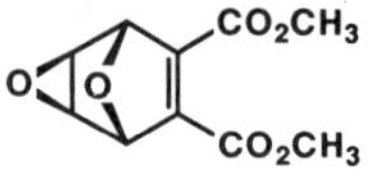 CO_2CH_3, O, CO_2CH_3	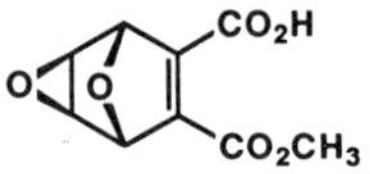 CO_2H, O, CO_2CH_3	Quant	77	43

TABLE VII. ENANTIOSELECTIVE HYDROLYSIS OF DIACYLATED *meso*-DIOLS

Substrate	Product	Yield (%)	ee (%)	Refs.
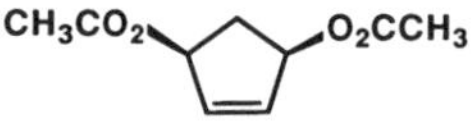 CH_3CO_2, O_2CCH_3	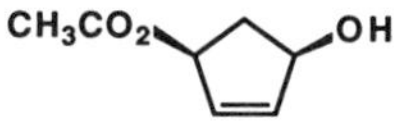 CH_3CO_2, OH	— 86	80.3 86[a]	38 39
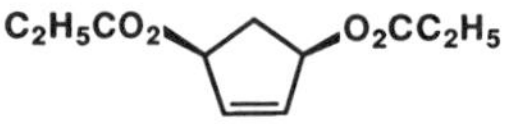 $C_2H_5CO_2$, $O_2CC_2H_5$	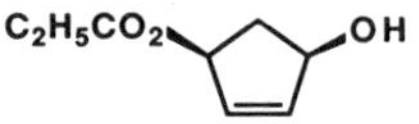 $C_2H_5CO_2$, OH	52	66	39
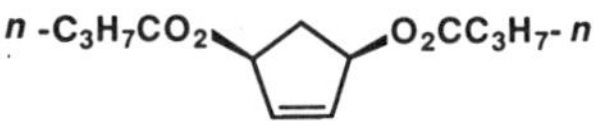 *n*-$C_3H_7CO_2$, $O_2CC_3H_7$-*n*	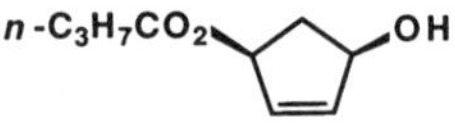 *n*-$C_3H_7CO_2$, OH	Trace	30	39
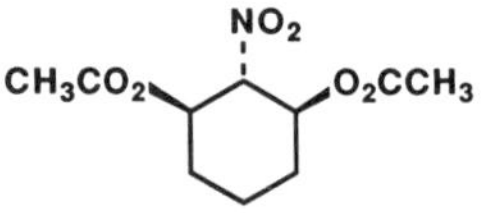 NO_2, CH_3CO_2, O_2CCH_3	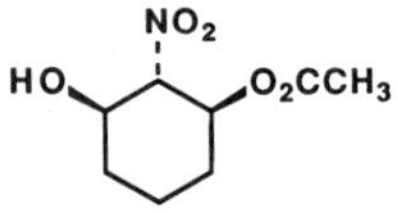 NO_2, HO, O_2CCH_3	89	98	40
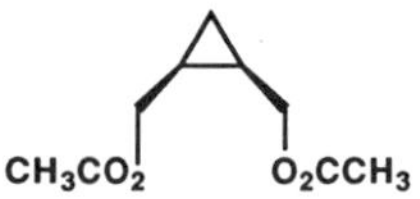 CH_3CO_2, O_2CCH_3	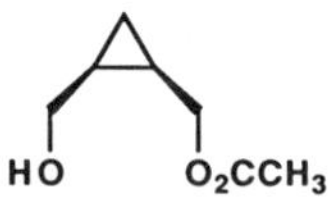 HO, O_2CCH_3	54	44	27[b]
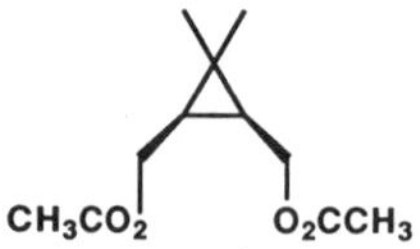 CH_3CO_2, O_2CCH_3	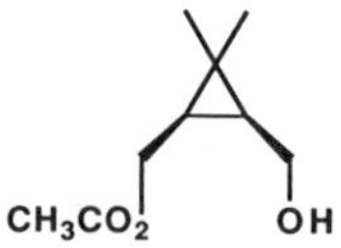 CH_3CO_2, OH	70	33	157
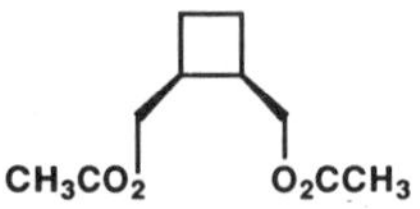 CH_3CO_2, O_2CCH_3	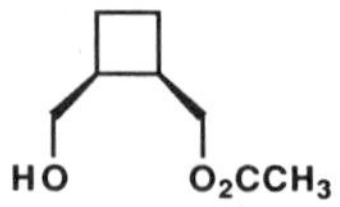 HO, O_2CCH_3	62 44	0 4	27[b] 157

TABLE VII. ENANTIOSELECTIVE HYDROLYSIS OF DIACYLATED *meso*-DIOLS (*Continued*)

Substrate	Product	Yield (%)	ee (%)	Refs.
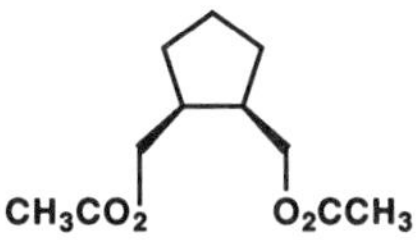 CH_3CO_2 O_2CCH_3	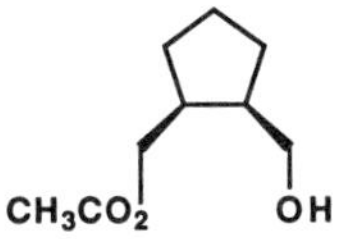 CH_3CO_2 OH	40	8	27[b]
O_2CCH_3 O_2CCH_3	O_2CCH_3 OH	—	12	31
		31	4	27[b]
O_2CCH_3 O_2CCH_3	OH O_2CCH_3	78	96	31
		43	40	27
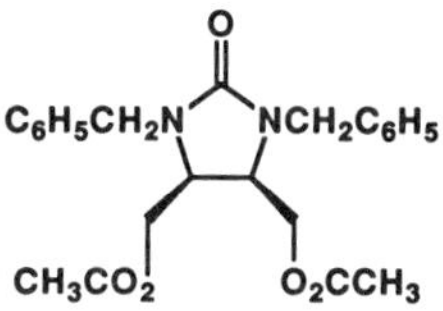 O $C_6H_5CH_2N$ $NCH_2C_6H_5$ CH_3CO_2 O_2CCH_3	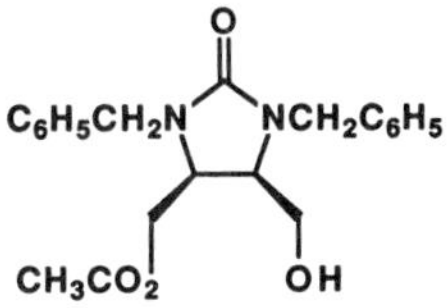 O $C_6H_5CH_2N$ $NCH_2C_6H_5$ CH_3CO_2 OH	70	90	32
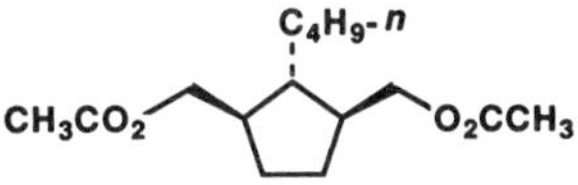 C_4H_9-*n* CH_3CO_2 O_2CCH_3	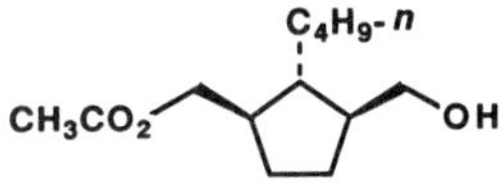 C_4H_9-*n* CH_3CO_2 OH	4	23	156
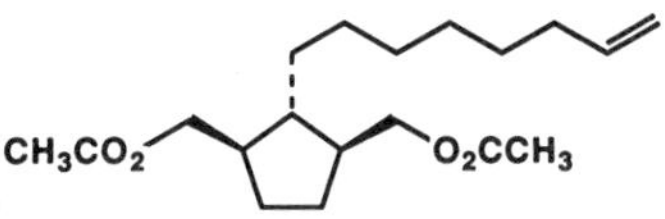 CH_3CO_2 O_2CCH_3	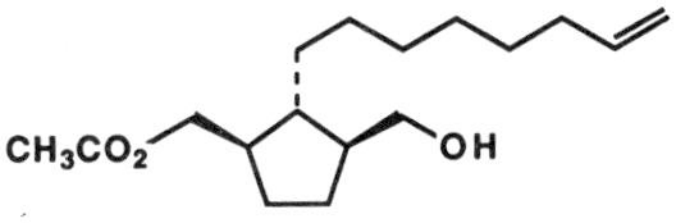 CH_3CO_2 OH	3	59	156

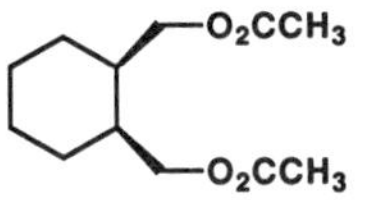

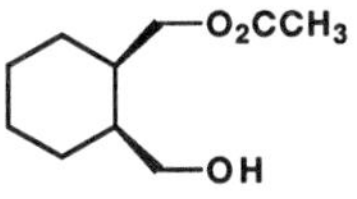

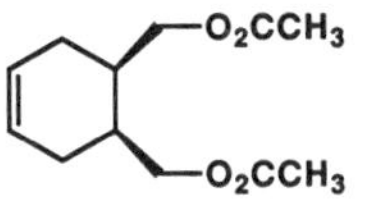

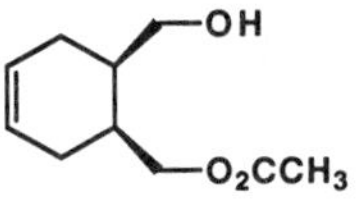

[a] The enantiomeric excess is evaluated after one recrystallization.
[b] The assignment of the absolute stereochemistry of the product is based on ref. 98.

TABLE VIII. KINETIC RESOLUTION OF RACEMIC 3-HYDROXY-3-METHYLALKANOIC ESTERS

Substrate	Fraction of Racemate Hydrolyzed	Recovered Ester			Recovered Acid				
		Configuration	Recovery (%)	ee (%)	Configuration	Recovery (%)	ee (%)	E[a]	Refs.
$C_2H_5COH(CH_3)CH_2CO_2CH_3$	0.88	(*S*)	12	98	(*R*)	51	13	4	45
n-$C_6H_{13}COH(CH_3)CH_2CO_2CH_3$	0.36	(*S*)	41	26	(*R*)	22	47	4	45
$(CH_3O)_2CHCH_2COH(CH_3)CH_2CO_2CH_3$	0.67	(*S*)	26	94	(*R*)	45	47	10	45
$(CH_3O)_2CHCH_2COH(CH_3)CH_2CO_2C_2H_5$	0.75	(*S*)	22	94	(*R*)	[b]	32	6	45
$C_6H_5CH_2OCH_2CH_2COH(CH_3)CH_2CO_2CH_3$	0.40	(*S*)	31	22	(*R*)	32	33	2	45
$C_6H_5CH_2OCH_2CH_2COH(CH_3)CH_2CO_2C_2H_5$	0.55	(*S*)	40	66	(*R*)	44	55	7	45
$CH_2{=}CHCH_2COH(CH_3)CH_2CO_2CH_3$	0.84	(*S*)	11	94	—	75	—	4	45
	0.51	(*S*)	44	49	(*R*)	43	48	5	45
	0.51	(*S*)	40	48	(*R*)	45	47	4	45
	0.50	(*S*)	46	44	—	47	44	4	45
$CH_2{=}CHCH_2COH(CH_3)CH_2CO_2C_2H_5$	0.50	(*S*)	51	50	—	—	—	5	45
	0.49	(*S*)	44	37	(*R*)	45	38	3	45

[a] The enantiomeric ratio (E) was calculated as described previously.[158]

[b] The methyl ester was recovered in low yield by methylation with dimethyl sulfate in aqueous solution.

TABLE IX. KINETIC RESOLUTION OF RACEMIC CYCLOPROPANECARBOXYLATES

Substrate	Conversion (%)	Product	Yield (%)	ee (%)	E	Refs.
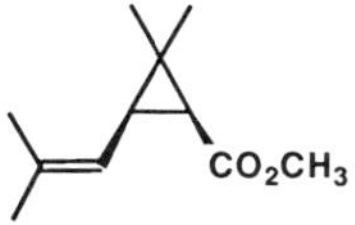 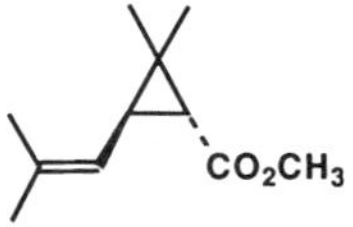 (±)- CO_2CH_3 + (±)- CO_2CH_3	50	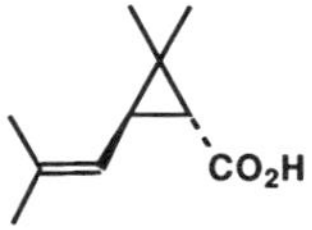 CO_2H	90	40	2.3	46
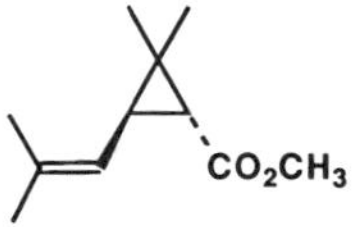 (±)- CO_2CH_3	50	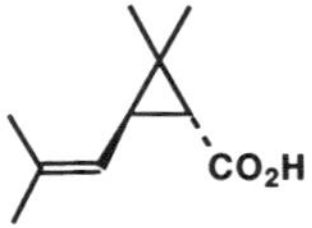 CO_2H	85	46	2.7	46
		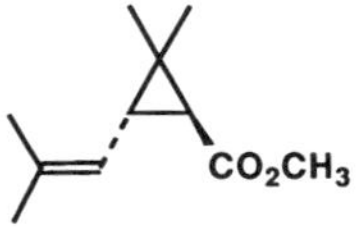 CO_2CH_3	80	40	2.3	46
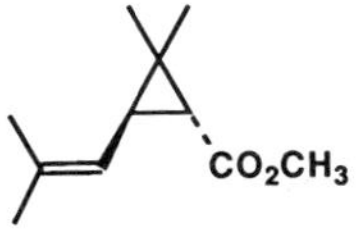 (±)- CO_2CH_3	30	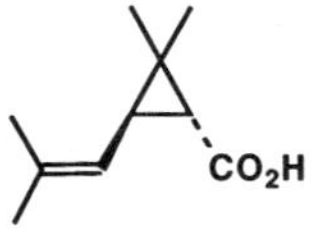 CO_2H	75	60	4.0	46

TABLE IX. KINETIC RESOLUTION OF RACEMIC CYCLOPROPANECARBOXYLATES (*Continued*)

Substrate	Conversion (%)	Product	Yield (%)	ee (%)	E	Refs.
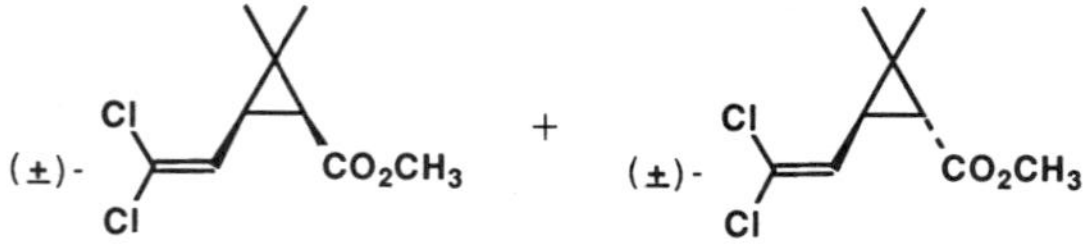 (±)- Cl, Cl; CO_2CH_3 + (±)- Cl, Cl; CO_2CH_3	50	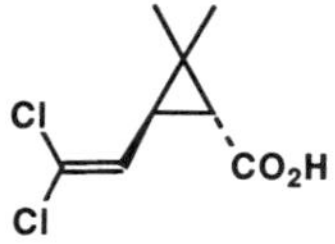 Cl, Cl; CO_2H	90	80	9.0	46
 (±)- CH_3O_2C CO_2CH_3 + (±)- CH_3O_2C CO_2CH_3	50	 HO_2C CO_2CH_3	85	60	4.0	46
		 HO_2C CO_2CH_3	85	60	4.0	46
 (±)- CH_3O_2C CO_2CH_3	50	 CH_3O_2C CO_2CH_3	90	50	3.0	46

TABLE X. KINETIC RESOLUTION OF *trans*-4-OXOCYCLOPENTANE-1,2-DICARBOXYLATE

Substrate	Product	Yield (%)	ee (%)	E	Ref.
(±)- 4-oxocyclopentane, CH_3O_2C / CO_2CH_3	4-oxocyclopentane, CH_3O_2C / CO_2H	44	95	39	48
	4-oxocyclopentane, CH_3O_2C / CO_2CH_3	45	95	39	

TABLE XI. KINETIC RESOLUTION OF *trans*-1,2-CYCLOHEXANEDICARBOXYLIC ESTERS

Substrate	Conversion (%)	ee (%) of Products		E	Refs.
(±)- cyclohexane, CO_2R / CO_2R		cyclohexane, CO_2R / CO_2R	cyclohexane, CO_2H / CO_2R		
R = CH_3	64	53	33	2.0	47
R = C_2H_5	67	83	45	2.6	47
R = n-C_3H_7	46	44	40	2.3	47

REFERENCES

[1] S. Hanessian, *Total Synthesis of Natural Products: The Chiron Approach*, Pergamon, Oxford, 1983.

[2] J. B. Jones, O. Perlman, and C. J. Sih, *Application of Biochemical Systems in Organic Chemistry*, Wiley-Interscience, New York, 1976.

[3] J. B. Jones, in *Enzymic and Non-enzymic Catalysis*, P. Dunnil, A. Wiseman, and N. Blakebrough, Eds., Ellis Horwood, New York, 1980, pp. 54–83.

[4] A. R. Battersby, *Chem. Br.*, **1984**, 611.

[5] C. J. Sih and C.-S. Chen, *Angew. Chem., Int. Ed. Engl.*, **23**, 570 (1984).

[6] G. M. Whitesides and C.-H. Wong, *Aldrichimica Acta*, **16**, 27 (1984).

[7] G. M. Whitesides and C.-H. Wong, *Angew. Chem., Int. Ed. Engl.* **24**, 617 (1985).

[8] J. B. Jones, *Tetrahedron*, **42**, 3351 (1986).

[9] J. B. Jones, in *Asymmetric Synthesis*, J. D. Morrison, Ed., Academic, New York, 1986, pp. 309–344.

[10] S. Butt and S. B. Roberts, *Chem. Br.*, **1987**, 127.

[11] K. Krisch, in *The Enzymes*, P. D. Boyer, Ed., Academic, New York, 1971, pp. 43–69.

[12] For the definition and usage of stereochemical terms, see J. Rétey and J. R. Robinson, *Stereospecificity in Organic Chemistry and Enzymology*, Verlag Chemie, Weinheim, 1982.

[13] S. G. Cohen and E. Khedouri, *J. Am. Chem. Soc.*, **83**, 1093 (1961).
[14] F.-C. Huang, L. F. H. Lee, R. S. D. Mittal, P. R. Ravikumar, J. A. Chan, C. J. Sih, E. Capsi, and C. R. Eck, *J. Am. Chem. Soc.*, **97**, 4144 (1975).
[15] M. Ohno, S. Kobayashi, T. Iimori, Y.-F. Wang, and T. Izawa, *J. Am. Chem. Soc.*, **103**, 2405 (1981).
[16] K. Adachi, S. Kobayashi, and M. Ohno, *Chimia*, **40**, 311 (1986).
[17] L. K. P. Lam, R. A. H. F. Hui, and J. B. Jones, *J. Org. Chem.*, **51**, 2047 (1986).
[18] F. Björkling, J. Boutelje, S. Gatenbeck, K. Hult, T. Norin, and P. Szmulik, *Tetrahedron*, **41**, 1347 (1985).
[19] H. Suemune, Y. Mizuhara, H. Akita, and K. Sakai, *Chem. Pharm. Bull.*, **34**, 3440 (1986).
[20] D. Breitgoff, K. Laumen, and M. P. Schneider, *J. Chem. Soc., Chem. Commun.*, **1986**, 1523.
[21] V. Kerscher and W. Kreiser, *Tetrahedron Lett.*, **28**, 531 (1987).
[22] Y. Nagao, M. Kume, R. C. Wakabayashi, T. Nakamura, and M. Ochiai, *The 13rd Symposium of Progress in Organic Reactions and Synthesis*, Tokushima, Japan, 1986, symposium paper, pp. 34–38.
[23] C.-S. Chen, Y. Fujimoto, and C. J. Sih, *J. Am. Chem. Soc.*, **103**, 3580 (1981).
[24] P. Mohr, N. Waespe-Sarĉeviĉ, C. Tamm, K. Gawronska, and J. K. Gawronski, *Helv. Chim. Acta*, **66**, 2501 (1983).
[25] G. Sabbioni, M. L. Shea, and J. B. Jones, *J. Chem. Soc., Chem. Commun.*, **1984**, 236.
[26] M. Schneider, N. Engel, P. Hönicke, G. Heinemann, and H. Görisch, *Angew. Chem., Int. Ed. Engl.*, **23,** 67 (1984).
[27] K. Laumen and M. Schneider, *Tetrahedron Lett.*, **26**, 2073 (1985).
[28] S. Kobayashi, K. Kamiyama, T. Iimori, and M. Ohno, *Tetrahedron Lett.*, **25**, 2557 (1984).
[29] H.-J. Gais and K. L. Lukas, *Angew. Chem., Int. Ed. Engl.*, **23**, 142 (1984).
[30] H.-J. Gais, K. L. Lukas, W. A. Ball, S. Braun, and H. J. Lindner, *Justus Liebigs Ann. Chem.*, **1986**, 687.
[31] G. Guanti, L. Banfi, E. Narisano, R. Riva, and S. Thea, *Tetrahedron Lett.*, **27**, 4639 (1986).
[32] Y.-F. Wang and C. J. Sih, *Tetrahedron Lett.*, **25**, 4999 (1984).
[33] S. Iriuchijima, K. Hasegawa, and G. Tsuchihashi, *Agric. Biol. Chem.*, **46**, 1907 (1982).
[34] N. Ohashi, K. Shimago, T. Ikeda, K. Ishizumi, Eur. Pat. Appl. EP 84,892 [*C.A.*, **99**, 211182t (1983)].
[35] J. B. Jones, R. S. Hinks, and P. G. Hultin, *Can. J. Chem.*, **63**, 452 (1985).
[36] M. Kurihara, K. Kamiyama, S. Kobayashi, and M. Ohno, *Tetrahedron Lett.*, **26**, 5831 (1985).
[37] F. Björkling, J. Boutelje, H. Hjalmarsson, K. Hult, and T. Norin, *J. Chem. Soc., Chem. Commun.*, **1987**, 1041.
[38] Y.-F. Wang, C.-S. Chen, G. Girdaukas, and C. J. Sih, *J. Am. Chem. Soc.*, **106**, 3695 (1984).
[39] K. Laumen and M. Schneider, *Tetrahedron Lett.*, **25**, 5875 (1984).
[40] D. Seebach and M. Eberle, *Chimia*, **40**, 315 (1986).
[41] R. Bloch, E. Guibe-Jampel, and G. Girard, *Tetrahedron Lett.*, **26**, 4087 (1985).
[42] M. Ohno, *Nucleosides and Nucleotides*, **4**, 21 (1985).
[43] Y. Ito, T. Shibata, M. Arita, H. Sawai, and M. Ohno, *J. Am. Chem. Soc.*, **103**, 6739 (1981).
[44] M. Arita, K. Adachi, Y. Ito, H. Sawai, and M. Ohno, *J. Am. Chem. Soc.*, **105**, 4049 (1983).
[45] W. K. Wilson, S. B. Baca, Y. J. Barber, T. J. Scallen, and C. J. Morrow, *J. Org. Chem.*, **48**, 3960 (1983).
[46] M. Schneider, N. Engel, and H. Boensmann, *Angew. Chem., Int. Ed. Engl.*, **23**, 64 (1984).
[47] F. Björkling, J. Boutelje, S. Gatenbeck, K. Hult, and T. Norin, *Appl. Microbiol. Biotechnol.*, **21**, 16 (1985).
[48] S. Tanaka, H. Ohaishi, H. Suemune, and K. Sakai, *The 107th Annual Meeting, Pharmaceutical Society of Japan*, 1987.
[49] S. Sicsic, M. Ikbal, and F. Le Goffic, *Tetrahedron Lett.*, **28**, 1887 (1987).
[50] P. Mohr, L. Rösslein, and C. Tamm, *Helv. Chim. Acta*, **70**, 142 (1987).
[51] H. Ritter and C. Siebel, *Makromol. Chem., Rapid Commun.*, **6**, 521 (1985).
[52] A. Hazato, T. Tanaka, T. Toru, N. Okamura, K. Bannai, S. Sugiura, K. Manabe, and S. Kurozumi, *Nippon Kagaku Kaishi*, **9**, 1390 (1983) [*C.A.*, **100**, 120720q (1984)].
[53] K. Okano, *Dissertation*, University of Tokyo, 1984.

[54] T. Ineyama, K. Izawa, Y. Nagao, H. Tanaka, and K. Soda, *The 54th Annual Meeting, Chemical Society of Japan*, 1987.

[55] J. A. Jongejan and J. A. Duine, *Tetrahedron Lett.*, **28**, 2767 (1987).

[56] Y.-F. Wang, C.-S. Chen, G. Girdaukas, and C. J. Sih, in *Enzymes in Organic Synthesis. Ciba Foundation Symposium 111*, Pitman Publishing, London, 1985, pp. 128–145.

[57] F. Björkling, J. Boutelje, S. Gatenbeck, H. Hult, T. Norin, and P. Szmulik, *Bioorg. Chem.*, **14**, 176 (1986).

[58] B. Cambou and A. M. Klibanov, *Biotechnol. Bioeng.*, **26**, 1449 (1984).

[59] K. Laumen, E. H. Reimerdes, and M. Schneider, *Tetrahedron Lett.*, **26**, 407 (1985).

[60] Microbiol Research Foundation, Japan Tokkyo Koho, JP 57,159,493 [82,159,493] [*C.A.*, **98**, 51923j (1983)].

[61] M. Ohno, S. Kobayashi, and K. Adachi, in *Enzymes as Catalysts in Organic Synthesis*, M. P. Schneider, Ed., D. Reidel Publishing, Dordrecht, 1986, pp. 123–142.

[61a] J. Zemlicka, L. E. Craine, M.-J. Heeg, and J. P. Oliver, *J. Org. Chem.*, **53**, 937 (1988).

[62] M. Ohno, in *Enzymes in Organic Synthesis, Ciba Foundation Symposium 111*, Pitman Publishing, London, 1985, pp. 171–187.

[63] S. Kobayashi, T. Iimori, T. Izawa, and M. Ohno, *J. Am. Chem. Soc.*, **103**, 2406 (1981).

[64] K. Okano, T. Izawa, and M. Ohno, *Tetrahedron Lett.*, **24**, 217 (1983).

[65] T. Iimori, Y. Takahashi, T. Izawa, S. Kobayashi, and M. Ohno, *J. Am. Chem. Soc.*, **105**, 1659 (1983).

[66] M. Ohno, in *Organic Synthesis: An Interdisciplinary Challenge*, J. Streith, H. Prinzbach, and G. Schill, Eds., Blackwell Scientific Publications, Oxford, 1985, pp. 189–204.

[67] K. Okano, Y. Kyotani, H. Ishihama, S. Kobayashi, and M. Ohno, *J. Am. Chem. Soc.*, **105**, 7186 (1983).

[68] Y.-F. Wang, T. Izawa, S. Kobayashi, and M. Ohno, *J. Am. Chem. Soc.*, **104**, 6465 (1982).

[69] M. Ohno, S. Kobayashi, T. Izawa, and Y.-F. Wang, *Nippon Kagaku Kaishi*, **1983**, 1299 [*C.A.*, **100**, 85460f (1984)].

[70] F. VanMiddlesworth, Y.-F. Wang, B.-N. Zhou, D. DiTullio, and C. J. Sih, *Tetrahedron Lett.*, **26**, 961 (1985).

[71] F. VanMiddlesworth, D. V. Patel, J. Donaubauer, P. Gannett, and C. J. Sih, *J. Am. Chem. Soc.*, **107**, 2996 (1985).

[72] P. Mohr, M. Tori, P. Grossen, P. Herold, and C. Tamm, *Helv. Chim. Acta*, **65**, 1412 (1982).

[73] P. Herold. P. Mohr, and C. Tamm, *Helv. Chim. Acta*, **66**, 744 (1983).

[74] T. Tschamber, N. Waespe-Sarĉeviĉ, and C. Tamm, *Helv. Chim. Acta*, **69**, 621 (1986).

[75] D. W. Brooks and J. T. Palmer, *Tetrahedron Lett.*, **24**, 3059 (1983).

[76] Y. Ito, M. Arita, K. Adachi, T. Shibata, H. Sawai, and M. Ohno, *Nucleic Acids Res., Symposium Ser.*, **10**, 45 (1981).

[77] M. Ohno, Y. Ito, M. Arita, T. Shibata, K. Adachi, and H. Sawai, *Tetrahedron*, **40**, 145 (1984).

[78] M. Arita, K. Adachi, Y. Ito, H. Sawai, and M. Ohno, *Nucleic Acids Res., Symposium Ser.*, **11**, 13 (1982).

[79] M. Arita, T. Okumoto, T. Saito, Y. Hoshino, K. Fukukawa, S. Shuto, M. Tsujino, H. Sakakibara, and M. Ohno, *Carbohydr. Res.*, **171**, 233 (1987).

[80] M. Arita, K. Adachi, H. Sawai, and M. Ohno, *Nucleic Acids Res., Symposium Ser.*, **12**, 25 (1983).

[81] J. H. Bateson, R. I. Hickling, P. M. Roberts, T. C. Smale, and R. Southgate, *J. Chem. Soc., Chem. Commun.*, **1980**, 1084.

[82] N. Tamura, Y. Kawano, Y. Matsushita, K. Yoshioka, and M. Ochiai, *Tetrahedron Lett.*, **27**, 3749 (1986).

[83] K. Kamiyama, S. Kobayashi, and M. Ohno, *Chem. Lett.*, **1987**, 29.

[84] H.-J. Gais, in *Enzymes as Catalysts in Organic Synthesis*, M. P. Schneider, Ed., D. Reidel Publishing, Dordrecht, 1986, pp. 97–122.

[85] H.-J. Gais, T. Lied, and K. L. Lukas, *Angew. Chem., Int. Ed. Engl.*, **23**, 511 (1984).

[86] H.-J. Gais, H. L. Lindner, T. Lied, K. L. Lukas, W. A. Ball, B. Rosenstock, and H. Sliwa, *Justus Liebigs Ann. Chem.*, **1986**, 1179.

[87] H.-J. Gais and T. Lied, *Angew. Chem., Int. Ed. Engl.*, **23**, 145 (1984).

[88] S. Takano, K. Tanigawa, and K. Ogasawara, *J. Chem. Soc., Chem. Commun.*, **1976**, 189.
[89] M. Harre, P. Raddatz, R. Walenta, and E. Winterfeldt, *Angew. Chem., Int. Ed. Engl.*, **21**, 480 (1982).
[90] M. Suzuki, A. Yonesawa, and R. Noyori, *J. Am. Chem. Soc.*, **107**, 3348 (1985).
[91] E. J. Corey, K. Niimura, Y. Konishi, S. Hashimoto, and Y. Hamada, *Tetrahedron Lett.*, **27**, 2199 (1986).
[92] F. Björkling, J. Boutelje, S. Gatenbeck, K. Hult, and T. Norin, *Tetrahedron Lett.*, **26**, 4957 (1985).
[93] D. Seebach and P. Knochel, *Helv. Chim. Acta*, **67**, 261 (1984).
[94] D. Seebach, G. Calderani, and P. Knochel, *Tetrahedron*, **41**, 4861 (1985).
[95] J. K. Whitesell and R. M. Lawrence, *Chimia*, **40**, 318 (1986).
[96] S. G. Cohen and E. Khedouri, *J. Am. Chem. Soc.*, **83**, 4228 (1961).
[97] S. G. Cohen, Y. Sprinzak, and E. Khedouri, *J. Am. Chem. Soc.*, **83**, 4225 (1961).
[98] W. Kasel, P. G. Hultin, and J. B. Jones, *J. Chem. Soc., Chem. Commun.*, **1985**, 1563.
[99] H. Hemmerle and H.-J. Gais, *Tetrahedron Lett.*, **28**, 3471 (1987).
[100] J. B. Jones and R. S. Hinks, *Can. J. Chem.*, **65**, 704 (1987).
[101] Y. Nagao, T. Nakayama, M. Ochiai, K. Fuji, and E. Fujita, *J. Chem. Soc., Chem. Commun.*, **1987**, 267.
[102] Y. Nagao, T. Nakamura, M. Kume, M. Ochiai, K. Fuji, and E. Fujita, *J. Chem. Soc., Chem. Commun.*, **1987**, 269.
[103] D. R. Deardorff, A. J. Matthews, D. S. McMeekin, and C. L. Craney, *Tetrahedron Lett.*, **27**, 1255 (1986).
[104] K. Laumen and M. Schneider, *J. Chem. Soc., Chem. Commun.*, **1986**, 1298.
[105] C. R. Johnson and T. D. Penning, *J. Am. Chem. Soc.*, **108**, 5655 (1986).
[106] H. Kotani, Y. Kuze, S. Uchida, T. Miyabe, T. Iimori, K. Okano, S. Kobayashi, and M. Ohno, *Agric. Biol. Chem.*, **47**, 1363 (1983).
[107] K. Yamamoto, H. Ando, and H. Chikamatsu, *J. Chem. Soc., Chem. Commun.*, **1987**, 334.
[108] S. Miura, S. Kurozumi, T. Toru, T. Tanaka, M. Kobayashi, S. Matsubara, and S. Ishimoto, *Tetrahedron*, **32**, 1893 (1976).
[109] B. I. Glänzer, K. Faber, and H. Griengl, *Tetrahedron Lett.*, **27**, 4293 (1986).
[110] W. E. Ladner and G. M. Whitesides, *J. Am. Chem. Soc.*, **106**, 7250 (1984).
[111] W. J. Marsheck and M. Miyano, *Biochim. Biophys. Acta*, **316**, 363 (1973).
[112] K. Mori and H. Akao, *Tetrahedron*, **36**, 91 (1980).
[113] M. Ohta and H. Tetsukawa, *Agric. Biol. Chem.*, **44**, 863 (1980).
[114] T. Oritani, M. Ichimura, Y. Hanyu, K. Yamashita, *Agric. Biol. Chem.*, **47**, 2613 (1983).
[115] M. Kasai, K. Kawai, M. Imuta, and H. Ziffer, *J. Org. Chem.*, **49**, 675 (1984).
[116] C.-Q. Han, D. DiTullio, Y.-F. Wang, and C. J. Sih, *J. Org. Chem.*, **51**, 1253 (1986).
[117] G. Kirchner, M. P. Scollar, and A. M. Klibanov, *J. Am. Chem. Soc.*, **107**, 7072 (1985).
[118] C.-S. Chen, S.-H. Wu, G. Girdaukas, and C. J. Sih, *J. Am. Chem. Soc.*, **109**, 2812 (1987).
[119] G. Fülling and C. J. Sih, *J. Am. Chem. Soc.*, **109**, 2845 (1987).
[120] H. Ohta, H. Tetsukawa, and N. Noto, *J. Org. Chem.*, **47**, 2400 (1982).
[121] B. Hadorn, F. Leuthardt, E. Ménard, and D. Vischer, *Helv. Chim. Acta*, **46**, 2003 (1963).
[122] C. Bally and F. Leuthardt, *Helv. Chim. Acta*, **53**, 732 (1970).
[123] A. J. Irwin and J. B. Jones, *J. Am. Chem. Soc.*, **99**, 556 (1977).
[124] J. B. Jones and K. P. Lok, *Can. J. Chem.*, **57**, 1025 (1979).
[125] I. J. Jakovac, H. B. Goodbrand, K. P. Lok, and J. B. Jones, *J. Am. Chem. Soc.*, **104**, 4659 (1982).
[126] A. J. Bridges, P. S. Raman, G. S. Y. Ng, and J. B. Jones, *J. Am. Chem. Soc.*, **106**, 1461 (1984).
[127] J. B. Jones and C. J. Francis, *Can. J. Chem.*, **62**, 2578 (1984).
[128] K. P. Lok, I. J. Jakovac, and J. B. Jones, *J. Am. Chem. Soc.*, **107**, 2521 (1985).
[129] V. Prelog, *Pure Appl. Chem.*, **9**, 119 (1964).
[130] W. P. Schneider and H. C. Murray, *J. Org. Chem.*, **38**, 397 (1973).
[131] D. W. Brooks, M. Wilson, and M. Webb, *J. Org. Chem.*, **52**, 2244 (1987).

[132] S. Terashima, S. Yamada, and M. Nara, *Tetrahedron Lett.*, **1977**, 1001.
[133] S. Terashima, M. Nara, and S. Yamada, *Tetrahedron Lett.*, **1978**, 1487.
[134] P. Schwartz and H. E. Carter, *Proc. Natl. Acad. Sci., USA*, **40**, 499 (1954).
[135] R. Altschul, P. Bernstein, and S. G. Cohen, *J. Am. Chem. Soc.*, **78**, 5091 (1956).
[136] Y. Kawasaki, J. Hiratake, Y. Yamamoto, and J. Oda, *J. Chem. Soc., Chem. Commun.*, **1984**, 779.
[137] T. Rosen and C. H. Heathcock, *J. Am. Chem. Soc.*, **107**, 3731 (1985).
[138] M. Oshima and T. Mukaiyama, *Chem. Lett.*, **1987**, 377.
[139] J. Hiratake, Y. Yamamoto, and J. Oda, *J. Chem. Soc., Chem. Commun.*, **1985**, 1717.
[140] J. Hiratake, M. Inagaki, Y. Yamamoto, and J. Oda, *J. Chem. Soc., Chem. Commun.*, **1987**, 1053
[141] Y. Nagao, T. Ikeda, M. Yagi, E. Fujita, and M. Shiro, *J. Am. Chem. Soc.*, **104**, 2079 (1982).
[142] Y. Nagao, T. Inoue, E. Fujita, S. Terada, and M. Shiro, *J. Org. Chem.*, **48**, 132 (1983).
[143] Y. Nagao, T. Inoue, E. Fujita, S. Terada, and M. Shiro, *Tetrahedron*, **40**, 1215 (1984).
[144] Y. Nagao, T. Ikeda, T. Inoue, M. Yagi, M. Shiro, and E. Fujita, *J. Org. Chem.*, **50**, 4072 (1985).
[145] K. Osakada, M. Obana, T. Ikariya, M. Saburi, and S. Yoshikawa, *Tetrahedron Lett.*, **22**, 4297 (1981).
[146] Y. Ishii, K. Osakada, T. Ikariya, M. Saburi, and S. Yoshikawa, *Chem. Lett.*, **1982**, 1179.
[147] J. Ichikawa, M. Asami, and T. Mukaiyama, *Chem. Lett.*, **1984**, 949.
[148] T. Mukaiyama, Y. Watanabe, and M. Shimizu, *Chem. Lett.*, **1984**, 401.
[149] T. Mukaiyama, I. Tomioka, and M. Shimizu, *Chem. Lett.*, **1984**, 49.
[150] S. Hatakeyama, K. Sakurai, and S. Takano, *J. Chem. Soc., Chem. Commun.*, **1985**, 1759.
[151] B. Häfele, D. Schröter, and V. Jäger, *Angew. Chem., Int. Ed. Engl.*, **25**, 87 (1986).
[152] B. Koppenhoefer, M. Walser, D. Schröter, B. Häfele, and V. Jäger, *Tetrahedron*, **43**, 2059 (1987).
[153] J. J. Horgan, J. K. Stoops, E. C. Webb, and B. Zerner, *Biochemistry*, **8**, 2000 (1969).
[154] M. Schneider, N. Engel, and H. Boensmann, *Angew. Chem., Int. Ed. Engl.*, **23**, 66 (1984).
[155] D. Häbich and W. Hartwig, *Tetrahedron Lett.*, **28**, 781 (1987).
[156] H. Suemune, K. Okano, H. Akita, and K. Sakai, *Chem. Pharm. Bull.*, **35**, 1741 (1987).
[157] G. Sabbioni and J. B. Jones, *J. Org. Chem.*, **52**, 4565 (1987).
[158] C.-S. Chen, Y. Fujimoto, G. Girdaukas, and C. J. Sih, *J. Am. Chem. Soc.*, **104**, 7294 (1982).

CHAPTER 2

THE ELECTROPHILIC SUBSTITUTION OF ALLYLSILANES AND VINYLSILANES

IAN FLEMING

University Chemical Laboratory, Cambridge, England

JACQUES DUNOGUÈS

Laboratoire de Chimie Organique et Organométallique, Université de Bordeaux I, Bordeaux, France

ROGER SMITHERS

Department of Chemistry, University of Malaya, Kuala Lumpur, Malaysia

CONTENTS

ACKNOWLEDGMENT

We thank Drs. Françoise Pisciotti and Françoise Simonin for their help in surveying the literature for the reactions of allylsilanes, vinylsilanes, and allenylsilanes.

INTRODUCTION

Allylsilanes and vinylsilanes usually react with electrophiles in the sense of Eqs. 1 and 2 to give substitution. These reactions are conveniently understood as the reactions of alkenes that have been significantly but only slightly modified by the presence of the silyl group. In both reactions, substitution is favored over addition, and both the site of attack and the site of the double bond in the product are usually determined by the site of the silyl group in the starting material.

$$R_3Si\text{-}CH_2CH{=}CH_2 \xrightarrow{E^+} CH_2{=}CHCH_2\text{-}E \quad \text{(Eq. 1)}$$

$$CH_2{=}CH\text{-}SiR_3 \xrightarrow{E^+} CH_2{=}CH\text{-}E \quad \text{(Eq. 2)}$$

In this review we discuss only the electrophilic substitution reactions of allylsilanes, vinylsilanes, and allenylsilanes. We further restrict ourselves to the reactions of tetraorganosilanes because they are synthetically the most interesting in the laboratory. However, although they are not included in the tables, allylsilanes and vinylsilanes that react by addition rather than substitution or that are not tetraorganosilanes are referred to occasionally in the text wherever their reactions illuminate the discussion. Many of the features of the reactions discussed here are shared by the reactions of arylsilanes, ethynylsilanes, propargylsilanes, cyclopropylsilanes, and cyclopropylmethylsilanes.[1] The methods by which allylsilanes and vinylsilanes are synthesized have been summarized in several places.[1–4]

Historically, the first electrophilic substitution of an allylsilane (Eq. 3) was carried out in 1948[5] and of a vinylsilane (Eq. 4) in 1954.[6] The first electrophilic substitution using a heteroatom electrophile was sulfonation (Eqs. 5 and 6) in the late 1960s.[7–9]

$$(CH_3)_3Si\text{-}CH_2CH{=}CH_2 \xrightarrow{HBr} (CH_3)_3Si\text{-}CH_2CH(Br)CH_3 \xrightarrow{heat} CH_2{=}CHCH_3 + (CH_3)_3SiBr \quad \text{(Eq. 3)}$$

$$CH_2{=}CH\text{-}Si(CH_3)_3 \xrightarrow{HI} I\text{-}CH_2CH_2\text{-}Si(CH_3)_3 \xrightarrow{OH^-} CH_2{=}CH_2 + (CH_3)_3SiOH \quad \text{(Eq. 4)}$$

$$(CH_3)_3Si\text{-}CH_2CH{=}CH\text{-}Si(CH_3)_3 \xrightarrow{ClSO_3Si(CH_3)_3} CH_2{=}CH\text{-}CH(Si(CH_3)_3)(SO_3Si(CH_3)_3) \quad \text{(Eq. 5)}$$

(70%)

$$CH_2{=}CH\text{-}Si(CH_3)_3 \xrightarrow{ClSO_3Si(CH_3)_3} CH_2{=}CH\text{-}SO_3Si(CH_3)_3 \quad \text{(Eq. 6)}$$

(70%)

The first carbon electrophile was used in 1970 (Eqs. 7 and 8),[10] and acid chlorides and other carbon electrophiles were introduced shortly afterwards.

$$(CH_3)_3Si\text{-}CH_2CH{=}CH_2 \xrightarrow[ZnCl_2]{ClCH_2OCH_3} CH_2{=}CHCH_2CH_2CH_2OCH_3 \quad \text{(Eq. 7)}$$

$$CH_2{=}CHSi(CH_3)_3 \xrightarrow[ZnCl_2]{ClCH_2OCH_3} CH_2{=}CHCH_2OCH_3 \quad \text{(Eq. 8)}$$

The first proof that an allylsilane reacted with allylic shift (Eq. 9) was published in 1956,[11] and the regiospecificity of reaction by a vinylsilane was first demonstrated (Eq. 10) in 1975.[12] Fluoride ion was introduced as a nucleophilic catalyst (Eq. 11) in 1978.[13] The stereoselectivity of vinylsilane reactions was first observed with bromination, which took place with inversion of configuration (Eq. 12),[14] and with protodesilylation, which took place with retention of configuration (Eq. 13).[15] The full stereospecificity was then shown in 1973 (Eq. 14).[16] The first investigations of the stereochemistry of the S_E2' reaction of allylsilanes were carried out with cyclic allylsilanes, which showed both *anti* (Eq. 15) and *syn* (Eq. 16) reactions.[17,18] Stereochemical studies of open-chain allylsilanes (Eqs. 17, 18)[19,20] also showed both *syn* and *anti* reactions. That the latter pattern is normal, and the former exceptional, was then established firmly in 1982 with a wide range of electrophiles (Eq. 19).[21–24] The first attempts to induce chirality in the products of electrophilic attack using chiral silicon[25,26] are illustrated in Eqs. 20 and 21, and chiral carbon has also been attached to silicon.[27]

4-i-C_3H_7-1-[$(CH_3)_3SiCH_2$]cyclohexene $\xrightarrow{CH_3CO_2H}$ 4-i-C_3H_7-1-methylenecyclohexane (Eq. 9)

4,4-dimethyl-1-$Si(CH_3)_3$-cyclohexene $\xrightarrow{CH_3COCl,\ AlCl_3}$ 4,4-dimethyl-1-$COCH_3$-cyclohexene

5,5-dimethyl-1-$Si(CH_3)_3$-cyclohexene $\xrightarrow{CH_3COCl,\ AlCl_3}$ 5,5-dimethyl-1-$COCH_3$-cyclohexene (Eq. 10)

$$(CH_3)_3SiCH_2CH{=}CH_2 + OHCCH_2CH_2C_6H_5 \xrightarrow[\text{THF, reflux}]{(n\text{-}C_4H_9)_4NF} CH_2{=}CHCH_2CH(OH)CH_2CH_2C_6H_5 \ (86\%) \quad \text{(Eq. 11)}$$

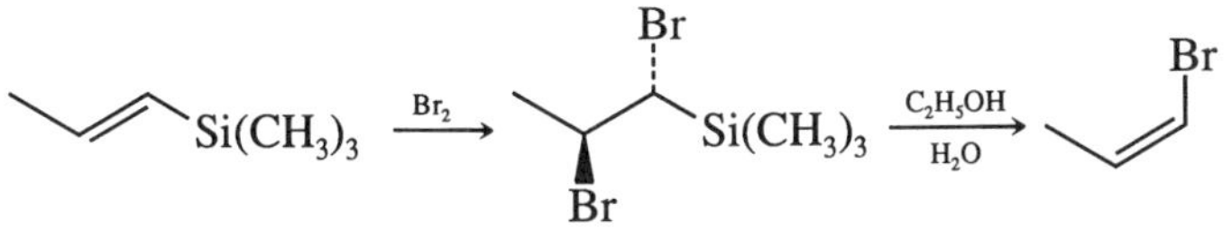

Br
Br
Si(CH3)3 Br2 → Si(CH3)3 C2H5OH / H2O → Br (Eq. 12) 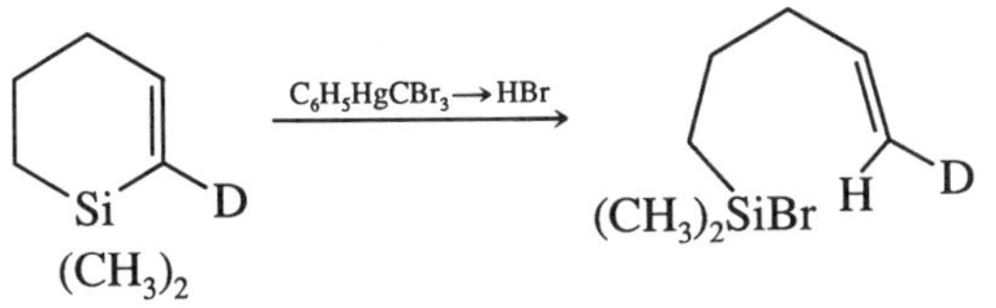
Br

C6H5HgCBr3 → HBr (Eq. 13) 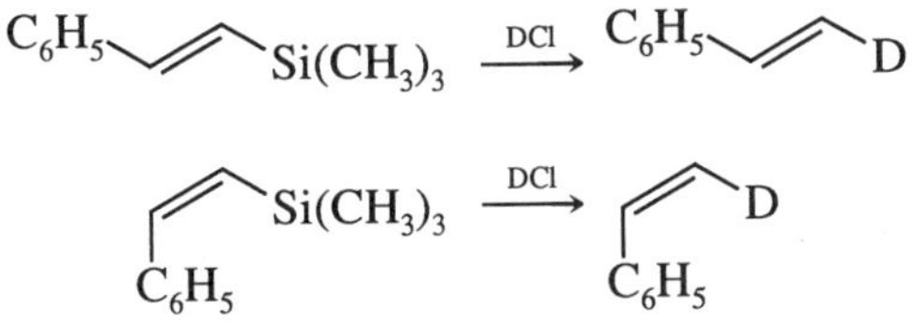
Si D (CH3)2 (CH3)2SiBr H D

C6H5 Si(CH3)3 DCl → C6H5 D
(Eq. 14) 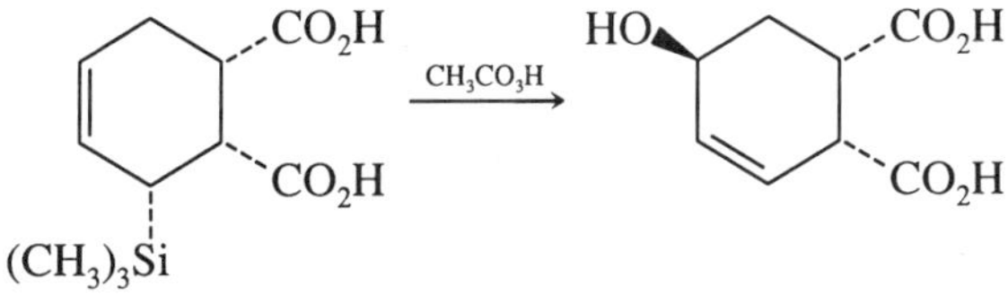
Si(CH3)3 DCl → D
C6H5 C6H5

CO2H CH3CO3H → HO CO2H (Eq. 15) 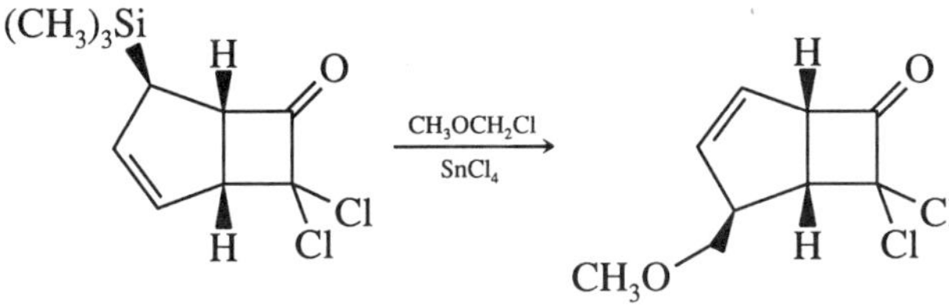
CO2H CO2H
(CH3)3Si

(CH3)3Si H O H O
CH3OCH2Cl / SnCl4 → (Eq. 16) 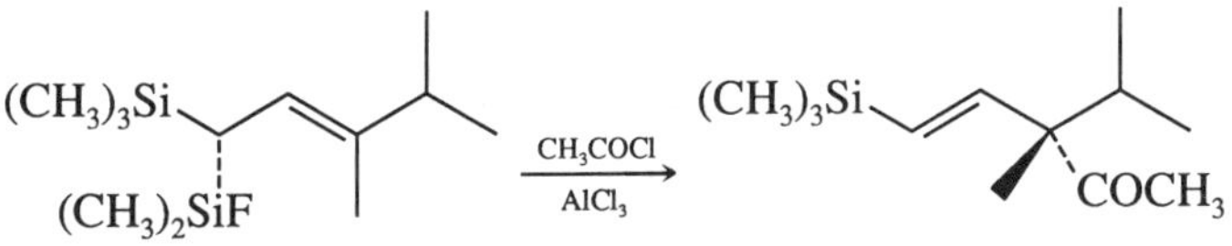
H Cl Cl H Cl Cl
CH3O

(CH3)3Si (CH3)3Si
CH3COCl / AlCl3 → (Eq. 17) 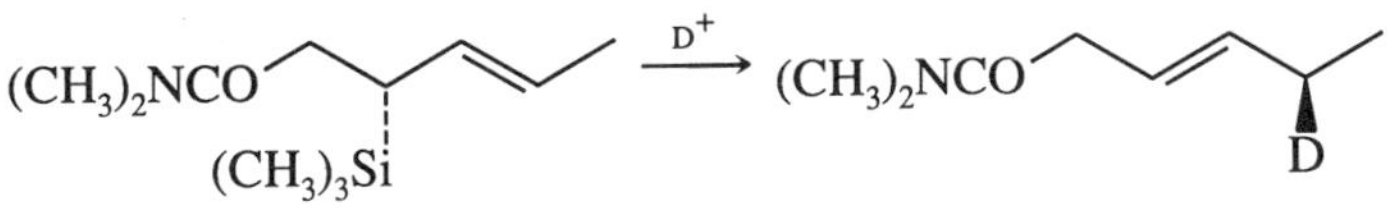
(CH3)2SiF COCH3

(CH3)2NCO D+ → (CH3)2NCO (Eq. 18) 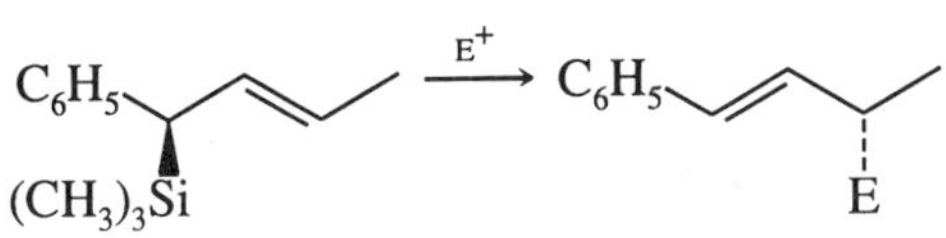
(CH3)3Si D

C6H5 E+ → C6H5 (Eq. 19)
(CH3)3Si E

E = t-C_4H_9, RCHOH, OH

C_6H_5, $C_{10}H_7$-α, Si, CH_3 —MCPBA→ 14% ee (Eq. 20)

CH_3, α-$C_{10}H_7$, Si, C_6H_5 —$C_6H_5CH(OCH_3)_2$, $BF_3 \cdot O(C_2H_5)_2$→ OCH_3, C_6H_5 5% ee (Eq. 21)

MECHANISM AND STEREOCHEMISTRY

The Simple Picture

At its most simple, the mechanism of the substitution reaction involves electrophilic attack on the π bond of the allylsilane **1** or vinylsilane **4** to generate cationic intermediates **2** or **5**, and a nucleophile then displaces the silyl group with the formation of an alkene (**3** or **6**). Since the loss of a silyl group is faster with oxygen or halogen nucleophiles than is the loss of a comparable proton,[28] the position of the double bond is determined by the atom to which the silicon is attached.

R_3Si, E^+ (**1**) → Nu^-, R_3Si, +, E (**2**) → E (**3**)

SiR_3, E^+ (**4**) → +, SiR_3, ^-Nu, E (**5**) → E (**6**)

The stepwise nature of the reaction is demonstrated by the protodesilylation of silanes **7** and **8**, both of which are simultaneously allylsilanes and vinylsilanes. Regardless of which silane is used, protodesilylation gives the same 4:1 mixture of the allylsilanes **10** and **11**.[29] There must, therefore, be a common intermediate, and the cation **9** is a likely candidate. Other evidence for a cationic intermediate is found in the reaction of allyltrimethylsilane with tetracyanoethylene, where the rate of the reaction is markedly dependent upon solvent polarity.[30]

$(CH_3)_3Si$–CH₂CH=CHCH₂–$Si(CH_3)_2C_6H_5$ **7** $\xrightarrow{H^+}$ $(CH_3)_3Si$–CH₂CH⁺CH₂CH₂–$Si(CH_3)_2C_6H_5$ **9** → CH₂=CHCH₂–$Si(CH_3)_2C_6H_5$ **10** + $(CH_3)_3Si$–CH₂CH=CH₂ **11**

$(CH_3)_3Si$–CH=CHCH₂CH₂–$Si(CH_3)_2C_6H_5$ **8** $\xrightarrow{H^+}$ **9**

10:11 4:1

The site of attack by the electrophile is controlled by the fact that cations **2** and **5** are stabilized by hyperconjugative overlap of the Si—C bonding orbital with the empty *p* orbital.[31–33] The degree of this stabilization is high: the silylethyl cation is calculated to be 38 kcal mol^{-1} more stable than the ethyl cation,[34] and experimentally the σ^+ value of the trimethylsilylmethyl group is -0.6,[31,35] comparable to an acetamido group and more negative than a methyl group (-0.3). Since this makes a trimethylsilylmethyl group comparable to two methyl groups, it is sometimes possible to achieve *anti*-Markovnikov attack at the more substituted end of a double bond, as in the deuterodesilylation of the allylsilane **12**. Only one deuterium is incorporated into the major product **13**, and that is at the methine position, showing that rapid and reversible attack at the less-substituted end of the double bond had not taken place. The product **14** of deuteronation at C-2 is isolated in this reaction, but it is a product of addition and is quite minor.[36] The regioselectivity in this reaction can also be understood by looking at the reactant side of the reaction coordinate: a MINDO-3 calculation[37] suggests that the coefficient on C-3 in the HOMO of 3-methyl-2-butenylsilane is slightly higher than that on C-2.

$(CH_3)_3Si$–CH₂CH=C(CH₃)₂ (C-2, C-3) **12** $\xrightarrow{CF_3CO_2D}$ CH₂=CH–CD(CH₃)₂ **13** (83%) + $(CH_3)_3Si$–CH₂CHD–C(CH₃)₂O_2CCF_3 **14** (17%)

However, the site of attack by the electrophile does not always lead directly to the silicon-stabilized cations **2** or **5**, and, even when it does, the silyl group is not quite always the electrofugal group. The problem is discussed in the section Scope and Limitations, since the overall reaction is not the electrophilic substitution of an allylsilane or a vinylsilane. However, one case in which the overall reaction is such a substitution belongs here because it represents merely an abnormal mechanism. The deuterodesilylation of the allylsilane **15** (in contrast to the deuterodesilylation **12** → **13**) gives the alkenes **18** and **20** in a ratio of 2:3.[38] The initial site of attack is now evidently the Markovnikov position C-2, giving at least in part the cation **16**, which then undergoes a 1,2 shift of hydride or deuteride to give the cations **17** and **19**, and hence the alkenes **18** and **20**. But for the choice of deuterium as the electrophile, this pathway would have been invisible, for the overall reaction with a proton appears quite normal. It is possible that other reactions of allylsilanes with this substitution pattern may similarly involve Markovnikov attack followed by migration of the electrophilic group. The migration step is well precedented for hydride, for alkyl groups, and for phenyl groups.[39]

$(CH_3)_3Si$ 2 C_6H_5 $\xrightarrow{D^+}$

15

$(CH_3)_3Si$ H D + C_6H_5

16

D shift → $(CH_3)_3Si$ + D C_6H_5 **17** → D C_6H_5 **18**

H shift → $(CH_3)_3Si$ D + C_6H_5 **19** → D C_6H_5 **20**

The Stereochemistry of Reactions of Allylsilanes

In most cases, open-chain allylsilanes react with electrophiles with *anti* stereoselectivity. The first example of this (Eq. 18)[20] was followed by others with deuterons,[40–42] protons,[43,44] and several other electrophiles (Eq. 19).[21–24,44] The simple explanation for the stereochemistry follows from the probable conformation of the allylsilane. The preferred conformation **21** has the small substituent H more or less eclipsing the double bond. The picture of a double bond as a pair of bent σ bonds is a useful alternative, because it emphasizes that there are regions of high electron population above and below the plane of the C=C double bond, and that the region of maximum electron repulsion does not necessarily lie along the line joining the two carbon atoms.[45] The large and electropositive silyl group encourages attack by the electrophile on the lower surface and only a 30° rotation is needed to get from **21** to **22**. The hyperconjugative overlap in the intermediate **22** is probably powerful enough for the configuration to be maintained, until the silyl group is lost in the second step **22** → **23**, with the result that the double bond produced in

23 is *trans*, and the overall reaction is stereoselectively *anti*. This picture conforms to the general rule for electrophilic attack on a double bond adjacent to a chiral center, if one assumes the silyl group to be the largest group.[46]

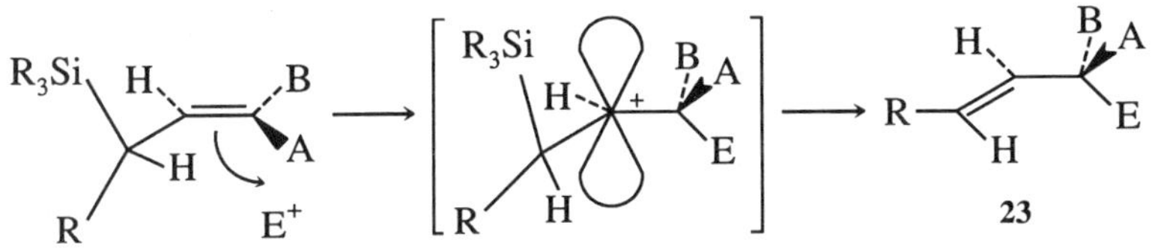

The precise conformation of any particular allylsilane, either in the ground state or at the time of reaction, is not, of course, known. Allylsilane itself (**21**, R = A = B = H) has a conformation in which the silyl group is tilted from the vertical by 12–14° (somewhat less than the 30° shown in **21**), a range in which calculations[47,48] and experiments[49,50] agree. At the transition state, the angle will again be different. Nevertheless, for the purpose of discussion, it is simple to use the conformation **21** to stand for all the subtle variations of conformation close to it. Furthermore, it is known that with large R groups, like phenyl and isopropyl, the three-bond coupling constant between the two hydrogens illustrated is at the maximum level (11–12 Hz) for coupling between hydrogens attached to adjacent trigonal and tetrahedral carbons, implying that the two hydrogens have a dihedral angle close to 180° in the ground state.[51]

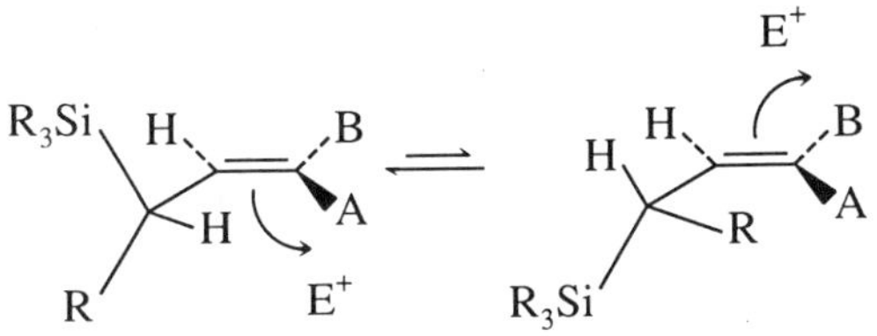

The simple picture of stereospecific *anti* attack with formation of a *trans* double bond is not always followed. In the first place, the difference in size between the substituents H and R in **21** is not always enough to guarantee only a *trans* double bond in the product; an appreciable proportion of the reaction can take place in the alternative conformation **24**, as shown by the protodesilylation in Eq. 22.[52] This problem is only serious when the substituent A in **21** is a hydrogen atom. This means that, for simple allylsilanes with a 1,2-disubstituted double bond, the *Z* isomer is likely to react with cleaner stereochemistry than the *E* isomer, as in the contrasting epoxidations of the allylsilanes (*E*)-**25** and (*Z*)-**25** with MCPBA (*m*-chloroperbenzoic acid). With the *E* isomer, it appears that more attack takes place from the conformation **24**, with the R group eclipsing the double bond, than it did in the protodesilylation in Eq. 22. In this conformation, the hydrogen is on the side of the double bond being attacked, and hydrogen offers less hindrance to the electrophile than the R group does in conformation **21**. Presumably the electrophile in epoxidation is larger than that in protodesilylation. With osmylation by osmium tetroxide in the presence of *N*-methylmorpholine *N*-oxide (NMMO) (Eq. 24), the electrophile is so much larger again that attack on conformation **24** is now the major pathway with the *E* isomer, and is even noticeable with the *Z* isomer (*Z*)-**25**.[53] In summary, there is a tradeoff in unfavorable interactions between the R group and the A group in conformation **24** on the one hand, and the R group and the incoming electrophile in conformation **21** on the other.

$(CH_3)_2NCO$–…–C_6H_5 (with $Si(CH_3)_3$) $\xrightarrow{HF}$ $(CH_3)_2NCO$–…–C_6H_5 (Eq. 22)

E:*Z* 88:12

$C_6H_5(CH_3)_2Si$ (*E*)-**25** $\xrightarrow[CD_2Cl_2,\ 0°]{MCPBA,\ NaHCO_3}$ epoxides (Eq. 23)

61:39

$C_6H_5(CH_3)_2Si$ (*Z*)-**25** $\xrightarrow[CD_2Cl_2,\ 0°]{MCPBA,\ NaHCO_3}$ epoxide > 95:5

(*E*)-**25** $\xrightarrow[\text{NMMO, } (CH_3)_2CO, \text{ room temp}]{OsO_4 \text{ cat.}}$ diols (OH, OH)

34:66

(Eq. 24)

(*Z*)-**25** $\xrightarrow[\text{NMMO, } (CH_3)_2CO, \text{ room temp}]{OsO_4 \text{ cat.}}$ diols (OH, OH)

78:22

A second problem is that the silyl group cannot be guaranteed to force electrophilic attack entirely *anti* to itself. The size of the other group R will clearly have an influence, not only on the relative concentration and reactivity of the conformations **21** and **24**, but also, when it does react in conformation **21**, on the hindrance it exerts toward the incoming electrophile relative to the hindrance exerted by the silyl group. The epoxidation of the allylsilane **26**, with either a phenyl or an isopropyl group on the chiral center, is more selective than the epoxidation of the allylsilane (*E*)-**25**, with a methyl group.[51] This implies that steric hindrance by the R group is more important in fixing the conformation as **21** than it is in hindering the attack of an electrophile on that conformation. Incidentally, it is not clear how much of the attack *anti* to a silyl group is caused by steric or electronic effects. Fortunately, the silyl group is large, and the steric and electronic effects are usually in the same direction.

$C_6H_5(CH_3)_2Si$ (R, H) **26** —MCPBA, CH_2Cl_2→ $C_6H_5(CH_3)_2Si$ (R, H, O) + $C_6H_5(CH_3)_2Si$ (R, H, O)

R = CH_3	61:39
R = C_6H_5	89:11
R = *i*-C_3H_7	95:5

Any groups on A or B in **21** that shield one or another surface of the double bond may override the preference for attack *anti* to the silyl group. This is seen in the protodesilylations of the allylsilanes **27** and **28**. The former reacts cleanly *anti*, because the lower surface is *anti* to both silyl groups and is unhindered, but the allylsilane **28** gives more *syn* reaction than *anti*.[41] Other *syn* stereoselective reactions (e.g., Eq. 16) are in bicyclo[3.2.0]heptanes and

$(CH_3)_3Si$, $Si(CH_3)_3$ **27** —D^+→ $Si(CH_3)_3$, D

$(CH_3)_3Si$, $Si(CH_3)_3$ **28** —D^+→ $Si(CH_3)_3$, D + $Si(CH_3)_3$, D

67:33

bicyclo[3.3.0]octanes, where the ring system might reasonably be responsible for overriding the normal preference for an *anti* reaction. Even in these systems, there are some reactions (e.g., Eq. 25)[54] in which the *anti* preference of the allylsilane controls the stereochemistry against the constraints of the ring system.[54,55]

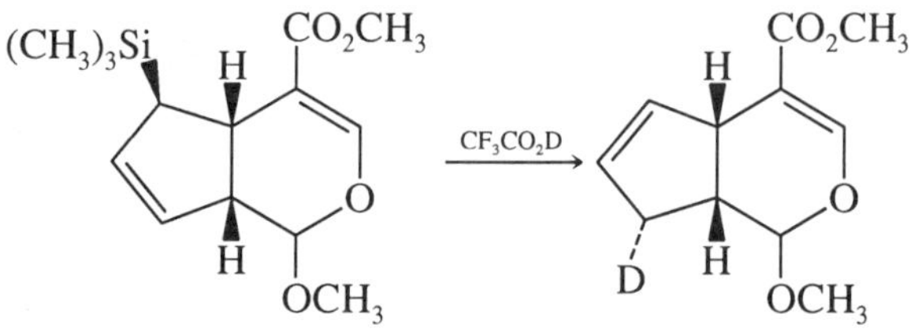

$(CH_3)_3Si$ CO_2CH_3 H H OCH_3 $\xrightarrow{CF_3CO_2D}$ CO_2CH_3 H D H OCH_3 (Eq. 25)

A different problem arises with 3,3-disubstituted allylsilanes. Here the conformation is as well controlled as it is in (*Z*)-allylsilanes like (*Z*)-**25**, but now the site of initial attack can be C-2, as it was with the allylsilane **15**. This causes a loss of stereochemical control. Thus the allylsilane **29** reacts with acid to give the alkene **30** in a normal *anti* reaction, because axial attack is favored by both the ring system and the orientation of the silyl group. However, in the diastereoisomeric allylsilane **31**, the preference for axial attack and control by the silyl group are in opposition, and protodesilylation now gives a mixture of **30** and **32**. Clean *anti* stereospecificity is lost because **31** is first protonated on C-2, as revealed by deuteration experiments.[44]

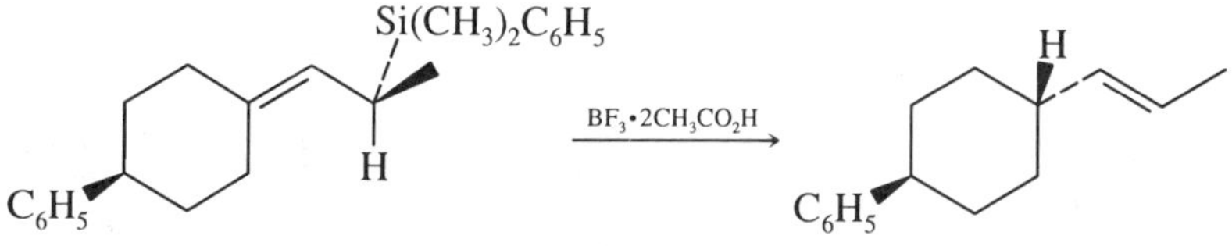

$Si(CH_3)_2C_6H_5$ H C_6H_5 **29** $\xrightarrow{BF_3\cdot 2CH_3CO_2H}$ H C_6H_5 **30** (Eq. 26)

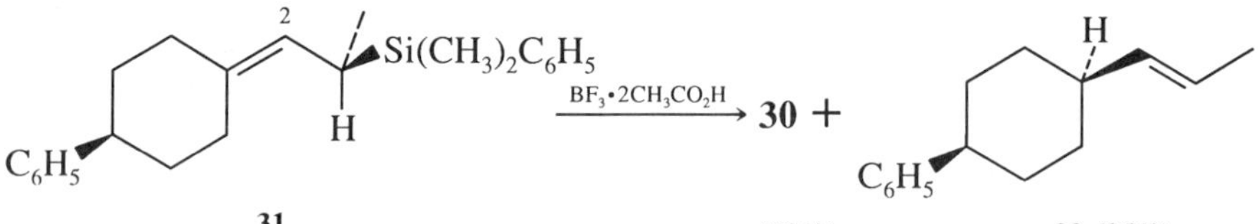

2 $Si(CH_3)_2C_6H_5$ H C_6H_5 **31** $\xrightarrow{BF_3\cdot 2CH_3CO_2H}$ **30** + H C_6H_5 (53%) **32** (36%)

(Eq. 27)

Finally there is the problem of the stereochemistry in Eq. 17; this clean *syn* stereospecific reaction, although the first to be studied in an open-chain allylsilane, is anomalous, since it is now known that acylation is normally *anti* stereospecific, both for allyltrimethylsilanes[21] and for allylfluorodimethylsilanes.[56] A possible explanation[21] is that the electrophile attacks *anti* to the trimethylsilyl group in the normal way, but that the fluorodimethylsilyl group is lost from the intermediate cation after a 60° rotation. In the protodesilylation of this allylsilane the proton mainly (70:30) attacks *anti* to the trimethylsilyl group, and only the trimethylsilyl group is lost.[43]

The Stereochemistry of Reactions of Vinylsilanes

The simple mechanism of electrophilic substitution is also consistent with the retention of stereochemistry usually observed in the electrophilic substitution of vinylsilanes. The initial attack takes place on the top (or bottom) face of the π bond, as in **33**, and the silyl group moves into position in order to stabilize the intermediate cation **37**.[16] The rotation from **34** to **37** is by a shorter path (60°) than the rotation (120°) from **34** to **35**. The drawing **34** represents a cation formed after the attack of the electrophile but before the rotation that brings the silyl group into place in **35**; this cation is probably not formed as such, since rotation will begin as the electrophile approaches. The cation **37**, however, is likely to be an intermediate, which, like the intermediate **22**, retains its configuration until the silyl group is lost **37** → **38**. The overall result is then retention of configuration.

E^+ SiR_3

33

+ E SiR_3 **34**

120° ccw

+ $^-$Nu SiR_3 E **35**

E **36**

60° cw

+ E SiR_3 Nu^- **37**

E **38**

As with allylsilanes, the simple mechanism is not always followed, and inversion of configuration is sometimes observed, notably with bromination and chlorination, which are discussed in detail in the section on halogen electrophiles.

The Less-Simple Possibilities

Addition before Substitution. The first complication that must be added to the simplified picture given above is the possibility that the intermediate cations **2** and **5** may capture a nucleophile before the silyl group is lost. If the capture of the nucleophile takes place *anti* to the silyl group, the stereochemical outcome is unaffected, both for allylsilanes and for vinylsilanes, because the subsequent elimination is stereospecifically *anti.* There is no evidence, in most of the common reactions, whether or not nucleophilic capture is a step on the reaction pathway. In some reactions, however, it is clearly taking place, as in the historically important examples cited in Eqs. 3 and 4. It is also the cause of the anomalous stereochemistry in the halogenation of vinylsilanes,[57] discussed in detail in the section on halogen electrophiles. The reaction of allylsilanes and vinylsilanes with peracids gives epoxides, and the overall electrophilic substitution has to be completed in a second step. The stereochemistry of these reactions is not necessarily anomalous. Nevertheless, they clearly have quite different mechanisms from the simple picture above, and the stepwise mechanism allows unusual stereochemical events to occur. These reactions are discussed in the section on oxygen electrophiles.

Nucleophilic Catalysis. A second complication that must be added to the simple picture is that a nucleophile may bond to silicon at any stage in the sequence. Thus any or all of the intermediates shown in the simple mechanisms may, at the time of reaction, have a nucleophile attached to silicon, which is then in a hypervalent state. Thus the removal of the silyl group **2** → **3** or **5** → **6** may take place in a stepwise manner by way of hypervalent intermediates. However, this is only a detail, about which nothing is known except that the stereochemistry of attack on the silicon takes place overall (Eq. 28) with inversion of configuration in the protodesilylation of an allylsilane, but with retention in mercuridesilylation.[58]

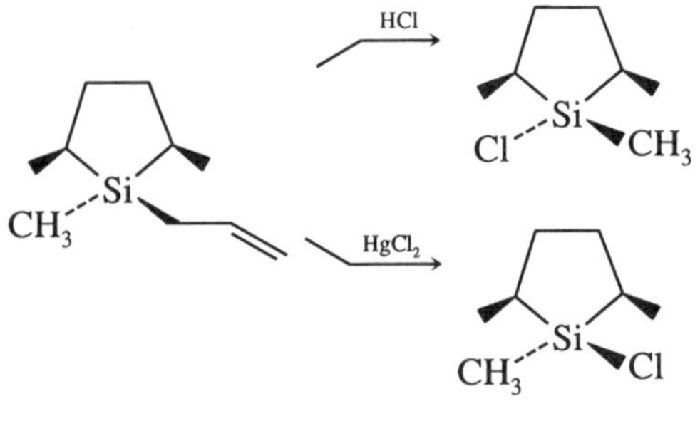

(Eq. 28)

A more substantial complication is that a nucleophile may bond to silicon before the electrophile attacks. This is what must be happening in the many reactions of allylsilanes and vinylsilanes catalyzed by nucleophiles, as in the reactions of the allylsilane **39** and the vinylsilanes **40** and **41** with benzaldehyde,[13,59] which are catalyzed by fluoride ion in the form of tetrabutylammonium fluoride (TBAF). It is clear that the electrophile in this type of reaction does not need to be as powerful as it is in the usual reactions: benzaldehyde reacts without the usual Lewis acid catalysis. The intermediates involved could be hypervalent silicon anions of the general type **42** or **44**. The hypervalent fluorotrimethylsilyl anion group will be more electron-donating than a trimethylsilyl group. The reactivity should be greater, and it is therefore not unreasonable that these intermediates react directly with electrophiles such as benzaldehyde. It is even possible that a hexacoordinate silyl anion, produced by complexation of the carbonyl group on the pentacoordinate species **42**, is an intermediate. Alternatively, the hypervalent silyl anions may fragment to give the free allyl or vinyl anions **43** or **45** before the attack on the

$(CH_3)_3Si$ (allyl) **39** + $OHCC_6H_5$ $\xrightarrow{\text{TBAF}}$ (homoallyl alcohol: OH, C_6H_5) (Eq. 29)

CF_3, SC_6H_5, $Si(CH_3)_3$ **40** + $OHCC_6H_5$ $\xrightarrow{\text{TBAF}}$ CF_3, SC_6H_5, C_6H_5, OH (Eq. 30)

CF_3, $Si(CH_3)_3$, SC_6H_5 **41** + $OHCC_6H_5$ $\xrightarrow{\text{TBAF}}$ HO, C_6H_5, CF_3, SC_6H_5 (Eq. 31)

R_3Si (allyl) $\overset{F^-}{\rightleftharpoons}$ $FR_3\bar{S}i$ (allyl) **42** $\rightleftharpoons$ R_3SiF + $^-$(allyl) **43** $\xrightarrow{E^+}$ E (allyl); **42** $\xrightarrow{E^+}$ E (allyl)

(vinyl)SiR_3 $\overset{F^-}{\rightleftharpoons}$ (vinyl)$\bar{S}iR_3F$ **44** $\rightleftharpoons$ (vinyl, H)$^-$ **45** + R_3SiF $\xrightarrow{E^+}$ (vinyl)E; **44** $\xrightarrow{E^+}$ (vinyl)E

electrophile. It is not known which of these pathways is followed. Whatever the intermediate, it retains the configuration of the vinyl group. In gas phase reactions, stable pentacoordinate species **42** and **44** can be detected, and so can free allyl and vinyl anions.[60] In related reactions of silyl enol ethers, there is evidence for free enolate ions[61] as well as for hexacoordinate intermediates.[62,63]

In the reactions in Eqs. 29–31, the fluoride ion is truly catalytic: fluoride ion is regenerated by the attack of the oxyanion derived from benzaldehyde on the trimethylsilyl fluoride released from the allylsilane. The silyl ether is hydrolyzed in the workup. However, the word catalysis is used rather freely throughout this chapter. It is applied to all those reactions in which an electrophile, typically a Lewis acid, participates in the activation of a substrate, or in which a nucleophile, typically fluoride ion, participates in the activation of a silyl group. It does not necessarily imply that only a catalytic quantity is needed, although in many cases that is all that is used. However, both with Lewis acids and with fluoride ion, a molar or even an excess of catalyst is often used, and it may or may not be unchanged in the course of the reaction.

One important consequence of nucleophilic catalysis is the loss of regiospecificity in the reactions of unsymmetrical allylsilanes. In the reaction of the allylsilane **46** with butyraldehyde, when Lewis acid catalysis is used,[64] only the alcohol **47** is produced, but with fluoride ion catalysis,[13] both alcohols **47** and **48** are produced. These results can be explained by either of the plausible mechanisms: clearly the free allyl anion can react at either end, but it is also known that the allylsilane **46** is allylically unstable (Eq. 33) in the presence of fluoride ion.[65] Hypervalent silyl anions are definitely involved in the reactions of vinylpentafluoro dianions such as **49**, which can be isolated before they are treated with electrophiles.[66–68] Because these compounds are not tetraorganosilanes, they are not included in the tables.

$(CH_3)_3Si$ (**46**) + $OHCC_3H_7$-*n* — $TiCl_4$, CH_2Cl_2 → **47** (OH, C_3H_7-*n*); — TBAF, THF → **47** (41%) + **48** (30%) (OH, C_3H_7-*n*) (Eq. 32)

$(CH_3)_3Si$ (**46**) ⇌ TBAF (cat.), THF → $Si(CH_3)_3$ (Eq. 33)

$$R\text{-CH=CH-}SiCl_3 \xrightarrow{KF} R\text{-CH=CH-}SiF_5^{2-} \xrightarrow{E^+} R\text{-CH=CH-}E$$

49

GENERAL FEATURES OF REACTIVITY

We discuss here those general features of the reactivity of allylsilanes and vinylsilanes that are not specific to any particular electrophile, and draw attention especially to the circumstances in which electrophilic substitution fails, either through failure of regiocontrol or through failure of the silyl group to be electrofugal enough.

Allylsilanes and Vinylsilanes Compared

The trimethylsilyl group is a σ donor and a π acceptor ($\sigma_I = -0.09$; $\sigma_R^\circ = +0.07$).[69] In consequence, its effect on the ground state of a double bond to which it is directly attached is uncertain, but some calculations suggest that it is overall a mild donor.[70] Therefore, as far as the reactant side of the reaction coordinate is concerned, a silyl substituent should have a small activating effect on reactivity. On the product side, the silyl group in a vinylsilane must move into conjugation (**34** → **37**) before it can stabilize the intermediate cation. Again, this should lead the silyl group to have only a small activating effect. In practice, the reactivity of vinylsilanes toward electrophilic attack appears to be very similar to that of the corresponding alkenes lacking the silyl group, although few direct comparisons have been made. The only quantitative work has been on the reactivity of arylsilanes, where it is well known that a silyl group activates the benzene ring to electrophilic attack *ipso* to the silyl group. This activation is high for protodesilylation (a factor of about 10^4) and for bromodesilylation, lower for sulfonation, and absent (or possibly even negative) for nitration.[71] Direct evidence for activation in vinylsilane reactions is more circumstantial. Thus the protodesilylation of (*E*)-vinylsilanes (Eq. 34)[72] takes place cleanly when a proper choice of acid is made. This implies that the protonation of the vinylsilane is significantly faster than the protonation of the product, which, had it taken place, would have caused stereochemical equilibration and double bond shifts. Another example of the greater reactivity of vinylsilanes is in the epoxidation of the diene **50**, where the silyl-substituted double bond is epoxidized about five times faster than the unsubstituted double bond.[73] Other dienes are also epoxidized selectively on the silyl-substituted double bond.[74,75] Furthermore, silylcycloalkenes can be acylated under conditions in which the corresponding cycloalkenes without the silyl group are unreactive.[76] However, the balance between vinylsilane reactivity and the reactivity of the corresponding alkene is a delicate one.

$$n\text{-}C_6H_{13}\text{-CH=C(Si(CH_3)_3)-}C_6H_{13}\text{-}n \xrightarrow{HI} n\text{-}C_6H_{13}\text{-CH=CH-}C_6H_{13}\text{-}n \quad \text{(Eq. 34)}$$

Z:*E* 94:6

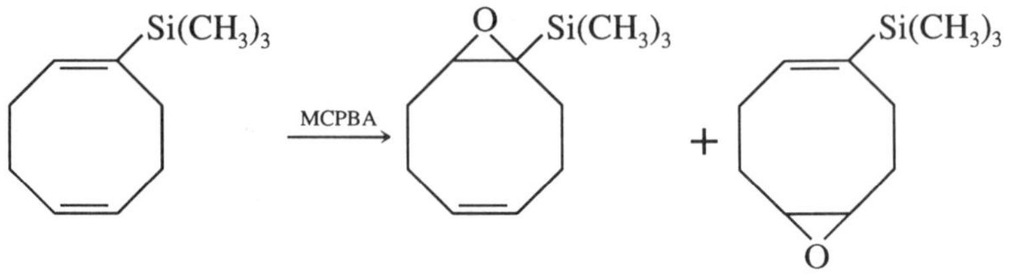

Si(CH3)3 MCPBA → Si(CH3)3 + Si(CH3)3

50 5:1

In contrast, the trimethylsilylmethyl group is both a σ and a π donor ($\sigma_I = -0.1; \sigma_R^\circ = -0.15$).[69] The hyperconjugative overlap of the Si–C bond with the C=C double bond of an allylsilane should raise the energy of the HOMO,[33,48,77,78] and there is evidence that it does.[70,78–81] Furthermore, only a 30° rotation or less about the C-1–C-2 bond is needed before the intermediate cation is fully stabilized. Allylsilanes therefore can be expected to be more reactive toward electrophiles than the corresponding simple alkenes, and this appears generally to be true. Allyltrimethylsilane reacts with the diphenylmethyl cation 30,700 times faster than propene,[82] and there are many reactions (e.g., Eq. 9) in which the double bond produced in the reaction might be expected to be reactive toward electrophiles, but is not attacked as fast as the original allylsilane double bond. However, there are problems when the alkene produced in the first step is more substituted than the double bond of the original allylsilane. This is the situation when allylsilanes with the silyl group at the more-substituted end of the allyl system react with electrophiles, as in Eq. 35, where the first-formed product **52** reacts further to give the chloride **53** to a substantial extent.[83]

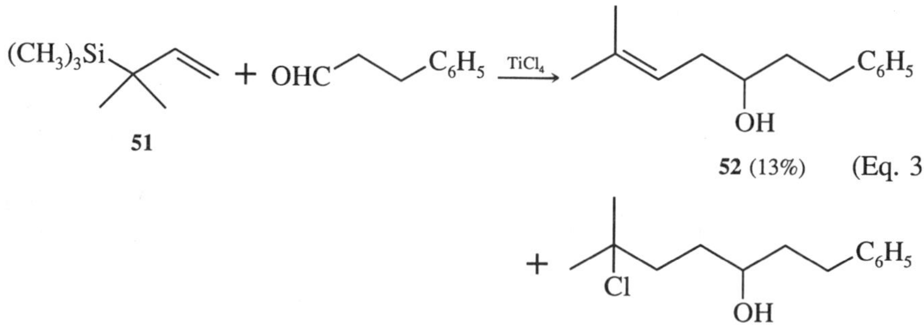

(CH3)3Si **51** + OHC C6H5 —TiCl4→ C6H5 OH **52** (13%) (Eq. 35)

+ Cl C6H5 OH **53** (76%)

Similarly, we can expect that allylsilanes will be more reactive than comparable vinylsilanes. Thus allyltrimethylsilane reacts with dichloroketene to give a cyclobutane, but vinyltrimethylsilane does not.[84,85] Proper comparisons using electrophilic substitution reactions have not been made, but the general impression given by the large body of reactions known for these two classes of compounds is that allylsilanes react under milder conditions or with weaker electrophiles than the corresponding vinylsilanes. Comparisons are readily made with systems that are simultaneously allylsilanes and vinylsilanes, like the compound **54**[86] and allenylsilanes **55** in general. These react with electrophiles as allylsilanes, not as vinylsilanes.

$(CH_3)_3Si$ / $(CH_3)_3Si$ **54** $\xrightarrow[AlCl_3]{CH_3COCl}$ $COCH_3$ / $(CH_3)_3Si$

R_3Si–=C= $\xrightarrow{E^+}$ HC≡C–E

55

Allylsilanes are also more reactive than vinylsilanes when nucleophilic catalysis is used, the usual pattern being that the rate is correlated with the stability of the anion that is departing. Thus indenylsilanes are very readily cleaved by alkali,[87] 3-phenylallyltrimethylsilane is cleaved by 0.64 M aqueous methanolic alkali at 40°,[88] and allyltrimethylsilane is cleaved by boiling methanolic potash.[5] On the other hand, vinylsilanes are not easily cleaved by aqueous alkali unless they are vinyltrichlorosilanes[89,90] or vinyltrifluorosilanes.[91]

As far as ground-state stability is concerned, vinylsilanes appear to be more stable than the corresponding allylsilanes, as shown by the equilibrium in Eq. 36.[92,93] No doubt this thermodynamic difference also contributes to the greater reactivity of allylsilanes. However, the balance is easily disturbed: the metallic nature of silicon makes 5-trimethylsilylcyclopentadiene[94] and the corresponding indene[93] the major isomers at equilibrium.

$(C_6H_5)_3Si$ $\underset{\text{quinoline}}{\overset{\text{heat}}{\rightleftharpoons}}$ $(C_6H_5)_3Si$ (2 kcal) (Eq. 36)

The Effect of the Substituents on the Silyl Group

In most work with allylsilanes and vinylsilanes the trimethylsilyl group is the obvious choice of silyl group. It is cheap and readily available, it contributes only a sharp singlet to NMR spectra, and the byproducts, usually hexamethyldisiloxane, are conveniently volatile. However, other groups are sometimes used, either because they are easier to introduce or because a different degree of reactivity is needed. Other reasons are that larger alkyl groups can decrease the inconvenient volatility of the simplest silanes, phenyl groups provide a chromophore as well as increasing the likelihood of getting crystalline derivatives, and three different groups on silicon can give rise to asymmetric induction. In general, the substituents on the silyl group not directly involved in the reaction have only a small effect on reactivity, at least by comparison with the effect of the same substituents bonded to carbon in ordinary organic chemistry. They do, however, have a noticeable effect: electron-donating groups increase the reactivity, and electron-withdrawing groups decrease it. Quantitative data on arylsilanes provide a guide to the relative reactivity of the corresponding vinylsilanes. Thus the relative rates for protodesilylation of *p*-silylanisoles (p-$R_3SiC_6H_4OCH_3$) are: R_3 = $(CH_3)_3$ (1000), $(C_2H_5)_3$ (490), (i-$C_3H_7)_3$ (55), $(CH_3)_2C_6H_5$ (330), $CH_3(C_6H_5)_2$ (74), and $(C_6H_5)_3$ (16), and the figures for bromodesilylation are very similar.[71] Hydrogen in place of methyl groups appears, spectroscopically at least, to be relatively electron-withdrawing.[95] Thus it appears that the trimethylsilyl group imparts the greatest reactivity of these common groups, but there is one piece of evidence to suggest that the pentamethyldisilyl group is a little more reactive: vinylpentamethyldisilane is epoxidized about twice as fast as vinyltrimethylsilane.[96]

The same trends are present in allylsilanes: whereas 2-methyl-3-trimethylsilylpropene is more than 1000 times more reactive toward diarylmethyl cations than isobutene itself, the corresponding allyltrichlorosilane is approximately 1000 times less reactive.[97] For this reason, substitution can be avoided with allyltrichlorosilanes, as in Eq. 37.[98] The rates of attack by allyltrialkylsilanes on the *p*-methoxydiphenylmethyl cation increase as the alkyl groups on the silicon atom are made larger in the series methyl < ethyl < *n*-butyl, with relative rates of 1:1.6:2.5, respectively. However, the most striking effect of having larger alkyl groups on silicon is that the silyl group may no longer be the electrofugal group, as discussed in the section on failure of the substitution reaction. Allyltriphenylsilane, on the other hand, is less reactive, with a relative rate of 0.02 on the same scale.[97] In agreement with this, the allylsilane **56** reacts with 95% selectivity for loss of the trimethylsilyl group.[99]

Cl_3Si + C_6H_6 $\xrightarrow{AlCl_3}$ Cl_3Si (C_6H_5) + Cl_3Si (C_6H_5) (Eq. 37)

91:9

$(CH_3)_3Si$ $Si(C_6H_5)_3$ **56** $\xrightarrow[AlCl_3]{CH_3COCl}$ $Si(C_6H_5)_3$ ($COCH_3$)

One complication that can arise when one of the groups on silicon is phenyl is cleavage of that group. Thus the phenyl group is cleaved from decyldimethylphenylsilane with tetrabutylammonium fluoride in tetrahydrofuran and dimethyl sulfoxide at 80° for 30 minutes.[100] However, this is only a problem with relatively unreactive allylsilanes or vinylsilanes.

In contrast to these trends, electron-withdrawing groups on silicon sometimes increase the rate of nucleophilic removal of the silyl group. Thus fluoride ion displaces a phenyldimethylsilyl but not a trimethylsilyl group from an unfunctionalized vinylsilane (Eq. 38).[101] However, this is not always the case: the cationic intermediate **9** loses a trimethylsilyl group to give **10** approximately four times faster than it loses a phenyldimethylsilyl group.[29] Similarly, allyltriphenylsilane reacts with 2-acetylnaphthoquinone to give the product **57**, whereas allyltrimethylsilane gives the normal electrophilic substitution product **58**.[102] Evidently the triphenylsilyl group is not displaced from the intermediate cation as fast as the trimethylsilyl group. However, in the acylative desilylation in Eq. 17, the fluorodimethylsilyl group is the electrofugal group, but in the corresponding protodesilylation the trimethylsilyl group is lost.[43] Clearly no generalizations can safely be made at this stage, except perhaps to say that when the groups on silicon are significantly hindered, as with *tert*-butyldimethylsilyl and triisopropylsilyl, it is normal to find that the silyl group is not the electrofugal group.

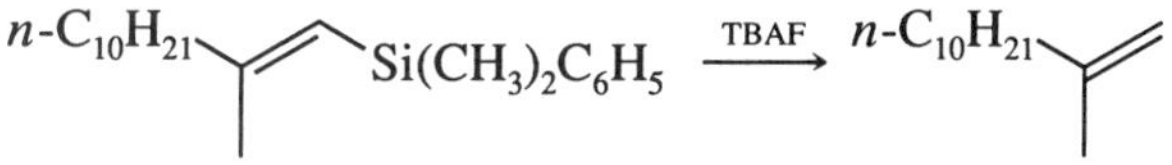

n-$C_{10}H_{21}$ ... $Si(CH_3)_2C_6H_5$ $\xrightarrow{\text{TBAF}}$ n-$C_{10}H_{21}$ ... (Eq. 38)

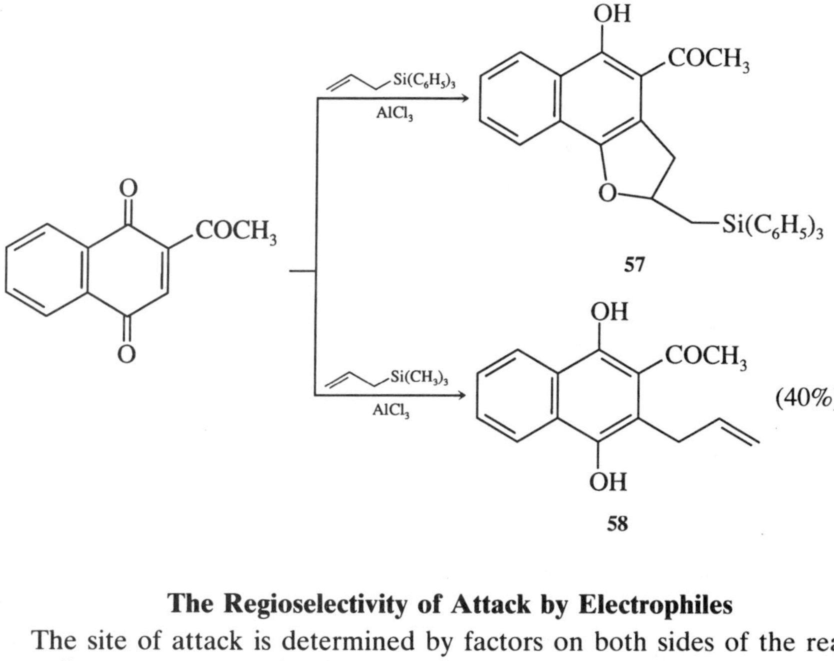

OH COCH₃ O COCH₃ O

$Si(C_6H_5)_3$ / $AlCl_3$ → OH, $COCH_3$, O, $Si(C_6H_5)_3$ (61%)

57

$Si(CH_3)_3$ / $AlCl_3$ → OH, $COCH_3$, OH (40%)

58

The Regioselectivity of Attack by Electrophiles

The site of attack is determined by factors on both sides of the reaction coordinate, as usual. The first consideration is the relative stability of the cations produced by attack at each end of the double bond, but it is likely that the ground-state polarization of the double bond influences regioselectivity to some extent, and there is some experimental support for the ground-state polarization of the HOMO. In the hydroboration of allylsilanes, for example, the boron commonly attaches itself to C-3 unless the substitution pattern polarizes the double bond in the opposite direction.

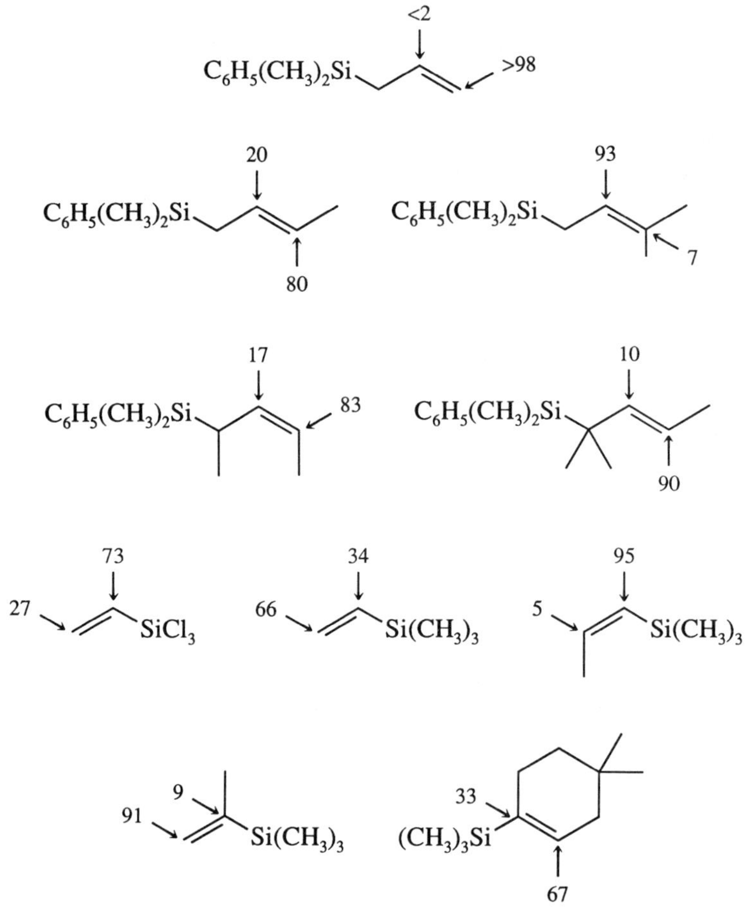

Hydroboration does not take place by way of a cationic intermediate, and the regioselectivity of attack is therefore likely to be affected more by the ground-state polarization of the HOMO than by product stability.[103] The arrows on the structures show the relative proportions of attack by boron at the two sites in the hydroboration of some allylsilanes and vinylsilanes with diborane.[104–111] Cycloaddition of allylsilanes to nitrones, which also takes place without the formation of cationic intermediates, shows strong regioselectivity in the same sense.[112] Furthermore, a silyl group consistently causes an upfield shift of 2–3 ppm in the ^{13}C resonance of C-3 relative to its position in the corresponding hydrocarbon, whereas C-2 is shifted downfield by about 1.5 ppm.[105,113]

Although ground-state polarization plays some part, it is more usual to assess the likelihood of clean electrophilic substitution in an allylsilane or vinylsilane by looking at the two possible cations that can be produced, one in which the silyl group is able to stabilize the cation by hyperconjugation and one in which it is not. Although the silyl group exerts a considerable directing effect, other factors can easily override this control. We have already mentioned how this can come about with 3,3-disubstituted allylsilanes such as allylsilane **15**, with which the overall reaction is an electrophilic substitution. However, the more usual result of electrophilic attack on the central carbon of an allylsilane is that the silyl group remains in the molecule, as in Eq. 39, where the electrophile attacks both the usual site, leading to the product **59**, and at C-2, leading to the lactone **60**.[114] With an even more powerful

$(CH_3)_3Si$ 2 — Cl, $CO_2C_2H_5$, SC_6H_5 / $SnCl_4$ →

$CO_2C_2H_5$, SC_6H_5 + C_6H_5S, O, O, $(CH_3)_3Si$ (Eq. 39)

59 **60**

donor, as in the enamine **61**, the nitrogen directs electrophilic attack by benzoyl chloride entirely to C-2, and the silyl group is ineffective. However, with cesium fluoride as catalyst and benzaldehyde as the electrophile, the silyl

$(CH_3)_3Si$ 1 2 3 N **61**

C_6H_5COCl → C_6H_5CO, $(CH_3)_3Si$, N

C_6H_5CHO / CsF → C_6H_5 O N + N, C_6H_5 OH

2:3

group directs attack away from C-2; there is still the usual loss of regiocontrol with respect to attack at C-1 and C-3.[115] The same problem can arise with vinylsilanes whenever electrophilic attack on the atom to which the silyl group is *not* attached would lead to a more stabilized cation than the usual β-silyl cation, as when the silyl group is attached to C-2 of a terminal alkene (Eq. 40).[116] The problem also occurs when the substituent attached to the same atom as the silyl group is a more powerfully cation-stabilizing group than the substituent attached to the other end of the double bond (Eqs. 41 and 42).

$(CH_3)_3Si$ COCl $\xrightarrow{AlCl_3}$ [$(CH_3)_3Si$ + O] ≡ O (Eq. 40)

$Si(CH_3)_3$ $\xrightarrow[AlCl_3]{CH_3COCl}$ $(CH_3)_3Si$ Cl $COCH_3$ (Eq. 41)

$Si(CH_3)_3$ SC_6H_5 $\xrightarrow[AlCl_3]{CH_3COCl}$ CH_3CO $Si(CH_3)_3$ SC_6H_5 (Eq. 42)

The phenyl and the phenylthio groups are more influential in directing electrophilic attack away from the silyl group than is the silyl group in directing attack to the atom to which it is bound.[117–119]

In cyclization reactions, the constraints of ring formation can be more powerful in determining regiochemistry than the presence of a silyl group. An example is the 6-*endo*-trig reaction in Eq. 43,[120] but it is not clear, in view of the low yield, to what extent the expected 5-*exo*-trig reaction has actually taken place.

CH_3O CH_3O $Si(CH_3)_3$ C_6H_{13}-*n* $\xrightarrow{TiCl_4}$ CH_3O C_6H_{13}-*n* (32%) (Eq. 43)

Failure of the Electrophilic Substitution Because the Silyl Group is not the Electrofugal Group

Sometimes, electrophilic substitution takes place but the silyl group is not the electrofugal group. Thus an allylsilane or vinylsilane that carries a strongly electrofugal group can lose that group rather than the silyl group (Eq. 44).[121] There are many examples with even better electrofugal groups such as tin,[122] boron,[123] copper,[124] aluminum,[125] magnesium,[126] and lithium.[127] These reactions are useful in the synthesis of allylsilanes and vinylsilanes, and they show how easily a silyl group can be carried through synthetic steps before it is used.

$$\mathrm{RCH{=}C[Si(CH_3)_2C_6H_5][Ge(C_2H_5)_3]} \xrightarrow{I_2} \mathrm{RCH{=}C[Si(CH_3)_2C_6H_5]I} \quad \text{(Eq. 44)}$$

A more delicate balance exists between a silyl group and a proton as an electrofugal group. The usual result is the loss of a trimethylsilyl group when the nucleophilic counterion is a halide or oxygen nucleophile. With nitrogen nucleophiles, a proton is removed as in the vinylsilane synthesis (Eq. 45).[128] More hindered silyl groups, like *tert*-butyldimethylsilyl and triisopropylsilyl, are not so readily lost, and are therefore frequently retained in the product.[129–131] However, there are occasions when even halide nucleophiles do not remove a trimethylsilyl group. In the reaction sequence in Eq. 46,[132] the first step is addition **62** → **63**. When this is followed by acid-catalyzed elimination **63** → **66**, the silyl group is not lost. Most probably the silyl group in the intermediate bromonium ion **65** is constrained by the presence of the adamantyl ring to lie in a pseudoequatorial position, and the C–H bond is then better conjugated with the π bond. The problem is overcome by treating the oxetane **63** with fluoride ion, which cleanly removes the silyl group to give the (Z)-vinyl bromide **64**. Evidently the balance of silicophilicity and basicity makes a fluoride ion more selective for silicon than a chloride ion.

$$\mathrm{CH_2{=}CHSi(CH_3)_3} \xrightarrow{Br_2} \mathrm{BrCH_2CH(Br)Si(CH_3)_3} \xrightarrow{(C_2H_5)_2NH} \mathrm{CH_2{=}C(Br)Si(CH_3)_3} \quad \text{(Eq. 45)}$$

This problem is unusual in open-chain silanes like **62**, but it is common with cyclic systems, where conformational constraints quite frequently prevent effective overlap of the Si–C bond and hence the easy loss of the silyl group. The allylsilane **67** undergoes a succession of normal protodesilylations, but it appears that the first step is an isomerization (**67** → **69**).[133] The problem may well be that the silyl group on C-4 is equatorial (the stereochemistry is not actually known), and it may be reluctant to become axial, even in the inter-

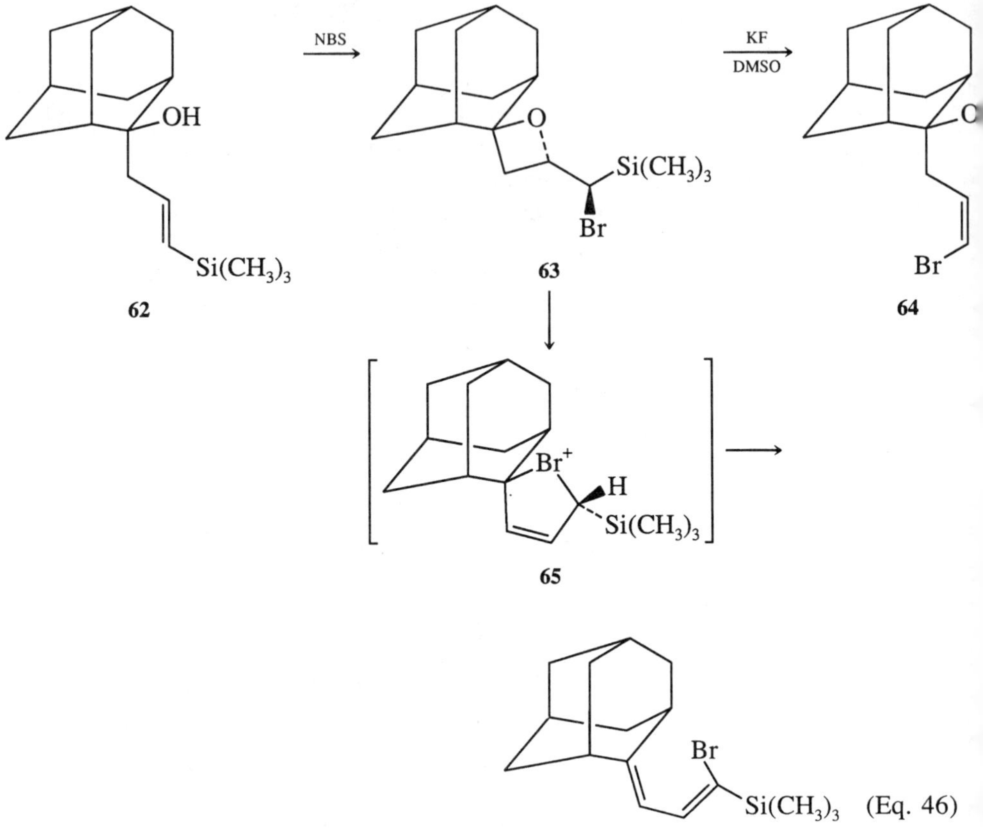

NBS

KF

DMSO

OH

Si(CH$_3$)$_3$

62

O

Si(CH$_3$)$_3$

Br

63

Br

64

Br$^+$

H

Si(CH$_3$)$_3$

65

Br

Si(CH$_3$)$_3$ (Eq. 46)

66

Si(CH$_3$)$_3$

Si(CH$_3$)$_3$

4

Si(CH$_3$)$_3$

Si(CH$_3$)$_3$

67

CH_3CO_2H

reflux

Si(CH$_3$)$_3$

Si(CH$_3$)$_3$

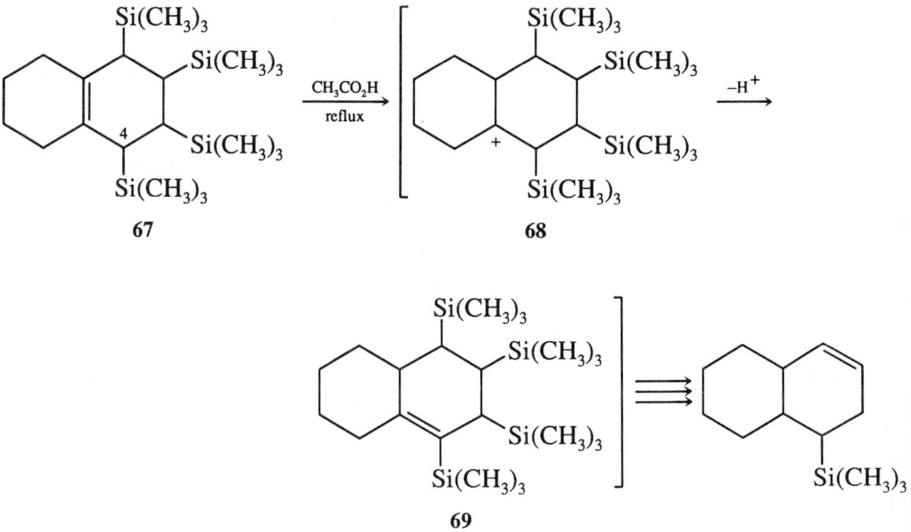

+

Si(CH$_3$)$_3$

Si(CH$_3$)$_3$

68

$-H^+$

Si(CH$_3$)$_3$

Si(CH$_3$)$_3$

Si(CH$_3$)$_3$

Si(CH$_3$)$_3$

69

Si(CH$_3$)$_3$

mediate cation **68**. If this is the case, the C–H bond on C-4 will be axial, and it is therefore the C–H bond that is broken. Similarly, epoxidation of the double allylsilane **70** followed by treatment with *p*-toluenesulfonic acid (TsOH) gives the dienes **71** and **72** with retention of one or both of the silyl groups.[134] The same problem is sometimes found when the silyl group is part of the ring: for example, the sulfonation of the cyclic allylsilane **73** is not entirely selective for the breaking of the Si–C bond.[135] Another manifestation of the same phenomenon is that the vinylsilane epoxide **74** opens with aqueous acid, but the diol **75** is unable to undergo the usual acid-catalyzed *anti* elimination step because the silyl and hydroxy groups are *cis*.[136]

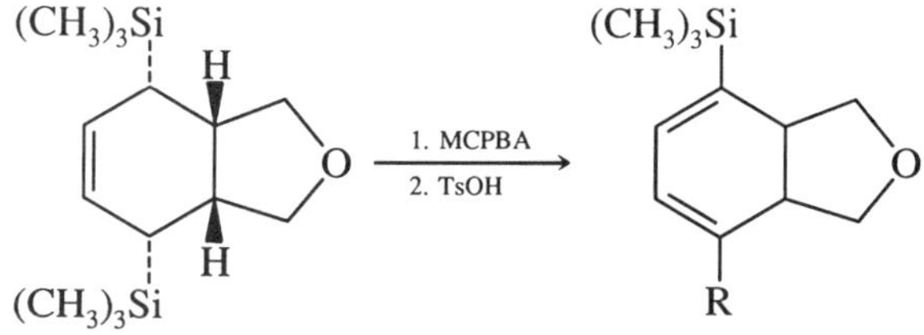

$(CH_3)_3Si$ H O $(CH_3)_3Si$ H **70** —1. MCPBA, 2. TsOH→ $(CH_3)_3Si$ O R

71 R = H
72 R = $Si(CH_3)_3$

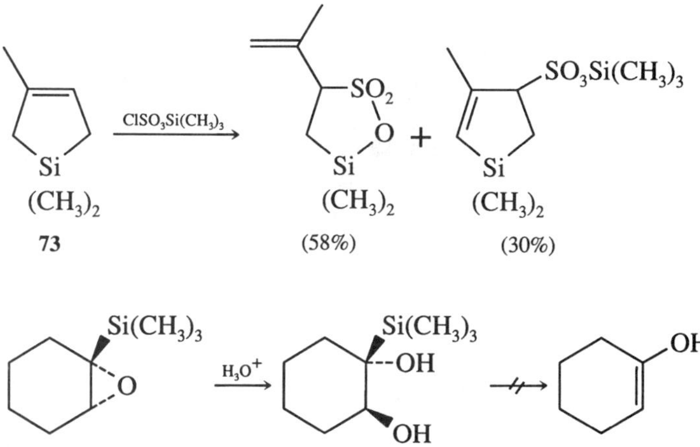

73 Si $(CH_3)_2$ —$ClSO_3Si(CH_3)_3$→ SO_2 O Si $(CH_3)_2$ (58%) + $SO_3Si(CH_3)_3$ Si $(CH_3)_2$ (30%)

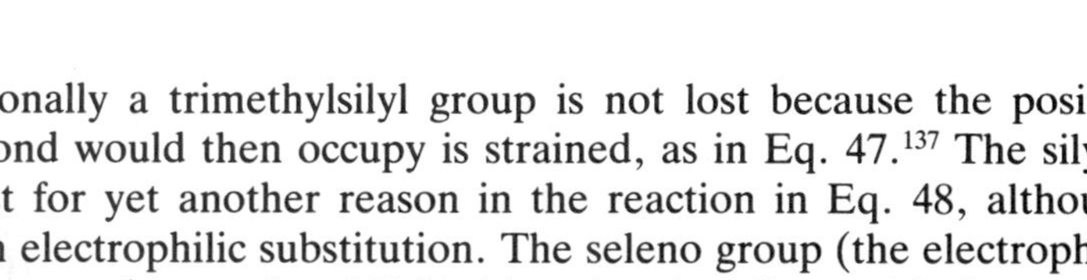

74 $Si(CH_3)_3$ O —H_3O^+→ **75** $Si(CH_3)_3$ OH OH —/→ OH

Occasionally a trimethylsilyl group is not lost because the position the double bond would then occupy is strained, as in Eq. 47.[137] The silyl group is not lost for yet another reason in the reaction in Eq. 48, although it is overall an electrophilic substitution. The seleno group (the electrophile) and the nitro group (the nucleophile) add to the vinylsilane with the usual regiochemistry. However, an oxidation step makes the seleno group the nucleofuge, and the cycloelimination **76** then leaves the silyl group in the molecule.[138]

OTs heat $Si(CH_3)_3$ → [+ $Si(CH_3)_3$] → $Si(CH_3)_3$ (25%) (Eq. 47)

$Si(CH_3)_3$ $\xrightarrow[2.\ AgNO_2]{1.\ C_6H_5SeCl}$ SeC_6H_5 $Si(CH_3)_3$ NO_2 $\xrightarrow{H_2O_2}$

^-O $\overset{+}{Se}$ C_6H_5 H $Si(CH_3)_3$ NO_2 **76** → O_2N $Si(CH_3)_3$ (Eq. 48)

Finally, there are a number of pericyclic reactions in which the attack is electrophilic in character but in which the cyclic transition state involves a hydrogen atom rather than a silyl group. Thus the allylsilanes in Eqs. 49–51 undergo Alder-ene reactions without transfer of the silyl group, even though in the latter cases the nucleophilic atom is oxygen.[139–141] This pathway may also be followed in some of those few Lewis acid catalyzed reactions in which a silyl group is not lost, such as those in Eqs. 249–251. The same selectivity for hydrogen atom transfer is found in the reactions of allylsilanes with singlet oxygen,[142,143] diethyl azodicarboxylate,[139] *N*-phenyltriazolinedione,[139,144] and *N*-sulfinylbenzenesulfonamide ($C_6H_5SO_2NSO$).[145] The silyl group in these reactions may well be activating the adjacent hydrogen.[146]

$(CH_3)_3Si$ + O O O $\xrightarrow{heat}$ $(CH_3)_3Si$ O O O (Eq. 49)

$(CH_3)_3Si$ + $C_2H_5O_2C$ (O) $CO_2C_2H_5$ $\xrightarrow{heat}$

$(CH_3)_3Si$ OH $CO_2C_2H_5$ $CO_2C_2H_5$ (Eq. 50)

$$(CH_3)_3Si\text{-}CH_2CH{=}CH_2 + F_3C\text{-}CO\text{-}CF_3 \xrightarrow[\text{no catalyst}]{100°} (CH_3)_3Si\text{-}CH{=}CH\text{-}CH_2\text{-}C(OH)(CF_3)_2 \quad (86\%) \qquad \text{(Eq. 51)}$$

Other Reactions of Allylsilanes and Vinylsilanes

Allylsilanes and vinylsilanes undergo reactions other than electrophilic substitution, and a few of these are relevant to this chapter.

Allylsilanes are remarkably stable (in the absence of fluoride ion) to allylic rearrangement in the sense of Eq. 33. The uncatalyzed [1,3]-sigmatropic shift takes place at 500° intramolecularly with inversion of configuration at the silyl group.[92] At higher temperatures (>580°), allyltrimethylsilane decomposes by pericyclic (Eq. 52) and radical pathways.[147] There is possibly one example of a 1,3 shift of a silyl group catalyzed by Pd(II),[148] and certainly one photochemical 1,3 shift.[149]

$$(CH_3)_2Si(CH_2CH{=}CH_2)(CH_2\text{-}H) \xrightarrow{\text{heat}} [(CH_3)_2Si{=}CH_2] + CH_3CH{=}CH_2 \qquad \text{(Eq. 52)}$$

The geometry of the double bond of an allylsilane can be equilibrated by catalysis with diisobutylaluminum hydride.[150] The corresponding equilibration of vinylsilane geometry is possible with *N*-bromosuccinimide in pyridine,[151] and photochemically in the case of styrylsilanes.[152]

In contrast to electrophiles and fluoride nucleophiles, radicals frequently attack allylsilanes and vinylsilanes to give addition (Eqs. 53–55).[6,153,154] Evidently the loss of a trialkylsilyl radical from the intermediate β-silylethyl radical is not as easy as is the loss of a trialkylstannyl radical from the corresponding tin compounds.[155] An alternative radical reaction is hydrogen replacement (Eqs. 56 and 57).[156,157] However, substitution can take place by a radical pathway when the chain-carrying step is a halogen abstraction (Eq. 58) because this creates a β-silylethyl halide **77**, which undergoes easy β elimination in the usual way.[158] A similar reaction is presumably involved in the *C*-allylation of the iodopyrimidone in Eq. 59,[159] and in the substitution reaction (Eq. 60), which was actually the first to establish carbon–carbon bond formation from an allylsilane.[160] In this type of reaction, vinyltrimethylsilane simply undergoes addition (Eq. 61), and shows little capacity for polymerization.[160]

$$C_2H_5CHO + CH_2{=}CH\text{-}Si(CH_3)_3 \xrightarrow{(CH_3CO_2)_2} C_2H_5CO\text{-}CH_2CH_2\text{-}Si(CH_3)_3 \qquad \text{(Eq. 53)}$$

$(CH_3)_3Si$ (allyl) $\xrightarrow{C_6H_5SH}$ $(CH_3)_3Si$–$CH_2CH_2CH_2$–SC_6H_5 (Eq. 54)

(vinyl)$Si(CH_3)_3$ $\xrightarrow{C_6H_5SH}$ C_6H_5S–CH_2CH_2–$Si(CH_3)_3$ (Eq. 55)

Si $(CH_3)_2$ $\xrightarrow{t\text{-}C_4H_9O_3CC_6H_5}$ $O_2CC_6H_5$ Si $(CH_3)_2$ (Eq. 56)

$Si(C_2H_5)_3$ $\xrightarrow{NBS}$

Br $Si(C_2H_5)_3$ + Br $Si(C_2H_5)_3$ (Eq. 57)

$(CH_3)_3Si$ + O O ClN $\longrightarrow$

[O O Cl $(CH_3)_3Si$ N] $\longrightarrow$ O O N (Eq. 58)

77

O CH_3N O N I CH_3 + $Si(CH_3)_3$ $\xrightarrow{h\nu}$ O CH_3N O N CH_3 (Eq. 59)

$(CH_3)_3Si$ + $BrCCl_3$ $\xrightarrow[80^\circ]{(C_6H_5CO_2)_2}$ CCl_3 (52%) (Eq. 60)

(vinyl)$Si(CH_3)_3$ + $BrCCl_3$ $\xrightarrow[80^\circ]{(C_6H_5CO_2)_2}$ Cl_3C–CH_2–CHBr–$Si(CH_3)_3$ (89%) (Eq. 61)

SCOPE AND LIMITATIONS

In this section we discuss, electrophile by electrophile (in order of the atomic number of the electrophilic atom), the reactions of allylsilanes, vinylsilanes, and allenylsilanes, describing the reaction conditions commonly used and identifying the special features associated with each electrophile. Where appropriate, nucleophile-catalyzed reactions are discussed after the corresponding acid-catalyzed reactions.

Protodesilylation and Deuterodesilylation

The proton is one of the most frequently used electrophiles in both allylsilane and vinylsilane chemistry. Protodesilylation is also frequently a major source of side products when other electrophilic substitutions are being attempted,[161–163] particularly those involving the use of molar proportions of Lewis acids[164–166] or of fluoride ions.[167,168]

Allylsilanes. It is usually possible to isolate the first-formed product of protodesilylation before it is isomerized by further protonations and deprotonations, because the silyl group has an activating effect on the double bond of the starting material. In acid-catalyzed reactions, it is simply a matter of finding a protic acid and reaction conditions strong enough for the first step but not for the second. The choice of acids is wide; in many cases it is idiosyncratic, but usually a brief trial of the commonly used acids is enough to find one that works. Clearly the choice of acid, of concentration, and of time and temperature are largely determined by the degree of substitution of the double bond in the allylsilane.

When the intermediate cation is tertiary, a fairly weak acid or mild conditions can be used. Thus hot acetic acid is enough to protodesilylate the allylsilane in Eq. 9, and cold dilute aqueous hydrochloric acid is adequate in a similar situation (Eq. 62).[169]

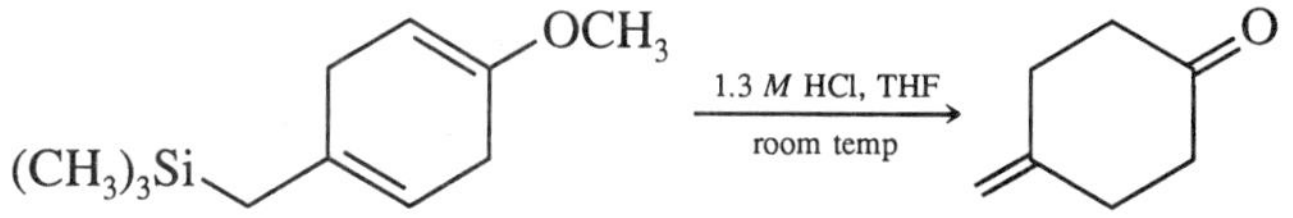

$(CH_3)_3Si$ ... OCH_3 $\xrightarrow[\text{room temp}]{1.3\ M\ \text{HCl, THF}}$ ... O (Eq. 62)

When the intermediate cation is secondary, several strong acids are effective, no single one of which can be said to be the most popular. The boron trifluoride–acetic acid complex in dichloromethane at room temperature (Eq. 63),[170,171] trifluoroacetic acid in dichloromethane at room temperature (Eq. 64),[172] and liquid hydrofluoric acid at −20° (Eq. 22),[20] have all been rec-

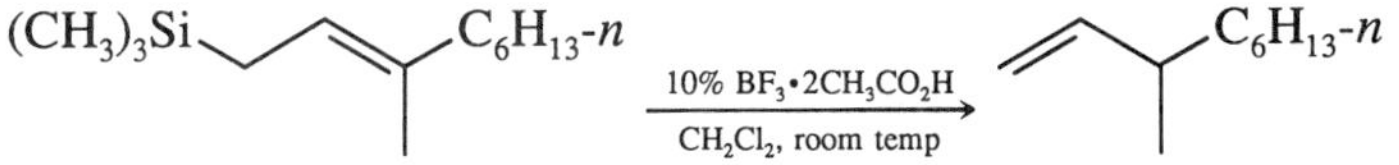

$(CH_3)_3Si$ ... C_6H_{13}-*n* $\xrightarrow[CH_2Cl_2,\ \text{room temp}]{10\%\ BF_3 \cdot 2CH_3CO_2H}$ ... C_6H_{13}-*n* (Eq. 63)

$C_6H_5(CH_3)_2Si$ —— 2% TFA, CCl_4 / room temp → (Eq. 64)

ommended. Other acids that have found favor in particular cases are *p*-toluenesulfonic acid in refluxing benzene,[134] concentrated sulfuric acid in methanol at room temperature,[55] methanesulfonic acid,[173] acetyl chloride in methanol,[174] fluoroboric acid etherate,[175] pyridinium trifluoromethanesulfonate,[176] and trifluoromethanesulfonic acid.[177] Difficulties in preserving the double bond in the product can be expected when the intermediate cation is secondary and the product is a 1,1-disubstituted alkene. So far, examples of this type of substitution have not been studied.

When the intermediate cation is destabilized by a neighboring carbonyl group, rather vigorous acidic conditions have to be used (Eq. 65).[178] In this

O, C_4H_9-*n*, $Si(CH_3)_2C_6H_5$ —— $BF_3 \cdot 2CH_3CO_2H$, CCl_4 / reflux → O, C_4H_9-*n* (Eq. 65)

78

example, it is possible that the ketone is enolized during protodesilylation, but in Eq. 66 enolization is impossible and the reaction still occurs.[179] A

$CON(C_2H_5)_2$, $Si(CH_3)_3$, O —— HCl, MeOH / reflux → $CON(C_2H_5)_2$, O (Eq. 66)

mechanistic detail is revealed in the protodesilylation in Eq. 67, in which there is no *endo–exo* isomerization of the neopentyl group in the allylsilanes **79** and **80** as the reaction proceeds. On the reasonable assumption that the cationic intermediates **81** and **82** are in rapid equilibrium, this result shows that they always lose a silyl group and never a proton.[180] The first step of the protodesilylation mechanism is therefore effectively irreversible.

Protodesilylation of allylsilanes takes place stereospecifically *anti*, as described in the section on mechanism. A different kind of stereocontrol is that illustrated by the examples in Eqs. 68 and 69, where the internal delivery of the proton in a chairlike transition state accounts for the relative configuration found for the major products **83** and **84** with 1,3- and 1,4-related stereocenters.[181,182]

79 $\xrightarrow{H_2SO_4}$ 81 ⇅ 82 $\xleftarrow{H_2SO_4}$ 80 → $HO_3SOSi(CH_3)_2$ (Eq. 67)

$\xrightarrow{BF_3 \cdot 2CH_3CO_2H}$ 83 >8:1 (Eq. 68)

$\xrightarrow{BF_3 \cdot 2CH_3CO_2H}$ 84 9:1 (Eq. 69)

Deuterodesilylations can be carried out with deuterated trifluoroacetic acid[29,41,44] and with deuterated *p*-toluenesulfonic acid.[55]

Protodesilylation is also possible using nucleophilic catalysis. The detailed mechanism of such protodesilylations is complex and far from fully understood.[183] Reagents that are used with unfunctionalized allylsilanes are methanolic potash,[28] potassium *tert*-butoxide in hexamethylphosphoramide (HMPA) at 60° for a few hours,[184] and cesium fluoride in dimethylformamide (DMF) or dimethyl sulfoxide (DMSO) at 100° for an hour.[185-187] The protonation step usually leads to the more-substituted alkene, regardless of the site of the silyl group, so that acid and nucleophile-catalyzed protodesilylations can be complementary (Eq. 70).[185]

Nucleophile-catalyzed desilylation is easier when there are anion-stabilizing groups like sulfur substituents present. The protonation takes place adjacent

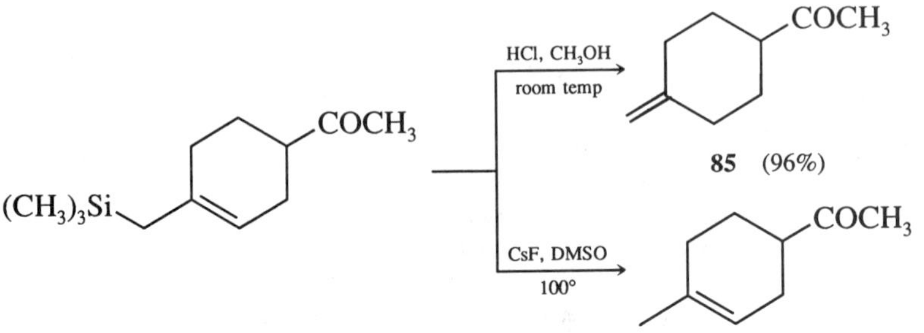

(Eq. 70)

to the sulfur substituent.[188,189] It is also easier when the nucleophile is built into the molecule, as in Eq. 71. The product in this example is set up for an oxy-Cope rearrangement.[190]

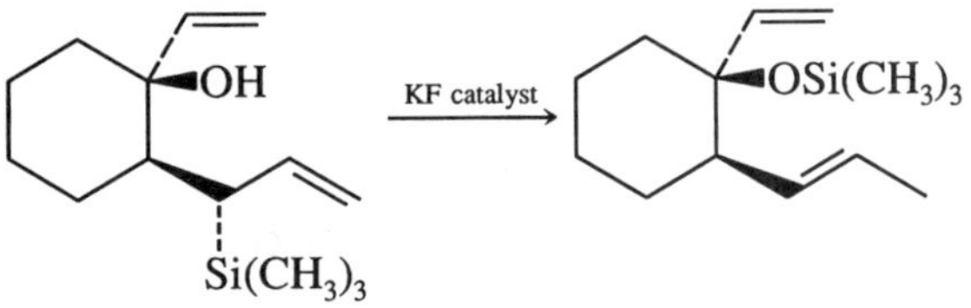

(Eq. 71)

In synthesis, acid-catalyzed protodesilylation is a powerful method for controlling the position of an isolated double bond. With allylsilanes, it is used in the synthesis of δ-terpineol,[174,185] dihydrojasmone (**78**),[178] some cytochalasins,[187] some carbaprostacyclins,[176] ϵ-muurolene and ϵ-cadinene,[185] and the Prelog–Djerassi lactone.[191] The protodesilylation of allylsilanes and vinylsilanes is also used as a method of silylating alcohols (Eq. 72).[177,192–195] The

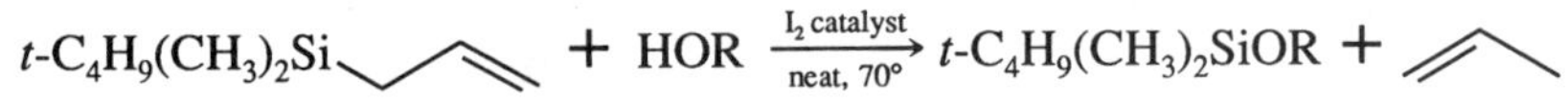

$$t\text{-}C_4H_9(CH_3)_2Si\text{CH}_2\text{CH=CH}_2 + HOR \xrightarrow[\text{neat, }70^\circ]{I_2\text{ catalyst}} t\text{-}C_4H_9(CH_3)_2SiOR + \text{CH}_2\text{=CHCH}_3 \qquad \text{(Eq. 72)}$$

method is particularly significant because it uses acidic conditions, in contrast to the usual methods of silylating alcohols. Protodesilylation of allylsilanes and vinylsilanes consumes protons, and it is implicit therefore that allylsilanes and vinylsilanes could be used to remove protons completely from acidic reaction media. This capacity does not appear to have been exploited. Finally, protodesilylation of allylsilanes is a method for converting tetraorganosilanes into triorganofluorosilanes, preparatory to carrying out the oxidative rearrangement that converts a silyl group into an alcohol.[196] Most of these reactions are not included in Tables I and II because the product usually isolated is the silyl ether or the silyl fluoride, not the propylene or ethylene that is the electrophilic substitution product relevant to this chapter.

The fluoride-catalyzed reaction in Eq. 70 is used in a synthesis of α-terpineol, just as the acid-catalyzed reaction is used in the synthesis of δ-terpineol.[185,186]

Vinylsilanes. As with allylsilanes the choice of acid to use for protodesilylation is a matter of judging the stability of the intermediate cation, but for vinylsilanes the choice of acid is more delicate, especially when the intermediate cation is tertiary. Thus the product **87** from the protodesilylation

$(CH_3)_3Si$ — TsH, CH_3CN, H_2O, room temp →

86 **87**

of the vinylsilane **86** is sensitive to acid, and a careful choice of conditions has to be made to avoid subsequent reactions. *p*-Toluenesulfinic acid (TsH) is effective in this case,[197] but *p*-toluenesulfonic acid in acetonitrile–water–tetrahydrofuran is effective for another exocyclic compound (Eq. 73), where the configuration and position of the double bond are maintained.[198] Hydriodic

$Si(CH_3)_3$ — TsOH, CH_3CN, THF, H_2O, reflux → (73%) (Eq. 73)

acid, in the form of iodine in wet benzene, is similarly effective in another delicate example (Eq. 74),[72] where a higher concentration of hydriodic acid causes isomerization of the product.

$Si(CH_3)_3$ C_5H_{11}-*n* — I_2, C_6H_6, H_2O → C_5H_{11}-*n* (Eq. 74)

When the intermediate cation is secondary, there is rarely much difficulty in finding a suitable acid. Hydriodic acid in benzene at room temperature is commonly used; the reaction is usually stereospecific, the reaction in Eq. 34 giving only 6% of *trans* product.[72] Fluoroboric acid in aqueous acetonitrile is recommended,[175] and other acids like hydrochloric acid in chloroform[199] and *p*-toluenesulfinic acid[197] are effective.

Deuterodesilylations can be carried out with iodine and deuterium oxide[72] and with deuterium chloride in acetonitrile.[16]

With unsubstituted vinylsilanes, when the intermediate cation is primary, relatively vigorous conditions have to be used. Tetravinylsilane, for example, needs hydrogen chloride and aluminum chloride at 85°.[200] In less vigorous conditions, addition is usually the first step (Eq. 4) because the subsequent elimination to give ethylene itself is comparatively slow. When the silyl group is on C-2 of a terminal alkene, addition sometimes takes place in the sense of Eq. 75, so that subsequent desilylative elimination is not possible.[6] Alternatively, the reaction is simply low-yielding (Eq. 76).[197] In the absence of a

$Si(CH_3)_3$ —(HCl, H_2O, 50°)→ Cl $Si(CH_3)_3$ (Eq. 75)

C_6H_{11} $Si(CH_3)_3$ **88** —(TsH, CH_3CN, H_2O, reflux)→ C_6H_{11} (5%) (Eq. 76)

good nucleophile for carbon, rearrangement takes place rather than simple addition, as in the synthesis of *tert*-butyldimethylsilyl trifluoromethanesulfonate (Eq. 77),[201] which is based on a similar reaction giving the corresponding

$Si(CH_3)_3$ —(HO_3SCF_3)→ [+ CH_3 $Si(CH_3)_2$ $^-OSO_2CF_3$] ⟶ $t\text{-}C_4H_9(CH_3)_2SiO_3SCF_3$ (66%) (Eq. 77)

sulfate.[202] The greater reactivity of terminal vinylsilanes relative to vinylsilanes like **88**, in which the silyl group is on C-2, is shown by the selective protodesilylation of the terminal silyl group in the vinylsilanes **89**.[203,204]

$(CH_3)_3Si$ R $Si(CH_3)_3$ **89** —(R = C_6H_5) CH_3CO_2H, reflux; (R = $n\text{-}C_6H_{13}$) CH_3CO_2H, H_2O, 110°→ $(CH_3)_3Si$ R

As with allylsilanes, the reaction is slow but still possible when the intermediate cation is conjugated with a carbonyl group (Eq. 78).[117] However, with the vinylsilane **90**, the deactivation by the ester groups is countered by activation by the amino group. In this case, methanol alone removes the silyl group.[205]

The protodesilylation of vinylsilanes nearly always takes place with preservation of the double bond and with a high degree of retention of configuration, as discussed in the section on mechanism. Exceptions to this rule are usually the result of the subsequent equilibration of the product olefin, but

$(CH_3)_3Si$ … $COCH_3$ $\xrightarrow[\text{reflux}]{\text{TsOH, } C_6H_6}$ … $COCH_3$ (Eq. 78)

$CO_2C_2H_5$, RNH, $Si(CH_3)_3$, $CO_2C_2H_5$ $\xrightarrow{CH_3OH}$ RNH, $CO_2C_2H_5$, $CO_2C_2H_5$

90

the stereoisomeric vinylsilanes (E)-**91** and (Z)-**91** equilibrate faster than they undergo protodesilylation.[197] Deuterodesilylation is therefore stereorandom. In this exceptional case, it appears that the loss of a proton **92** → **91** is faster than the loss of the silyl group **92** → **93**.

n-C_6H_{13} $Si(CH_3)_3$ (E)-**91** ⇌ (TsH, CH_3CN) n-C_6H_{13} (+) $Si(CH_3)_3$ **92** → n-C_6H_{13} **93**

$Si(CH_3)_3$ n-C_6H_{13} (Z)-**91**

Nucleophile-catalyzed protodesilylation of vinylsilanes is difficult unless the putative vinyl anion is stabilized. When there is no anion stabilization, trimethylsilyl groups can be removed with sodium ethoxide in dimethyl sulfoxide at 130°,[206] but a phenyldimethylsilyl group can be removed with tetrabutylammonium fluoride in dimethyl sulfoxide at 80° (Eq. 38). A proton is the only electrophile that can be used in this reaction.[101] However, anion stabilization makes desilylation much easier. A halogen substituent, for example (Eqs. 79 and 80), allows warm sodium methoxide in methanol to work, and

Br, $Si(CH_3)_3$ $\xrightarrow[40°]{NaOCH_3,\ CH_3OH}$ Br (74%) (Eq. 79)

$Si(CH_3)_3$, Br $\xrightarrow[40°]{NaOCH_3,\ CH_3OH}$ Br (84%) (Eq. 80)

configuration is cleanly maintained.[207] Cyano groups help the protodesilylation even more, so that sodium methoxide in methanol at 0° is now powerful enough.[208]

A suitably placed hydroxy group helps to remove the silyl group from vinylsilanes. There are two ways in which such hydroxy groups can be used. In one, the hydroxy group is probably functioning as a means of delivering fluoride ion intramolecularly (Eq. 81).[209] In the other, the alkoxide ion itself is an intramolecular nucleophile (Eq. 82).[210] In both cases, the reaction fails

C_6H_5, OH, $Si(CH_3)_3$ $\xrightarrow[80°]{(CH_3)_4NF,\ CH_3CN}$ C_6H_5, OH (65%) (Eq. 81)

$(CH_3)_3Si$, OH $\xrightarrow[30°]{NaH,\ HMPA,\ THF}$ OH (88%) (Eq. 82)

without the hydroxy group. These reactions are useful because the bulk of the silyl group is often helpful in controlling stereochemistry on both the double bond and the neighboring chiral center. However, the double-bond geometry is not preserved when the intermediate anion is phenyl-substituted (Eq. 83).[211]

OH, C_6H_5, $Si(CH_3)_2C_6H_5$ $\xrightarrow[0.5\ h]{NaH\ cat.,\ HMPA}$ OH, C_6H_5 (83%) (Eq. 83)

E:*Z* 89:11

The acid-catalyzed protodesilylation of vinylsilanes is used in the synthesis of some insect pheromones[212,213] and of β-agarofuran **87**,[214] and the base-catalyzed removal of a silyl group is used in a synthesis of the sex pheromone of the codling moth[215] and in syntheses of vertinolide, eldanolide, and protomycinolide IV.[126,216,217] Base-catalyzed reactions of the type in Eq. 82 are used in syntheses of (+)-blastimycinone, corynomycolic acid, brevicomin, and isoavenaciolide.[218–221] The regiocontrol of both allylsilane and vinylsilane protodesilylations is combined in the sequence shown in Eq. 84: the sulfoxide

C_6H_5, C_6H_5SO, $Si(CH_3)_3$ $\xrightarrow{heat}$ C_6H_5, $Si(CH_3)_3$ **94** (83%) + C_6H_5, $Si(CH_3)_3$ **95** (17%) $\xrightarrow[C_6H_6]{HI}$ C_6H_5 (79%) (Eq. 84)

undergoes thermal elimination in both directions although more toward the silyl group than away from it. However, both the vinylsilane **94** and the allylsilane **95** undergo protodesilylation to place the double bond specifically at the end of the chain.[222]

Allenylsilanes. There are few reports on the reaction of allenylsilanes with protons. The allenylsilanes **96** and **97** are sulfonic acid derivatives, and the sulfonic acid group is the source of the protons. Allenylsilane **96** behaves as an allylsilane, but **97** is anomalous in that a silyl shift takes place in addition

$(CH_3)_3Si$, $(CH_3)_3Si$ =C= $SO_3Si(CH_3)_3$, C_4H_9-*t* (**96**) $\xrightarrow[\text{2. } NaHCO_3]{\text{1. } H_2O}$ $(CH_3)_3Si—C≡C—$ SO_3Na, C_4H_9-*t*

$(CH_3)_3Si$, HO_3S =C= $Si(CH_3)_3$ (**97**) $\xrightarrow{H_2O}$ O_2S, O, $Si(CH_3)_3$

to other steps.[223] Silyl shifts in allenylsilane reactions are quite common, as in Eqs. 193 and 246. The diastereoisomeric allenylsilanes **98** and **99** undergo protodesilylation stereospecifically *anti*, but the yields are not high.[38]

98 (C_6H_5, C=C, CH_3, $Si(CH_3)_2C_6H_5$) $\xrightarrow[CH_2Cl_2,\ 0°]{BF_3\cdot 2CH_3CO_2H}$ (C_6H_5, H, C≡C, CH_3) (46%)

99 (C_6H_5, C=C, $Si(CH_3)_2C_6H_5$, CH_3) $\xrightarrow[CH_2Cl_2,\ 0°]{BF_3\cdot 2CH_3CO_2H}$ (C_6H_5, H, C≡CCH_3) (46%)

Nucleophilic removal of a silyl group from an allene appears to be easier than from a vinylsilane; the products are mixtures of allenes and acetylenes (Eq. 85).[224]

$(CH_3)_3Si$ C= $\xrightarrow{KOH,\ CH_3OH}$ C= + HC≡C

56:44

(Eq. 85)

Carbon Electrophiles

The Choice of Lewis Acid or Nucleophilic Catalyst. Most uncharged carbon electrophiles are not powerful enough to react with allylsilanes or vinylsilanes on their own; they usually need catalysis by Lewis acids or nucleophiles. The choice of Lewis acid is often an empirical one, but some guidance can be found in the sections that follow. There is no simple order of Lewis acid strength, since it is dependent in part upon whether the Lewis acid is complexing with nitrogen, oxygen, halogen, or sulfur. However, it is commonly accepted that aluminum chloride and titanium tetrachloride are powerful Lewis acids, that boron trifluoride etherate and stannic chloride are moderately powerful, and that zinc halides and silyl halides are at the weak end of the range. It is also commonly found that, for any given reaction, one Lewis acid often works much better than any other, and it is not always the same Lewis acid for any given electrophile, allylsilane, or vinylsilane. It is always wise to try several Lewis acids if the first one does not succeed.[225] One of the most common side reactions is protodesilylation. One solution to this problem is to use trimethylsilyl trifluoromethanesulfonate, which often works in catalytic amounts, unlike most Lewis acids, which have to be used stoichiometrically. Another solution is to use alkylaluminum chlorides, since they are inherently proton-free.[226] It is also possible to add dialkyltin dichloride as a proton scavenger before performing the reaction with the carbon electrophile.[227a] However, these devices do not always solve the problem, because the offending reaction may be a metal-for-silicon exchange rather than protodesilylation. Another occasional problem is equilibration of alkene geometry by the Lewis acid before the reaction takes place.[228]

The choice of nucleophilic catalyst is relatively easy because of the high selectivity of fluoride ion for attack at silicon rather than at carbon or hydrogen. However, the choice of the fluoride ion source is still a matter for experiment, and protodesilylation is again a common problem, particularly since most fluoride ion sources are extremely difficult to dry thoroughly. Tetrabutylammonium fluoride (TBAF) in tetrahydrofuran is notorious in this respect,[227b] but there are occasions when it works only if it is slightly wet.[227a] For a really dry fluoride ion source, trisdimethylaminosulfonium difluorotrimethylsilicate (TASF) is recommended,[61] but for most purposes tetrabutylammonium fluoride dried with molecular sieves is effectively proton-free.[162] Benzyltrimethylammonium fluoride is another useful fluoride ion source, as are cesium fluoride and other alkali metal fluorides.

A few electrophiles are strong enough to react directly without needing catalysis, notably chlorosulfonyl isocyanate (CSI) and tetracyanoethylene (TCNE), which are uncharged, and a few metal-stabilized cationic electrophiles.

Carbon Electrophiles Needing No Catalysis

Allylsilanes. Chlorosulfonyl isocyanate is apt to give cycloaddition products with simple alkenes, and does so with allylsilanes (Eq. 86). However,

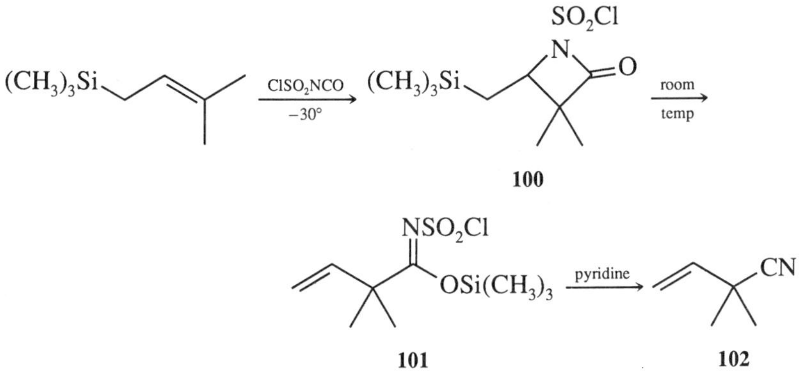

(Eq. 86)

substitution is favored by the presence of the silyl group, so that merely warming the intermediate **100** gives its isomer **101**, which can be converted to the nitrile **102**.[229] In some cases the intermediate lactam is not observed, and aqueous workup simply gives the primary amide. Chlorosulfonyl isocyanate is used as the electrophile in the key step (Eq. 87) in a synthesis of loganin.[55]

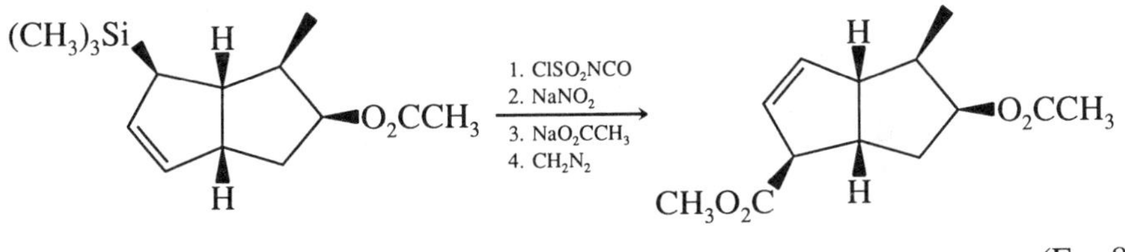

(Eq. 87)

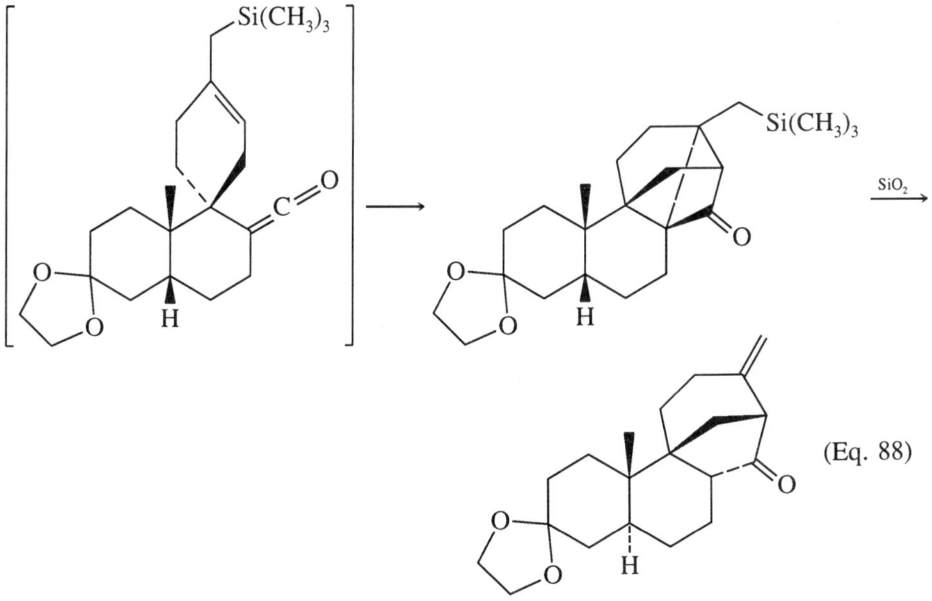

(Eq. 88)

Dichloroketene usually reacts with allylsilanes to give cycloaddition products,[55,84] but in one case the cycloaddition product with a dialkylketene, formed in an intramolecular reaction (Eq. 88), is so strained that it opens in the presence of silica gel. The overall result is an electrophilic substitution on the allylsilane group by an acylating agent.[230]

Tetracyanoethylene reacts with allyltrimethylsilane to give a mixture of cycloaddition and substitution products (Eq. 89), the proportions of which are very solvent-dependent.[30]

Solvent	Cycloadduct	Substitution product
CH_2Cl_2	(> 95%)	(< 5%)
$CH_3CO_2C_2H_5$	(66%)	(33%)
CH_3CN	(13%)	(87%)

(Eq. 89)

Allylsilanes react with nitrones in a cycloaddition reaction (Eq. 90), and reduction of the cycloadduct **103** followed by acid- or base-catalyzed elimination of the silyl and hydroxyl groups gives homoallylamines.[231,232]

103 (94%) → 1. $LiAlH_4$; 2. H^+ or OH^- → (72%)

(Eq. 90)

The dithienium cation **104**,[233] the cobalt-stabilized propargylic cation **105**,[234] and iron-stabilized cyclohexadienyl cations like **106**[235–237] react with allylsilanes with high levels of regiocontrol in both partners. The metal group can be removed from the products by oxidation.

$(CH_3)_3Si$

104 (79%)

C_6H_5 $Co_2(CO)_6$ CH **105** (91%)

$Fe(CO)_3$ CO_2CH_3 **106** (95%)

Vinylsilanes. As with allylsilanes, vinylsilanes react with chlorosulfonyl isocyanate to give cycloaddition products that can be induced to open during workup (Eq. 91).[238] They also undergo cycloaddition with aldehyde-derived

C_6H_5 $Si(CH_3)_3$ $\xrightarrow{CSI}$ C_6H_5 $Si(CH_3)_3$ N SO_2Cl =O $\xrightarrow{H_2O}$ C_6H_5 $CONH_2$

(Eq. 91)

nitrones, and the cycloadducts can be opened in an acid-catalyzed elimination of the silyl group with breaking of the N–O bond to give eventually α,β-unsaturated aldehydes (Eq. 92).[239] Alternatively, styryltrimethylsilane gives a regioisomeric cycloadduct that can be reduced, and the product converted stereospecifically into the *cis* or the *trans* allylamine (Eq. 93).[232]

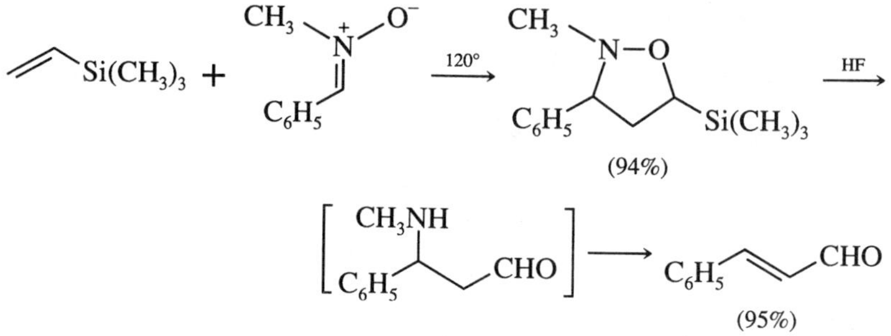

(Eq. 92)

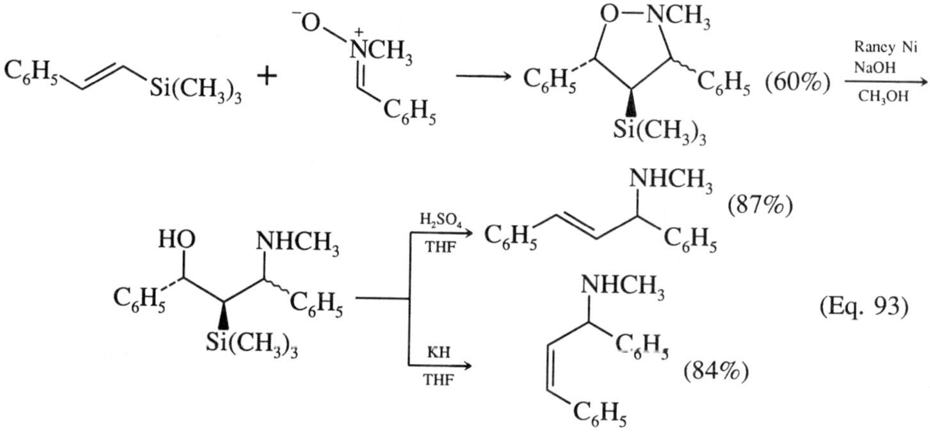

(Eq. 93)

Simple iminium ions react with both allylsilanes and vinylsilanes. Although they can sometimes be isolated, and would then properly belong in this section, they are more frequently generated in situ. For this reason, they are all, with the exception of the nitrones shown above, discussed in the section on iminium ions, regardless of whether the iminium ion needs catalysis for its generation or not.

Alkyl Halides, Alcohols, Ethers, Esters, Nitroalkanes, Alkenes, and Arenes. Allylsilanes and vinylsilanes react not only with metal-stabilized carbonium ions but also with carbonium ions generated in situ from suitable precursors. Most of the substrates are therefore of the kind associated with S_N1 behavior, but the electrophile may not always be a free cation.

Allylsilanes. Tertiary alkyl halides react stereospecifically and regiospecifically in the presence of titanium tetrachloride (Eq. 94),[21] and the reaction

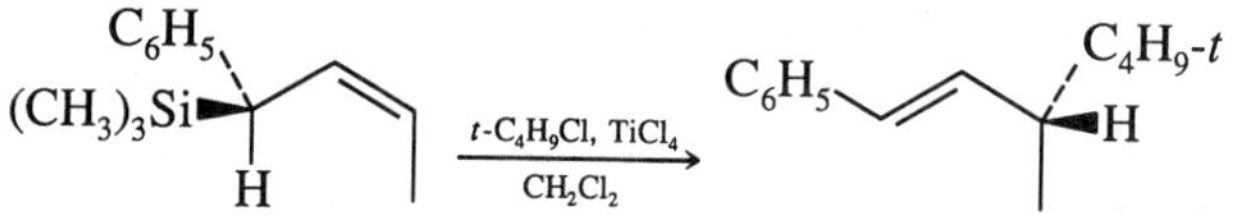

(Eq. 94)

even works when two quaternary centers are bonded (Eq. 95).[170] Secondary benzylic or allylic halides or ethers react similarly.[240] Unsymmetrical allylsi-

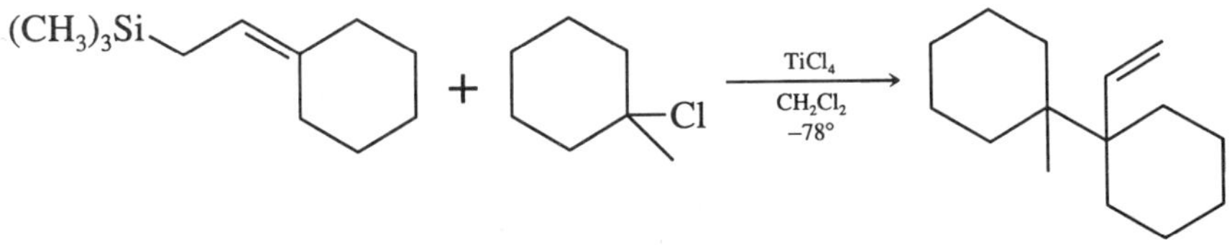

$(CH_3)_3Si$ + Cl $\xrightarrow[CH_2Cl_2,\ -78°]{TiCl_4}$ (98%)

(Eq. 95)

lanes and unsymmetrical allyl halides and ethers show complete regiocontrol with respect to the allylsilane-derived portion of the product, but not usually with respect to the allyl halide or ether portion.[241] The deuterium label in Eq. 96, however, shows that the integrity of the allylic system of the electro-

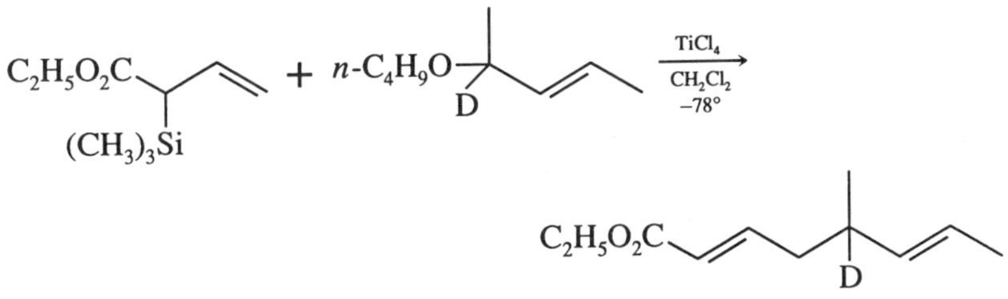

$C_2H_5O_2C$, $(CH_3)_3Si$ + n-C_4H_9O, D $\xrightarrow[CH_2Cl_2,\ -78°]{TiCl_4}$ $C_2H_5O_2C$, D (60%)

(Eq. 96)

phile is retained in this case and that a free allylic cation is not involved.[240] Unsymmetrical allylic lactones (Eq. 97) react mainly at the less-substituted end of the allylic system.[242] The catalyst in this case is Meerwein's salt, rather than the usual Lewis acid.

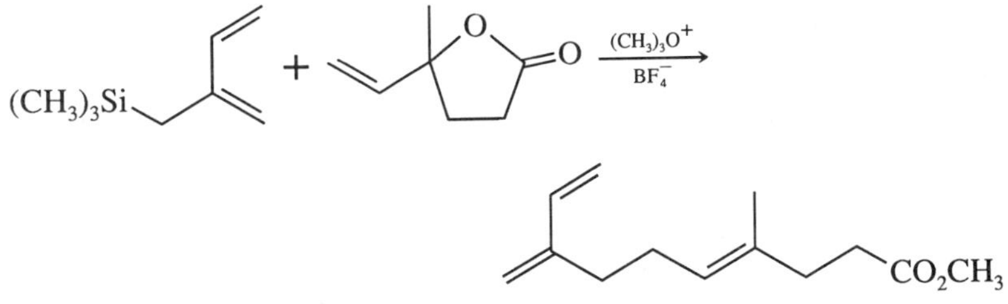

$(CH_3)_3Si$ + O, =O $\xrightarrow[BF_4^-]{(CH_3)_3O^+}$ CO_2CH_3

(Eq. 97)

π-Allylpalladium cationic complexes are presumably involved in the palladium(0)-catalyzed reactions of the allylic acetate in Eq. 98.[243] Some poly-

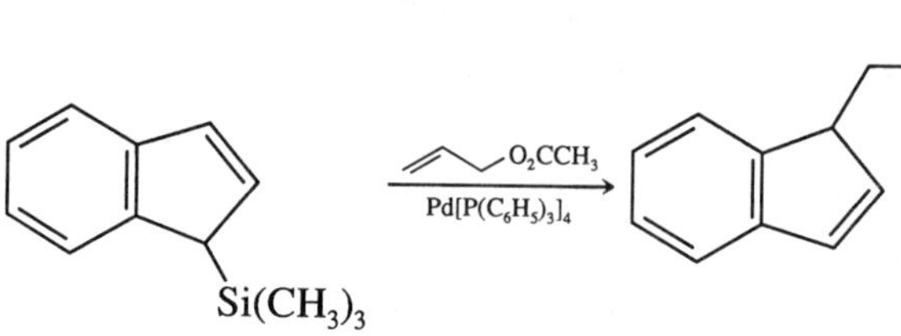

(Eq. 98)

halogenated primary alkyl halides react with allylsilanes with iron or ruthenium carbonyl catalysts.[244] Tertiary allylic and benzylic nitroalkanes react in the presence of stannic chloride with displacement of the nitro group (Eq. 99).[245]

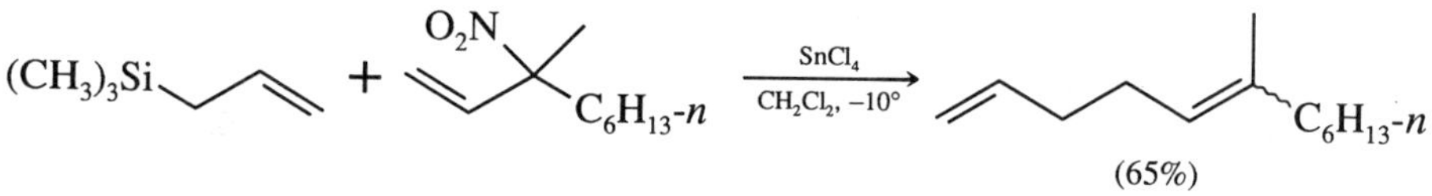

(Eq. 99)

Alcohols, in contrast to ethers and esters, usually react with allylsilanes (and vinylsilanes) in the presence of Lewis acids to induce protodesilylation and the formation of silyl ethers. However, tertiary benzyl and cumyl alcohols react with allylsilanes by alkylation (Eq. 100), although there is some difficulty

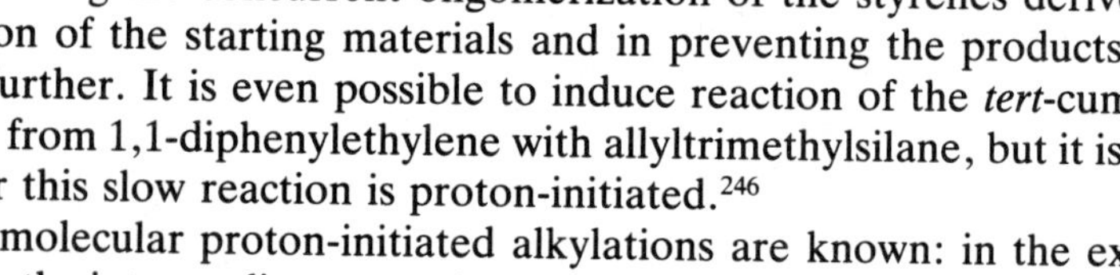

(Eq. 100)

in preventing the concurrent oligomerization of the styrenes derived by dehydration of the starting materials and in preventing the products from reacting further. It is even possible to induce reaction of the *tert*-cumyl cation derived from 1,1-diphenylethylene with allyltrimethylsilane, but it is not clear whether this slow reaction is proton-initiated.[246]

Intramolecular proton-initiated alkylations are known: in the example in Eq. 101, the intermediate secondary alkyl halide **107** reacts with the allylsilane

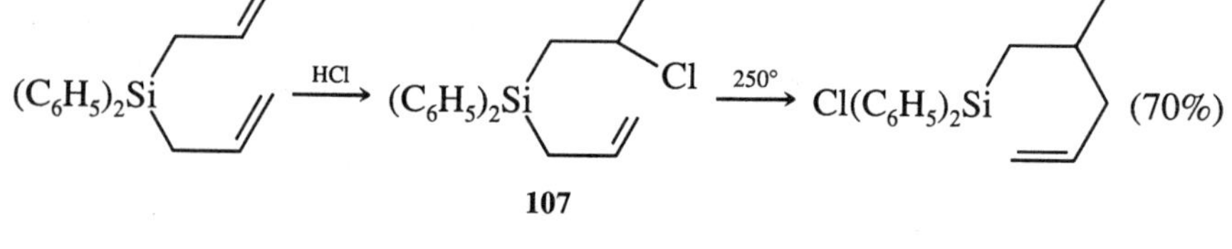

(Eq. 101)

group intramolecularly without catalysis,[247] but intramolecular reaction is much easier when the cationic center is tertiary (Eq. 102).[248] In reactions like

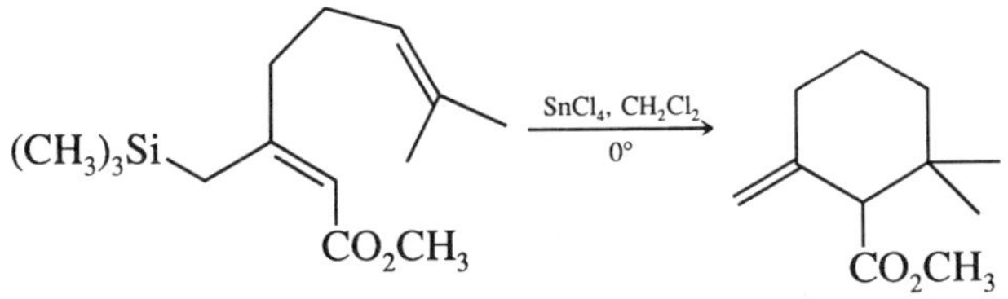

$(CH_3)_3Si$ ... CO_2CH_3 $\xrightarrow[0°]{SnCl_4,\ CH_2Cl_2}$ ... CO_2CH_3 (Eq. 102)

these where the group initiating cyclization is an alkene, there is a problem in ensuring that it is attacked by the electrophilic catalyst faster than the allylsilane group is attacked directly. In the example in Eq. 102, the methoxycarbonyl group lowers the reactivity of the allylsilane. The problem does not arise in the example in Eq. 103, where the allylsilane unit is embedded

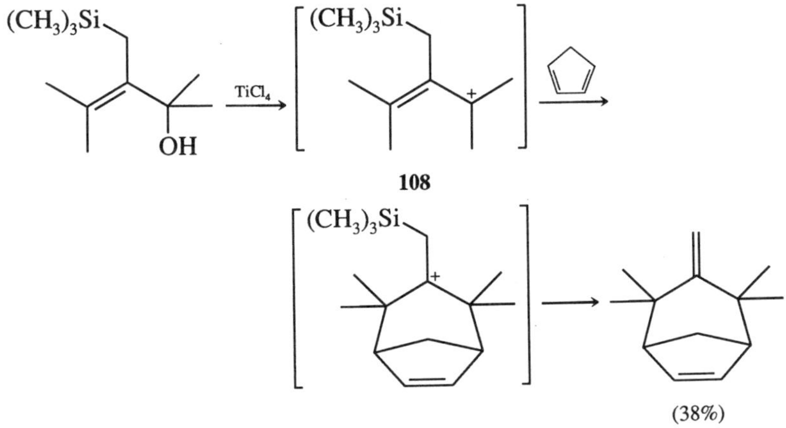

$(CH_3)_3Si$ ... OH $\xrightarrow{TiCl_4}$ [$(CH_3)_3Si$ ... $^+$] **108** $\longrightarrow$ [$(CH_3)_3Si$ ... $^+$] $\longrightarrow$ ... (38%) (Eq. 103)

in the cationic unit **108** that initiates the cyclization.[249] The problem of chemoselectivity is more generally solved in reactions like that in Eq. 104, where

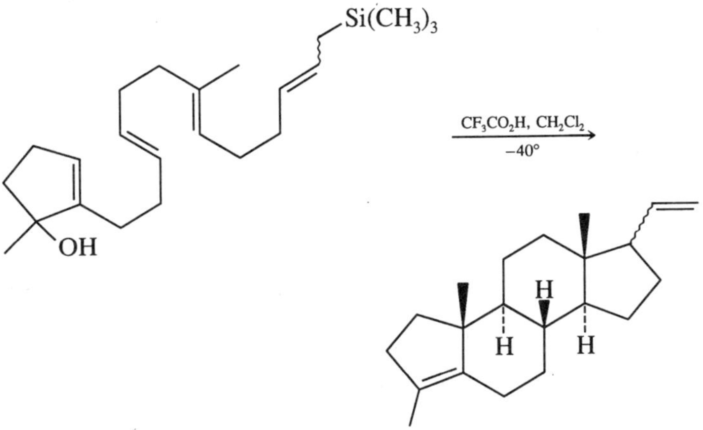

$Si(CH_3)_3$ ... OH $\xrightarrow[-40°]{CF_3CO_2H,\ CH_2Cl_2}$ ... H H H (83%) (Eq. 104)

an allylsilane is a strikingly effective terminating group for polyene cyclization. Here, acid leads to ionization of the tertiary allylic alcohol much faster than it protodesilylates the allylsilane group.[250]

4-Chloro-β-lactams react with allylsilanes (Eq. 232), presumably by way of an iminium ion. However, as mentioned earlier, iminium ions are available from such a variety of sources that all their reactions with allylsilanes and vinylsilanes are discussed later in a special section, regardless of the functional group from which the ions are derived.

Oxygen- and sulfur-stabilized carbocations are usually generated from acetals (q.v.) but they are also available from alkoxymethyl halides and from alkyl- and arylthiomethyl halides. Thus α-alkoxyalkyl halides (Eqs. 7 and 105)[10,251] and α-thioalkyl halides (Eq. 106)[252,253] react with allylsilanes in the

$(CH_3)_3Si$ + $ClCH_2OCH_2C_6H_5$ $\xrightarrow[CH_3CN,\ room\ temp]{(CH_3)_3SiI\ (cat.)}$ $OCH_2C_6H_5$ (78%) (Eq. 105)

$(CH_3)_3Si$ ($CO_2C_2H_5$) + $ClCH_2SC_6H_5$ $\xrightarrow[-78°]{TiCl_4,\ CH_2Cl_2}$ $C_2H_5O_2C$ $CH_2SC_6H_5$ (52%) (Eq. 106)

$(CH_3)_3Si$ + OBn, BnO, OBn, OBn, X, O **109** $\xrightarrow{X = Cl:\ (CH_3)_3SiO_3SCF_3\ cat.,\ CH_3CN;\ X = F:\ BF_3\cdot O(C_2H_5)_2\ CH_2Cl_2,\ 0°}$

Bn = $CH_2C_6H_5$

OBn, BnO, OBn, OBn, O (Eq. 107)

α:β >10:1 (81%) from X = Cl
α:β >20:1 (95%) from X = F

presence of a Lewis acid. Similarly anomeric sugar chlorides[254] and fluorides[255] **109** can be used to prepare *C*-glycosides (Eq. 107). Two equivalents of an allylsilane react with dichloromethyl methyl ether.[256]

The reactions of primary alkyl bromides **110** that are attached to tetrahydrofuran and tetrahydropyran rings are more complicated. On treatment with silver tetrafluoroborate, they give mixtures of products, of which one (**113**) is derived from the cation **111**, and the other (**114**) from a rearranged cation **112** (Eq. 108). On the whole, the tetrahydrofurans react primarily by

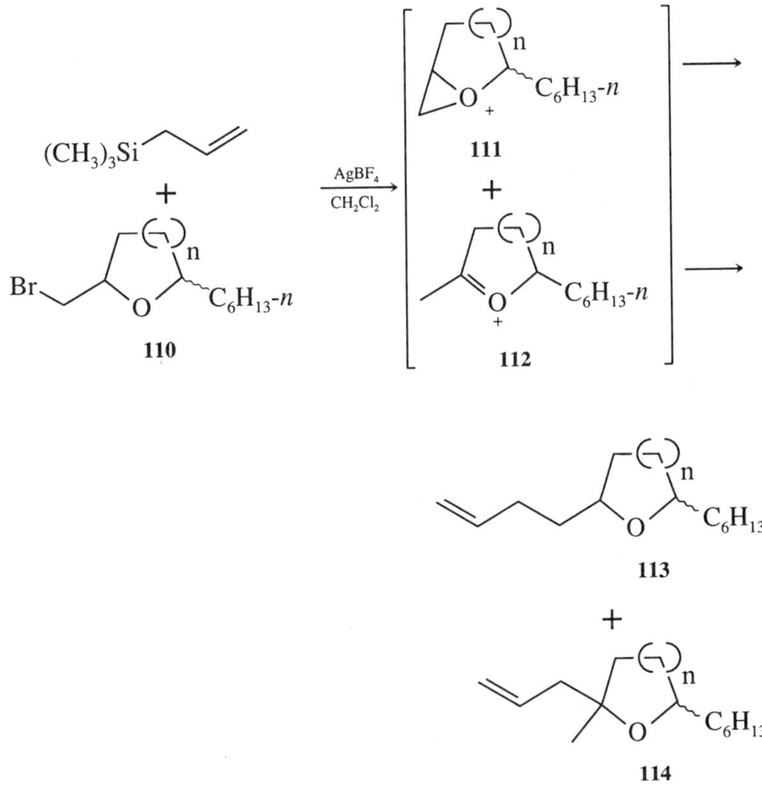

(Eq. 108)

n = 1 **113**:**114** 79:21 (76%)
n = 2 **113**:**114** 0:100 (88%)

the former pathway, and the tetrahydropyrans by the latter. Open-chain β-alkoxyalkyl halides do not give clean reactions.[257]

Nucleophile-catalyzed alkylations are restricted to allylsilanes likely to give well-stabilized anions and to the most reactive primary alkyl halides (Eqs. 109 and 110).[258,259] However, intramolecular removal of the silyl group makes

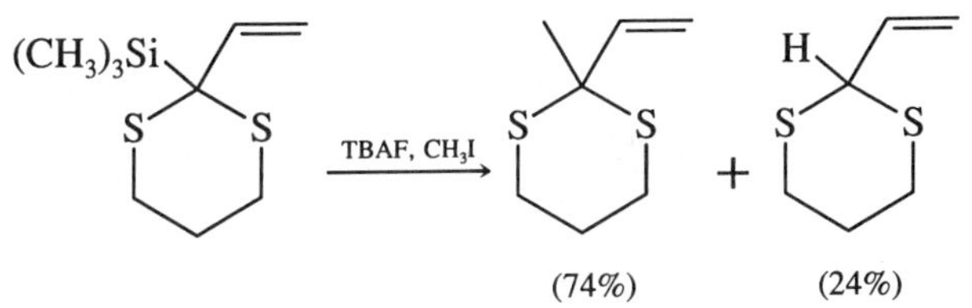

(Eq. 109)

(74%) (24%)

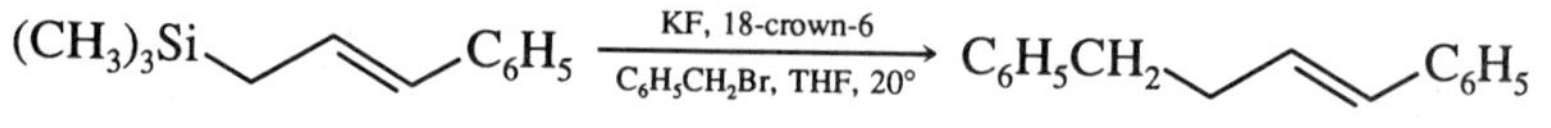

(Eq. 110)

it possible to alkylate some otherwise unfunctionalized allylsilanes (Eqs. 111 and 112).[260,190]

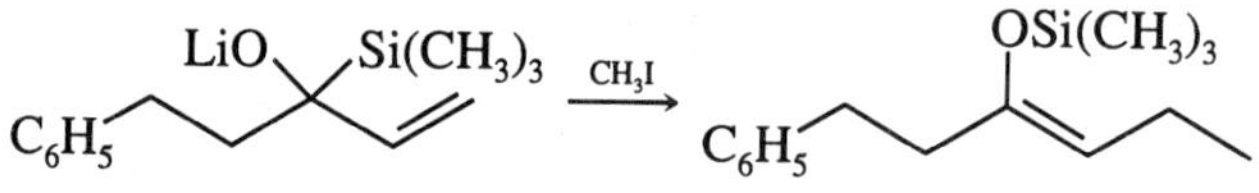

(83%) (Eq. 111)

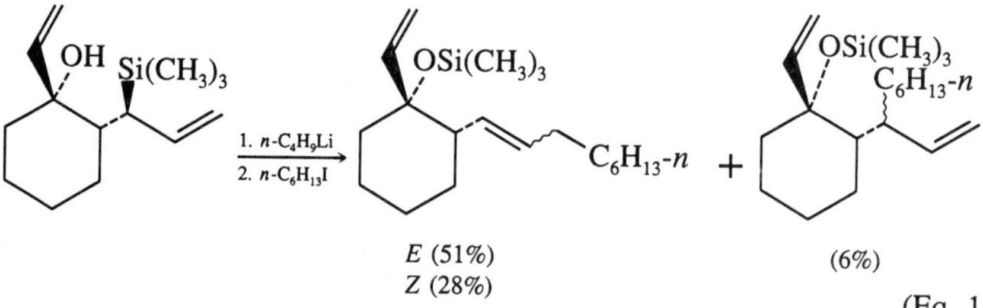

E (51%)
Z (28%)
(6%)
(Eq. 112)

Arenes are not usually electrophilic enough to react with allylsilanes, but they can become so on irradiation. Allylsilanes react with *o*- or *p*-dicyanobenzene on irradiation in acetonitrile; the cyano group is displaced regiospecifically, but the allyl group is incorporated with loss of regiocontrol (Eq. 113). Most probably, the reaction involves electron transfer from the allylsilane to the photo-excited arene, followed by the loss of the silyl group from the allylsilane radical cation; coupling of the two components and loss of cyanide ion complete the sequence.[261]

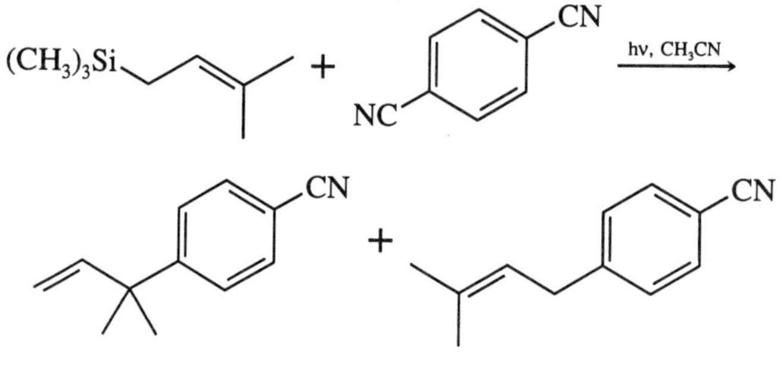

3:2 (91%) (Eq. 113)

The alkylation in Eq. 97 is the key step in a synthesis of sinensal,[242] and the cycloalkylation of allylsilanes is used in syntheses of steroids (Eq. 104),[250,262,263] of limonene (Eq. 114),[264] of 3-hydroxylabdadienoic acid,[265] of albicanyl acetate and isodrimenin,[266] and of trixagol (from the product of Eq. 102).[248]

$TiCl_4$, $C_6H_5NHCH_3$ / CH_2Cl_2 (79%) (Eq. 114)

Vinylsilanes. Alkylation of a vinylsilane would make the product more reactive than the starting material. Accordingly, useful alkylations are largely restricted to intramolecular reactions like those in Eqs. 115–118 where this

$(C_2H_5)_3N$, CH_2Cl_2, $-25°$ (85%) (Eq. 115)

$(i\text{-}C_4H_9)_2AlCl$ (Eq. 116)

$AlCl_3$ (Eq. 117)

$SbCl_5$, CH_2Cl_2 / $-78°$ (51%) (Eq. 118)

problem does not arise.[267-270] The reaction in Eq. 118 is remarkable because the attack on the vinylsilane group is from a secondary center, and the overall stereochemistry at the vinylsilane double bond is, perforce, inversion of configuration. It appears that a vinylsilane group, like an allylsilane group, is likely to be an effective terminator for polyene cyclization.

Vinylsilanes react well with methoxymethyl chloride (Eq. 8) and with dichloromethyl methyl ether[228] in the presence of titanium tetrachloride; in the latter case, an aqueous workup gives the α,β-unsaturated aldehyde (Eq. 119).[256] The reaction is not stereospecific, the thermodynamically more stable

C_2H_5 C_2H_5 $Si(CH_3)_3$ + Cl_2CHOCH_3 $\xrightarrow[\text{2. } H_2O]{\text{1. } TiCl_4,\ CH_2Cl_2,\ -78°}$ C_2H_5 C_2H_5 CHO (73%) (Eq. 119)

product (usually E) being isolated.[256,271] The reaction works on cyclopentenylsilanes[272] and on 1-silylbutadienes.[273]

The reaction in Eq. 115 is a synthesis of (E)-γ-bisabolene,[267] and the reaction of a vinylsilane with dichloromethyl methyl ether is used in a synthesis of nuciferal.[256]

Allenylsilanes. The allenylsilane **115** reacts as a vinylsilane with benzyl chloride in the presence of aluminum chloride, an unusual pattern of reactivity for allenylsilanes.[274]

$[(CH_3)_3Si]_2C{=}C{=}C[Si(CH_3)_3]_2$ (**115**) $\xrightarrow[AlCl_3]{ClCH_2C_6H_5}$ $[(CH_3)_3Si]_2C{=}C{=}C[Si(CH_3)_3](CH_2C_6H_5)$

Epoxides, Oxetanes, and Episulfonium Salts

Allylsilanes. Ethylene oxide reacts regiospecifically (Eq. 120)[170] and stereospecifically (Eq. 121)[24] with allylsilanes in the presence of titanium tetra-

$(CH_3)_3Si$ … C_6H_{13}-n $\xrightarrow[TiCl_4,\ CH_2Cl_2,\ -78°]{\text{ethylene oxide}}$ OH … C_6H_{13}-n (93%) (Eq. 120)

$(CH_3)_3Si$ (cyclopentenyl), 22–25% ee $\xrightarrow[TiCl_4,\ CH_2Cl_2,\ -78°]{\text{ethylene oxide}}$ OH, 24% ee (93%) (Eq. 121)

chloride. Oxetane reacts similarly (Eq. 122).[275] However, propylene oxide,

$(CH_3)_3Si$ ... + oxetane $\xrightarrow{TiCl_4}$ ... OH (85%)

(Eq. 122)

and presumably other higher oxides, give uncharacterized mixtures of several products. It seems likely that Lewis acid catalyzed rearrangement of the oxide to propionaldehyde is competing with the direct reaction, and the propionaldehyde then reacts with the allylsilane.[276] This problem is avoided when the reaction is intramolecular and 5-, 6-, and 7-membered rings are made (Eqs. 123–125).[277,265,278] In the last example, the choice of Lewis acid is critical; the

$(CH_3)_3Si$ ... $\xrightarrow[-95^\circ]{TiCl_4,\ CH_2Cl_2}$ HO ... (55%) (Eq. 123)

cis:*trans* 4:1

$(CH_3)_3Si$ ... CO_2CH_3 $\xrightarrow[-56^\circ]{BF_3 \cdot O(C_2H_5)_2}$ OH ... CO_2CH_3 (80%) (Eq. 124)

$(CH_3)_3Si$ ... O $\xrightarrow[-60^\circ]{BF_3 \cdot O(C_2H_5)_2}$ ... OH (71%) (Eq. 125)

usual Lewis acid, titanium tetrachloride, gives the chlorohydrin from the epoxide and causes protodesilylation of the allylsilane.

Alkenes react stereospecifically with sulfenyl chlorides to give β-chlorosulfides by way of episulfonium ions. The β-chlorosulfides **116** and **117** react with allyltrimethylsilane in the presence of a Lewis acid (Eqs. 126 and 127)

$\xrightarrow{C_6H_5SCl}$ C_6H_5S ... Cl **116** $\xrightarrow[ZnBr_2,\ CH_3NO_2,\ 20^\circ]{\text{allyl-}Si(CH_3)_3}$ C_6H_5S ... (92%) (Eq. 126)

C_6H_5SCl → C_6H_5S ... Cl **117** → ($Si(CH_3)_3$ allylsilane; $ZnBr_2$, CH_3NO_2, 20°) → C_6H_5S ... (40%)

(Eq. 127)

to give the products of *anti* carbosulfenylation of the alkene, presumably by way of the same episulfonium ions. With unsymmetrical alkenes, the allyl group is introduced at the more-substituted end of the alkene (Markovnikov), completely so with styrene and isobutylene, and predominantly so (55:35) with propene.[279] Sulfenylation of the allylsilane is a minor pathway in these reactions, but selenenylation is the only pathway in attempts to achieve the corresponding carboselenenylation of alkenes.[279]

Reactions of allylsilanes with epoxides assisted by nucleophiles are found in one special case. The alkoxide **118** is available from the reaction of vinylmagnesium bromide and the appropriate acylsilane. It appears to be in equilibrium with its isomer **119**, which is the result of a Brook rearrangement,

C_6H_5, C_2H_5, $(CH_3)_3Si$, OMgBr **118** ⇌ C_6H_5, C_2H_5, $(CH_3)_3SiO$, MgBr **119** → (1. $CuC{\equiv}CSi(CH_3)_3$; 2. epoxide C_4H_9-*n*) → C_6H_5, C_2H_5, $(CH_3)_3SiO$, C_4H_9-*n*, OH (66%)

(Eq. 128)

and this species, as its cuprate, reacts with epoxides (Eq. 128).[280] This sequence is clearly related to the alkylation in Eq. 111.

The alkylation in Eq. 124 is the key step in a synthesis of karahana ester.[265]

Vinylsilanes. The vinylsilane **120** reacts intramolecularly with the epoxide group in the presence of titanium tetrachloride to give a cyclopentanol (Eq. 129).[120]

O, $Si(CH_3)_3$, C_6H_{13}-*n* **120** → ($TiCl_4$) → OH, C_6H_{13}-*n* (54%)

(Eq. 129)

Aldehydes and Ketones. Aldehydes and ketones are among the most frequently used electrophiles with allylsilanes, in both Lewis acid catalyzed and nucleophile-catalyzed reactions. The high level of regiocontrol and stereocontrol available makes this a good method for preparing homoallylic alcohols, which are versatile intermediates in organic synthesis. Vinylsilanes and allenylsilanes are much less often used.

Allylsilanes. Very electrophilic ketones like hexafluoroacetone react with allylsilanes without catalysis in an ene reaction (Eq. 51),[141] but a wide variety of aldehydes and ketones react in the presence of Lewis acids to give substitution. Typically, titanium tetrachloride is used at $-78°$ for a few minutes, but reactions are frequently carried out at 0°, or at room temperature when the conditions do not need to be mild. Several other Lewis acids, notably aluminum chloride, ethylaluminum dichloride, and stannic chloride are used. The reaction is regiospecific (Eqs. 130 and 131),[64] and stereospecifically *anti* with the formation of a *trans* double bond, as usual (Eq. 132).[21]

$(CH_3)_3Si$ ... C_6H_5 + $OHCC_3H_7$-*n* —($TiCl_4$, CH_2Cl_2, room temp)→ OH, C_3H_7-*n*, C_6H_5 (87%)

(Eq. 130)

$(CH_3)_3Si$, C_6H_5 + $OHCC_3H_7$-*n* —($TiCl_4$, CH_2Cl_2, room temp)→ C_6H_5 ... OH, C_3H_7-*n* (54%)

(Eq. 131)

C_6H_5, $(CH_3)_3Si$, H (85% ee) + CH_2O —($TiCl_4$, CH_2Cl_2, $-78°$)→ C_6H_5 ... OH, H (86% ee)

(Eq. 132)

Frequently, and especially when aluminum chloride is the Lewis acid, the first-formed product reacts with another molecule of aldehyde giving 4-chlorotetrahydropyrans (Eq. 133). It is possible to use two different aldehydes in succession, and hence get unsymmetrical chlorotetrahydropyrans.[281]

When the allylsilane has an extra double bond extending the conjugation, reaction takes place at the terminus of the diene system remote from the silyl group (Eq. 134).[282]

$(CH_3)_3Si\text{-}CH_2CH{=}CH_2 + OHCC_6H_5 \xrightarrow[CH_2Cl_2]{AlCl_3} [CH_2{=}CHCH_2CH(OH)C_6H_5] \xrightarrow[C_6H_5CHO]{AlCl_3}$ 4-Cl-2,6-$(C_6H_5)_2$-tetrahydropyran (72%) (Eq. 133)

$(CH_3)_3Si\text{-}CH_2CH{=}CHCH{=}CH_2 + OHCC_2H_5 \xrightarrow[-40^\circ]{TiCl_4,\ CH_2Cl_2} CH_2{=}CHCH{=}CHCH_2CH(OH)C_2H_5$ (79%)

(Eq. 134)

When the product alkene is highly substituted, there are problems from its further reaction, as in Eq. 35, where the trisubstituted double bond in the first-formed product **52** is evidently close in nucleophilicity to the double bond in the allylsilane **51**.[83] Note also the lower yield in Eq. 131 than in Eq. 130, which probably stems from the same cause. Comparable reactions with acetals (q.v.) are less troublesome. There is also a problem when the hydroxy group in the product is highly activated, since it can ionize when the conditions are harsh and hence undergo further attack by the counterion or by the allylsilane (Eq. 135).[246]

$(CH_3)_3Si\text{-}CH_2CH{=}CH_2 + OC(C_6H_5)_2 \xrightarrow[reflux]{TiCl_4,\ CHCl_3} (CH_2{=}CHCH_2)_2C(C_6H_5)_2$ (25%) (Eq. 135)

Allylsilanes with suitably placed ester groups in the molecule give lactones (Eqs. 136–138) as a consequence of the cyclization of the first-formed prod-

$(CH_3)_3SiCH(CO_2C_2H_5)C(CH_3){=}CH_2 + O{=}C(CH_3)_2 \xrightarrow[room\ temp]{TiCl_4,\ CH_2Cl_2}$

$C_2H_5O_2CCH{=}C(CH_3)CH_2C(CH_3)_2OH + C_2H_5O_2CCH{=}C(CH_3)CH_2C(CH_3)_2Cl +$ 4,6,6-trimethyl-5,6-dihydro-2-pyranone

76:14:10

(Eq. 136)

$(CH_3)_3Si$ ... $CO_2Si(CH_3)_3$ + O= ... $\xrightarrow[\text{2. HCl, } H_2O]{\text{1. } TiCl_4, CH_2Cl_2 \text{, } -78° \text{ to room temp}}$

(80%) (Eq. 137)

$(CH_3)_3Si$ ... $CO_2C_2H_5$ + $OHCC_4H_9$-*n* $\xrightarrow[0\text{–}25°]{TiCl_4, CH_2Cl_2}$

C_4H_9-*n* (25%) (Eq. 138)

ucts.[252,283–285] The reaction in Eq. 136 is notable in ensuring attack γ to the carbonyl group (d^4 reactivity), in contrast to the usual kinetically controlled reactions of lithium enolates (d^2 reactivity). The ester in Eq. 139, however, appears to react with unusual regiocontrol; the allylsilane first rearranges to the silyl dienol ether **121**, which then reacts with the aldehyde.[286] This reaction

$(CH_3)_3Si$ ... CO_2CH_3 + $OHCC_3H_7$-*n* $\xrightarrow[-78°]{TiCl_4, CH_2Cl_2}$

[$(CH_3)_3SiO$ ... OCH_3] **121** ⟶ *n*-C_3H_7 ... (49%) (Eq. 139)

is therefore not strictly the electrophilic substitution of an allylsilane, and examples of it are not included in the tables. In contrast, the allylsilane **122** attacks aldehydes directly (Eq. 140).[287]

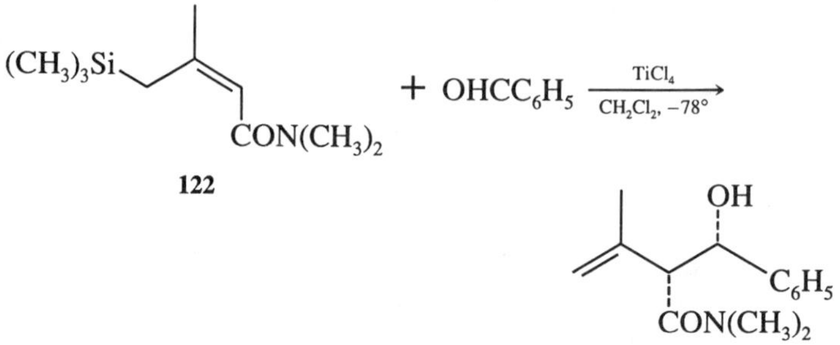

$(CH_3)_3Si$ ... $CON(CH_3)_2$ (122) + $OHCC_6H_5$ $\xrightarrow[CH_2Cl_2,\ -78°]{TiCl_4}$ OH, C_6H_5, $CON(CH_3)_2$ (62%) (Eq. 140)

syn:*anti* 97:3

Aldehydes give somewhat better yields than ketones, and in aldehydo-ketones (Eqs. 141 and 142) it is the aldehyde that reacts.[288,289]

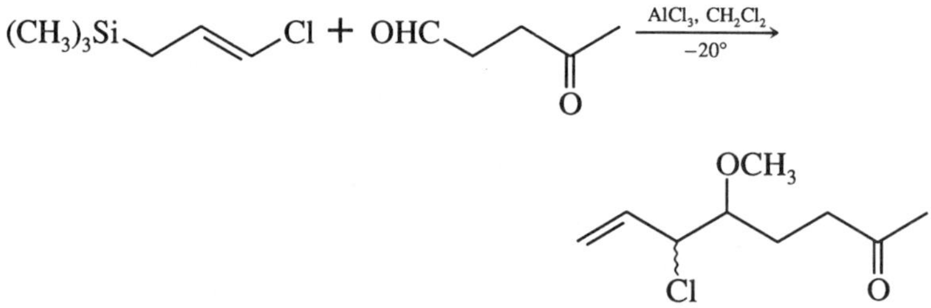

$(CH_3)_3Si$ ... Cl + OHC ... O $\xrightarrow[-20°]{AlCl_3,\ CH_2Cl_2}$ OCH_3, Cl, O (Eq. 141)

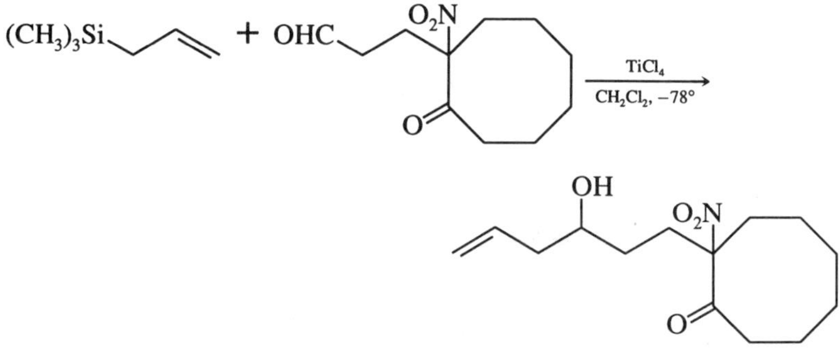

(24%)

$(CH_3)_3Si$ ... + OHC ... O_2N ... O $\xrightarrow[CH_2Cl_2,\ -78°]{TiCl_4}$ OH, O_2N, O (Eq. 142)

(100%)

Intramolecular reactions with a wide variety of structures work well (Eqs. 143–146).[290–292,258]

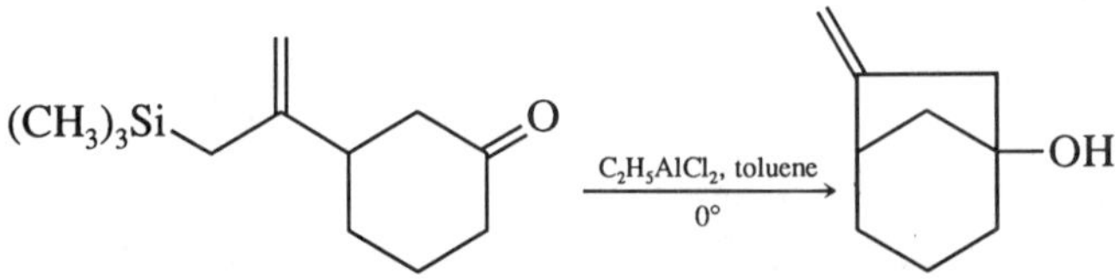

$(CH_3)_3Si$ ... O $\xrightarrow[0°]{C_2H_5AlCl_2,\ toluene}$ OH (91%) (Eq. 143)

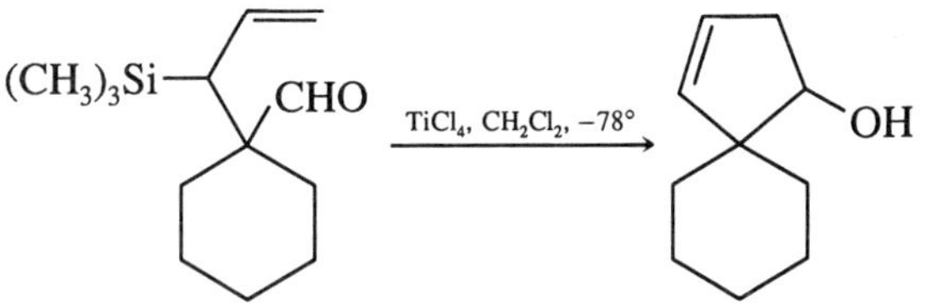

$(CH_3)_3Si$ ... CHO $\xrightarrow{TiCl_4,\ CH_2Cl_2,\ -78°}$ ... OH (67%) (Eq. 144)

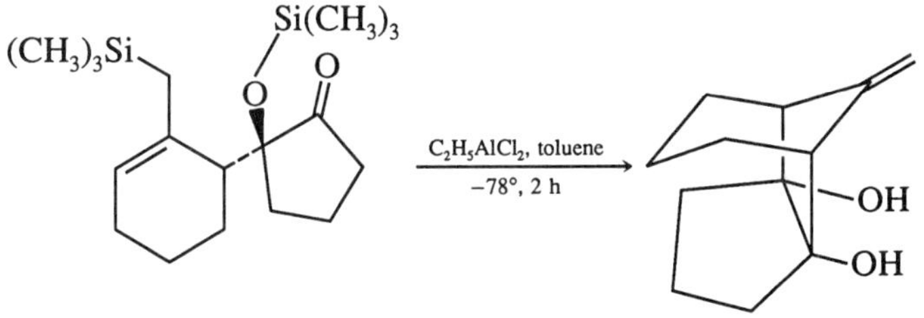

$(CH_3)_3Si$ ... $Si(CH_3)_3$ ... O ... O $\xrightarrow[-78°,\ 2\ h]{C_2H_5AlCl_2,\ toluene}$ ... OH ... OH (85%) (Eq. 145)

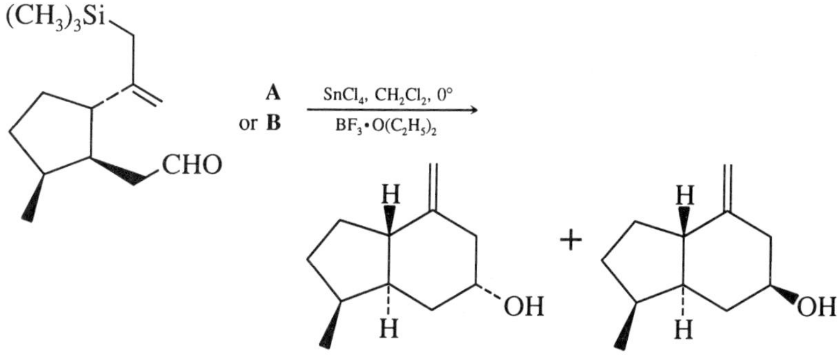

$(CH_3)_3Si$ ... CHO **A** or **B** $\xrightarrow[BF_3\cdot O(C_2H_5)_2]{SnCl_4,\ CH_2Cl_2,\ 0°}$ (Eq. 146)

123 + **124**

A 59:41
B 85:15

When the allylsilane has a substituent at C-3, the reaction with aldehydes gives largely the product **125** in which the substituents are *syn* on the carbon chain (*erythro* in the alternative nomenclature) (see also Eq. 140). The (*E*)-allylsilane is highly selective in this sense (> 95:5) (Eq. 147), but the (*Z*)-

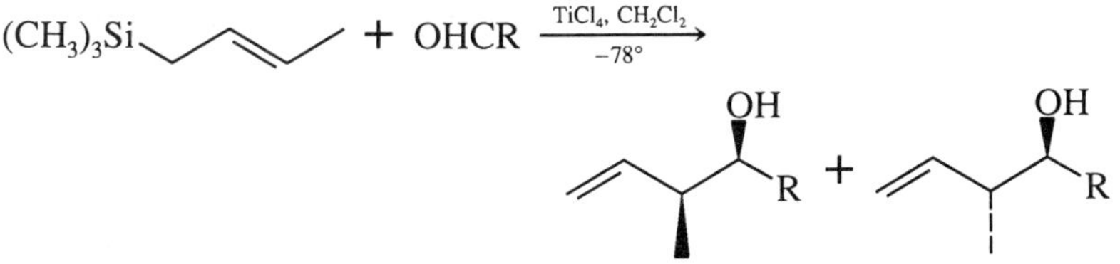

$(CH_3)_3Si$ + OHCR $\xrightarrow[-78°]{TiCl_4,\ CH_2Cl_2}$ (Eq. 147)

125 >95:5 **126**
(*syn* = *erythro*) (*anti* = *threo*)

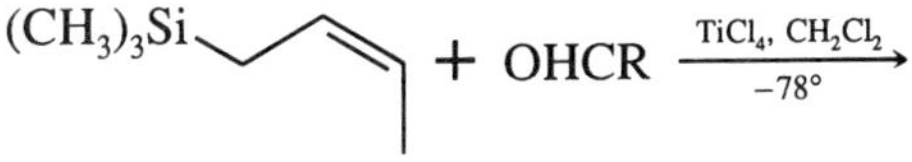

$(CH_3)_3Si$ + OHCR $\xrightarrow[-78°]{TiCl_4,\ CH_2Cl_2}$ **125** + **126** (Eq. 148)
65:35

allylsilane is much less so (~65:35) (Eq. 148).[293] When the allylsilane is, in addition, chiral and resolved, the usual *anti* selectivity with respect to the allylsilane stereochemistry is essentially complete, and *syn* (*erythro*) selectivity with respect to the substituents remains (Eq. 149); it is even high with the (*Z*)-allylsilane when the aldehyde is heavily substituted (Eq. 150).[22]

C_6H_5, $(CH_3)_3Si$, H + $OHCC_4H_9$-*t* $\xrightarrow[-78°]{TiCl_4,\ CH_2Cl_2}$

OH, C_6H_5, C_4H_9-*t* (47%) (Eq. 149)

>99:1

C_6H_5, $(CH_3)_3Si$, H + $OHCC_4H_9$-*t* $\xrightarrow[-78°]{TiCl_4,\ CH_2Cl_2}$

OH, C_6H_5, C_4H_9-*t* (27%) (Eq. 150)

>99:1

To explain these results, one school of thought prefers antiperiplanar transition states **A–D**, and suggests that the transition states **A** and **C** are more favored than the transition states **B** and **D** because of the steric interactions in the latter. Another school of thought prefers synclinal transition states **E–L**, and suggests that the transition states **E** or **F** and **I** or **J** are favored over **G**, **H**, **K**, and **L** because of interactions from the Lewis acid bonded to the oxygen *anti* to the R group in the aldehyde. Neither the antiperiplanar nor the synclinal pictures explains well why (*E*)-allylsilanes are more *syn* (*erythro*) selective than (*Z*)-allylsilanes—the opposite result would seem to be more likely.

X_nM, O^+, H, H, R, SiR_3 → *syn* (= *erythro*)

A

MX_n, ^+O, H, R, H, SiR_3 → *anti* (= *threo*)

B

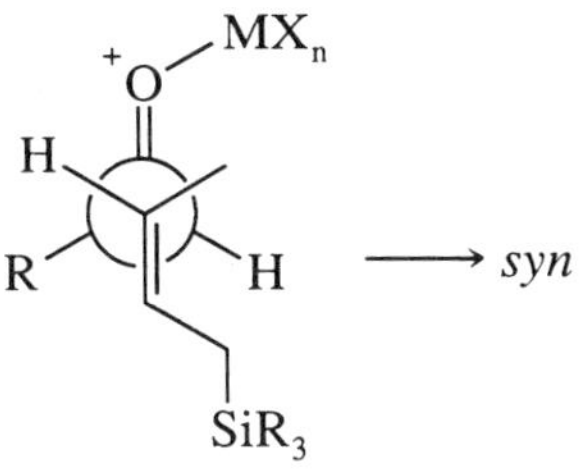

C $\longrightarrow$ *syn*

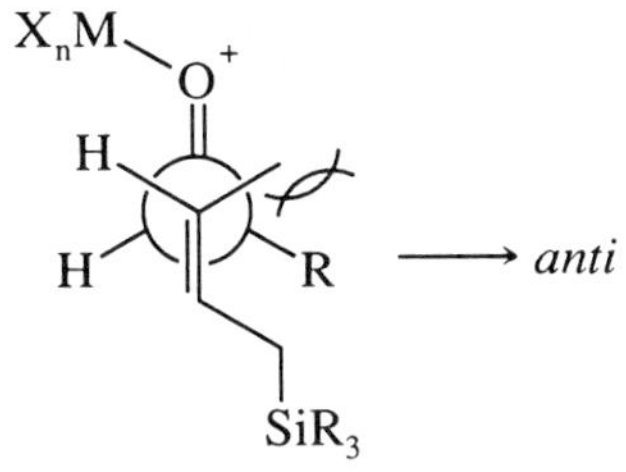

D $\longrightarrow$ *anti*

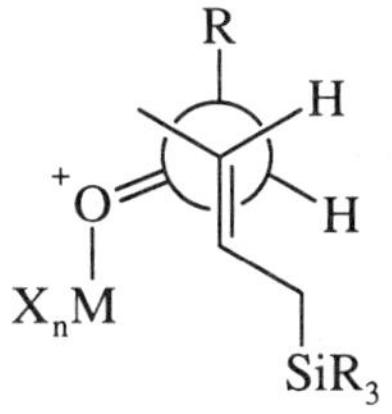

E or F $\longrightarrow$ *syn*

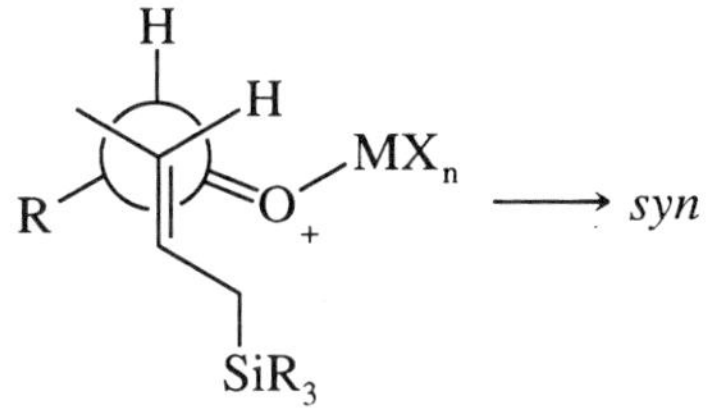

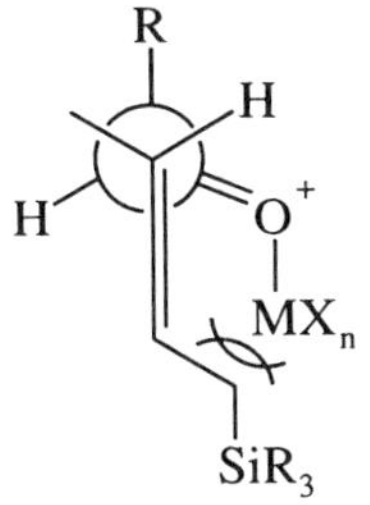

G or H $\longrightarrow$ *anti*

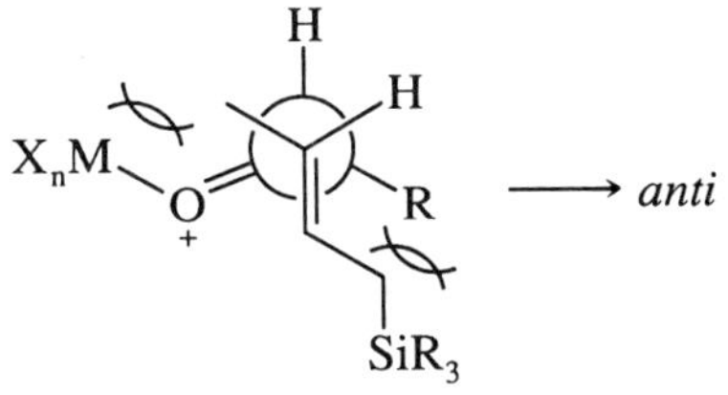

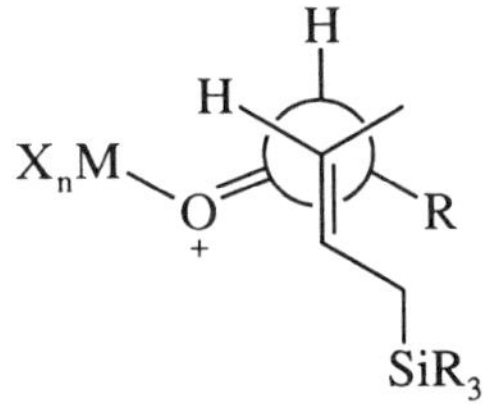

I or J $\longrightarrow$ *syn*

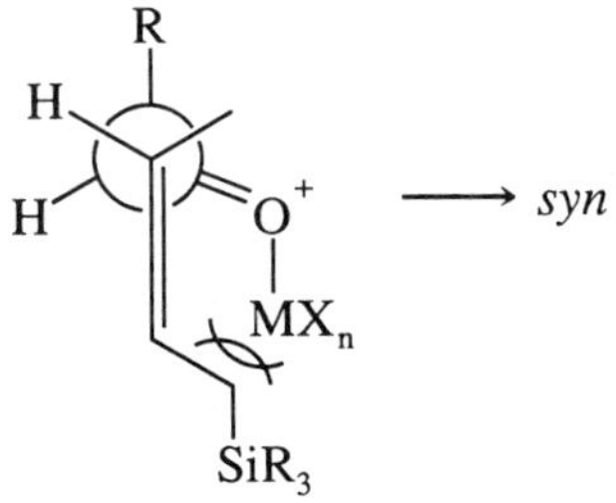

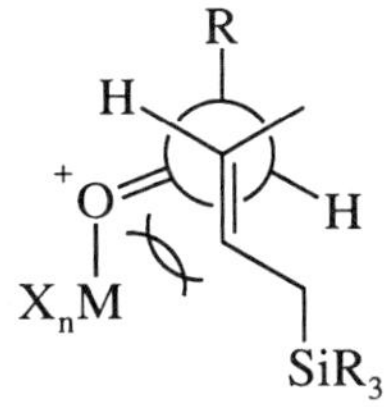

K or L $\longrightarrow$ *anti*

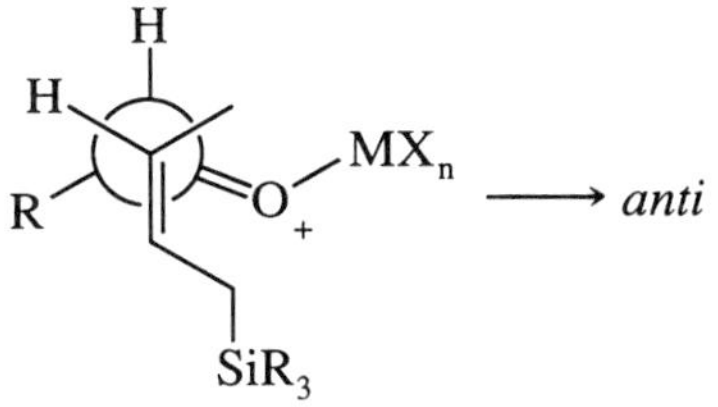

Furthermore, the available evidence is inconclusive. In a cyclic reaction (Eq. 151) designed to test whether synclinal or antiperiplanar transition states are preferred, the synclinal reaction is favored. Significantly, the nature of the Lewis acid substantially affects the proportions of the reaction that follow

H O⁺ X_nM $Si(CH_3)_3$ + X_nM O⁺ H $Si(CH_3)_3$ $\xrightarrow{MX_n}$

synclinal *antiperiplanar*

H OH + HO H (Eq. 151)

127 **128**

MX_n	**127:128**
$SnCl_4$	49:51
$(C_2H_5)_2AlCl$	66:34
$FeCl_3$	70:30
$BF_3{\cdot}O(C_2H_5)_2$	80:20

the synclinal and the antiperiplanar pathways,[294] just as it does in the cyclization of Eq. 146.[258] However, in the cyclizations in Eq. 152, the *Z* and *E*

CHO SC_6H_5 $Si(CH_3)_3$ $\xrightarrow{CF_3CO_2H,\ CH_2Cl_2,\ 0°}$ OH SC_6H_5 (54%) + **130** (<2%)

129

(Eq. 152)

CHO C_6H_5S $(CH_3)_3Si$ $\xrightarrow{CF_3CO_2H,\ CH_2Cl_2,\ 0°}$ OH SC_6H_5 (52%) + **129** (<7%)

130

isomers give different products, the *E* isomer giving the *cis* product and the *Z* isomer the *trans*.[295] This is consistent with the synclinal transition states **E** or **F** for the former (the antiperiplanar transition state **A** being impossible) and either the antiperiplanar transition state **D** or the synclinal transition state **K** for the latter. But this time, high selectivity is only observed with catalysis by a protic acid; Lewis acids give mixtures of the stereoisomeric products, and any explanation cannot now depend upon the size of the Lewis acid. Finally, in the 5-*endo-trig* cyclization (Eq. 153), in which both isomers give

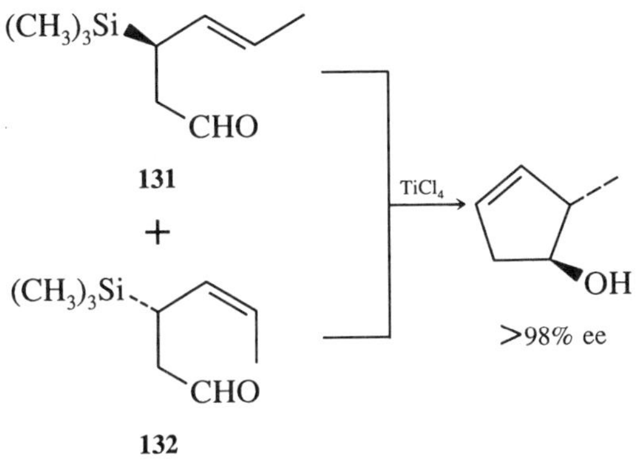

$(CH_3)_3Si$ ... CHO **131** + $(CH_3)_3Si$ ... CHO **132** $\xrightarrow{TiCl_4}$ ... OH (59%) >98% ee (Eq. 153)

131:**132** 83:17

the same product, the stereochemical result is only consistent with the antiperiplanar transition state **A** for the major *E* isomer **131** and the synclinal transition state **I** for the minor *Z* isomer **132**.[296] Clearly this is still a confused area. A coherent picture of all the factors that control the stereochemistry of reactions between two trigonal carbon atoms has yet to emerge.

When the aldehyde or ketone contains a chiral center, diastereoface selectivity in the attack on the carbonyl group is possible. Menthyl pyruvate, for example, is moderately selective in its reaction with allyltrimethylsilane (up to 55% diastereoisomeric excess),[297] but better designed chiral auxiliaries give better control, phenylmenthyl glyoxylate giving 80% diastereoisomeric excess.[298] Similarly, when the chiral center is not functional, as with 2-phenylpropionaldehyde, only the normal moderate degree of selectivity is found (up to 76% diastereoisomeric excess).[299] However, when the chiral center has an ether group α or β to the carbonyl group, very high diastereoselectivity is observed (Eqs. 154–156), consistent with attack by the allylsilane from the less hindered side of a ring made up by chelation of the Lewis acid between the ether and carbonyl oxygens.[299,300] Some Lewis acids cause dehydration or debenzylation of the product, and the precise order of mixing the reagents is important.[301] Boron trifluoride, which has only one coordination site, gives low and opposite stereoselectivity in the reaction in Eq. 154.[299] In the sugar aldehyde in Eq. 157, the choice of Lewis acid determines the diastereoselectivity, titanium tetrachloride giving the product **133** expected from chelation control, and boron trifluoride the product **134** of normal Felkin–Anh con-

$(CH_3)_3Si$ + OHC–CH(OCH$_2$C$_6$H$_5$)CH$_3$, $SnCl_4$, −78° → syn:anti 45:1 (81%) (Eq. 154)

$(CH_3)_3Si$ + OHC–CH(CH$_3$)CH$_2$OCH$_2$C$_6$H$_5$, $SnCl_4$, −78° → syn:anti 12:1 (92%) (Eq. 155)

$(CH_3)_3Si$ + OHC–CH$_2$CH(OCH$_2$C$_6$H$_5$)CH$_3$, $TiCl_4$ or $SnCl_4$ → anti:syn 95:5 (85%) (Eq. 156)

syn:*anti* 45:1

syn:*anti* 12:1

anti:*syn* 95:5

$(CH_3)_3Si$ + OHC (furanoside, OCH$_3$)

$TiCl_4$, CH_2Cl_2, −78° → **133**:**134** 20:1 (89%)

133

$BF_3{\cdot}O(C_2H_5)_2$, CH_2Cl_2, −78° → **134**:**133** 90:1 (80%)

134

(Eq. 157)

trol.[302] Surprisingly, this is not the case in the reaction in Eq. 156, where boron trifluoride gives the same result as the other Lewis acids.[301]

Nucleophilic displacement of the silyl group from allylsilanes works well for protodesilylation (q.v.), but only a limited range of other electrophiles can be used, among which aldehydes and ketones are the best. Protodesilylation is only a minor pathway in most cases, in spite of the difficulties of preparing reagents and solvents completely free of proton sources. The reactions are usually carried out with tetrabutylammonium fluoride in tetrahydrofuran at room temperature or at reflux, but benzyltrimethylammonium fluoride,[303] although little used as yet, may be better.[304] Other fluoride ion sources and solvents are cesium fluoride in dimethylformamide,[115,305] potassium fluoride with 18-crown-6 in tetrahydrofuran,[259] and either potassium *tert*-butoxide or tris(dimethylamino)sulfonium difluorotrimethylsilicate in polar aprotic solvents.[306] The first-formed product is usually the silyl ether, which is hydrolyzed to an alcohol during workup or, in more hindered cases, deliberately hydrolyzed in a second step. The reactions are sometimes highly regioselective (Eqs. 158 and 159), but not regiospecific (see Eq. 32), because

$(CH_3)_3Si$ F F **135** or $(CH_3)_3Si$ F F **136** + $OHCC_6H_5$ → 1. TASF, room temp; 2. HCl, H_2O → OH C_6H_5 F F

(100% from **135**)
(58% from **136**) (Eq. 158)

$(CH_3)_3Si$ + O → TBAF, THF, reflux → OH (90%) (Eq. 159)

selectivity depends upon the ambident character of the intermediate anion.[306,167] When the intermediate anion is stabilized, the conditions can be somewhat milder (Eqs. 160 and 161).[258,307] In the reaction in Eq. 161, the ambident intermediate is attacked at the end of the allyl system away from the silyl group. However, when the disilylated allyl group in the starting material also has a highly hindering group at one end, as in **137**, the ambident intermediate can be attacked at the silicon-substituted end, and the product then undergoes an easy Peterson elimination (Eq. 162).[227a]

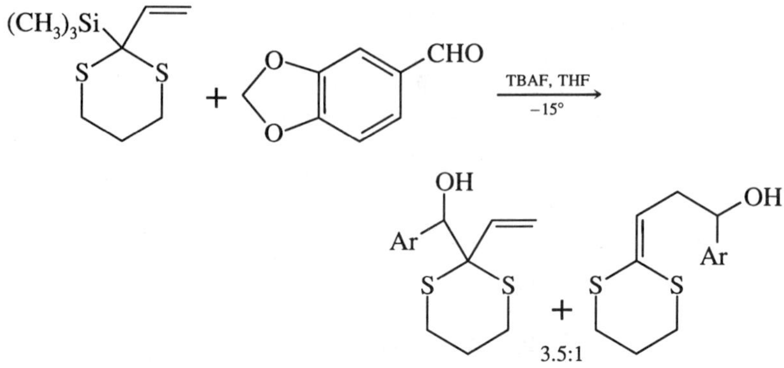

(Eq. 160)

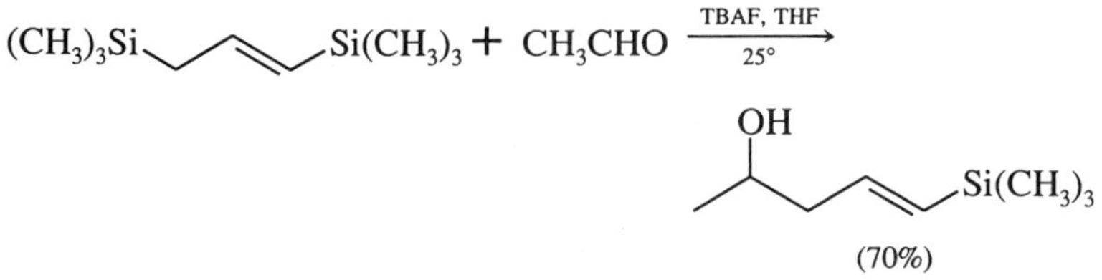

(Eq. 161)

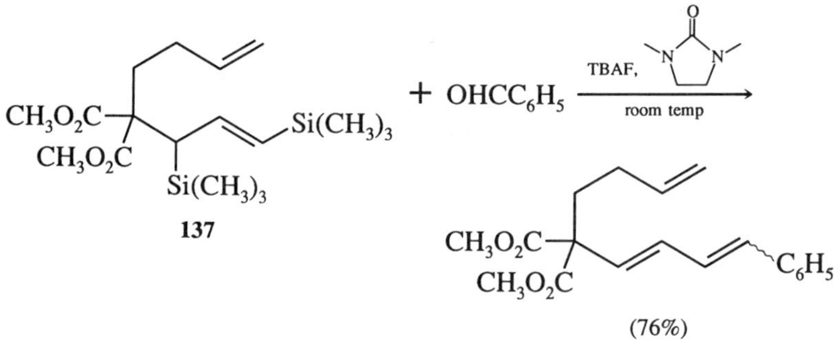

(Eq. 162)

The base-catalyzed reaction also works intramolecularly (Eqs. 163 and 164).[258,166] The stereochemistry in Eq. 163 is opposite to that found for the reaction catalyzed by Lewis acids (Eq. 146). The byproduct in Eq. 164 is evidence for some competing protodesilylation.

The reaction of allylsilanes with aldehydes is used in syntheses of tagetol (ipsenol) (**138**),[174,308,309] muscone,[310] and sesbanimide A.[311] The reaction in Eq.

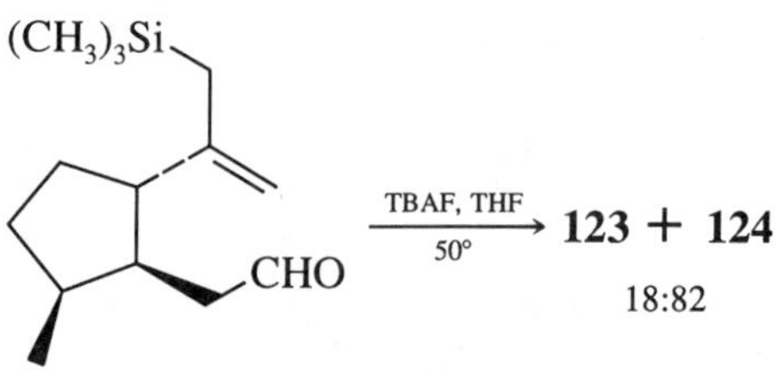

(Eq. 163)

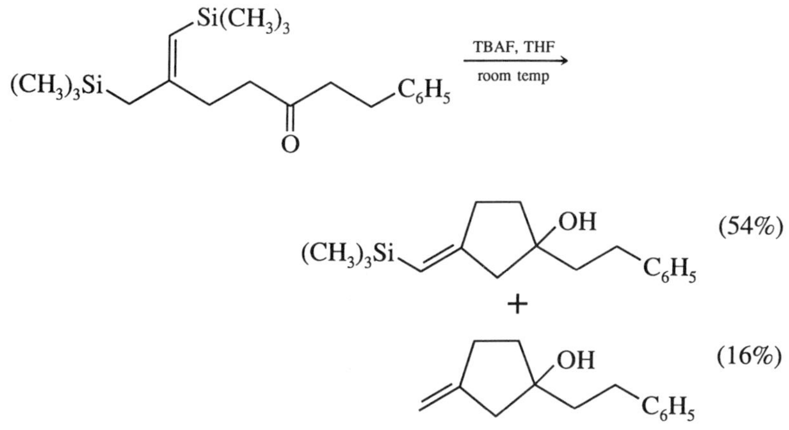

TBAF, THF, room temp

(CH3)3Si ... OH ... C6H5 (54%)

+ (Eq. 164)

OH ... C6H5 (16%)

159 is used in the synthesis of vetispirene,[167] and the reaction in Eq. 165 is used in a synthesis of the Prelog–Djerassi lactone.[312]

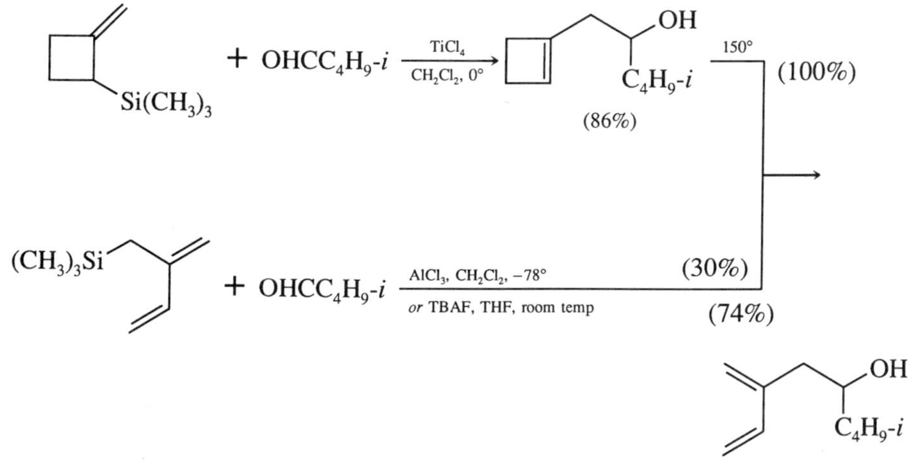

Si(CH3)3 + OHCC4H9-i —TiCl4, CH2Cl2, 0°→ OH C4H9-i (86%) —150°→ (100%)

(CH3)3Si + OHCC4H9-i —AlCl3, CH2Cl2, −78° or TBAF, THF, room temp→ (30%) (74%)

OH C4H9-i

138

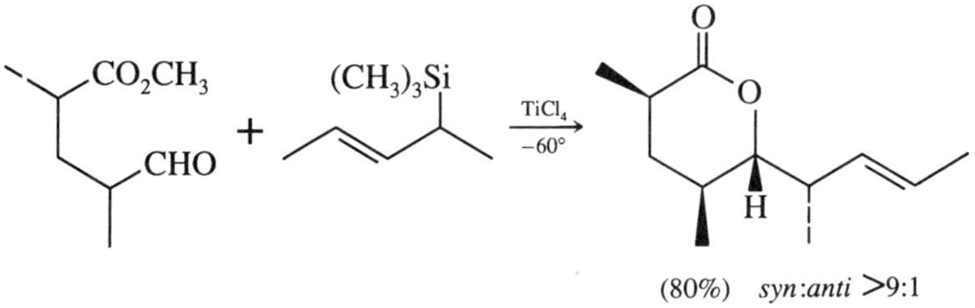

CO2CH3 CHO + (CH3)3Si —TiCl4, −60°→ O O H (Eq. 165)

(80%) *syn*:*anti* >9:1

Vinylsilanes. Vinylsilanes react with chloral in the presence of Lewis acids (Eq. 166), but there are very few other examples of this type of reaction.[313]

$(CH_3)_3Si\diagup\!\!=\!\!\diagdown Si(CH_3)_3 + OHCCCl_3 \xrightarrow[\text{room temp}]{AlCl_3,\ CH_2Cl_2} (CH_3)_3SiCH{=}CHCH(OH)CCl_3$

(Eq. 166)

Nucleophilic catalysis works when an anion-stabilizing group is present (Eq. 167), but not otherwise.[314] The usual byproduct in this reaction is that derived from protodesilylation, which is a more severe problem with ketone substrates than with aldehydes. The reaction is stereospecific (Eqs. 30 and 31), and can be used in ring synthesis (Eq. 168).[101]

$n\text{-}C_7H_{15}CH{=}C(CN)Si(CH_3)_3 + OHCCH_3 \xrightarrow[-20°\text{-room temp}]{\text{TBAF, THF}} n\text{-}C_7H_{15}CH{=}C(CN)CH(OH)CH_3$ (70%) (Eq. 167)

$OHC(CH_2)_4CH{=}C(SC_6H_5)Si(CH_3)_3 \xrightarrow[25°]{\text{TBAF, THF}}$ 2-(SC_6H_5)-cyclohex-2-en-1-ol (60%) (Eq. 168)

Allenylsilanes. Allenylsilanes react with aldehydes and ketones in the presence of titanium tetrachloride. Intermediate β-silylvinyl chlorides are sometimes observed, but these can be converted to the acetylene with potassium fluoride in dimethyl sulfoxide. With aldehydes, the reaction is mod-

$(CH_3)_3SiCH{=}C{=}CHCH_3 + OHC\text{-}C_6H_{11} \xrightarrow[CH_2Cl_2]{TiCl_4}$

$(CH_3)_3SiCH{=}C(Cl)CH(CH_3)CH(OH)C_6H_{11} \xrightarrow[25°]{\text{KF, DMSO}} HC{\equiv}CCH(CH_3)CH(OH)C_6H_{11}$ (81%) (Eq. 169)

syn:*anti* 4:1

erately selective in favor of *syn* products (Eq. 169), in the same sense but not to the same degree as the corresponding reaction with allylsilanes (Eq. 147).[315]

α,β-Unsaturated Carbonyl Compounds and α,β-Unsaturated Nitriles

Allylsilanes. Allylsilanes react with α,β-unsaturated aldehydes by direct attack at the carbonyl group,[316] but they react with α,β-unsaturated ketones (Eq. 170) by conjugate addition.[317,318] The reaction is usually known as the

$(CH_3)_3Si$ + C_6H_5 COCH$_3$ —($TiCl_4$, CH_2Cl_2, −40°)→ C_6H_5 COCH$_3$ (80%)

(Eq. 170)

Sakurai reaction. Occasionally, the intermediate cation is trapped intramolecularly before the silyl group is lost (Eq. 171).[319,320]

$(CH_3)_3Si$ + CH_3CO —($TiCl_4$, CH_2Cl_2, −30°)→

CH_3CO + $(CH_3)_3Si$ COCH$_3$ (Eq. 171)

trans (75%)
cis (8%)

(17%)

The reaction works well with methyl vinyl ketone[170,317] and other open-chain enones, acylcycloalkenes,[290,319,321] α-alkylidenecycloalkanones,[319,322,323] cycloalkenones,[317,322,324–326] and bicyclic enones.[317,327] There appears occasionally to be some difficulty with cyclopentenones, notably 3-substituted cyclopentenones,[328] but the presence of a 2-phenylthio substituent helps these reactions (Eq. 172)[329]

Cl, $(CH_3)_3Si$ + C_6H_5S, O —($C_2H_5AlCl_2$, CH_2Cl_2, −20°)→ Cl, C_6H_5S, O (63%)

(Eq. 172)

If the reaction mixture is not worked up, the intermediate is a titanium enolate. Electrophiles can be added directly to this intermediate to establish two carbon–carbon bonds in one pot (Eq. 173).[330]

$$(CH_3)_3Si\text{CH}_2CH{=}CH_2 \xrightarrow[\text{2. } C_6H_5CHO]{\text{1. } C_6H_5CH{=}CHCOCH_3,\ TiCl_4} CH_2{=}CHCH_2CH(C_6H_5)CH(COCH_3)CH(OH)C_6H_5 \quad (64\%)$$

(Eq. 173)

α,β-Unsaturated esters do not react with allylsilanes with Lewis acid catalysis. This limitation is overcome by using either an α,β-unsaturated acyl cyanide (Eq. 174)[331] or an α,β-unsaturated acylsilane (Eq. 175).[332] Both the acyl cyanides and the acylsilanes are versatile products that can be converted into acids and esters. The hexenoyl cyanides undergo an intramolecular ene reaction to give cyclohexenones,[333] and acylsilanes can be converted into aldehydes, which eliminates the difficulty of α,β-unsaturated aldehydes themselves reacting by attack at the carbonyl group. When there are two conjugated double bonds, as with sorbyl cyanide, addition takes place almost equally at the β and δ positions.[331]

$$(CH_3)_3SiCH_2C(CH_3){=}CH_2 + CH_3CH{=}CHCOCN \xrightarrow[-30^\circ]{TiCl_4,\ CH_2Cl_2} CH_2{=}C(CH_3)CH_2CH(CH_3)CH_2COCN \quad (95\%)$$

(Eq. 174)

$$(CH_3)_3SiCH_2CH{=}C(CH_3)_2 + (CH_3)_2C{=}CHCOSi(CH_3)_3 \xrightarrow[-78^\circ]{TiCl_4,\ CH_2Cl_2} CH_2{=}CHC(CH_3)_2C(CH_3)_2CH_2COSi(CH_3)_3 \quad (59\%)$$

(Eq. 175)

Silylpentadienes give some conjugate addition at the terminus of the diene system, but the Diels–Alder reaction is now a competing process (Eq. 176).[282]

Stereochemically, the reaction is moderately well controlled in open-chain systems by a neighboring chiral center (Eq. 177),[299] and is unusual with 4-substituted cyclohexenones and cycloheptenones in giving largely the *cis* product (Eq. 178), in contrast to cuprate reactions, which give largely the *trans* product.[326] With 5-substituted cyclohexenones and cycloheptenones, allylsilanes give the *trans* product, as usual for nucleophilic attack on these systems, but the selectivity (> 98:2) is higher than that for comparable cuprate reactions (~ 83:17).[326,334]

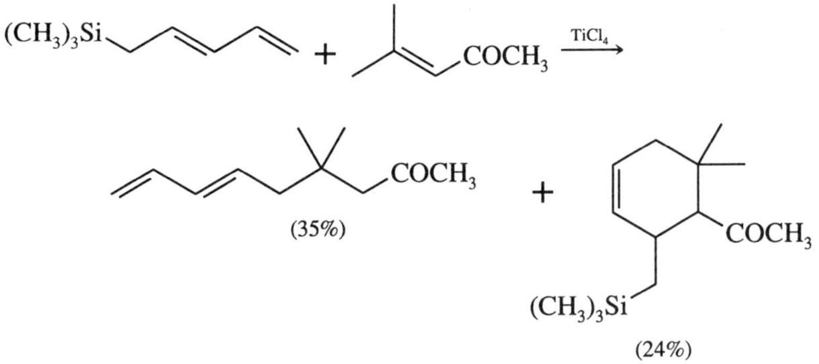

$(CH_3)_3Si$ + COCH$_3$ $\xrightarrow{TiCl_4}$

COCH$_3$ (35%) + COCH$_3$ $(CH_3)_3Si$ (24%) (Eq. 176)

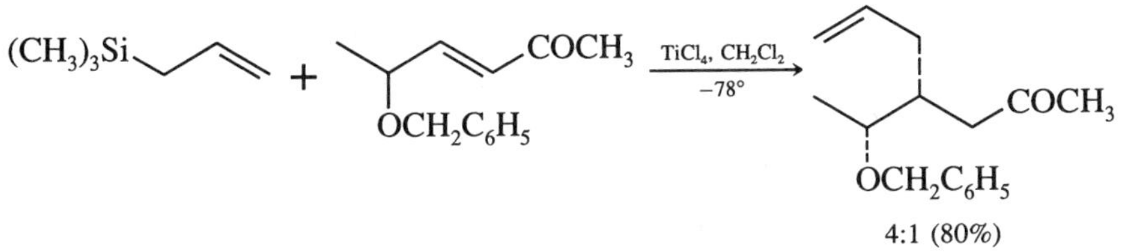

$(CH_3)_3Si$ + COCH$_3$ OCH$_2$C$_6$H$_5$ $\xrightarrow[-78°]{TiCl_4,\ CH_2Cl_2}$ COCH$_3$ OCH$_2$C$_6$H$_5$ 4:1 (80%)

(Eq. 177)

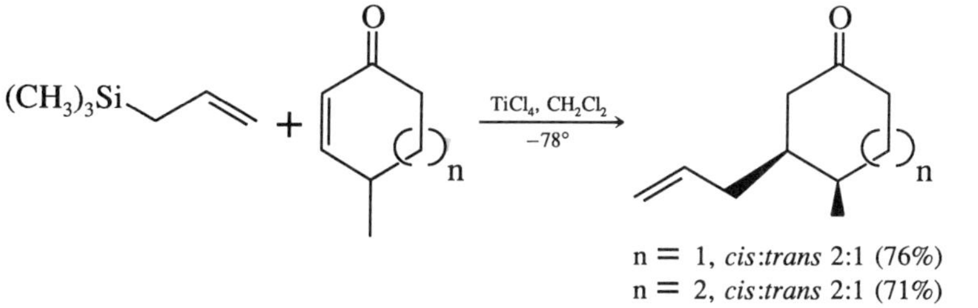

$(CH_3)_3Si$ + O n $\xrightarrow[-78°]{TiCl_4,\ CH_2Cl_2}$ O n (Eq. 178)

n = 1, *cis*:*trans* 2:1 (76%)
n = 2, *cis*:*trans* 2:1 (71%)

The intramolecular Sakurai reaction shows many remarkable features. The reaction is very sensitive to the choice of Lewis acid. Titanium tetrachloride can be used (Eq. 179),[335] but it quite frequently causes protodesilylation instead.[168,226] Ethylaluminum dichloride avoids this problem (Eq. 180),[226] but it is not as powerful a Lewis acid. As with the reactions between allylsilanes and aldehydes, it is not clear whether antiperiplanar or synclinal transition

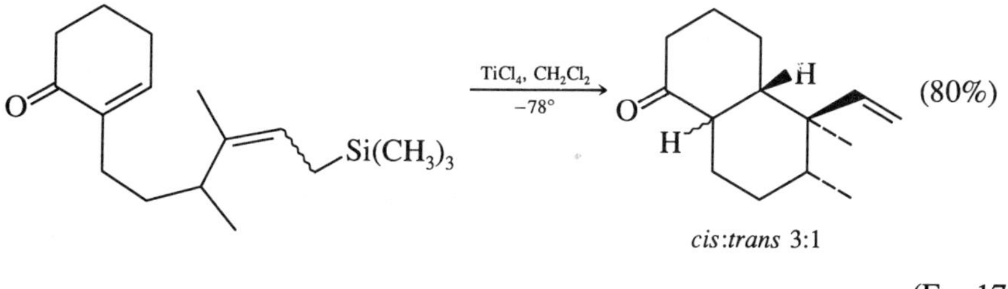

O Si(CH$_3$)$_3$ $\xrightarrow[-78°]{TiCl_4,\ CH_2Cl_2}$ O H H (80%)

cis:*trans* 3:1

(Eq. 179)

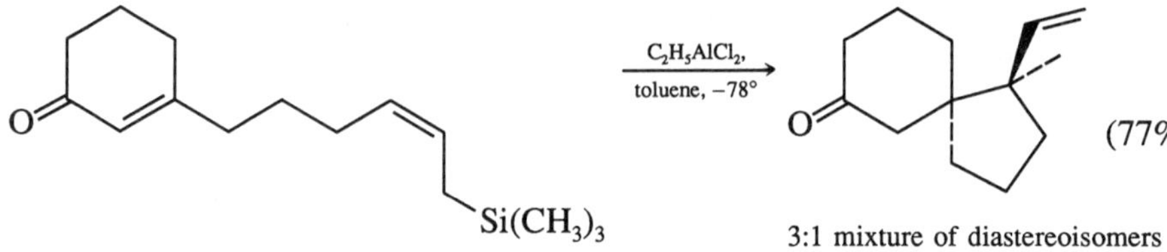

$C_2H_5AlCl_2$, toluene, −78°

Si(CH$_3$)$_3$

(77%)

3:1 mixture of diastereoisomers

(Eq. 180)

states are inherently preferred. In the reaction in Eq. 179, and in several others,[333,336] the transition state appears to be antiperiplanar, but in the reaction in Eq. 180 it is synclinal.

The conjugate addition is thwarted in the reaction in Eq. 181,[337] because a four-membered ring would have been formed, but in the reaction in Eq. 182 two quaternary centers are bonded efficiently.[337] In these two examples,

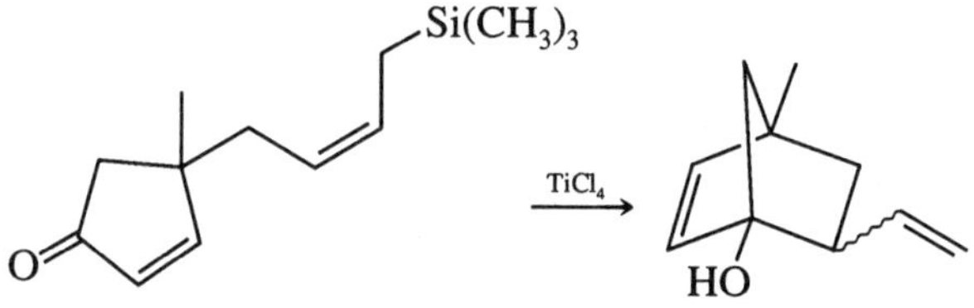

Si(CH$_3$)$_3$

$TiCl_4$

HO

(45%) (Eq. 181)

139

Si(CH$_3$)$_3$ $TiCl_4$

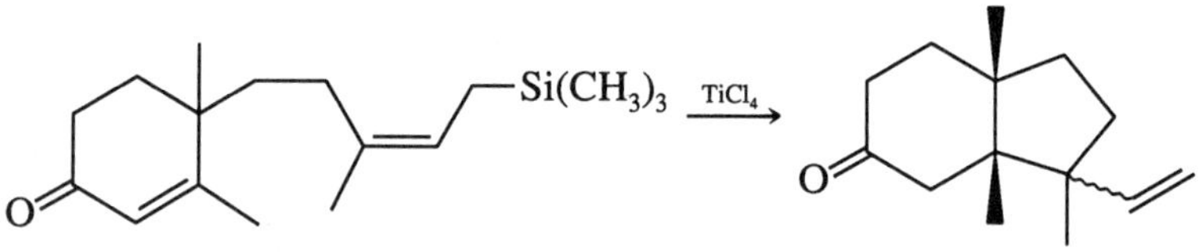

(78%)

(Eq. 182)

ethylaluminum dichloride is not powerful enough to induce reaction, but fortunately titanium tetrachloride works well. The allylsilanes in Eqs. 183 and 184 react at the terminus of the doubly conjugated dienone systems.[338,339]

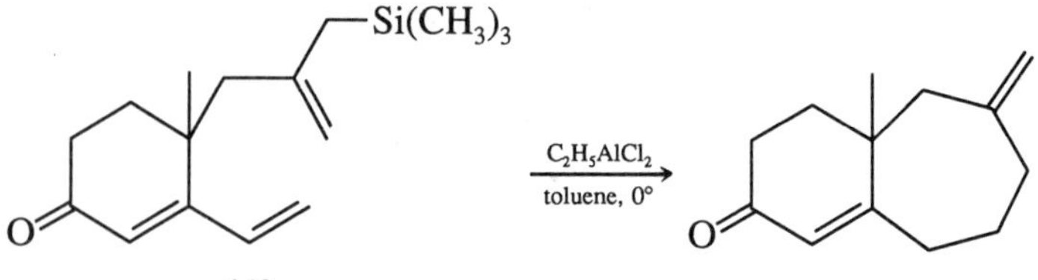

Si(CH$_3$)$_3$

$C_2H_5AlCl_2$, toluene, 0°

(55%) (Eq. 183)

140

$Si(CH_3)_3$

O= **141** $\xrightarrow[\text{toluene, 0°}]{C_2H_5AlCl_2}$ O= (60%)

(Eq. 184)

The Sakurai reaction can be catalyzed with fluoride ion (Eq. 185) but, in contrast to the Lewis acid catalyzed reaction (Eq. 170), it gives mixtures of

$(CH_3)_3Si$ + C_6H_5 $COCH_3$ $\xrightarrow[\text{reflux}]{\text{TBAF, THF}}$

C_6H_5 $COCH_3$ (24%)

+ (Eq. 185)

OH C_6H_5 (50%)

products because some direct attack at the carbonyl group takes place as well as conjugate attack.[13] However, unlike the Lewis acid catalyzed reaction, it works with α,β-unsaturated esters, amides, and nitriles (Eq. 186).[162] In in-

$(CH_3)_3Si$ + C_6H_5 Z $\xrightarrow[\text{room temp}]{\text{TBAF, DMF, HMPA}}$ C_6H_5 Z

Z = CO_2CH_3 (90%)
Z = $CON(C_2H_5)_2$ (80%)
Z = CN (65%)

(Eq. 186)

tramolecular reactions, nucleophilic catalysis frequently shows different selectivity from the Lewis acid catalyzed reaction, as in the reactions in Eqs. 187 and 188,[337,338] which can be compared with the reactions in Eqs. 181 and 184. Clearly the differences in these pairs of reactions stem from the ability of the allylsilane to react in the presence of fluoride ion without allylic shift. However, it is not obvious why the fluoride-catalyzed reaction in Eq. 189 should give such different products from those in Eq. 183.[339] The intramolecularly assisted removal of the silyl group, already seen in Eqs. 111 and 128, also works when α,β-unsaturated carbonyl compounds are the electrophiles and copper catalysis is used.[280]

139 $\xrightarrow[\text{room temp}]{\text{TBAF, DMF, HMPA}}$ (85%) (Eq. 187)

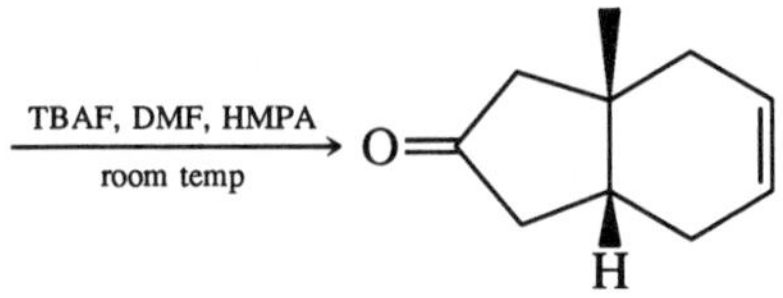

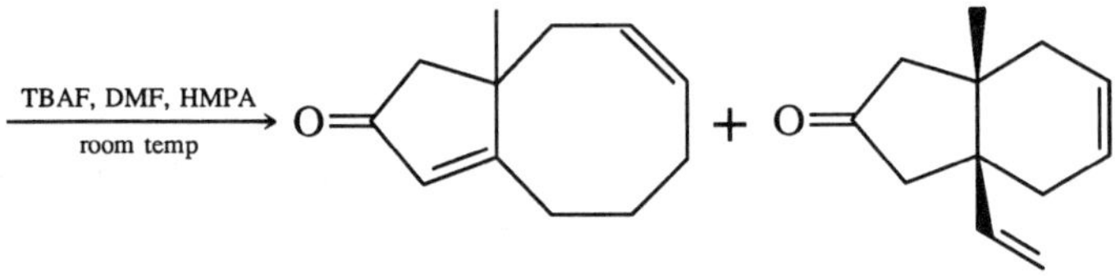

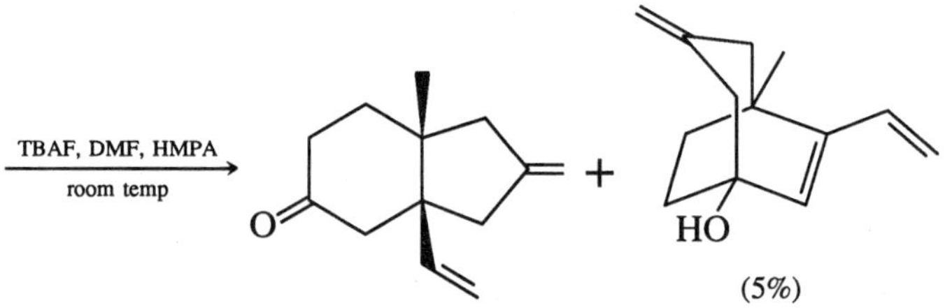

141 $\xrightarrow[\text{room temp}]{\text{TBAF, DMF, HMPA}}$ + (Eq. 188)

(33%) (23%)

140 $\xrightarrow[\text{room temp}]{\text{TBAF, DMF, HMPA}}$ + (Eq. 189)

HO

(5%)

(45%)

The Sakurai reaction is used in syntheses of lycopodine,[340] nootkatone,[316,323] ptilocaulin,[334] hirsutene,[341] and fawcettimine.[342]

Vinylsilanes. The only known intermolecular reaction (Eq. 190) involves a vinylsilane that is also an allylsilane.[320] Unusually, it reacts effectively as a vinylsilane; perhaps the intermediate cation loses the silyl group with intramolecular participation of the oxygen of the carbonyl group.

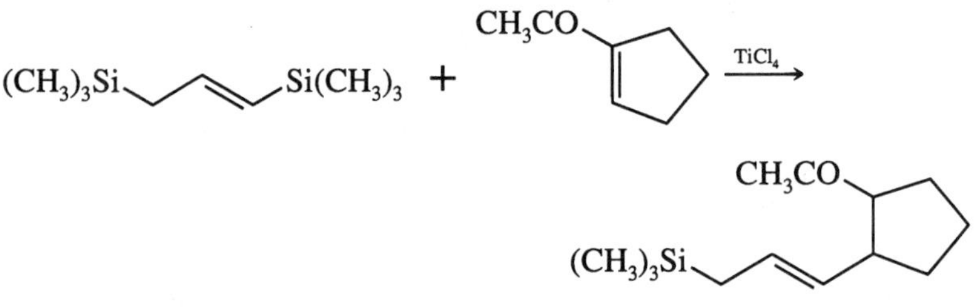

$(CH_3)_3Si$ $Si(CH_3)_3$ + CH_3CO $\xrightarrow{TiCl_4}$

CH_3CO $(CH_3)_3Si$ (Eq. 190)

(87%)

The silicon-controlled Nazarov reaction (Eq. 191), although pericyclic in nature, is formally the intramolecular reaction of a vinylsilane with an enone. Lewis acids are needed and ferric chloride has been recommended,[343,344] although its oxidizing power sometimes makes it unsuitable.[345] Other Lewis acids do work, notably boron trifluoride etherate and stannic chloride.[345,131] The reaction is used in a synthesis of capnelline.[345]

$(CH_3)_3Si$ $\xrightarrow[20°]{FeCl_3,\ CH_2Cl_2}$ (84%) (Eq. 191)

Allenylsilanes. Allenyltrimethylsilane reacts with acetylcyclopentene in the presence of titanium tetrachloride to give 1-acetyl-2-propargylcyclopentane in low yield, but the reaction with α,β-unsaturated acyl cyanides is better (Eq. 192).[331] Allenylsilanes with more substituents, especially 1-alkylallenylsilanes, react in a different way (Eq. 193): the intermediate cation **142**, instead of simply losing the silyl group, rearranges and then cyclizes to give the product **143** of cyclopentene annulation.[346] The overall result is not an electrophilic substitution, and these reactions are not included in the tables.

$(CH_3)_3Si$ C + COCN $\xrightarrow[-30°]{TiCl_4,\ CH_2Cl_2}$ HC≡C COCN (60%)

(Eq. 192)

$(CH_3)_3Si$ C + O $\xrightarrow[-78°]{TiCl_4}$ [$(CH_3)_3Si$ + $OTiCl_3$] ⟶

142

$(CH_3)_3Si$ O (85%) (Eq. 193)

143

Quinones

Allylsilanes react with quinones in the presence of titanium tetrachloride very much as they do with α,β-unsaturated ketones, the usual products coming from attack at the C=C double bond (Eq. 194),[347] but occasionally attack at the C=O double bond is seen (Eq. 195).[83,347]

The reaction of methyl α-silyl-3-butenoate with a 2-acetylnaphthoquinone is used in syntheses of nanaomycin A and deoxyfrenolicin.[348]

$(CH_3)_3Si$; , $TiCl_4$; CH_2Cl_2, −78° ; OH ; OH ; (58%) (Eq. (194)

CH_3O ; OCH_3, $TiCl_4$; CH_2Cl_2, −78° ; O ; CH_3O ; OCH_3 ; OH ; (81%) (Eq. 195)

α,β-Unsaturated Nitro and Nitroso Compounds

Allylsilanes react with nitroalkenes in the presence of aluminum chloride or titanium tetrachloride. The product is a nitronic acid derivative, but this is comparatively unstable and it is best treated immediately with a reducing agent: titanium trichloride gives the ketone by a Nef-type reaction (Eq. 196),[349] and, when the nitroalkene is β-nitrostyrene, zinc gives the corresponding nitrile (Eq. 197).[350]

$(CH_3)_3Si$; n-$C_{10}H_{21}$, NO_2 ; $AlCl_3$, −20° ; n-$C_{10}H_{21}$; O^- ; N^+ ; OH ; $TiCl_3$; $C_{10}H_{21}$-n ; O ; (50%) (Eq. 196)

$TiCl_4$, −15° ; C_6H_5, NO_2 ; C_6H_5 ; $OTiCl_3$; N^+ ; $OSi(CH_3)_3$; Zn ; C_6H_5 ; CN (Eq. 197)

Allylsilanes also react with nitrosoalkenes, generated in situ, to give cycloadducts; an acid-catalyzed step opens the ring to complete an electrophilic substitution reaction (Eq. 198).[351]

$(CH_3)_3Si$ + HON=C($CO_2C_2H_5$)CH_2Br $\xrightarrow[CH_2Cl_2]{Na_2CO_3}$

$\xrightarrow{HClO_4}$ $CO_2C_2H_5$ (Eq. 198)

(53%)

Acetals and Ketals

Allylsilanes. Acetals and ketals are exceptionally good electrophiles for allylsilanes. The reaction is typically catalyzed by stoichiometric amounts of titanium tetrachloride in dichloromethane at −78° and is complete in a few minutes (Eqs. 199 and 200).[352,83] Ketals require somewhat longer times than aldehyde acetals. Other Lewis acids commonly used are boron trifluoride etherate, stannic chloride, and, in catalytic amounts, trimethylsilyl trifluoromethanesulfonate,[353,354] trimethylsilyl iodide,[355] triphenylmethyl perchlorate,[356] and montmorillonite clays.[357] Hemiacetals also react cleanly, with replacement of the hydroxy group (Eq. 201).[358]

$(CH_3)_3Si$ + CH_3O–CH(OCH_3)–C_4H_9-*n* $\xrightarrow[-78°]{TiCl_4,\ CH_2Cl_2}$

OCH_3, C_4H_9-*n* (80%) (Eq. 199)

$(CH_3)_3Si$ + CH_3O–CH(OCH_3)–C_4H_9-*n* $\xrightarrow[-78°]{TiCl_4,\ CH_2Cl_2}$

OCH_3, C_4H_9-*n* (83%) (Eq. 200)

$(CH_3)_3Si$ + HO, O $\xrightarrow[CH_2Cl_2]{BF_3 \cdot O(C_2H_5)_2}$ O (57%)

(Eq. 201)

The reactions with acetals and ketals are often cleaner than with the corresponding aldehydes and ketones, as in Eqs. 200, 202, and 203,[83,252,284] which can be compared with Eqs. 35, 136, and 138, respectively, in each of which further reaction occurs.

$(CH_3)_3Si$ ... $CO_2C_2H_5$ + CH_3O / CH_3O $\xrightarrow[CH_2Cl_2,\ -78^\circ]{(CH_3)_3SiOSO_2CF_3}$ $C_2H_5O_2C$ ... OCH_3

(83%) E:Z, 86:14

(Eq. 202)

$(CH_3)_3Si$ ... $CO_2C_2H_5$ + CH_3O ... OCH_3 ... C_4H_9-n $\xrightarrow[0^\circ]{TiCl_4,\ CH_2Cl_2}$ $C_2H_5O_2C$ ... OCH_3 ... C_4H_9-n

(85%)

(Eq. 203)

Nevertheless, some reactions do go beyond the first-formed product (Eq. 204).[227a] The corresponding reaction with aldehydes is not clean with Lewis acids, although it is with nucleophilic catalysis (Eq. 162), which is actually the best way to do this type of reaction.

CH_3O_2C / CH_3O_2C ... $Si(CH_3)_3$ ... $Si(CH_3)_3$ + CH_3O ... OCH_3 ... C_6H_5 $\xrightarrow[(n\text{-}C_4H_9)_2SnCl_2,\ \text{room temp}]{(CH_3)_3SiOSO_2CF_3}$

CH_3O_2C / CH_3O_2C ... C_6H_5 (Eq. 204)

(38%)

Since trimethylsilyl iodide and silyl ethers can be prepared by the reaction of iodine with allyltrimethylsilane, and since acetals can be prepared from silyl ethers and aldehydes in the presence of trimethylsilyl iodide, it is possible to combine all three reactions (Eq. 205).[359]

$(CH_3)_3Si$ ... + CH_3OH + $OHCC_6H_5$ $\xrightarrow[CH_2Cl_2,\ 40^\circ]{I_2\ \text{cat.}}$ ... OCH_3 ... C_6H_5 (89%)

(Eq. 205)

Chemoselectivity in the acetal **144** involves exclusive attack at the acetal group, but chemoselectivity in acetal **145** depends upon the Lewis acid used (Eqs. 206 and 207): with aluminum chloride, attack takes place at the ketone group, but with titanium tetrachloride, both functional groups are attacked.[360] However, in the corresponding α-phenylthioalkyl halides (Eq. 208), attack takes place only on the thioalkyl halide group,[361] and with sugar aldehydes reaction takes place at the aldehyde group rather than at the anomeric position.[302]

$(CH_3)_3Si$ + CH_3O, CH_3O, $COCH_3$ → ($TiCl_4$, CH_2Cl_2, −15°) CH_3O, $COCH_3$

144 (60%)

(Eq. 206)

$(CH_3)_3Si$ + CH_3O, OCH_3, $COCH_3$ **145**

→ ($TiCl_4$, CH_2Cl_2, 0°) OCH_3, HO (45%)

→ ($AlCl_3$, CH_2Cl_2, −15° to 0°) CH_3O, OCH_3, HO (80%)

(Eq. 207)

$(CH_3)_3Si$

→ (C_6H_5S, Cl, $COCH_3$, $SnCl_4$) C_6H_5S, $COCH_3$ (90%)

→ (C_6H_5S, Cl, $COCH_3$, $SnCl_4$) C_6H_5S, $COCH_3$ (39%)

(Eq. 208)

Allylsilanes with a second double bond react at the terminus of the pentadienyl unit (Eq. 209),[362] just as they do with aldehydes.

$(CH_3)_3Si$ + C_2H_5O, C_2H_5O, C_4H_9-*i* → ($TiCl_4$, CH_2Cl_2, −78°) C_2H_5O, C_4H_9-*i*

(65%)

(Eq. 209)

Cyclizations with loss of an alkoxy group are possible (Eqs. 210 and 211).[363,364] Displacement of the methoxyethoxy group is a general phenomenon with this type of mixed acetal (Eq. 212), and presumably occurs because the Lewis acid is complexed to the two oxygen atoms of the glycol ether group.[364] Monothioacetals also react selectively with cleavage of the C–S acetal bond when stannic chloride is used as the Lewis acid (Eq. 213).[365]

$Si(CH_3)_3$; CH_3O OCH_3 $\xrightarrow[\text{room temp}]{SnCl_4,\ CCl_4}$ OCH_3 (72%) (Eq. 210)

$(CH_3)_3Si$; CH_3O O O C_6H_{13}-*n* $\xrightarrow[-78\text{ to }-50^\circ]{TiCl_4,\ CH_2Cl_2}$ O C_6H_{13}-*n* (96%) (Eq. 211)

$(CH_3)_3Si$ — $CH_3O(CH_2)_2OCH_2O(CH_2)_2C_6H_5$, $TiCl_4$, CH_2Cl_2, −20° (93%) (Eq. 212); $SnCl_4$, CH_2Cl_2, $CH_3SCH_2O(CH_2)_2C_6H_5$, room temp (40%) (Eq. 213) → $CH_2O(CH_2)_2C_6H_5$

When there is a substituent at C-3 of the allylsilane, mixtures of diastereoisomers are produced. With aliphatic acetals, the *syn* isomer is substantially the major product, whatever the geometry of the double bond in the allylsilane. With aromatic acetals the double bond geometry affects the stereochemistry, and the degree of stereospecificity is highest when the acetal has an electron-withdrawing group conjugated to it (Eq. 214).[366] Merely changing

146 (CH_3O, CH_3O, CN) $\xrightarrow[BF_3\cdot O(C_2H_5)_2]{(CH_3)_3Si\text{(}E\text{-crotyl)}}$ **147** (OCH_3, CN) + **148** (OCH_3, C[N]) 80:20

146 $\xrightarrow[BF_3\cdot O(C_2H_5)_2]{(CH_3)_3Si\text{(}Z\text{-crotyl)}}$ **147** + **148** 20:80

(Eq. 214)

the *para* substituent in **146** from cyano to methoxy removes the stereospecificity altogether, and the products are then formed in a 45:55 ratio from both allylsilanes.[336]

Allylsilanes attack cyclohexanone ketals axially (93:7),[353] and attack 2-phenylpropionaldehyde acetal with a degree of Cram selectivity that is influenced by the choice of Lewis acid, stannic chloride giving the highest ratio (3.5:1).[367] With β-alkoxyacetals, chelation control is not possible as it is with the corresponding aldehydes (Eq. 156). Stereoselectivity is in the opposite sense (Eq. 215) and depends upon the choice of Lewis acid, but even at its best (2.8:1), it is not high.[367]

$(CH_3)_3Si$ + CH_3O (OCH_3, $OCH_2C_6H_5$) $\xrightarrow[Ti(OC_3H_7\text{-}i)_4]{TiCl_4}$ (OCH_3, $OCH_2C_6H_5$)

syn:*anti* 2.8:1

(Eq. 215)

Anomeric sugar acetals and acetates react with attack on the ring (Eq. 216),[254] but the corresponding halides (Eq. 107) react somewhat faster and need less catalyst. The reaction shows good to excellent stereoselectivity for α (axial) attack because of the anomeric effect.[368,369] It appears that the better the leaving group, the more polar the solvent, and the more powerful the Lewis acid, the better the stereoselectivity, the combination of *p*-nitrobenzoates and boron trifluoride etherate being notably effective.[370] Anchimeric participation by a 3-acyl group does not take place.[368,369]

$(CH_3)_3Si$ + (OR, RO, OR, CH_3O, O, OR) $\xrightarrow[CH_3CN]{(CH_3)_3SiOSO_2CF_3\ (cat.)}$

R = $CH_2C_6H_5$

(OR, RO, OR, O, OR)

α:β, 10:1
(86%)

(Eq. 216)

The acetals of optically active 1,2- and 1,3-glycols show striking stereoselectivity, giving homoallylic ethers in high diastereoisomeric excess, especially when titanium tetrachloride–titanium tetraisopropoxide mixtures are used as the Lewis acid (Eq. 217).[225,371] In this series, in contrast to the sugar series above, an earlier S_N2-like transition state (**149**) is more likely than a transition state with a free oxocarbonium ion. The ether product can be converted to the free alcohol by oxidation and β elimination. Even noncyclic acetals made from aldehydes and optically active 1-phenylethanol give homoallylic ethers with high (80%) diastereoselectivity.[372]

$(CH_3)_3Si$ + [dioxane acetal, C_4H_9-*i*] $\xrightarrow{TiCl_4,\ Ti(OC_3H_7\text{-}i)_4}$ [HO… homoallylic ether, C_4H_9-*i*] (Eq. 217)

[Structure 149: Nu, *i*-C_4H_9, O, O, MX_n]

149

The reaction of allylsilanes with sugar esters is a key step in synthetic work directed toward the synthesis of palytoxin;[369] the reaction in Eq. 217 creates one of the chiral centers in a synthesis of myoporone,[225] and a similar reaction is used in a synthesis of calcitrol.[373]

Nucleophile-catalyzed reactions do not occur with acetals.

Vinylsilanes. In the reaction between acetals and vinylsilanes, the first-formed allylic ether usually reacts further (Eqs. 218 and 219).[374] The problem

C_6H_5CH=CH$Si(CH_3)_3$ $\xrightarrow[CH_2Cl_2,\ -78\ \text{to}\ -20°]{C_6H_5CH(OC_2H_5)_2,\ MoCl_5}$ C_6H_5CH=CHCH(C_6H_5)CH=CHC_6H_5 (62%) (Eq. 218)

C_6H_5CH=CH$Si(CH_3)_3$ $\xrightarrow[\text{room temp}]{C_2H_5O\text{–acetal}\ (OC_2H_5),\ TiCl_4}$ C_6H_5CH=CH…Cl (46%) (Eq. 219)

is partly solved by using monothioacetals, although the products are allylic thioethers and the reaction is not stereospecific (Eq. 220).[375] Secondary reactions are less of a problem in intramolecular reactions (Eq. 221), although

$C_6H_5CH{=}CHSi(CH_3)_3$ (Z or E isomer) + $C_2H_5OCH(SC_6H_5)CH_3$ $\xrightarrow[-78 \text{ to } -20^\circ]{MoCl_5,\ CH_2Cl_2}$ $C_6H_5CH{=}CHCH(SC_6H_5)CH_3$ (E only) (42%)

(Eq. 220)

$\xrightarrow[10^\circ,\ 12 \text{ h } Z \text{ isomer};\ 24 \text{ h } E \text{ isomer}]{ZnBr_2,\ CCl_4}$ (Eq. 221)

Si(CH$_3$)$_3$, CH$_3$O, OCH$_3$ → OCH$_3$

(70% from Z isomer)
(55% from E isomer)

even here it is necessary to choose Lewis acid and solvent with great care to avoid decomposition of the product almost as fast as it is formed.[376] The Z isomer reacts approximately twice as fast as the E isomer, presumably because it is following the favored pathway with retention of configuration. The cyclization in Eq. 43 appears anomalous, because the product isolated is not the result of attack at the silicon-bearing carbon. It seems likely that in this case the major product, an allylic ether, has decomposed, leaving a low yield of the homoallylic ether. Endocylic attack on an acetal is also successful (Eqs. 222 and 223), and the double bond exocyclic to the ring is set up stereospecifically with retention of configuration.[377] Sometimes the subsequent reaction of the first-formed product is desired, as in the arene synthesis in Eq. 224.[378]

$\xrightarrow[-15^\circ]{SnCl_4,\ CH_2Cl_2}$ (Eq. 222)

C$_4$H$_9$-n, Si(CH$_3$)$_3$, O, O, OCH$_3$ → C$_4$H$_9$-n, O

> 98% E (57%)

$\xrightarrow[-15^\circ]{SnCl_4,\ CH_2Cl_2}$ (Eq. 223)

C$_4$H$_9$-n, Si(CH$_3$)$_3$, O, O, OCH$_3$ → C$_4$H$_9$-n, O

> 98% Z (71%)

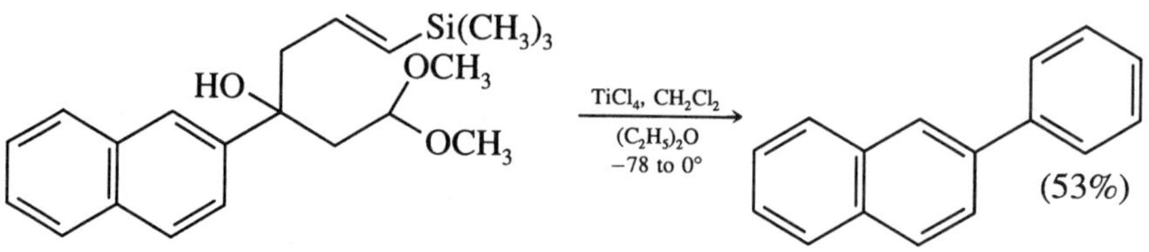

$TiCl_4$, CH_2Cl_2, $(C_2H_5)_2O$, −78 to 0°

(53%)

(Eq. 224)

When the silyl group is on C-2 of a terminal double bond, the intramolecular reaction still works to make medium-sized rings, but the regiochemistry, as usual with this type of vinylsilane, is no longer determined by the silyl group (Eq. 225).[379]

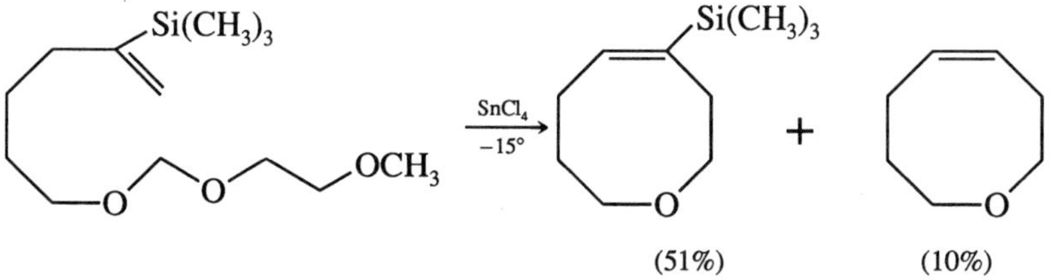

$SnCl_4$, −15°

(51%) (10%)

(Eq. 225)

Dithioacetals react comparatively cleanly in intramolecular reactions; the choice of Lewis acid is critical, dimethyl(methylthio)sulfonium fluoroborate having a remarkably selective thiophilicity. Even so, the product is unstable with respect to allylic shift of the thio group (Eq. 226).[380]

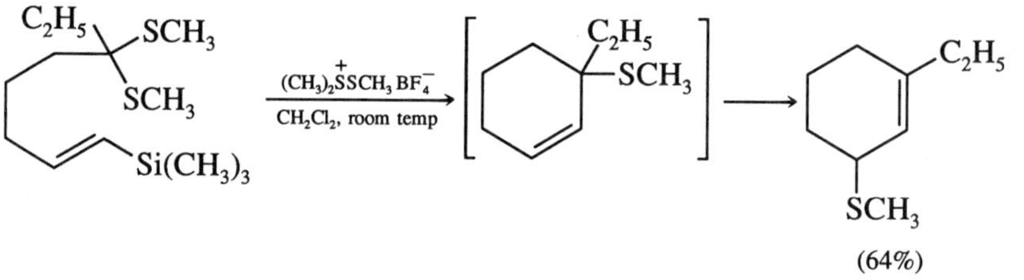

$(CH_3)_2\overset{+}{S}SCH_3\,BF_4^-$, CH_2Cl_2, room temp

(64%)

(Eq. 226)

Allenylsilanes. Allenylsilanes react with acetals in the presence of titanium tetrachloride to give addition, but the substitution reaction can be completed by treatment with fluoride ion (Eq. 227).[381]

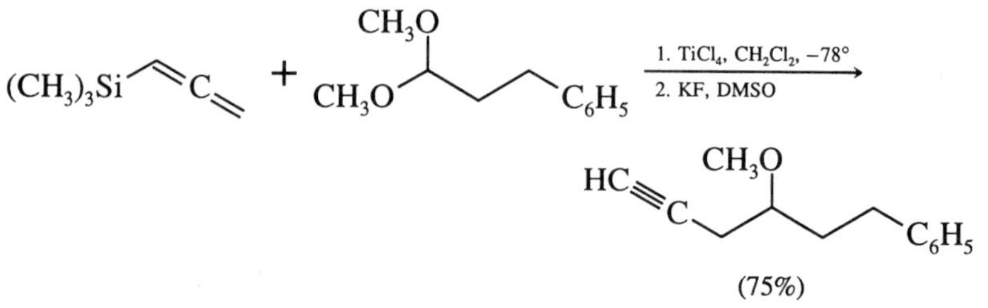

1. $TiCl_4$, CH_2Cl_2, −78°
2. KF, DMSO

(75%)

(Eq. 227)

α,β-Unsaturated Acetals

Allylsilanes react with α,β-unsaturated acetals in the presence of Lewis acids. The products, which are allylic ethers, frequently react again (Eq. 228).[382] The secondary reaction can be prevented by using milder Lewis acids like the mixture of titanium tetrachloride and titanium tetraisopropoxide[382] or catalytic amounts of trimethylsilyl trifluoromethanesulfonate (Eq. 229).[353]

$(CH_3)_3Si$ + CH_3O, CH_3O; $TiCl_4$, CH_2Cl_2, −78° ; [OCH_3] → (61%) (Eq. 228)

$(CH_3)_3Si$ + CH_3O, CH_3O, C_6H_5; A $TiCl_4$, $Ti(OC_3H_7\text{-}i)_4$ or B $(CH_3)_3SiO_3SCF_3$; RO, C_6H_5 (Eq. 229)

A, R = C_3H_7-*i* (81%)
B, R = CH_3 (78%)

Glycal 3-acetates are vinylogous acetals, and the cationic electrophile generated from them has the same substitution pattern as the electrophile derived from α,β-unsaturated acetals; they react well with allylsilanes to introduce the allyl group axially at the anomeric position (Eq. 230).[383,384]

$(CH_3)_3Si$ + OAc, OAc, OAc, O; $TiCl_4$, CH_2Cl_2, −78° or $BF_3\cdot O(C_2H_5)_2$, −50° ; OAc, OAc, O (94%) α:β 16:1

(Eq. 230)

In all these reactions, there is evidently a strong preference for the allylsilane to react with the electrophilic species at the carbon atom having the oxygen substituent.

The reaction in Eq. 230 is used in the synthesis of okadaic acid.[384]

Iminium Cations

Iminium ions are almost certainly the effective electrophiles in several reactions of allylsilanes and vinylsilanes. Except for the nitrone cycloaddition reactions discussed in the section on carbon electrophiles needing no catalysis, they are all grouped together here, regardless of the way in which they have been generated.

Allylsilanes. Allylsilanes react with acyliminium ions (Eq. 231),[385] generated in situ in a variety of ways. The reaction can be used with β-lactams (Eq. 232)[386] and in more conjugated systems (Eq. 233), where the selectivity

$(CH_3)_3Si$ + C_2H_5O … C_4H_9-*n* —($TiCl_4$, CH_2Cl_2, 25°)→ … C_4H_9-*n*

(Eq. 231)

p-$O_2NC_6H_4CO_2$ … CO_2CH_3 + $CO_2C_6H_4NO_2$-*p* … $Si(CH_3)_3$ —($AgBF_4$, CH_2Cl_2, −60 to 0°)→ *p*-$O_2NC_6H_4CO_2$ … $CO_2C_6H_4NO_2$-*p* … CO_2CH_3 (69%)

(Eq. 232)

$(CH_3)_3Si$ + OH … $CO_2C_2H_5$ —($SnCl_4$, CH_2Cl_2, −78°)→ … $CO_2C_2H_5$ (90%)

(Eq. 233)

is again for attack α to the heteroatom.[387] The reaction is especially powerful in ring synthesis, where it is possible to use protic acids as well as Lewis acids and to make 5-*endo*-trig reactions work (Eqs. 234 and 235).[388,389]

OH Si(CH_3)$_3$ $\xrightarrow[0°]{CF_3CO_2H,\ CH_2Cl_2}$ H H N O N O (72%) (Eq. 234)

Si(CH_3)$_3$ C_2H_5O C_6H_5 N $CO_2C_2H_5$ $\xrightarrow[\text{room temp}]{HCO_2H}$ C_6H_5 N $CO_2C_2H_5$ (81%) (Eq. 235)

The double bond produced in the reaction of an allylsilane with an acyliminium ion is itself available for a second carbon–carbon bond-forming step when a second acyliminium ion precursor group is present (Eq. 236).[390]

(CH_3)$_3$Si + CH_3O N O OCH_3 $\xrightarrow{TiCl_4,\ CH_2Cl_2}$ Cl N O (77%)

(Eq. 236)

Simple iminium ions, generated in situ from the salts of secondary amines and aqueous formaldehyde, react cleanly with allylsilanes (Eq. 237). The corresponding reaction with primary amine salts and formaldehyde does not stop simply at substitution; the first-formed product reacts intramolecularly with another iminium ion (Eq. 238).[391] This reaction, and the similar one in Eq. 236, revealingly illustrate the different ways in which an allylsilane and a simple alkene react with electrophiles: the former by substitution and the

(CH_3)$_3$Si + CH_2O + $CH_3NHCH_2C_6H_5$ $\xrightarrow[50°]{H_2O}$ $NCH_2C_6H_5$ CH_3 (76%)

(Eq. 237)

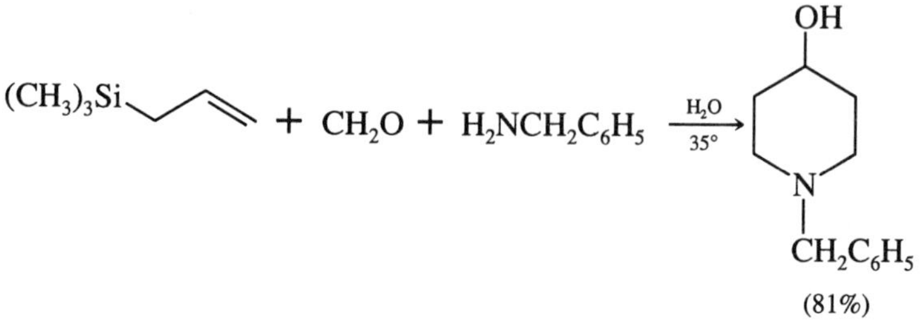

(Eq. 238)

(81%)

latter by addition. The same type of reaction taking place intramolecularly can be used to make six-, seven-, and eight-membered rings (Eq. 239).[392]

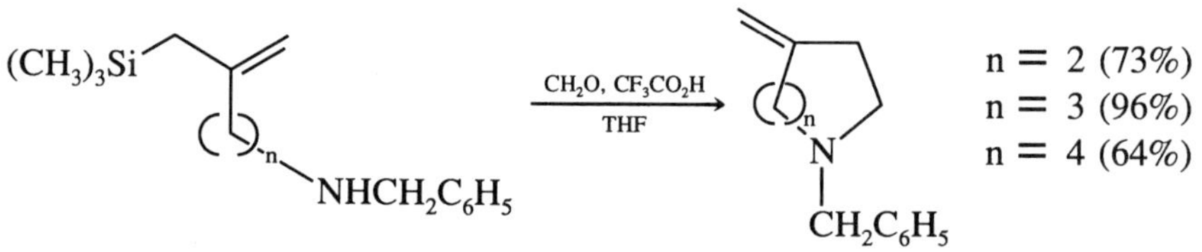

n = 2 (73%)
n = 3 (96%)
n = 4 (64%)

(Eq. 239)

The iminium ion **150** is not as reactive as the acyliminium ions or the simple iminium ions in the reactions above, but it can be induced to react photochemically (Eq. 240). The reaction takes place by electron transfer, loss of the silyl group, and radical coupling, and the placing of the silyl group does not therefore determine the regiochemistry of the allylic coupling.[393] This reaction can also be used intramolecularly, where it works in one case that is 5-*endo*-trig at both ends.[394]

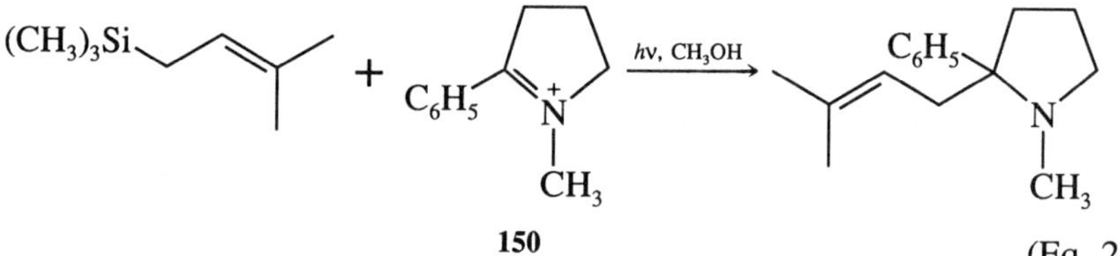

150

(Eq. 240)

The only nucleophile-catalyzed reaction is of allyltrimethylsilane with pyridine *N*-oxides. The products are 2-prop-1′-enylpyridines, the double bond having moved into conjugation (Eq. 241).[395]

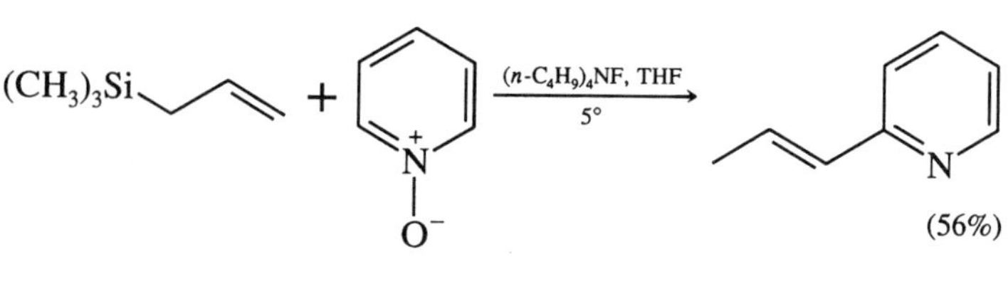

(Eq. 241)

(56%)

The cyclization in Eq. 234 is used in syntheses of isoretronecanol and epilupinine,[388] and a similar reaction is used in a synthesis of mesembrine.[396]

Vinylsilanes. Vinylsilanes react intramolecularly with iminium ions (Eq. 242)[397] and acyliminium ions (Eq. 243).[398] The reactions are sometimes very

C_6H_5 ... $Si(CH_3)_3$ —(CF_3CO_2H, CH_3CN, 82°)→ C_6H_5 ... N H (90%) (Eq. 242)

O ... N ... HO ... $Si(CH_3)_3$ ... Br —(CF_3CO_2H, 25°)→ O ... N ... Br (63%) (Eq. 243)

sensitive to the geometry of the vinylsilane, the reaction in Eq. 242, for example, failing with the *E* isomer.[397] However, in the reaction in Eq. 244, the vinylsilane geometry is scrambled by the rapidity of the aza-Cope rearrangements **151** ⇌ **152** ⇌ **153**, with the result that the geometry of the vinylsilane double bond does not affect the reaction. Strictly speaking, the cyclization step **152** → **154** is an allylsilane reaction, not a vinylsilane reaction. The reaction of the *E* isomer shown in Eq. 245 is stereospecific, with the *Z* isomer giving the alternative geometrical isomer of the product, thus providing

C_3H_7-*n* ... N H ... $Si(CH_3)_3$ —($(CH_2O)_n$, CH_3CN, RSO_3H, reflux)→

[C_3H_7-*n* ... $\overset{+}{N}$ ... $Si(CH_3)_3$ **151** ⇌ C_3H_7-*n* ... N^+ ... $Si(CH_3)_3$ **152** ⇌ C_3H_7-*n* ... $\overset{+}{N}$ ... $Si(CH_3)_3$ **153**]

152 → C_3H_7-*n* ... N **154** (90%) (Eq. 244)

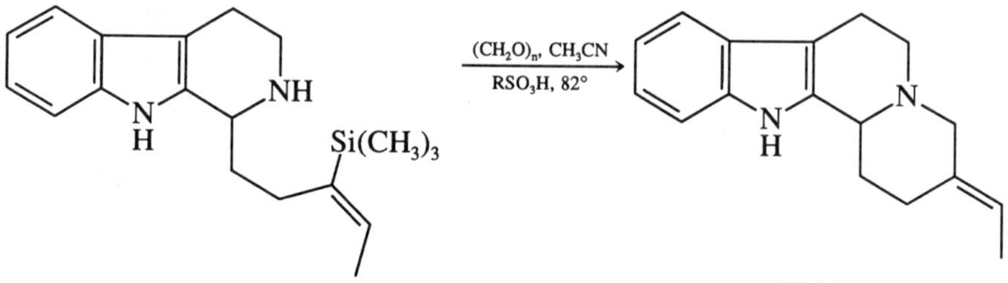

$(CH_2O)_n$, CH_3CN, RSO_3H, 82°

(83%)

(Eq. 245)

an elegant method for controlling the geometry of an exocyclic double bond.[399] Intramolecular reactions of a vinylsilane with an iminium ion are used in syntheses of elaeokanine[398] and epielwesine[397] and of three of the pumiliotoxins.[400,401]

Allenylsilanes. Allenyltrimethylsilane reacts with the acyliminium ion derived from **155** to give both the substitution product **156** and the annelation product **157** (Eq. 246). Substitution is completely supressed when a *tert*-butyldimethylsilyl group is used in place of the trimethylsilyl group.[129]

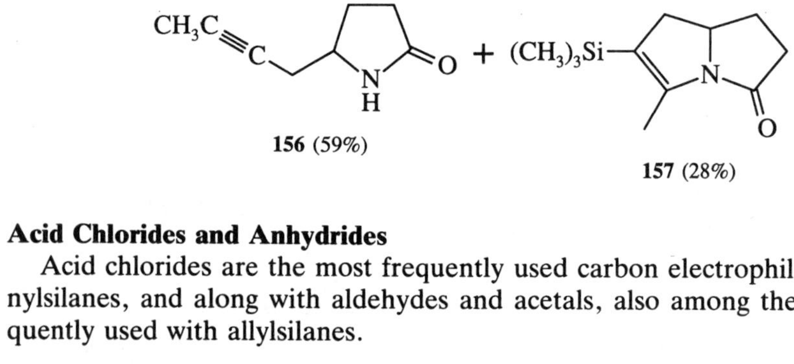

$TiCl_4$, CH_2Cl_2

155

156 (59%)

157 (28%)

(Eq. 246)

Acid Chlorides and Anhydrides

Acid chlorides are the most frequently used carbon electrophiles with vinylsilanes, and along with aldehydes and acetals, also among the most frequently used with allylsilanes.

Allylsilanes. The reaction of allylsilanes with acid chlorides is usually catalyzed with aluminum chloride or titanium tetrachloride in dichloromethane at −78° for a few minutes. Other Lewis acids used include stannic chloride at 0°,[55] zinc chloride at room temperature,[402] and indium and gallium chlorides at 40–60°.[403] The reaction with anhydrides is catalyzed with boron trifluoride etherate,[404] with titanium tetrachloride,[319] or even with a protic acid.[405] The reaction is regiospecific (Eq. 247)[406] and stereospecifically *anti* (Eq. 248),[21] although one special example (Eq. 17) discussed earlier is ster-

$(CH_3)_3Si$ + ClCO $\xrightarrow[-60°]{AlCl_3,\ CH_2Cl_2}$ (90%)

(Eq. 247)

C_6H_5, $(CH_3)_3Si$, H + $ClCOCH_3$ $\xrightarrow[-78°]{AlCl_3,\ CH_2Cl_2}$ C_6H_5, $COCH_3$, H

85% ee 53% ee

(Eq. 248)

eospecifically *syn*.[19] The products of these reactions are β,γ-unsaturated ketones, which are very susceptible to enolization and hence to loss of configuration. When 3,3-disubstituted allylsilanes are used, the products are not susceptible to enolization, and stereochemistry should be well controlled. However, with this type of allylsilane there are frequently problems in controlling the site of attack by the electrophile (Eqs. 249 and 250).[38,407] Protodesilylation is also a source of byproducts, and the products of protodesilylation sometimes react with the electrophile. The reactions in Eq. 251 show such interfering protodesilylation as well as loss of regiocontrol in the site of attack on the allyl system.[408]

$(CH_3)_3Si$ + Cl(C=O)R $\xrightarrow[CH_2Cl_2]{TiCl_4}$ + $(CH_3)_3Si$

R = CH_3	100:0
R = C_6H_5	75:25
R = 2-furyl	50:50

(Eq. 249)

$C_6H_5(CH_3)_2Si$ + $ClCOCH_3$ $\xrightarrow[CH_2Cl_2,\ -78°]{TiCl_4,\ Ti(OC_3H_7\text{-}i)_4}$

$C_6H_5(CH_3)_2Si$, $COCH_3$ (65%)

(Eq. 250)

$(CH_3)_3Si$–[cyclohexenyl]–C_6H_5 + $ClCOCH_3$ $\xrightarrow[-25^\circ]{AlCl_3}$ [R, $COCH_3$, C_6H_5 cyclohexene] + [$COCH_3$, C_6H_5 cyclohexene]

R = $Si(CH_3)_3$ (19%) (32%)
R = H (32%)

(Eq. 251)

Protodesilylation may not always be the cause of this problem. The byproduct **159** in the reaction in Eq. 252 looks like the result of protodesilylation followed by acylation. However, when acylation is deliberately carried out on the product of protodesilylation, there is very little of the product **159** in the complex mixture produced. It seems likely that the competing reaction is not protodesilylation but direct attack of titanium tetrachloride on the allylsilane, in other words a metal-for-silicon exchange (q.v.), the allyltitanium species then reacting with the acid chloride. Consistent with this explanation is the observation that the same reaction as that in Eq. 252 can be carried out by premixing the acid chloride and the titanium tetrachoride before adding the allylsilane; using this protocol, the product is **158** (77%), uncontaminated with the regioisomer **159**.[409] This protocol is generally recommended, as it is also for the reaction of vinylsilanes with acid chlorides.[76]

$C_6H_5(CH_3)_2Si$–[allyl]–C_6H_5 + $ClCOCH_3$ $\xrightarrow[-78^\circ]{TiCl_4,\ CH_2Cl_2}$

[$COCH_3$, C_6H_5] + CH_3CO–[allyl]–C_6H_5 (Eq. 252)

158 (62%) **159** (26%)

Pentadienylsilanes react with acid chlorides with less regiocontrol than with other carbon electrophiles: there is substantial attack at the γ position and at the terminus (Eq. 253, compare Eq. 134). Attack at the γ position is the major reaction when the ε position is substituted.[410]

$(CH_3)_3Si$–[pentadienyl, γ, ε] + $ClCOC_4H_9$-*t* $\xrightarrow[-78^\circ]{TiCl_4,\ CH_2Cl_2}$

[dienyl]–COC_4H_9-*t* + [divinyl]–COC_4H_9-*t* (Eq. 253)

(34%) (25%)

α-Functionalized allylsilanes are useful because of the extra functionality that is attached to the double bond in the product. Ester groups allow an effective γ acylation of crotonate systems. The products isolated are mixtures of the γ-acyl α,β-unsaturated ester and the corresponding α-pyrone, formed by cyclization.[252,411] Phenylthio groups make the product an (*E*)-vinyl sulfide (Eq. 254).[412] Silyloxy groups are versatile because the first-formed products

$$(CH_3)_3Si\text{-}CH(SC_6H_5)CH{=}CHCH_3 + ClCOC_3H_7\text{-}n \xrightarrow[-78^\circ]{AlCl_3} C_6H_5SCH{=}CHCH(CH_3)COC_3H_7\text{-}n \quad (75\%)$$

(Eq. 254)

are silyl (or titanium) enol ethers. These can be worked up directly to give 1,4-ketoaldehydes (Eq. 255),[413] or they can be treated with a second electrophile (Eq. 256).[330]

$$(CH_3)_3Si\text{-}CH(OSi(CH_3)_2C_4H_9\text{-}t)CH{=}CH_2 + ClCOC_4H_9\text{-}t \xrightarrow[-78^\circ]{TiCl_4} OHC\text{-}CH_2CH_2\text{-}COC_4H_9\text{-}t \quad (75\%)$$

(Eq. 255)

$$\xrightarrow[2.\ CH_3CH(OC_2H_5)_2]{1.\ TiCl_4,\ -78^\circ} OHC\text{-}CH(CH(CH_3)OC_2H_5)CH_2\text{-}COC_4H_9\text{-}t \quad (53\%)$$

(Eq. 256)

The acylation of allylsilanes is used in syntheses of artemesia ketone (Eq. 247),[406] myrcenone,[308] tagetone,[414] and dihydrojasmone.[415]

Vinylsilanes. Acid chlorides are the best carbon electrophiles for vinylsilanes, presumably because the acyl group that is introduced lowers the nucleophilicity of the double bond of the product relative to that of the starting material. With other carbon electrophiles, the nucleophilicity of the double bond is less reduced, or even enhanced, and electrophilic reaction cannot then be stopped as easily at the first-formed product. The reaction is generally carried out with aluminum chloride in dichloromethane at 0°, but other temperatures between −78° and room temperature are common. Titanium tetrachloride and stannic chloride are also used. The reaction is stereospecific with retention of configuration (Eq. 257),[271] unless subsequent isomerization of the double bond catalyzed by the Lewis acid, or by addition of hydrogen chloride and subsequent elimination occurs.[416,417] The addition of hydrogen chloride to the double bond of the product is a major source of byproducts in some cases. Treatment with base during workup converts these compounds into the α,β-unsaturated ketones. Occasionally there are advantages to an

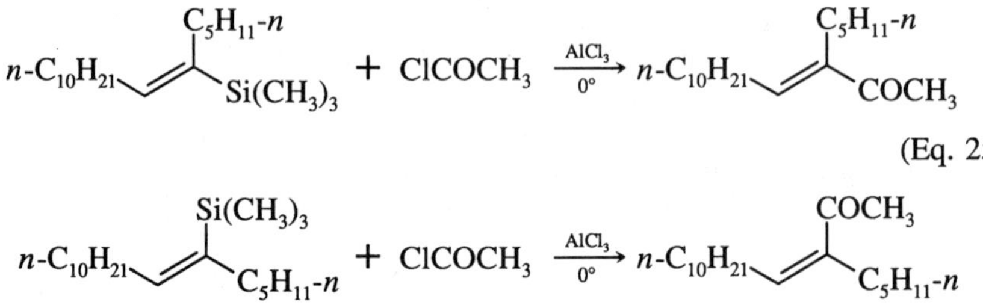

(Eq. 257)

alternative workup, in which hydrogen chloride is deliberately passed through the reaction mixture, so that β-chloroketones are formed.[117]

The regiocontrol in attack on unsymmetrical vinylsilanes is good (Eq. 10) unless the asymmetry is considerable (Eq. 258[418] and Eqs. 40–42).[296,117–119]

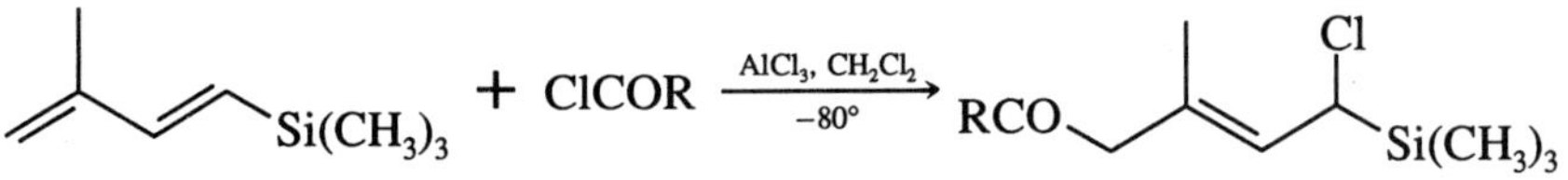

(Eq. 258)

The reaction works well intramolecularly, both in 5-*exo*-trig (Eq. 259)[417] and in 5-*endo*-trig reactions (Eq. 260).[419] Even four-membered rings can be formed, but there are substantial amounts of byproduct from attack at the end of the vinyl group away from the silyl substituent (Eq. 261).[116] When the

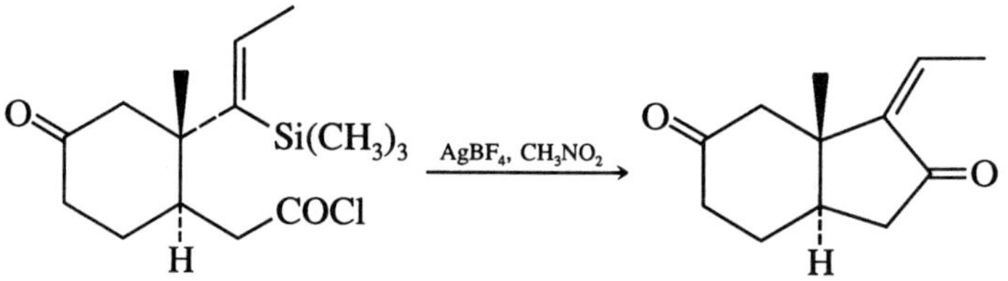

(71%) (Eq. 259)

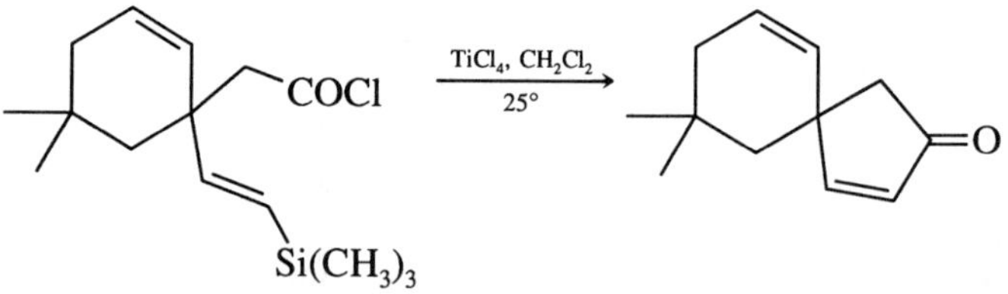

(82%) (Eq. 260)

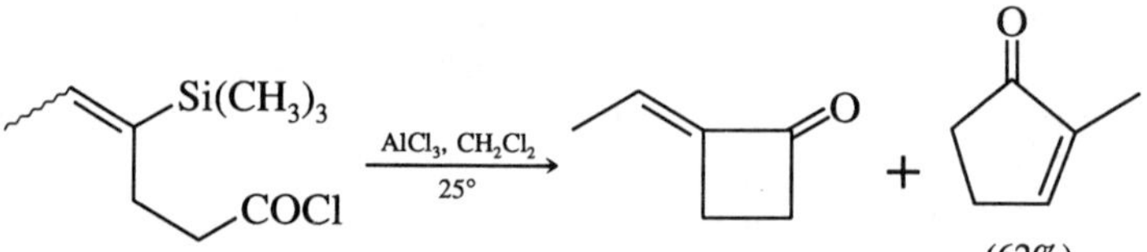

mixture of *Z* and *E*

E only (38%)

(62%)

(Eq. 261)

acid chloride is presented with two similar double bonds, one of which is a vinylsilane and the other is not, it reacts with the vinylsilane (Eq. 262), but when one of the double bonds can give a tertiary cation, it is that double bond which reacts (Eq. 263).[420] In this example, the silyl group is lost from the other double bond by protodesilylation.

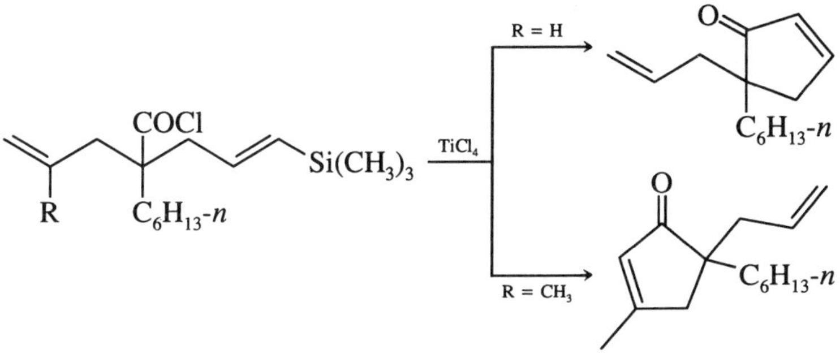

COCl Si$(CH_3)_3$ R C_6H_{13}-*n* $TiCl_4$ R = H (69%) (Eq. 262) C_6H_{13}-*n* R = CH_3 C_6H_{13}-*n* (50%) (Eq. 263)

The acylation of vinylsilanes is particularly useful when an α,β-unsaturated acid chloride is used, because the first-formed product **160** can undergo a Nazarov cyclization in situ. There are two ways of doing this reaction to make bicyclic ketones: to use vinyltrimethylsilane and a cyclic α,β-unsaturated acid chloride (Eq. 264),[421] or to use a cyclic vinylsilane and an open-chain α,β-unsaturated acid chloride (Eq. 265).[76] The intermediate dienone can be isolated in some cases (Eq. 265). When the vinylsilane carries a phenylthio group, the reaction is still well-behaved, but different products are obtained from 2-(phenylthio)vinyltrimethylsilane (**161**) and from 1-(phenylthio)vinyltrimeth-

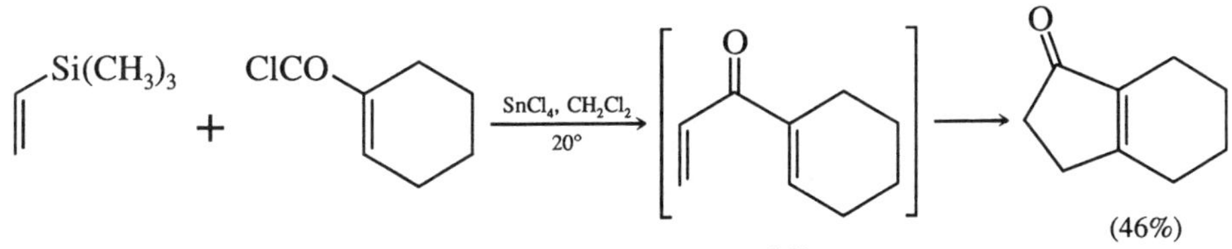

Si$(CH_3)_3$ + ClCO $\xrightarrow{SnCl_4,\ CH_2Cl_2,\ 20°}$ [**160**] → (46%) (Eq. 264)

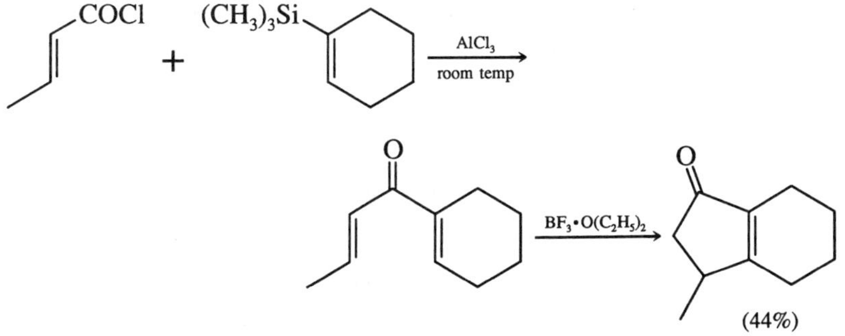

COCl + $(CH_3)_3$Si $\xrightarrow[\text{room temp}]{AlCl_3}$

$\xrightarrow{BF_3 \cdot O(C_2H_5)_2}$ (44%) (Eq. 265)

ylsilane (**164**). With the former (Eq. 266), the silyl and the phenylthio groups reinforce each other, clean substitution takes place, and the first-formed product is the dienone **162**. Nazarov cyclization and shift of the phenylthio group then leads to the α-phenylthioketone **163**.[119] With 1-(phenylthio)vinylsilane

$Si(CH_3)_3$ ClCO

$AlCl_3$
$Cl(CH_2)_2Cl$
50°

C_6H_5S

161

O SC_6H_5

(Eq. 266)

C_6H_5S H

162 **163** (40%)

(**164**), however, the phenylthio group controls the regiochemistry of attack (Eq. 267). Now the intermediate dienone **165** retains the silyl group, and the Nazarov cyclization **165** → **166** is another example of a vinylsilane reacting intramolecularly with an enone (q.v.).[119]

ClCO

$AgBF_4$
$Cl(CH_2)_2Cl$
−20°

C_6H_5S $Si(CH_3)_3$

164

O O

(Eq. 267)

C_6H_5S $Si(CH_3)_3$ C_6H_5S

165 **166** (35%)

The acylation of vinylsilanes is used in syntheses of naginata and isoegoma ketones,[422] of hirsutene,[119] and of grandisol.[423]

Allenylsilanes. Allenylsilanes react with acid chlorides, but there is insufficient work to make any generalizations.[274,424]

Nitriles

Allylsilanes react with nitriles in the presence of boron trichloride to give β,γ-unsaturated ketones (Eq. 268).[425]

$(CH_3)_3Si$–$CH_2CH{=}CHCH_3$ + $NCCCl_3$ $\xrightarrow{BCl_3,\ CH_2Cl_2}$ $CH_2{=}CHCH(CH_3)COCCl_3$ (Eq. 268)

(79%)

Orthoesters

Allylsilanes. Two equivalents of allyltrimethylsilane react with ethyl orthoformate (Eq. 269)[352] and with dichloromethyl methyl ether[273] in the presence of titanium tetrachloride. The allylsilane **167** reacts with methyl orthoformate to give the aldehyde **168**. This is the only reaction in which this allylsilane appears to react directly with an electrophile rather than first rearranging to a silyl dienol ether (see Eq. 139). However, the balance is a very delicate one, since the corresponding pyrrolidino enamine is normal, the electrophile attacking at the γ position.[285]

$(CH_3)_3SiCH_2CH{=}CH_2$ + $(C_2H_5O)_3CH$ $\xrightarrow[\text{room temp}]{TiCl_4,\ CH_2Cl_2}$ $(CH_2{=}CHCH_2)_2CHOC_2H_5$ (24%) (Eq. 269)

167 + $(CH_3O)_3CH$ $\xrightarrow[-78°]{TiCl_4,\ CH_2Cl_2}$ **168** (50%)

Vinylsilanes. 2-Ethoxy-1,3-dithiolan reacts nonstereospecifically with (Z)- or (E)-β-trimethylsilylstyrene in the presence of boron trifluoride etherate, to give (E)-2-styryl-1,3-dithiolan, but the yield is only 20%.[375] Formylation is more usually carried out using dichloromethyl methyl ether (Eq. 119).

Carbon Electrophiles at the Oxidation State of Carbon Dioxide

The most important electrophile in this class is chlorosulfonyl isocyanate, which was discussed earlier because it does not need catalysis. The only other reactions in this class are the fluoride-catalyzed reaction between phenyl cyanate and the allylsilane **169**, which is used in a synthesis of hinesol and β-vetivone,[167] and the reaction in Eq. 270, in which a chlorocarbonate is, most unusually, an acylating agent on carbon.[426]

$Si(CH_3)_3$ + C_6H_5OCN $\xrightarrow[\text{reflux}]{\text{TBAF, THF}}$ CN (69%)

169

$(CH_3)_3Si$ ... O–COCl $\xrightarrow[0^\circ]{SnCl_4,\ CH_2Cl_2}$ (57%) (Eq. 270)

Nitrogen Electrophiles

Allylsilanes. Triazolinediones and diethyl azodicarboxylate react with allylsilanes in an ene reaction with loss of a proton rather than of the silyl group.[139,142] The loss of the silyl group is encouraged by polar solvents and lower temperatures (Eq. 271), presumably because ionic intermediates are involved to some extent in place of the more concerted ene pathway.[427] The nitrosation of allylsilanes is reported to give addition products, which are not very tractable.[428] The subsequent elimination step is only rarely possible.[429,430] Nitration of allylsilanes is straightforward in simple cases (Eq. 272).[431]

$(CH_3)_3Si$ + NC_6H_5 $\xrightarrow{25^\circ}$

R NC_6H_5 in benzene, R = H (9%); in CH_3CN, R = H (85%) + $(CH_3)_3Si$ NC_6H_5

R = H or $Si(CH_3)_3$

(Eq. 271)

$(CH_3)_3Si$ $\xrightarrow[CH_2Cl_2,\ -78^\circ]{NO_2^+\ BF_4^-}$ NO_2 (75%) (Eq. 272)

The radical-chain reaction described earlier (Eq. 58) achieves overall the effect of an electrophilic substitution with a nitrogen electrophile.

Vinylsilanes. 1-Trimethylsilylcycloalkenes are too unreactive toward mild nitrating reagents for the synthesis of the corresponding nitroalkenes.[432] Nitrosation gives only addition products,[428] but trimethylsilyl azide and vinylsilanes give substitution as the overall reaction by way of cycloaddition, loss of nitrogen, and a silyl shift from carbon to nitrogen (Eq. 273).[433]

$$C_6H_5CH{=}CHSi(CH_3)_3 + N_3Si(CH_3)_3 \xrightarrow{\text{reflux}} C_6H_5CH{=}CHN[Si(CH_3)_3]_2 \quad (55\%) \qquad \text{(Eq. 273)}$$

Allenylsilanes. Allenylsilanes also react with triazolinediones (Eq. 274), but without loss of the silyl group.[145]

$$(CH_3)_3SiCH{=}C{=}C(CH_3)_2 + \text{N=N triazolinedione } (NC_6H_5) \longrightarrow (CH_3)_3SiC{\equiv}C{-}C(CH_3)_2{-}N(\text{urazole, HN}, NC_6H_5) \qquad \text{(Eq. 274)}$$

Oxygen Electrophiles

Allylsilanes

Peracid Epoxidation. Allylsilanes have long been known to react rapidly with peracids,[434,435] but in the early work the products were not characterized. The cyclic allylsilane **170** is an exception; here the epoxide **171** is easily isolated, and the overall sequence is completed by hydroxide ion.[436] Presumably the epoxide is relatively stable because the C–O bond and the Si–C bond in **171** are badly aligned for the elimination step **171** → **172**. The allylsilane

$$\mathbf{170}\ (\text{Si}(C_6H_5)_2) \xrightarrow[30^\circ]{HO_3CC_6H_4NO_2\text{-}p,\ (C_2H_5)_2O} \mathbf{171}\ (95\%)\ (\text{Si}(C_6H_5)_2) \xrightarrow[\text{reflux}]{NaOH,\ H_2O} (C_6H_5)_2SiOH \text{ (OH) } \mathbf{172}$$

oxide **173** is also isolable for similar reasons, but it is unstable, hence the low yield; on standing, it rearranges to the silyl ether of the alcohol **174**. In practice, it is usual not to try to isolate intermediate oxides, but to proceed directly to the allylic alcohol by treatment with acid or fluoride ion. Thus, if the intermediate epoxide **173** is not isolated, the overall yield of **174** is much improved (73%).[55] Similarly, in the acidic conditions used with the allylsilane in Eq. 15, the allylic alcohol is produced directly.[134]

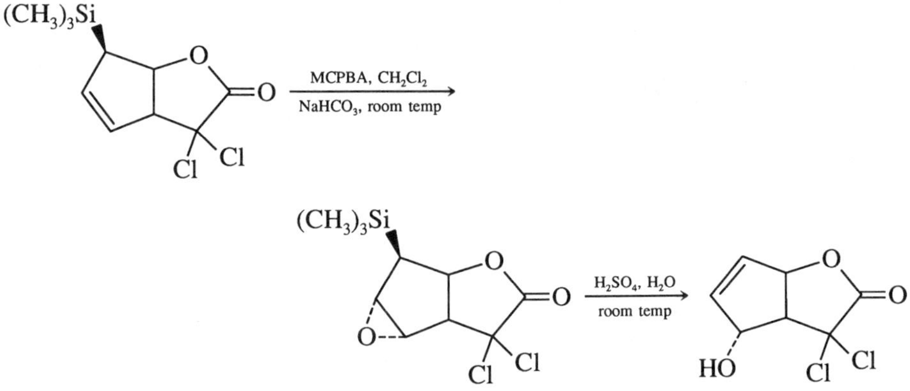

$(CH_3)_3Si$ — MCPBA, CH_2Cl_2 / $NaHCO_3$, room temp → $(CH_3)_3Si$ — H_2SO_4, H_2O / room temp → HO … Cl Cl

173 (18%) **174** (73%)

When the double bond of the allylsilane has a *trans* configuration, the diastereoselectivity in the epoxidation step is less than in most other electrophilic reactions, as discussed earlier (Eq. 23) and illustrated in Eq. 275. The desilylative opening of the epoxide is cleanly *anti*, with the result that the overall reaction (Eq. 275) gives a mixture of two products, with opposite configurations at the carbinol carbon and the double bond.[437,53] The overall reaction is stereospecifically *anti* for both products, but the fact that a mixture is obtained reduces the synthetic usefulness of this reaction. The overall conversion of an allylsilane to an allylic alcohol, whether achieved by epoxidation or by osmylation (see below), is a powerful transformation in organic synthesis.

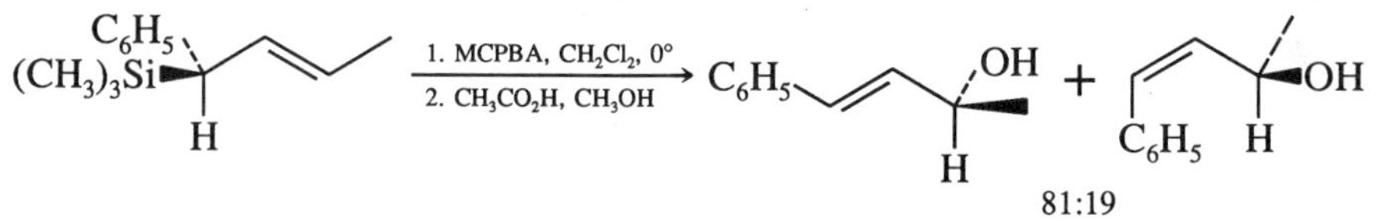

C_6H_5, $(CH_3)_3Si$, H — 1. MCPBA, CH_2Cl_2, 0° / 2. CH_3CO_2H, CH_3OH → C_6H_5 … OH, H + … OH, C_6H_5 H

81:19

(Eq. 275)

However, allylsilane epoxidations are not completely reliable for the synthesis of allylic alcohols.[438] Uncharacterizable mixtures are occasionally found, and a number of recognizable side reactions can occur. When acid is used on the epoxide **171**, the reaction only goes as far as the chlorohydrin.[436] The mixture of stereoisomeric allylsilane epoxides **175** gives the rearranged allylic alcohol **176** as the major product, as well as an unusual byproduct **177**.[408] The chlorohydrin **178** reacts with base, presumably to give the epoxide **179**. How-

$(CH_3)_3Si$ — **175** — H_2SO_4, H_2O, room temp → [OH, C_6H_5] → HO, C_6H_5 **176** (49%); OH, C_6H_5 **177** (28%)

ever, this is unstable and spontaneously undergoes an unexpected rearrangement **179** → **180**.[438] Epoxidation of the allylsilane **181** is also anomalous. The

$(CH_3)_3Si$, OH, Cl **178** — LiH, t-C_4H_9OH, reflux → [$(CH_3)_3Si$, O] **179** → $(CH_3)_3Si$, CHO **180** (40%)

first-formed epoxide **182** is not observed; it opens and captures a nucleophile β to the silyl group, but only after a rearrangement of the silyl group. The final elimination step **183** → **184** disguises these curious events because the overall reaction and the high level of *anti* stereoselectivity are normal.[437] Clearly the cleavage of the C–O bond β to the silyl group in the epoxides **179** and **182** is not concerted with the loss of the silyl group, and in each case cationic rearrangements take place more rapidly than desilylation.

$(CH_3)_3Si$ **181** 22–25% ee — MCPBA, $NaHCO_3$ → [$(CH_3)_3Si$, O] **182** →

$Si(CH_3)_3$, m-$ClC_6H_4CO_2$, OH **183** (78%) — TBAF, THF, room temp → OH **184** 19% ee (36%)

The epoxidation of allylsilanes is used in syntheses of dihydronepetalactone[439] and sarkomycin.[440]

Other Oxygen Electrophiles. Osmium tetroxide reacts with allylsilanes stoichiometrically or catalytically, and the products can be converted to allylic alcohols in three ways: by a Peterson elimination (Eq. 276),[53] by acylation and treatment with fluoride ion (Eq. 277),[44] or by treatment with acid (Eq. 278).[441] The overall result is an electrophilic substitution similar to that achieved by epoxidation. Osmium tetroxide is stereochemically a more demanding electrophile than a peracid, as shown by the clean attack *exo* to the bicyclic system in **185** (Eq. 279), whereas the corresponding epoxidation sequence gives a mixture of both stereoisomers at the carbinol carbon of **186**.[442] However, the greater steric demand of osmium tetroxide makes it less selective

$C_6H_5(CH_3)_2Si$ H

OsO_4 cat. / $O(CH_2CH_2)_2NCH_3O$, $(CH_3)_2CO$, room temp

OH, OH, $C_6H_5(CH_3)_2Si$ H

KOC_4H_9-*t* / DMSO, room temp

OH (Eq. 276)

$C_6H_5(CH_3)_2Si$ H C_6H_5

OsO_4 / C_5H_5N room temp

OH OH $C_6H_5(CH_3)_2Si$ H C_6H_5 (94%)

1. $(CH_3CO)_2O$ $(C_2H_5)_3N$ DMAP 2. TBAF, THF room temp

OH C_6H_5 (80%)

(Eq. 277)

O_2CCH_3 CO_2CH_3 $Si(CH_3)_3$

OsO_4 cat. / $O(CH_2CH_2)_2NCH_3O$ *t*-C_4H_9OH, $(CH_3)_2CO$ H_2O, room temp

O_2CCH_3 HO CO_2CH_3 HO $Si(CH_3)_3$ (96%)

TsOH, C_6H_6 / reflux

O_2CCH_3 HO CO_2CH_3 (98%) (Eq. 278)

$$\mathbf{185} \xrightarrow[\text{3. } SOCl_2,\ C_5H_5N \quad \text{4. } (i\text{-}C_4H_9)_2AlH]{\text{1. } OsO_4 \quad \text{2. } (CH_3CO)_2O,\ C_5H_5N} \mathbf{186}\ (59\%)$$

(Eq. 279)

than epoxidation in reactions (Eq. 24) where the stereocontrol stems from the chiral center carrying the silyl group.

Hydrogen peroxide in the presence of fluoride ion converts the allylsilane in Eq. 280 into an allylic alcohol.[24] Although overall an electrophilic substitution, this reaction is of a quite different kind from the others in this section. It is not dependent upon the substrate being an allylsilane but requires a leaving group attached to the silyl group. The peroxy anion attacks the silyl group, displacing the leaving group, and a rearrangement with retention of configuration establishes the C–O bond at the same atom as that to which the silyl group was attached.

$$(C_2H_5O)_2CH_3Si\text{-cyclopentenyl} \xrightarrow{H_2O_2,\ KF,\ DMF} HO\text{-cyclopentenyl} \quad \text{(Eq. 280)}$$

22–25% ee → 16% ee

Palladium(II) catalyzed reactions with allylsilanes in the presence of oxygen, which lead to allylic alcohols and α,β-unsaturated carbonyl compounds, are discussed in the section on metal electrophiles. A similar reaction with rhodium catalysts and oxygen is a good method for converting allylsilanes into β-silylenones.[443] Singlet oxygen[142] leads to an ene reaction without loss of the silyl group, and molecular oxygen in the presence of peroxides also leaves the silyl group in the molecule.

Electrochemical oxidation of allylsilanes in hydroxylic media results in the replacement of the silyl group by an oxygen nucleophile, but without regiocontrol (Eq. 281).[444]

$$(CH_3)_3SiCH_2CH{=}CHCH_2C_6H_5 \xrightarrow[CH_3OH]{\text{electrode}} CH_3OCH_2CH{=}CHCH_2C_6H_5 + CH_2{=}CHCH(OCH_3)CH_2C_6H_5$$

37:63 (76%)

(Eq. 281)

The reaction in Eq. 278 is used in a synthesis of shikimic acid.[441]

Vinylsilanes

Peracid Epoxidation. Vinylsilanes are easily epoxidized, usually with one equivalent of a buffered peracid to avoid subsequent epoxide opening.[143] Proper comparisons have not been made, but in general vinylsilanes react with peracids at much the same rate as the corresponding alkenes without the silyl group,[96] or even a little faster (see **50**).[73,74]

Vinylsilane epoxides are usually stable under the reaction conditions, and a second step has to be carried out to complete the electrophilic substitution reaction. In dilute aqueous acid, for example, the regioisomeric epoxides **187** and **192** are opened stereospecifically, and the diols **188** and **193** can be isolated. In more concentrated acid the silyl group and the β-hydroxy group undergo elimination to give the enols **190** and **195**, respectively, thus completing what amounts to an electrophilic substitution.[445–447] Alternatively, the diols can rearrange by silyl shift to give α-silyl carbonyl compounds **189** and **194**, which subsequently lose their silyl groups to give the same enols **190** and **195**. The latter pathway is detected when *tert*-butyldimethylsilyl groups are used, because the α-silylcarbonyl compounds can then be isolated.[448] In either case, the enols **190** and **195** give the aldehyde **191** and the ketone **196**, re-

$Si(CH_3)_3$; C_6H_{13}-*n* — **187** —(H_2SO_4, H_2O; THF, 0° to room temp)→ HO, OH, $Si(CH_3)_3$, C_6H_{13}-*n* — **188** (76%) → [*n*-C_6H_{13}, O, H, $Si(CH_3)_3$] **189**

188 → [*n*-C_6H_{13}, OH] **190** → *n*-C_6H_{13}, CHO **191**

n-C_6H_{13}, O, $Si(CH_3)_3$ — **192** —(H_2SO_4, H_2O; THF, 0° to room temp)→ *n*-C_6H_{13}, OH, $(CH_3)_3Si$ OH — **193** (83%) → [*n*-C_6H_{13}, $Si(CH_3)_3$, O] **194**

193 → [*n*-C_6H_{13}, OH] **195** → *n*-C_6H_{13}, O **196**

spectively, as the final products. The overall conversion of a vinylsilane to an aldehyde or ketone[445] has found much use in synthesis because the position of the silyl group determines the position of the carbonyl group. The conditions of the acid-catalyzed hydrolysis are noticeably milder when a ketone group is nearby (Eq. 282), since it can participate in the opening of the epoxide. In this case, the synthesis of the epoxide **197** and the opening step can easily be combined in one operation, simply by not buffering the peracid used in the epoxidation.[449]

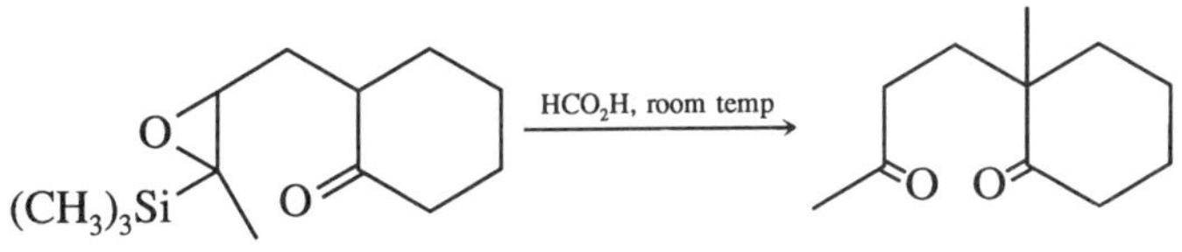

(Eq. 282)

Any nucleophile that can open the epoxide usually attacks at the carbon atom α to the silyl group,[73,143,450–452] creating a β-silyl alcohol that can undergo elimination. The net result is that the atom which replaces the silyl group arrives as a *nucleophile*, as in Eqs. 283–286, yet the overall result is the same as that of an electrophilic substitution. This reaction, therefore, is different from most of the others in this chapter, although the sequence of events is essentially the same as that in halodesilylation. Because the opening of the epoxide is stereospecific (inversion of configuration at C-1) and the elimination

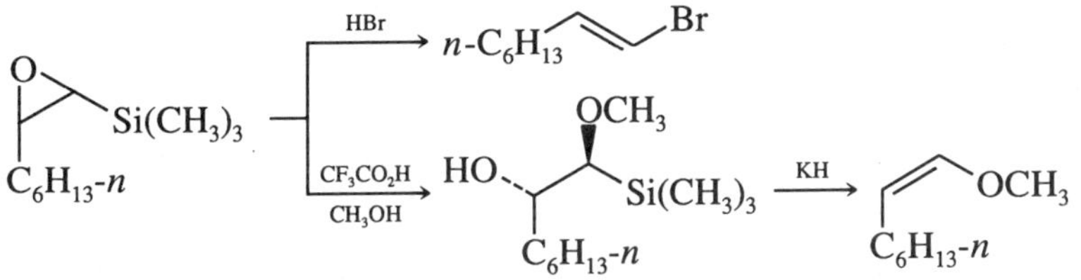

(Eq. 283)

(Eq. 284)

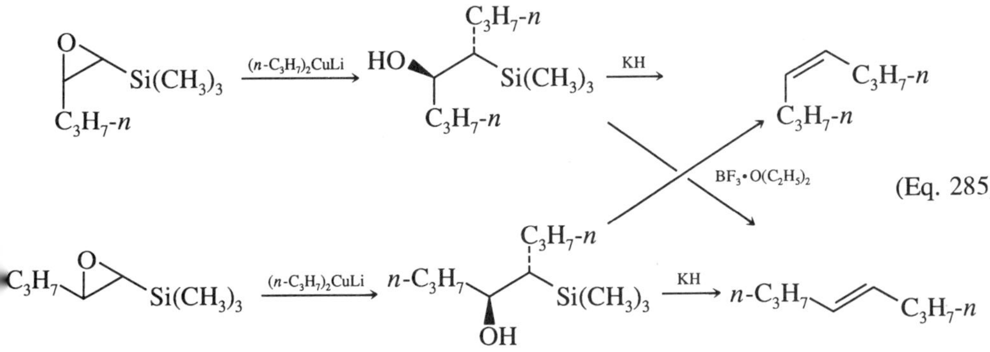

(Eq. 285)

(Eq. 286)

of the silyl and hydroxyl groups is also stereospecific (*syn* with base, *anti* with acid, Eqs. 285 and 286), the overall reaction is stereospecific, and the result can be either retention or inversion of configuration about the double bond.[451,453] Examples of all these reactions are included in the tables under the heading "Oxygen Electrophiles," even though the nucleophilic group that eventually replaces the silyl group can be halogen (Eq. 283), oxygen (Eq. 284), nitrogen, sulfur, or even carbon (Eqs. 285 and 286).

Some rather more complicated reactions take place when epoxysilanes are treated with Lewis acids[454] or are pyrolyzed.[455,456] Rearrangement to an α-silyl ketone or aldehyde, or a derivative thereof, occurs in some reactions (Eqs. 287–290), which are remarkable for producing mainly the carbonyl compound

(Eq. 287): epoxysilane with $Si(CH_3)_3$ and C_3H_7-n $\xrightarrow[\text{room temp}]{MgBr_2,\ (C_2H_5)_2O}$ n-$C_3H_7COCH_2Si(CH_3)_3$ (93%)

(Eq. 288): same epoxysilane $\xrightarrow{600°}$ n-$C_3H_7C(OSi(CH_3)_3)=CH_2$ (57%)

(Eq. 289): n-C_3H_7 epoxide with $Si(CH_3)_3$ $\xrightarrow[\text{room temp}]{MgBr_2,\ (C_2H_5)_2O}$ n-$C_3H_7CH=C(CH_3)OSi(CH_3)_3$ (82%)

(Eq. 290): same epoxide $\xrightarrow{600°}$ n-$C_3H_7CH=CHCH_2OSi(CH_3)_3$ (50%) + $C_2H_5CH=CHCH_2CH_2OSi(CH_3)_3$ (26%)

or a derivative regioisomeric with that obtained by simple hydrolysis. In contrast, the β,β-disubstituted vinylsilane epoxides **198** and **199** open without nucleophilic participation; with boron trifluoride etherate, the carbonyl derivative is produced with the normal regioselectivity, and the rearrangement takes place with retention of configuration.[457] There are further complications in that skeletal rearrangement can take place (Eq. 291),[454] and rearrangement to an α-silylaldehyde can precede the attack of a Grignard reagent (Eq. 292).[453]

198: C_2H_5/CH_3 epoxide with $Si(CH_3)_2C_6H_5$ $\xrightarrow[CH_2Cl_2,\ -78°]{BF_3\cdot O(C_2H_5)_2}$ $C_2H_5(CH_3)C=CHOSi(CH_3)_2C_6H_5$ (69%)

199: CH_3/C_2H_5 epoxide with $Si(CH_3)_2C_6H_5$ $\xrightarrow[CH_2Cl_2,\ -78°]{BF_3\cdot O(C_2H_5)_2}$ $CH_3(C_2H_5)C=CHOSi(CH_3)_2C_6H_5$ (68%)

$\xrightarrow[\text{room temp}]{MgBr_2,\ (C_2H_5)_2O}$ (84%) (Eq. 291)

$Si(CH_3)_3$ $OSi(CH_3)_3$

$n\text{-}C_6H_{13}$ $Si(CH_3)_3$ $\xrightarrow[(C_2H_5)_2O,\ \text{room temp}]{C_2H_5MgBr}$

$n\text{-}C_6H_{13}$ C_2H_5 OH $Si(CH_3)_3$ (84%)

$\xrightarrow[CH_2Cl_2,\ \text{room temp}]{BF_3\cdot O(C_2H_5)_2}$ $n\text{-}C_6H_{13}$ C_2H_5 (90%)

(Eq. 292)

$\xrightarrow[\text{room temp}]{KH,\ THF}$ $n\text{-}C_6H_{13}$ C_2H_5 (90%)

Another way of getting ketones from epoxysilanes involves the reduction of the epoxysilane, followed by oxidation and hydrolytic removal of the silyl group (Eq. 293).[143] The product is again the ketone in which the carbonyl group is produced at the carbon that did *not* bear the silyl group. This sequence is so far from being a simple electrophilic substitution that examples are not included in the tables.

$(CH_3)_3Si$ O $\xrightarrow[2.\ Na_2Cr_2O_7]{1.\ LiAlH_4}$ O (70%) (Eq. 293)

With cyclic vinylsilane epoxides, the opening process leads to a β-hydroxysilane in which the silyl group and the hydroxyl group are *cis* and are unable to undergo *anti* elimination except in large rings.[458] Thus the reaction of trimethylsilylcyclohexene oxide (**200**) in methanolic acid stops with the formation of the glycol monoether **201**. However, *syn* Peterson elimination is still possible (Eq. 294).[73] When the nucleophile is a hydroxy group, the Peterson elimination occurs in low yield because the oxyanion geminal to the

$Si(CH_3)_3$ O **200** $\xrightarrow{H_2SO_4,\ CH_3OH}$ $Si(CH_3)_3$ OCH_3 OH **201** (78%) $\xrightarrow[\text{room temp}]{NaH,\ DMF}$ OCH_3

(Eq. 294)

silyl group can remove it in a Brook rearrangement.[452] However, this reaction can be made the major pathway by using sodium hydride in ether, **207** → **208** → **209**, and it is a good, stereospecific synthesis of silyl enol ethers.[459] Cyclic vinylsilane epoxides like **202** do not simply open: rearrangement is encouraged in this system by the stabilization of the cation **203**, and a furan is produced. The electrophilic substitution is thus thwarted.[460]

H^+ $Si(CH_3)_3$ **202** $\xrightarrow[CH_3OH,\ 50°]{H_2SO_4,\ H_2O}$ [$(CH_3)_3Si$ HO **203**] ⟹ (92%)

The vinylsilane **204** is also an allylsilane, but it reacts abnormally in both capacities. The epoxide **205** is formed and opens as usual for a 2,2-disubstituted vinylsilane epoxide, but one of the silyl groups migrates (**205** → **206**) faster than either is lost.[166] The rearrangement of the silyl group is similar to the hydride shift **179** → **180** and to the silyl shifts **188** → **189** and **193** → **194**. Although the silyl group α to the carbonyl group in **206** is inherently unstable with respect to solvolytic displacement, it nevertheless survives in this case.

$(CH_3)_3Si$–$CH_2C(R)$=$CHSi(CH_3)_3$ **204** $\xrightarrow[-78°\ to\ -20°]{MCPBA}$ [$(CH_3)_3Si$ ... $Si(CH_3)_3$, H^+ **205**] ⟶

$R = (CH_2)_2COCH_2SO_2C_6H_5$

$(CH_3)_3Si$ ... $Si(CH_3)_3$ (OH) $\xrightarrow{-H^+}$ $(CH_3)_3SiCH_2C(R)(Si(CH_3)_3)CHO$ **206**

The epoxidation of vinylsilanes to make ketones or aldehydes is used in syntheses of dihydrojasmone,[461] gymnomitrol,[462] three pseudo-senoxydenes,[463] and quinghaosu.[464]

Other Oxygen Electrophiles. Vinylsilanes react with osmium tetroxide, which can be used catalytically in the presence of trimethylamine *N*-oxide (Eqs. 295 and 296). The diols are then converted stereospecifically into silyl enol ethers by treatment with sodium hydride, which induces simultaneously

C_6H_{13}-n vinyl–$Si(CH_3)_3$ $\xrightarrow[t\text{-}C_4H_9OH,\ \text{reflux}]{OsO_4,\ (CH_3)_3NO}$ HO, OH, $Si(CH_3)_3$, C_6H_{13}-n (**207**) $\xrightarrow[(C_2H_5)_2O,\ \text{room temp}]{NaH}$ [HO, O^-, $Si(CH_3)_3$, C_6H_{13}-n (**208**)] — C_6H_{13}-n, $OSi(CH_3)_3$ (Eq. 295)

209 > 99% *Z* (73%)

$n\text{-}C_6H_{13}$–CH=CH–$Si(CH_3)_3$ $\xrightarrow[t\text{-}C_4H_9OH,\ \text{reflux}]{OsO_4,\ (CH_3)_3NO}$ $n\text{-}C_6H_{13}$, OH, OH, $Si(CH_3)_3$ $\xrightarrow[\text{room temp}]{NaH,\ (C_2H_5)_2O}$

$n\text{-}C_6H_{13}$–CH=CH–$OSi(CH_3)_3$ (Eq. 296)

(65%) 99% *E*

a Brook rearrangement and a β elimination **208** → **209**.[459] The use of sodium hydride in ether is critical in making this reaction clean.

Ozonization of vinylsilanes forms α-hydroxycarbonyl compounds (e.g., Eq. 297), or masked versions thereof, but the reaction is not simply an electrophilic substitution.[465] The reaction of singlet oxygen with allylsilanes and vinylsilanes has been mentioned earlier.[142,143] The products are allylic alcohols, but the silyl group is retained. *tert*-Butyl hydroperoxide with copper(I) chloride reacts with allylsilanes to give allylic hydroperoxides, again without loss of the silyl group.[466] None of these reactions is included in the tables.

C_6H_{11}, $Si(CH_3)_3$ $\xrightarrow[2.\ (CH_3)_2S,\ 20°]{1.\ O_3,\ CH_3OH,\ CH_2Cl_2,\ 20°}$ C_6H_{11}, O, OH (Eq. 297)

(73%)

Allenylsilanes

The epoxidation of the allenylsilane **210** does not give substitution: the allene oxide–cyclopropanone rearrangement **211** → **212** takes place, and the internal nucleophile captures the latter intermediate.[467] The allene oxide–cyclopropanone manifold is accessible in another way from vinylsilanes, namely, by epoxidation of the vinylsilane **213** followed by fluoride-catalyzed elimination (Eq. 298).[468] These reactions are not included in the tables.

$(CH_3)_3Si$ $\xrightarrow[C_6H_5CN]{H_2O_2}$

OH

210

$(CH_3)_3Si$ OH → $(CH_3)_3Si$ OH → $(CH_3)_3Si$ O O

211 **212**

Cl Si$(CH_3)_3$ C_4H_9-*t* $\xrightarrow[\text{NaO}_2\text{CCH}_3,\ \text{CH}_2\text{Cl}_2,\ \text{reflux}]{CH_3CO_3H}$ Cl O Si$(CH_3)_3$ C_4H_9-*t* $\xrightarrow[\text{diglyme, room temp}]{CsF}$ [O C_4H_9-*t*] (Eq. 298)

213

Phosphorus Electrophiles

The only phosphorus electrophile to have been used on allylsilanes is phosphorus pentachloride; it gives a mixture of products (Eq. 299),[469] presumably derived by subsequent reactions of the product of electrophilic attack.

$(CH_3)_3Si$ $\xrightarrow[\text{2. SO}_2]{\text{1. PCl}_5,\ -20° \text{ to } 20°}$

$POCl_2$ + $POCl_2$ + Cl $POCl_2$ (Eq. 299)

Sulfur Electrophiles

Arenesulfenyl chlorides add to allylsilanes[55,134] and vinylsilanes,[421] but the net substitution reaction can be completed with fluoride ion (Eqs. 300 and 301). Benzenesulfenyl tetrafluoroborate, however, gives the substitution product directly (Eq. 300).[134] Benzenesulfenyl chloride reacts with a polymeric vinylsilane to give substitution.[470]

Methanesulfonyl chloride and benzenesulfonyl chloride react with allylsilanes in the presence of copper(I) chloride. Although electrophilic substitution

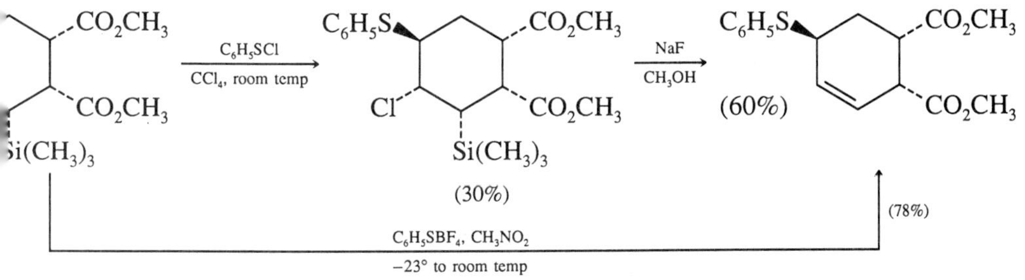

(Eq. 300)

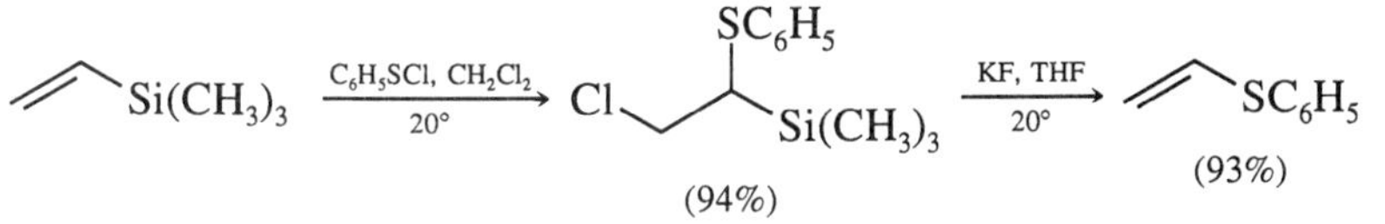

(Eq. 301)

is the overall result, it is possible that the reaction involves a radical chain.[471] The more powerful electrophile trimethylsilyloxysulfonyl chloride reacts readily with allylsilanes, vinylsilanes, and allenylsilanes (Eqs. 302–304).[8,223]

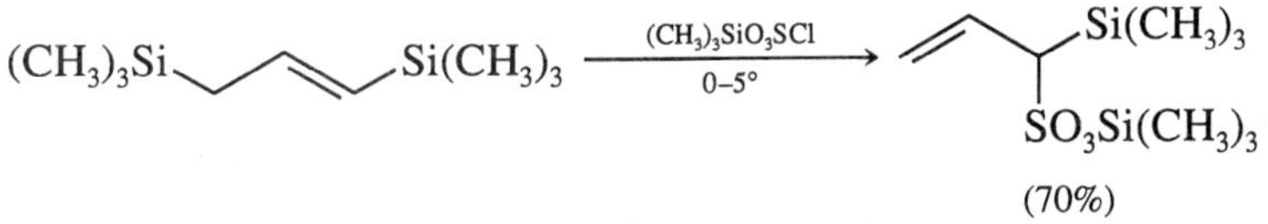

(Eq. 302)

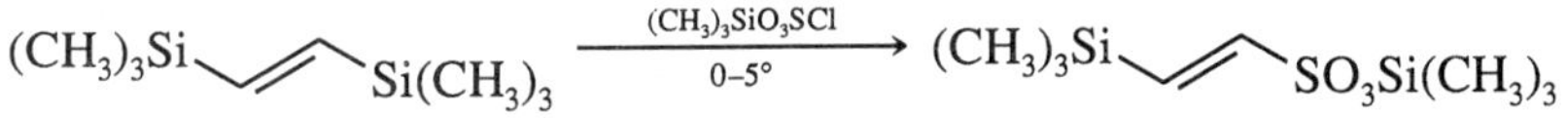

(Eq. 303)

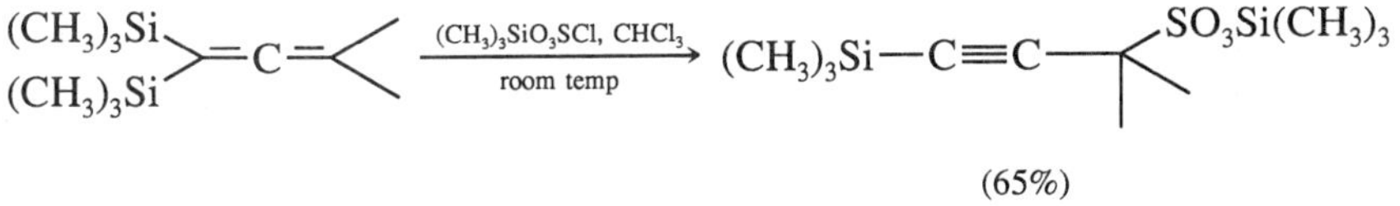

(Eq. 304)

Sulfenylation, assisted intramolecularly by a neighboring alkoxide, is also possible (Eq. 305).[260]

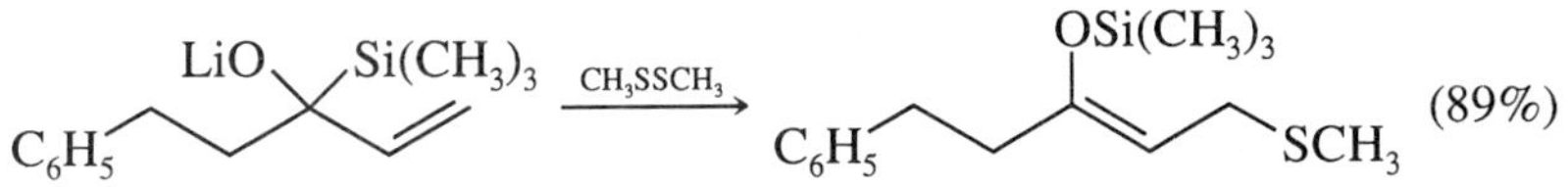

(Eq. 305)

The sulfur electrophile *N*-sulfinylbenzenesulfonamide ($C_6H_5SO_2NSO$) has been discussed earlier; it reacts with allylsilanes in an ene reaction without removing the silyl group.[145]

Selenium Electrophiles

Benzeneselenenyl chloride adds to allylsilanes, and a catalytic amount of tin(II) chloride or chromatography on florisil completes the substitution process (Eq. 306).[472] However, the allyl selenides produced are unstable with respect to allylic rearrangement. In consequence, the only product isolated from unsymmetrical allylsilanes has the selenenyl group at the less-substituted end of the allyl system, regardless of which allylsilane (**214** or **215**) one starts with (Eq. 307).[472,473] This problem is much less severe with allyl sulfides than with allyl selenides.[472] The allyl sulfides and selenides are particularly useful in synthesis because they can be easily oxidized to sulfoxides and selenoxides, and hence to allylically rearranged alcohols.

$(CH_3)_3Si$ [allyl] $\xrightarrow[-78°]{C_6H_5SeCl,\ CH_2Cl_2}$ $(CH_3)_3Si$–[CH(Cl)]–SeC_6H_5 $\xrightarrow[\text{or florisil}]{SnCl_2,\ CH_2Cl_2,\ 0°}$ [allyl]SeC_6H_5 (90%) (Eq. 306)

214: $(CH_3)_3Si$ ... O_2CCH_3 $\xrightarrow{1.\ C_6H_5SeCl,\ CH_2Cl_2,\ -78°;\ 2.\ SnCl_2,\ CH_2Cl_2,\ 0°}$ [C_6H_5Se ... O_2CCH_3] → (91%) C_6H_5Se ... O_2CCH_3

215: $(CH_3)_3Si$ ... O_2CCH_3 $\xrightarrow{1.\ C_6H_5SeCl;\ 2.\ SnCl_2}$ (80%) C_6H_5Se ... O_2CCH_3 (Eq. 307)

Selenium dioxide alone, or benzeneseleninic anhydride catalyzed by boron trifluoride etherate, reacts with allylsilanes. The simple substitution products such as **216** are not observed because they undergo rapid [2,3]-sigmatropic shift. In the selenium dioxide reaction, a mixture of enal and acylsilane is isolated (Eq. 308).[474] In the benzeneseleninic anhydride reaction, the product

$(CH_3)_3Si$ ... C_6H_5 $\xrightarrow[\text{reflux}]{SeO_2,\ THF}$ R–C(=O)– ... C_6H_5 (Eq. 308)

R = $Si(CH_3)_3$ (13%)
R = H (27%)

is an allylic alcohol with the hydroxy group on the carbon atom to which the silyl group was originally attached (Eq. 309).[405] This method avoids the two-step sequence of sulfenylation followed by oxidative rearrangement.

$(CH_3)_3Si$, CH_3CO_2 → [$(C_6H_5SeO)_2O$, CH_2Cl_2, $BF_3 \cdot O(C_2H_5)_2$, 20°] → [O^-, $^+SeC_6H_5$, CH_3CO_2 **216**] → HO, CH_3CO_2 (Eq. 309)

Halogen Electrophiles

Allylsilanes. Bromine and chlorine initially add to allylsilanes, typically at −70°. The subsequent elimination step is so fast that it is not usual to isolate the intermediate, although it is possible with 3-silacyclopentenes[36] or with allylsilanes having hindered silyl groups.[475] The usual product isolated is simply the allyl bromide or chloride. Occasionally, addition of another mole of bromine or chlorine to the allyl halide takes place,[5] but this can usually be avoided by using only one equivalent of halogen.

1,4-Disilyl-3-butenes **217** react with one equivalent of bromine or with *N*-bromosuccinimide (NBS) to give 1,3-dienes, which can be trapped by the 1,4 addition of bromine (Eq. 310)[476] or by dienophiles.[477] The initial product of bromodesilylation loses trimethylsilyl bromide in a very fast β elimination.

$(CH_3)_3Si$, $COCH_3$, $(CH_3)_3Si$ **217** → [Br_2, CuBr, propylene oxide, THF, −78°] → [$COCH_3$] → Br, $COCH_3$, Br (41%) (Eq. 310)

A similar reaction is the iodination of the disilylcyclohexadiene (Eq. 311), which is mainly of interest as a source of iodotrimethylsilane.[478]

$(CH_3)_3Si$ ... $Si(CH_3)_3$ $\xrightarrow[0°]{I_2,\ \text{aprotic solvent}}$ benzene + $2(CH_3)_3SiI$ (Eq. 311)

3-Silylcyclohexenes, however, are not always well behaved, presumably because the silyl group is in an equatorial position much of the time. The allylsilane **218** reacts with bromine or with NBS, but the first-formed product is aromatized, and to some extent retains the silyl group.[408] Similarly the bis(allylsilane) **70** gives silicon-containing products on bromination,[134] in contrast to the analogous open-chain bis(allylsilane) **217**.

$(CH_3)_3Si$, C_6H_5 **218** $\xrightarrow[\text{or } \mathbf{B}\ \text{NBS},\ CCl_4,\ \text{reflux}]{\mathbf{A}\ Br_2,\ C_6H_{12},\ \text{room temp}}$ C_6H_5 + $(CH_3)_3Si$, C_6H_5

A 72:28
B 58:42

Iodosobenzene reacts with allylsilanes in the presence of boron trifluoride etherate. An allyl–iodine(III) intermediate is probably produced; in the presence of an internal nucleophile this is displaced, achieving overall an umpolung of allylsilane reactivity (Eq. 312).[479] In the absence of an internal nucleophile, the intermediate is attacked by another equivalent of iodosobenzene, and the product is an enal (Eq. 313).[480]

$(CH_3)_3Si$, OR, C_6H_{13}-*n*

R = H: $\xrightarrow[\text{dioxane, room temp}]{C_6H_5IO,\ BF_3\cdot O(C_2H_5)_2}$ O, C_6H_{13}-*n* (65%) (Eq. 312)

R = $COCH_3$: $\xrightarrow[\text{dioxane, room temp}]{C_6H_5IO,\ BF_3\cdot O(C_2H_5)_2}$ CHO, O_2CCH_3, C_6H_{13}-*n* (72%) (Eq. 313)

Vinylsilanes. The reaction of bromine or chlorine with vinylsilanes is more complicated. Addition is almost always the first step, and, although the intermediates can be isolated, it is usual to add fluoride ion or sodium methoxide to complete the electrophilic substitution reaction (Eq. 314).[481] If an amine base is used, the silyl group is retained (Eq. 45).[128,482] The complication is in the stereochemistry: these electrophiles are anomalous in giving overall inversion of configuration (Eqs. 11 and 314). The distinctive feature about bromination and chlorination is that the electrophile becomes bonded simultaneously to both carbon atoms of the original vinyl group to give a bridged halonium ion **220**. Furthermore, with the nucleophilic halide ion as the counterion, this intermediate is opened to give the product **221** or **222** of *anti* addition. Whether the incoming nucleophile attacks β to the silyl group to give **221** or α to it to give **222** does not affect the argument; in most cases it is not known which takes place, and obviously it depends upon the substituents at the α and β carbons. The intermediate dihalide is usually isolable in these reactions. The desilylhalogenation is then brought about with alkoxide or fluoride ion, and is an *anti* stereospecific process taking place in the conformations **223** or **224**. Both of these pathways lead to the vinyl bromide or chloride **225** that is the product of inversion of configuration.[14,481]

C_4H_9-*n* … $Si(CH_3)_3$ $\xrightarrow[-78°]{Br_2,\ CH_2Cl_2}$ Br … Br … $Si(CH_3)_3$, C_4H_9-*n* $\xrightarrow[\text{room temp}]{NaOCH_3,\ CH_3OH}$ Br … C_4H_9-*n*, 99% *E* (Eq. 314)

X—X, SiR_3 **219** ⟶ **220** (a, X^+, b, X^-, SiR_3) → a: **221** ⇌ (rotation) **223**; b: **222** ⇌ (rotation) **224** → anti elim. → **225**

Bromination with retention of configuration is possible by using cyanogen bromide and a Lewis acid.[271] Presumably the absence of bromide ion gives the intermediate **220** a longer life, and hence more opportunity to open unimolecularly to an intermediate similar to **37**. A further complication with halogenation is found in the special case of β-silylstyrenes **226**. Here, the addition of bromine to the double bond is stereospecifically *syn*, as shown by the isolation of the crystalline intermediate **227a** from the addition of bromine to the styrene **226a**.[483] The subsequent desilylbromination **227** → **228** is *anti*, as usual, and the overall result is retention of configuration.[483,484] This seemingly unusual behavior is also found with styrylboranes.[485] Presumably the *syn* addition takes place, both with the silanes and the boranes, because the intermediate bromonium ion **229** can open, more or less completely, to a cation **230** stabilized by both the metal–carbon bond and overlap with the benzene ring. Such an intermediate can be expected to capture the nucleophile from the direction *anti* to the metal group to give **231**, the result of *syn* addition.

C_6H_5–CH=CH–SiR_3 —(*syn* addition) Br_2, $CHCl_3$, 0° (R = C_6H_5) or Br_2, CS_2, −100° (R = CH_3)→ C_6H_5–CHBr–CHBr–SiR_3 —(*anti* elimination) DMSO, 25° (R = C_6H_5); CH_3CN, room temp (R = CH_3)→ C_6H_5–CH=CH–Br

226a (R = C_6H_5)
226b (R = CH_3)

227a (R = C_6H_5)
227b (R = CH_3)

228

(M) = SiR_3 or BR_2

[**229** ⇌ **230** ⇌ **231**] →

Another anomaly is in the bromination and chlorination of vinylsilanes that have large substituents like *tert*-butyl on the double bond. With the *cis* isomer (Eq. 315), the *anti* addition of bromine is normal, but a problem arises in the elimination step, because the silyl group and the halide can be *anti* only in a conformation having a gauche interaction of the silyl and *tert*-butyl groups. With sodium methoxide, *anti* elimination takes place from a lower-energy conformation, in which a proton and the bromine are antiperiplanar and the product retains the silyl group. The silyl group can be removed with fluoride ion, but with *syn* stereochemistry in a conformation that avoids the gauche interaction of the silyl group and the *tert*-butyl group (Eq. 316). The corresponding *trans* vinylsilane is normal in giving *anti* addition and *anti* elimination (Eq. 317), but only when the reaction is carried out in carbon tetrachloride. In the usual solvent, dichloromethane, the major product is contaminated with a product **234** from rearrangement and further bromination (Eq. 318).[57]

Br_2, CH_2Cl_2, −78° (96%); $NaOCH_3$, CH_3OH, reflux (Eq. 315); KF, DMSO, room temp (70%)

232 (*cis*:*trans* 99:1)

(Eq. 316)

Br_2, CCl_4 → **233**; $NaOCH_3$, CH_3OH, room temp → **232** (31%) *cis*:*trans* 87:13 (Eq. 317)

Br_2, CH_2Cl_2, −78° → **233** + **234** (Eq. 318)

Bromination of cyclic vinylsilanes takes place perforce with retention of configuration (Eq. 319),[486] but it is not easy to get high yields because the elimination step is *syn* and therefore slow. Elimination is best achieved with

Br_2, CH_2Cl_2, −78° to room temp (30%) (Eq. 319)

fluoride ion.[487] With a medium-sized ring, both the *syn* and the *anti* elimination steps are slow enough to allow a transannular hydride shift to take place with the formation of an anomalous product (Eq. 320).[488]

$$\text{cyclooctenyl-}Si(CH_3)_3 \xrightarrow[-78^\circ \text{ to room temp}]{Br_2,\ CH_2Cl_2} \text{Br-cyclooctene} \quad (70\%) \qquad \text{(Eq. 320)}$$

Overall inversion of configuration is also found in iodination with iodine chloride,[489,490] and with iodine and silver trifluoroacetate.[491] Iodination with iodine itself[271,491–493] usually takes place with retention of configuration (Eq. 321),[271] either because the intermediate iodonium ion is not susceptible to attack by iodide ion to give a vicinal diiodide or because the vicinal diiodide always reverts to the iodonium ion. Terminal vinylsilanes are exceptional in allowing the addition step to take place. Both the *cis* and *trans* vinylsilanes **235** and **236** give the *cis* vinyl iodide **237**,[490] and *trans* 1,2-bis(trimethylsilyl)ethylene gives a mixture of *cis*- and *trans*-2-iodovinyltrimethylsilane.[416] However, in the presence of Lewis acids such as aluminum chloride, *trans* vinylsilanes give the *trans* vinyl iodide, presumably because the chloride ion removes the silyl group before the vicinal diiodide is formed.[494]

$$C_6H_{11}CH{=}C(C_2H_5)Si(CH_3)_3 \xrightarrow[\text{room temp}]{I_2,\ CH_2Cl_2} C_6H_{11}CH{=}C(C_2H_5)I \qquad \text{(Eq. 321)}$$

$$\underset{\mathbf{235}}{n\text{-}C_8H_{17}CH{=}CHSi(CH_3)_3} \xrightarrow[\text{room temp}]{I_2,\ CH_2Cl_2} (65\%) \rightarrow \underset{\mathbf{237}}{n\text{-}C_8H_{17}CH{=}CHI}$$

$$\underset{\mathbf{236}}{n\text{-}C_8H_{17}CH{=}CHSi(CH_3)_3} \xrightarrow{I_2,\ CH_2Cl_2} (40\%) \rightarrow \mathbf{237}$$

When iodination with iodine is carried out in the presence of other nucleophiles like azide ion, intermediates can be isolated, but they are not easily susceptible to the subsequent loss of the silyl group which is necessary to achieve an electrophilic substitution reaction.[495,496]

Iodosobenzene reacts stereospecifically with some vinylsilanes in the presence of Meerwein's salt (Eq. 322), except that acetylene formation takes place when there is a hydrogen *anti* to the iodonium group. The vinyliodonium ion products **238** can be converted with a wide variety of nucleophiles, mostly cuprates, into vinyl halides, cyanides, sulfides, and nitro compounds.[497] When boron trifluoride etherate is used in place of Meerwein's salt, acetylene formation is the only reaction (Eq. 323), presumably taking place by β elimination of the first-formed iodonium ion.[498]

$$\mathbf{235} \xrightarrow[\text{CH}_2\text{Cl}_2,\ 25°]{\text{C}_6\text{H}_5\text{IO},\ (\text{C}_2\text{H}_5)_3\text{O}^+\text{BF}_4^-} n\text{-C}_8\text{H}_{17}\text{CH=CH}\overset{+}{\text{I}}\text{C}_6\text{H}_5\ \ \text{BF}_4^- \quad \mathbf{238}\ (72\%) \qquad \text{(Eq. 322)}$$

$$\mathbf{235} \xrightarrow[\text{CH}_2\text{Cl}_2,\ \text{room temp; 2. NaOH, room temp}]{\text{1. C}_6\text{H}_5\text{IO},\ \text{BF}_3\cdot\text{O(C}_2\text{H}_5)_2} (51\%)\ \ n\text{-C}_8\text{H}_{17}\text{C}\equiv\text{CH} \qquad \text{(Eq. 323)}$$

$$\mathbf{236} \xrightarrow[\text{or C}_6\text{H}_5\text{IO},\ (\text{C}_2\text{H}_5)_3\text{O}^+\text{BF}_4^-,\ \text{CH}_2\text{Cl}_2,\ 25°\ (90\%)]{\text{C}_6\text{H}_5\text{IO},\ \text{BF}_3\cdot\text{O(C}_2\text{H}_5)_2,\ 25°} (99\%)\ \ n\text{-C}_8\text{H}_{17}\text{C}\equiv\text{CH}$$

The chlorination of a vinylsilane is used in the synthesis of mycorrhizin A,[499] and iodination in syntheses of 2-tricosene and other pheromones.[494,500]

Metal Electrophiles

Replacement of one metal bonded to carbon by another is a common reaction, but replacement of a silyl group by another metal is comparatively rare. Allylsilanes and vinylsilanes do not in general react with butyllithium by metal exchange as the corresponding tin compounds do, although there is one report of such a reaction.[501] The reactions that take place instead are deprotonation with allylsilanes[115,132,502–505] and addition with vinylsilanes.[506–509] Allylsilanes and vinylsilanes do not usually react with metal halides, presumably because there is a limited range of metals for which such a reaction is thermodynamically favorable. In special cases, however, it is possible for a carbon–silicon bond to be replaced by a carbon–metal bond (Eq. 128),[280] and the intervention of such exchange in allylsilane chemistry has already been referred to as a cause of failure in some Lewis acid catalyzed reactions. Certainly Lewis acids react (Eq. 324 and 325),[510,511] although in Eq. 324 the allyl fragment is not identified and it could conceivably have been removed by protodesilylation or fluorodesilylation. Lewis acid attack followed by a β elimination is the most likely explanation for the reaction in Eq. 326, and initial attack by the metal followed by attack of the allylstannane on the epoxide explains the anomalous regiochemistry in the reaction in Eq. 327.[258] The reaction of allyltrimethylsilane with tungsten hexachloride (Eq. 328) is probably the first step in the activation of that reagent as a metathesis catalyst.[512]

$$(\text{C}_6\text{H}_5)_2\text{Si(OC}_2\text{H}_5)\text{CH}_2\text{CH=CH}_2 \xrightarrow[25°]{\text{SbF}_5} (\text{C}_6\text{H}_5)_2\text{SiF}_2\ (100\%) \qquad \text{(Eq. 324)}$$

$(CH_3)_3Si$, $(CH_3)_3Si$ [5,5-bis(trimethylsilyl)cyclopentadiene] $\xrightarrow[0°]{BI_3}$ I_2B–C_5H_4–BI_2 (48%) (Eq. 325)

$(CH_3)_3Si$ [2-(cyclohexenyl)-1,3-dithiane] $\xrightarrow{SnCl_4}$ [S, S, $SnCl_3$] $\xrightarrow[-HCl]{-SnCl_2}$ S, S

(Eq. 326)

$(CH_3)_3Si$, S, S $\xrightarrow[CH_2Cl_2,\ -78°]{SnCl_4,\ \text{ethylene oxide}}$ HO, S, S (30%) (Eq. 327)

$(CH_3)_3Si$~CH=CH$_2$ $\xrightarrow{WCl_6}$ [CH$_2$=CH~WCl_5] (Eq. 328)

Palladation. Allylsilanes react with palladium(II) salts with the formation of π-allylpalladium complexes (Eq. 329).[513] The reaction is stereospecific (Eqs.

$(CH_3)_3Si$~CH=CH$_2$ $\xrightarrow[50°]{Li_2PdCl_4,\ CH_3OH}$ (π-allyl $PdCl)_2$ (32%) (Eq. 329)

330 and 331) with palladium attacking *anti* to the silyl group.[23] The substituents on the allyl framework of the isolated products are in the W configuration regardless of the configuration of the original double bond, but diastereofacial integrity is maintained.

C_6H_5, $(CH_3)_3Si$, H **239** *or* $(CH_3)_3Si$, H, C_6H_5 **240** $\xrightarrow[0°]{Li_2PdCl_4,\ CH_3OH}$ C_6H_5~(π-allyl) $PdCl)_2$

(84%) from **239**
(76%) from **240**

(Eq. 330)

C_6H_5 $(CH_3)_3Si$ H **241** or $(CH_3)_3Si$ H C_6H_5 **242** $\xrightarrow[0°]{Li_2PdCl_4,\ CH_3OH}$ C_6H_5 $PdCl)_2$

(100%) from **241**
(79%) from **242**

(Eq. 331)

Palladium(II) bistrifluoroacetate and oxygen react with allylsilanes to give allylic alcohols (Eq. 332).[514] When the allylsilane carries an electron-withdrawing group, the oxidation goes further, giving an α,β-unsaturated aldehyde or ketone (Eq. 333).[514]

$(CH_3)_3Si$ C_9H_{19}-*n* $\xrightarrow[O_2,\ h\nu]{Pd(O_2CCF_3)_2,\ (CH_3)_2CO}$ C_9H_{19}-*n* (OH) + HO C_9H_{19}-*n* (Eq. 332)

1:1

$(CH_3)_3Si$ *p*-$CH_3C_6H_4SO_2$ $\xrightarrow[O_2,\ h\nu]{Pd(O_2CCF_3)_2,\ (CH_3)_2CO}$

p-$CH_3C_6H_4SO_2$ (O) (58%) (Eq. 333)

When the allylsilane is also an allylic acetate, a reaction with palladium(0) complexes leads to the trimethylenemethane complex **243**, which adds to electrophilic alkenes (Eq. 334).[515] Although the overall result is partly an electrophilic substitution, these reactions are not included in the tables because the first step is presumably nucleophilic attack by Pd(0) on the allylic acetate part of the molecule. This subject has been reviewed.[516]

CH_3CO_2 $(CH_3)_3Si$ $\xrightarrow{PdL_4}$ [$\overset{+}{Pd}L_2$] **243** $\xrightarrow{CO_2CH_3}$ CO_2CH_3

(Eq. 334)

Vinylsilanes react with palladium(II) salts to give electrophilic substitution in the usual way (Eq. 335).[517] The intermediate **244** is not isolated, but its subsequent reactions identify it.

$$C_6H_5\text{CH=CH}Si(CH_3)_3 \xrightarrow{PdCl_2} [C_6H_5\text{CH=CH}PdCl]\ \mathbf{244}$$

$$\mathbf{244} \xrightarrow[\text{room temp}]{CH_3OH,\ CuCl_2} C_6H_5\text{CH=CHCH=CH}C_6H_5\ (52\%)$$

$$\mathbf{244} \xrightarrow[CuCl_2,\ \text{room temp}]{CH_2{=}CH_2,\ CH_3CN} C_6H_5\text{CH=CH}(CH_2)_2Cl\ (30\%)$$

$$\mathbf{244} \xrightarrow{CH_2{=}CHCO_2CH_3} C_6H_5\text{CH=CHCH=CH}CO_2CH_3\ (35\%)$$

(Eq. 335)

Vinylsilanes also react (Eq. 336) with phenylpalladium cations, prepared from phenyldiazonium cations and a Pd(0) complex.[518] The main pathway appears to be *syn* addition of the benzene ring and the palladium to the double bond, followed by competing *syn* and *anti* eliminations of the silyl and palladium groups. The regiochemistry is not as well controlled as with most electrophilic substitutions because the usual cationic intermediates are not involved.

$$\underset{Z \text{ or } E}{C_6H_5\text{CH=CH}Si(CH_3)_3} \xrightarrow[Pd(C_6H_5CH{=}CHCOCH{=}CHC_6H_5)_2]{C_6H_5N_2^+BF_4^-} C_6H_5\text{CH=CH}C_6H_5 + (C_6H_5)_2C{=}CH_2$$

4:1 (97% from Z)
2:1 (98% from E)

(Eq. 336)

Heck reactions on vinylsilanes give styrenes (Eq. 337).[519] Remarkably, when silver nitrate is included in the reaction mixture the silyl group remains in the molecule (Eq. 338).[520]

$$CH_2{=}CHSi(CH_3)_3 + IC_6H_4OCH_3\text{-}p \xrightarrow[(o\text{-}CH_3C_6H_4)_3P,\ (C_2H_5)_3N,\ 70^\circ]{Pd(O_2CCH_3)_2,\ DMF} CH_2{=}CHC_6H_4OCH_3\text{-}p\ (60\%)$$

(Eq. 337)

$$CH_2{=}CHSi(CH_3)_3 + IC_6H_4OCH_3\text{-}p \xrightarrow[AgNO_3,\ 50^\circ]{Pd(O_2CCH_3)_2,\ (C_2H_5)_3N} p\text{-}CH_3OC_6H_4\text{CH=CH}Si(CH_3)_3\ (67\%)$$

(Eq. 338)

Mercuration. Allylsilanes[521,522] and 2,2-disubstituted vinylsilanes[214] undergo mercuridesilylation (Eqs. 339 and 340), but less reactive vinylsilanes give mainly addition products (Eqs. 341 and 342).[523,521]

$(CH_3)_3Si$ (allyl) $\xrightarrow[\text{room temp}]{Hg(O_2CCH_3)_2,\ CH_3CO_2H}$ (allyl) HgO_2CCH_3 (40%)

(Eq. 339)

$(CH_3)_3Si$ $\xrightarrow[CH_3CO_2H,\ \text{room temp}]{Hg(O_2CCH_3)_2,\ THF,\ H_2O}$ CH_3CO_2Hg (O)

(Eq. 340)

$Si(CH_3)_3$ $\xrightarrow[\text{2. NaOH, } H_2O \text{ 3. NaCl, } H_2O]{\text{1. } Hg(O_2CCH_3)_2,\ H_2O}$ HO, HgCl, $Si(CH_3)_3$ (88%) (Eq. 341)

$Si(CH_3)_3$ $\xrightarrow[H_2O,\ \text{room temp; 2. } NaBH_4,\ NaOH,\ \text{room temp}]{\text{1. } Hg(O_2CCH_3)_2,\ THF}$

HO, $Si(CH_3)_3$ (44%) + HO (19%) + OH (7%) (Eq. 342)

Thallation. Thallium(III) salts, typically thallium tris(trifluoroacetate) (TTFA), react with allylsilanes to give allylthallium(II) intermediates. These are now *electrophilic*, and an umpolung of reactivity has been achieved, similar to that found in the reaction of allylsilanes with iodosobenzene. The nucleophile that attacks these species is always included in the reaction mixture, so that the intermediates are not detected. Nucleophiles that work include nitriles (Eq. 343),[524] alcohols (Eq. 344), carboxylic acids (Eq. 345),[525] and aromatic

$(CH_3)_3Si$ (allyl) $\xrightarrow[0°]{TTFA,\ CH_3CN,\ -20°,\ 2\ h}$ (allyl) $NHCOCH_3$ (48%) (Eq. 343)

$\xrightarrow[0°]{TTFA,\ C_2H_5OH}$ (allyl) OC_2H_5 (86%) (Eq. 344)

$\xrightarrow[\text{room temp}]{Tl(O_2CCH_3)_3,\ CH_3CO_2H}$ (allyl) O_2CCH_3 (81%) (Eq. 345)

$\xrightarrow[\text{room temp}]{TTFA,\ C_6H_5OCH_3}$ (allyl) $C_6H_4OCH_3$ (88%) *o:p* 1:1.4

(Eq. 346)

rings (Eq. 346).[526] Because the allylthallium intermediates are not stable with respect to allylic transposition, unsymmetrical allylsilanes give mixtures of products (Eq. 347).[525]

$$(CH_3)_3Si\text{CH}_2\text{CH}{=}\text{CH}C_6H_5 \xrightarrow[0^\circ]{\text{TTFA, }CH_3OH} \text{CH}_2{=}\text{CHCH(OCH}_3)C_6H_5 + CH_3O\text{CH}_2\text{CH}{=}\text{CH}C_6H_5 \quad 75{:}25 \quad (88\%)$$

(Eq. 347)

COMPARISON WITH OTHER METHODS

The electrophiles that attack allylsilanes and vinylsilanes are, in general, those that attack ordinary alkenes. However, silicon is more electropositive than carbon and is therefore a metal with respect to carbon. Thus the electrophilic substitution reactions of allylsilanes and vinylsilanes have many of the features of the electrophilic substitution reactions of more conventional allyl–metal and vinyl–metal compounds. The difference is that the electrophile has to be more powerful for the silanes than for other metal derivatives, and there is a continuity of behavior, with allyltin and vinyltin reagents, for example, falling between their silicon and lithium counterparts in reactivity.

Allylsilanes and vinylsilanes are useful in synthesis, not only because they often react with electrophiles in a well-behaved and reasonably predictable way, but also because they have enough stability to survive many reaction conditions. Unlike other metal derivatives, they are stable to air and water, they can usually be purified by distillation, crystallization, or chromatography, and they can be stored indefinitely. Unlike other allyl–metal compounds, allylsilanes are stable with respect to 1,3-allylic rearrangement, with the result that regiocontrol is more reliable with allylsilanes than with other allyl–metal compounds. Although some regiostable allylstannanes and allylboranes are effective reagents, neither is as widely reliable as allylsilanes for regiocontrol. Allylic stability also confers upon allylsilanes a high degree of configurational stability when the silyl group is attached to a chiral center.

The stability of allylsilanes and vinylsilanes relative to other metal compounds makes it possible to carry these groups through several stages of a synthesis without the silicon interfering.[55] As long as powerful electrophiles are not used, the group will survive until it is used in the reaction for which it was designed. Because allylsilanes and vinylsilanes are only a little more reactive than the corresponding alkenes, it is easy to predict what reaction conditions allylsilanes and vinylsilanes will survive: if the corresponding alkene is likely to retain its stereochemistry and its position in the molecule, there is a fair probability that the allylsilane or vinylsilane will survive.

The fundamental silicon-containing starting materials, such as chlorotrimethylsilane, are abundant and relatively cheap. Tetraorganosilanes are rarely toxic. Reactions can be followed easily by ^{1}H NMR spectroscopy, even in solvents like ether, because of distinctive changes in the position of resonance of the methyl groups attached to silicon. The silicon-containing byproducts of the reactions, typically hexamethyldisiloxane, are volatile, easily removed, and, as far as we know, harmless. Furthermore, hexamethyldisiloxane can be converted back to chlorotrimethylsilane if need be. Finally, silicon–carbon bonds can be made by a wide variety of methods. All these advantages more than compensate for the need to use a relatively powerful electrophile at the point when the substitution reaction is to be carried out. Furthermore, although the metallic character of silicon is weak, allylsilanes and vinylsilanes are usually more reactive than their simple alkene counterparts. Either for this reason or because protons are not released in most allylsilane and vinylsilane reactions, silicon-containing alkenes usually give better yields than the simple alkenes in similar reactions.

In general, there is a strong parallel between the reactions of allylsilanes and the reactions of silyl enol ethers. The latter are somewhat more reactive, but the electrophiles that react with silyl enol ethers[527] are almost exactly the same electrophiles that react with allylsilanes. The outcome too is similar, except that silyl enol ethers give a carbonyl group in the product where allylsilanes give a carbon–carbon double bond. In some cases, it is particularly useful to introduce a future carbonyl group in masked form by an allylsilane reaction, and to release the carbonyl group later by ozonolysis.[322]

EXPERIMENTAL PROCEDURES

***trans*-1-Phenyl-4(*E*)-(1-propenyl)cyclohexane [Regiospecific Acid-Catalyzed Protodesilylation of an Allylsilane]** (Eq. 26).[44] The allylsilane **29** (200 mg, 0.6 mmol) and the boron trifluoride acetic acid complex (0.08 mL, 40% BF_3 in CH_3CO_2H) were stirred in 5 mL of dichloromethane at 0° for 20 minutes and then at 20° for 40 minutes. Sodium bicarbonate solution was added, the aqueous layer was extracted with hexane, and the organic layers were combined, dried ($MgSO_4$), and evaporated. The residue was crystallized from methanol to give 110 mg (92%) of the alkene **30**, mp 43–44°. IR (CCl_4) 985 (*trans* CH=CH) cm^{-1}; ^{1}H NMR (CCl_4, 250 MHz) δ 0.9–2.6 (m, 10*H*), 1.6 (d, J = 5 Hz, 3*H*), 5.15–5.45 (m, 2*H*), 7.12 (s, 5*H*); ^{13}C NMR ($CDCl_3$) δ 18.0 (q), 33.4 (t), 34.0 (t), 40.4 (d), 44.1 (d), 122.4 (d), 125.8 (d), 126.8 (d), 128.3 (d), 137.2 (d), 147.2 (d).

(*Z*)-7-Tetradecene [Stereospecific Acid-Catalyzed Protodesilylation of a Vinylsilane] (Eq. 34).[72,528] Constant boiling hydriodic acid (0.04 mL) was added to a solution of (*E*)-7-trimethylsilyl-7-tetradecene (134 mg, 0.5 mmol) in 1 mL of benzene and the mixture was stirred at room temperature for 15 minutes. Alkali treatment and chromatography gave 89.5 mg (91%) of (*Z*)-7-tetradecene; ^{1}H NMR δ 0.9 (t, J = 6 Hz, 6*H*), 1.1–1.8 (m, 16*H*), 1.8–2.4

(m, 4H), 5.20 (t, J = 5 Hz, 2H); identified also by comparison of its derived epoxide with an authentic sample. GLC analysis indicated that the product was a 94:6 mixture with the E isomer.

(*Z*)-7-Deutero-7-tetradecene [Stereospecific Acid-Catalyzed Deuterodesilylation of a Vinylsilane].[72,528] Deuterium oxide (0.1 mL, 99% d_2) and iodine (25.4 mg, 0.1 mmol) were added to (E)-7-trimethylsilyl-7-tetradecene (268 mg, 1 mmol) in 2 mL of benzene and the mixture refluxed for 1.5 hours. Sodium hydroxide solution was added, the aqueous layer was extracted three times with ether, the ether layers were combined, dried (Na_2SO_4), and concentrated. Chromatography on silica gel eluting with hexane gave 177 mg (90%) of (Z)-7-deutero-7-tetradecene, which was 89% monodeuterated, as determined by mass spectrometry.

(3*S*,4*S*)-4-Methyl-5-hexen-3-ol [Intramolecularly Assisted Base-Catalyzed Protodesilylation of a Vinylsilane] (Eq. 82).[210] Sodium hydride (0.33 g of a 50% dispersion in mineral oil, 6.44 mmol) was added to a solution of (3S,4R)-4-methyl-5-trimethylsilyl-5-hexen-3-ol (1.20 g, 6.45 mmol) in 3 mL of dry hexamethylphosphoramide. The mixture was stirred at 30° for 2 hours, then slowly poured into 15 mL of aqueous 3 M hydrochloric acid at 0°, and stirring continued for half an hour. The mixture was extracted with a 1:1 mixture of light petroleum and ether (3 × 40 mL), and the combined organic extracts were dried ($MgSO_4$) and evaporated. The residue was chromatographed on silica gel eluting with a mixture of light petroleum and ether to give 0.65 g (88%) of (3S,4S)-4-methyl-5-hexen-3-ol $[\alpha]_D^{25}$ −45.0° (c 0.95, $CHCl_3$). ^{1}H NMR (CCl_4) δ 0.92 (t, J = 7 Hz, 3H), 1.20–1.80 (m, 3H), 1.98–2.41 (m, 1H), 3.27 (dt, J = 4.5 and 8 Hz, 1H), 4.83–5.20 (m, 2H), 5.72 (ddd, J = 7, 9, and 16 Hz).

2,2-Dimethyl-3-butenonitrile [Reaction of an Allylsilane with Chlorosulfonyl Isocyanate] (Eq. 86).[37] Chlorosulfonyl isocyanate (4.8 g, 34 mmol) was added dropwise to a solution of 3,3-dimethylallyltrimethylsilane (4.95 g, 35 mmol) in 10 mL of carbon tetrachloride at 0°. At this stage, the β-lactam cycloadduct (**100**) was present; IR (CCl_4) 1815 (C=O), 1410, 1175 (SO_2) cm^{-1}; ^{1}H NMR (CCl_4) δ 0 (s, 9H), 1.2 (s, 3H), 1.25 (m, 2H), 1.3 (s, 3H), 4.0 (dd, J = 4 and 11 Hz, 1H). The solution was brought to room temperature for 1 hour and the solvent was removed to give N-chlorosulfonyl-2,2-dimethyl-1-trimethylsilyloxy-3-butenimine (**101**); IR (neat) 3080 (=CH), 1640 (C=C), 1540 (C=N), 1370, 1170–1190 (SO_2), 1255, 845 [$Si(CH_3)_3$] cm^{-1}; ^{1}H NMR δ 0.48 (s, 9H), 1.32 (s, 6H), 5.1 (dd, J = 17 and 1 Hz, 1H), 5.2 (dd, J = 9 and 1 Hz, 1H), 5.8 (dd, J = 17 and 9 Hz, 1H). This was dissolved in 30 mL of ether, and pyridine (3 g, 38 mmol) was added dropwise at 0°. The pyridine–sulfur trioxide complex was removed by filtration and washed twice with ether. The combined ethereal layers were washed twice with ammonium chloride solution at 0°, dried ($MgSO_4$), and evaporated. The residue was chromatographed on silica gel eluting with dichloromethane to give 2.5 g (77%) of the nitrile **102**; IR (neat) 3080 (=CH_2), 2240 (CN), 1640 (C=C), 990, 930 (CH=CH_2) cm^{-1}, ^1H NMR δ 1.25 (s, 6*H*), 5.1 (dd, J = 17 and 1 Hz, 1*H*), 5.2 (dd, J = 9 and 1 Hz, 1*H*), 5.7 (dd, J = 17 and 9 Hz, 1*H*).

3-Methyl-3-vinyl-1-nonanol [Reaction of an Allylsilane with Ethylene Oxide] (Eq. 120).[170,276] Titanium tetrachloride (0.6 mL, 5.5 mmol) in 5 mL of dichloromethane was cooled in a dry-ice–acetone bath and transferred by syringe to a stirred solution of 3-methyl-1-trimethylsilyl-2-nonene (1.06 g, 5 mmol) and ethylene oxide (0.26 g, 6 mmol) in 8 mL of dichloromethane under nitrogen in a dry-ice–acetone bath, and kept for 1 hour. The mixture was poured into 25 mL of sodium bicarbonate solution and extracted with ether (3 × 50 mL), and the organic phases were dried ($MgSO_4$) and evaporated under reduced pressure. The residue was chromatographed on silica gel to give 0.86 g (93%) of the alcohol; IR (film) 3350 (OH), 3080 (=CH), 1640 (C=C) cm^{-1}; ^1H NMR ($CDCl_3$) δ 0.89 (s, 3*H*), 1.00 (s, 3*H*), 1.1–1.71 (m, 12*H*), 2.01 (s, 1*H*, OH), 3.63 (t, J = 7 Hz, 2*H*) 4.79–5.08 (m, 2*H*), 5.75 (dd, J = 11.5 and 17.5 Hz, 1*H*).

4(*RS*)-Methyl-5(*SR*)-phenylthio-1-hexene [Carbosulfenylation of an Alkene] (Eq. 126).[279,529] Benzenesulfenyl chloride (2.93 g, 20 mmol) in 20 mL of dichloromethane was added dropwise to (*Z*)-2-butene (1.40 g, 25 mmol) in 20 mL of dichloromethane at −78° under argon and then allowed to warm to room temperature. Solvent was evaporated to give the crude adduct **116**; ^1H NMR ($CDCl_3$, 60 MHz) δ 1–2.5 (m, 8*H*), 3.1–3.46 (m, 1*H*), 3.8–4.12 (m, 1*H*), 7.15–7.5 (m, 5*H*). The crude adduct (0.41 g, 2 mmol) and allyltrimethylsilane (0.25 g, 2.2 mmol) were dissolved in 4 mL of nitromethane, anhydrous sublimed zinc bromide (0.113 g, 0.5 mmol) was added, and the mixture was stirred for 16 hours at room temperature. The yellow solution was poured into water, extracted with 10 mL of dichloromethane, and the organic layer was dried ($MgSO_4$) and concentrated. The residue was chromatographed (silica gel, 1% ether in light petroleum) on a radial chromatotron to give 0.37 g (92%) of the adduct; IR (film) 3085 (=CH), 1638 (C=C), 1580 (Ar) cm^{-1}; ^1H NMR ($CDCl_3$) δ 0.98 (d, J = 7 Hz, 3*H*), 1.21 (d, J = 6 Hz, 3*H*), 1.74–1.92 (m, 1*H*), 1.96–2.30 (m, 2*H*), 3.32 (dq, J = 4 and 6 Hz, 1*H*), 5.00 (br s, 1*H*), 5.07 (d, J = 5 Hz, 1*H*), 5.64–5.92 (m, 1*H*), 7.14–7.48 (m, 5*H*); ^{13}C NMR ($CDCl_3$) δ 14.85, 15.79, 36.82, 39.07, 47.46, 116.18, 126.43, 128.78, 131.46, 135.97, 136.96; mass spectrum, m/z (rel. intensity) 206 (12), 165 (5), 97 (9), 96 (37), 56 (5), 55 (100).

(*E*)-1-Phenyl-1-hepten-4-ol [Regioselective Allylation of an Aldehyde Catalyzed by Lewis Acid] (Eq. 131).[64] Titanium tetrachloride (0.19 g, 1 mmol) was added dropwise from a syringe to a solution of *n*-butyraldehyde (0.144 g, 2 mmol) in 3 mL of dichloromethane at room temperature under nitrogen, and the mixture was stirred for 5 minutes. 1-Phenylallyltrimethylsilane (0.38

g, 2 mmol) was added rapidly and the mixture was stirred for 30 seconds at room temperature. Water was added and the mixture extracted with ether. The organic layer was washed with water, dried (Na_2SO_4), and evaporated. The residue was chromatographed on silica gel to give 0.205 g (54%) of (*E*)-1-phenyl-1-hepten-4-ol.

6-Phenyl-1-hexen-4-ol [Fluoride Ion Catalyzed Allylation of an Aldehyde] (Eq. 11).[13] Tetrabutylammonium fluoride (26 mg, 0.1 mmol) and 4A molecular sieves (50 mg) in 5 mL of dry tetrahydrofuran were added to a stirred mixture of 3-phenylpropanal (264 mg, 2 mmol) and allyltrimethylsilane (229 mg, 2 mmol) in 5 mL of tetrahydrofuran at room temperature under argon, and the resulting pale yellow mixture was refluxed for 4 hours. Methanol and hydrochloric acid were added. The solvents were evaporated, and the residue was purified by TLC to give 303 mg (86%) of the alcohol.

(*E*)-2-(1-Hydroxyethyl)dec-2-enenitrile [Fluoride Ion Catalyzed Reaction of a Vinylsilane with an Aldehyde] (Eq. 167).[314,530] A dry, 30 mL, three-necked flask equipped with a magnetic stirrer, a septum, and a dropping funnel was charged with (*E*)-2-trimethylsilyldec-2-enenitrile (223 mg, 1 mmol) and a solution of acetaldehyde (53 mg, 1.2 mmol) in 5 mL of tetrahydrofuran under argon. The mixture was cooled to −20°, and a solution of tetrabutylammonium fluoride in tetrahydrofuran (5.5 mL, 0.2 *M*) was added dropwise slowly with stirring. After half an hour, the mixture was allowed to warm slowly to room temperature and stirring continued for half an hour. Hexane (50 mL) was added and the mixture was washed with saturated sodium chloride solution (3 × 15 mL). The organic layer was dried ($MgSO_4$) and concentrated on a rotary evaporator, and the residue was distilled (kugelrohr) at 140° (26 mm) to give 137 mg (70%) of (*E*)-2-(hydroxyethyl)dec-2-enenitrile; IR (film) 3410 (OH), 2210 (CN), 1635 (C=C) cm^{-1}; ^1H NMR ($CDCl_3$) δ 0.85 (t, 3*H*), 1.1–1.6 (m, 10*H*), 1.44 (d, J = 6.7 Hz, 3*H*), 2.34 (dt, J = 7.7 and 7 Hz, 2*H*), 4.39 (q, J = 6.7 Hz, 1*H*), 6.43 (t, J = 7.7 Hz, 1*H*); ^{13}C NMR ($CDCl_3$) δ 68.5, 119.7, 147.4, and 116.0 ($^3J_{CN\text{-}H3}$ = 10 Hz).

4-Phenyl-6-hepten-2-one [Lewis Acid Catalyzed Intermolecular Sakurai Reaction] (Eq. 170).[318] Titanium tetrachloride (22 mL, 0.2 mol) was added slowly by syringe to a stirred solution of benzalacetone (29.2 g, 0.2 mol) in 300 mL of dichloromethane kept at −40° (dry-ice–methanol). After 5 minutes allyltrimethylsilane (30.2 g, 0.26 mol) in dichloromethane was added dropwise with stirring over 30 minutes, and the mixture stirred at −40° for a further 30 minutes. Water (400 mL) and 500 mL of ether were added and the combined organic layers were washed with saturated sodium bicarbonate solution and brine, dried (Na_2SO_4), and evaporated. The residue was distilled to give 29.2–30 g of ketone (78–80%) bp 69–71° (0.2 mm) n_D^{20} 1.5156; IR (film) 1710 (CO), 1630 (C=C) cm^{-1}; ^1H NMR ($CDCl_3$) δ 1.97 (s, 3*H*), 2.35 (t, J = 7.5

Hz, 2*H*), 2.72 (d, J = 7.5 Hz, 2*H*), 3.27 (quintet, J = 7.5 Hz, 1*H*), 4.8–5.1 (m, 2*H*), 5.4–5.9 (m, 1*H*), 7.0–7.4 (m, 5*H*).

1-Vinylspiro[4.5]decan-7-one [Lewis Acid Catalyzed Intramolecular Sakurai Reaction] (Eq. 180).[226] (*Z*)-3-(6-Trimethylsilyl-4-hexenyl)-2-cyclohexenone (165 mg, 0.66 mmol) in 30 mL of dry toluene was added to ethylaluminum dichloride (0.1 mL, 0.73 mmol) at 0° and the mixture was stirred for 30 minutes. Ice water (10 mL) was added, the aqueous phase was separated and extracted with ether (2 × 30 mL), and the combined organic layers were dried (Na_2SO_4) and evaporated. The residue was distilled (kugelrohr) at 60° (0.08 mm) to give 90.5 mg (77%) of the spirocyclic ketone as a 3:1 mixture of diastereoisomers. The major isomer had IR (film) 1715 (CO), 920 (=CH) cm^{-1}; ^1H NMR ($CDCl_3$, 90 MHz) δ 1.1–2.5 (m, 15*H*), 5.1 (m, 2*H*), 5.7 (m, 1*H*); ^{13}C NMR δ 21.4 (t), 23.5 (t), 29.7 (t), 35.8 (t), 41.2 (t), 47.0 (t), 50.2 (s), 54.4 (d), 116.4 (t), 138.5 (d), 212.4 (s).

Methyl 3-Phenyl-5-hexenoate [Fluoride Ion Catalyzed Sakurai Reaction with an α,β-Unsaturated Ester] (Eq. 186).[162] Tetrabutylammonium fluoride (20 mg) was kept at room temperature under high vacuum (0.1 mm) for 30 minutes. The flask was then flushed with nitrogen, 2 mL of dry dimethylformamide was added, and the solution was transferred by syringe onto 4A molecular sieves (300 mg, flame dried for 5 minutes under high vacuum) and stirred for 30 minutes. A flask containing 4A molecular sieves (100 mg) was flame dried under vacuum for 5 minutes and kept under nitrogen. The solution of anhydrous tetrabutylammonium fluoride in dimethylformamide was added, followed by methyl cinnamate (10 mg, 0.62 mmol) in 1 mL of dimethylformamide. Freshly distilled allyltrimethylsilane (212 mg, 1.8 mmol) and hexamethylphosphoramide (331 mg, 1.8 mmol) in 2 mL of dimethylformamide was then added dropwise at room temperature. Coloration occurred immediately. After 10 minutes, analysis by TLC revealed that the reaction was complete. Methanolic hydrochloric acid (1 mL, 1 M) was added, and the mixture was diluted with 20 mL of water. Workup and column chromatography gave 101 mg (79%) of product; IR (film) 2960 (CH), 1740 (C=O), 990, 905, 752 ($CH{=}CH_2$ and C_6H_5) cm^{-1}; ^1H NMR ($CDCl_3$) δ 2.1–2.5 (m, 5*H*), 3.0–3.2 (m, 1*H*), 3.45 (s, 3*H*), 4.7–5.0 (m, 2*H*), 5.25–5.8 (m, 1*H*); mass spectrum m/z 173 (M − 41).

***cis*-3a,4,5,6,7,7a-Hexahydro-1*H*-inden-1-one [Silicon Controlled Nazarov Cyclization]** (Eq. 191).[343,531] Anhydrous ferric chloride (337 mg, 2.08 mmol) was added in one portion to a solution of (*E*)-1-(1-cyclohexenyl)-3-trimethylsilyl-2-propen-1-one (456 mg, 1.98 mmol) in 27 mL of dichloromethane and the mixture stirred magnetically for 4 hours at 0°, by which time the starting material had been consumed. Water (27 mL) was added, the mixture was diluted with 23 mL of ether, and the aqueous phase was separated and washed with ether (2 × 27 mL). The individual organic extracts were washed with 16 mL of water and brine (3 × 16 mL), combined, dried (K_2CO_3), and concentrated. The residue was purified by flash chromatography on silica gel (eluting with hexane–ethyl acetate 3:1) followed by distillation, bp 100° (0.08 mm), to give 226 mg (84%) of the indenone; IR ($CHCl_3$) 3010, 2940, 2860 (CH), 1705 (C=O), 1585 (C=C) cm^{-1}; ^1H NMR ($CDCl_3$, 360 MHz) δ 1.0–2.0 (m, 8*H*), 2.40 (q, *J* = 6.2 Hz, 1*H*), 2.94–3.0 (m, 1*H*), 6.14 (dd, *J* = 5.7 and 1.1 Hz, 1*H*), 7.64 (dd, *J* = 5.7 and 2.8 Hz, 1*H*); mass spectrum, *m/z* (rel. intensity) 135 (21), 118 (10), 108 (40), 107 (100), 95 (54), 94 (22), 93 (38), 91 (17), 81 (28), 80 (37), 79 (87), 77 (34), 67 (32), 53 (36).

(5β)-6α-Vinyl-1-azabicyclo[3.3.0]octan-2-one [Cyclization of an Allylsilane on an Acyliminium Ion] (Eq. 234).[388] A solution of (*Z*)-5-hydroxy-*N*-(5-trimethylsilyl-3-pentenyl)pyrrolidone (140 mg, 0.6 mmol) in 2 mL of dichloromethane was added dropwise to a solution of trifluoroacetic acid (0.2 mL, 2.6 mmol) in 10 mL of dichloromethane under nitrogen and the mixture was stirred for 1 hour at 0°. The mixture was diluted with 20 mL of dichloromethane, washed with saturated sodium bicarbonate solution (2 × 25 mL), dried (K_2CO_3), and concentrated. The residue was chromatographed to give 64 mg (72%) of the product; IR ($CHCl_3$) 1675 (C=O), 995, 925 ($CH{=}CH_2$) cm^{-1}; ^1H NMR ($CDCl_3$) δ 1.75–2.10 (m, 3*H*), 2.32 (m, 1*H*), 2.55–2.7 (m, 2*H*), 3.07 (m, 1*H*), 3.61 (dt, 1*H*), 4.03 (br q, 1*H*), 5.0–5.15 (m, 1*H*), 5.5–5.7 (m, 1*H*); ^{13}C NMR ($CDCl_3$) δ 19.9, 31.7, 33.2, 39.1, 42.8, 63.2, 116.0, 134.5, 173.9. The stereochemistry of the product was established by nuclear Overhauser difference experiments showing, among other effects, enhancement of intensity from the allylic methine to the bridgehead hydrogen.

(±)-Deplancheine [Stereospecific Intramolecular Reaction of a Vinylsilane with an Iminium Ion] (Eq. 245).[399] A solution of (*E*)-2,3,4,9-tetrahydro-1-[3-(trimethylsilyl)-3-pentenyl]-1*H*-pyrido[3,4-*b*]indole (390 mg, 1.24 mmol) and *d*-camphorsulfonic acid (290 mg, 1.16 mmol) in 25 mL of acetonitrile was stirred with paraformaldehyde (1.08 g, 36 mmol) for 2 hours at 82°. The excess of paraformaldehyde was removed by filtration through a plug of glass wool. A basic workup using dichloromethane and drying (Na_2SO_4), concentration, chromatography on silica gel (eluting with chloroform-2-propanol–ammonium hydroxide solution 20:1:0.1), and sublimation at 140° (0.3 mm) gave 260 mg (83%) of (±)-deplancheine mp 143°; UV (CH_3OH) nm max 235, 273, 283, 290; IR ($CHCl_3$) 3440 (NH), 2840–2700 (Bohlmann bands) cm^{-1}; ^1H NMR ($CDCl_3$, 250 MHz)) δ 1.5–1.65 (m, 1*H*), 1.63 (d, *J* = 6.8 Hz, 3*H*), 1.93–2.05 (br t, 1*H*), 2.1–2.2 (m, 1*H*), 2.6–2.85 (m, 3*H*), 2.95–3.1 (m, 3*H*), 3.3–3.45 (m, 3*H*), 5.43 (br q, *J* = 6.8 Hz, 1*H*), 7.04–7.48 (m, 4*H*), 7.70 (br s, 1*H*); ^{13}C NMR ($CDCl_3$) δ 12.9, 21.8, 26.1, 30.5, 53.1, 60.4, 63.7, 108.6, 110.9, 118.4, 119.5, 119.6, 121.5, 127.7, 134.3, 134.9, 136.3; mass spectrum (CI, isobutane), *m/z* (rel. intensity) 253 (100). GLC analysis (25 m

SE-30 glass capillary) showed that the product was a 95:5 mixture with the *Z* isomer.

Artemesia Ketone [Acylation of an Unsymmetrical Allylsilane] (Eq. 247).[37,406] A mixture of senecioyl chloride (6.52 g, 55 mmol) and aluminum chloride (6.68 g, 50 mmol) in 50 mL of dichloromethane was added over 30 minutes to a stirred solution of 3,3-dimethylallyltrimethylsilane (7.1 g, 50 mmol) in 100 mL of dichloromethane at −60°. The mixture was stirred for an additional 15 minutes at −60° and then poured slowly into a mixture of ice and ammonium chloride cooled to −35°. The layers were separated and the aqueous layer was extracted with ether (2 × 150 mL). The combined organic layers were washed with sodium bicarbonate solution and ammonium chloride solution, dried (Na_2SO_4), and concentrated. The residue was distilled to give 6.84 g (90%) of artemesia ketone, bp 87° (200 mm); IR (neat) 3080 (CH), 1690 (C═O), 1630 (C═C) cm^{-1}; ^1H NMR (CCl_4) δ 1.13 (s, 6*H*), 1.85 (d, *J* = 1.5 Hz, 3*H*), 2.08 (d, *J* = 1.5 Hz, 3*H*), 4.90 (m, 1*H*), 5.90 (m, 1*H*), 6.13 (m, 1*H*).

1-Acetyl-4,4-dimethylcyclohexene [Regiospecific Acetylation of a Vinylsilane] (Eq. 10).[117] 4,4-Dimethyl-1-trimethylsilylcyclohexene (364 mg, 2 mmol) in 100 mL of dichloromethane was added over 6 hours to a stirred mixture of acetyl chloride (0.47 g, 10 mmol) and aluminum chloride (0.8 g, 10 mmol) in 20 mL of dichloromethane at 0°. The mixture was stirred for an additional 15 minutes and poured into 150 mL of sodium bicarbonate solution. The aqueous layer was extracted with dichloromethane (3 × 20 mL) and the combined organic layers were washed with sodium bicarbonate solution (2 × 50 mL), dried (Na_2SO_4), and evaporated. The residue was purified by TLC on silica gel eluting with dichloromethane to give 234 mg (77%) of the unsaturated ketone; IR (CCl_4) 1670 (C═O), 1640 (C═C) cm^{-1}; ^1H NMR (CCl_4) δ 0.88 (s, 6*H*), 1.4–2.2 (m, 9*H* including a singlet at 2.14), 6.6–6.76 (m, 1*H*); semicarbazone mp 219–220°.

4-Phenyl-3-buten-2-ol [Stereospecific Epoxidation of an Allylsilane and Desilylative Opening of an Allylsilane Oxide] (Eq. 275).[437] (*R*)(*E*)-1-Phenyl-1-trimethylsilyl-2-butene (0.622 g, 3.04 mmol, 81% ee) in 10 mL of dichloromethane was mixed with sodium bicarbonate (0.252 g, 2.99 mmol) and *m*-chloroperbenzoic acid (0.72 g, 80%, 3.34 mmol) in 15 mL of dichloromethane and stirred at −78°. The mixture was stirred at 0° for 1 hour, and the solvent was removed under reduced pressure. Methanol (12 mL) and 2 mL of acetic acid were added, and the solution was washed with 20% sodium hydroxide solution (4 × 50 mL) and with water. The organic layer was dried ($MgSO_4$) and concentrated, and the residue was purified by TLC on silica gel eluting with chloroform to give 0.441 g (98%) of an 81:19 mixture of the *E* and *Z* isomers of the allylic alcohols. These were separated as their silyl ethers by GLC (silicone DC550) to give the pure *E* isomer $[\alpha]_D^{20}$ −11.8° (*c* 2.4, C_6H_6); ^{1}H NMR (CCl_4) δ 1.31 (d, *J* = 6 Hz, 3*H*), 1.35 (s, 1*H*), 4.38 (quintet, *J* = 6 Hz, 1*H*), 6.12 (dd, *J* = 6 and 16 Hz, 1*H*), 6.48 (d, *J* = 16 Hz, 1*H*), 7.0–7.4 (m, 5*H*); and the pure *Z* isomer; ^{1}H NMR (CCl_4) δ 1.30 (d, *J* = 6 Hz, 3*H*), 1.39 (s, 1*H*), 4.63 (dq, *J* = 9 and 6 Hz, 1*H*), 5.58 (dd, *J* = 9 and 12 Hz, 1*H*), 6.38 (d, *J* = 12 Hz, 1*H*), 6.95–7.35 (m, 5*H*). The enantiomeric excess of each isomer (73 and 72%, respectively) was determined using a chiral europium shift reagent after hydrogenation of the separated alcohols.

(*Z*)-1-(Trimethylsilyloxy)-1-octene [Osmylation of a Vinylsilane and its Stereospecific Conversion to a Silyl Enol Ether] (Eq. 295).[459] A solution of osmium tetroxide in *tert*-butanol (0.3 mL, 2.5% w/v) was added to a solution of (*Z*)-1-trimethylsilyloctene (1.38 g, 7.5 mmol, 99% *Z*) and trimethylamine *N*-oxide dihydrate (1.125 g, 10.2 mmol) in 0.6 mL of pyridine, 4 mL of water, and 15 mL of *tert*-butyl alcohol, and the mixture was refluxed for 24 hours. Aqueous sodium bisulfite solution (20 mL, 20%) was added at room temperature, most of the *tert*-butyl alcohol was removed on a rotary evaporator at room temperature, and the residue was saturated with sodium chloride and extracted five times with ether. The ether was washed with brine, dried ($MgSO_4$), and concentrated, and the residue was distilled (kugelrohr) at 95–100° to give 1.14 g (70%) of the diol **207**; IR (film) 3400 (OH), 2950 (CH), 1250 ($SiCH_3$) cm^{-1}; ^{1}H NMR ($CDCl_3$) δ 0.08 (s, 9*H*), 0.8–1.5 (m, 13*H*), 1.99 (br s, 2*H*, OH's), 3.40 (d, *J* = 3 Hz, 1*H*), 3.80 (m, 1*H*). Sodium hydride (144 mg of a 50% dispersion in oil) was stirred under argon successively with two portions of petroleum ether and the liquid was removed by pipet. Anhydrous ether (15 mL) was added, followed by a solution of the diol (218 mg, 1 mmol) in 5 mL of ether, and the mixture was stirred overnight. Saturated sodium bicarbonate solution (5 mL) was added and the aqueous layer was extracted with ether. The combined organic layers were washed with water, dried ($MgSO_4$), and concentrated, and the residue was distilled (kugelrohr) at 60–65° to give 146 mg (73%) of the silyl enol ether; IR (film) 2950 (=CH), 1650 (C=C), 1250 ($SiCH_3$), 1085, 840 (CH=CH) cm^{-1}; ^{1}H NMR ($CDCl_3$) δ 0.13 (s, 9*H*), 0.6–2.3 (m, 13*H*), 4.3–4.65 (m, 1*H*), 6.13 (m, 1*H*). GLC analysis showed the presence of octanal (1.5%) and the *E* isomer (0.5%).

(*E*)-2-Bromo-2-heptene [Stereospecific Bromodesilylation of a Vinylsilane] (Eq. 314).[481] Bromine (5.0 g, 30 mmol) in 15 mL of dichloromethane was slowly added to a stirred solution of (*Z*)-2-trimethylsilyl-2-heptene (4.3 g, 25 mmol) in 25 mL of dichloromethane at −78° under nitrogen. Methanol (25 mL) and sodium sulfite (0.5 g) were added and the mixture was stirred vigorously until the red-orange color became light yellow. The still cold (−78°) mixture was quickly poured into 10% sodium sulfite solution and shaken until all the color had disappeared. The aqueous layer was thoroughly extracted with pentane, and the combined organic layers were dried (Na_2SO_4). The solvents were removed on a rotary evaporator in the dark at room temperature to give 8.2 g (98%) of the crude dibromide. Freshly prepared sodium methoxide in methanol (37.5 mL, 1 *M*, 37.5 mmol) was cooled to 0° and stirred with a solution of the crude dibromide in 10 mL of methanol at 0° for 1 hour. The mixture was allowed to warm to room temperature, stirred for a further 2 hours, mixed with water and pentane, and separated. The aqueous layer was thoroughly extracted with pentane and the combined organic layers were dried (Na_2SO_4). The solvents were removed by distillation at atmospheric pressure in the presence of solid sodium carbonate, and the residue was distilled, again in the presence of sodium carbonate, to give 4.1 g (91%) of the bromide, bp 63–66° (16 mm); IR (neat) 2975, 2875 (CH), 846 (=CH) cm^{-1}; ^1H NMR (CCl_4, 60 MHz) δ 0.91 (br t, J = 6 Hz, 3H), 1.33 (m, 4H), 2.0 (br q, J = 7.5 Hz, 2H), 2.18 (d, J = 1.5 Hz, 3H), 5.8 (tq, J = 7.5 and 1.5 Hz, 1H). GLC analysis (20% SF-96 on Chromosorb Q) showed that the product was a 99:1 mixture with the *Z* isomer.

3-Ethoxy-1-propene [Umpolung of Allylsilane Reactivity Using a Thallium(III) Salt] (Eq. 344).[525] Allyltrimethylsilane (0.114 g, 1 mmol) was added dropwise with stirring to a solution of thallium tris(trifluoroacetate) (0.543–-0.65 g, 1–1.2 mmol) in 5.8 mL of ethanol under nitrogen, and the mixture was stirred at 0° for 30 minutes. The mixture was poured into aqueous sodium bicarbonate solution and extracted with ether. The ether layer was washed with water and brine and dried, and the solution was analyzed by GLC after addition of an internal standard. The yield was 86%. A pure sample of the allylic ether was isolated by GLC, bp 65–67°; ^1H NMR ($CDCl_3$) δ 1.22 (t, J = 7 Hz, 3H), 3.51 (q, J = 7 Hz, 2H), 3.97 (d, J = 6 Hz, 2H), 5.1–5.4 (m, 2H), 5.7–6.2 (m, 1H).

TABULAR SURVEY

We have attempted to cover the literature thoroughly up to the end of 1986, but a few entries from 1987 are also included. References to additional papers published through December 1988 are listed, table by table, in the addenda to the tables.

The tables are arranged, as is the text, in order of the atomic number of the electrophilic atom, and, for carbon electrophiles, in order of increasing oxidation state at the electrophilic carbon. For each type of electrophile there are separate tables for allylsilanes, vinylsilanes, and, when there are any, allenylsilanes. Those compounds that are both allylsilanes and vinylsilanes are listed under allylsilanes. Within each table, allylsilanes and vinylsilanes are cited first in order of the number of connected carbon atoms in the allyl or vinyl portion of the molecule, ignoring groups connected only by heteroatoms. Second, they are ordered by the number of hydrogen atoms attached to those carbon atoms; this is opposite to the usual practice in *Organic Reactions,* but has the effect here of putting the most simple allylsilanes and vinylsilanes first in the table. Third, they are ordered by the atomic number of any heteroatoms attached to the carbon chain; fourth, with linear chains ahead of branched; fifth, with open-chain before cyclic molecules; sixth, by the size of the other groups on the silicon atom; and seventh, with intermolecular reactions ahead of intramolecular reactions. For each allylsilane or vinylsilane, the reactions are listed first, where applicable, with aldehyde and acetal electrophiles ahead of ketone and ketal electrophiles; second, where applicable, by the number of connected carbon atoms in the electrophile; third, by the molecular weight of the electrophile; and fourth, with Lewis acid catalyzed reactions ahead of nucleophile catalyzed reactions. Occasionally, a subtable departs from these priorities in order to save space. Numbers in parentheses are yields, and a dash indicates that no yield is reported. Where isolated yields and yields determined by GLC are reported, we only give the former. Numbers without parentheses are ratios of products.

A high proportion of the papers reporting electrophilic substitution reactions of allylsilanes and vinylsilanes are preliminary publications, with the result that many reactions that are described as taking place, for example, at −78°, are in fact allowed to warm to room temperature before workup. It is not always possible, therefore, to be thorough in identifying the temperature at which reactions actually take place. Similarly, many reactions are carried out by mixing the reagents at low temperatures, and then keeping them at higher temperatures. In cases where the length of time at the lower temperature is not reported, we record only the higher temperature and the time that the mixture is held at this temperature. Also, the solvent, commonly dichloromethane, is often used but not mentioned in the text; the absence of any mention of a solvent in the tables cannot be assumed to mean that there was none. Finally, workup may be more or less severe, in some cases, for example, cleaving silyl groups from oxygen atoms and in some cases not. We largely omit details of workup.

The following abbreviations are used in the tables:

Ac	acetyl
acac	acetylacetonate
AIBN	azobis(isobutyronitrile)
Bn	benzyl
Bz	benzoyl
CSA	camphorsulfonic acid
CSI	chlorosulfonyl isocyanate
DBN	1,5-diazabicyclo[4.3.0]nonane
DBU	1,5-diazabicyclo[5.4.0]undecane
DCC	dicyclohexylcarbodiimide
DDQ	dichlorodicyanobenzoquinone
de	diastereomeric excess
DMAP	4-*N*,*N*-dimethylaminopyridine
DME	1,2-dimethoxyethane
DMF	*N*,*N*-dimethylformamide

DMPU	*N*,*N'*-dimethylpropyleneurea
DMSO	dimethyl sulfoxide
ee	enantiomeric excess
GLC	gas–liquid chromatography
HMPA	hexamethylphosphoramide
MCPBA	*m*-chloroperbenzoic acid
NBS	*N*-bromosuccinimide
NMMO	*N*-methylmorpholine *N*-oxide
NMR	nuclear magnetic resonance
PCC	pyridinium chlorochromate
Py	pyridine
TASF	tris(dimethylamino)sulfonium difluorotrimethylsilicate
TBAF	tetrabutylammonium fluoride
TFA	trifluoroacetic acid
THF	tetrahydrofuran
THP	tetrahydropyranyl
Ts	*p*-toluenesulfonyl

TABLE I. PROTODESILYLATION AND DEUTERODESILYLATION OF ALLYLSILANES

Reactant	Conditions	Product(s) and Yield(s)	Refs.
C_3			
$(CH_3)_3SiCH_2CH{=}CH_2$	HCl, reflux, 24 h	$CH_2{=}CHCH_3$ (15) + $(CH_3)_3SiCH_2CH{=}CH_2$ (48)	5
	1. HBr, −78° 2. > 40°	$CH_2{=}CHCH_3$ (73)	5
	H_2SO_4, −20°	" (48)	5
	1. HI, 0° 2. 80°, 30 min	" (55)	5
	CH_3MgI, H_2O[a]	" (—)	532
	TsOH, ROH, CH_3CN, 70–80°, 1.5–8 h (R = primary or secondary alkyl, aryl, and acyl)	" (—) + $ROSi(CH_3)_3$ (83–94)[b]	192
	CF_3SO_3H, CH_2Cl_2, 1 h	" (—) + $(CH_3)_3SiO_3SCF_3$ (85)[c]	177
	KF, C_2H_5OH, reflux, 6 d	" (—)[d]	533
$C_6H_5CH{=}NCH[Si(C_2H_5)_3]CH{=}CH_2$	CsF, solvent, H_2O, 1 h	$C_6H_5CH{=}NCH{=}CHCH_3$ (70–85)	534
	Solvent — Moles of H_2O	E:Z	
	CH_3CN — 0.06	5:10	
	" — 0.03	4:10	
	" — 0.3	1:1	
	" + 18-crown-6 — 0.06	9:10	
	THF + 18-crown-6 — 0.03	6:1	
	" + " — 0.5	1:1	
	CsF, CH_3CN, D_2O, 1 h	$C_6H_5CH{=}NCH{=}CHCH_2D$ (—)	534
(E)-$(CH_3)_3SiCH_2CH{=}CHSiC_6H_5(CH_3)_2$ or (E)-$C_6H_5(CH_3)_2SiCH_2CH{=}CHSi(CH_3)_3$[e]	Acid, $CDCl_3$, 27°	$(CH_3)_3SiCH_2CH{=}CH_2$ I + $CH_2{=}CHCH_2SiC_6H_5(CH_3)_2$ II	29
	TFA, 5 min ($t_{1/2}$ < 1 min)	I:II 1:3.7 (—)	
	BF_3·2AcOH, 5 min ($t_{1/2}$ < 1 min)	1:4.3 (—)	
	Cl_2CHCO_2H ($t_{1/2}$ 10 min)	1:3.5 (—)	
	$ClCH_2CO_2H$ ($t_{1/2}$ 10.5 h)	1:3.5 (—)	
	CH_3CO_2H, 5 d	No reaction	

(*E*)-$(CH_3)_3SiCH_2CH{=}CHSiC_6H_5(CH_3)_2$	CF_3CO_2D, $CDCl_3$, 5 min	$(CH_3)_3SiCH_2CH{=}CHD$ + $CH_2{=}CHCHDSiC_6H_5(CH_3)_2$ 1:3.7 (—)	29
(*E*)-$C_6H_5(CH_3)_2SiCH_2CH{=}CHSi(CH_3)_3$	", ", "	$(CH_3)_3SiCHDCH{=}CH_2$ + $CDH{=}CHCH_2SiC_6H_5(CH_3)_2$ 1:3.6 (—)	29
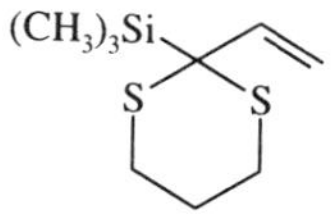	Acid treatment	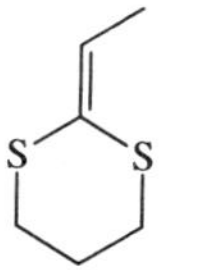 (—)	257
	TBAF, THF, H_2O	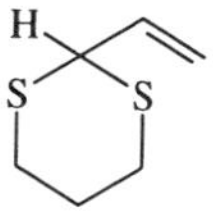 (—)	257
C_4			
(*Z*)-$(CH_3)_3SiCH_2CH{=}CHCH_3$	TFA, 60°, 3 h	$CH_2{=}CHC_2H_5$ (100)	184
	KOC_4H_9-*t*, HMPA, 60°, 3 h	$CH_2{=}CHC_2H_5$ (10) + (*Z*)-$CH_3CH{=}CHCH_3$ (65)	184
(*Z*)-$(CH_3)_3SiCH_2CH{=}CHCH_2Si(CH_3)_3$	TFA, 60°, 3 h	$CH_2{=}CH(CH_2)_2Si(CH_3)_3$ (100)	184
	KOC_4H_9-*t*, HMPA, 60°, 3 h	$CH_2{=}CHC_2H_5$ (10) + (*Z*)-$CH_3CH{=}CHCH_3$ (70)	184
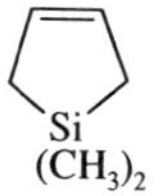	CH_3SO_3H, 70–80°, 13 h	$CH_2{=}CH(CH_2)_2Si(CH_3)_2X$ X = O_3SCH_3 (90)	173
	TsOH, 70–80°, 13 h	X = OTs (90)	173
	$BF_3{\cdot}2AcOH$, Et_2O, 0°, 0.5 h	X = F (90)	173, 171
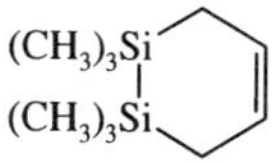	TFA, 60°, 3 h	No reaction	184
	KOC_4H_9-*t*, HMPA, 60°, 3 h	X = $OSi(CH_3)_2(CH_2)_2CH{=}CH_2$ (50)	184
	RCO_2H	$RCO_2Si(CH_3)_2Si(CH_3)_2(CH_2)_2CH{=}CH_2$	535
	R = H, 8–10°, 3 h	(70)	
	R = CH_3, 120–140°, 6 h	(64)	
	R = C_2H_5, 120–140°, 6 h	(60)	

TABLE I. PROTODESILYLATION AND DEUTERODESILYLATION OF ALLYLSILANES (*Continued*)

Reactant	Conditions	Product(s) and Yield(s)	Refs.
C_5			
$(CH_3)_3SiCH_2C(CH_3)=CHCH_3$	TFA, 60°, 3 h	$(CH_3)_2CHCH=CH_2^f$ (100)	184
	KOC_4H_9-*t*, 60°, 3 h	$(CH_3)_2CHCH=CH_2^f$ (5) + $(CH_3)_2C=CHCH_3$ (85)	184
$(CH_3)_3SiCH_2CH=C(CH_3)_2$	CF_3CO_2D, $CDCl_3$, 3.5 h	$CH_2=CHCD(CH_3)_2$ + $(CH_3)_3SiCH_2CHDC(CH_3)_2O_2CCF_3$ 83:17 (—)	36
1,1,3-trimethylsilacyclopent-3-ene [structure: ring Si$(CH_3)_2$]	HCl, CH_3CO_2H, reflux, 6 h	$[(CH_3)_2C=CHCH_2Si(CH_3)_2]_2O$ (76)	173
	$BF_3{\cdot}2AcOH$, $(C_2H_5)_2O$, 0°, 0.5 h	$CH_2=C(CH_3)(CH_2)_2Si(CH_3)_2F$ (90) + $(CH_3)_2C=CHCH_2Si(CH_3)_2F$ (trace) I	171
	$BF_3{\cdot}2AcOH$, CCl_4, 0°, 0.5 h	$CH_2=CHCH(CH_3)CH_2Si(CH_3)_2F$ (60) + I (32)	171
$(CH_3)_3Si$-cyclopentadiene [structure]	TFA, 60°, 3h	cyclopentadiene [structure] (100)	184
	KOC_4H_9-*n*, 60°, 3 h	" (100)	184
C_6			
$(CH_3)_3SiCH_2C(CH_3)=C(CH_3)CH_2Si(CH_3)_3$ *E* + *Z*[g]	CH_3CO_2H, reflux, 24 h	$CH_2=C(CH_3)CH(CH_3)CH_2Si(CH_3)_3$ (—)	536
$CH_3CH=CHCH(Si(CH_3)_3)CH_2CON(CH_3)_2$ [structure]	HF, −20°, 30 min	$CH_3CH_2CH=CHCH_2CON(CH_3)_2$ [structure] *E*:*Z* 94:6 (85)	52

Substrate	Conditions	Product(s) and Yield(s) (%)	Refs.
$(CH_3)_2HSi$ / $(CH_3)_2HSi$ (cyclohexene) + $(CH_3)_2HSi$ / $SiH(CH_3)_2$ (cyclohexene)	HCl, C_6H_6, CH_3CO_2H, reflux, 48 h	$(CH_3)_2HSi$ (cyclohexene) (50–60) + $(CH_3)_2HSi$ (cyclohexene) (—) + cyclohexene (—)	537
$(CH_3)_3Si$ / $Si(CH_3)_3$ (cyclohexene)	TFA, $CHCl_3$	$(CH_3)_3Si$ (cyclohexene) (—)	41
	CF_3CO_2D	$Si(CH_3)_3$ / D (cyclohexene) + $Si(CH_3)_3$ / D (cyclohexene) (—) 70:30	41
$(CH_3)_3Si$ / $(CH_3)_3Si$ (cyclohexene) **I**	TFA, $CHCl_3$	Cyclohexene (—)	41
	CF_3CO_2D	(D) / (D) / (D) (cyclohexene) (—)	41
	" (0.95 eq)	$(CH_3)_3Si$ / (D) / (D) (cyclohexene) + **I** (20) 67:13 (—)	41

TABLE I. PROTODESILYLATION AND DEUTERODESILYLATION OF ALLYLSILANES (*Continued*)

Reactant	Conditions	Product(s) and Yield(s)	Refs.
$(CH_3)_3Si$, $Si(CH_3)_3$, $Fe(CO)_3$ + $(CH_3)_3Si$, $Si(CH_3)_3$, $Fe(CO)_3$ + $(CH_3)_3Si$, $Si(CH_3)_3$, $Fe(CO)_3$ 8:4:1		$Si(CH_3)_3$, $Fe(CO)_3$ I + $Fe(CO)_3$ II	538
	TFA (10 eq), CH_2Cl_2, 10 min[h]	I:II 48:4 (—)	
	", ", 20 min	I (72) + II (15)	
	", ", 40 min[h]	I:II 71:29 (—)	
	", ", 2 d[h]	I:II 0:100 (—)	
	CCl_3CO_2H, CH_2Cl_2, reflux, 70 min	II (35)	
	CF_3CO_2D, CH_2Cl_2, 3 h	$(d_1–d_3)$-I + $(d_1–d_4)$-II 3:1 (—)	
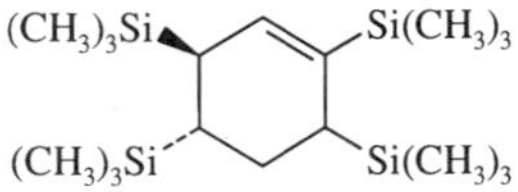 $(CH_3)_3Si$, $Si(CH_3)_3$, $(CH_3)_3Si$, $Si(CH_3)_3$ (mainly *trans*)	CH_3CO_2H, reflux, 16 h	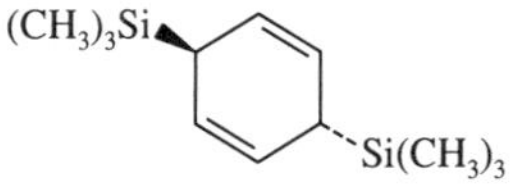 $(CH_3)_3Si$, $Si(CH_3)_3$ (mainly *trans*) (100)	539
C_7			
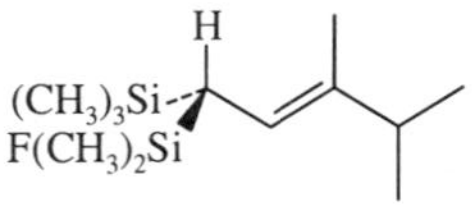 H, $(CH_3)_3Si$, $F(CH_3)_2Si$ 90% ee	TFA, 0°, 4.5 h	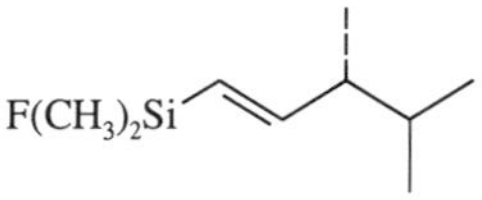 $F(CH_3)_2Si$ *E*:*Z* 97:3 (41) 44% ee	43
	$(CH_3)_3O^+BF_4^-$, CH_3OH, CD_2Cl_2, 2.5 h	" 40% ee (31)	43
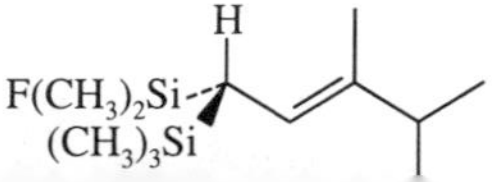 H, $F(CH_3)_2Si$, $(CH_3)_3Si$	CF_3CO_2D, CD_2Cl_2, 0°, 3 h	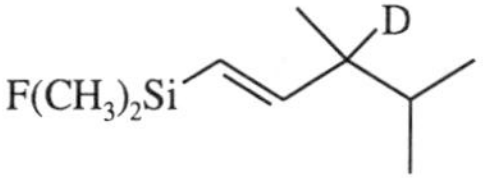 $F(CH_3)_2Si$, D (—)	43

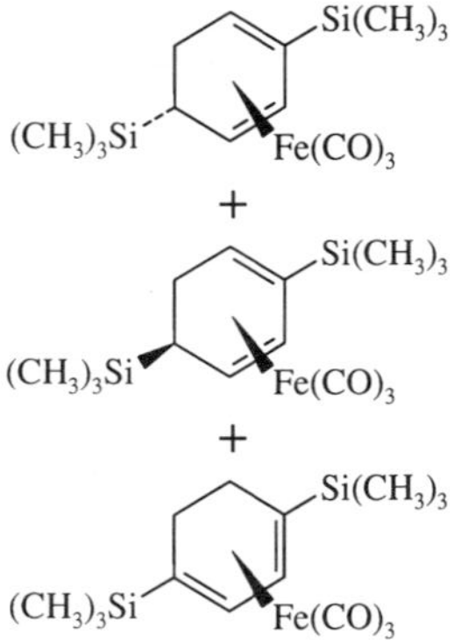

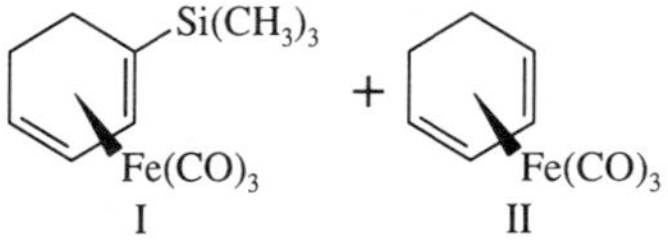

Reactant	Conditions	Product(s) and Yield(s) (%)	Refs.
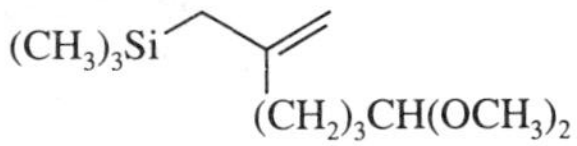 $(CH_3)_3Si$ … $(CH_2)_3CH(OCH_3)_2$	HCl, CH_3OH, 15 min	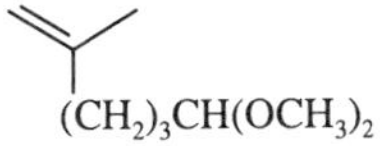 $(CH_2)_3CH(OCH_3)_2$ (76)	363
$C_6H_5(CH_3)_2Si$	TFA, CCl_4, 20 h	I ($R^1 = R^2 = H$) (98)	172
	CF_3CO_2D, CCl_4, 2.5 h	I ($R^1 = H, R^2 = D$) + I ($R^1 = D, R^2 = H$) 52:48 (—)	38
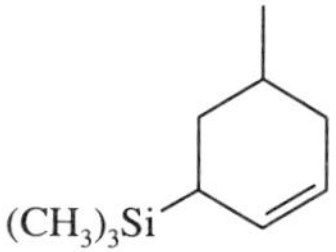 $(CH_3)_3Si$ *cis:trans* 88:12	CF_3CO_2D, $CHCl_3$ or CH_2Cl_2	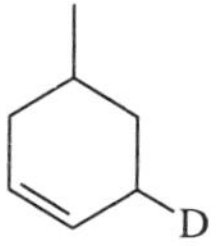 D *cis:trans* 16:84 (—)	42
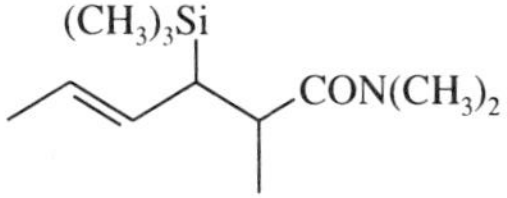 $(CH_3)_3Si$ … $CON(CH_3)_2$ *erythro*	HF, −20°, 30 min	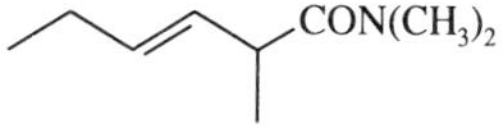 $CON(CH_3)_2$ (88)	52
erythro + *threo*		(99)	
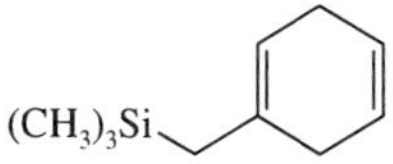 $(CH_3)_3Si$	HCl, THF	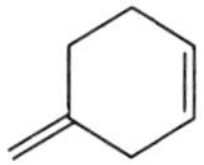 (71)	169
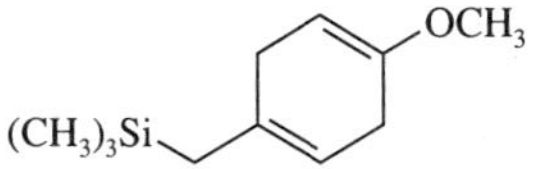 $(CH_3)_3Si$ … OCH_3	HCl, THF, H_2, 20 h	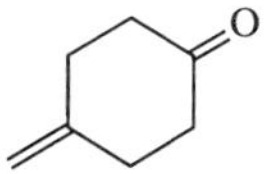 O (80)	169

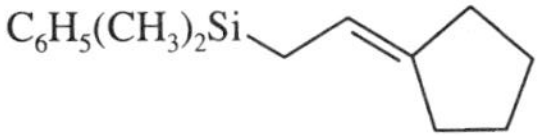

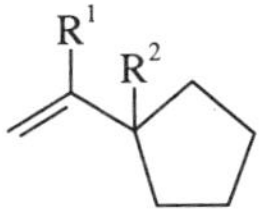

TABLE I. PROTODESILYLATION AND DEUTERODESILYLATION OF ALLYLSILANES (*Continued*)

Reactant	Conditions	Product(s) and Yield(s)	Refs.
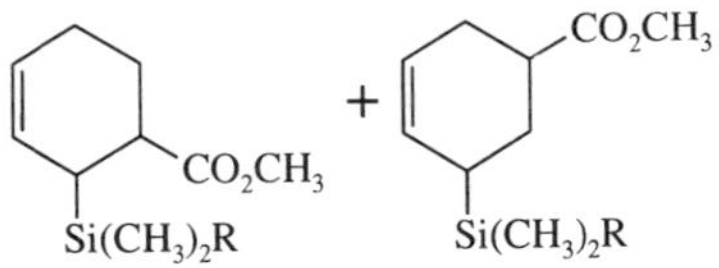 [structure: cyclohexene with CO_2CH_3 and $Si(CH_3)_2R$] + [structure: cyclohexene with CO_2CH_3 and $Si(CH_3)_2R$]	TsOH, C_6H_6, reflux,	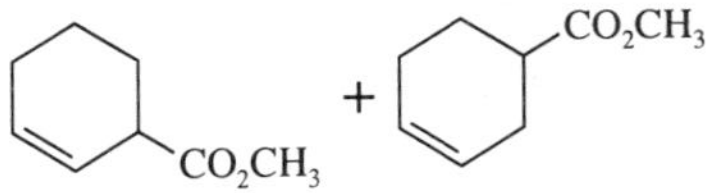 [structure: cyclohexene with CO_2CH_3] + [structure: cyclohexene with CO_2CH_3]	134
R = CH_3	1 h	(87)	
R = $Si(CH_3)_3$	0.5 h	(99)	
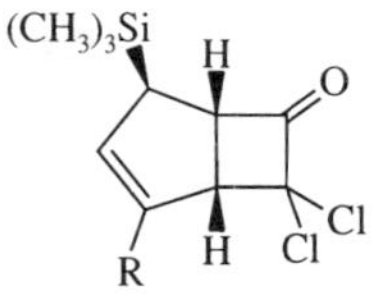 [structure: bicyclic dichlorocyclobutanone with $(CH_3)_3Si$, R]		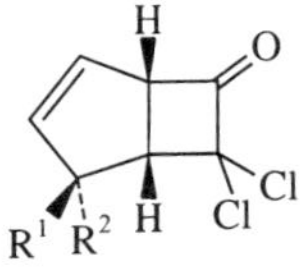 [structure: bicyclic dichlorocyclobutanone with R^1, R^2] R^1 / R^2	
R = H	H_2SO_4, CH_3OH, 26 h	H / H (69)	55
	TsOD, C_6H_{12}, reflux, 4 d	D / H (51)	55
R = $Si(CH_3)_3$	TsOH, C_6H_{12}, reflux, 5 d	$Si(CH_3)_3$ / H (18.5)	540
	TsOD, C_6H_{12}, reflux, 7 d	$Si(CH_3)_3$ / D (10.5)	540
	H_2SO_4, CH_3OH, 24 h	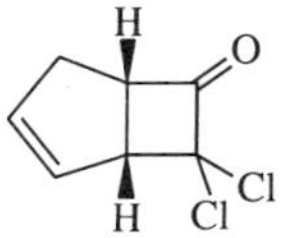 [structure: bicyclic dichlorocyclobutanone] (37)	540
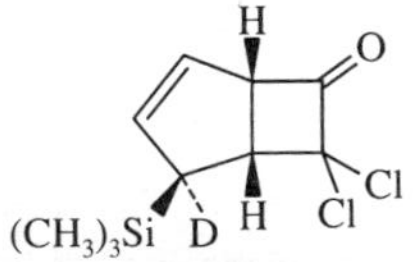 [structure: bicyclic dichlorocyclobutanone with $(CH_3)_3Si$, D]	H_2SO_4, CH_3OH, 24 h	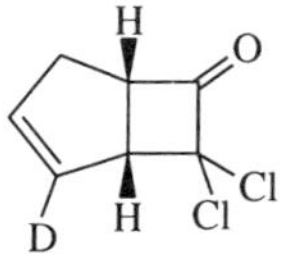 [structure: bicyclic dichlorocyclobutanone with D] (—)	540

C_8

Reactant	Conditions	Product(s) and Yield(s) (%)	Refs.
[structure: allylsilane with $Si(CH_3)_3$, OR^2, R^1]	$BF_3 \cdot 2AcOH$, CH_2Cl_2	[structures: two diastereomeric dienes with OR^2, R^1]	182
R^1 / R^2			
CH_3 / H		66:34 (—)	
n-C_4H_9 / "		75:25 (—)	
t-C_4H_9 / "		87:13 (—)	
$CH_2{=}CHC(CH_3)_2$ / "		90:10 (—)	
CH_3 / $COCH_3$		33:67 (—)	
n-C_4H_9 / "		33:67 (—)	
t-C_4H_9 / "		34:66 (—)	
$CH_2{=}CHC(CH_3)_2$ / "		34:66 (—)	
(Z,E)-$(CH_3)_3SiCH_2CH{=}CH(CH_2)_2$-$CH{=}CHCH_3$	TFA, 60°, 3 h	$CH_2{=}CH(CH_2)_3CH{=}CHCH_3$ $E{:}Z$ 1:1 I (100)	184
	KOC_4H_9-t, HMPA, 60°, 3 h	(Z,E)-$CH_3CH{=}CH(CH_2)_2CH{=}CHCH_3$ (65) + (E)-I (—)	184
(Z)-$(CH_3)_3SiCH_2$-$CH{=}CHCH_2CH(CH_3)CH{=}CH_2$	TFA	$CH_2{=}CH(CH_2)_2CH(CH_3)CH{=}CH_2$ (100)	184
	KOC_4H_9-t, HMPA, 60°, 3 h	(Z)-$CH_3CH{=}CHCH_2CH(CH_3)CH{=}CH_2$ (95) + I (5)	184
[structure: $(CH_3)_3Si$-methylcycloheptene] + [structure] *trans* 70% + *cis* 17%	CF_3CO_2D	[structure: D-methylcycloheptane] + [structure: D-methylcycloheptene] (—) *cis* 70% + *trans* 18% *cis* 9% + *trans* 3%	541

TABLE I. PROTODESILYLATION AND DEUTERODESILYLATION OF ALLYLSILANES (*Continued*)

Reactant	Conditions	Product(s) and Yield(s)	Refs.
2-Methyl-3-(trimethylsilyl)cycloheptene, $(CH_3)_3Si$; *trans* 10% + *cis* 3%			
$R(CH_3)_2SiCH_2CH=C_6H_{10}$ (cyclohexylidene)		1-R-1-vinylcyclohexane	
R = CH_3	$BF_3 \cdot 2AcOH$, $CHCl_3$, 5 min	R = H (99)	170
R = C_6H_5	TFA, CCl_4, 20 h	R = H (95)	172
R = C_6H_5	CF_3CO_2D, CCl_4, 20 h	R = D (—)	38
$[i\text{-}C_3H_7CH{=}CHC(OLi)[Si(CH_3)_3]CH{=}CH_2]^{i}$	$HC{\equiv}CC_3H_7\text{-}n$, THF	$i\text{-}C_3H_7CH{=}CHC(OSi(CH_3)_3){=}CHCH_3$ (81)	260
$C_6H_5(CH_3)_2SiCH(R^1)C(R^2){=}C_5H_8$ (cyclopentylidene)		$R^1CH{=}C(R^2)C_5H_9$ (cyclopentyl)	
R^1 R^2		R^1 R^2	
CH_3 H	TFA, CCl_4, 20 h	CH_3 H (99)	172
H CH_3	", ", "	H CH_3 (92)	172
$(CH_3)_3SiC(CH_3)_2C(=CHCH_3)COCH_3$	$BF_3 \cdot 2AcOH$, $CDCl_3$, 70°, 15 min	$(CH_3)_2C{=}C(C_2H_5)COCH_3$ (77)	542

Reactant	Conditions	Product(s) and Yield(s) (%)	Refs.
$R(CH_3)_2Si$, $R(CH_3)_2Si$ (cyclooctene) I ($R = CH_3$)	CH_3CO_2H, reflux, 24 h	$R(CH_3)_2Si$ (cyclooctene) II ($R = CH_3$) (70)	537
I ($R = H$) + $H(CH_3)_2Si$, $H(CH_3)_2Si$ (cyclooctene)	CH_3CO_2H, reflux, 48 h	II ($R = H$) (50–60) + cyclooctene (—)	537
R^1, R^2, R^3, CO_2H, $Si(CH_3)_3$	$BF_3{\cdot}2AcOH$, CH_2Cl_2, 25°, 5 min	R^1, H, R^2, R^3, CO_2H	
R^1 / R^2 / R^3			
H / CH_3 / CH_3		(85)	181
CH_3 / H / CH_3		(100)	181
H / CH_3 / $CH_2{=}C(CH_3)$		(90)	543
$(CH_3)_3Si$ (cyclopropyl)cyclopentene	TsOH, C_6H_6, reflux, 30 min	TsO (cyclopentylidene) (87)	544
$(CH_3)_3Si$, R^1, R^2, R^3		R^1, R^2, R^3	169

TABLE I. PROTODESILYLATION AND DEUTERODESILYLATION OF ALLYLSILANES (*Continued*)

Reactant	Conditions	Product(s) and Yield(s)	Refs.
R^1 R^2 R^3			
CH_3 H H	HCl, THF, CH_3OH, 20 h	(95)	
H H CO_2CH_3	", ", ", 24 h	(84)	
H CO_2CH_3 CO_2CH_3	HI, C_6H_6, 24 h	(75)	
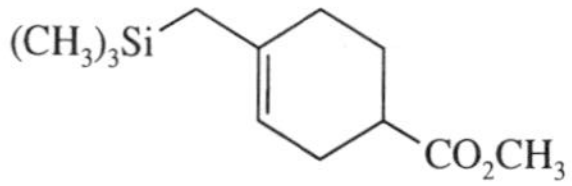 $(CH_3)_3Si$ CO_2CH_3	CH_3COCl, CH_3OH, 0°, 5 min	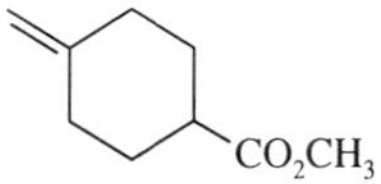 CO_2CH_3 (88)	174
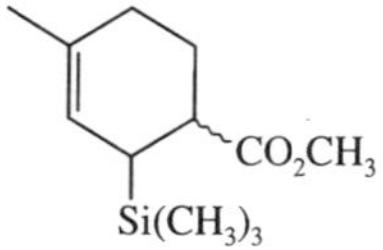 CO_2CH_3 $Si(CH_3)_3$	TsOH, C_6H_6, reflux, 3 h	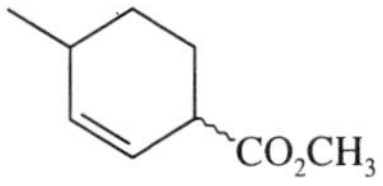 CO_2CH_3 (86)	134
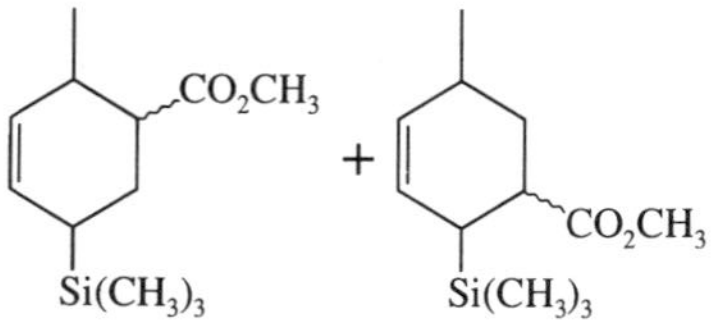 CO_2CH_3 $Si(CH_3)_3$ + CO_2CH_3 $Si(CH_3)_3$	TsOH, C_6H_6, reflux, 2 h	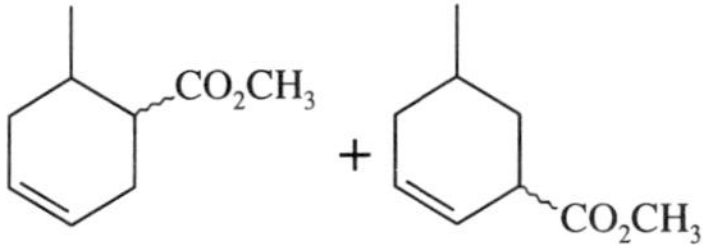 CO_2CH_3 + CO_2CH_3 (97)	134
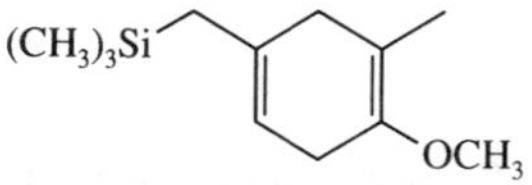 $(CH_3)_3Si$ OCH_3	HCl, THF, CH_3OH, H_2O, 30 h	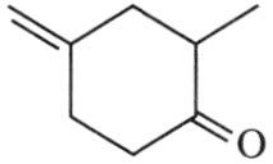 O (60)	545

Reactant	Conditions	Product(s) and Yield(s) (%)	Refs.
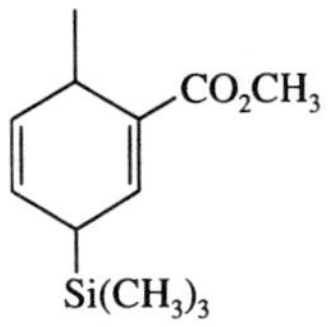 CO_2CH_3, $Si(CH_3)_3$ I	TsOH, C_6H_6, reflux, 1 h	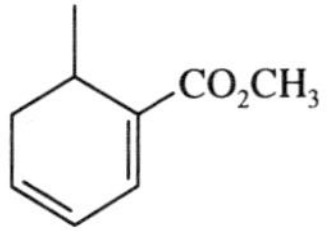 CO_2CH_3 (95)	134
I 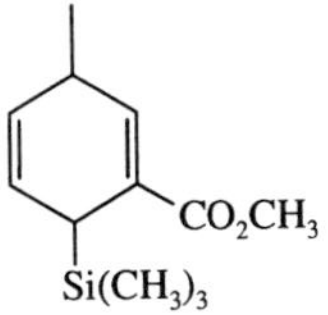 + CO_2CH_3, $Si(CH_3)_3$	1. TsOH, C_6H_6, reflux, 1.5 h 2. DDQ	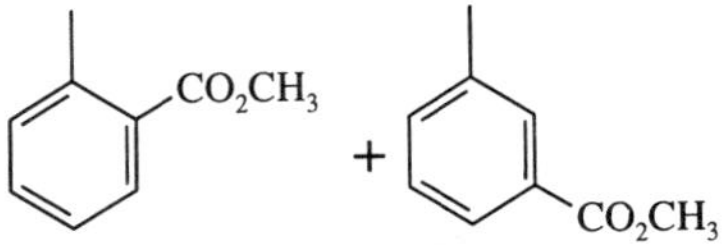 CO_2CH_3 + CO_2CH_3 (94)	134
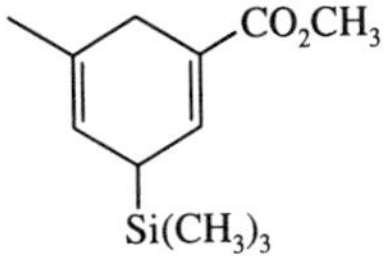 CO_2CH_3, $Si(CH_3)_3$	TsOH, C_6H_6, reflux, 1 h	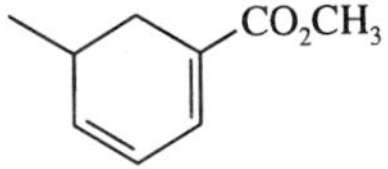 CO_2CH_3 (79)	134
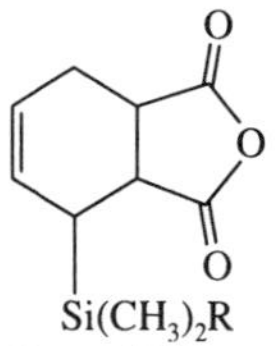 O, O, O, $Si(CH_3)_2R$ R = CH_3 R = $Si(CH_3)_3$	TsOH, C_6H_6, reflux, 2 h	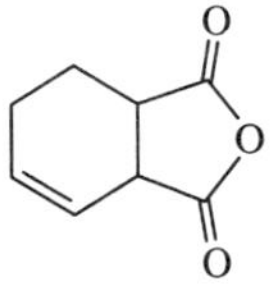 O, O, O (76) (90)	134
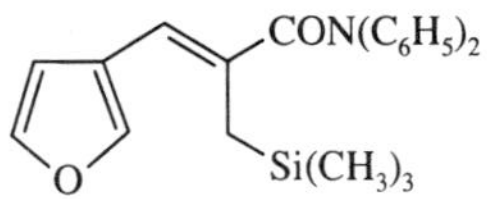 O, $CON(C_6H_5)_2$, $Si(CH_3)_3$	HCl, CH_3OH, reflux, 2 h	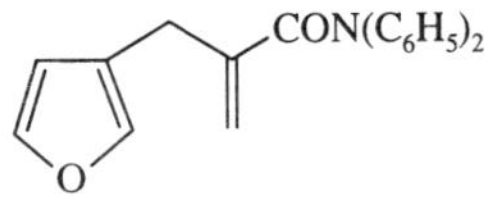 O, $CON(C_6H_5)_2$ (91)	179

TABLE I. PROTODESILYLATION AND DEUTERODESILYLATION OF ALLYLSILANES (*Continued*)

Reactant	Conditions	Product(s) and Yield(s)	Refs.
C_9			
(E)-$(CH_3)_3SiCH_2CH{=}CHC_6H_{13}$-$n$	$BF_3 \cdot 2AcOH$, $CHCl_3$, 5 min	$CH_2{=}CHC_7H_{15}$-n (97)	170
$(CH_3)_3SiCH_2CH{=}C{=}CHC_5H_{11}$-$n$	$BF_3 \cdot 2AcOH$, CH_2Cl_2, −40°, 30 min	(Z)-$CH_2{=}CHCH{=}CHC_5H_{11}$-$n$ (85)	546
$(CH_3)_3SiCH_2C(C_5H_{11}$-$n){=}C{=}CH_2$	$BF_3 \cdot 2AcOH$, CH_2Cl_2, −78°, 30 min	$CH_2{=}C(C_5H_{11}$-$n)CH{=}CH_2$ (80)	546
$(CH_3)_3Si$, R^1, R^2		R^1, R^2	547
R^1 R^2		R^1 R^2	
H CH_3	$BF_3 \cdot 2AcOH$ (0.5 eq), −25°	H CH_3 (85)	
CH_3 $Si(CH_3)_3$	", "	CH_3 $Si(CH_3)_3$ (90)	
" "	$BF_3 \cdot 2AcOH$ (1 eq)	H CH_3 (85)	
$(CH_3)_3Si$, $COCH_3$	HCl, CH_3OH, 15 min	$COCH_3$ I (96)	185
	KF, DMSO, 120°, 12 h	I (8) + $COCH_3$ II (52)	186, 185
	KF, DMF, 120°, 25 h	I (7) + II (41)	186
	CsF, DMSO, 140°, 0.2 h	I (8) + II (47)	186
	", ", 100°, 0.5 h	I (7) + II (44)	186
	", ", 70°, 23 h	I (9) + II (46)	186

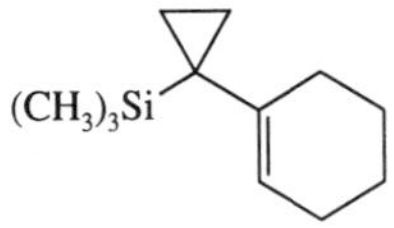 $(CH_3)_3Si$ (cyclopropyl–cyclohexenyl)	TsOH, C_6H_6, reflux, 30 min	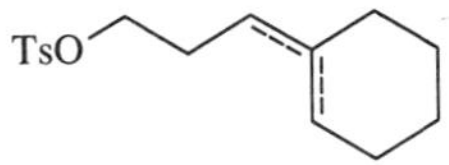 TsO– (—)	544
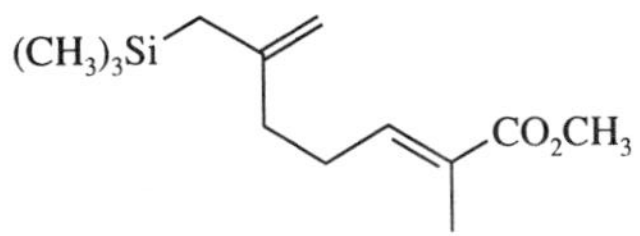 $(CH_3)_3Si$–…CO_2CH_3	$BF_3 \cdot O(C_2H_5)_2$	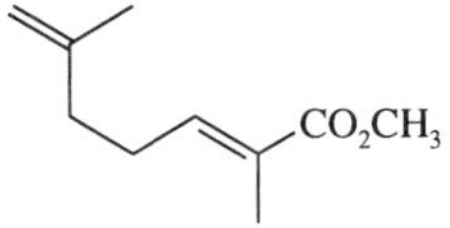 CO_2CH_3 (—)	168
	TBAF	(—)	168
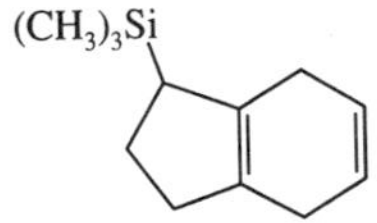 $(CH_3)_3Si$ (tetrahydroindanyl)	HCl, THF	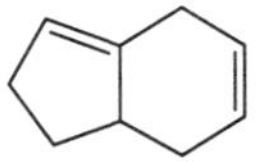 (66) + other isomers (33)	169
(*E*)-$R_3SiCH_2CH{=}CHC_6H_5$	$NaOCH_3$, CH_3OH, 50°, 20 h	(*E*)-$CH_3CH{=}CHC_6H_5$ + $CH_2{=}CHCH_2C_6H_5$	
R = CH_3		(60–65) (40–35)	183, 88
R = C_2H_5		(65–69) (35–30) + (*Z*)-isomer (0.4)	183
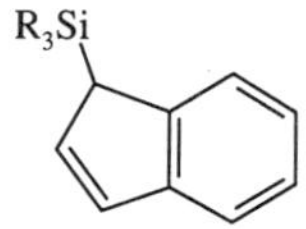 R_3Si (indenyl)	NaOH, H_2O, 100°, 12 h		
R = CH_3		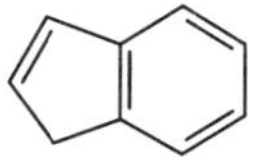 (55)	548, 87
R = C_2H_5		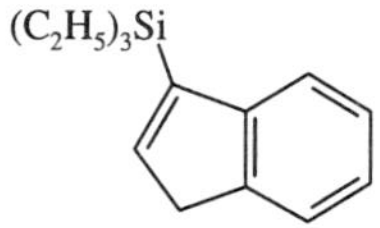 $(C_2H_5)_3Si$ (75)	548

TABLE I. PROTODESILYLATION AND DEUTERODESILYLATION OF ALLYLSILANES (*Continued*)

Reactant	Conditions	Product(s) and Yield(s)	Refs.
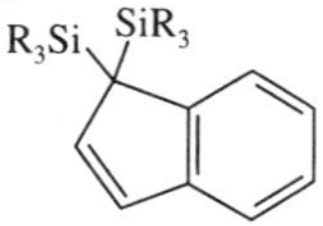 1,1-bis(R_3Si)indene	C_2H_5OH, 6 h	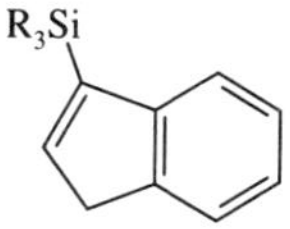 3-(R_3Si)indene	548
R = CH_3		(20)	
R = C_2H_5		(7)	
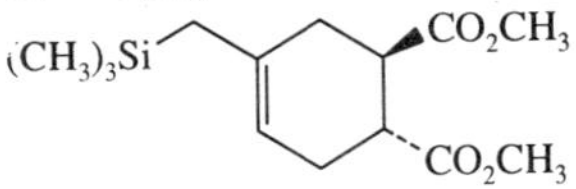 $(CH_3)_3SiCH_2$-cyclohexene with CO_2CH_3, CO_2CH_3	TsOH, C_6H_6, reflux, 2 h	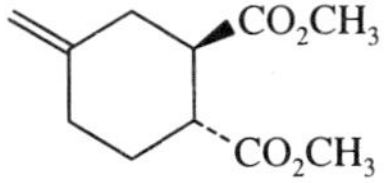 methylenecyclohexane with CO_2CH_3, CO_2CH_3 (100)	549
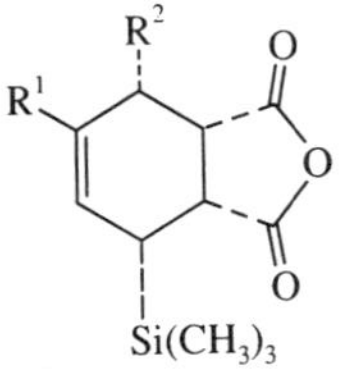 R^1, R^2, $Si(CH_3)_3$-substituted tetrahydrophthalic anhydride	TsOH, C_6H_6, reflux, 2 h	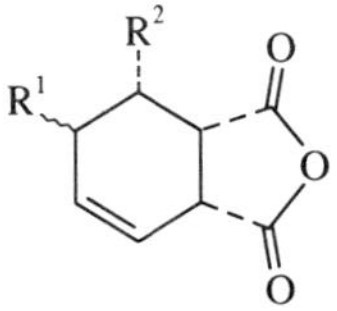 R^1, R^2-substituted tetrahydrophthalic anhydride	134
R^1 R^2			
H CH_3		(84)	
CH_3 H		(65)	
$[(CH_3)_3SiCH_2CH{=}CHC_6H_4R\text{-}p]^j$	HI, C_6H_6	$CH_2{=}CHCH_2C_6H_4R\text{-}p$	222
R = H		(79)	
OCH_3		(68)	
Cl		(86)	
CH_3		(93)	
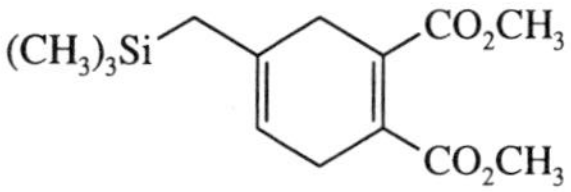 $(CH_3)_3SiCH_2$-cyclohexadiene with CO_2CH_3, CO_2CH_3	TsOH	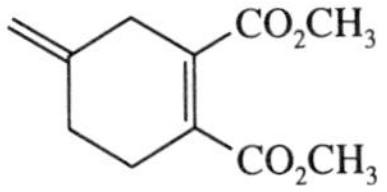 methylenecyclohexene with CO_2CH_3, CO_2CH_3 (98)	549

Reactant	Conditions	Product(s) and Yield(s) (%)	Refs.
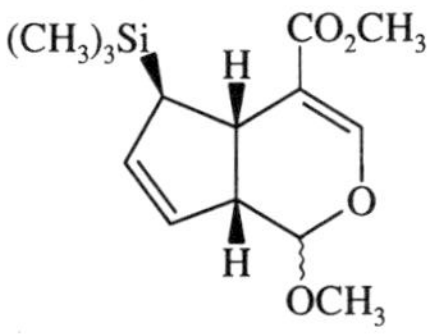 (structure: $(CH_3)_3Si$, H, CO_2CH_3, H, OCH_3)		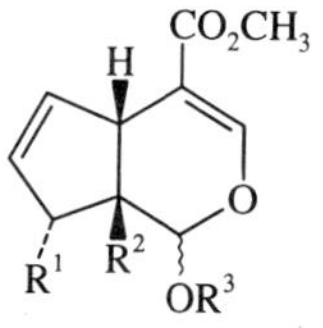 (structure: H, CO_2CH_3, R^1, R^2, OR^3)	54
		R^1 / R^2 / R^3	
	TsOH, C_6H_6, 80°, 1.5 h	H / H / CH_3 (60)	
	TsOD, ", ", "	D / " / " (—)	
	", ", CD_3OD, 80°, 1.5 h	" / H / CD_3 (—)	
	CF_3CO_2D, CH_2Cl_2, 0–25°, 16 h	" / D / CH_3 (—)	
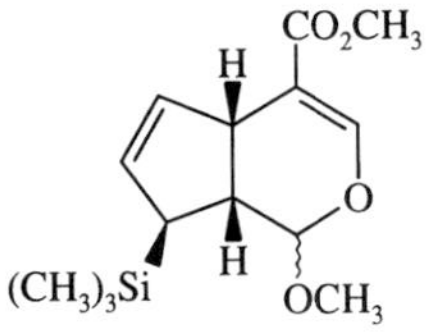 (structure: CO_2CH_3, H, H, $(CH_3)_3Si$, OCH_3)		"	54
		R^1 / R^2 / R^3	
	TsOH, C_6H_6, 80°, 1.5 h	H / H / CH_3 (65)	
	TsOD, ", ", "	D / " / " (—)	
	", ", CD_3OD, 80°, 1.5 h	" / H / CD_3 (—)	
	CF_3CO_2D, CH_2Cl_2, 0–25°, 16 h	" / D / CH_3 (—)	
C_{10}			
$R(CH_3)_2SiCH_2CH{=}C(CH_3)(CH_2)_3C_3H_7$-$i$[k]	$BF_3 \cdot 2AcOH$, 0°, 0.5 h	$CH_2{=}CHCH(CH_3)(CH_2)_3C_3H_7$-$i$	550
R = CH_3		(85)	
R = i-C_3H_7		(80)	
$(CH_3)_3SiCH_2CH{=}C(CH_3)C_6H_{13}$-$n$	$BF_3 \cdot 2AcOH$, $CHCl_3$, 5 min	$CH_2{=}CHCH(CH_3)C_6H_{13}$-$n$ (93)	170
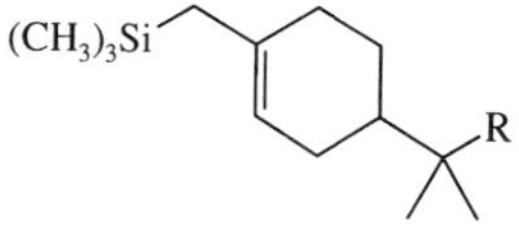 (structure: $(CH_3)_3Si$, R)		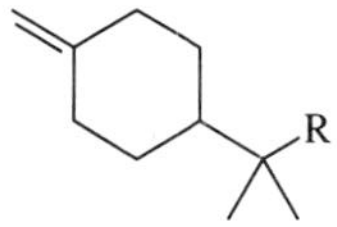 (structure: R)	

TABLE I. PROTODESILYLATION AND DEUTERODESILYLATION OF ALLYLSILANES (*Continued*)

Reactant	Conditions	Product(s) and Yield(s)	Refs.
R = H	CH_3CO_2H, reflux (rapid) room temp (slow)	(100)	11
R = OH	CH_3COCl, CH_3OH, 0°, 15 min	(92)	174
$(CH_3)_3SiCH_2C(CH_3){=}CH(CH_2)_3$- $C(CH_3){=}CH_2$	TFA, $(C_2H_5)_2O$, 35°	$CH_2{=}C(CH_3)(CH_2)_4C(CH_3){=}CH_2$ (90) + I $(CH_3)_2C{=}CH(CH_2)_3C(CH_3){=}CH_2$ (10) II	184
	KOC_4H_9-*t*, HMPA, 60°, 3 h	I (10) + II (80)	184
$(CH_3)_3SiCH_2C(C_5H_{11}$-*n*$){=}C{=}CHCH_3$	$BF_3{\cdot}2AcOH$, CH_2Cl_2, −78°, 30 min	$CH_2{=}C(C_5H_{11}$-*n*$)CH{=}CHCH_3$ *E*:*Z* 76:24 (100)	546
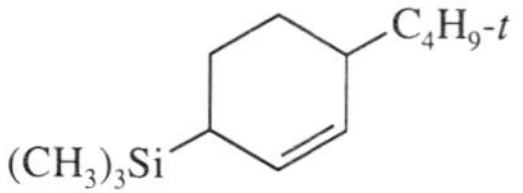 3-(trimethylsilyl)-5-*tert*-butylcyclohexene, $(CH_3)_3Si$, C_4H_9-*t* *cis*:*trans* 54:46	CF_3CO_2D, $CHCl_3$	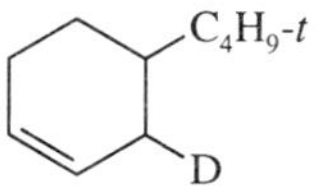 3-deuterio-4-*tert*-butylcyclohexene, D, C_4H_9-*t* (—) *cis*:*trans* 22:78	551, 41
$[(CH_3)_3Si]_2CHCH{=}C(CH_3)$- $(CH_2)_2CH{=}C(CH_3)_2$	$BF_3{\cdot}2AcOH$	$CH_2{=}CHCH(CH_3)(CH_2)_2CH{=}C(CH_3)_2$ (80)	547
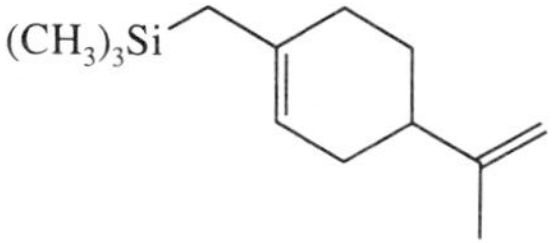 1-[(trimethylsilyl)methyl]-4-isopropenylcyclohexene, $(CH_3)_3Si$	HCl, CH_3OH, 15 min	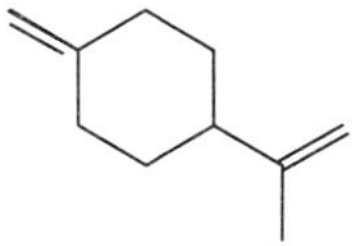 1-methylene-4-isopropenylcyclohexane (78)	185
	CsF, DMSO, 100°, 30 min	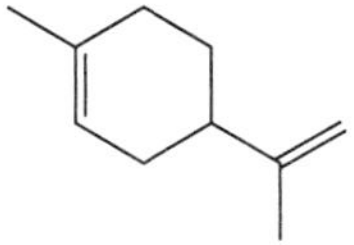 1-methyl-4-isopropenylcyclohexene (46)	185

Reactant	Conditions	Product(s) and Yield(s) (%)	Refs.
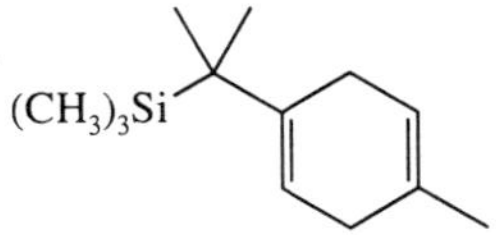 $(CH_3)_3Si$ [structure]	HCl, THF, CH_3OH, 20 h	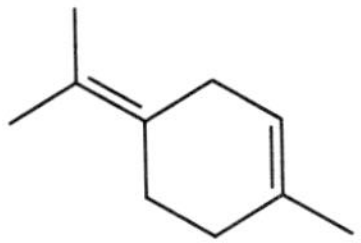 [structure] (80)	169
$(CH_3)_3SiCH_2C(=CHCO_2H)(CH_2)_2CH{=}C(CH_3)_2$	$NaOCH_3$, CH_3OH, reflux	$CH_2{=}C(CH_2CO_2H)(CH_2)_2CH{=}C(CH_3)_2$ (—)	283
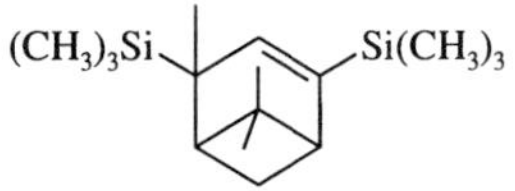 $(CH_3)_3Si$ [structure] $Si(CH_3)_3$	$BF_3{\cdot}2AcOH$ (0.5 eq)	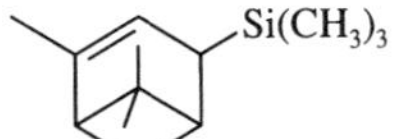 [structure] $Si(CH_3)_3$ (85)	547
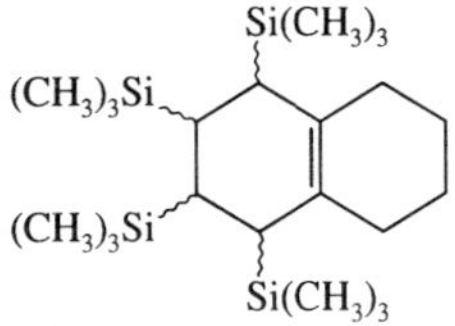 $Si(CH_3)_3$, $(CH_3)_3Si$, $(CH_3)_3Si$, $Si(CH_3)_3$ [structure]	CH_3CO_2H, reflux, 48 h	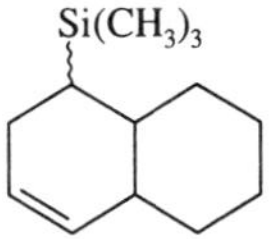 $Si(CH_3)_3$ [structure] (65)	133
$(CH_3)_3SiCH_2CH{=}C(CH_3)C_6H_5$	CF_3CO_2D, $BF_3{\cdot}O(C_2H_5)_2$, CCl_4, 2 min	$CH_2{=}CDCH(CH_3)C_6H_5$ (45) + $CH_2{=}CHCD(CH_3)C_6H_5$ (30)	38
C_{11}			
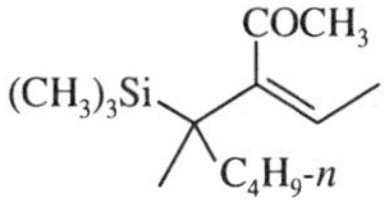 $(CH_3)_3Si$, $COCH_3$, C_4H_9-*n* [structure]	$BF_3{\cdot}2AcOH$, $CDCl_3$, 70°, 15 min	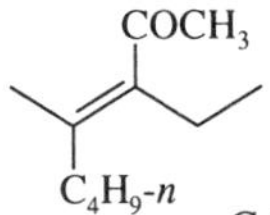 $COCH_3$, C_4H_9-*n* [structure] (98)	542
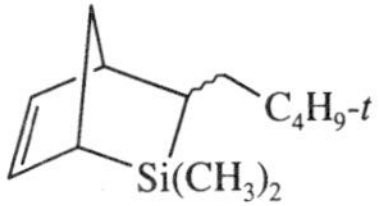 [structure] C_4H_9-*t*, $Si(CH_3)_2$		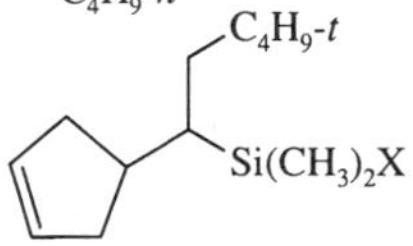 C_4H_9-*t*, $Si(CH_3)_2X$ [structure]	180
	HCl, H_2O, 2 h	X = OH (95)	
	", CH_3OH	X = OCH_3 (72)	
	1. H_2SO_4, hexane 2. NH_4F	X = F (80)	

TABLE I. PROTODESILYLATION AND DEUTERODESILYLATION OF ALLYLSILANES (*Continued*)

Reactant	Conditions	Product(s) and Yield(s)	Refs.
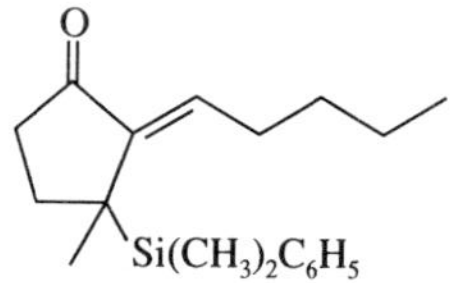 $Si(CH_3)_2C_6H_5$	$BF_3 \cdot 2AcOH$, CCl_4, reflux, 1 h	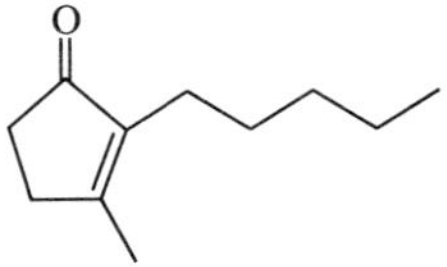 (42)	178
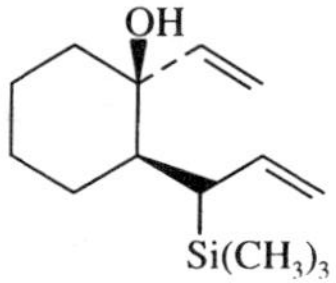 OH, $Si(CH_3)_3$	KH (0.1 eq), THF	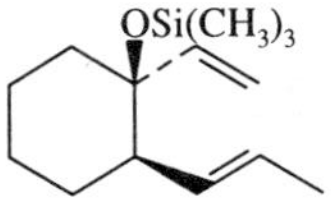 $OSi(CH_3)_3$ (89)	261
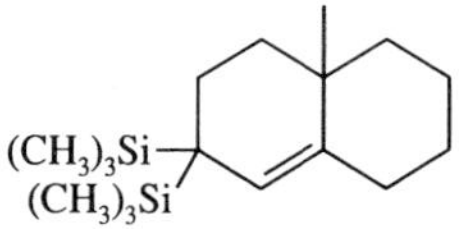 $(CH_3)_3Si$, $(CH_3)_3Si$	$BF_3 \cdot 2AcOH$ (0.5 eq), −25°	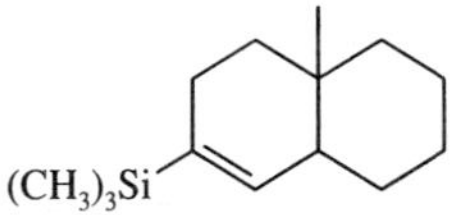 $(CH_3)_3Si$ (95)	547
$C_6H_5(CH_3)_2SiCH_2CH{=}CHCH(CH_3)C_6H_5$	$BF_3 \cdot 2AcOH$, CH_2Cl_2, 0°, 4 h	$CH_2{=}CHCH_2CH(CH_3)C_6H_5$ (75)	409
	$BF_3 \cdot 2AcOD$, CH_2Cl_2, ", "	$CH_2{=}CHCHDCH(CH_3)C_6H_5$ (71) 98% d_1, diastereoisomers 52:48	409
$[C_6H_5(CH_2)_2C(OLi)[Si(CH_3)_3]CH{=}CH_2]^{l}$	$HC{\equiv}CC_3H_7$-*n*, THF	$C_6H_5(CH_2)_2C(OSi(CH_3)_3){=}CHCH_3$ (79)	260
C_{12}			
$C_6H_5(CH_3)_2SiCH_2C(CH_3){=}CHCH$- $(CH_3)C_6H_5$	$BF_3 \cdot 2AcOH$, CH_2Cl_2, 0°, 4 h	$CH_2{=}C(CH_3)CH_2CH(CH_3)C_6H_5$ (85)	409
	$BF_3 \cdot 2AcOD$, CH_2Cl_2, 0°, 4 h	$CH_2{=}C(CH_3)CHDCH(CH_3)C_6H_5$ (92) 94% d_1, diastereoisomers 80:20	409

Reactant	Conditions	Product(s) (yield)	Refs.
$C_6H_5(CH_3)_2SiCH_2CH{=}C(CH_3)$- $CH(CH_3)C_6H_5$	$BF_3 \cdot 2AcOH$, CH_2Cl_2, 0°, 4 h	C_6H_5 + C_6H_5 56:44 (75)	409
$(CH_3)_3Si$, C_6H_5	$AlCl_3$ or $TiCl_4$	C_6H_5 + C_6H_5 (—)	408
$(CH_3)_3Si$, $Si(CH_3)_3$	CCl_4,[m] 72 h	$(CH_3)_3Si$, R R = H (15) + R = $Si(CH_3)_3$ (70) + 1,2-bis(trimethylsilyl)acenaphthene (5) and 1,5-bis(trimethylsilyl)acenaphthylene (6)	552
C_{13}			
H, $Si(CH_3)_3$, $(CH_2)_4CO_2C_2H_5$, H	PyH^+ $CF_3SO_3^-$, CH_2Cl_2, 0°, 2 h	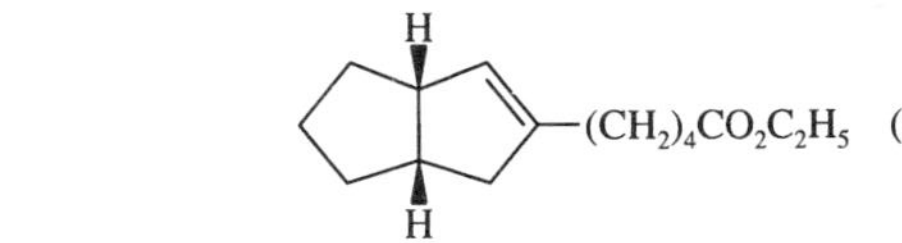 H, $(CH_2)_4CO_2C_2H_5$ (87), H	176
$C_6H_5(CH_3)_2SiCH_2C(CH_3){=}C(CH_3)$- $CH(CH_3)C_6H_5$	$BF_3 \cdot 2AcOH$, CH_2Cl_2, 0°, 4 h	C_6H_5 + C_6H_5 60:40 (86)	409
(E)-$C_6H_5(CH_2)_2CH{=}CHCH(Si(CH_3)_3)CH_2CON(CH_3)_2$	HF, −20°, 30 min	$C_6H_5(CH_2)_3CH{=}CHCH_2CON(CH_3)_2$ E:Z 88:12 (76)	52
	HF, Py, 4 h	" E:Z 83:17 (95)	52

TABLE I. PROTODESILYLATION AND DEUTERODESILYLATION OF ALLYLSILANES (*Continued*)

Reactant	Conditions	Product(s) and Yield(s)	Refs.
C_{14}			
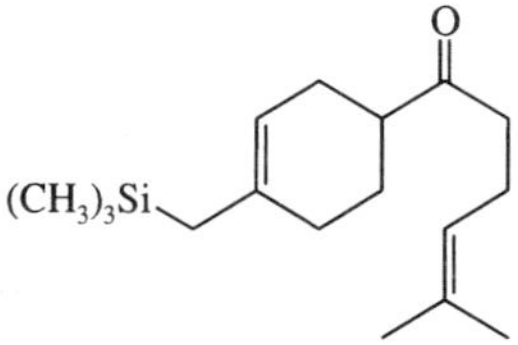 $(CH_3)_3Si$ (structure)		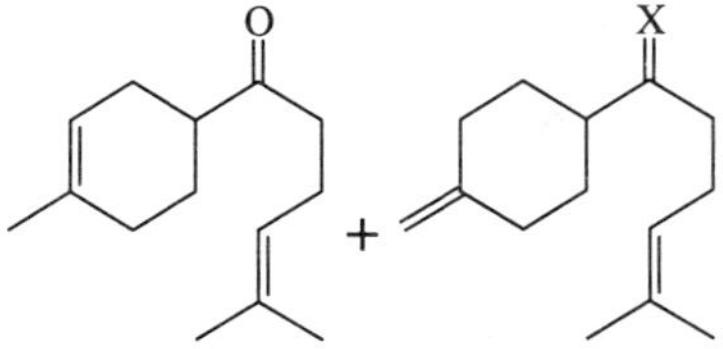 I + II (structures)	
	CsF, DMSO, 100°, 1 h	I (79) + II X = O (—)	186, 185
	1. $(CH_3)_3SiCH_2MgCl$, $(C_2H_5)_2O$, reflux, 1 h 2. HCl, CH_3OH, 15 h	II X = CH_2 (84)	185
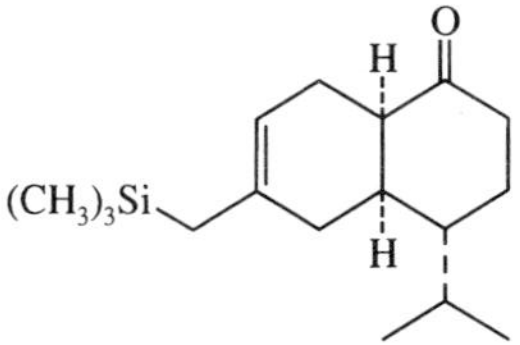 $(CH_3)_3Si$ (structure)		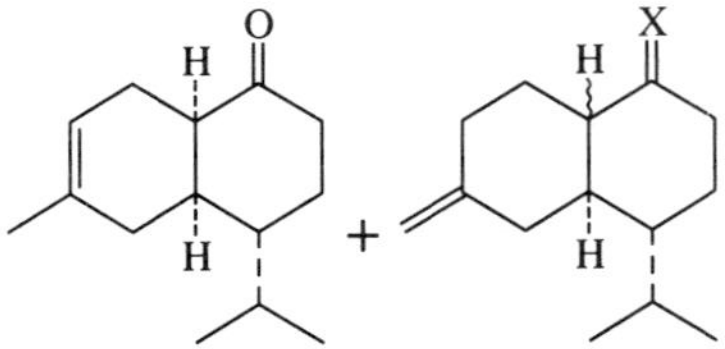 I + II (structures)	
	CsF, DMSO, 100°, 1 h	I (52) + II X = O, $H_{10\alpha}$ (8)	186, 185
	1. $(CH_3)_3SiCH_2MgCl$, $(C_2H_5)_2O$, reflux, 1 h 2. HCl, CH_3OH, 15 h	II X = CH_2, $H_{10\alpha}$ (25) + II X = CH_2, $H_{10\beta}$ (38)	185
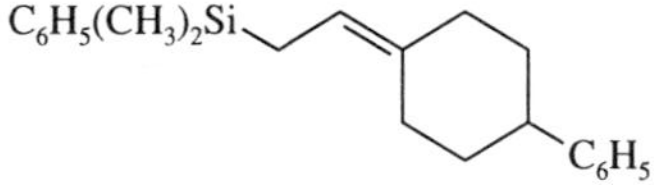 $C_6H_5(CH_3)_2Si$ (structure), C_6H_5	$BF_3 \cdot 2AcOH$, CH_2Cl_2, 0°, 20 min; 20°, 40 min	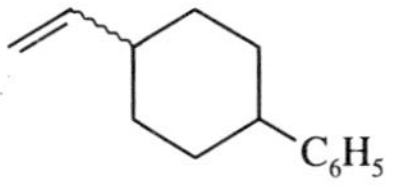 (structure) C_6H_5 *trans* (74) + *cis* (15)	44

Substrate	Conditions	Product(s) and Yield(s) (%)	Refs.
C_{15}			
(±)-$C_6H_5(CH_3)_2Si$ (allylsilane, 4-phenylcyclohexylidene)	$BF_3 \cdot 2AcOH$, CH_2Cl_2, 0°, 20 min; 20°, 40 min	I $R^1 = R^2 = H$ (92)	44
	CF_3CO_2D, CH_2Cl_2, 20°, 2.5 h	I $R^1 = H$, $R^2 = D$ (78)	44
(±)-$C_6H_5(CH_3)_2Si$ (allylsilane, 4-phenylcyclohexylidene, diastereomer)	$BF_3 \cdot 2AcOH$, CH_2Cl_2, 0°, 20 min; 20°, 40 min	II $R^1 = R^2 = H$ (36) + I $R^1 = R^2 = H$ (53)	44
	CF_3CO_2D, CH_2Cl_2, 20°, 2.5 h	I $R^1 = H$, $R^2 = D$ $(21)^n$ + I $R^1 = D$, $R^2 = H$ $(21)^n$ + II $R^1 = H$, $R^2 = D$ (42)	44
$C_6H_5CH(Si(CH_3)_3)CH=CHC_6H_5$ ≃ 94% ee	CF_3CO_2D, 20 h	$C_6H_5CH=CHCHDC_6H_5$ *E*:*Z* 85:15 (77) (*E*)-isomer ≃ 96% ee	40
C_{17}			
$C_6H_5(CH_3)_2SiCH(C_6H_5)CH=C(CH_3)CH_2R$ and $C_6H_5(CH_3)_2SiCH(C_6H_5)CH=C(CH_3)CH_2R$ (stereoisomers)	$BF_3 \cdot 2AcOH$, CH_2Cl_2, 5°, 25 min	$C_6H_5CH=CHCH(CH_3)CH_2R$ I + $C_6H_5CH=CHCH(CH_3)CH_2R$ II; I:II 83:17 (86)	190

TABLE I. PROTODESILYLATION AND DEUTERODESILYLATION OF ALLYLSILANES (*Continued*)

Reactant	Conditions	Product(s) and Yield(s)	Refs.
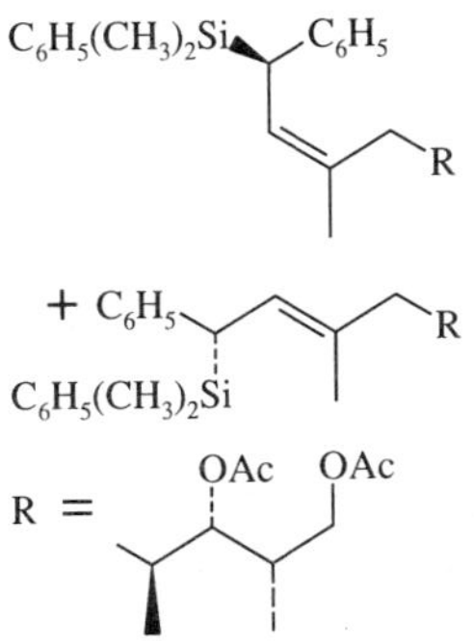 $C_6H_5(CH_3)_2Si$, C_6H_5 (structure, R) + C_6H_5, $C_6H_5(CH_3)_2Si$ (structure, R); R = (structure, OAc, OAc)	$BF_3 \cdot 2AcOH$, CH_2Cl_2, 5°, 25 min	I + II I:II 20:80 (85)	190
C_{21}			
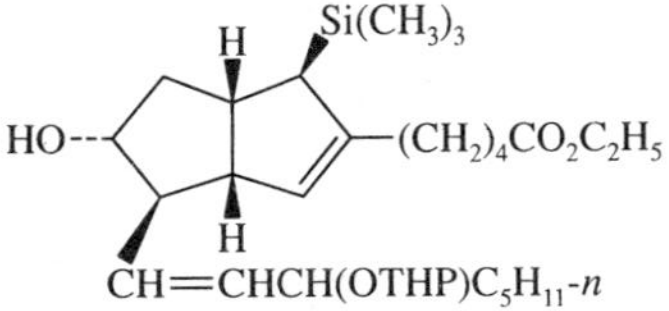 HO (bicyclic structure, H, $Si(CH_3)_3$, H) $-(CH_2)_4CO_2C_2H_5$, $CH{=}CHCH(OTHP)C_5H_{11}$-*n*	PyH^+ $CF_3SO_3^-$, CH_2Cl_2, 0°, 2 h	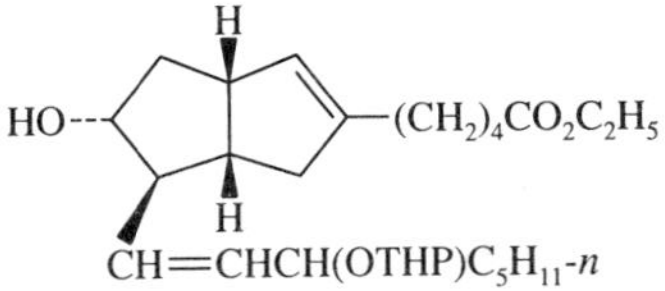 HO (bicyclic structure, H, H) $-(CH_2)_4CO_2C_2H_5$ (49), $CH{=}CHCH(OTHP)C_5H_{11}$-*n*	176

C_{26}

Reactant	Conditions	Product(s) (Yield, %)	Refs.
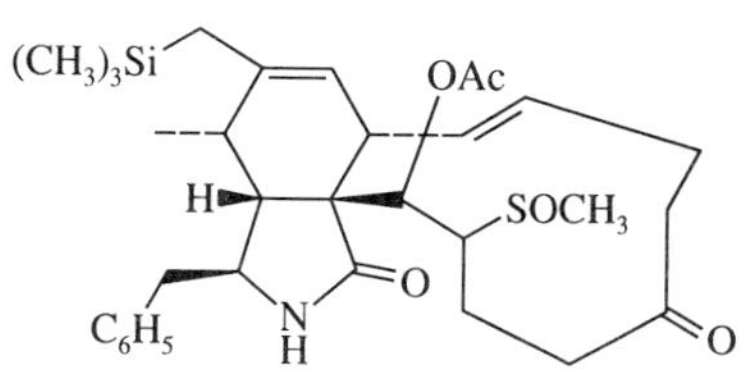 (structure: $(CH_3)_3Si$, OAc, $SOCH_3$, C_6H_5, NH, O)	1. CsF, DMSO, 20°, 1.5 h 2. Xylene, $CaCO_3$, 140°	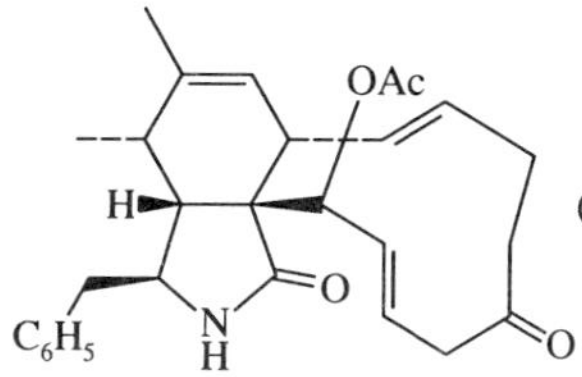 (structure: OAc, C_6H_5, NH, O) (57) 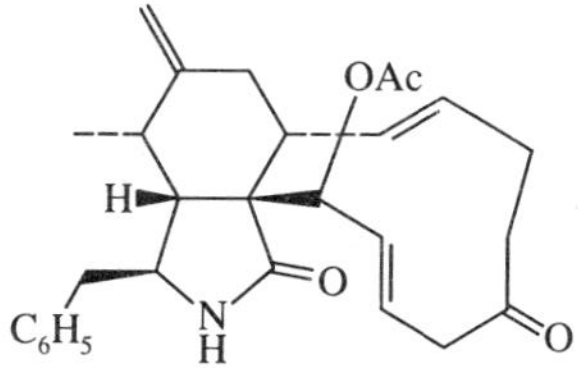 + (structure: OAc, C_6H_5, NH, O) (6) I	187
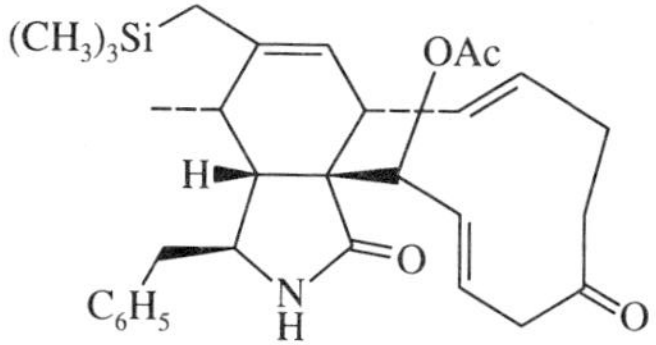 (structure: $(CH_3)_3Si$, OAc, C_6H_5, NH, O)	$BF_3·2AcOH$, CH_2Cl_2, 0°	I (80)	187

TABLE I. PROTODESILYLATION AND DEUTERODESILYLATION OF ALLYLSILANES (*Continued*)

Reactant	Conditions	Product(s) and Yield(s)	Refs.
C_{27}			
THPO, HO, $(CH_3)_3Si$ SC_6H_5	KF, 18-crown-6 (cat.), CH_3OH, reflux, 13 h	HO, SC_6H_5 (85)	188

[a] These are the conditions of workup of a reaction of $Cl_3SiCH_2CH{=}CH_2$ with CH_3MgI.

[b] This reaction is principally a synthesis of the silyl ethers, see also refs. 177 and 193–195.

[c] This reaction is principally a synthesis of trimethylsilyl trifluoromethanesulfonate.

[d] No product is reported but the starting material is described as "35% cleaved."

[e] Both isomers give the same results.

[f] There are two possible explanations for this result, both typographical. The more probable is that the starting material was $(CH_3)_3SiCH_2CH{=}C(CH_3)_2$, not as stated. The alternative is that the product is really $CH_2{=}C(CH_3)C_2H_5$.

[g] The *Z* isomer reacts faster.

[h] This is a selection of three from 11 runs over different periods.

[i] This intermediate is prepared by the addition of vinyllithium to 5-methyl-1-trimethylsilyl-2-pentenone.

[j] These allylsilanes are formed, mixed with the corresponding vinylsilanes, in 93–100% yield, by pyrolysis of the 2-phenylsulfoxides.

[k] This allylsilane is prepared by selective hydrogenation (H_2, nickel boride) of geranyltrimethylsilane.

[l] This intermediate is prepared by the addition of vinyllithium to 3-phenyl-1-trimethylsilylpropanone.

[m] No acid is added.

[n] These compounds are also deuterated (up to d_4) in the cyclohexane ring.

TABLE II. PROTODESILYLATION AND DEUTERODESILYLATION OF VINYLSILANES

Reactant	Conditions	Product(s) and Yield(s)	Refs.
C_2			
$(CH_3)_3SiCH{=}CH_2$	1. HBr, $(C_6H_5CO_2)_2$ 2. 0.1 *N* NaOH, H_2O	$CH_2{=}CH_2$ (—)	5, 553
	1. HI, 4 h 2. 0.1 *N* NaOH, H_2O	" (—)	6
	H_2SO_4, 40°, 4 h; 20°, 6 h	" (50)	553
	TFA, 120°	" (100)	184
	KOC_4H_9-*t*, HMPA, 60°, 3 h	No reaction	184
$(C_2H_5)_3SiCH{=}CH_2$	HCl, CCl_4, 0–5°	$(C_2H_5)_3Si(CH_2)_2Cl$ (47)	554
$[(CH_3)_3Si]_3CSi(CH_3)_2CH{=}CH_2$	TFA, 3 d	$[(CH_3)_3Si]_3CSi(CH_3)_2O_2CCF_3$ (—)	555
1,1,3,3-tetramethyl-1,3-disilacyclopent-4-ene [ring: $(CH_3)_2Si$–CH_2–$Si(CH_3)_2$–CH=CH–]	H_2SO_4, CH_3OD, reflux, 91 h	(*Z*)-$(CH_3)_2Si(CH{=}CHD)CH_2Si(CH_3)_2OCH_3$ (11)	556
	hν, ROX	(*E*)-$(CH_3)_2Si(CH{=}CHX)CH_2Si(CH_3)_2OR$	556
	R: CH_3, X: H	(25)	
	R: ", X: " + xylene	(48)	
	R: ", X: D	(—)	
	R: ", X: " + xylene	(—)	
	R: CF_3CH_2, X: H	(30)	
	R: *t*-C_4H_9, X: "	(24)	
	R: ", X: " + xylene	(20)	
	R: ", X: D	(—)	
	R: ", X: " + xylene	(—)	
$[(CH_3)_3Si]_2C{=}C[Si(CH_3)_3]_2$	MCPBA, CH_2Cl_2, reflux	$[(CH_3)_3Si]_2C{=}CHSi(CH_3)_3$ (—) + $[(CH_3)_3Si]_2C\text{—}C[Si(CH_3)_3]_2$ (epoxide, O bridging the two C) (major)	557
$(CF_2{=}CF)_4Si$	1. KOH, H_2O 2. Br_2, 80°, 4 h	$CF_2BrCHFBr$ (90)	91

TABLE II. PROTODESILYLATION AND DEUTERODESILYLATION OF VINYLSILANES (*Continued*)

Reactant	Conditions	Product(s) and Yield(s)	Refs.
$(CH_3)_3SiCH{=}CBrOC_2H_5$	H_2O, 0°	$CH_3CO_2C_2H_5$ (—)	558
C_3			
$(CH_3)_3SiC(CH_3){=}CH_2$	HCl, H_2O, 50°, 17 h	$(CH_3)_3SiC(CH_3)_2Cl$ (35)	6
$(CH_3)_3SiCH{=}CHCH_2Si(CH_3)_2C_6H_5$	see Table I		
$(CH_3)_3SiCH_2CH{=}CHSi(CH_3)_2C_6H_5$	" " "		
$(CH_3)_3SiC(CH_2OH){=}CH_2$	KF, DMSO, 150°, 2 h	$CH_2{=}CHCH_2OH$ (>90)	209
(*Z*)-$(CH_3)_3SiCH{=}C[Si(CH_3)_2C_6H_5]CH_2OH$	TBAF, THF	(*E*)-$(CH_3)_3SiCH{=}CHCH_2OH$ (—)	559
C_4			
(*E*)-$(CH_3)_3SiCH{=}CH(CH_2)_2Si(CH_3)_3$	TFA, 60°, 3 h	$CH_2{=}CHC_2H_5$ (56) + $C_2H_5CH{=}CHSi(CH_3)_3$ (14)[a]	184
	KOC$_4$H$_9$-*t*, HMPA, 60°, 3 h	No reaction	184
$(CH_3)_3SiCCl{=}C(CH_3)_2$	$(CH_3)_4NF$, H_2O (1 eq), diglyme, 75 h	$HClC{=}C(CH_3)_2$ (62)	560
(*Z*)-$(CH_3)_3SiCH{=}C[Si(CH_3)_2C_6H_5](CH_2)_2OH$	TBAF, THF	(*E*)-$(CH_3)_3SiCH{=}CH(CH_2)_2OH$ (—)	213
$(CH_3)_3SiCX{=}CHCH(OTHP)CH_3$	$NaOCH_3$, CH_3OH, 40°, 4 h	$XCH{=}CHCH(OTHP)CH_3$	207
E X = Br		*E* (82)	
Z X = Br		*Z* (75)	
E X = I		*E* (68)	
$(CH_3)_3SiC(CO_2CH_3){=}C(CO_2CH_3)NHR$	CH_3OH, reflux, 1 h	$CH_3O_2CCH{=}C(CO_2CH_3)NHR$	205
R = $Si(CH_3)_3$		R = H (90)	
R = C_3H_7-*n*		R = C_3H_7-*n* (90)	
R = C_4H_9-*t*		R = C_4H_9-*t* (88)	
C_5			
$C_6H_5(CH_3)_2SiCH{=}C(CH_3)(CH_2)_2OCH_2C_6H_5$	TBAF, THF, 80°, 1 h	$CH_2{=}C(CH_3)(CH_2)_2OCH_2C_6H_5$ (80)	101, 100
[$(CH_3)_3SiCH{=}CBrC_3H_7$-*n*][b]	HBr, pentane, 0°, 15–30 min	$CH_2{=}CBrC_3H_7$-*n*	561

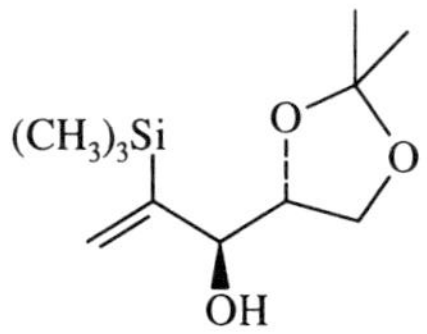

Reactant	Conditions	Product(s) and Yield(s) (%)	Refs.
$(CH_3)_3Si$ … OH	NaH, HMPA	… OH (—)	562
$(CH_3)_3Si$ … OH	NaH, HMPA	… OH (—)	562
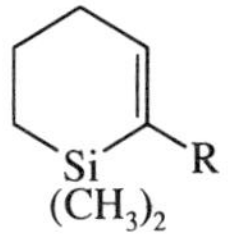 Si $(CH_3)_2$, R R = H	HBr, C_6H_6, reflux, 30 min	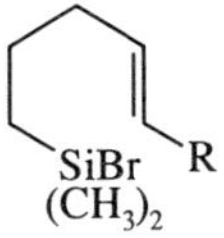 SiBr $(CH_3)_2$ R R = H (100) I	15
R = H	$C_6H_5HgCBr_3$, C_6H_6, reflux, 2 h	I R = H (15) + Br, Br, Si $(CH_3)_2$, R R = H (50) II	15
R = D	", ", ", 4 h	I R = D (—) + II R = D (—)	15
 $(CH_3)_3Si$, $SO_2C_6H_4CH_3$-*p*	$NaOCH_3$, CH_3OH, 25°, 30 min	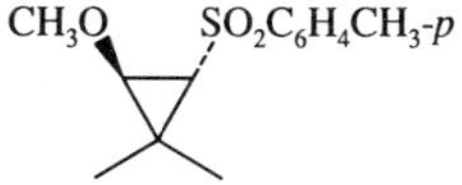 CH_3O, $SO_2C_6H_4CH_3$-*p* (98)	563
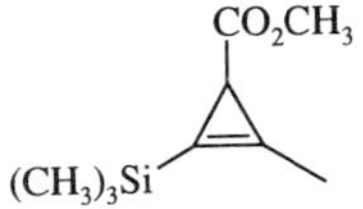 CO_2CH_3, $(CH_3)_3Si$	NaOD, D_2O, 1–2 d	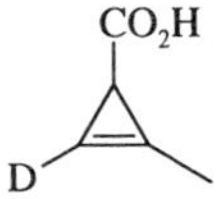 CO_2H, D (68)	564

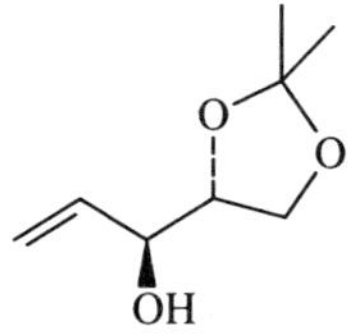

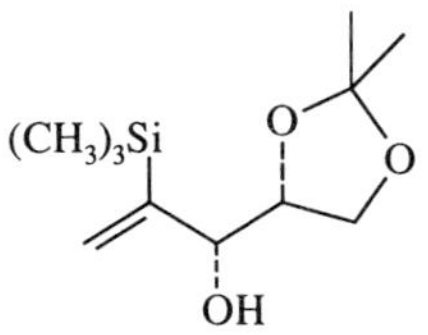

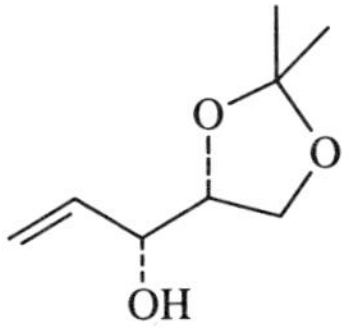

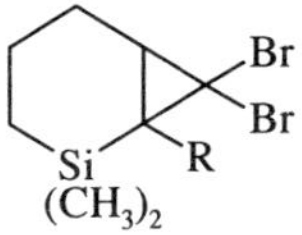

TABLE II. PROTODESILYLATION AND DEUTERODESILYLATION OF VINYLSILANES (*Continued*)

Reactant	Conditions	Product(s) and Yield(s)	Refs.
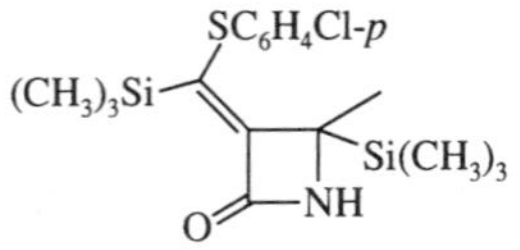 $(CH_3)_3Si$–C(SC_6H_4Cl-*p*)= 3-methyl-3-$Si(CH_3)_3$-azetidin-2-one (structure)	TBAF (2 eq), CH_3CO_2H, THF	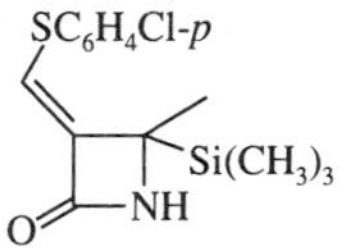 SC_6H_4Cl-*p* CH= 3-methyl-3-$Si(CH_3)_3$-azetidin-2-one (structure) (80)	565
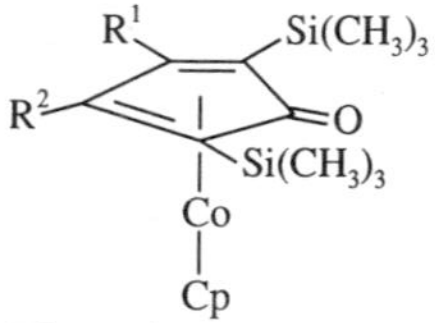 R^1, R^2, $Si(CH_3)_3$, $Si(CH_3)_3$ cyclopentadienone–Co–Cp complex (structure)	$C_6H_5CH_2\overset{+}{N}(CH_3)_3$ F^- (2 eq), THF, C_2H_5OH, reflux	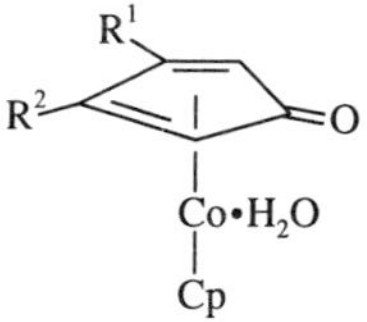 R^1, R^2 cyclopentadienone–$Co\cdot H_2O$–Cp complex (structure)	566
R^1 R^2			
H H		(70–80)	
-$(CH_2)_n$ n = 2, 3, and 4		(70–80)	
C_6			
$[(CH_3)_3SiCH{=}CHC_4H_9$-*n*$]$[c]	HI, C_6H_6	$CH_2{=}CHC_4H_9$-*n* (Good)	222
(*E*)-$C_6H_5(CH_3)_2SiC(C_2H_5){=}CHC_2H_5$	HI, toluene, 5 min	(*Z*)-$C_2H_5CH{=}CHC_2H_5$ (—)	567
$[(CH_3)_3SiCH{=}CBrC_4H_9$-*n*$]$[b]	HBr, pentane, 0°, 15–30 min	$CH_2{=}CBrC_4H_9$-*n* (94)	561
$(CH_3)_3SiCX{=}CHR$		CHX=CHR	207
X R			
E Br *n*-C_4H_9	$NaOCH_3$, CH_3OH, 40°, 4 h	*E* (76)	
" I "	", ", THF, 40°, 4 h	" (75)	
Z " "	", ", 40°, 4 h	*Z* (69)	
E Br *t*-C_4H_9	", ", ", "	*E* (93)	
" " $CH_2CH(OTHP)C_2H_5$	", ", ", "	" (78)	
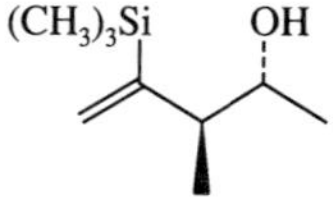 $(CH_3)_3Si$, OH (structure)	NaH, HMPA, THF, 30°, 2 h	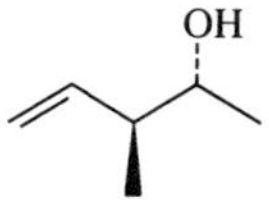 OH (structure) (40)	210

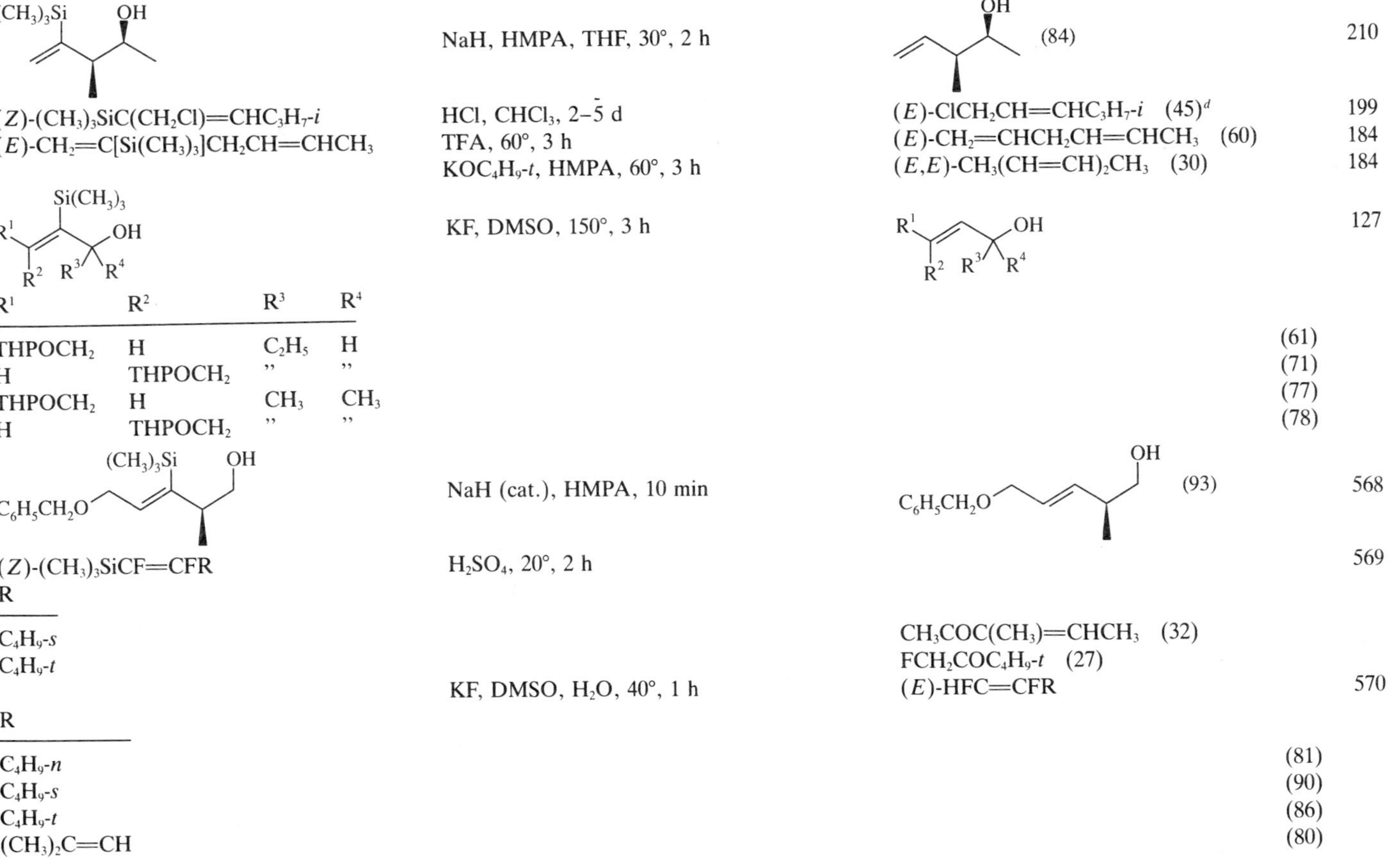

Reactant	Conditions	Product(s) and Yield(s) (%)	Refs.
(structure: $(CH_3)_3Si$, OH)	NaH, HMPA, THF, 30°, 2 h	(structure: OH) (84)	210
(Z)-$(CH_3)_3SiC(CH_2Cl)$=CHC_3H_7-*i*	HCl, $CHCl_3$, 2–5 d	(E)-$ClCH_2CH$=CHC_3H_7-*i* (45)[d]	199
(E)-CH_2=$C[Si(CH_3)_3]CH_2CH$=$CHCH_3$	TFA, 60°, 3 h	(E)-CH_2=$CHCH_2CH$=$CHCH_3$ (60)	184
	KOC_4H_9-*t*, HMPA, 60°, 3 h	(E,E)-$CH_3(CH$=$CH)_2CH_3$ (30)	184
(structure: R^1, R^2, $Si(CH_3)_3$, OH, R^3, R^4)	KF, DMSO, 150°, 3 h	(structure: R^1, R^2, OH, R^3, R^4)	127

R^1	R^2	R^3	R^4	Yield
$THPOCH_2$	H	C_2H_5	H	(61)
H	$THPOCH_2$	"	"	(71)
$THPOCH_2$	H	CH_3	CH_3	(77)
H	$THPOCH_2$	"	"	(78)

Reactant	Conditions	Product(s) and Yield(s) (%)	Refs.
(structure: $C_6H_5CH_2O$, $(CH_3)_3Si$, OH)	NaH (cat.), HMPA, 10 min	(structure: $C_6H_5CH_2O$, OH) (93)	568
(Z)-$(CH_3)_3SiCF$=CFR	H_2SO_4, 20°, 2 h		569
R = C_4H_9-*s*		$CH_3COC(CH_3)$=$CHCH_3$ (32)	
R = C_4H_9-*t*		$FCH_2COC_4H_9$-*t* (27)	
	KF, DMSO, H_2O, 40°, 1 h	(E)-HFC=CFR	570

R	Yield
C_4H_9-*n*	(81)
C_4H_9-*s*	(90)
C_4H_9-*t*	(86)
$(CH_3)_2C$=CH	(80)

TABLE II. PROTODESILYLATION AND DEUTERODESILYLATION OF VINYLSILANES (*Continued*)

Reactant	Conditions	Product(s) and Yield(s)	Refs.
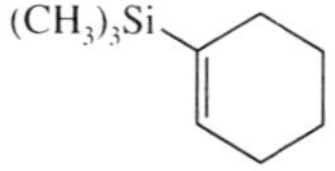 $(CH_3)_3Si$ (cyclohexenyl structure)	TsH, CH_3CN, H_2O, reflux, 3 h	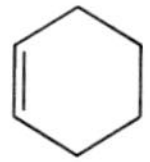 (cyclohexene structure) (81)	197
(*E*)-$(CH_3)_3SiC(CN)$=CHC_3H_7-*n*	NaOH, CH_3OH, $(C_2H_5)_2O$, 0°, 1 h	(*Z*)-NCCH=CHC_3H_7-*n* (65)	208
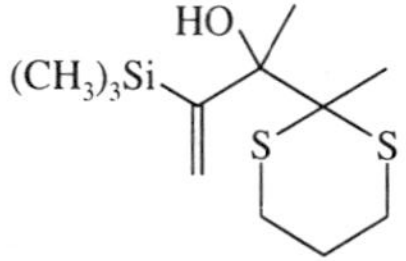 $(CH_3)_3Si$, HO (dithiane structure)	1. KH, THF, 10 min 2. H_2O	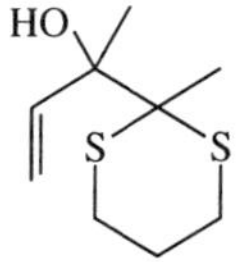 HO (dithiane structure) (26)	571
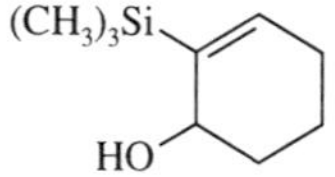 $(CH_3)_3Si$, HO (cyclohexenol structure)	TBAF, CH_3CN	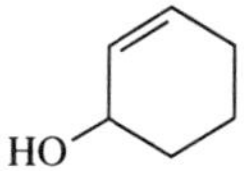 HO (cyclohexenol structure) (97)	572
$(CH_3)_3Si$ (oxepin structure)		R (oxepin structure)	74
		R	
	$(C_2H_5)_4NF$, CH_3CN	H (100)	
	", CD_3CN	D:H 85:15 (100)	
	", CD_3SOCD_3	" <15:85 (100)	
	", ", D_2O	" 69:31 (100)	
	TBAF, CH_3CN	H (100)	
	KF, 18-crown-6, CD_3CN	D:H 79:21 (100)	
$R^1_3SiC(CO_2CH_3)$=$C(CO_2CH_3)C{\equiv}CNR^2C_6H_5$		CH_3O_2CCH=$C(CO_2CH_3)CH_2CONR^2C_6H_5$	573

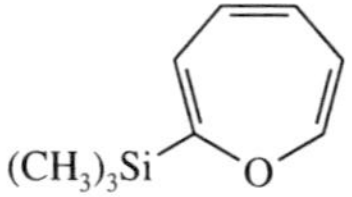

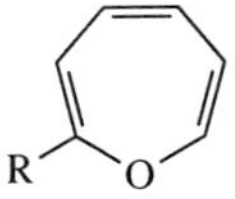

Reactant	Conditions	Product(s) and Yield(s) (%)	Refs.
R^1 R^2			
CH_3 CH_3	SiO_2, $CHCl_3$, $(C_2H_5)_2O$	*E* (23)	
" C_6H_5	", ", "	*E* (21)	
C_6H_5 CH_3	HCl, $(CH_3)_2CO$, H_2O, 2 h	*E* + *Z* (—)	
" C_6H_5	SiO_2, $CHCl_3$, $(C_2H_5)_2O$	*E* (43) + *Z* (28)	
$[CH_3C(CO_2C_2H_5)_2C(Si(CH_3)_3){=}CHSO_2C_6H_5]$[e]	CH_3CO_2H, THF	(*Z*)-$CH_3C(CO_2C_2H_5)_2CH{=}CHSO_2C_6H_5$ (80)	574
C_7			
$C_2H_5CH{=}C[Si(CH_3)_3]CHOHC_2H_5$	TBAF, CH_3CN	$C_2H_5CH{=}CHCHOHC_2H_5$	572
E		*Z* (100)	
Z		*E* (98)	
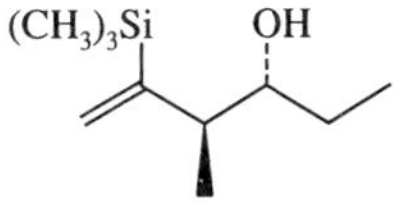 $(CH_3)_3Si$ OH	NaH, HMPA, THF, 30°, 6 h	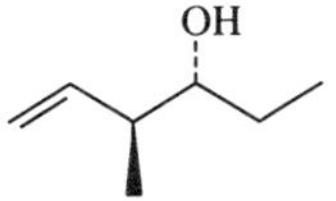 OH (90)	219, 210, 575
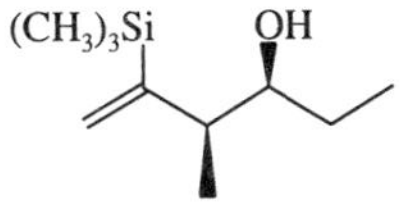 $(CH_3)_3Si$ OH	NaH, HMPA, THF, 30°, 3 h	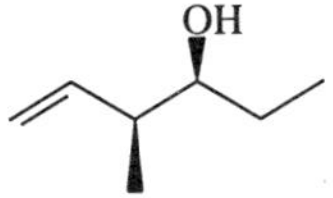 OH (86)	219, 210, 575
(*E*)-$C_6H_5(CH_3)_2SiC(i\text{-}C_3H_7CHOH){=}CHCH_3$	TBAF, DMSO, 1 h	(*Z*)-*i*-$C_3H_7CHOHCH{=}CHCH_3$ (56)	576
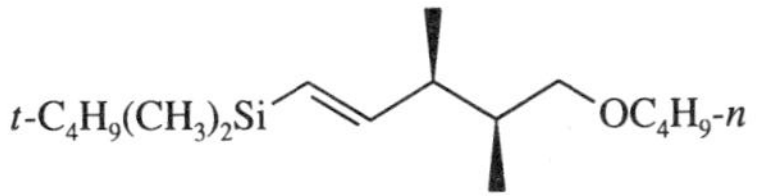 *t*-$C_4H_9(CH_3)_2Si$ … OC_4H_9-*n*	HBF_4, CH_3CN, 55°, 1.5 h	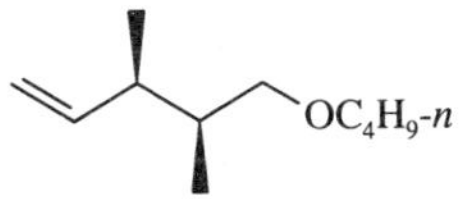 OC_4H_9-*n* (94)	175
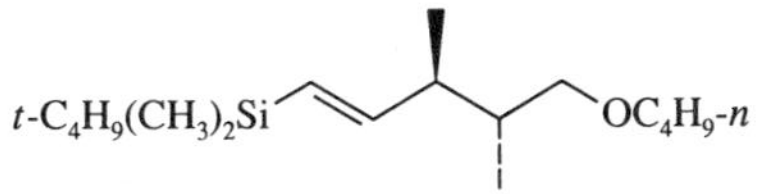 *t*-$C_4H_9(CH_3)_2Si$ … OC_4H_9-*n*	", ", ", "	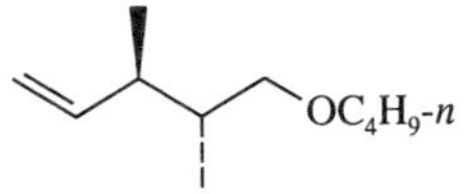 OC_4H_9-*n* (96)	175
$[(CH_3)_3SiCH{=}CBrC_5H_{11}\text{-}n]$[b]	HBr, pentane, 0°, 15–30 min	$CH_2{=}CBrC_5H_{11}$-*n* (80)	561
$[(CH_3)_3SiCH{=}CBr(CH_2)_2C_3H_7\text{-}i]$[b]	", ", ", "	$CH_2{=}CBr(CH_2)_2C_3H_7$-*i* (76)	561

TABLE II. PROTODESILYLATION AND DEUTERODESILYLATION OF VINYLSILANES (*Continued*)

Reactant	Conditions	Product(s) and Yield(s)	Refs.
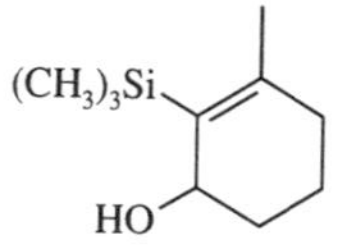 2-$(CH_3)_3Si$-3-methylcyclohex-2-en-1-ol (structure: $(CH_3)_3Si$, HO)	TBAF, CH_3CN	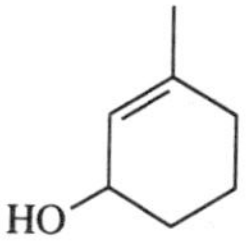 3-methylcyclohex-2-en-1-ol (structure: HO) (51)	572
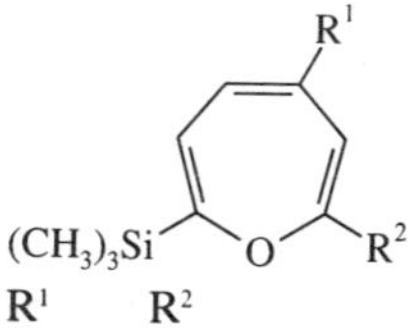 2-$(CH_3)_3Si$-oxepin (structure: $(CH_3)_3Si$, O, R^1, R^2)	$(C_2H_5)_4NF$, CD_3CN	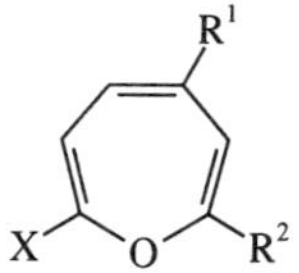 oxepin (structure: X, O, R^1, R^2)	74
R^1 R^2		X	
CH_3 H		D:H 81:19 (100)	
H CH_3		D:H 88:12 (100)	
$[(CH_3)_3SiC(CH_3COC(CH_3)CO_2C_2H_5)=CHSO_2C_6H_5]$[e]	CH_3CO_2H, THF	(*Z*)-$CH_3COC(CH_3)(CO_2C_2H_5)CH=CH(C_6H_5SO_2)$ (73)	574
C_8			
$(CH_3)_3SiCH=CHC_6H_{13}$-*n*		$RCH=CHC_6H_{13}$-*n*	
		R	
	TsH, CH_3CN, H_2O, reflux, 3 h	H (85)	197
E	HI, C_6H_6, 4 h	" (92)	72
"	I_2, C_6H_6, D_2O, reflux, 3 h	*E* D 90% D_1 (97)	72
"	TsD, CH_3CN, D_2O, reflux	*E*:*Z* 60:40 D (—)	197
Z	HI, C_6H_6, 4 h	H (97)	72
"	I_2, C_6H_6, D_2O, reflux, 15 min	*Z* D (—)[f]	197
"	", ", ", ", 3 h	*Z* D (100)	72
"	TsD, CH_3CN, D_2O, reflux, 15 min	*E*:*Z* 60:40 D (10)	197

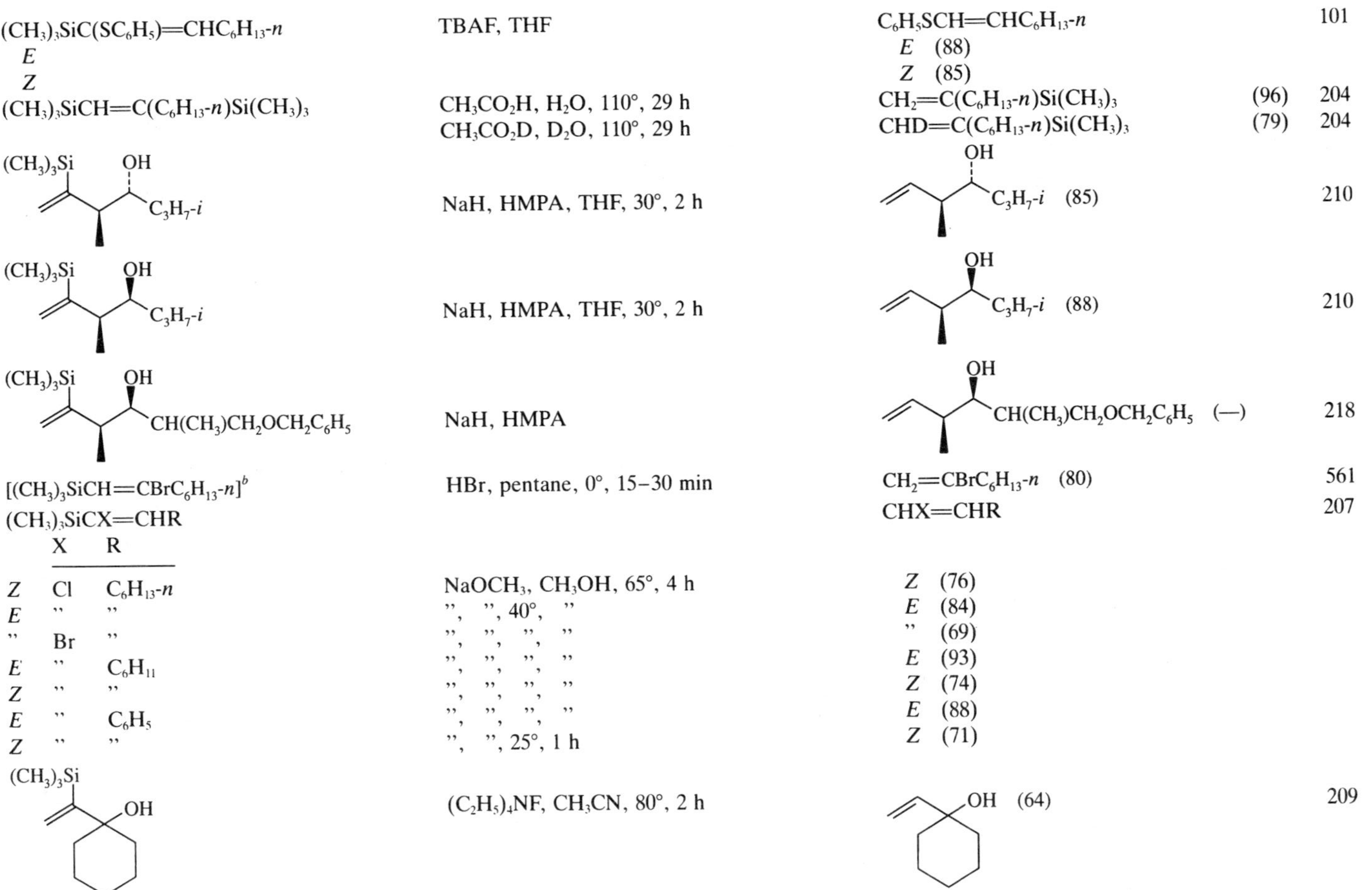

Reactant	Conditions	Product(s) and Yield(s) (%)	Refs.
$(CH_3)_3SiC(SC_6H_5){=}CHC_6H_{13}$-*n*	TBAF, THF	$C_6H_5SCH{=}CHC_6H_{13}$-*n*	101
E		*E* (88)	
Z		*Z* (85)	
$(CH_3)_3SiCH{=}C(C_6H_{13}$-*n*$)Si(CH_3)_3$	CH_3CO_2H, H_2O, 110°, 29 h	$CH_2{=}C(C_6H_{13}$-*n*$)Si(CH_3)_3$ (96)	204
	CH_3CO_2D, D_2O, 110°, 29 h	$CHD{=}C(C_6H_{13}$-*n*$)Si(CH_3)_3$ (79)	204
$(CH_3)_3Si$, OH, C_3H_7-*i*	NaH, HMPA, THF, 30°, 2 h	OH, C_3H_7-*i* (85)	210
$(CH_3)_3Si$, OH, C_3H_7-*i*	NaH, HMPA, THF, 30°, 2 h	OH, C_3H_7-*i* (88)	210
$(CH_3)_3Si$, OH, $CH(CH_3)CH_2OCH_2C_6H_5$	NaH, HMPA	OH, $CH(CH_3)CH_2OCH_2C_6H_5$ (—)	218
$[(CH_3)_3SiCH{=}CBrC_6H_{13}$-*n*$]^b$	HBr, pentane, 0°, 15–30 min	$CH_2{=}CBrC_6H_{13}$-*n* (80)	561
$(CH_3)_3SiCX{=}CHR$		$CHX{=}CHR$	207
X R			
Z Cl C_6H_{13}-*n*	$NaOCH_3$, CH_3OH, 65°, 4 h	*Z* (76)	
E " "	", ", 40°, "	*E* (84)	
" Br "	", ", ", "	" (69)	
E " C_6H_{11}	", ", ", "	*E* (93)	
Z " "	", ", ", "	*Z* (74)	
E " C_6H_5	", ", ", "	*E* (88)	
Z " "	", ", 25°, 1 h	*Z* (71)	
$(CH_3)_3Si$, OH	$(C_2H_5)_4NF$, CH_3CN, 80°, 2 h	OH (64)	209

TABLE II. PROTODESILYLATION AND DEUTERODESILYLATION OF VINYLSILANES (*Continued*)

Reactant	Conditions	Product(s) and Yield(s)	Refs.
(*E*)-$(CH_3)_3SiC(CN)=CHC_5H_{11}$-*n*	NaOH, CH_3OH, $(C_2H_5)_2O$, 0°, 1 h	(*Z*)-$NCCH=CHC_5H_{11}$-*n* (85)	208
1-$(CH_3)_3Si$-2-methyl-5-acetylcyclopentene ($COCH_3$)	K_2CO_3, CH_3OH, 9 h	1-methyl-5-acetylcyclopentene ($COCH_3$) (68)	346
$[(CH_3)_3SiCR=CHSO_2C_6H_5]^e$	CH_3CO_2H, THF	$RCH=CHSO_2C_6H_5$	574
R = $CH_2=CHCH_2C(CO_2C_2H_5)_2$		*E*:*Z* 1:5 (93)	
R = 2-oxo-1-($CO_2C_2H_5$)cyclopentyl		" 1:8 (85)	
R = 1-methyl-2,5-dioxocyclopentyl (CH_3)		*Z* (51)	
(*E*)-$(CH_3)_3Si$CH= pyrrolizidinone (H, R)	TsH, CH_3CN, H_2O, reflux, 18 h	methylene pyrrolizidinone (H, R)	
R = H		(—)	577, 578
R = O_2CCH_3		(81)	578

Reactant	Conditions	Product(s) and Yield(s) (%)	Refs.
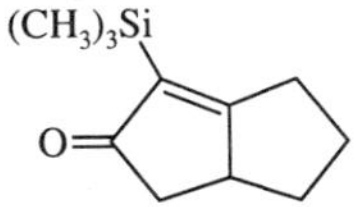 $(CH_3)_3Si$ (structure), O=		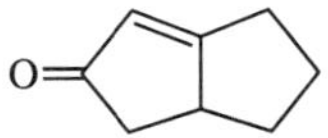 O= (structure)	579
	HI, C_6H_6	(45)	
	NaOH, CH_3OH, $(C_2H_5)_2O$	(20)	
	TBAF, THF	(45)	
(E)-$(CH_3)_3SiCH{=}CHC_6H_5$	H_2SO_4, 0°, 40 min	Polystyrene (90)	6
	HCl, CS_2, −100°, 1 h	$CH_2{=}CHC_6H_5$ + $CH_3CHClC_6H_5$ 1:1 (—)	16
	DCl or DBr, CH_3CN, reflux	$CHD{=}CHC_6H_5$ >96% E (—)	16
(Z)-$(CH_3)_3SiCH{=}CHC_6H_5$	DCl or DBr, CH_3CN, reflux	$CHD{=}CHC_6H_5$ >96% Z (—)	16
$(CH_3)_3SiCH{=}C(C_6H_5)Si(CH_3)_3$	CH_3CO_2H, prolonged reflux	$CH_2{=}C(C_6H_5)Si(CH_3)_3$ (100)	203
(Z)-$(CH_3)_3SiCF{=}CFC_6H_5$	H_2SO_4, 20°, 2 h	$FCH_2COC_6H_5$ (13)	569
	KF, DMSO, H_2O, 40°, 1 h	(E)-$FCH{=}CFC_6H_5$ (76)	570
$[(CH_3)_3Si]_2C{=}C(C_6H_5)Si(CH_3)_3$	", reflux, 24 h	$(CH_3)_3SiCH{=}C(C_6H_5)Si(CH_3)_3$ (80) + $CH_2{=}C(C_6H_5)Si(CH_3)_3$ (5–10)	203
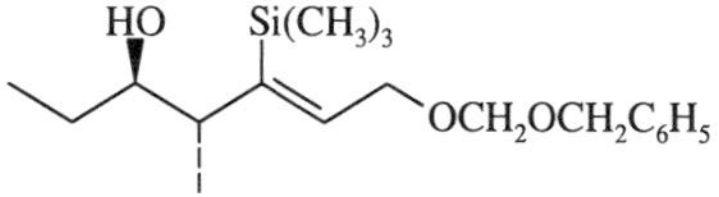 HO, $Si(CH_3)_3$, $OCH_2OCH_2C_6H_5$ (structure)	NaH, HMPA, 30 min	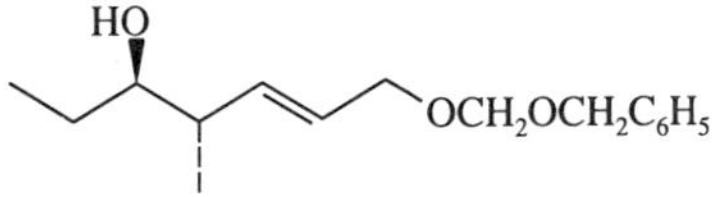 HO, $OCH_2OCH_2C_6H_5$ (structure) (96)	217
C_9			
(Z)-$(CH_3)_3SiCF{=}CFC_7H_{15}$-$n$	HBr, CH_3CN, 20°, 1 h	$CHF{=}CFC_7H_{15}$-n $E:Z$ 98:2 (80)	569
	H_2SO_4, 20°, 2 h	$FCH_2COC_7H_{15}$-n (66)	569
	KF, DMSO, H_2O, 40°, 1 h	(E)-$CHF{=}CFC_7H_{15}$-n (85)	570
$(CH_3)_3SiC(CH_2C_6H_{11}){=}CH_2$	TsH, CH_3CN, H_2O, reflux, 18 h	$C_6H_{11}CH_2CH{=}CH_2$ (<5)	197
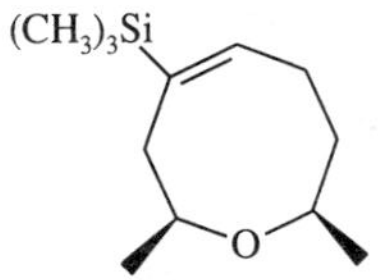 $(CH_3)_3Si$, O (structure)	HCl, $(C_2H_5)_2O$, 25°	 O (structure) (—)	379

TABLE II. PROTODESILYLATION AND DEUTERODESILYLATION OF VINYLSILANES (*Continued*)

Reactant	Conditions	Product(s) and Yield(s)	Refs.
(*E*)-$C_6H_5(CH_3)_2SiC{=}CHCH_3$ (1-hydroxycyclohexyl)	TBAF, DMSO, 1 h	*Z* 1-(CH=CHCH$_3$)cyclohexanol (64)	576
(*Z*)-$(CH_3)_3SiC(CH_2Cl){=}CHR$	HCl, $CHCl_3$, 2–5 d	(*E*)-$ClCH_2CH{=}CHR$	199
R = C_6H_{11}		(62)	
R = C_6H_5		(>90)	
$(CH_3)_3Si$, OH, $OCH_2C_6H_5$, dioxolane	NaH, HMPA	OH, $OCH_2C_6H_5$, dioxolane (—)	220
THPO, $(CH_3)_3Si$, OH (cyclohexyl)	KF, DMSO, 150°, 3 h	THPO, OH (cyclohexyl) (78)	127
(*E*)-$(CH_3)_3SiC(CN){=}CHR$	NaOH, CH_3OH, $(C_2H_5)_2O$, 0°, 1 h	(*Z*)-$NCCH{=}CHR$	208
R			
C_6H_{11}		(85)	
C_6H_5		(97)	
$C_6H_4OCH_3$-*p*		(92)	
[$(CH_3)_3SiCH{=}CBrCH_2$-(1,4-dioxaspiro[4.5]decan-8-yl)][b]	HBr, pentane, 0°, 15–30 min	$CH_2{=}CBrCH_2$-(dioxaspiro) (83)	561

Reactant	Conditions	Product(s) and Yield(s) (%)	Refs.
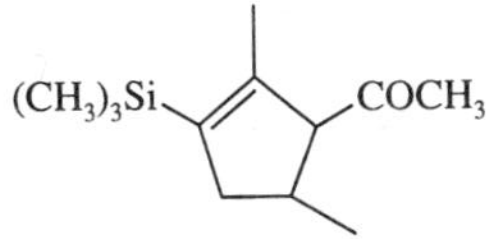 $(CH_3)_3Si$, $COCH_3$ (cyclopentene)	K_2CO_3, CH_3OH, 9 h	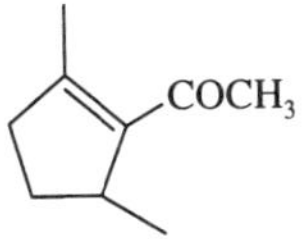 $COCH_3$ (64)	346
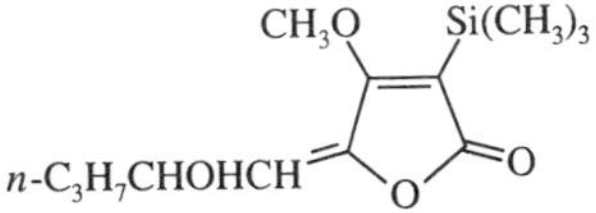 CH_3O, $Si(CH_3)_3$, n-$C_3H_7CHOHCH$=, O, O (furanone)	HCl, CH_3OH, 23°, 7 d	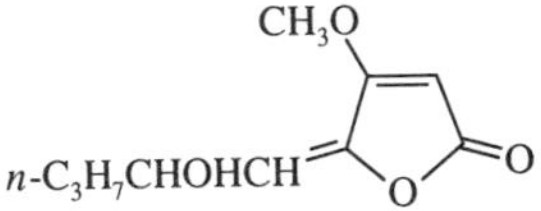 CH_3O, n-$C_3H_7CHOHCH$=, O, O (80)	580
$(CH_3)_3SiCH{=}CHCH_2C_6H_4R$-$p$[g]	HI, C_6H_6	$CH_2{=}CHCH_2C_6H_4R$-p (68–93)	222
(E)-$C_6H_5(CH_3)_2SiC(C_6H_5){=}CHCH_2OH$	NaH (cat.), HMPA, 0.5 h	$C_6H_5CH{=}CHCH_2OH$ $E:Z$ 50:50 (80)	211
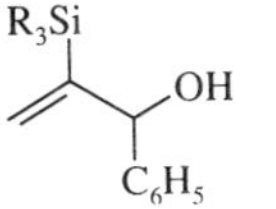 R_3Si, OH, C_6H_5		$CH_2{=}CHCHOHC_6H_5$	209
R = CH_3	$(CH_3)_4NF$, CH_3CN, 80°, 3 h	(65)	
R = C_6H_5	", ", ", 2 h	(80)	
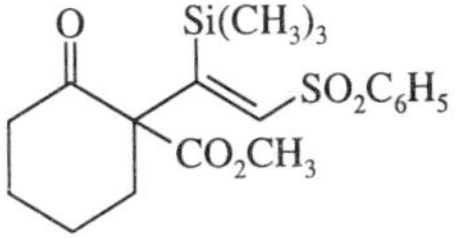 O, $Si(CH_3)_3$, $SO_2C_6H_5$, CO_2CH_3 (cyclohexanone)	CH_3CO_2H, THF	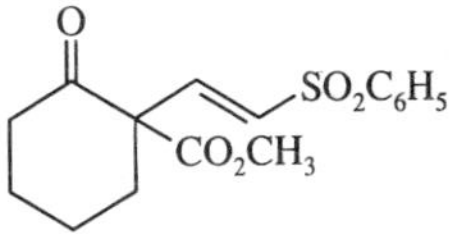 O, $SO_2C_6H_5$, CO_2CH_3 (79)	574
C_{10}			
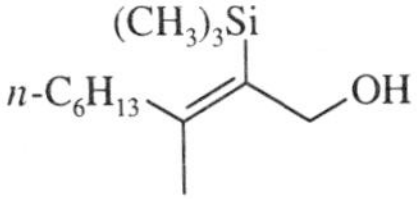 $(CH_3)_3Si$, n-C_6H_{13}, OH	TBAF, DMSO, 70°	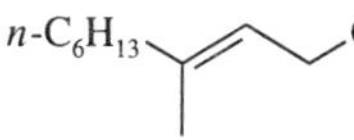 n-C_6H_{13}, OH (—)	493, 581
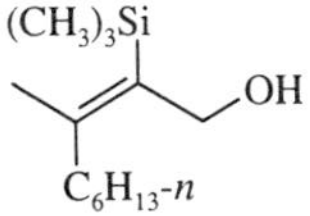 $(CH_3)_3Si$, OH, C_6H_{13}-n	TBAF, DMSO, 70°	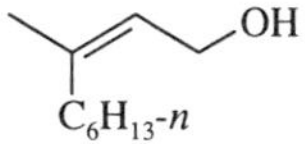 OH (—), C_6H_{13}-n	581

TABLE II. PROTODESILYLATION AND DEUTERODESILYLATION OF VINYLSILANES (*Continued*)

Reactant	Conditions	Product(s) and Yield(s)	Refs.
$(CH_3)_3SiC(C_4H_9\text{-}n){=}C(C_4H_9\text{-}n)Si(CH_3)_3$	$NaOC_2H_5$, DMSO, 130°, 15 min	$n\text{-}C_4H_9CH{=}CRC_4H_9\text{-}n$	206
E		*Z* R = $Si(CH_3)_3$ (6) + *E* R = H (90)	
Z		*Z* R = $Si(CH_3)_3$ (53) + *E* R = H (19) + *Z* R = H (28)	
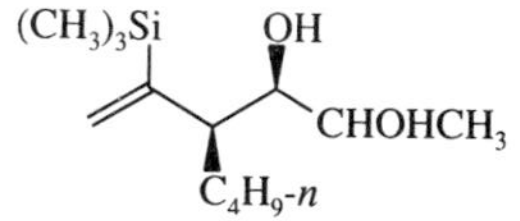 $(CH_3)_3Si$, OH, $CHOHCH_3$, $C_4H_9\text{-}n$	KH, HMPA, 10 h	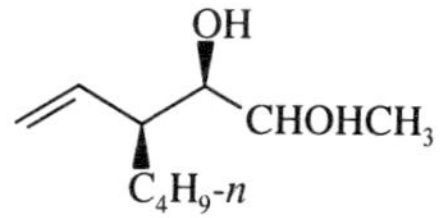 OH, $CHOHCH_3$, $C_4H_9\text{-}n$ (75)[h]	218
(*E*)-$(CH_3)_3SiC(CH_2CN){=}CHC_6H_{13}\text{-}n$	I_2, H_2O	(*Z*)-$NCCH_2CH{=}CHC_6H_{13}\text{-}n$ (—)	582
(*E*)-$(CH_3)_3SiC(CN){=}CHCH(C_2H_5)C_4H_9\text{-}n$	NaOH, CH_3OH, $(C_2H_5)_2O$, 0°, 1 h	(*Z*)-$NCCH{=}CHCH(C_2H_5)C_4H_9\text{-}n$ (91)	208
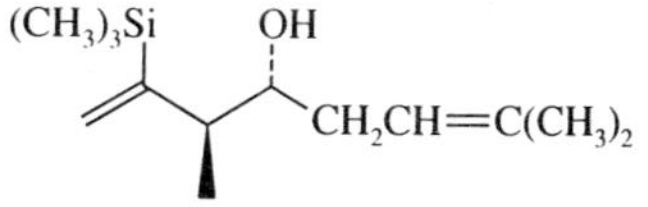 $(CH_3)_3Si$, OH, $CH_2CH{=}C(CH_3)_2$	NaH, HMPA, 10 min	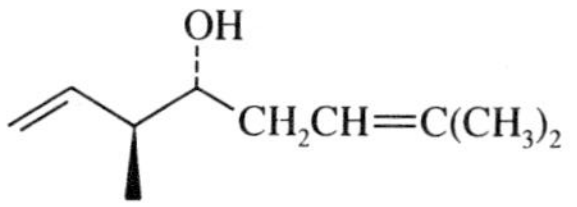 OH, $CH_2CH{=}C(CH_3)_2$ (91)	216
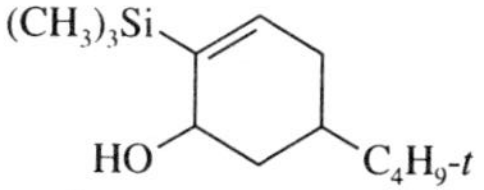 $(CH_3)_3Si$, HO, $C_4H_9\text{-}t$	TBAF, CH_3CN	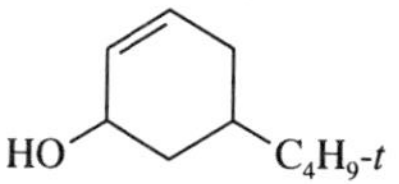 HO, $C_4H_9\text{-}t$ (96)	572
$(\dot{C}H_3)_3SiCH{=}CHCH_2CH{=}CH(CH_2)_3CH{=}CH_2$		$CH_2{=}CHCH_2CH{=}CH(CH_2)_3CH{=}CH_2$	
E, *E*	TFA, 60°, 3 h	(100)	184
"	$KOC_4H_9\text{-}t$, HMPA, 60°, 3 h	(95)	184, 215
Z, *E*	$MoCl_5$, C_6H_6, 40–50°, 2–3 h	(41)	164
$(CH_3)_3Si$ + $(CH_3)_3Si$	TsOH, CH_3CN, THF, H_2O, reflux, 1 h	+	198
92:8		9:1 (73)	
9:91		1:9 (75)	

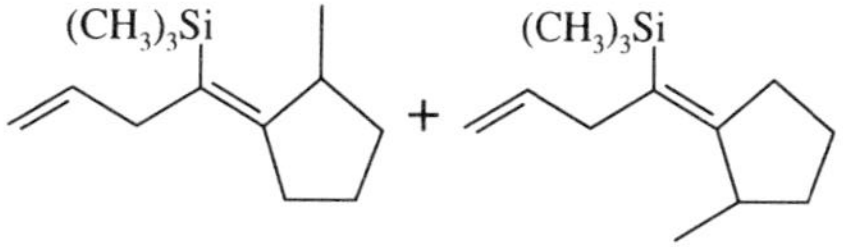

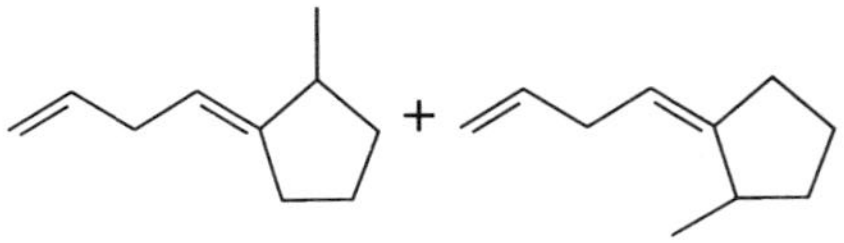

Reactant	Conditions	Product(s) and Yield(s) (%)	Refs.
$CH_2{=}C[Si(CH_3)_3]COH(CH_3)$- $(CH_2)_2CH{=}C(CH_3)_2$	$(C_2H_5)_4NF$, CH_3CN, 80°, 2 h	$CH_2{=}CHCOH(CH_3)(CH_2)_2CH{=}C(CH_3)_2$ (52)	209
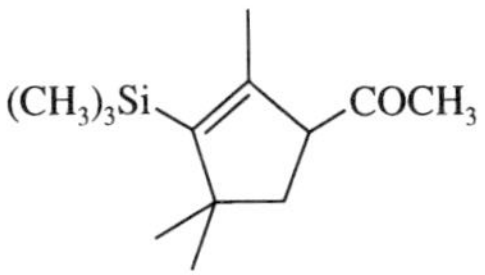 $(CH_3)_3Si$, $COCH_3$	HF, CH_3CN, 48 h	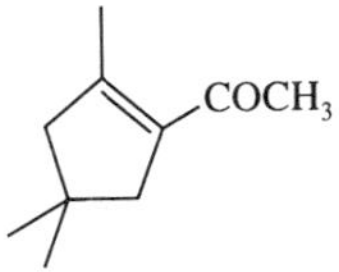 $COCH_3$ (66)	346
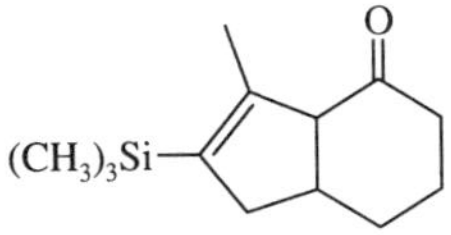 $(CH_3)_3Si$—, O	K_2CO_3, CH_3OH, 9 h	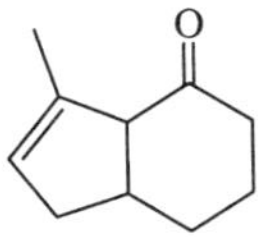 O (86)	346
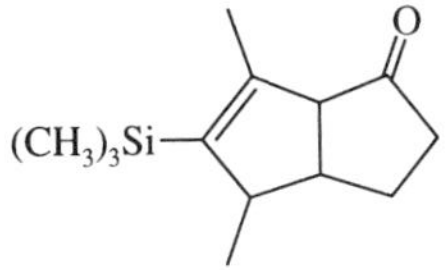 $(CH_3)_3Si$—, O	HF, CH_3CN, 48 h	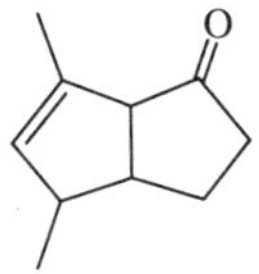 O (92)	346
(E)-$C_6H_5(CH_3)_2SiC(C_6H_5){=}CHCHOHCH_3$	NaH (cat.), HMPA, 0.5 h	$C_6H_5CH{=}CHCHOHCH_3$ $E:Z$ 89:11 (83)	211
(E)-$C_6H_5(CH_3)_2SiC(CHOHC_6H_5){=}CHCH_3$	TBAF, DMSO, 1 h	(Z)-$C_6H_5CHOHCH{=}CHCH_3$ (81)	576
$(CH_3)_3SiC(CHOHC_6H_5){=}CHCH_2OTHP$	KF, DMSO, 150°, 3 h	$C_6H_5CHOHCH{=}CHCH_2OTHP$ (75)	127
C_{11}			
$(CH_3)_3SiC(C_5H_{11}$-$n){=}C(C_2H_5)_2$	I_2, C_6H_6	n-$C_5H_{11}CH{=}C(C_2H_5)_2$ (90)	72
	HI (0.5 eq), C_6H_6	" + n-$C_6H_{13}C(C_2H_5){=}CHCH_3$ (—)	72
$(CH_3)_3SiC(CH_2CH{=}CH_2){=}CHC_6H_{13}$-$n$	HI, C_6H_6, 1 h	$CH_2{=}CHCH_2CH{=}CHC_6H_{13}$-$n$ (100)	583
$C_6H_5(CH_3)_2SiC{=}CHC_4H_9$-$n$ $(CH_3O)_2CH(CH_2)_4$	TBAF, DMSO, THF, 80°, 15 min	$(CH_3O)_2CH(CH_2)_4CH{=}CHC_4H_9$-$n$ (91)	100
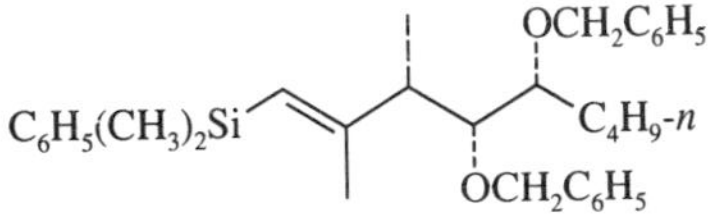 $C_6H_5(CH_3)_2Si$, $OCH_2C_6H_5$, C_4H_9-n, $OCH_2C_6H_5$	TBAF, DMSO, THF, 100°, 2 h	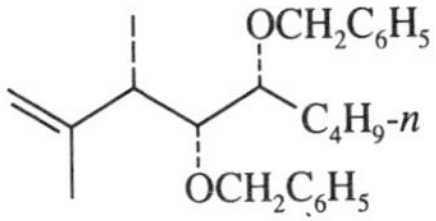 $OCH_2C_6H_5$, C_4H_9-n, $OCH_2C_6H_5$ (95)	100, 101

TABLE II. PROTODESILYLATION AND DEUTERODESILYLATION OF VINYLSILANES (*Continued*)

Reactant	Conditions	Product(s) and Yield(s)	Refs.
$[(CH_3)_3SiCH{=}CBr(CH_2)_8CO_2CH_3]$[b]	HBr, pentane, 0°, 15–30 min	$CH_2{=}CBr(CH_2)_8CO_2CH_3$ (61)	561
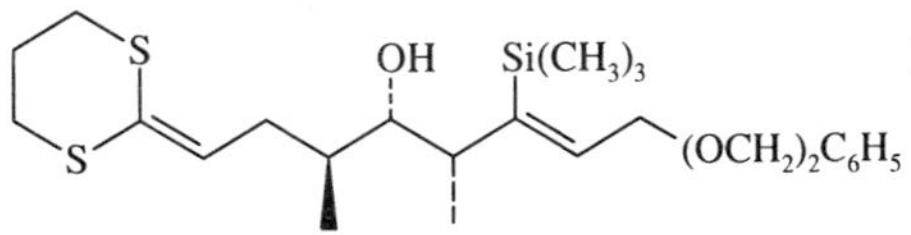 S, S, OH, $Si(CH_3)_3$, $(OCH_2)_2C_6H_5$	NaH, HMPA, THF, 0°	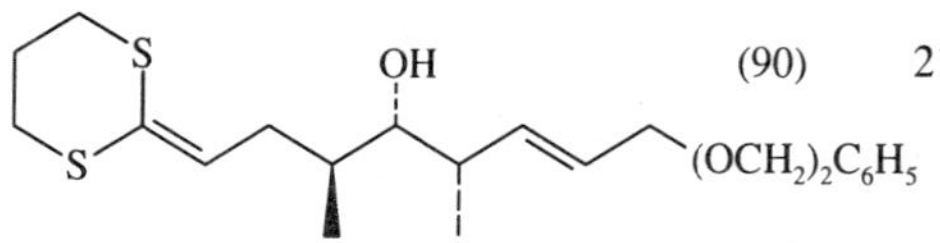 S, S, OH, $(OCH_2)_2C_6H_5$ (90)	217
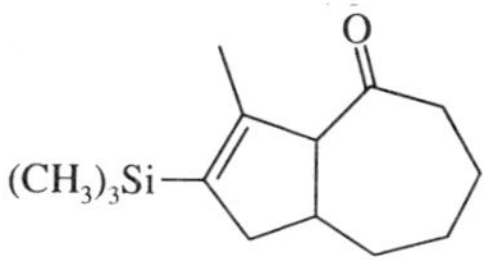 $(CH_3)_3Si$, O	K_2CO_3, CH_3OH, 9 h	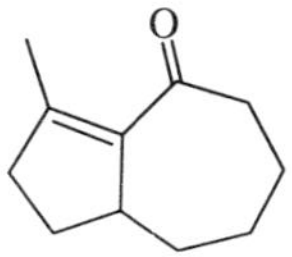 O (91)	346
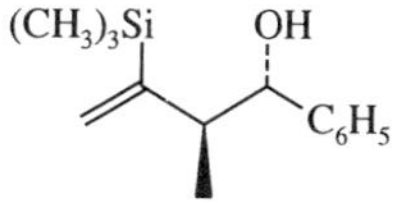 $(CH_3)_3Si$, OH, C_6H_5	NaH, HMPA, THF, 30°, 2 h	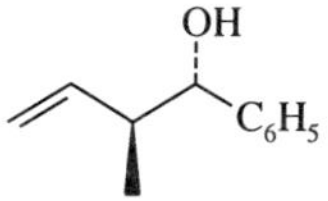 OH, C_6H_5 (93)	210
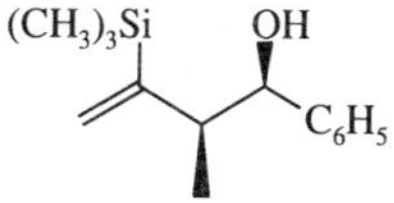 $(CH_3)_3Si$, OH, C_6H_5	NaH, HMPA, THF, 30°, 2 h	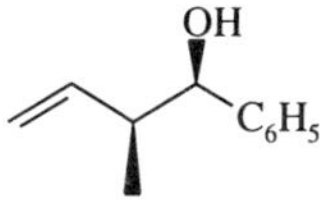 OH, C_6H_5 (90)	210
(E)-$(CH_3)_3SiCH{=}C[Si(CH_3)_3]CH(CH_2C_6H_5)CH_3$	CF_3CO_2D, CH_2Cl_2, 25°, 1 h	(E)-$CHD{=}C[Si(CH_3)_3]CH(CH_3)CH_2C_6H_5$ (99)	584
(E)-$(CH_3)_3SiC(CN){=}CH(CH_2)_2C_6H_5$	NaOH, CH_3OH, $(C_2H_5)_2O$, 0°, 1 h	(Z)-$NCCH{=}CH(CH_2)_2C_6H_5$ (95)	208
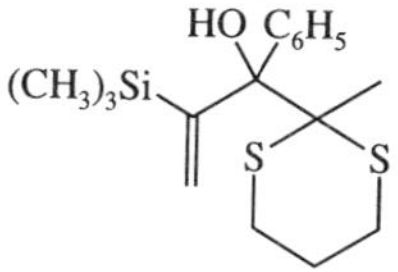 $(CH_3)_3Si$, HO, C_6H_5, S, S	1. KH, HMPA 2. H_2O	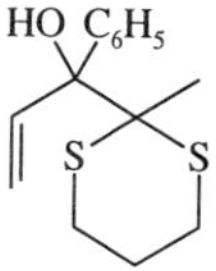 HO, C_6H_5, S, S (—)	585

 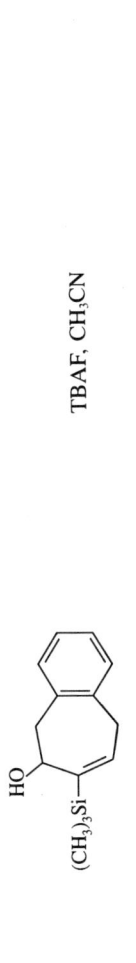	TBAF, CH_3CN	HO (benzocycloheptenol structure) (60)	572
C_{12}			
(*Z*)-$R(CH_3)_2SiCH{=}CHC_{10}H_{21}$-*n*		$CH_2{=}CHC_{10}H_{21}$-*n* I + $HO(CH_3)_2SiCH{=}CHC_{10}H_{21}$-*n* II + $O[(CH_3)_2SiCH{=}CHC_{10}H_{21}$-*n*$]_2$ III	100
R			
$CH_2{=}CHCH_2$	TBAF, THF, 25°, 5–30 min	I (trace) + II (22–68) + III (68–16)	
"	", DMSO, THF, 80°, 30 min	I (92)	
$C_6H_5(CH_3)_2Si$	", THF, 25°, 30 min	I (trace) + II (67) + III (9)	
"	", DMSO, THF, 80°, 30 min	I (100)	
(*Z*)-$R^1(CH_3)_2SiCR^2{=}CHR^3$		(*E*)-$R^2CH{=}CHR^3$	
R^1 R^2 R^3			
C_6H_5 *n*-$C_{10}H_{21}$ D	TBAF, DMSO, 80°, 30 min	(80)	100, 101
CH_3 *n*-C_4H_9 $(CH_2)_6OH$	I_2, C_6H_6, H_2O	(>84)[i]	461
CH_3 *n*-C_4H_9 $(CH_2)_6OTHP$	", ", ", reflux, 2 h	(82)[i]	213
(*E*)-$(CH_3)_3SiCBr{=}CHC_{10}H_{21}$-*n*	TBAF, HMPA, THF, 80°, 0.3 h	(*E*)-$BrCH{=}CHC_{10}H_{21}$-*n* (95)	100
HO, $(CH_3)_3Si$ (cyclododecenol structure) *E*	TBAF, CH_3CN, heat	HO (cyclododecenol structure) *Z* (96)	572
Z		*E* (96)	
$C_6H_5(CH_3)_2Si$, $CH_2{=}CHCH_2$ (cyclononene structure)	TBAF, HMPA, THF, 80°, 2 h	$CH_2{=}CHCH_2$ (cyclononene structure) (68)	100

TABLE II. PROTODESILYLATION AND DEUTERODESILYLATION OF VINYLSILANES (*Continued*)

Reactant	Conditions	Product(s) and Yield(s)	Refs.
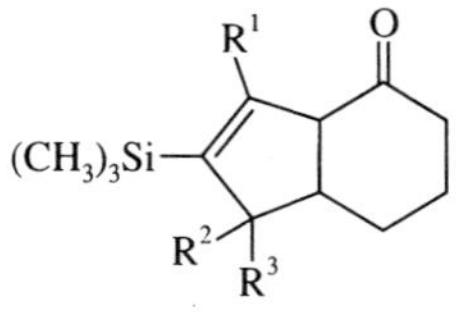 $(CH_3)_3Si$ (structure; R^1, R^2, R^3)		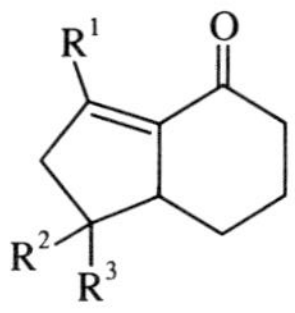 (structure; R^1, R^2, R^3, O)	346
R^1 / R^2 / R^3			
CH_3 / C_2H_5 / H	HF, CH_3CN, 48 h	(88–91)	
" / CH_3 / CH_3	", ", "	(94)	
i-C_3H_7 / H / H	K_2CO_3, CH_3OH, 9 h	(75)	
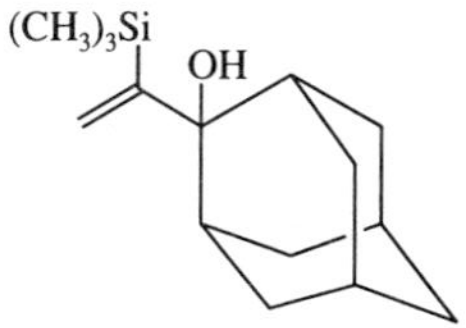 $(CH_3)_3Si$, OH (structure)	1. KH, HMPA, 0°, 1 min 2. H_2O	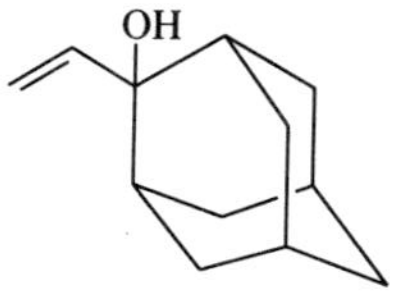 OH (structure) (37)	585
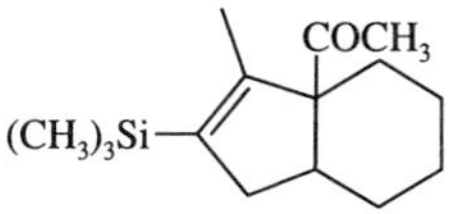 $(CH_3)_3Si$, $COCH_3$ (structure)	HF, CH_3CN, 48 h	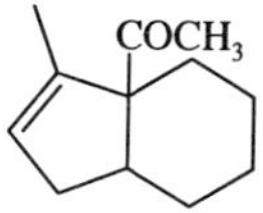 $COCH_3$ (structure) (96)	346
$(CH_3)_3SiCH{=}CHC(CH_3)_2COC_6H_5$	HI, C_6H_6	$CH_2{=}CHC(CH_3)_2COC_6H_5$ (75)	586
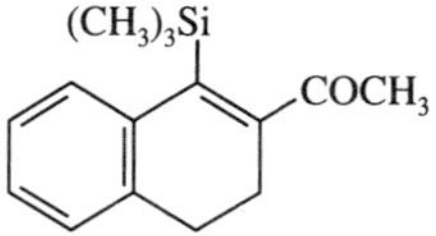 $(CH_3)_3Si$, $COCH_3$ (structure)	TsOH, C_6H_6, reflux, 2 h	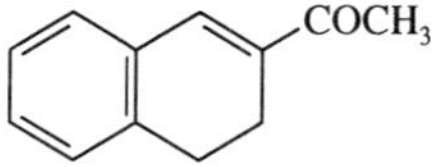 $COCH_3$ (structure) (96)	117

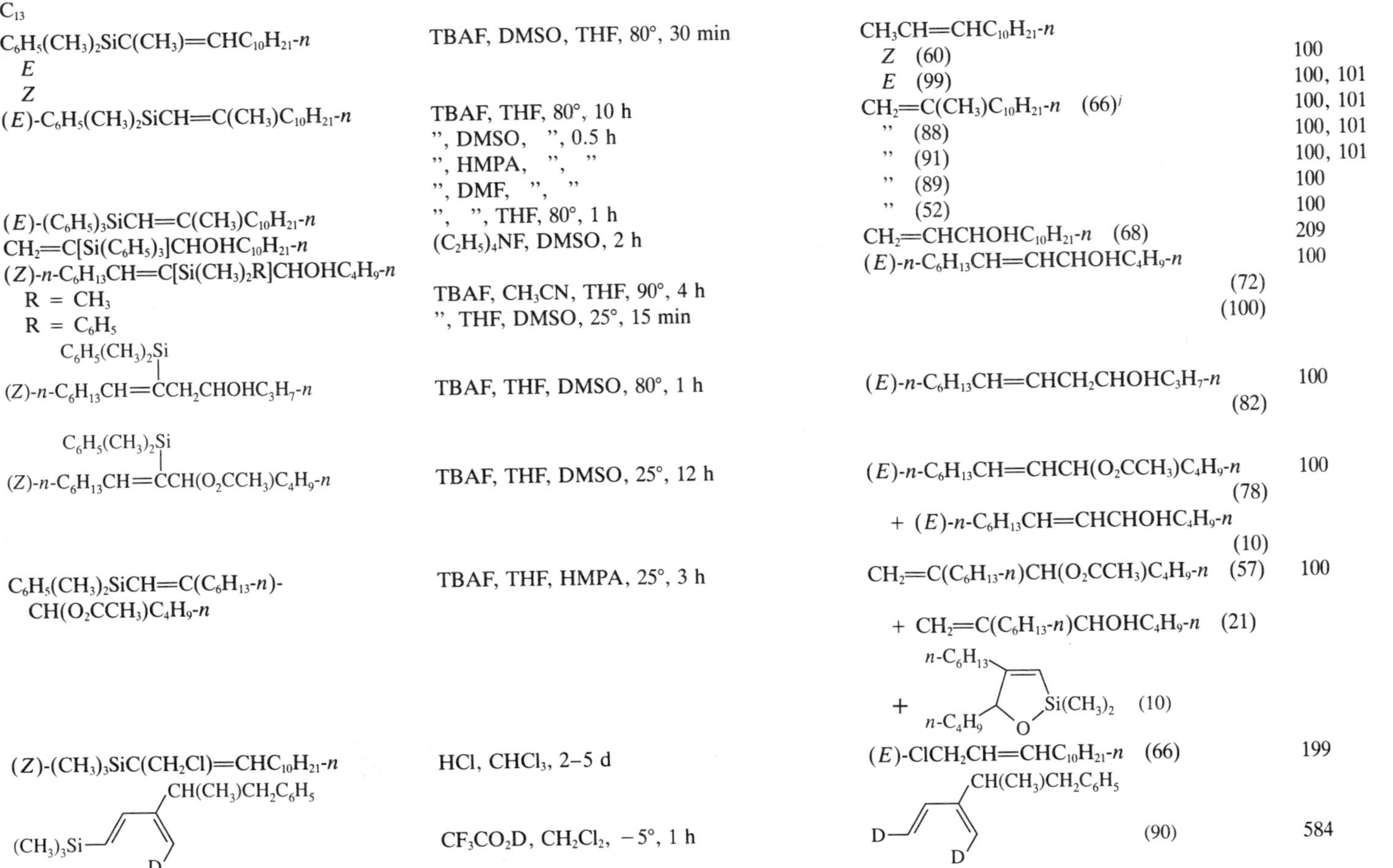

Reactant	Conditions	Product(s) and Yield(s) (%)	Refs.
C_{13}			
$C_6H_5(CH_3)_2SiC(CH_3){=}CHC_{10}H_{21}$-*n*	TBAF, DMSO, THF, 80°, 30 min	$CH_3CH{=}CHC_{10}H_{21}$-*n*	
E		*Z* (60)	100
Z		*E* (99)	100, 101
(*E*)-$C_6H_5(CH_3)_2SiCH{=}C(CH_3)C_{10}H_{21}$-*n*	TBAF, THF, 80°, 10 h	$CH_2{=}C(CH_3)C_{10}H_{21}$-*n* (66)[j]	100, 101
	", DMSO, ", 0.5 h	" (88)	100, 101
	", HMPA, ", "	" (91)	100, 101
	", DMF, ", "	" (89)	100
(*E*)-$(C_6H_5)_3SiCH{=}C(CH_3)C_{10}H_{21}$-*n*	", ", THF, 80°, 1 h	" (52)	100
$CH_2{=}C[Si(C_6H_5)_3]CHOHC_{10}H_{21}$-*n*	$(C_2H_5)_4NF$, DMSO, 2 h	$CH_2{=}CHCHOHC_{10}H_{21}$-*n* (68)	209
(*Z*)-*n*-$C_6H_{13}CH{=}C[Si(CH_3)_2R]CHOHC_4H_9$-*n*		(*E*)-*n*-$C_6H_{13}CH{=}CHCHOHC_4H_9$-*n*	100
R = CH_3	TBAF, CH_3CN, THF, 90°, 4 h	(72)	
R = C_6H_5	", THF, DMSO, 25°, 15 min	(100)	
(Z)-*n*-$C_6H_{13}CH{=}C[Si(CH_3)_2C_6H_5]CH_2CHOHC_3H_7$-*n*	TBAF, THF, DMSO, 80°, 1 h	(*E*)-*n*-$C_6H_{13}CH{=}CHCH_2CHOHC_3H_7$-*n* (82)	100
(Z)-*n*-$C_6H_{13}CH{=}C[Si(CH_3)_2C_6H_5]CH(O_2CCH_3)C_4H_9$-*n*	TBAF, THF, DMSO, 25°, 12 h	(*E*)-*n*-$C_6H_{13}CH{=}CHCH(O_2CCH_3)C_4H_9$-*n* (78) + (*E*)-*n*-$C_6H_{13}CH{=}CHCHOHC_4H_9$-*n* (10)	100
$C_6H_5(CH_3)_2SiCH{=}C(C_6H_{13}$-*n*)-$CH(O_2CCH_3)C_4H_9$-*n*	TBAF, THF, HMPA, 25°, 3 h	$CH_2{=}C(C_6H_{13}$-*n*)$CH(O_2CCH_3)C_4H_9$-*n* (57) + $CH_2{=}C(C_6H_{13}$-*n*)$CHOHC_4H_9$-*n* (21) + 2,2-dimethyl-4-(*n*-C_6H_{13})-5-(*n*-C_4H_9)-oxasilole [$Si(CH_3)_2$, O ring] (10)	100
(*Z*)-$(CH_3)_3SiC(CH_2Cl){=}CHC_{10}H_{21}$-*n*	HCl, $CHCl_3$, 2–5 d	(*E*)-$ClCH_2CH{=}CHC_{10}H_{21}$-*n* (66)	199
$(CH_3)_3Si$–CH=CH–C[$CH(CH_3)CH_2C_6H_5$]=CHD	CF_3CO_2D, CH_2Cl_2, −5°, 1 h	DCH=CH–C[$CH(CH_3)CH_2C_6H_5$]=CHD (90)	584

TABLE II. PROTODESILYLATION AND DEUTERODESILYLATION OF VINYLSILANES (*Continued*)

Reactant	Conditions	Product(s) and Yield(s)	Refs.
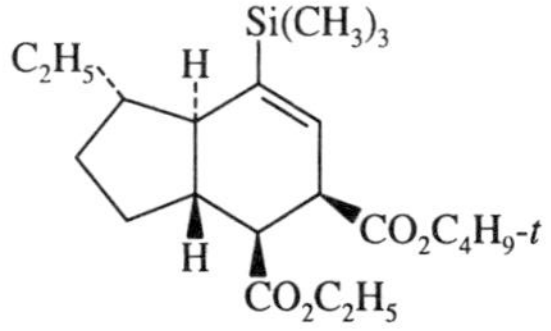 C_2H_5, H, $Si(CH_3)_3$, H, $CO_2C_4H_9$-*t*, $CO_2C_2H_5$	HF, CH_2Cl_2, 15 min	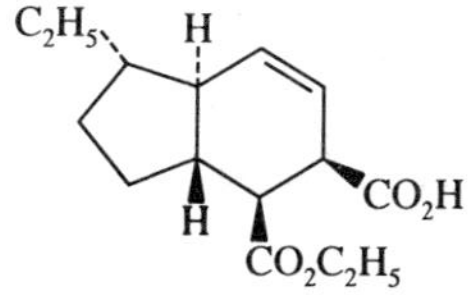 C_2H_5, H, H, CO_2H, $CO_2C_2H_5$ (92)	587
C_{14}			
(*E*)-$(CH_3)_3SiC(C_6H_{13}$-*n*$)=CHC_6H_{13}$-*n*	HI, C_6H_6, 15 min	(*Z*)-*n*-$C_6H_{13}CH=CHC_6H_{13}$-*n* (91)	72
	I_2, C_6H_6, D_2O, reflux, 1 h	(*Z*)-*n*-$C_6H_{13}CD=CHC_6H_{13}$-*n* 89% D_1 (90)	72
$C_6H_5(CH_3)_2SiCH=C(CH=CH_2)C_{10}H_{21}$-*n*	TBAF, DMF, THF, 25°, 0.5 h	$CH_2=C(CH=CH_2)C_{10}H_{21}$-*n* (82)	100
	", THF, 80°, 0.5 h	" (89)	100
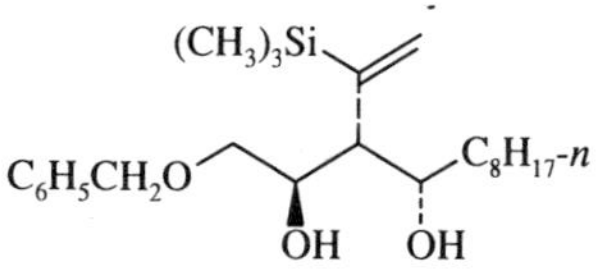 $(CH_3)_3Si$, $C_6H_5CH_2O$, C_8H_{17}-*n*, OH, OH	NaH, HMPA[k]	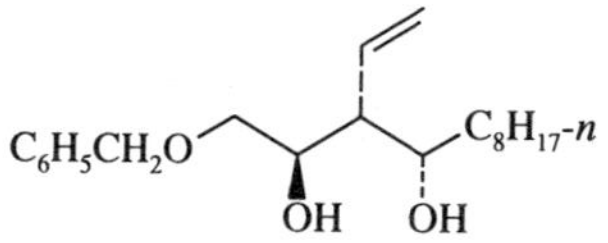 $C_6H_5CH_2O$, C_8H_{17}-*n*, OH, OH (—)	221
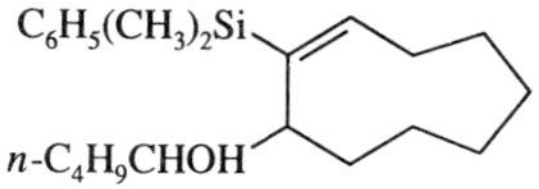 $C_6H_5(CH_3)_2Si$, *n*-C_4H_9CHOH	TBAF, THF, DMSO, 80°, 1 h	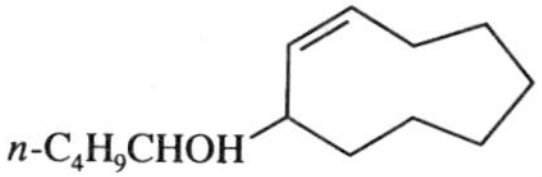 *n*-C_4H_9CHOH (93)	100
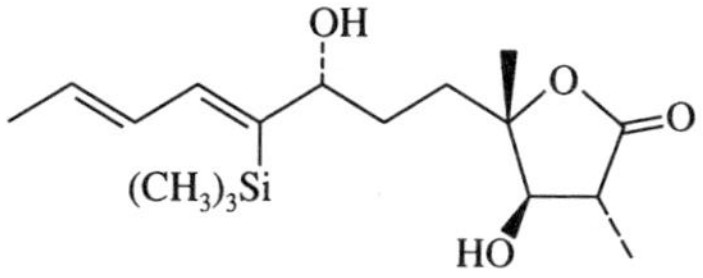 OH, O, O, $(CH_3)_3Si$, HO	TBAF, THF, 1 h, 45°, 4 h	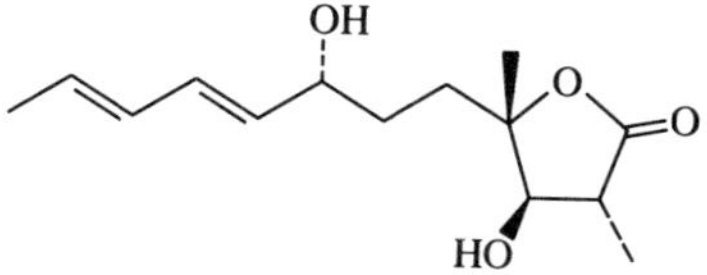 OH, O, O, HO (73)	126

Reactant	Conditions	Product(s) and Yield(s) (%)	Refs.
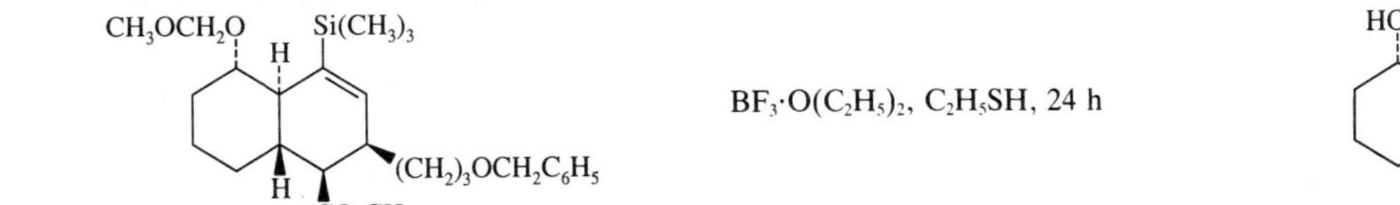	$BF_3 \cdot O(C_2H_5)_2$, C_2H_5SH, 24 h	(84)	587
$(CH_3)_3SiC(C_6H_5){=}CHC_6H_5$	HBF_4, CH_3CN, 25°,	$C_6H_5CH{=}CHC_6H_5$	588
E	22 min	*E*:*Z* 0.8:8 (9)	
”	124 min	” 11:35 (48)	
”			
Z	25 min	” 4:0 (5)	
”	122 min	” 28:0.3 (35)	
C_{15}			
$C_6H_5(CH_3)_2SiCH{=}C(CH_2CH{=}CH_2)C_{10}H_{21}$-*n*	TBAF, THF, 80°, 2 h	$CH_2{=}C(CH{=}CHCH_3)C_{10}H_{21}$-*n* (89)	101
	”, DMSO, ”, ”	” (100)	101
	”, HMPA, 25°, 1 h	” (88)	101
	TsH, CH_3CN, 10 h	(92)	197
	”, ”, reflux, 1.5 h	(92)	214
	TBAF, THF, DMSO, 80°, 0.5 h	(85)	100, 101
	K_2CO_3, CH_3OH, 9 h	COC_6H_5 + COC_6H_5 89:11 (68)	346

TABLE II. PROTODESILYLATION AND DEUTERODESILYLATION OF VINYLSILANES (*Continued*)

Reactant	Conditions	Product(s) and Yield(s)	Refs.
$(C_6H_5)_3SiC(=CH_2)C(C_6H_5)_2OH$	$(C_2H_5)_4NF$, DMSO, 2 h	$CH_2=CHC(C_6H_5)_2OH$ (65)	209
C_{16}			
(Z)-$(CH_3)_3SiC(=CH(CH_2)_4OTHP)CH_2CH_2CH=C(n\text{-}C_3H_7)_2$	I_2, C_6H_6, reflux, 2 h	(*E*)-$(n\text{-}C_3H_7)_2C=CH(CH_2)_2$-$CH=CH(CH_2)_4OTHP$ (83)	213
$(n\text{-}C_3H_7)_2C=C[Si(CH_3)_3]$-$(CH_2)_2C\equiv C(CH_2)_2OTHP$	1. I_2, C_6H_6, reflux, 2 h 2. H_3O^+	$(n\text{-}C_3H_7)_2C=CH(CH_2)_2C\equiv C(CH_2)_2OH$ (76)	589
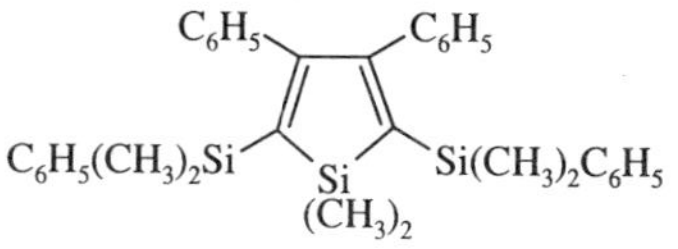 3,4-Diphenyl-1,1-dimethyl-2,5-bis[$Si(CH_3)_2C_6H_5$]silole	HCl, $(C_2H_5)_2O$, 7 h	$CH_2=C(C_6H_5)C(C_6H_5)=CH_2$ (84)	590
C_{19}			
$(CH_3)_3SiC(=CHC_{10}H_{21}\text{-}n)(CH_2)_4C_3H_7\text{-}i$ *E*:*Z* 87:13	HI, CH_2Cl_2, 30 min	$i\text{-}C_3H_7(CH_2)_4CH=CHC_{10}H_{21}\text{-}n$ (100) *E*:*Z* 13:87	212
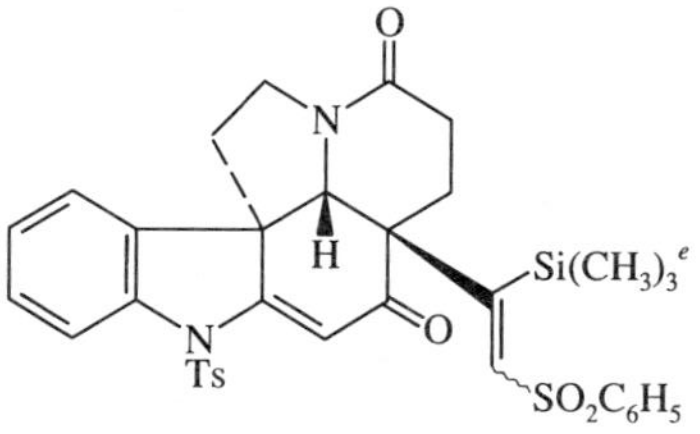 [structure: N-Ts indole-fused pentacyclic lactam ketone bearing $C(Si(CH_3)_3)=CHSO_2C_6H_5$][e]	CH_3CO_2H, THF	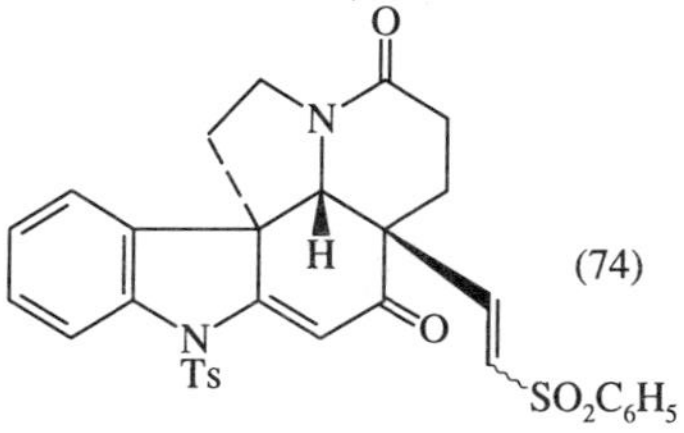 [structure: same skeleton bearing $CH=CHSO_2C_6H_5$] (74)	574
C_{21}			
$C_6H_5(CH_3)_2SiC(C_8H_{17}\text{-}n)=CHC_{11}H_{23}\text{-}n$ *E*:*Z* >95:5	TBAF, THF, HMPA, 90–100°, 1.5 h	$n\text{-}C_8H_{17}CH=CHC_{11}H_{23}\text{-}n$ (58) *E*:*Z* <2:98	576

C_{33}			
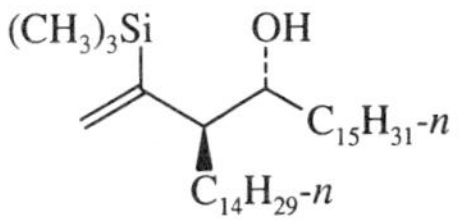 $(CH_3)_3Si$, OH, $C_{15}H_{31}$-*n*, $C_{14}H_{29}$-*n*	NaH, THF, HMPA, 30°, 1 h	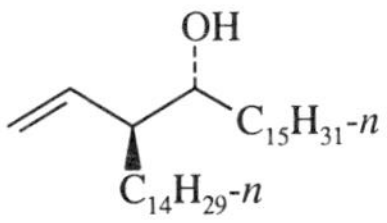 OH, $C_{15}H_{31}$-*n*, $C_{14}H_{29}$-*n* (86)	219

[a] No explanation is offered for the formation of this anomalous product. The probable structure is $CH_2{=}CH(CH_2)_2Si(CH_3)_3$.

[b] This vinylsilane is produced in situ by the addition of HBr to the corresponding acetylene $(CH_3)_3SiC{\equiv}CR$.

[c] This vinylsilane is formed as the major product, mixed with the corresponding allylsilane, by pyrolysis of the 2-phenyl sulfoxide.

[d] The yield by NMR is 75%.

[e] This compound is prepared in situ by conjugate addition of the appropriate enolate to $(CH_3)_3SiC{\equiv}CSO_2C_6H_5$ using NaH in THF at −80°.

[f] All of the starting material is consumed.

[g] See Table I for the full entry.

[h] The yield is of the derived diacetate.

[i] This product is isolated as its acetate, *n*-$C_4H_9CH{=}CH(CH_2)_6O_2CCH_3$.

[j] Starting material (28%) is also recovered.

[k] The reagents are not given in this paper, but this research group always uses the reagents quoted (see refs. 216, 217, and 258).

TABLE III. PROTODESILYLATION OF ALLENYLSILANES

Reactant	Conditions	Product(s) and Yield(s)	Refs.
C_3			
$(CH_3)_3SiC[SO_3Si(CH_3)_3]{=}C{=}CH_2$	H_2O, $(C_2H_5)_2O$	$HO_3SCH{=}C{=}CH_2$ + $HO_3SC{\equiv}CCH_3$ } (40) + $(CH_3)_3SiC(SO_3H){=}C{=}CH_2$ (60)	591
	Na_2CO_3, H_2O, $(C_2H_5)_2O$	$NaO_3SCH{=}C{=}CH_2$ (40) + $(CH_3)_3SiC(SO_3Na){=}C{=}CH_2$ (60)	591
$[(CH_3)_3Si]_2C{=}C{=}C[Si(CH_3)_3]_2$	TFA (1 eq), CCl_4, 0°	$[(CH_3)_3Si]_2CHC{\equiv}CSi(CH_3)_3$ (66) + $(CH_3)_3SiCH{=}C{=}CHSi(CH_3)_3$ (7) + $(CH_3)_3SiC{\equiv}CCH_2Si(CH_3)_3$ (7)	592
	TFA (2 eq), CCl_4, 0°, 10 min	$(CH_3)_3SiC{\equiv}CCH_2Si(CH_3)_3$ (85)	593, 592
C_4			
$(CH_3)_3SiCH{=}C{=}CHCH_2N(C_2H_5)_2$ + $(CH_3)_3SiC{\equiv}C(CH_2)_2N(C_2H_5)_2$	KOH, H_2O, reflux, 2 h	$CH_2{=}C{=}CHCH_2N(C_2H_5)_2$ + $HC{\equiv}C(CH_2)_2N(C_2H_5)_2$ (66)	594
$(CH_3)_3SiCBr{=}C{=}CHCH_2Br$	KOH, CH_3OH, −5°, 4 h	$BrC{\equiv}CCH{=}CH_2$ (—)	595
C_5			
$(CH_3)_3SiC(SO_3H){=}C{=}C(CH_3)CH_2Si(CH_3)_3$	$CHCl_3$, 30–35°, 2 d	$Si(CH_3)_3$ O_2S O (—)	223
C_6			
$(CH_3)_3SiCH{=}C{=}CHC_3H_7$-*n* + $(CH_3)_3SiC{\equiv}CC_4H_9$-*n*	KOH, CH_3OH, reflux, 1.5 h	$CH_2{=}C{=}CHC_3H_7$-*n* + $HC{\equiv}CCH_2C_3H_7$-*n* (—)	224
C_7			
$[(CH_3)_3Si]_2C{=}C{=}C(C_4H_9\text{-}t)SO_3Si(CH_3)_3$	1. H_2O 2. $NaHCO_3$	$(CH_3)_3SiC{\equiv}CCH(C_4H_9\text{-}t)SO_3Na$ (—)	424

Reactant	Conditions	Product(s) and Yield(s) (%)	Refs.
C_8			
$(CH_3)_3SiCH{=}C{=}CHC_5H_{11}$-*n* + $(CH_3)_3SiC{\equiv}CCH_2C_5H_{11}$-*n*	KOH, CH_3OH, reflux, 1.5 h	$CH_2{=}C{=}CHC_5H_{11}$-*n* + $HC{\equiv}CCH_2C_5H_{11}$-*n* 22:78 (—)	224
C_9			
$[(CH_3)_3Si]_2C{=}C{=}C(C_6H_5)Si(CH_3)_3$	HCl, CH_3OH	$(CH_3)_3SiCH{=}C{=}CHC_6H_5$ (65)	596
C_{10}			
$(CH_3)_3SiC(C_6H_4OCH_3$-*p*$){=}C{=}CHCH_3$	CsF, CH_3CN, 8 h	*p*-$CH_3OC_6H_4CH{=}C{=}CHCH_3$ (98)	597
C_{14}			
$(CH_3)_3SiC(C_6H_4OCH_3$-*p*$){=}C{=}C(CH_2)_5$	CsF, CH_3CN, 8 h	*p*-$CH_3OC_6H_4CH{=}C{=}C(CH_2)_5$ (99)	597
C_{15}			
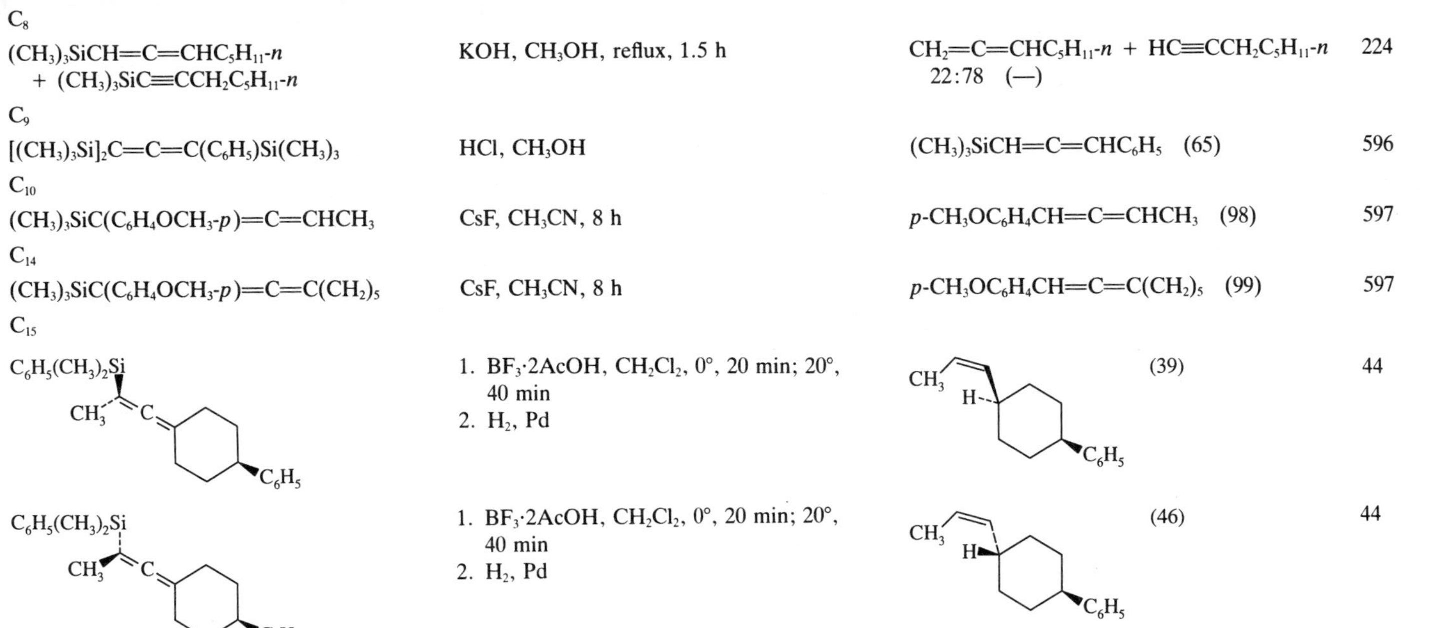	1. $BF_3{\cdot}2AcOH$, CH_2Cl_2, 0°, 20 min; 20°, 40 min 2. H_2, Pd	CH_3 H C_6H_5 (39)	44
$C_6H_5(CH_3)_2Si$ CH_3 C C_6H_5	1. $BF_3{\cdot}2AcOH$, CH_2Cl_2, 0°, 20 min; 20°, 40 min 2. H_2, Pd	CH_3 H C_6H_5 (46)	44

TABLE IV. Allylsilanes with Carbon Electrophiles Needing no Catalysis

Reactant	Conditions	Product(s) and Yield(s)	Refs.
C_3			
$(CH_3)_3SiCH_2CH{=}CH_2$	1,3-dithian-2-ylium (S⁺) BF_4^-, CH_2Cl_2, 1 h	$CH_2{=}CHCH_2$-(1,3-dithian-2-yl) (78)	519
	1. CSI, CCl_4, 0°, 0.5 h 2. Py, 0°	$CH_2{=}CHCH_2CN$ (64)	8
	$R^1R^2\overset{+}{C}C{\equiv}CH[Co_2(CO)_6]BF_4^-$, CH_2Cl_2, 0°	$R^1R^2C(CH_2CH{=}CH_2)C{\equiv}CH[Co_2(CO)_6]$	234
	R^1 R^2		
	H H	(83)	
	CH_3 CH_3	(70)	
	C_6H_5 C_6H_5	(92)	
	1. $RC{\equiv}\overset{+}{N}{-}O^-$, reflux, 12 h 2. Raney Ni or $LiAlH_4$, $TiCl_4$ 3. H^+ or base	$CH_2{=}CHCH_2CHRNH_2$ R = CH_3 and/or C_2H_5 (—)	231
	1. $RC{\equiv}\overset{+}{N}{-}O^-$, toluene, 25°, 16 h 2. Raney Ni, $B(OH)_3$, CH_3OH, H_2O, 2.5 h 3. $BF_3{\cdot}O(C_2H_5)_2$, CH_2Cl_2, 0°, 1 h	$CH_2{=}CHCH_2COR$ R = C_4H_9-*t* (53) R = C_6H_5 (57)	598
	1. $RCH{=}N^+(O^-)C_6H_5$, 120°, 40 h 2. Raney Ni or $LiAlH_4$, $TiCl_4$ 3. H^+ or base	$CH_2{=}CHCH_2CHRNHC_6H_5$	231
	R = *n*-C_3H_7	(72)	
	R = C_6H_5	(68)	
	$CH_2{=}CH\overset{+}{C}RC{\equiv}CH[Co_2(CO)_6]BF_4^-$, CH_2Cl_2, −78°	$CH_2{=}CH(CH_2)_2CH{=}CRC{\equiv}CH[Co_2(CO)_6]$	599
	R = H	(81)	
	R = CH_3	(76)	

Reactant	Conditions	Product(s) and Yield(s) (%)	Refs.
$(NC)_2C{=}C(CN)_2$		$CH_2{=}CHCH_2C(CN)_2CH(CN)_2$ + [cyclobutane: NC, CN, CN, CN; $(CH_3)_3Si$]	30
	CH_3CN, 50°, 1–3 d	87:13 (high)	
	$CH_3CO_2C_2H_5$	33:67 (—)	
	CH_2Cl_2	<5:95 (—)	
	C_6H_6	<5:95 (—)	
[cyclohexadienyl: R^1, R^2, R^3, $+$]—$Fe(CO)_3$ PF_6^-		$CH_2{=}CHCH_2$–[cyclohexadiene: R^1, R^2, R^3]—$Fe(CO)_3$	
R^1 R^2 R^3			
H H H	CH_3CN, 60°, 8 h	(52)	235
" " "	CH_2Cl_2, reflux	(78)	237
" CH_3 "	CH_2Cl_2, 40°, 36 h	(64)	235
OCH_3 H "	CH_2Cl_2, 40°, 20 h	(77)	235, 237
H OCH_3 "	CH_2Cl_2, 40°, 20 h	(77)	236
" " "	$Cl(CH_2)_2Cl$, reflux	(82)	237
—$(CH_2)_4$— "	CH_2Cl_2, 40°, 23 h	(98)	235
H CH_3 CO_2CH_3	CH_2Cl_2, 40°, 1 h	(80)	235, 237
$C_6H_5CH{=}CH$–[1,3-dithiolanylium: $\overset{+}{S}$, S] BF_4^-	CH_2Cl_2, 0°	$CH_2{=}CHCH_2CH(C_6H_5)CH{=}$[1,3-dithiolane: S, S] (86)	600
$C_6H_5CH{=}CH$–[1,3-dithianylium: $\overset{+}{S}$, S] BF_4^-	KF, CH_2Cl_2, 0°	$CH_2{=}CHCH_2CH(C_6H_5)CH{=}$[1,3-dithiane: S, S] (72)	600

TABLE IV. ALLYLSILANES WITH CARBON ELECTROPHILES NEEDING NO CATALYSIS (*Continued*)

Reactant	Conditions	Product(s) and Yield(s)	Refs.
$R_2(CH_2{=}CHCH_2)SiCH_2CHClCH_3$	250°	$R_2ClSiCH_2CH(CH_3)CH_2CH{=}CH_2$ + $R_2ClSiCH_2CH{=}CH_2$	247
R = CH_3		5:95 (—)	
R = C_6H_5		70:30 (—)	
C_4			
$(CH_3)_3SiCH(CH_3)CH{=}CH_2$	1. $[RC{\equiv}\overset{+}{N}{-}O^-]$, toluene, 25°, 16 h 2. H_2, Raney Ni, $B(OH)_3$, CH_3OH, H_2O, 2.5 h 3. $BF_3{\cdot}O(C_2H_5)_2$, CH_2Cl_2, 0°, 1 h	$CH_3CH{=}CHCH_2COR$	598
	R = t-C_4H_9	E:Z 1:2 (—)	
	R = C_6H_5	E:Z 1:1.5 (—)	
$(CH_3)_3SiCH_2CH{=}CHCH_2Si(CH_3)_3$	1. CSI, CCl_4, 60°, 1 h 2. Py, 0°	$CH_2{=}CHCH(CN)CH_2Si(CH_3)_3$ (72)	229
R($CH_3CHClCH_2$)Si(cyclopent-3-enyl) 	250°	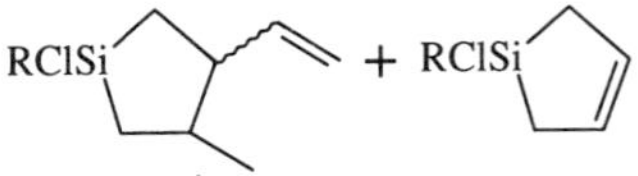 RClSi(3-methyl-4-vinylcyclopentyl) + RClSi(cyclopent-3-enyl)	247
R = CH_3		50:50 (—)	
R = C_6H_5		95:5 (—)	
C_5			
$(CH_3)_3SiCH(CH_3)CH{=}CHCH_3$	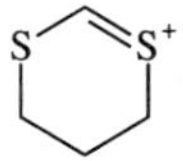 1,3-dithian-1-ium-2-ylidene BF_4^- ($S^+BF_4^-$), CH_2Cl_2, reflux, 30 min	$CH_3CH{=}CHCHCH_3$ 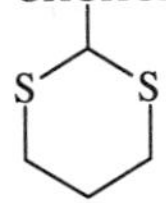 (2-(1,3-dithianyl)) (72)	519
$(CH_3)_3SiCH_2CH{=}C(CH_3)_2$	1. CSI, CCl_4, 1h 2. Py, 0°	$CH_2{=}CHC(CH_3)_2CN$ (80)	229
	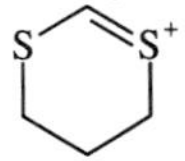 1,3-dithian-1-ium-2-ylidene BF_4^- ($S^+BF_4^-$), CH_2Cl_2, 4 h	$CH_2{=}CHC(CH_3)_2$  (2-(1,3-dithianyl)) (79)	519

Allylsilane	Electrophile, Conditions	Product(s) and Yield(s) (%)	Refs.
	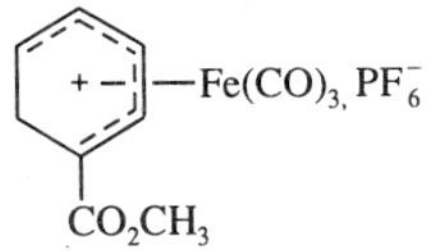 $-Fe(CO)_3, PF_6^-$; CO_2CH_3	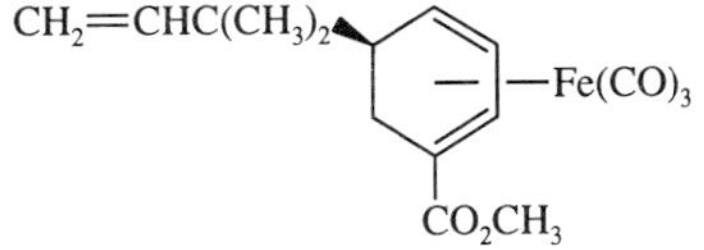 $CH_2{=}CHC(CH_3)_2$ … $-Fe(CO)_3$; CO_2CH_3 (95)	237
	$R^1R^2\overset{+}{C}C{\equiv}CH[Co_2(CO)_6]BF_4^-$, CH_2Cl_2, 0°	$CH_2{=}CHC(CH_3)_2CR^1R^2C{\equiv}CH[Co_2(CO)_6]$	234
	R^1 / R^2		
	H / H	(95)	
	CH_3 / CH_3	(0)	
	C_6H_5 / H	(91)	
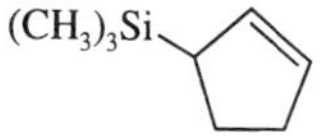 $(CH_3)_3Si$	$R^1R^2\overset{+}{C}C{\equiv}CH[Co_2(CO)_6]BF_4^-$, CH_2Cl_2	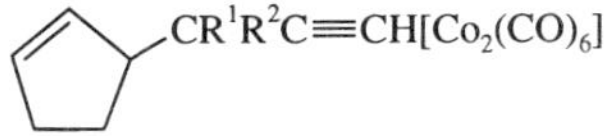 $CR^1R^2C{\equiv}CH[Co_2(CO)_6]$	234
	R^1 / R^2		
	H / H / 0°	(79)	
	CH_3 / CH_3 / "	(92)	
	C_6H_5 / H / "	(97)	
	$CH_2{=}CH$ / H / −78°	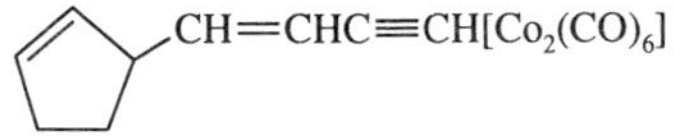 $CH{=}CHC{\equiv}CH[Co_2(CO)_6]$ (80)	599
C_7			
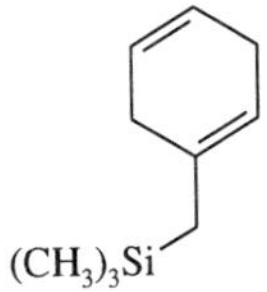 $(CH_3)_3Si$	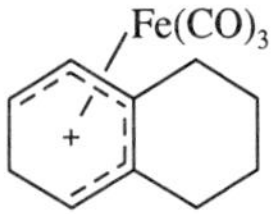 $Fe(CO)_3$; PF_6^-, CH_2Cl_2, reflux, 14 h	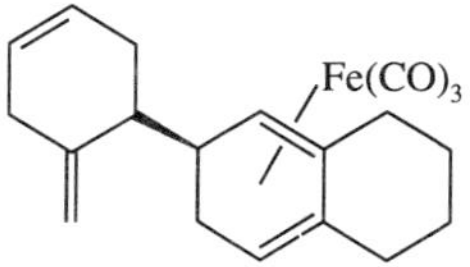 $Fe(CO)_3$ (98)	237
C_9			
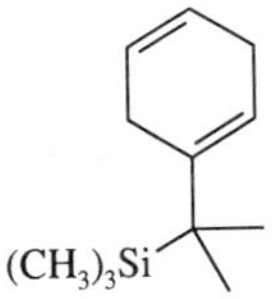 $(CH_3)_3Si$	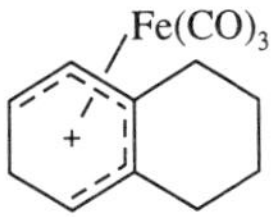 $Fe(CO)_3$; PF_6^-, CH_2Cl_2, reflux, 20 h	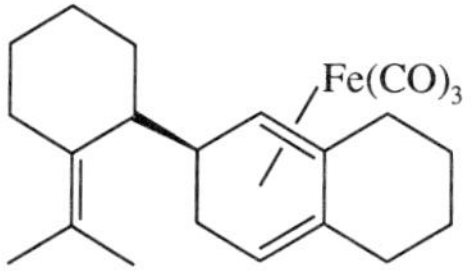 $Fe(CO)_3$ (86)	237

TABLE IV. ALLYLSILANES WITH CARBON ELECTROPHILES NEEDING NO CATALYSIS (*Continued*)

Reactant	Conditions	Product(s) and Yield(s)	Refs.
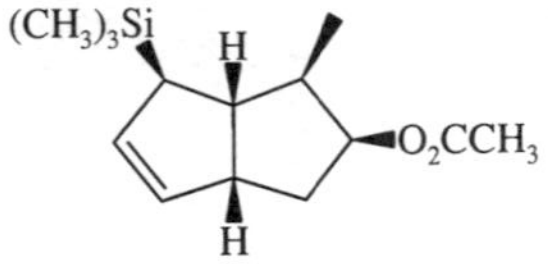 $(CH_3)_3Si$, H, H, O_2CCH_3	1. CSI, CCl_4, 2.5 h 2. HCl, $(CH_3)_2CO$, H_2O, 3.5 h	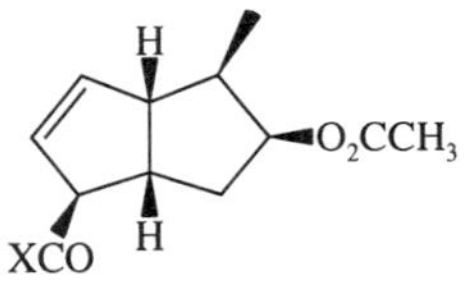 H, H, O_2CCH_3, XCO; X = NH_2 (55)	55
	1. CSI, CCl_4, 2.5 h 2. $NaNO_2$, CH_3CO_2H, $(CH_3CO)_2O$, 16 h 3. NaO_2CCH_3, H_2O, 0°, 2 h 4. CH_2N_2, $(C_2H_5)_2O$, 0.5 h	" X = OCH_3 (61)	
$(CH_3)_3SiCH_2CH{=}CHC_6H_5$	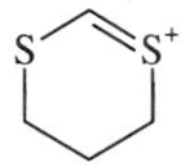 S, S^+ BF_4^-, CH_2Cl_2, reflux, 18 h	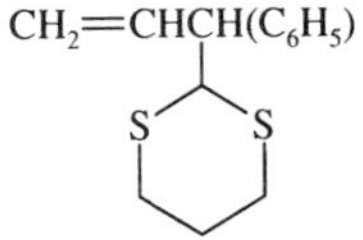 $CH_2{=}CHCH(C_6H_5)$, S, S (60)	519
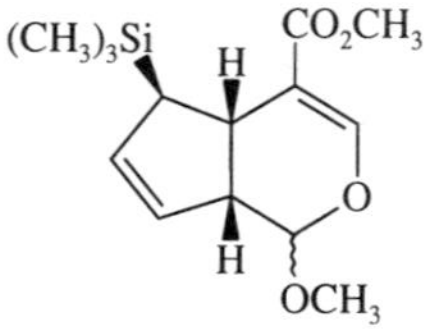 $(CH_3)_3Si$, H, H, CO_2CH_3, O, OCH_3	1. CSI 2. H_3O^+	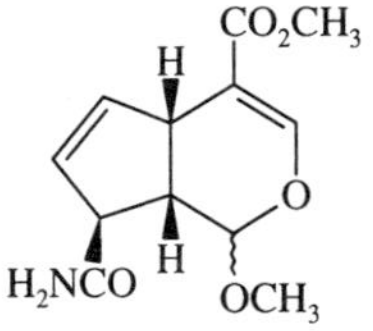 H, H, CO_2CH_3, O, H_2NCO, OCH_3 (30)	54
C_{11}			
$C_6H_5(CH_3)_2SiCH_2CH{=}CHCH(CH_3)C_6H_5$	1. CSI, CCl_4, 0°, 30 min 2. DMF, 50°, 10 min	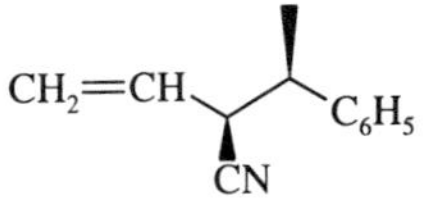 $CH_2{=}CH$, C_6H_5, CN (79)	409
C_{12}			
$C_6H_5(CH_3)_2SiCH_2C(CH_3){=}CHCH(CH_3)C_6H_5$ *E*:*Z* 74:26	1. CSI, CCl_4, 0°, 30 min 2. DMF, 50°, 10 min	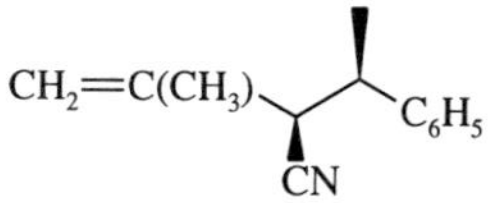 $CH_2{=}C(CH_3)$, C_6H_5, CN (82)	409

Reactant	Conditions	Product(s) and Yield(s) (%)	Refs.
$C_6H_5(CH_3)_2SiCH_2CH{=}C(CH_3)CH(CH_3)C_6H_5$ $E:Z$ 64:36	1. CSI, CCl_4, 0°, 30 min 2. DMF, 50°, 10 min	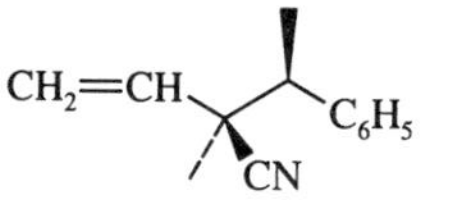 $CH_2{=}CH$, C_6H_5, CN (40)[a]	409
C_{13}			
$C_6H_5(CH_3)_2SiCH_2C(CH_3){=}C(CH_3)(CH(CH_3)C_6H_5)$ ($C_6H_5CHCH_3$ branch drawn below the $=CCH_3$ carbon) $E:Z$ 74:26	1. CSI, CCl_4, 0°, 30 min 2. DMF, 50°, 10 min	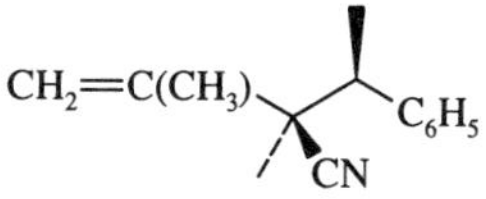 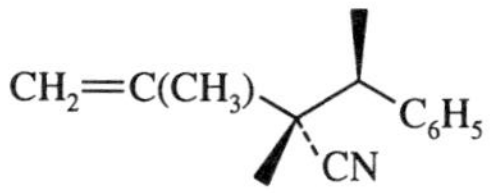 $CH_2{=}C(CH_3)$, C_6H_5, CN + $CH_2{=}C(CH_3)$, C_6H_5, CN 68:32 (81)[a]	409
C_{18}			
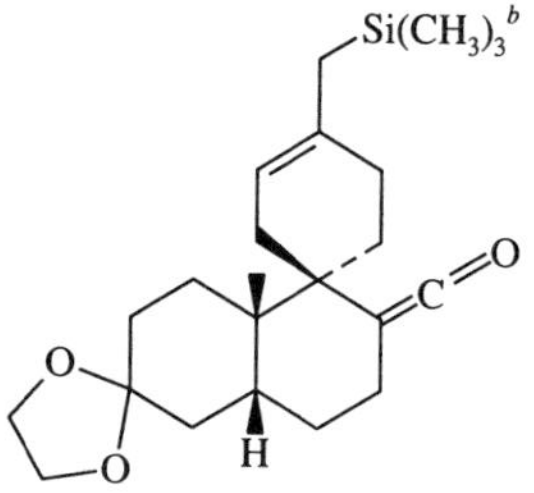 $Si(CH_3)_3$[b], C=O, O, O, H	SiO_2 gel, $(C_2H_5)_2O$, petroleum ether, 62 h	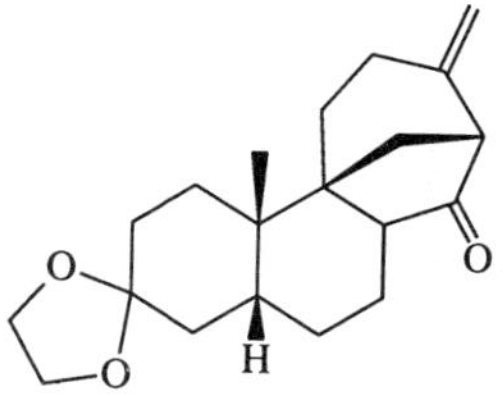 O, O, O, H (60)	230

[a] The relative stereochemistry is tentatively assigned, by analogy, but not proved.
[b] This ketene is prepared by a photo-Wolff rearrangement in ether at $-73°$, but not isolated.

TABLE V. VINYLSILANES WITH CARBON ELECTROPHILES NEEDING NO CATALYSIS

Reactant	Conditions	Product(s) and Yield(s)	Refs.
C_2			
$(CH_3)_3SiCH{=}CH_2$	1. $RCH{=}N^+(O^-)CH_3$, C_6H_6, reflux, 12–24 h 2. HF, CH_3CN, 0.5 h	(*E*)-RCH=CHCHO	239
	R		
	$C_2H_5O_2C$	(37)	
	2-furyl	(49)	
	C_6H_5	(89)	
	n-C_6H_{13}	(37)	
	2,5-bis($C_6H_5CH_2O$)phenyl-substituted 2-phenyl-4-methyl-1,3-dioxolan-4-yl (structure)	(63)	
C_8			
$C_6H_5CH{=}CHSi(CH_3)_3$	1. CSI, CCl_4, 0°, 0.5 h 2. HCl, $(CH_3)_2CO$, H_2O	$C_6H_5CH{=}CHCONH_2$ (63)	238

Reactant	Conditions	Product(s) and Yield(s) (%)	Refs.
C_{28}			
1,2,3,4-Tetraphenyl-1-sila-1,1-dimethylcyclopent-3-ene (C_6H_5 ×4, Si$(CH_3)_2$)	CSI, 75°, 24 h	No reaction	238
2,3,4,5-Tetraphenyl-1,1-dimethylsilole (C_6H_5 ×4, Si$(CH_3)_2$)	1. CSI, $CDCl_3$, 0°, 2 h 2. C_6H_5SH, −40°	β-Lactam-fused silacyclopentene (C_6H_5 ×4, NH, C=O, Si$(CH_3)_2$) (43)	238
	CSI, $(C_2H_5)_2O$, 12 h	Oxasilepin imine (C_6H_5 ×4, =NSO_2Cl, O, Si$(CH_3)_2$) (43)	238

TABLE VI. ALLYLSILANES WITH ALKYL HALIDES, ALCOHOLS, ETHERS, NITRO COMPOUNDS, ALKENES, AND ARENES

Reactant	Conditions	Product(s) and Yield(s)	Refs.
C_3			
$(CH_3)_3SiCH_2CH{=}CH_2$	$ClCH_2OCH_2C{\equiv}CH$, Zn	$CH_2{=}CH(CH_2)_2OCH_2C{\equiv}CH$ (46) + $(CH_3)_3SiCH_2CHCl(CH_2)_2OCH_2C{\equiv}CH$ (14)	601
	$(XCH_2)_2O$, $ZnCl_2$, $(C_2H_5)_2O$, 30°	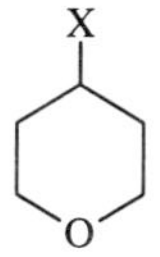 X = Cl (20); X = Br (26)	602
	Cl_2CHOCH_3, $TiCl_4$	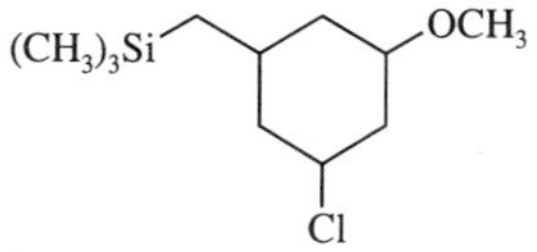 (—)	273
	$ClCH_2OCH_2C_6H_5$, $(CH_3)_3SiI$ (cat.), CH_3CN, 2.5 h	$CH_2{=}CH(CH_2)_2OCH_2C_6H_5$ (78)	251
	$ClCH(SC_6H_5)CO_2C_2H_5$, $SnCl_4$, CH_2Cl_2, 0°, 20 min	$CH_2{=}CHCH_2CH(SC_6H_5)CO_2C_2H_5$ (87)	361
	$ICFHCF_2Cl$, $Fe_3(CO)_{12}$, 60°, 3 h	$CH_2{=}CHCH_2CFHCF_2Cl$ (75)	244, 603
	$BrCFClCF_2Br$, ", ", 6 h	$CH_2{=}CHCH_2CFClCF_2Br$ (50)	603
	", $Ru_3(CO)_{12}$, ", "	" (59)	244, 603
	$ICFClCF_2Cl$, $h\nu$, 0°, 3 h	$CH_2{=}CHCH_2CFClCF_2Cl$ (65)	244
	", AIBN, 80°, 8 h	" (40)	244
	", $Fe_3(CO)_{12}$, 60°, 4 h	" (85)	244, 603
	", $Ru_3(CO)_{12}$, ", "	" (65)	603
	$I(CF_2)_2Br$, $Fe_3(CO)_{12}$, 60°, 6 h	$CH_2{=}CHCH_2(CF_2)_2Br$ (85)	244, 603
	", ", $H_2N(CH_2)_2OH$, 60°, 3 h	" (82)	603
	", $Ru_3(CO)_{12}$, ", "	" (67)	603

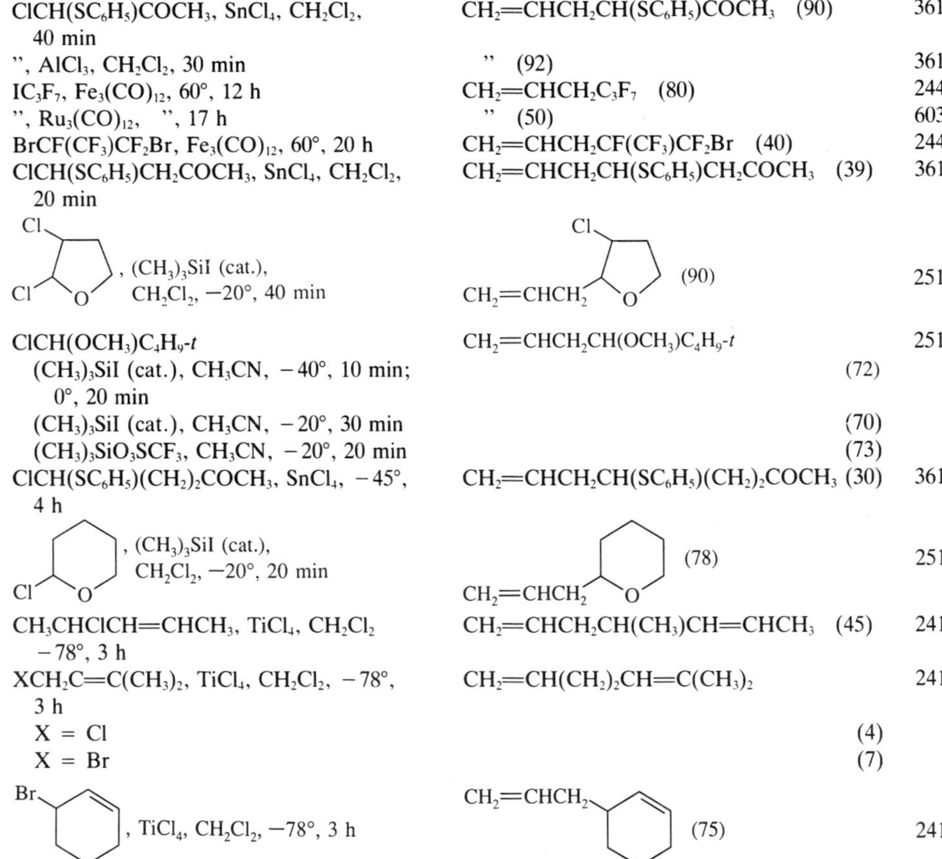

Reactant, Conditions	Product(s) and Yield(s) (%)	Refs.
$ClCH(SC_6H_5)COCH_3$, $SnCl_4$, CH_2Cl_2, 40 min	$CH_2{=}CHCH_2CH(SC_6H_5)COCH_3$ (90)	361
", $AlCl_3$, CH_2Cl_2, 30 min	" (92)	361
IC_3F_7, $Fe_3(CO)_{12}$, 60°, 12 h	$CH_2{=}CHCH_2C_3F_7$ (80)	244, 603
", $Ru_3(CO)_{12}$, ", 17 h	" (50)	603
$BrCF(CF_3)CF_2Br$, $Fe_3(CO)_{12}$, 60°, 20 h	$CH_2{=}CHCH_2CF(CF_3)CF_2Br$ (40)	244, 603
$ClCH(SC_6H_5)CH_2COCH_3$, $SnCl_4$, CH_2Cl_2, 20 min	$CH_2{=}CHCH_2CH(SC_6H_5)CH_2COCH_3$ (39)	361
2,3-dichlorotetrahydrofuran, $(CH_3)_3SiI$ (cat.), CH_2Cl_2, −20°, 40 min	2-allyl-3-chlorotetrahydrofuran (90)	251
$ClCH(OCH_3)C_4H_9$-*t*	$CH_2{=}CHCH_2CH(OCH_3)C_4H_9$-*t*	251
$(CH_3)_3SiI$ (cat.), CH_3CN, −40°, 10 min; 0°, 20 min	(72)	
$(CH_3)_3SiI$ (cat.), CH_3CN, −20°, 30 min	(70)	
$(CH_3)_3SiO_3SCF_3$, CH_3CN, −20°, 20 min	(73)	
$ClCH(SC_6H_5)(CH_2)_2COCH_3$, $SnCl_4$, −45°, 4 h	$CH_2{=}CHCH_2CH(SC_6H_5)(CH_2)_2COCH_3$ (30)	361
2-chlorotetrahydropyran, $(CH_3)_3SiI$ (cat.), CH_2Cl_2, −20°, 20 min	2-allyltetrahydropyran (78)	251
$CH_3CHClCH{=}CHCH_3$, $TiCl_4$, CH_2Cl_2 −78°, 3 h	$CH_2{=}CHCH_2CH(CH_3)CH{=}CHCH_3$ (45)	241
$XCH_2C{=}C(CH_3)_2$, $TiCl_4$, CH_2Cl_2, −78°, 3 h	$CH_2{=}CH(CH_2)_2CH{=}C(CH_3)_2$	241
X = Cl	(4)	
X = Br	(7)	
3-bromocyclohexene, $TiCl_4$, CH_2Cl_2, −78°, 3 h	3-allylcyclohexene (75)	241

TABLE VI. ALLYLSILANES WITH ALKYL HALIDES, ALCOHOLS, ETHERS, NITRO COMPOUNDS, ALKENES, AND ARENES (*Continued*)

Reactant	Conditions	Product(s) and Yield(s)	Refs.
	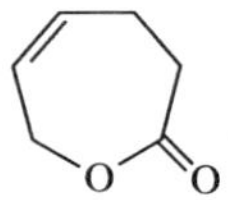 [structure], $(CH_3)_3OBF_4$, 94 h	(*Z*)-$CH_2{=}CH(CH_2)_2CH{=}CH(CH_2)_2CO_2CH_3$ (87)	242
	[structure], KF, 18-crown-6, THF, 25°, 4 d	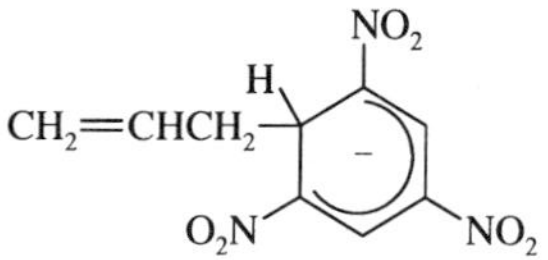 [structure] K^+ (45)	604
	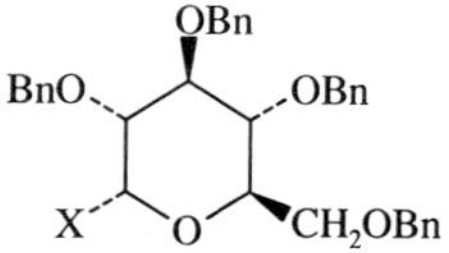 [structure]	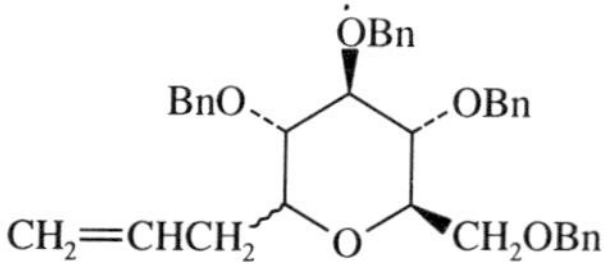 [structure]	
	X = F $BF_3{\cdot}O(C_2H_5)_2$, CH_2Cl_2	α:β >20:1 (95)	255
	X = Cl $(CH_3)_3SiO_3SCF_3$, CH_3CN, 6.5 h	" 10:1 (81)	254
	" $(CH_3)_3SiI$, CH_3CN, 16 h	" " (75)	254
	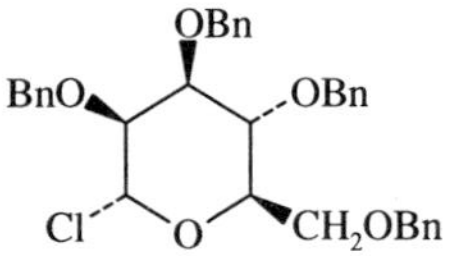 [structure], $(CH_3)_3SiO_3SCF_3$, CH_3CN, 4 h	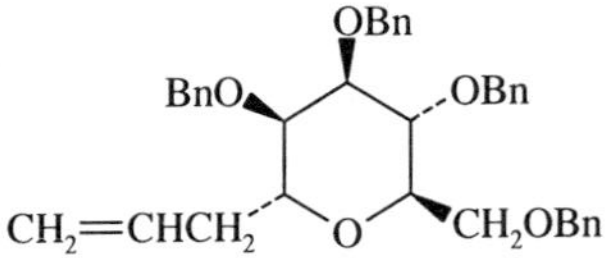 [structure] (80)	254
	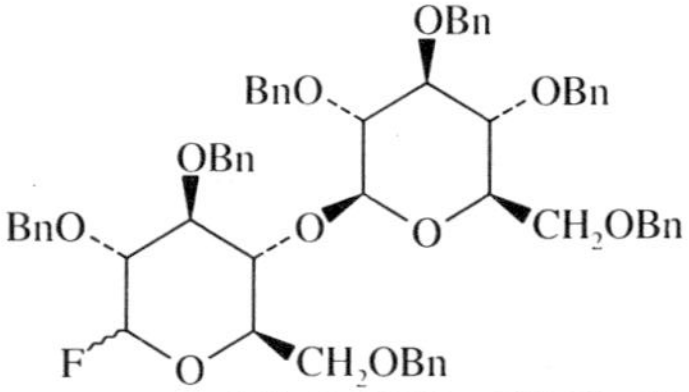 [structure] $BF_3{\cdot}O(C_2H_5)_2$, CH_2Cl_2, 0°	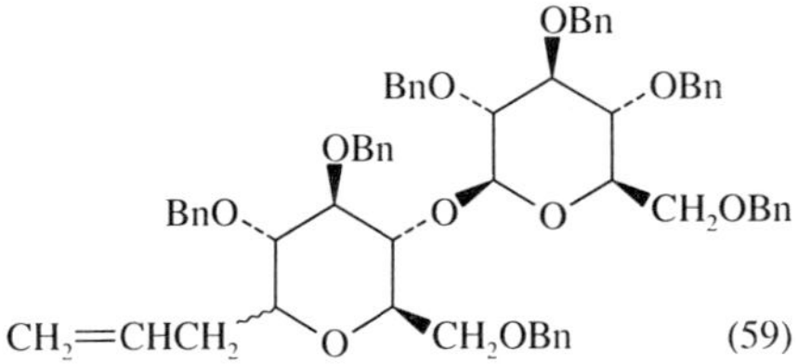 [structure] (59)	255

Reactant and Conditions	Product(s) and Yield(s) (%)	Refs.
$XCH_2C_6H_5$, $TiCl_4$, CH_2Cl_2, −78°, 3 h	$CH_2{=}CH(CH_2)_2C_6H_5$	241
X = Cl	(trace)	
Br	(14)	
I	(3)	
$R^1CH{=}CR^2$–(2-R^3-γ-butyrolactone-2-yl) I, $(CH_3)_3OBF_4$, 94 h	$CH_2{=}CHCH_2CHR^1CR^2{=}CR^3(CH_2)_2CO_2CH_3$	242
R^1 / R^2 / R^3		
H / H / H	(100)	
CH_3 / " / "	(85)	
H / CH_3 / "	(81)	
" / " / CH_3	(74)	
I(R^1 = R^3 = H, R^2 = CH_3)$(C_2H_5)_3OBF_4$	$CH_2{=}C(CH_2)_2CCH_3$ ‖ $C_2H_5O_2C(CH_2)_2CH$ (61)	605
n-$C_4H_9CH(OR)CH{=}CH_2$, $TiCl_4$, CH_2Cl_2, −78°, 3 h	$CH_2{=}CHCH_2CH(C_4H_9\text{-}n)CH{=}CH_2$ + $CH_2{=}CH(CH_2)_2CH{=}CHC_4H_9\text{-}n$	241
R = CH_3	21:79 (67)	
R = $Si(CH_3)_3$	30:70 (60)	
$CH_2{=}CHC(CH_3)(NO_2)(CH_2)_2CO_2CH_3$, $SnCl_4$, CH_2Cl_2, −10°, 10 min	$CH_2{=}CH(CH_2)_2CH$ ‖ $CH_3O_2C(CH_2)_2CCH_3$ (70)	245
n-$C_4H_9CH(OCH_3)CH{=}CHCH_3$, $TiCl_4$, CH_2Cl_2, −78°, 3 h	$CH_2{=}CHCH_2CH(C_4H_9\text{-}n)CH{=}CHCH_3$ + $CH_2{=}CHCH_2CH(CH_3)CH{=}CHC_4H_9\text{-}n$ 35:65 (75)	241
$C_6H_5CHOHCH_3$, BF_3, CH_2Cl_2, 0°	$CH_2{=}CHCH_2CH(CH_3)C_6H_5$ (22)	246
$NCC_6H_4CN\text{-}p$, $h\nu$, CH_3CN, 30–40 h	$CH_2{=}CHCH_2C_6H_4CN\text{-}p$ (66)	261
$NCC_6H_4CN\text{-}o$, $h\nu$, CH_3CN, 30–40 h	$CH_2{=}CHCH_2C_6H_4CN\text{-}o$ (67)	261
$BrCH_2$–(2-oxabicyclo[3.3.0]octan-3-yl), $AgBF_4$, CH_2Cl_2, 19 h	3-CH_3-3-($CH_2{=}CHCH_2$)-2-oxabicyclo[3.3.0]octane (50) + $CH_2{=}CH(CH_2)_2$–(2-oxabicyclo[3.3.0]octan-3-yl) (4)	257

TABLE VI. ALLYLSILANES WITH ALKYL HALIDES, ALCOHOLS, ETHERS, NITRO COMPOUNDS, ALKENES, AND ARENES (*Continued*)

Reactant	Conditions	Product(s) and Yield(s)	Refs.
	$ClCH(SC_6H_5)COC_6H_{13}$-*n*, $SnCl_4$ CH_2Cl_2, 20 min	$CH_2{=}CHCH_2CH(SC_6H_5)COC_6H_{13}$-*n* (91)	361
	IC_8F_{17}, $Ru_3(CO)_{12}$, 60°, 19 h	$CH_2{=}CHCH_2C_8F_{17}$ (71)	244, 603
	$ArC(CH_3)_2OH$, BF_3, CH_2Cl_2, 0°	$CH_2{=}CHCH_2C(CH_3)_2Ar$	246
	Ar		
	C_6H_5	(87)	
	m-HOC_6H_4	(79)	
	p-HOC_6H_4	(50)	
	p-(*i*-C_3H_7)C_6H_4	(90)	
	$C_6H_5C(CH_3)_2NO_2$, $SnCl_4$, CH_2Cl_2, −10°, 30 min	$CH_2{=}CHCH_2C(CH_3)_2C_6H_5$ (40)	245
	$BrCH_2$ [structure], $AgBF_4$, CH_2Cl_2, 22 h	CH_3, $CH_2{=}CHCH_2$ [structure] (13) + $CH_2{=}CH(CH_2)_2$ [structure] (69)	257
	$(CO)_3Cr$ [structure] O_2CCH_3, $TiCl_4$, CH_2Cl_2, 0°, 1 h	$(CO)_3Cr$ [structure] $CH_2CH{=}CH_2$ (80)	606
	[structure] OH, OH, CF_3SO_3H, CH_3CN, 1 h	[structure] OH, $CH_2CH{=}CH_2$ (18)	194

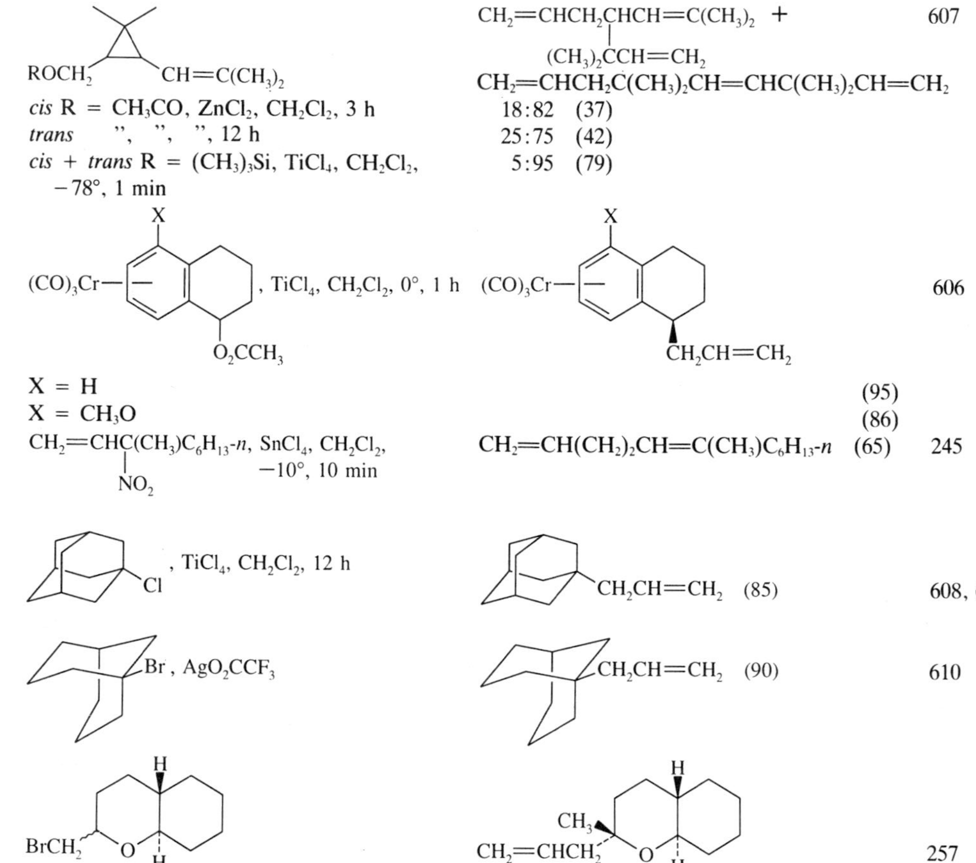

Reactant and Conditions	Product(s) and Yield(s) (%)	Refs.
$ROCH_2$ (2,2-dimethylcyclopropyl) $CH{=}C(CH_3)_2$	$CH_2{=}CHCH_2CHCH{=}C(CH_3)_2$ [branch $(CH_3)_2CCH{=}CH_2$] + $CH_2{=}CHCH_2C(CH_3)_2CH{=}CHC(CH_3)_2CH{=}CH_2$	607
cis R = CH_3CO, $ZnCl_2$, CH_2Cl_2, 3 h	18:82 (37)	
trans ", ", ", 12 h	25:75 (42)	
cis + *trans* R = $(CH_3)_3Si$, $TiCl_4$, CH_2Cl_2, −78°, 1 min	5:95 (79)	
$(CO)_3Cr$–(X-tetrahydronaphthalene, O_2CCH_3), $TiCl_4$, CH_2Cl_2, 0°, 1 h	$(CO)_3Cr$–(X-tetrahydronaphthalene, $CH_2CH{=}CH_2$)	606
X = H	(95)	
X = CH_3O	(86)	
$CH_2{=}CHC(CH_3)(NO_2)C_6H_{13}$-*n*, $SnCl_4$, CH_2Cl_2, −10°, 10 min	$CH_2{=}CH(CH_2)_2CH{=}C(CH_3)C_6H_{13}$-*n* (65)	245
(adamantyl)Cl, $TiCl_4$, CH_2Cl_2, 12 h	(adamantyl)$CH_2CH{=}CH_2$ (85)	608, 609
(bicyclic)Br, AgO_2CCF_3	(bicyclic)$CH_2CH{=}CH_2$ (90)	610
$BrCH_2$ (octahydrochromene, H, H)	$CH_2{=}CHCH_2$, CH_3 (octahydrochromene, H, H)	257

TABLE VI. ALLYLSILANES WITH ALKYL HALIDES, ALCOHOLS, ETHERS, NITRO COMPOUNDS, ALKENES, AND ARENES (*Continued*)

Reactant	Conditions	Product(s) and Yield(s)	Refs.
	α-$BrCH_2$, $AgBF_4$, CH_2Cl_2, 44 h	(41)	
	β-$BrCH_2$, ", ", 90 h	(73)	
	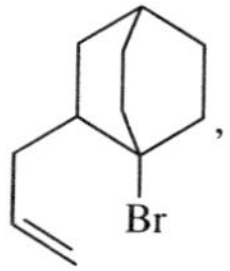 [structure: allyl bicyclo[2.2.2]octyl bromide, Br], $AgBF_4$, CH_2Cl_2, 0°	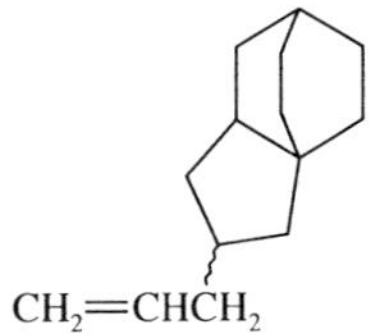 [structure: tricyclic product bearing CH_2=CHCH$_2$] (71)	611
	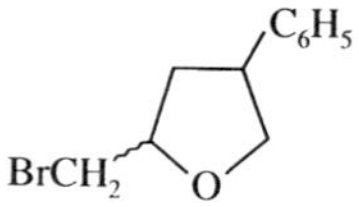 [structure: $BrCH_2$-tetrahydrofuran bearing C_6H_5], $AgBF_4$, CH_2Cl_2, 24 h	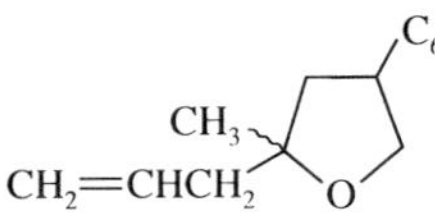 [structure: CH_3, CH_2=CHCH$_2$-tetrahydrofuran bearing C_6H_5] (71)	257
	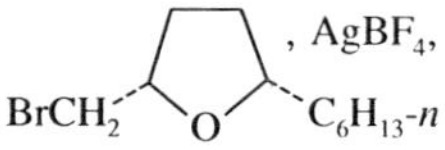 [structure: $BrCH_2$-tetrahydrofuran-C_6H_{13}-*n*], $AgBF_4$, CH_2Cl_2, 21 h	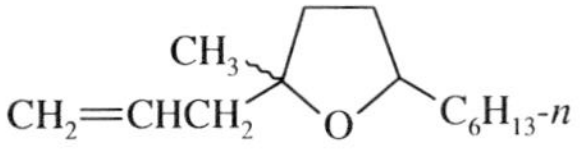 [structure: CH_3, CH_2=CHCH$_2$-tetrahydrofuran-C_6H_{13}-*n*] (14) 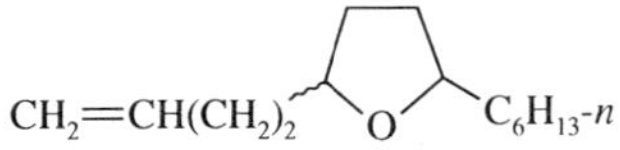 + [structure: CH_2=CH(CH_2)$_2$-tetrahydrofuran-C_6H_{13}-*n*] (14)	257
	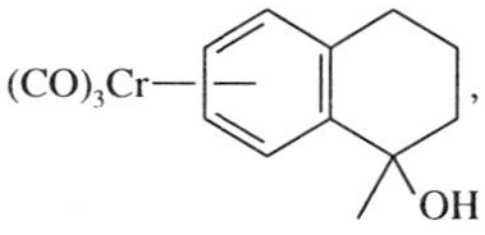 [structure: $(CO)_3Cr$-tetralin bearing CH_3, OH], $TiCl_4$, CH_2Cl_2, 0°, 1 h	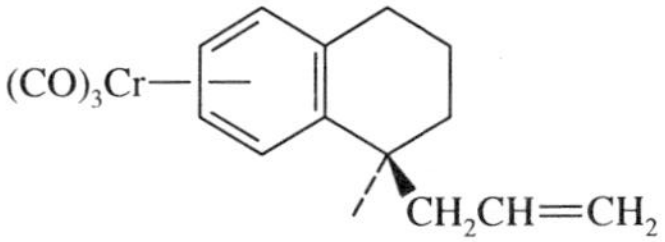 [structure: $(CO)_3Cr$-tetralin bearing CH_3, CH_2CH=CH_2] (99)	606
	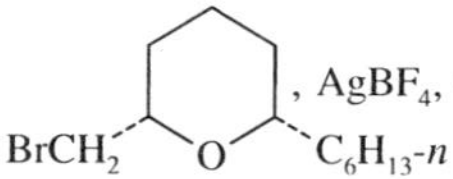 [structure: $BrCH_2$-tetrahydropyran-C_6H_{13}-*n*], $AgBF_4$, CH_2Cl_2, 42 h	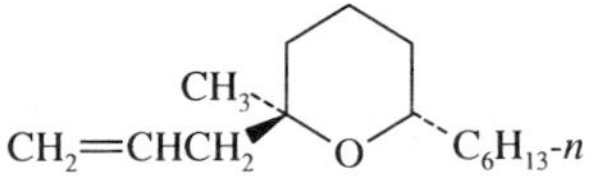 [structure: CH_3, CH_2=CHCH$_2$-tetrahydropyran-C_6H_{13}-*n*] (88)	257

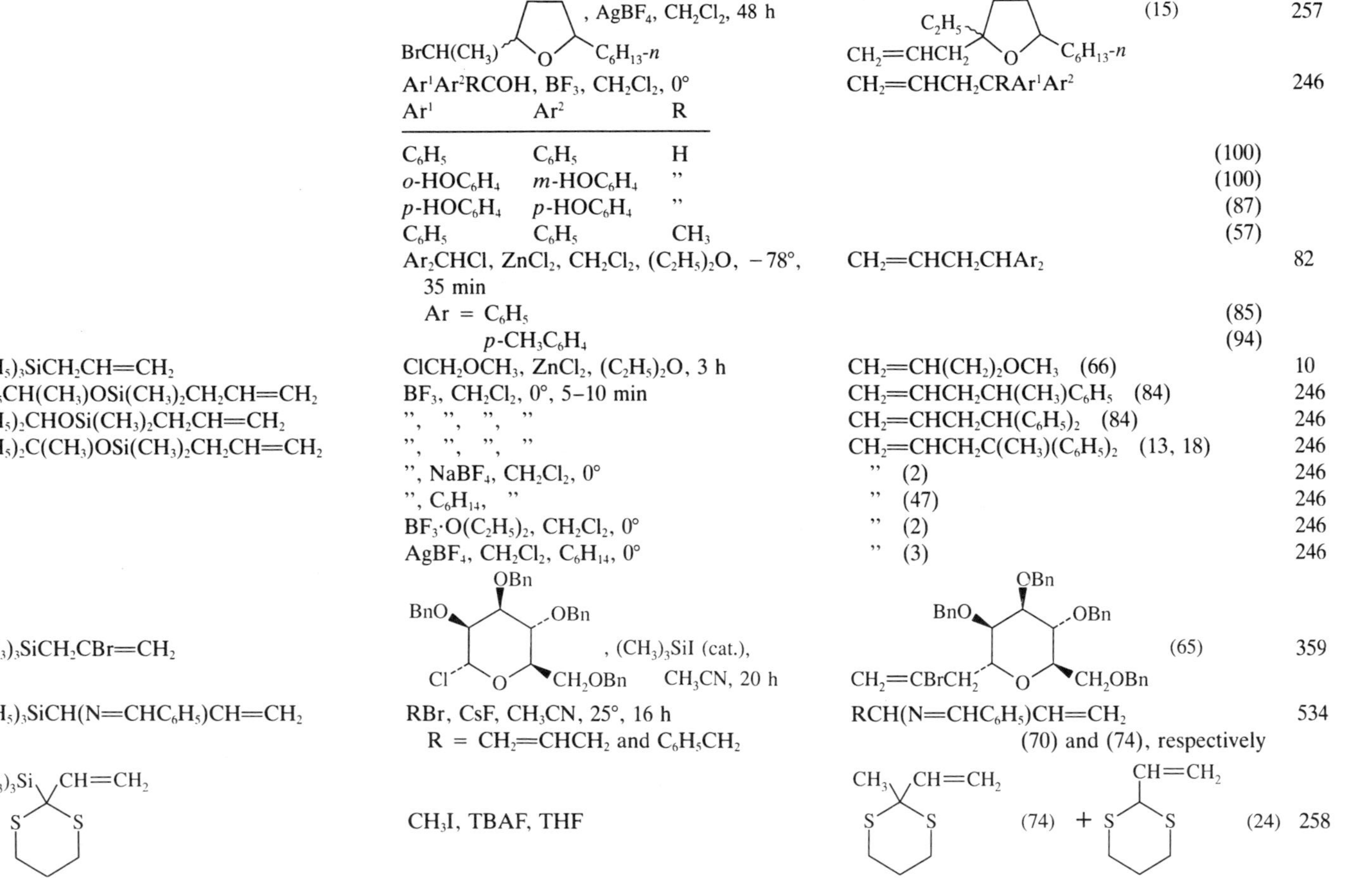

Reactant	Conditions	Product(s) and Yield(s) (%)	Refs.
	$BrCH(CH_3)$–(tetrahydrofuran)–C_6H_{13}-n, $AgBF_4$, CH_2Cl_2, 48 h	C_2H_5, $CH_2{=}CHCH_2$–(tetrahydrofuran)–C_6H_{13}-n (15)	257
	Ar^1Ar^2RCOH, BF_3, CH_2Cl_2, 0°	$CH_2{=}CHCH_2CRAr^1Ar^2$	246
	Ar^1 / Ar^2 / R		
	C_6H_5 / C_6H_5 / H	(100)	
	o-HOC_6H_4 / m-HOC_6H_4 / "	(100)	
	p-HOC_6H_4 / p-HOC_6H_4 / "	(87)	
	C_6H_5 / C_6H_5 / CH_3	(57)	
	Ar_2CHCl, $ZnCl_2$, CH_2Cl_2, $(C_2H_5)_2O$, −78°, 35 min	$CH_2{=}CHCH_2CHAr_2$	82
	Ar = C_6H_5	(85)	
	p-$CH_3C_6H_4$	(94)	
$(C_2H_5)_3SiCH_2CH{=}CH_2$	$ClCH_2OCH_3$, $ZnCl_2$, $(C_2H_5)_2O$, 3 h	$CH_2{=}CH(CH_2)_2OCH_3$ (66)	10
$C_6H_5CH(CH_3)OSi(CH_3)_2CH_2CH{=}CH_2$	BF_3, CH_2Cl_2, 0°, 5–10 min	$CH_2{=}CHCH_2CH(CH_3)C_6H_5$ (84)	246
$(C_6H_5)_2CHOSi(CH_3)_2CH_2CH{=}CH_2$	", ", ", "	$CH_2{=}CHCH_2CH(C_6H_5)_2$ (84)	246
$(C_6H_5)_2C(CH_3)OSi(CH_3)_2CH_2CH{=}CH_2$	", ", ", "	$CH_2{=}CHCH_2C(CH_3)(C_6H_5)_2$ (13, 18)	246
	", $NaBF_4$, CH_2Cl_2, 0°	" (2)	246
	", C_6H_{14}, "	" (47)	246
	$BF_3{\cdot}O(C_2H_5)_2$, CH_2Cl_2, 0°	" (2)	246
	$AgBF_4$, CH_2Cl_2, C_6H_{14}, 0°	" (3)	246
$(CH_3)_3SiCH_2CBr{=}CH_2$	(BnO, OBn, OBn, CH_2OBn, Cl-pyranose), $(CH_3)_3SiI$ (cat.), CH_3CN, 20 h	(BnO, OBn, OBn, CH_2OBn, $CH_2{=}CBrCH_2$-pyranose) (65)	359
$(C_2H_5)_3SiCH(N{=}CHC_6H_5)CH{=}CH_2$	RBr, CsF, CH_3CN, 25°, 16 h; R = $CH_2{=}CHCH_2$ and $C_6H_5CH_2$	$RCH(N{=}CHC_6H_5)CH{=}CH_2$ (70) and (74), respectively	534
2-$(CH_3)_3Si$-2-$CH{=}CH_2$-1,3-dithiane	CH_3I, TBAF, THF	2-CH_3-2-$CH{=}CH_2$-1,3-dithiane (74) + 2-$CH{=}CH_2$-1,3-dithiane (24)	258

TABLE VI. Allylsilanes with Alkyl Halides, Alcohols, Ethers, Nitro Compounds, Alkenes, and Arenes (*Continued*)

Reactant	Conditions	Product(s) and Yield(s)	Refs.
C_4			
$(CH_3)_3SiCH_2CH{=}CHCH_3$	$ClCH_2OCH_2C_6H_5$, $(CH_3)_3SiO_3SCF_3$ (cat.), CH_3CN, 3 h	$CH_2{=}CHCH(CH_3)CH_2OCH_2C_6H_5$ (72)	251
	$ClCH(OCH_3)C_4H_9$-*i*, $(CH_3)_3SiO_3SCF_3$ (cat.), CH_3CN, −20°, 15 min	$CH_2{=}CHCH(CH_3)CH(OCH_3)C_4H_9$-*i* (80)	251
	p-NCC_6H_4CN, $h\nu$, CH_3CN, 30–40 h	$CH_2{=}CHCH(CH_3)C_6H_4CN$-*p* + $CH_3CH{=}CHCH_2C_6H_4CN$-*p* 3:2 (80)	261
	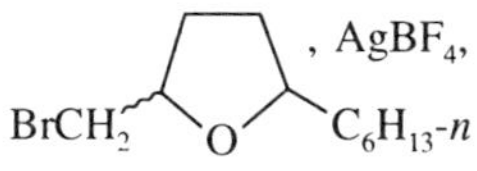 2-($BrCH_2$)-5-(C_6H_{13}-*n*)tetrahydrofuran, $AgBF_4$, CH_2Cl_2, 44 h	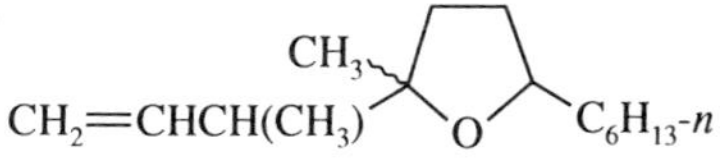 2-CH_3-2-[$CH_2{=}CHCH(CH_3)$]-5-(C_6H_{13}-*n*)tetrahydrofuran (16) 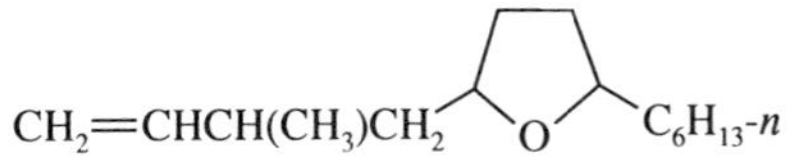 + 2-[$CH_2{=}CHCH(CH_3)CH_2$]-5-(C_6H_{13}-*n*)tetrahydrofuran (60)	257
	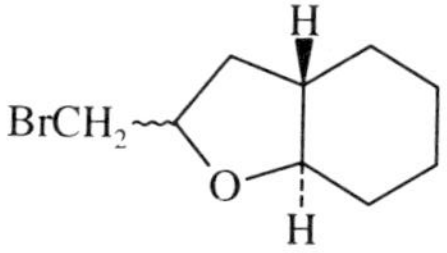 2-($BrCH_2$)-cis-hexahydrobenzofuran, $AgBF_4$, CH_2Cl_2, 45 h	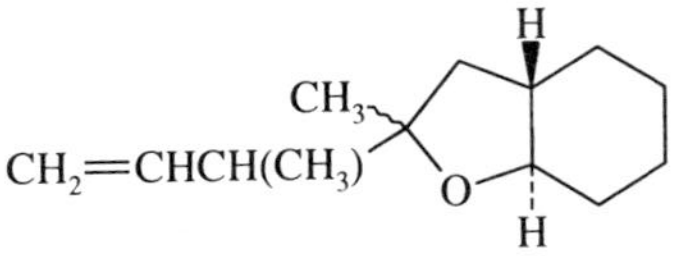 2-CH_3-2-[$CH_2{=}CHCH(CH_3)$]-cis-hexahydrobenzofuran (20) 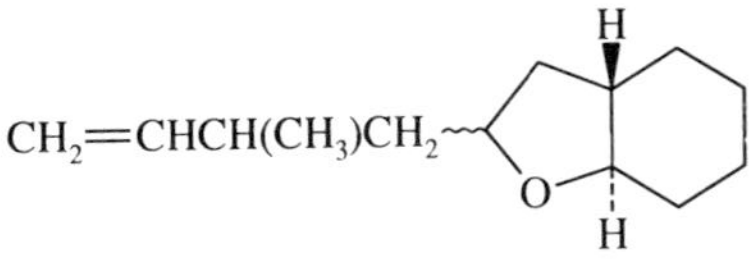 + 2-[$CH_2{=}CHCH(CH_3)CH_2$]-cis-hexahydrobenzofuran (58)	257
	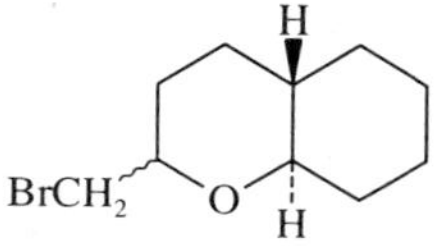 2-($BrCH_2$)-cis-octahydrochromene, $AgBF_4$, CH_2Cl_2, 70 h	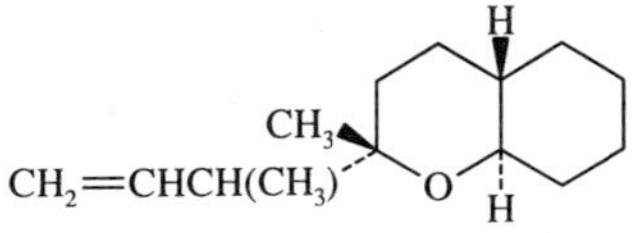 2-CH_3-2-[$CH_2{=}CHCH(CH_3)$]-cis-octahydrochromene (97)	257

(E)-$(CH_3)_3SiCH_2CH{=}CHCH_3$ $E:Z$ 94:6	$ClCH(OCH_3)C_6H_5$, $(CH_3)_3SiO_3SCF_3$ (cat.), CH_2Cl_2	$CH_2{=}CHCH(CH_3)CH(OCH_3)C_6H_5$ *syn* : *anti* 84:16 (53)	366
(Z)-$(CH_3)_3SiCH_2CH{=}CHCH_3$ $E:Z$ 11:89	", ", "	*syn* : *anti* 29:71 (80)	366
$(CH_3)_2SiHCH(CH_3)CH{=}CH_2$	$ICFClCF_2Cl$, $Ru_3(CO)_{12}$, 60°, 3 h	$CH_3CH{=}CHCH_2CFClCF_2Cl$ (82)	244
	NCC_6H_4CN-*p*, $h\nu$, CH_3CN, 30–40 h	$CH_2{=}CHCH(CH_3)C_6H_4CN$-*p* + $CH_3CH{=}CHCH_2C_6H_4CN$-*p* 3:2 (80)	261
$(CH_3)_3SiCH_2C(CH_3){=}CH_2$	$ClCH_2OCH_2C_6H_5$, $(CH_3)_3SiI$ (cat.), CH_3CN, 4.5 h	$CH_2{=}C(CH_3)(CH_2)_2OCH_2C_6H_5$ (76)	251
	$ClCH(OCH_3)C_4H_9$-*i*, $(CH_3)_3SiI$ (cat.), CH_3CN, −40°, 15 min; 0°, 1 h	$CH_2{=}C(CH_3)CH_2CH(OCH_3)C_4H_9$-*i* (71)	251
	$ClCH(OCH_3)C_4H_9$-*i*, $(CH_3)_3SiO_3SCF_3$ (cat.), CH_3CN, −20°, 25 min	" (68)	251
	$CH_2{=}CH$, O, O, $(CH_3)_3OBF_4$, CH_2Cl_2, 94 h	$CH_2{=}C(CH_3)(CH_2)_2$-$CH{=}CH(CH_2)_2CO_2CH_3$ (90)	242
	NCC_6H_4CN-*p*, $h\nu$, CH_3CN, 30–40 h	$CH_2{=}C(CH_3)CH_2C_6H_4CN$-*p* (80)	261
	NCC_6H_4CN-*o*, $h\nu$, CH_3CN, 30–40 h	$CH_2{=}C(CH_3)CH_2C_6H_4CN$-*o* (44)	261
$(CH_3)_3SiCH(CO_2C_2H_5)CH{=}CH_2$	$CH_3CH{=}CHCHXCH_3$, $TiCl_4$, CH_2Cl_2, −78°	(E,E)-$CH_3CH{=}CHCH(CH_3)$-$CH_2CH{=}CHCO_2C_2H_5$	240
	X = OCH_3 15 min	(64)	
	Cl 5 min	(65)	
	OC_4H_9-*n* 30 min	(60)	
	Br 10 min	(81)	
	$OSi(CH_3)_3$ 3 min	(20)	
	$CH_3CH{=}CHCD(CH_3)OC_4H_9$-*n*, $TiCl_4$, CH_2Cl_2, −78°	(E,E)-$CH_3CH{=}CHCD(CH_3)$-$CH_2CH{=}CHCO_2C_2H_5$ (—)	240
	X, $TiCl_4$, CH_2Cl_2, −78°	(E)- $CH_2CH{=}CHCO_2C_2H_5$	240
	X = OC_4H_9-*n*, 15 min	(62)	
	Br	(88)	

TABLE VI. ALLYLSILANES WITH ALKYL HALIDES, ALCOHOLS, ETHERS, NITRO COMPOUNDS, ALKENES, AND ARENES (*Continued*)

Reactant	Conditions	Product(s) and Yield(s)	Refs.
	$C_6H_5CH(CH_3)X$, $TiCl_4$, CH_2Cl_2, −78°	(*E*)-$C_6H_5CH(CH_3)CH_2CH{=}CHCO_2C_2H_5$	240
	X = OCH_3 15 min	(92)	
	OC_4H_9-*n* 30 min	(68)	
	Br 10 min	(87)	
	(cyclododecenyl)-OCH_3, $TiCl_4$, CH_2Cl_2, −78°, 5 min	(cyclododecenyl)-$CH_2CH{=}CHCO_2C_2H_5$ (73)	240
$C_6H_5(CH_3)_2SiCH(CO_2CH_3)CH{=}CH_2$	$ClCH(SC_6H_5)CH_3$, $TiCl_4$, CH_2Cl_2, 0°, 2.5 h	$CH_3CH(SC_6H_5)CH_2CH{=}CHCO_2CH_3$ (63)	411
C_5			
$(CH_3)_3SiCH_2CH{=}C(CH_3)_2$	$ClCH(SC_6H_5)CH_2CO_2C_2H_5$, $SnCl_4$, CH_2Cl_2, −78°, 3 h	$CH_2{=}CHC(CH_3)_2CH(SC_6H_5)CH_2CO_2C_2H_5$ (20)	361
	$ClCH(SC_6H_5)CH_2COCH_3$, $SnCl_4$, CH_2Cl_2, −78°, 3 h	$CH_2{=}CHC(CH_3)_2CH(SC_6H_5)CH_2COCH_3$ (20)	361
	$CH_3CH{=}CHCHClCH_3$, $TiCl_4$, CH_2Cl_2, −78°, 3 h	$CH_2{=}CHC(CH_3)_2CH(CH_3)CH{=}CHCH_3$ (78)	241
	$(CH_3)_2C{=}CHCH_2X$, $TiCl_4$, CH_2Cl_2, −78°, 3 h	$CH_2{=}CHC(CH_3)_2CH_2CH{=}C(CH_3)_2$	241
	X = Cl	(10)	
	Br	(12)	
	$CH_2{=}CH$-(γ-butyrolactone, 5-yl), $(CH_3)_3OBF_4$, 94 h	$CH_2{=}CHC(CH_3)_2CH_2$-$CH{=}CH(CH_2)_2CO_2CH_3$ (93)	242
	NCC_6H_4CN-*p*, *hν*, CH_3CN, 30–40 h	$CH_2{=}CHC(CH_3)_2C_6H_4CN$-*p* + $(CH_3)_2C{=}CHCH_2C_6H_4CN$-*p* 3:2 (91)	261
	$CH_2{=}CHCH(C_4H_9\text{-}n)OCH_3$, $TiCl_4$, CH_2Cl_2, −78°, 3 h	$CH_2{=}CHC(CH_3)_2CH_2CH{=}CHC_4H_9$-*n* + $CH_2{=}CHC(CH_3)_2CH(C_4H_9\text{-}n)CH{=}CH_2$ 70:30 (33)	241

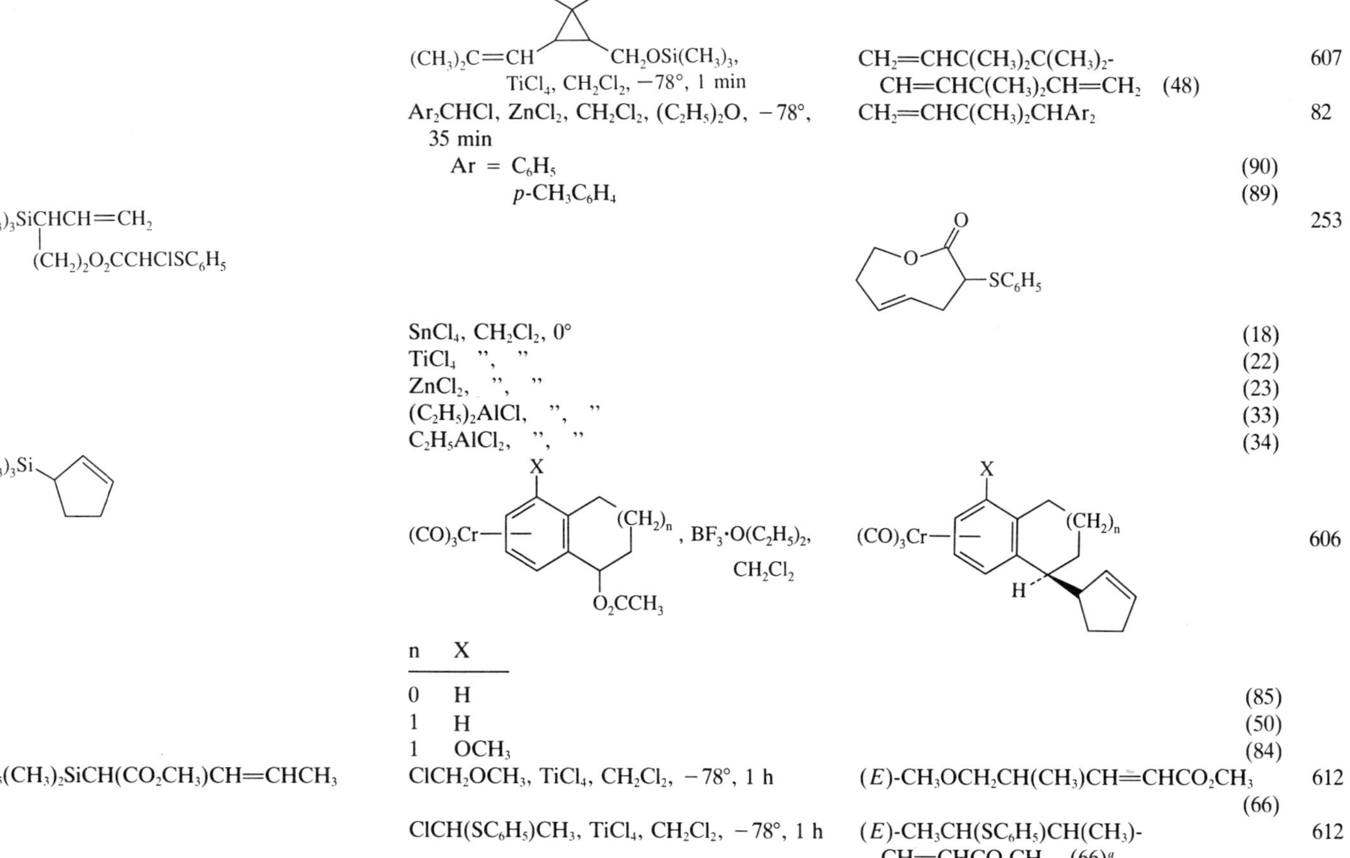

Reactant	Conditions	Product(s) and Yield(s) (%)	Refs.
	$(CH_3)_2C{=}CH$–(2,2-dimethylcyclopropyl)–$CH_2OSi(CH_3)_3$, $TiCl_4$, CH_2Cl_2, $-78°$, 1 min	$CH_2{=}CHC(CH_3)_2C(CH_3)_2$-$CH{=}CHC(CH_3)_2CH{=}CH_2$ (48)	607
	Ar_2CHCl, $ZnCl_2$, CH_2Cl_2, $(C_2H_5)_2O$, $-78°$, 35 min	$CH_2{=}CHC(CH_3)_2CHAr_2$	82
	Ar = C_6H_5	(90)	
	p-$CH_3C_6H_4$	(89)	
$(CH_3)_3SiCHCH{=}CH_2$ with $(CH_2)_2O_2CCHClSC_6H_5$		3-(SC_6H_5)-oxacyclonon-5-en-2-one	253
	$SnCl_4$, CH_2Cl_2, 0°	(18)	
	$TiCl_4$ ", "	(22)	
	$ZnCl_2$, ", "	(23)	
	$(C_2H_5)_2AlCl$, ", "	(33)	
	$C_2H_5AlCl_2$, ", "	(34)	
$(CH_3)_3Si$–(2-cyclopentenyl)	$(CO)_3Cr$–(X-substituted arene with $(CH_2)_n$ ring bearing O_2CCH_3), $BF_3{\cdot}O(C_2H_5)_2$, CH_2Cl_2	$(CO)_3Cr$–(X-substituted arene with $(CH_2)_n$ ring bearing 2-cyclopentenyl, H)	606
	n X		
	0 H	(85)	
	1 H	(50)	
	1 OCH_3	(84)	
$C_6H_5(CH_3)_2SiCH(CO_2CH_3)CH{=}CHCH_3$	$ClCH_2OCH_3$, $TiCl_4$, CH_2Cl_2, $-78°$, 1 h	(E)-$CH_3OCH_2CH(CH_3)CH{=}CHCO_2CH_3$ (66)	612
	$ClCH(SC_6H_5)CH_3$, $TiCl_4$, CH_2Cl_2, $-78°$, 1 h	(E)-$CH_3CH(SC_6H_5)CH(CH_3)$-$CH{=}CHCO_2CH_3$ (66)[a]	612

TABLE VI. ALLYLSILANES WITH ALKYL HALIDES, ALCOHOLS, ETHERS, NITRO COMPOUNDS, ALKENES, AND ARENES (*Continued*)

Reactant	Conditions	Product(s) and Yield(s)	Refs.
$(CH_3)_3SiCH(CO_2C_2H_5)C(CH_3)=CH_2$	$ClCH_2SC_6H_5$, $TiCl_4$, CH_2Cl_2, −78°, 15 min; 0°, 24 h	$C_6H_5S(CH_2)_2C(CH_3)=CHCO_2C_2H_5$ (52) *E*:*Z* 44:56	252
$C_6H_5(CH_3)_2SiC(CH_3)(CO_2CH_3)CH=CH_2$	$ClCH_2OCH_3$, $TiCl_4$, CH_2Cl_2, −78°, 1 h	(*E*)-$CH_3O(CH_2)_2CH=C(CH_3)CO_2CH_3$ (77)	612
	$ClCH(SC_6H_5)CH_3$, $TiCl_4$, CH_2Cl_2, −78°, 1 h	(*E*)-$CH_3CH(SC_6H_5)CH_2CH=C(CH_3)CO_2CH_3$ (73)	612
$(CH_3)_3SiCH(CH_2CF_3)CH=CH_2$	$ICFClCF_2Cl$, $Fe_3(CO)_{12}$, 60°, 2 h	$CF_3CH_2CH=CHCH_2CFClCF_2Cl$ (85)	244
$(CH_3)_3SiCH_2C(CH=CH_2)=CH_2$	$ClCH(OR^1)R^2$, CH_3CN	$CH_2=C(CH=CH_2)CHR^1(OR^2)$	251
	R^1 — R^2		
	i-C_3H_7 — H — $(CH_3)_3SiO_3SCF_3$ (cat.), −20°, 1 h	(74)	
	CH_3 — *i*-C_4H_9 — $(CH_3)_3SiO_3SCF_3$ (cat.), −20°, 10 min	(86)	
	" — " — $(CH_3)_3SiI$ (cat.), −40°, 10 min; 0°, 30 min	(80)	
	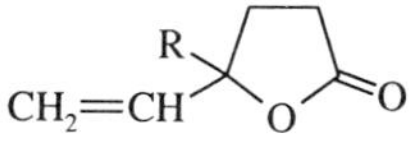, $(CH_3)_3OBF_4$	(*E*)-$CH_2=C(CH=CH_2)(CH_2)_2CH=CR(CH_2)_2CO_2CH_3$	242
	R = H, 94 h	(74)	
	R = CH_3, 71 h	(65)	
C_6			
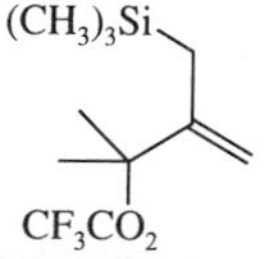	, $ZnCl_2$, $C_2H_5N(C_3H_7\text{-}i)_2$, CH_2Cl_2, 0°	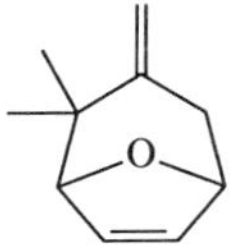 (—)	613

Reactant	Conditions	Product(s) and Yield(s) (%)	Refs.
	 cyclopentadiene, $ZnCl_2$, $C_2H_5N(C_3H_7\text{-}i)_2$, CH_3CN, 0°	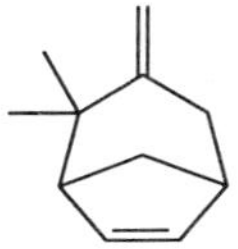 (45)	613
$(CH_3)_3SiCHCH{=}CH_2$ / $(CH_2)_3O_2CCHClSC_6H_5$	$C_2H_5AlCl_2$, CH_2Cl_2, 0°	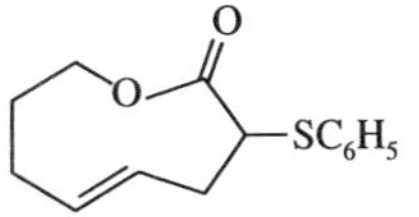 SC_6H_5 (48)	253
(*E*)-$(CH_3)_3SiCH(CO_2C_2H_5)$-$C(CH_3){=}CHCH_3$	$ClCH_2SC_6H_5$, $TiCl_4$, CH_2Cl_2, −78°, 15 min; 0°, 24 h	$C_6H_5S(CH_2)_2CH(CH_3)C(CH_3){=}CHCO_2C_2H_5$ (20) *E*:*Z* 54:46	252
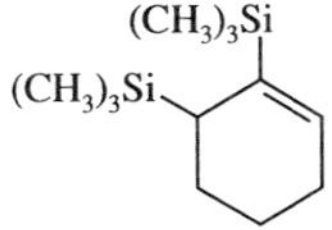 $(CH_3)_3Si$, $(CH_3)_3Si$	$ClCH_2OCH_3$, $AlCl_3$, CH_2Cl_2, −20°, 1.5 h	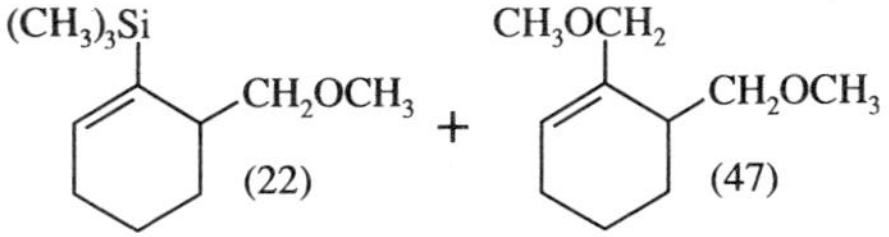 $(CH_3)_3Si$, CH_2OCH_3 (22) + CH_3OCH_2, CH_2OCH_3 (47)	86
C_7			
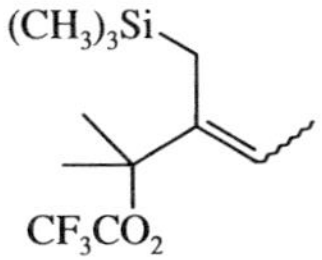 $(CH_3)_3Si$, CF_3CO_2	 furan, $ZnCl_2$, $C_2H_5N(C_3H_7\text{-}i)_2$, CH_2Cl_2, 0°	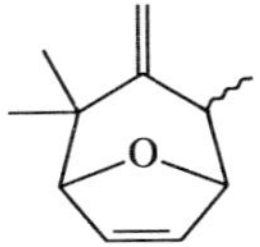 O (—)	613
	 cyclopentadiene, $ZnCl_2$, $C_2H_5N(C_3H_7\text{-}i)_2$, CH_3CN, 0°	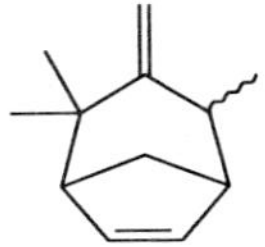 (60)	613
$(CH_3)_3SiCHCH{=}CH_2$ / $(CH_2)_4O_2CCHClSC_6H_5$	$C_2H_5AlCl_2$, CH_2Cl_2, 0°	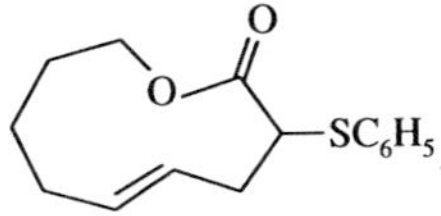 SC_6H_5 (48)	253

TABLE VI. ALLYLSILANES WITH ALKYL HALIDES, ALCOHOLS, ETHERS, NITRO COMPOUNDS, ALKENES, AND ARENES (*Continued*)

Reactant	Conditions	Product(s) and Yield(s)	Refs.
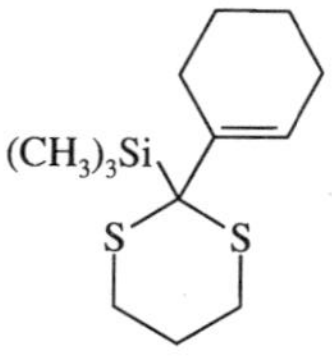 $(CH_3)_3Si$ (structure: 2-(1-cyclohexenyl)-1,3-dithiane)	$ClCH_2OCH_3$, TlF, CH_2Cl_2, 40°	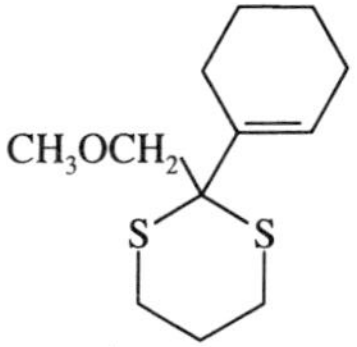 CH_3OCH_2 (structure: 2-(1-cyclohexenyl)-1,3-dithiane) (10)	258
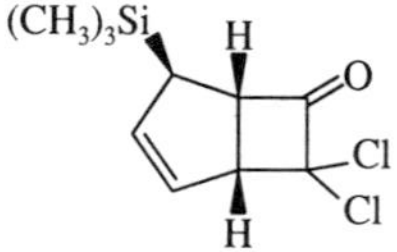 $(CH_3)_3Si$ (structure: bicyclic dichloro ketone, H, O, Cl, Cl)	$ClCH_2OCH_3$, $SnCl_4$, CH_2Cl_2, reflux, 60 h	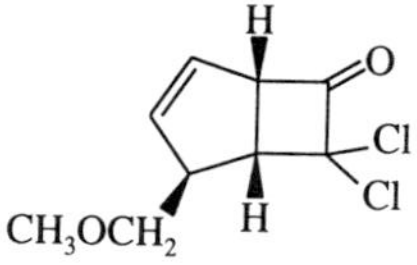 CH_3OCH_2 (structure: bicyclic dichloro ketone, H, O, Cl, Cl) (78)	55
C_8			
$(CH_3)_3SiCH_2CH{=}CHC_5H_{11}$-*n*	$ClCH(SC_6H_5)CH_2CO_2C_2H_5$, $SnCl_4$, CH_2Cl_2, 20 min	$CH_2{=}CHCH(C_5H_{11}$-*n*$)CH(SC_6H_5)$-$CH_2CO_2C_2H_5$ (78)	361
	$ClCH(SC_6H_5)CH_2COCH_3$, $SnCl_4$, CH_2Cl_2, 20 min	$CH_2{=}CHCH(C_5H_{11}$-*n*$)CH(SC_6H_5)$-CH_2COCH_3 (80)	361
	$ClCH(SC_6H_5)CH_2COC_5H_{11}$-*n*, $SnCl_4$, CH_2Cl_2, 20 min	$CH_2{=}CHCH(C_5H_{11}$-*n*$)CH(SC_6H_5)$-$CH_2COC_5H_{11}$-*n* (91)	361
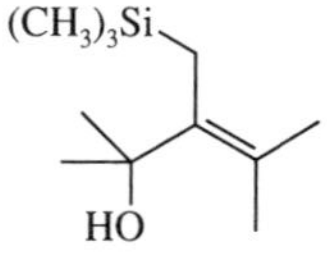 $(CH_3)_3Si$ (structure), HO	 (cyclopentadiene), $TiCl_4$, $CH_3NHC_6H_5$, CH_2Cl_2, −20°, 3 h	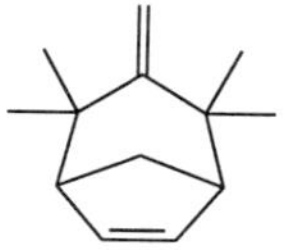 (structure: bicyclic diene) (38) 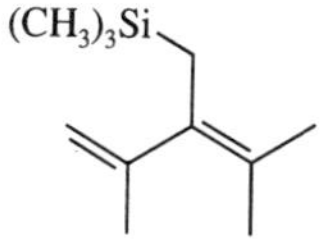 + $(CH_3)_3Si$ (structure: diene) (12)	249

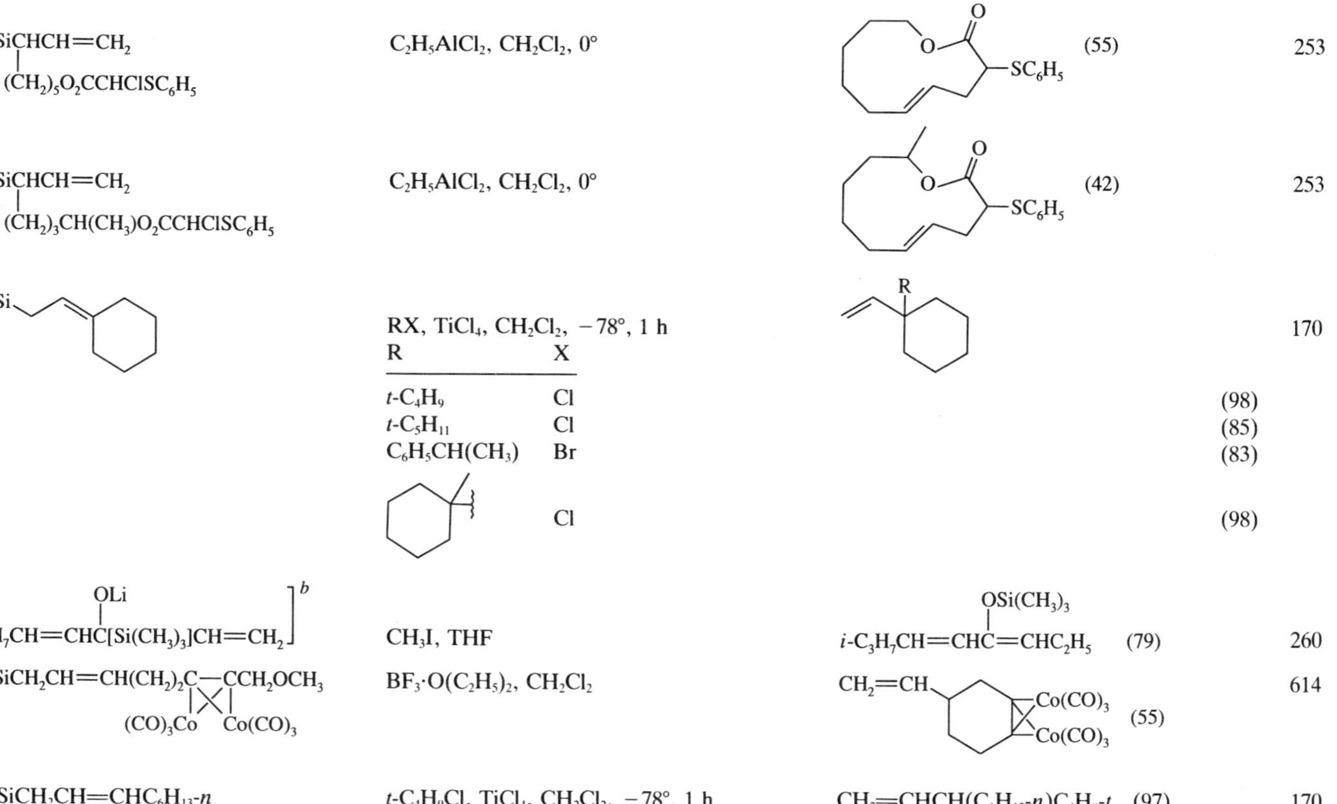

Reactant	Conditions	Product(s) and Yield(s) (%)	Refs.
$(CH_3)_3SiCH(CH=CH_2)(CH_2)_5O_2CCHClSC_6H_5$	$C_2H_5AlCl_2$, CH_2Cl_2, 0°	(55)	253
$(CH_3)_3SiCH(CH=CH_2)(CH_2)_3CH(CH_3)O_2CCHClSC_6H_5$	$C_2H_5AlCl_2$, CH_2Cl_2, 0°	(42)	253
$(CH_3)_3Si$	RX, $TiCl_4$, CH_2Cl_2, −78°, 1 h	R	170
	R: t-C_4H_9, X: Cl	(98)	
	R: t-C_5H_{11}, X: Cl	(85)	
	R: $C_6H_5CH(CH_3)$, X: Br	(83)	
	X: Cl	(98)	
[i-$C_3H_7CH=CHC(OLi)[Si(CH_3)_3]CH=CH_2$] [b]	CH_3I, THF	i-$C_3H_7CH=CHC(OSi(CH_3)_3)=CHC_2H_5$ (79)	260
$(CH_3)_3SiCH_2CH=CH(CH_2)_2C{-}CCH_2OCH_3$ ((CO)$_3$Co Co(CO)$_3$)	$BF_3{\cdot}O(C_2H_5)_2$, CH_2Cl_2	$CH_2=CH$ Co(CO)$_3$ Co(CO)$_3$ (55)	614
C_9			
$(CH_3)_3SiCH_2CH=CHC_6H_{13}$-$n$	t-C_4H_9Cl, $TiCl_4$, CH_2Cl_2, −78°, 1 h	$CH_2=CHCH(C_6H_{13}$-$n)C_4H_9$-t (97)	170

TABLE VI. ALLYLSILANES WITH ALKYL HALIDES, ALCOHOLS, ETHERS, NITRO COMPOUNDS, ALKENES, AND ARENES (*Continued*)

Reactant	Conditions	Product(s) and Yield(s)	Refs.
$(CH_3)_3SiCH_2CH{=}CH(CH_2)_3C{-}CCH_2OCH_3$ with $(CO)_3Co$, $Co(CO)_3$ bridging the C–C	$BF_3{\cdot}O(C_2H_5)_2$, CH_2Cl_2	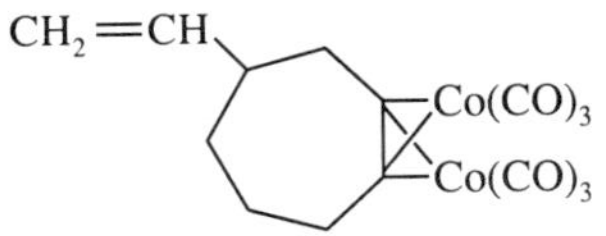 $CH_2{=}CH$-cycloheptyne $Co_2(CO)_6$ complex (structure) (55)	614
$(CH_3)_3SiCH_2CH{=}CHC_6H_5$	$C_6H_5CH_2Br$, KF, 18-crown-6, THF, 12 h	$C_6H_5(CH_2)_2CH{=}CHC_6H_5$ (40)	259
	$C_6H_5C(CH_3)_2OH$, BF_3, CH_2Cl_2, 0°	$CH_2{=}CHCH(C_6H_5)C(CH_3)_2C_6H_5$ (—)	246
	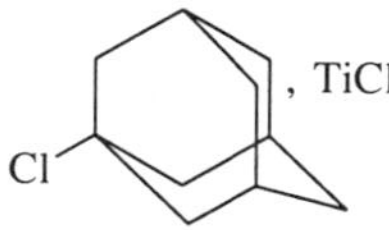 1-Chloroadamantane (structure), $TiCl_4$, CH_2Cl_2, 30 min	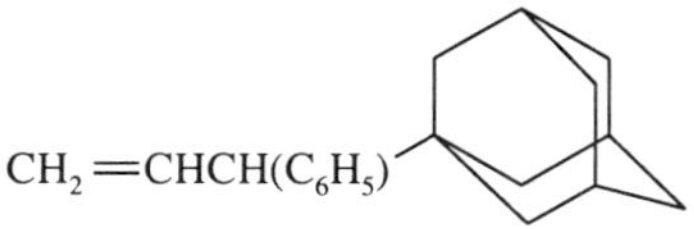 $CH_2{=}CHCH(C_6H_5)$-1-adamantyl (structure) (48)	608
	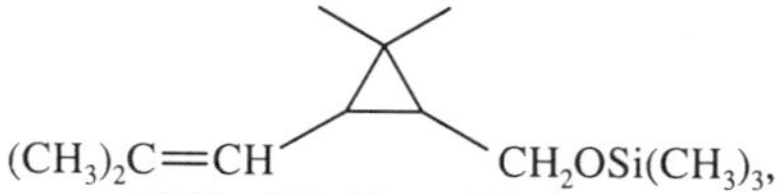 $(CH_3)_2C{=}CH$–(2,2-dimethylcyclopropyl)–$CH_2OSi(CH_3)_3$ (structure), $TiCl_4$, CH_2Cl_2, −78°, 1 min	$CH_2{=}CHCH(C_6H_5)C(CH_3)_2CH{=}CHC(CH_3)_2CH{=}CH_2$ (83)	607
	$(C_6H_5)_2CHOH$, BF_3, CH_2Cl_2, 0°	$CH_2{=}CHCH(C_6H_5)CH(C_6H_5)_2$ (—)	246
$(CH_3)_3SiCH_2CH{=}CHC_6H_4OCH_3$-*p*	$C_2H_5CHXC_6H_4OCH_3$-*p*, CH_2Cl_2, −78°	$CH_2{=}CHCH(C_6H_4OCH_3$-*p*$)CH(C_2H_5)C_6H_4OCH_3$-*p*	615
	X = OCH_3: $TiCl_4$, 30 min	(76)[a]	
	X = ": $BF_3{\cdot}O(C_2H_5)_2$, 2 h	(63)[a]	
	X = Cl: $TiCl_4$, 15 min	(72)[a]	
	X = ": $BF_3{\cdot}O(C_2H_5)_2$, 2 h	(70)[a]	

Reactant	Conditions	Product(s) and Yield(s) (%)	Refs.
$(CH_3)_3SiCH_2CH{=}CHC_6H_3(NO_2)(OCH_3)$-2,4	$RCH(OCH_3)C_6H_4OCH_3$-*p*, $TiCl_4$, CH_2Cl_2, −78°, 1 h	$CH_2{=}CHCHCH(C_6H_4OCH_3\text{-}p)R$ with $C_6H_4(NO_2)(OCH_3)$—2,4	615
	R = CH_3	*RR*:*RS* 3:1 (79)	
	R = C_2H_5	*RR*:*RS* 7:1 (70)	
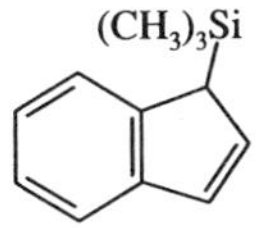 $(CH_3)_3Si$	$CH_2{=}CHCH_2O_2CCH_3$, $Pd[P(C_6H_5)_3]_4$	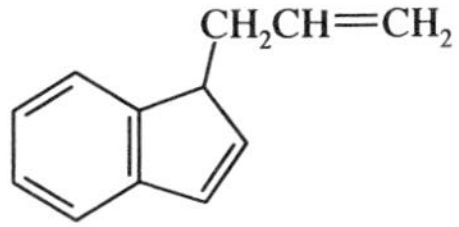 $CH_2CH{=}CH_2$ (—)	243
	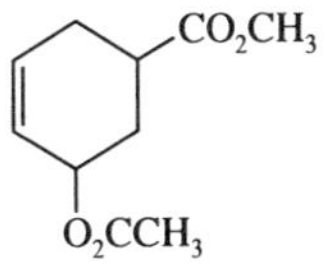 CO_2CH_3, O_2CCH_3, $Pd[P(C_6H_5)_3]_4$	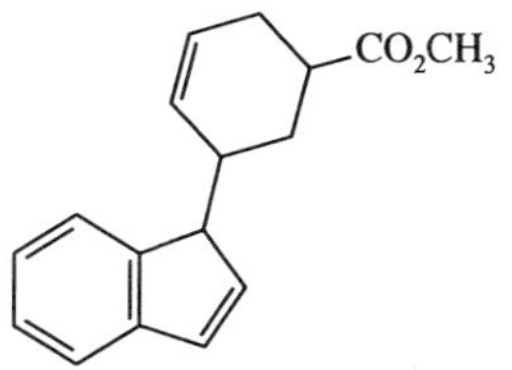 CO_2CH_3 (—)	243
	$C_6H_5CH{=}CHCH(O_2CCH_3)CN$, $Pd[P(C_6H_5)_3]_4$	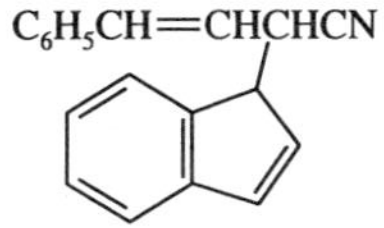 $C_6H_5CH{=}CHCHCN$ (—)	616
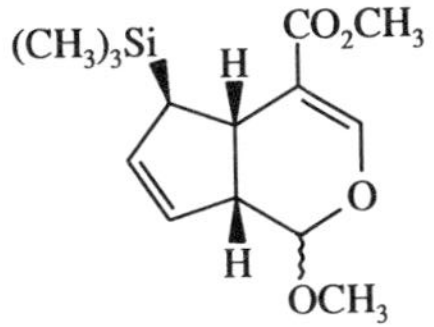 $(CH_3)_3Si$, H, CO_2CH_3, O, H, OCH_3	$ClCH_2OCH_3$, $SnCl_4$, CH_2Cl_2, −78°, 1 h	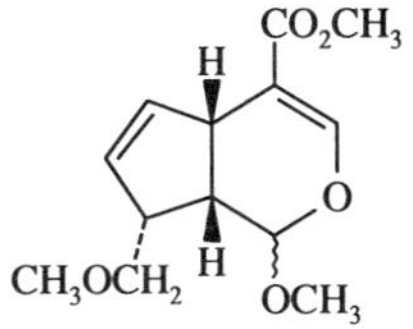 H, CO_2CH_3, O, CH_3OCH_2, H, OCH_3 (45)	54
C_{10}			
$(CH_3)_3SiCH_2CH{=}C(CH_3)C_6H_{13}$-*n*	*t*-C_4H_9Cl, $TiCl_4$, CH_2Cl_2, −78°, 1 h	$CH_2{=}CHC(CH_3)(C_4H_9\text{-}t)C_6H_{13}$-*n* (97)	170
$(CH_3)_3SiCH_2CH{=}C(CH_3)$-$(CH_2)_2CH{=}C(CH_3)_2$	NCC_6H_4CN-*p*, *hν*, CH_3CN, 30–40 h	$CH_2{=}CHC(CH_3)C_6H_4CN$-*p* with $(CH_2)_2CH{=}C(CH_3)_2$ + $CH_3C{=}CHCH_2C_6H_4CN$-*p* with $(CH_2)_2CH{=}C(CH_3)_2$ 1:1 (54)	261

TABLE VI. ALLYLSILANES WITH ALKYL HALIDES, ALCOHOLS, ETHERS, NITRO COMPOUNDS, ALKENES, AND ARENES (*Continued*)

Reactant	Conditions	Product(s) and Yield(s)	Refs.
$(CH_3)_3SiC(CH_3)CH{=}CH_2$ with $(CH_2)_2CH{=}C(CH_3)_2$ on the marked carbon	$CH_2{=}C(CH_3)$-(γ-butyrolactone), $(CH_3)_3OBF_4$ 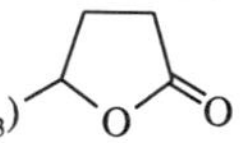	$CH_3C{=}CH(CH_2)_2C(CH_3){=}CH(CH_2)_2CO_2CH_3$ with $(CH_2)_2CH{=}C(CH_3)_2$ on the marked carbon (82)	605
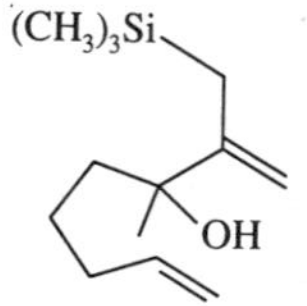 $(CH_3)_3Si$, OH (structure)	$TiCl_4$, $C_6H_5NHCH_3$, CH_2Cl_2, $-15°$	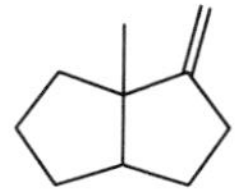 (structure) (15)	617
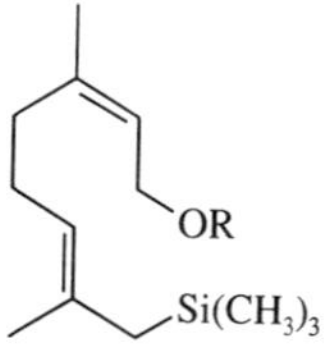 OR, $Si(CH_3)_3$ (structure)	R = H, $TiCl_4$, $C_6H_5NHCH_3$, CH_2Cl_2, $-23°$, 1 h	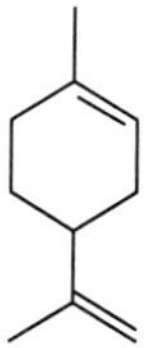 (structure) (77)	264
	R = $COCH_3$, $CH_3Al(O_2CCF_3)_2$, hexane, 25°, 17 h	" (28)	264
$(CH_3)_3Si$, CO_2R (structure)		CO_2R (structure) I + CO_2R (structure) II	
R: H	$TiCl_4$, CH_2Cl_2, 0°, 4 h	1:8 (—)	248
"	$BF_3{\cdot}O(C_2H_5)_2$, CH_3CO_2H, CH_2Cl_2, 0°, 4 h	1.5:1 (—)	248

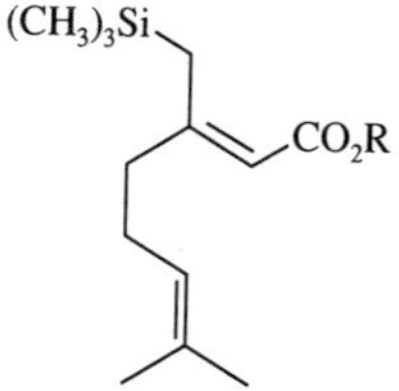

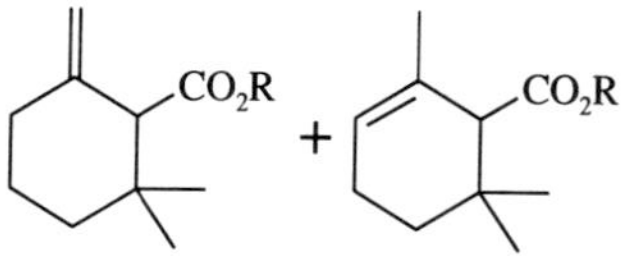

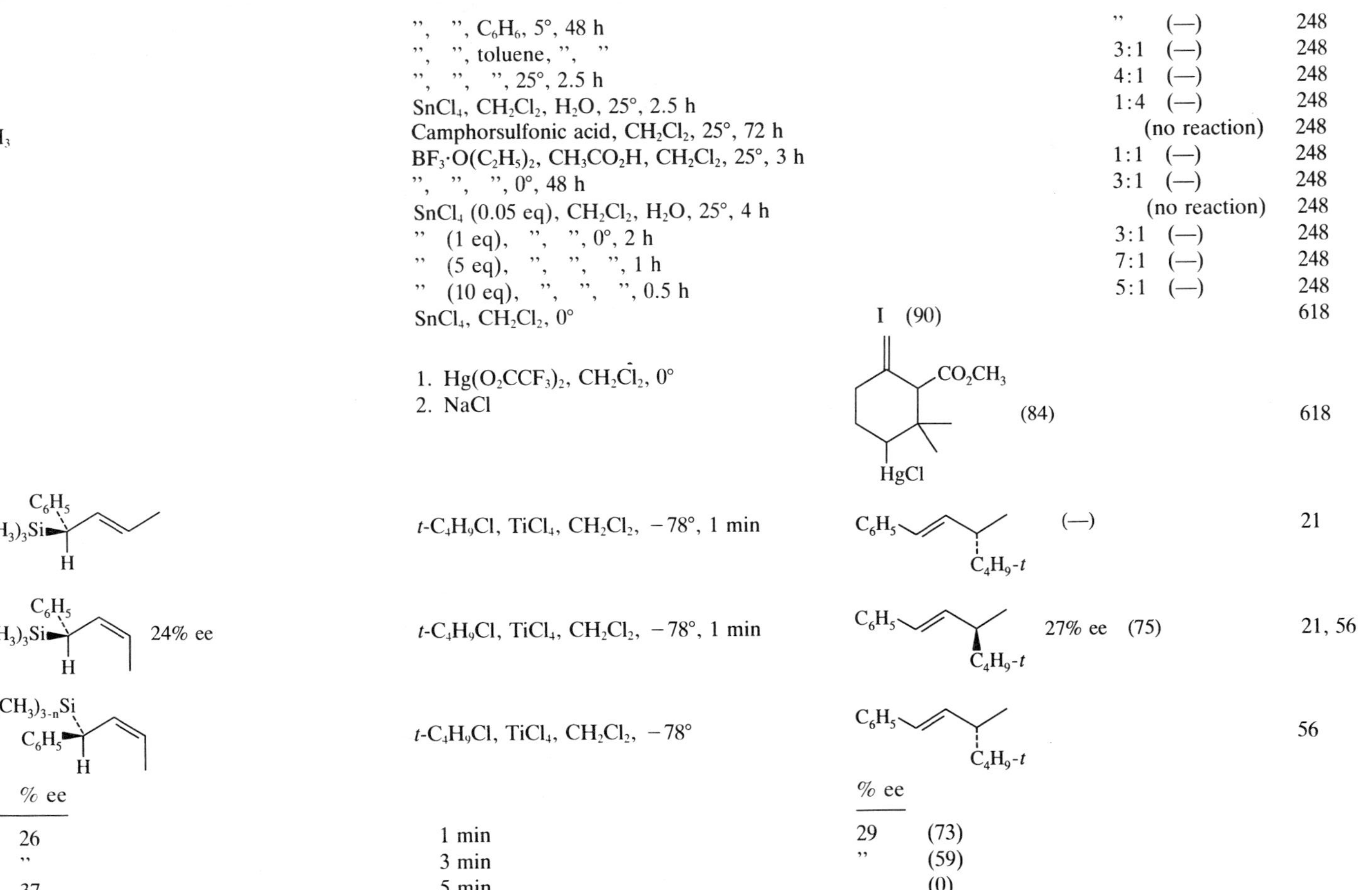

”	”, ”, C_6H_6, 5°, 48 h	” (—)	248
”	”, ”, toluene, ”, ”	3:1 (—)	248
”	”, ”, ”, 25°, 2.5 h	4:1 (—)	248
”	$SnCl_4$, CH_2Cl_2, H_2O, 25°, 2.5 h	1:4 (—)	248
CH_3	Camphorsulfonic acid, CH_2Cl_2, 25°, 72 h	(no reaction)	248
”	$BF_3{\cdot}O(C_2H_5)_2$, CH_3CO_2H, CH_2Cl_2, 25°, 3 h	1:1 (—)	248
”	”, ”, ”, 0°, 48 h	3:1 (—)	248
”	$SnCl_4$ (0.05 eq), CH_2Cl_2, H_2O, 25°, 4 h	(no reaction)	248
”	” (1 eq), ”, ”, 0°, 2 h	3:1 (—)	248
”	” (5 eq), ”, ”, ”, 1 h	7:1 (—)	248
”	” (10 eq), ”, ”, ”, 0.5 h	5:1 (—)	248
”	$SnCl_4$, CH_2Cl_2, 0°	I (90)	618
” ”	1. $Hg(O_2CCF_3)_2$, CH_2Cl_2, 0° 2. NaCl	CO_2CH_3, HgCl (84)	618
C_6H_5, $(CH_3)_3Si$, H	t-C_4H_9Cl, $TiCl_4$, CH_2Cl_2, −78°, 1 min	C_6H_5, C_4H_9-t (—)	21
C_6H_5, $(CH_3)_3Si$, H 24% ee	t-C_4H_9Cl, $TiCl_4$, CH_2Cl_2, −78°, 1 min	C_6H_5, C_4H_9-t 27% ee (75)	21, 56
$F_n(CH_3)_{3-n}Si$, C_6H_5, H	t-C_4H_9Cl, $TiCl_4$, CH_2Cl_2, −78°	C_6H_5, C_4H_9-t	56
n % ee		% ee	
1 26	1 min	29 (73)	
2 ”	3 min	” (59)	
3 37	5 min	(0)	

TABLE VI. ALLYLSILANES WITH ALKYL HALIDES, ALCOHOLS, ETHERS, NITRO COMPOUNDS, ALKENES, AND ARENES (*Continued*)

Reactant	Conditions	Product(s) and Yield(s)	Refs.
C_{11}			
[structure: 1-vinyl-2-[1-(trimethylsilyl)allyl]cyclohexanol; OH, H, $(CH_3)_3Si$]	1. n-C_4H_9Li, $-78°$, 10 min 2. RX, 0°	[structures I + II; $OSi(CH_3)_3$, H, R]	190
	R / X: CH_3 / I	(E)- + (Z)-I 86 + II (5)	
	R / X: n-C_6H_{13} / Br	(E)-I (27)[c] + (Z)-I (39)[c] + II (3)	
	R / X: n-C_6H_{13} / Br	(E)-I (51)[c] + (Z)-I (28)[c] + II (—) $E:Z$ 93:7	
$[C_6H_5(CH_2)_2C(OLi)[Si(CH_3)_3]CH{=}CH_2]$[d]	1. RX, THF 2. HCl, CH_3OH	$C_6H_5(CH_2)_2CO(CH_2)_2R$	260
	R / X: CH_3 / I	(83)	
	R / X: C_2H_5 / "	(86)	
	R / X: $CH_2{=}CHCH_2$ / Br	(62)	
C_{12}			
[structure: $(CH_3)_3Si$, OH]	$TiCl_4$, $C_6H_5NHCH_3$, CH_2Cl_2, $-15°$	[structure] (70)	617
$[C_6H_5(CH_2)_2C(OLi)[Si(CH_3)_3]C(CH_3){=}CH_2]$[e]	1. C_2H_5I, THF 2. HCl, CH_3OH	$C_6H_5(CH_2)_2COCH(CH_3)C_3H_7$-$n$ (64)	260

Reactant	Conditions	Product(s) and Yield(s) (%)	Refs.
C_{13}			
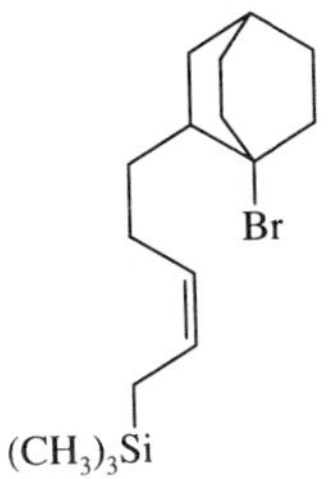 Br; $(CH_3)_3Si$	$AgBF_4$, CH_2Cl_2, 0°, 1 h	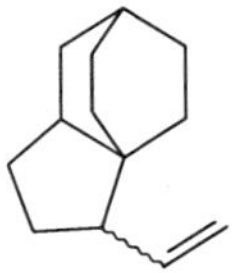 (75)	611
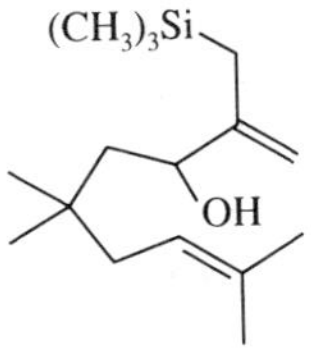 $(CH_3)_3Si$; OH	$TiCl_4$, $C_6H_5NHCH_3$, CH_2Cl_2, −15°	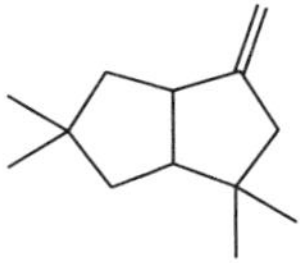 (55) 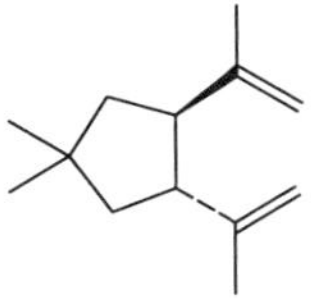 + (25)	617
C_{14}			
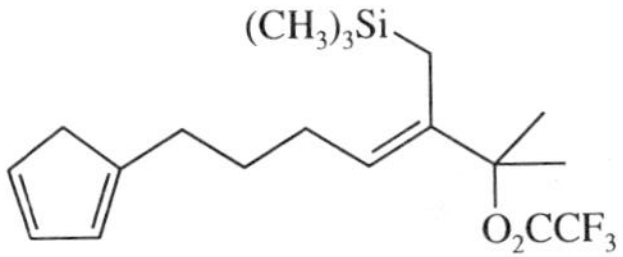 $(CH_3)_3Si$; O_2CCF_3	$ZnCl_2$, Al_2O_3, CH_2Cl_2, pentane, −30°	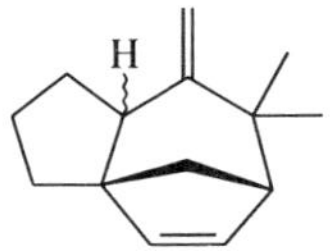 H; H_α (7) + H_β (9)	619
C_{15}			
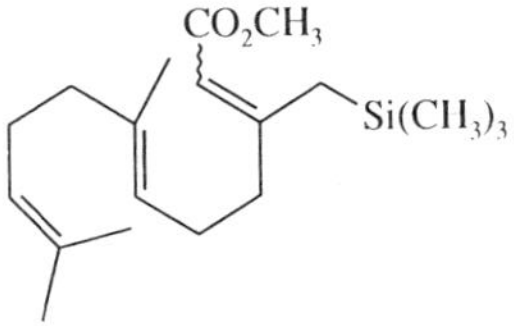 CO_2CH_3; $Si(CH_3)_3$	$SnCl_4$, CH_2Cl_2, −56°	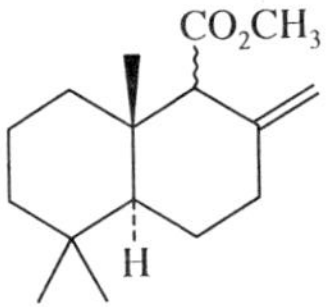 CO_2CH_3; H (95) CO_2CH_3 α:β 1:4	266

TABLE VI. Allylsilanes with Alkyl Halides, Alcohols, Ethers, Nitro Compounds, Alkenes, and Arenes (*Continued*)

Reactant	Conditions	Product(s) and Yield(s)	Refs.
	$Hg(O_2CCF_3)_2$, CH_3NO_2, $-20°$, 1 h	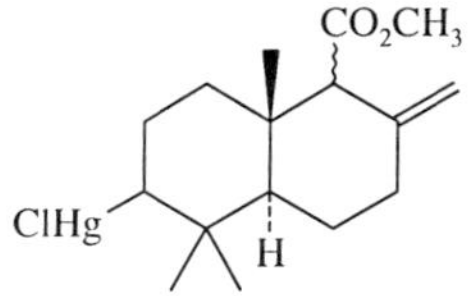 CO_2CH_3, ClHg, H (60)	266
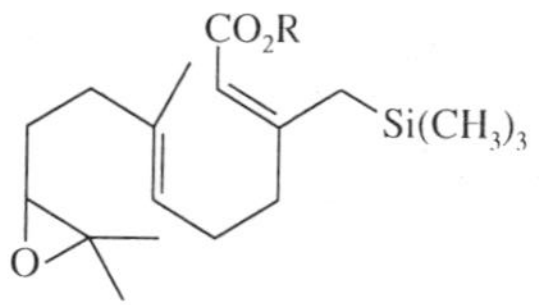 CO_2R, $Si(CH_3)_3$, O	$SnCl_4$, CH_2Cl_2, 0°, 1–4 h	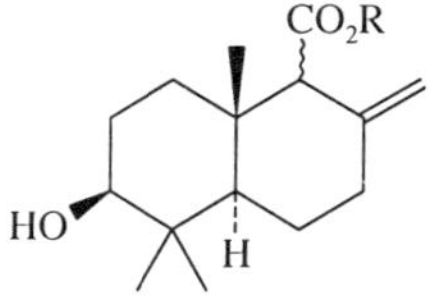 CO_2R, HO, H	265
R		$\alpha:\beta$	
CH_3		2:3 (55)	
C_3H_7-*i*		1:3 (51)	
$CH_2C_4H_9$-*t*		2:5 (62)	
$CH(C_4H_9$-*t*$)_2$		1:3 (60)	
"	$SnCl_4$, CH_3CO_2H, CH_2Cl_2, 0°	1:5 (55)	
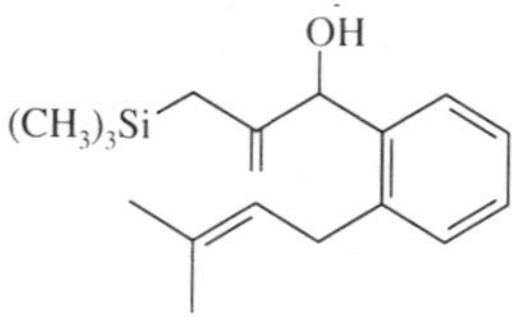 $(CH_3)_3Si$, OH	$TiCl_4$, $C_6H_5NHCH_3$, CH_2Cl_2, $-15°$	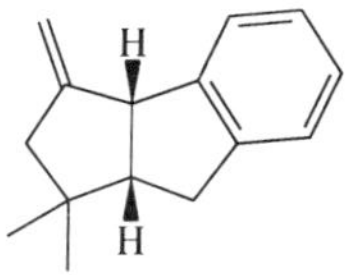 H, H (45)	617
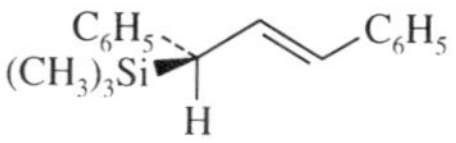 C_6H_5, $(CH_3)_3Si$, H, C_6H_5	*t*-C_4H_9Cl, $TiCl_4$, CH_2Cl_2, 0°, 1 h	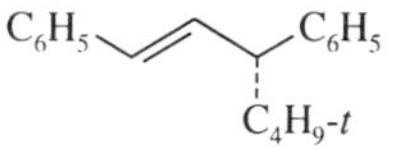 C_6H_5, C_6H_5, C_4H_9-*t* (—)	21
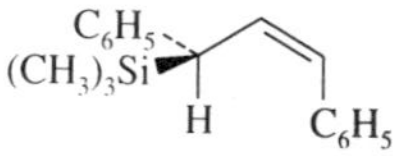 C_6H_5, $(CH_3)_3Si$, H, C_6H_5	*t*-C_4H_9Cl, $TiCl_4$, CH_2Cl_2, 0°, 1 h	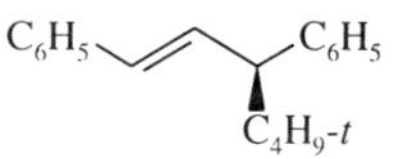 C_6H_5, C_6H_5, C_4H_9-*t* (—)	21

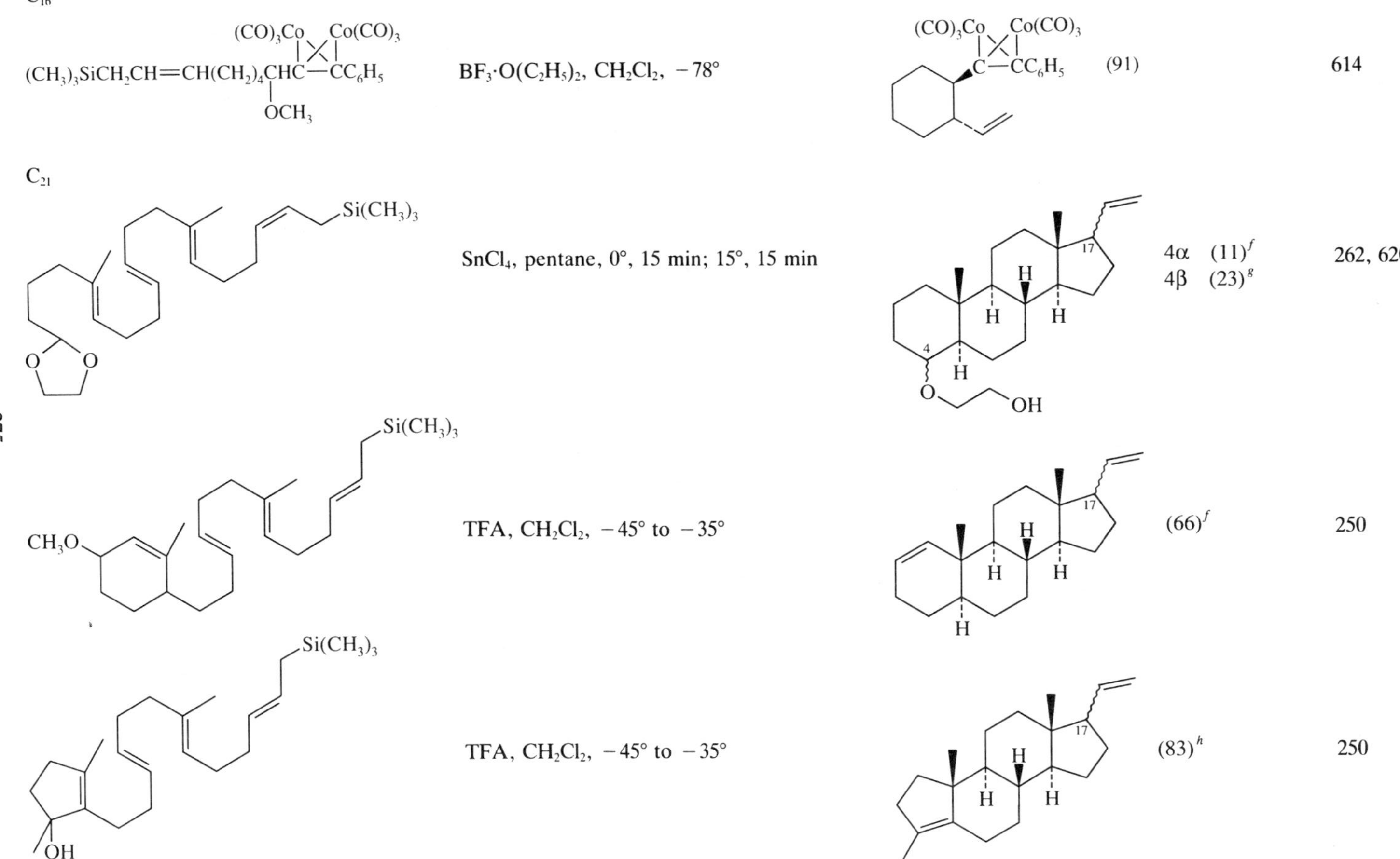

C_{16}
$(CO)_3Co$ $Co(CO)_3$
$(CH_3)_3SiCH_2CH{=}CH(CH_2)_4CHC{-}CC_6H_5$
OCH_3
$BF_3{\cdot}O(C_2H_5)_2$, CH_2Cl_2, −78°
$(CO)_3Co$ $Co(CO)_3$
$C{-}CC_6H_5$
(91)
614
C_{21}
$Si(CH_3)_3$
O O
$SnCl_4$, pentane, 0°, 15 min; 15°, 15 min
17
H
H H
4
H
O
OH
4α (11)[f]
4β (23)[g]
262, 620
$Si(CH_3)_3$
CH_3O
TFA, CH_2Cl_2, −45° to −35°
17
H
H H
H
(66)[f]
250
$Si(CH_3)_3$
OH
TFA, CH_2Cl_2, −45° to −35°
17
H
H H
(83)[h]
250

TABLE VI. ALLYLSILANES WITH ALKYL HALIDES, ALCOHOLS, ETHERS, NITRO COMPOUNDS, ALKENES, AND ARENES (*Continued*)

Reactant	Conditions	Product(s) and Yield(s)	Refs.
C_{22}			
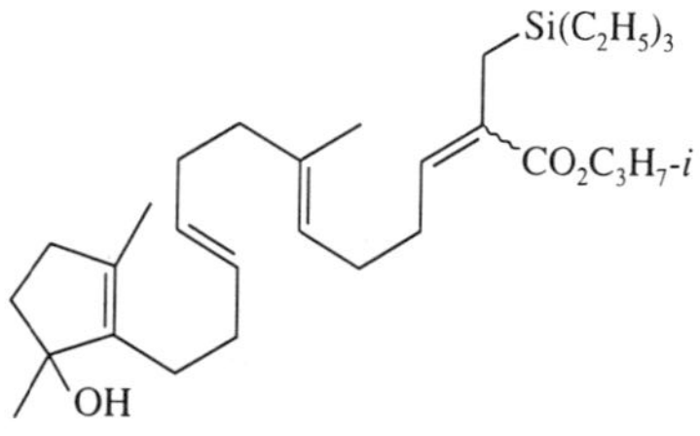 $Si(C_2H_5)_3$, $CO_2C_3H_7$-*i*, OH	TFA, CH_2Cl_2, $-20°$, 3.5 h	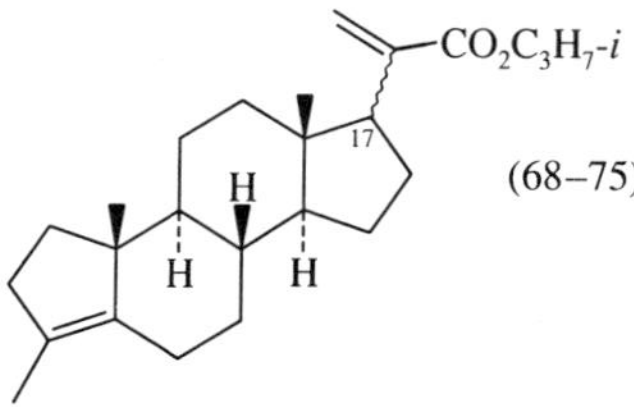 $CO_2C_3H_7$-*i*, 17, H, H, H (68–75)[f]	263

[a] The diastereoisomers are produced in a ratio of 1:1.
[b] This intermediate is prepared by addition of vinyllithium to 5-methyl-1-trimethylsilyl-2-pentenone.
[c] The product is isolated as the alcohol by treating the silyl ether with HCl in CH_3OH for 10 min.
[d] This intermediate is prepared by addition of vinyllithium to 3-phenyl-1-trimethylsilylpropanone.
[e] This intermediate is prepared by addition of 2-propenyllithium to 3-phenyl-1-trimethylsilylpropanone.
[f] This compound is a 1:1 mixture of diastereoisomers at C-17.
[g] This compound is a 2:3 mixture of diastereoisomers at C-17.
[h] This isomer is a 47:53 mixture of diastereoisomers at C-17.

TABLE VII. VINYLSILANES WITH ALKYL HALIDES, ALCOHOLS, AND ALKENES

Reactant	Conditions	Product(s) and Yield(s)	Refs.
C_2			
$(CH_3)_3SiCH{=}CH_2$	CH_3OCH_2Cl, $ZnCl_2$, 150°, 48°	$CH_3OCH_2CH{=}CH_2$ (—)	10
	", $AlCl_3$, CH_2Cl_2, 0°, 1.25 h	" (42)	228
(E)-$(CH_3)_3SiCH{=}CHSi(CH_3)_3$	CH_3OCH_2Cl, $AlCl_3$, CH_2Cl_2, 0°, 1.25 h	(E)-$CH_3OCH_2CH{=}CHSi(CH_3)_3$ (70)	228
	" (2 eq), ", ", ", "	(E)-$CH_3OCH_2CH{=}CHCH_2OCH_3$ (73)	228
	1. CH_3OCHCl_2, $AlCl_3$, −40° 2. H_2O	$(CH_3)_3SiCH{=}CHCHO$ (20)	228
$[(CH_3)_3Si]_2C{=}CH(CH_2)_2Br$	$(i$-$C_4H_9)_2AlCl$	1-$(CH_3)_3Si$-cyclobutene (61)	268
C_6			
(E)-$(CH_3)_3SiCH{=}CHC_4H_9$-n	CH_3OCHCl_2, $TiCl_4$, CH_2Cl_2, −90°, 30 min	(E)-$OHCCH{=}CHC_4H_9$-n (76)	621
(E)-$(CH_3)_3SiC(C_2H_5){=}CHC_2H_5$	CH_3OCHCl_2, $TiCl_4$, CH_2Cl_2, −78°	(E)-$OHCC(C_2H_5){=}CHC_2H_5$ (73)	256
(E)-$(CH_3)_2Si(CH_2Cl)CH{=}CHC_4H_9$-$t$	1. $AlCl_3$, CH_2Cl_2, 10 min 2. CH_3MgBr, $(C_2H_5)_2O$, reflux, 2 h	(E)-$(CH_3)_3SiCH_2CH{=}CHC_4H_9$-$t$ + 1-$(CH_3)_3Si$-2-C_4H_9-t-cyclopropane 87:13 (65)	622
	1. $AlCl_3$, CH_2Cl_2, 10 min 2. CH_3OH, $(C_2H_5)_3N$ 3. H_2O_2, KF, DMF	(E)-$HOCH_2CH{=}CHC_4H_9$-t + 1-HO-2-C_4H_9-t-cyclopropane 87:13 (58)	622
(Z)-$(CH_3)_3SiCF{=}CFC_4H_9$-n	s-C_4H_9OH, H_2SO_4, −20°, 2 h	t-$C_4H_9CHFCOC_4H_9$-n (23)	569
	s-C_4H_9Cl, $AlCl_3$, CH_2Cl_2, −50°, 1 h	(E)-s-$C_4H_9CF{=}CFC_4H_9$-n (20–30)	570
	t-C_4H_9Cl, ", ", −5°, "	(E)-t-$C_4H_9CF{=}CFC_4H_9$-n (80)	570

TABLE VII. Vinylsilanes with Alkyl Halides, Alcohols, and Alkenes (*Continued*)

Reactant	Conditions	Product(s) and Yield(s)	Refs.
	OH, H_2SO_4, −20°, 2 h	$CHFCOC_4H_9$-*n* (91)	569
	, H_2SO_4	" (26)	569
C_7			
(*E*)-$(CH_3)_3SiCH{=}CHC_5H_{11}$-*n*	CH_3OCHCl_2, $TiCl_4$, CH_2Cl_2, − 90°, 30 min	(*E*)-$OHCCH{=}CHC_5H_{11}$-*n* (85)	621
(*E*,*E*)-$(CH_3)_3Si(CH{=}CH)_2C_3H_7$-*n*	CH_3OCHCl_2, $TiCl_4$, CH_2Cl_2, −70°, 20 min	(*E*,*E*)-$OHC(CH{=}CH)_2C_3H_7$-*n* (71)	273
$(CH_3)_3SiCF{=}CFC_5H_{11}$-*n*	*t*-C_4H_9OH, H_2SO_4, −20°, 2 h	*t*-$C_4H_9CHFCOC_5H_{11}$-*n* (76)	569
C_8			
(*Z*)-$(CH_3)_3SiC(CH_3){=}CHC_5H_{11}$-*n*	CH_3OCHCl_2, $TiCl_4$, CH_2Cl_2, −78°, 1 h	(*E*)-$OHCC(CH_3){=}CHC_5H_{11}$-*n* (71)	256
(*E*)-$(CH_3)_3SiC(C_3H_7$-*n*$){=}CHC_3H_7$-*n*	", ", ", ", "	(*E*)-$OHCC(C_3H_7$-*n*$){=}CHC_3H_7$-*n* (79)	256
(*E*)-$(CH_3)_2SiCH{=}CHC_6H_{13}$-*n* CH_2Cl	1. $AlCl_3$, CH_2Cl_2, 10 min 2. CH_3MgBr, $(C_2H_5)_2O$, reflux, 2 h	(*E*)-$(CH_3)_3SiCH_2CH{=}CHC_6H_{13}$-*n* + C_6H_{13}-*n* $(CH_3)_3Si$ 78:22 (58)	622
(*Z*)-$(CH_3)_2SiCH{=}CHC_6H_{13}$-*n* CH_2Cl	1. $AlCl_3$, CH_2Cl_2, 10 min 2. CH_3MgBr, $(C_2H_5)_2O$, reflux, 2 h	(*Z*)-$(CH_3)_3SiCH_2CH{=}CHC_6H_{13}$-*n* (89)	622
	1. $AlCl_3$, CH_2Cl_2, 10 min 2. CH_3OH, $(C_2H_5)_3N$ 3. H_2O_2, KF, DMF	(*Z*)-$HOCH_2CH{=}CHC_6H_{13}$-*n* (48)	622
$(CH_3)_3SiC(C_6H_{13}$-*n*$){=}CH_2$ CH_2Cl	1. $AlCl_3$, CH_2Cl_2, 10 min 2. CH_3MgBr, $(C_2H_5)_2O$, reflux, 2 h	$(CH_3)_3Si$ *n*-C_6H_{13} (44)	622

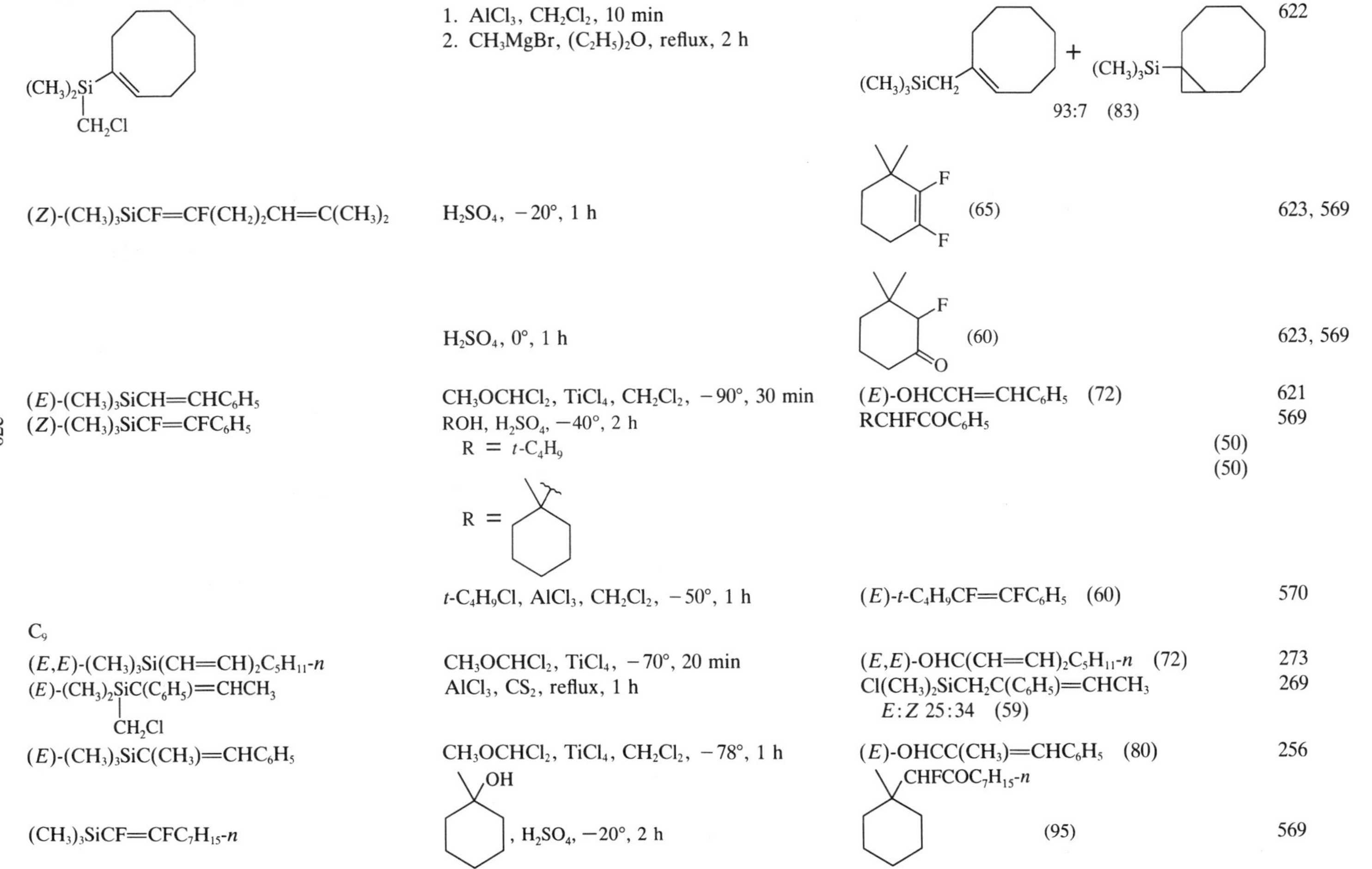

Reactant	Conditions	Product(s) and Yield(s) (%)	Refs.
$(CH_3)_2Si(CH_2Cl)$-cyclooctenyl	1. $AlCl_3$, CH_2Cl_2, 10 min 2. CH_3MgBr, $(C_2H_5)_2O$, reflux, 2 h	$(CH_3)_3SiCH_2$-cyclooctene + $(CH_3)_3Si$-bicyclo[6.1.0]nonane 93:7 (83)	622
(Z)-$(CH_3)_3SiCF{=}CF(CH_2)_2CH{=}C(CH_3)_2$	H_2SO_4, $-20°$, 1 h	1,2-difluoro-3,3-dimethylcyclohexene (65)	623, 569
	H_2SO_4, 0°, 1 h	2-fluoro-3,3-dimethylcyclohexanone (60)	623, 569
(E)-$(CH_3)_3SiCH{=}CHC_6H_5$	CH_3OCHCl_2, $TiCl_4$, CH_2Cl_2, $-90°$, 30 min	(E)-$OHCCH{=}CHC_6H_5$ (72)	621
(Z)-$(CH_3)_3SiCF{=}CFC_6H_5$	ROH, H_2SO_4, $-40°$, 2 h	$RCHFCOC_6H_5$	569
	R = t-C_4H_9	(50)	
	R = 1-methylcyclohexyl	(50)	
	t-C_4H_9Cl, $AlCl_3$, CH_2Cl_2, $-50°$, 1 h	(E)-t-$C_4H_9CF{=}CFC_6H_5$ (60)	570
C_9			
(E,E)-$(CH_3)_3Si(CH{=}CH)_2C_5H_{11}$-$n$	CH_3OCHCl_2, $TiCl_4$, $-70°$, 20 min	(E,E)-$OHC(CH{=}CH)_2C_5H_{11}$-n (72)	273
(E)-$(CH_3)_2Si(CH_2Cl)C(C_6H_5){=}CHCH_3$	$AlCl_3$, CS_2, reflux, 1 h	$Cl(CH_3)_2SiCH_2C(C_6H_5){=}CHCH_3$ $E{:}Z$ 25:34 (59)	269
(E)-$(CH_3)_3SiC(CH_3){=}CHC_6H_5$	CH_3OCHCl_2, $TiCl_4$, CH_2Cl_2, $-78°$, 1 h	(E)-$OHCC(CH_3){=}CHC_6H_5$ (80)	256
$(CH_3)_3SiCF{=}CFC_7H_{15}$-$n$	1-methylcyclohexanol, H_2SO_4, $-20°$, 2 h	1-methylcyclohexyl-$CHFCOC_7H_{15}$-n (95)	569

TABLE VII. VINYLSILANES WITH ALKYL HALIDES, ALCOHOLS, AND ALKENES (*Continued*)

Reactant	Conditions	Product(s) and Yield(s)	Refs.
C_{10}			
(*E*)-$(CH_3)_2SiC(C_4H_9\text{-}n)=CHC_4H_9\text{-}n$ with CH_2Cl on Si	1. $AlCl_3$, CH_2Cl_2, 10 min 2. CH_3MgBr, $(C_2H_5)_2O$, reflux, 2 h	(*E*)-$(CH_3)_3SiCH_2C(C_4H_9\text{-}n)=CHC_4H_9\text{-}n$ 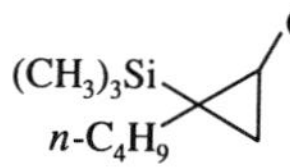 + $(CH_3)_3Si$, n-C_4H_9, C_4H_9-n cyclopropane 77:23 (93)	622
	1. $AlCl_3$, CH_2Cl_2, 10 min 2. CH_3OH, $(C_2H_5)_3N$ 3. H_2O_2, KF, DMF	$HOCH_2C(C_4H_9\text{-}n)=CHC_4H_9\text{-}n$ *E*:*Z* 92:8 (43)	622
(*E* or *Z*)-$(CH_3)_3SiC(C_2H_5)=CHC_6H_{11}$	CH_3OCHCl_2, $AlCl_3$, CH_2Cl_2, 0°	(*E*)-$OHCC(C_2H_5)=CHC_6H_{11}$ (90)	271
(*E*,*E*)-$(CH_3)_3Si(CH=CH)_2C_6H_5$	CH_3OCHCl_2, $TiCl_4$, CH_2Cl_2, −70°, 20 min	(*E*,*E*)-$OHC(CH=CH)_2C_6H_5$ (50)	273
C_{11}			
(*E*)-$(CH_3)_3SiCH=CH(CH_2)_8CO_2C_2H_5$	CH_3OCHCl_2, $TiCl_4$, CH_2Cl_2, −60°, 4 h	(*E*)-$OHCCH=CH(CH_2)_8CO_2C_2H_5$ (83)	256
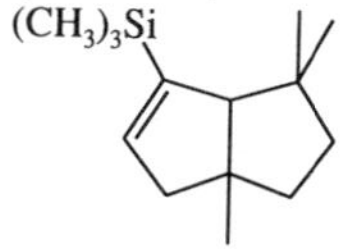 $(CH_3)_3Si$-substituted bicyclic structure	CH_3OCHCl_2, $AlCl_3$	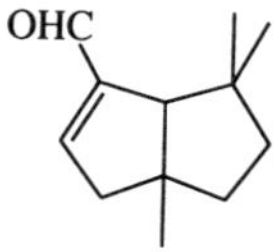 OHC-substituted bicyclic structure (50)	272
(*E*,*E*)-$(CH_3)_3SiCH=C(CH_3)CH=CHC_6H_5$	CH_3OCHCl_2, $TiCl_4$, CH_2Cl_2, −70°, 20 min	$OHCCH=C(CH_3)CH=CHC_6H_5$ *E*,*E*:*Z*,*E* 3:1 (68)	273
C_{14}			
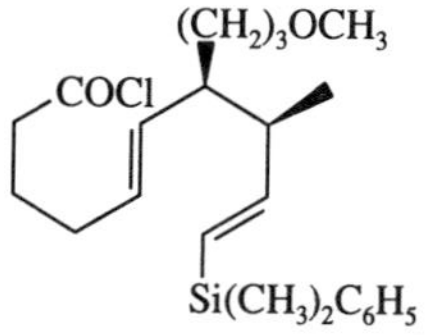 COCl, $(CH_2)_3OCH_3$, $Si(CH_3)_2C_6H_5$ structure	1. $SbCl_5$, CH_2Cl_2, −78°, 10 min 2. $NaHCO_3$, H_2O	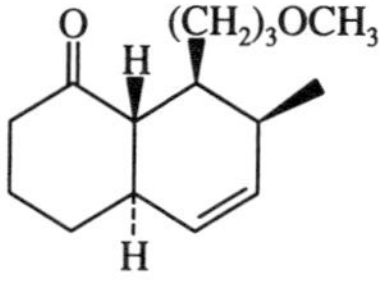 O, H, $(CH_2)_3OCH_3$, H bicyclic structure (51)	270

Reactant	Conditions	Product(s) and Yield(s) (%)	Refs.
(*E*)-$(CH_3)_3SiC(CH_3){=}CH(CH_2)_2CH(CH_3)C_6H_4CH_3$-*p*	CH_3OCHCl_2, $TiCl_4$, CH_2Cl_2, $-60°$, 4 h	(*E*)-$OHCC(CH_3){=}CH(CH_2)_2CH(CH_3)C_6H_4CH_3$-*p* (48)	256
C_{15} 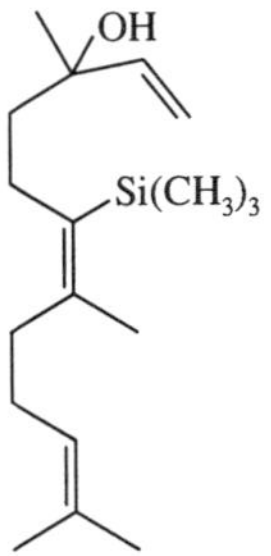 [structure: OH, Si(CH₃)₃]	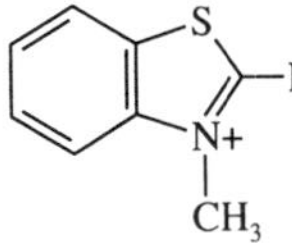 3-methyl-2-fluorobenzothiazolium $^-O_3SCF_3$, $(C_2H_5)_3N$, CH_2Cl_2, $-25°$, 6 h	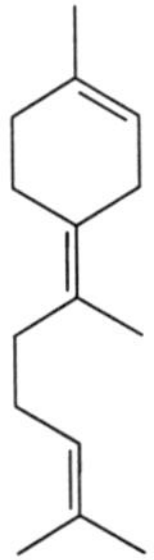 [structure] *E*:*Z* 97.5:2.5 (85)	267
	$TiCl_4$, $C_6H_5NHCH_3$, CH_2Cl_2, $-23°$, 1 h	" *E*:*Z* 99.6:0.4 (91)	267

TABLE VIII. ALLENYLSILANES WITH ALKYL HALIDES

Reactant	Conditions	Product(s) and Yield(s)	Refs.
C_3			
$[(CH_3)_3Si]_2C{=}C{=}C[Si(CH_3)_3]_2$	$ClCH_2C_6H_5$, $AlCl_3$	$[(CH_3)_3Si]_2C{=}C{=}C(CH_2C_6H_5)Si(CH_3)_3$ (—)	274

TABLE IX. Allylsilanes with Epoxides, Oxetanes, and Episulfonium Salts

Reactant	Conditions	Product(s) and Yield(s)	Refs.
C_3			
$(CH_3)_3SiCH_2CH{=}CH_2$	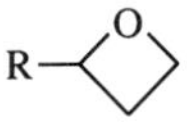 R-(oxetane), $TiCl_4$, CH_2Cl_2, $-100°$ to room temp overnight	$CH_2{=}CHCH_2CHR(CH_2)_2OH$	275
	R = H	(69)	
	R = CH_3	(20)	
	1. $R^1R^2C{=}CHR^3$, C_6H_5SCl, CH_3NO_2 2. Allylsilane, $ZnBr_2$ (cat.), 20°, 16 h		279
	R^1 / R^2 / R^3		
	CH_3 / H / H	$CH_2{=}CHCH_2CH(CH_3)CH_2SC_6H_5$ (55) + $CH_2{=}CH(CH_2)_2CH(CH_3)SC_6H_5$ (35)	
	H / CH_3 / CH_3 *E*	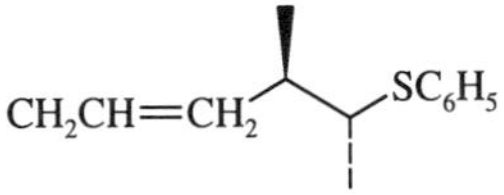 $CH_2CH{=}CH_2$–CH(CH_3)–CH(CH_3)SC_6H_5 (40)	
	" / " / " *Z*	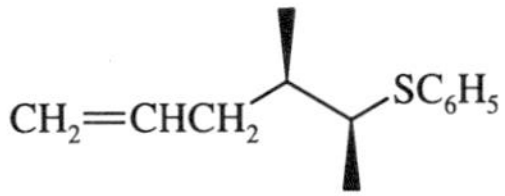 $CH_2{=}CHCH_2$–CH(CH_3)–CH(CH_3)SC_6H_5 (92)	
	CH_3 / " / H	$CH_2{=}CHCH_2C(CH_3)_2CH_2SC_6H_5$ (85)	
	H / —$(CH_2)_n$—	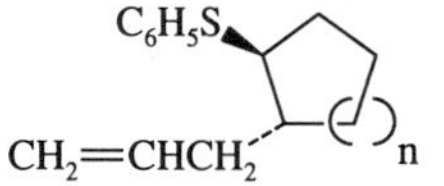 2-(C_6H_5S)-1-($CH_2{=}CHCH_2$)cycloalkane: n = 1 (91), n = 2 (78), n = 3 (50)	
	C_6H_5 / H / H	$CH_2{=}CHCH_2CH(C_6H_5)CH_2SC_6H_5$ (74)	

TABLE IX. Allylsilanes with Epoxides, Oxetanes, and Episulfonium Salts (*Continued*)

Reactant	Conditions	Product(s) and Yield(s)	Refs.
	1. $R^1CH{=}CHR^2$, C_6H_5SCl, CH_2Cl_2 2. Allylsilane, $ZnBr_2$ (cat.), 20°, 3 h		279
	R^1 = C_2H_5O, R^2 = H	$CH_2{=}CHCH_2CH(OC_2H_5)CH_2SC_6H_5$ (83)	
	R^1R^2 = —$O(CH_2)_3$—	3-(C_6H_5S)-2-($CH_2{=}CHCH_2$)tetrahydropyran (73)	
	1. $R^1R^2C{=}CHR^3$, $CH_3C_6H_4SCl$-*p*, CH_2Cl_2 2. Allylsilane, $TiCl_4$, −60°, 0.5 h	$CH_2{=}CHCH_2CR^1R^2CHR^3SC_6H_4CH_3$-*p*	624
	R^1 = CH_3, R^2 = CH_3, R^3 = H	(65)	
	R^1 = CH_3O, R^2 = H, R^3 = "	(85)	
	R^1 = C_2H_5O, R^2 = CH_3, R^3 = "	(90)	
	R^1 = ", R^2 = H, R^3 = CH_3	(73)	
	R^1 = H, R^2R^3 = —$O(CH_2)_4$—	(80)	
	R^1 = CH_3, R^2R^3 = —$O(CH_2)_3$—	(75)	
	R^1 = CH_3O, R^2R^3 = —$(CH_2)_4$—	(75)	
	R^1 = C_6H_5, R^2 = H, R^3 = H	(83)	
C_5			
$(CH_3)_3SiCH_2CH{=}C(CH_3)_2$	$(CH_2)_3O$, $TiCl_4$, CH_2Cl_2, −100° to room temp overnight	$CH_2{=}CHC(CH_3)_2(CH_2)_3OH$ (85)	275
3-$(CH_3)_3Si$-cyclopentene, 22–25% ee	$(CH_2)_2O$, $TiCl_4$, CH_2Cl_2, −78°, 1 min	3-[$(CH_2)_2OH$]cyclopentene, 24% ee (93)	24
3-$(CH_3)_3Si$-cyclopentene	$(CH_2)_3O$, $TiCl_4$, CH_2Cl_2, −100° to room temp overnight	3-[$(CH_2)_3OH$]cyclopentene (55)	275

Reactant	Conditions	Product(s) and Yield(s) (%)	Refs.
C_6			
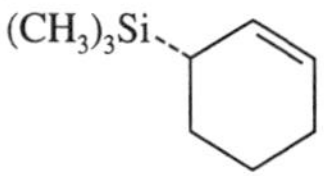 $(CH_3)_3Si$ >1% ee	$(CH_2)_2O$, $TiCl_4$, CH_2Cl_2, $-78°$, 1 min	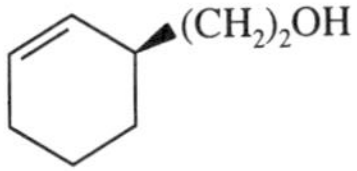 $(CH_2)_2OH$ 1.6% ee (70)	24
C_7			
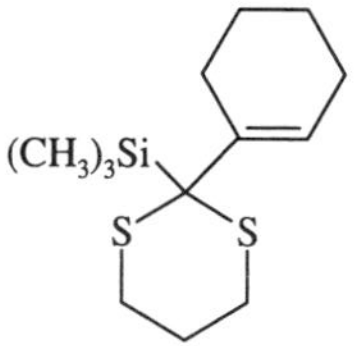 $(CH_3)_3Si$ S S	$(CH_2)_2O$, $SnCl_4$, CH_2Cl_2, $-78°$	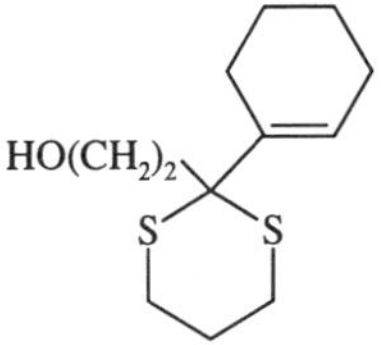 $HO(CH_2)_2$ S S (30)	258
C_8			
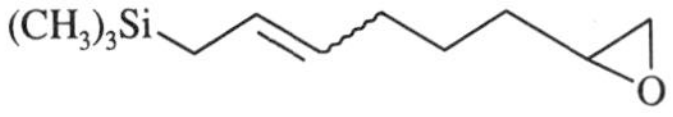 $(CH_3)_3Si$ O	$TiCl_4$, CH_2Cl_2, $-95°$	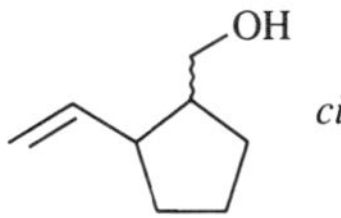 OH *cis*:*trans* 4:1 (55)	277
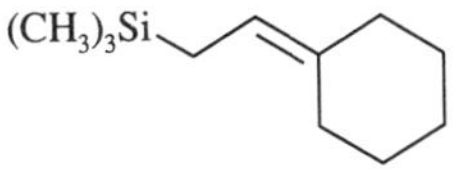 $(CH_3)_3Si$	$(CH_2)_2O$, $TiCl_4$, CH_2Cl_2, $-78°$, 1 h	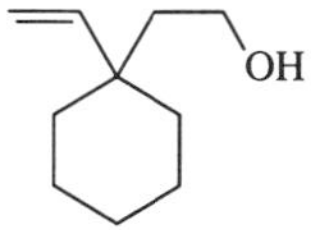 OH (87)	170
	$(CH_2)_3O$, $TiCl_4$, CH_2Cl_2, $-100°$ to room temp overnight	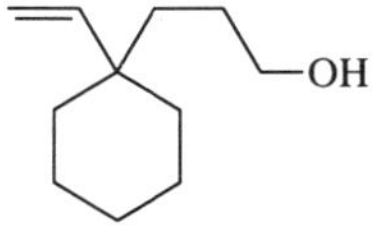 OH (86)	275
C_9			
$(CH_3)_3SiCH_2CH{=}CHC_6H_{13}$-*n*	$(CH_2)_2O$, $TiCl_4$, CH_2Cl_2, $-78°$, 1 h	$CH_2{=}CHCH(C_6H_{13}$-*n*$)(CH_2)_2OH$ (92)	170
(*E*)-$(CH_3)_3SiCH_2CH{=}CHC_6H_5$	$(CH_2)_3O$, $TiCl_4$, CH_2Cl_2, $-100°$ to room temp, overnight	$CH_2{=}CHCH(C_6H_5)(CH_2)_3OH$ (85)	275

TABLE IX. ALLYLSILANES WITH EPOXIDES, OXETANES, AND EPISULFONIUM SALTS (*Continued*)

Reactant	Conditions	Product(s) and Yield(s)	Refs.
C_{10}			
$(CH_3)_3SiCH_2CH{=}C(CH_3)C_6H_{13}$-*n*	$(CH_2)_2O$, $TiCl_4$, CH_2Cl_2, −78°, 1 h	$CH_2{=}CHC(CH_3)(C_6H_{13}$-*n*$)(CH_2)_2OH$ (93)	170
t-$C_4H_9(CH_3)_2SiCOH(C_7H_{15}$-*n*$)CH{=}CH_2$	1. *n*-C_4H_9Li, −70°, 20 min 2. $CuC{\equiv}CSi(CH_3)_3$, THF, −70°, 20 min 3. 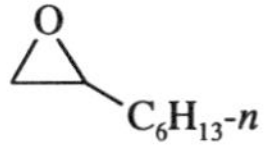 (epoxide) C_6H_{13}-*n*	*t*-$C_4H_9(CH_3)_2SiO$-(*n*-$C_7H_{15})C{=}CH(CH_2)_2CHOHC_6H_{13}$-*n* (74)	280
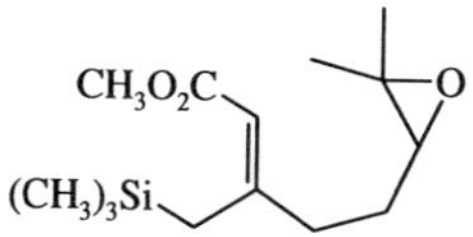 CH_3O_2C, $(CH_3)_3Si$ (structure)	$BF_3{\cdot}O(C_2H_5)_2$ (2 eq), CH_2Cl_2, −56°, 4 h	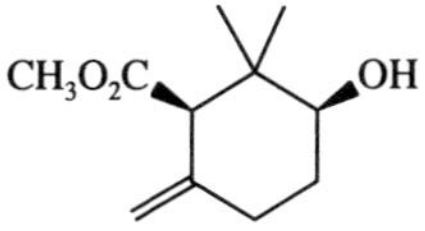 CH_3O_2C, OH (structure) (80)	265
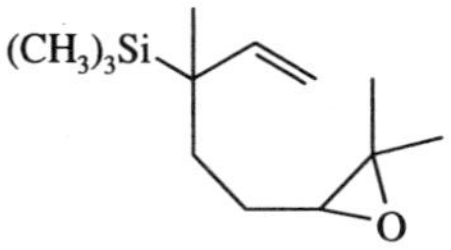 $(CH_3)_3Si$, O (structure)	$BF_3{\cdot}O(C_2H_5)_2$, −60°, 0.5 h	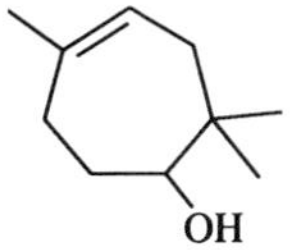 OH (structure) (71)	278
	$TiCl_4$	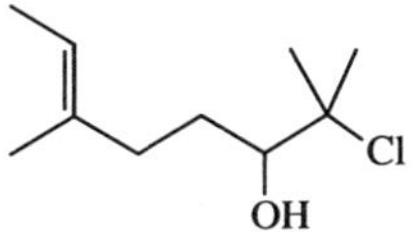 Cl, OH (structure) (—)	278
C_{11}			
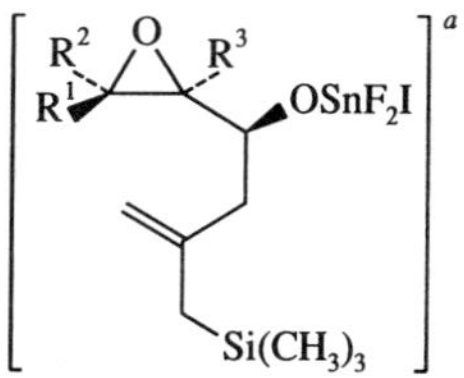 [R^2, O, R^3, R^1, $OSnF_2I$, $Si(CH_3)_3$ (structure)][a]	THF	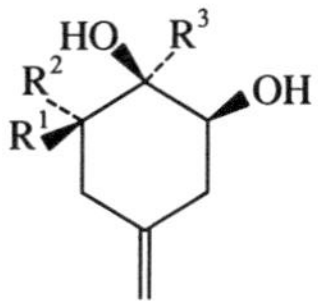 HO, R^3, R^2, R^1, OH (structure) + diastereoisomer	625

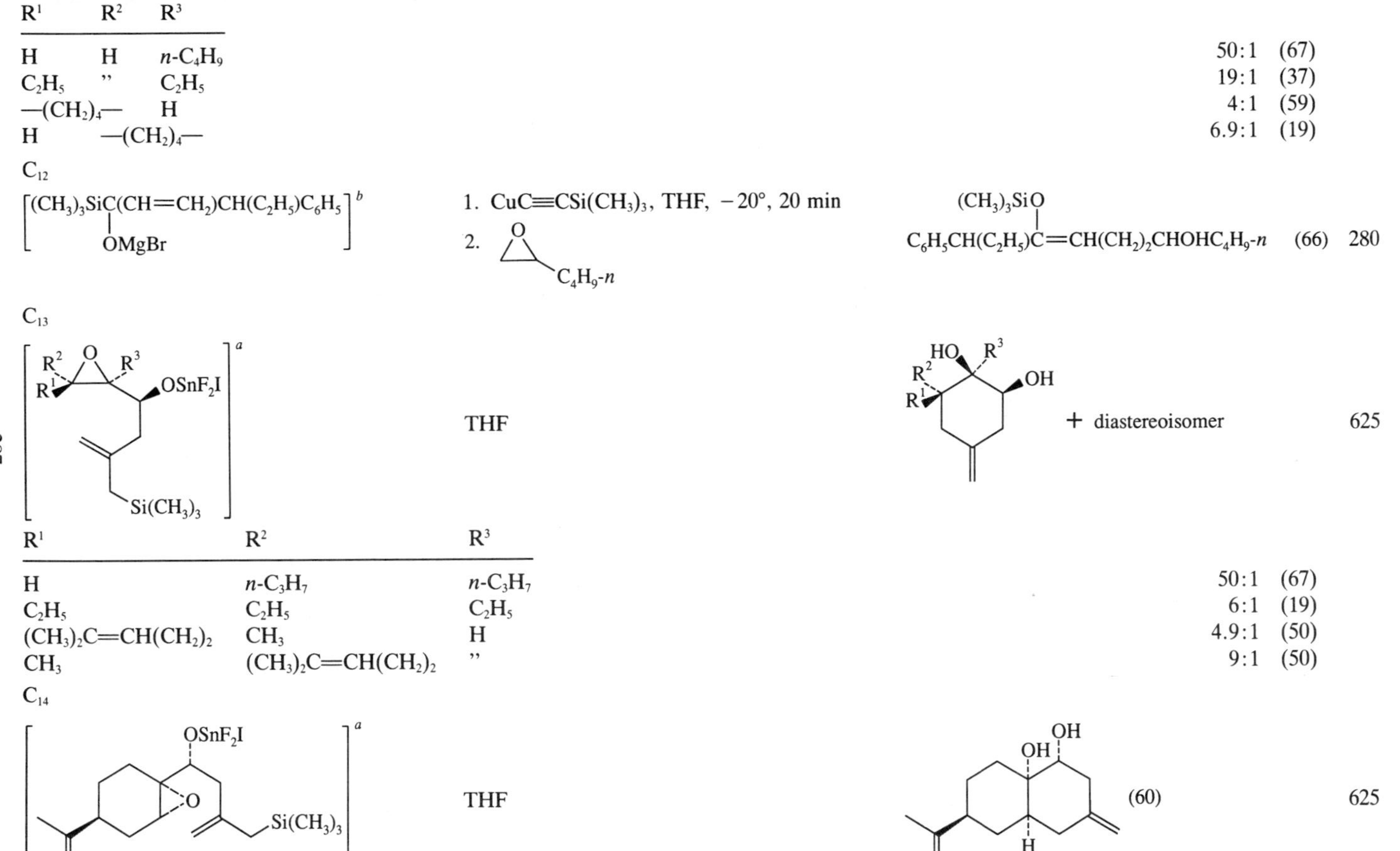

R^1	R^2	R^3		
H	H	n-C_4H_9	50:1	(67)
C_2H_5	"	C_2H_5	19:1	(37)
—$(CH_2)_4$—		H	4:1	(59)
H	—$(CH_2)_4$—		6.9:1	(19)

C_{12}

Reactant	Conditions	Product(s) and Yield(s) (%)	Refs.
[$(CH_3)_3SiC(OMgBr)(CH{=}CH_2)CH(C_2H_5)C_6H_5$] [b]	1. $CuC{\equiv}CSi(CH_3)_3$, THF, −20°, 20 min 2. 1,2-epoxyhexane (C_4H_9-n)	$C_6H_5CH(C_2H_5)C(OSi(CH_3)_3){=}CH(CH_2)_2CHOHC_4H_9$-$n$ (66)	280

C_{13}

Reactant	Conditions	Product(s) and Yield(s) (%)	Refs.
[epoxide, $OSnF_2I$, $Si(CH_3)_3$] [a]	THF	+ diastereoisomer	625

R^1	R^2	R^3		
H	n-C_3H_7	n-C_3H_7	50:1	(67)
C_2H_5	C_2H_5	C_2H_5	6:1	(19)
$(CH_3)_2C{=}CH(CH_2)_2$	CH_3	H	4.9:1	(50)
CH_3	$(CH_3)_2C{=}CH(CH_2)_2$	"	9:1	(50)

C_{14}

Reactant	Conditions	Product(s) and Yield(s) (%)	Refs.
[$OSnF_2I$, O, $Si(CH_3)_3$] [a]	THF	(60)	625

TABLE IX. ALLYLSILANES WITH EPOXIDES, OXETANES, AND EPISULFONIUM SALTS (*Continued*)

Reactant	Conditions	Product(s) and Yield(s)	Refs.
C_{15}			
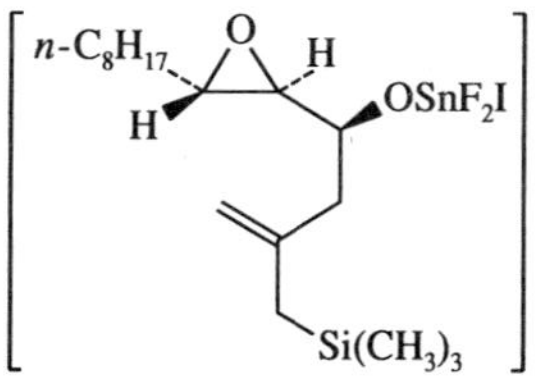 [n-C_8H_{17} epoxide, $OSnF_2I$, $Si(CH_3)_3$ structure][a]	THF	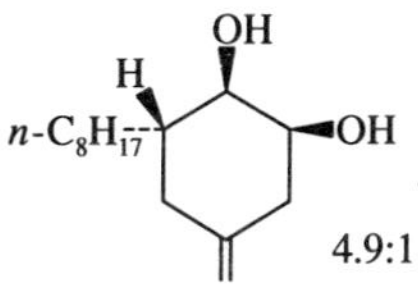 n-C_8H_{17}, H, OH, OH cyclohexane structure + diastereoisomer (53) 4.9:1	625

[a] This presumed intermediate is prepared by reaction of $(CH_3)_3SiCH_2C(CH_2I){=}CH_2$ with the appropriate aldehyde in the presence of SnF_2.

[b] This intermediate is prepared by the addition of vinylmagnesium bromide to the acylsilane.

TABLE X. VINYLSILANES WITH EPOXIDES

Reactant	Conditions	Product(s) and Yield(s)	Refs.
C_{14}			
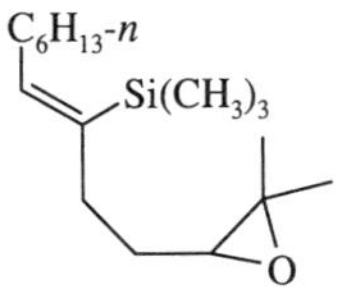 C_6H_{13}-*n*, $Si(CH_3)_3$, O	$TiCl_4$	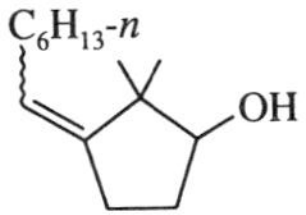 C_6H_{13}-*n*, OH (54)	120

TABLE XI. ALLYLSILANES WITH ALDEHYDES AND KETONES

Reactant	Conditions	Product(s) and Yield(s)	Refs.
C_3			
$(CH_3)_3SiCH_2CH{=}CH_2$	CH_3CHO, $AlCl_3$, CH_2Cl_2, 2 h	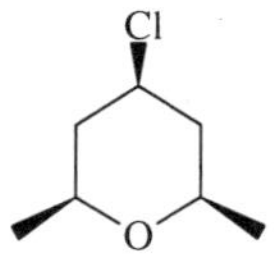 Cl (56) O	281
	1. CH_3CHO, $TiCl_4$, CH_2Cl_2, 1 min 2. $CH_3CH{=}CHCHO$, 4 h	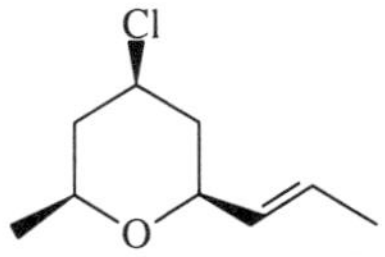 Cl (56) O	281
	Cl_3CCHO, $GaCl_3$, CH_2Cl_2, 100–110°, 24 h	$CH_2{=}CHCH_2CHOHCCl_3$ (55)	313
	C_2H_5CHO, $AlCl_3$, CH_2Cl_2, −25°, 2 h	$CH_2{=}CHCH_2CHOHC_2H_5$ (40–45)	626
	$RCH_2CH(OCH_2C_6H_5)CHO$	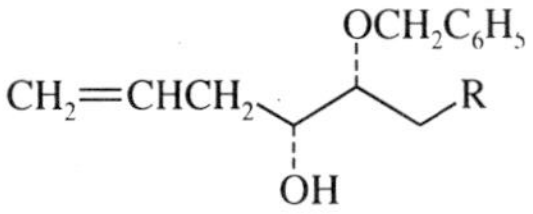 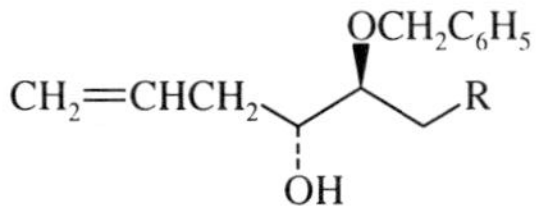 $OCH_2C_6H_5$ / $CH_2{=}CHCH_2$ … R / OH (I) + $OCH_2C_6H_5$ / $CH_2{=}CHCH_2$ … R / OH (II)	
	R / Conditions	I:II	
	H — $SnCl_4$, −78°, 20 min	35:1 (94)	299
	" — $BF_3{\cdot}O(C_2H_5)_2$, −78°, 20 min	1:1.5 (50)	299
	" — $TiCl_4$	(0)[a]	627

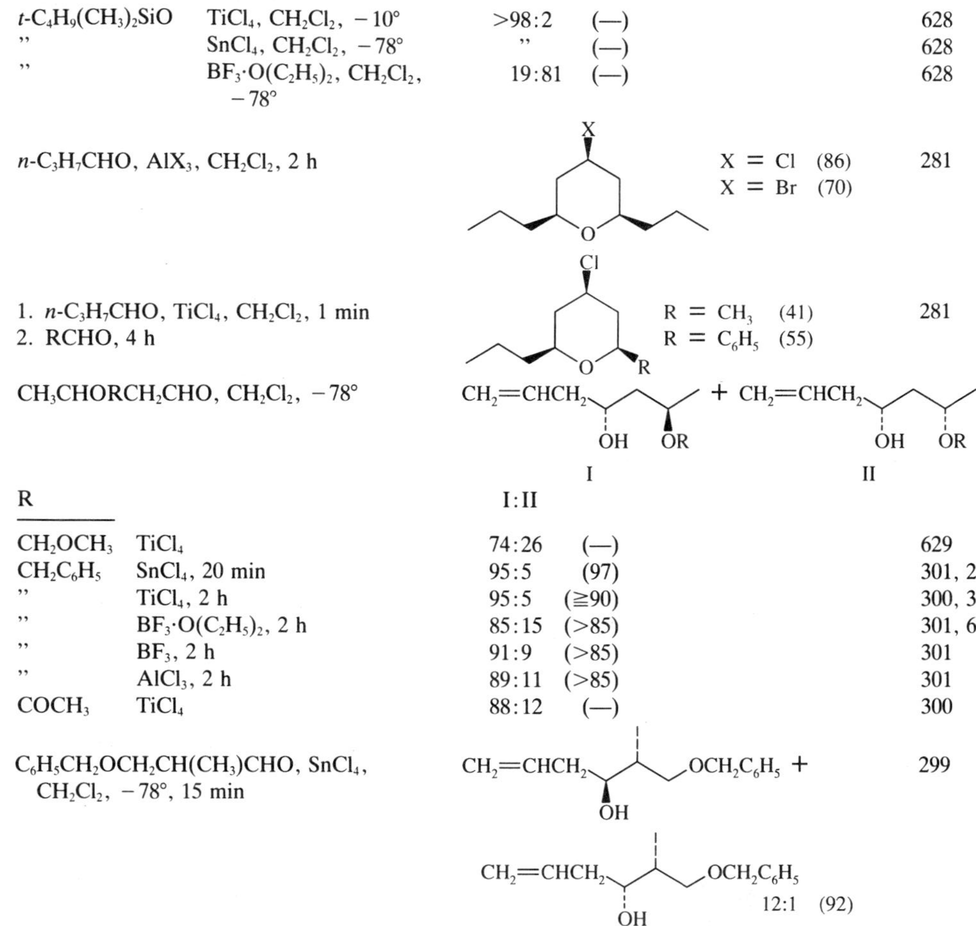

	Conditions	Product(s) and Yield(s) (%)	Refs.
t-$C_4H_9(CH_3)_2SiO$	$TiCl_4$, CH_2Cl_2, $-10°$	>98:2 (—)	628
"	$SnCl_4$, CH_2Cl_2, $-78°$	" (—)	628
"	$BF_3 \cdot O(C_2H_5)_2$, CH_2Cl_2, $-78°$	19:81 (—)	628
	n-C_3H_7CHO, AlX_3, CH_2Cl_2, 2 h	X = Cl (86); X = Br (70)	281
	1. n-C_3H_7CHO, $TiCl_4$, CH_2Cl_2, 1 min 2. RCHO, 4 h	R = CH_3 (41); R = C_6H_5 (55)	281
	$CH_3CHORCH_2CHO$, CH_2Cl_2, $-78°$	$CH_2{=}CHCH_2$ ... OH OR (I) + $CH_2{=}CHCH_2$ ... OH OR (II)	
R		I:II	
CH_2OCH_3	$TiCl_4$	74:26 (—)	629
$CH_2C_6H_5$	$SnCl_4$, 20 min	95:5 (97)	301, 299
"	$TiCl_4$, 2 h	95:5 (≧90)	300, 301
"	$BF_3 \cdot O(C_2H_5)_2$, 2 h	85:15 (>85)	301, 627
"	BF_3, 2 h	91:9 (>85)	301
"	$AlCl_3$, 2 h	89:11 (>85)	301
$COCH_3$	$TiCl_4$	88:12 (—)	300
	$C_6H_5CH_2OCH_2CH(CH_3)CHO$, $SnCl_4$, CH_2Cl_2, $-78°$, 15 min	$CH_2{=}CHCH_2$ ... OH ... $OCH_2C_6H_5$ + $CH_2{=}CHCH_2$ ... OH ... $OCH_2C_6H_5$ 12:1 (92)	299

TABLE XI. ALLYLSILANES WITH ALDEHYDES AND KETONES (*Continued*)

Reactant	Conditions	Product(s) and Yield(s)	Refs.
	n-C_3H_7CHO, $TiCl_4$, CH_2Cl_2, 30 s	$CH_2{=}CHCH_2CHOHC_3H_7$-n (87)	64
	", TBAF, THF, reflux, 15 h	" (83)[b]	13
	i-C_3H_7CHO, $TiCl_4$, CH_2Cl_2, 10 min	$CH_2{=}CHCH_2CHOHC_3H_7$-i (54)	64
	", $AlCl_3$, CH_2Cl_2, −25°, 2 h	" (40–45)	626
	n-C_4H_9CHO, TBAF, THF, reflux, 15 h	$CH_2{=}CHCH_2CHOHC_4H_9$-n (92)[b]	13
	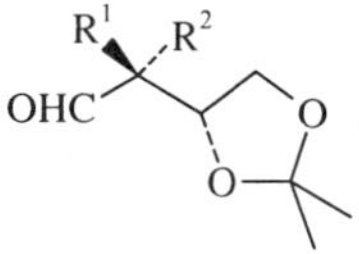 OHC–C(R¹)(R²)–(2,2-dimethyl-1,3-dioxolan-4-yl), $MgBr_2$, CH_2Cl_2, −20°	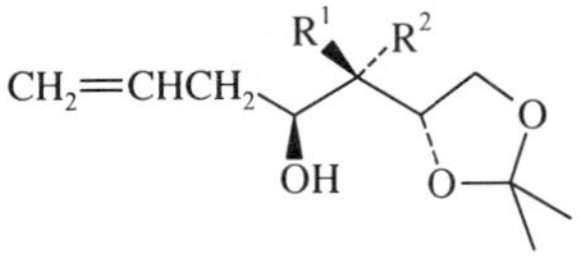 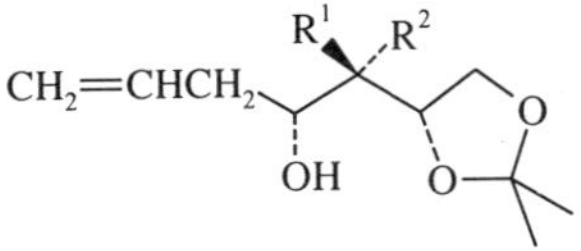 $CH_2{=}CHCH_2$–CH(OH)–C(R¹)(R²)–(2,2-dimethyl-1,3-dioxolan-4-yl) I + $CH_2{=}CHCH_2$–CH(OH)–C(R¹)(R²)–(2,2-dimethyl-1,3-dioxolan-4-yl) II	630
	R¹ = $C_6H_5CH_2O$, R² = H	I:II >98:2 (95)	
	R¹ = H, R² = $C_6H_5CH_2O$	I:II 2:98 (78)	
	OHC–(methyl 2,3-O-isopropylidene furanoside, OCH_3)	$CH_2{=}CHCH_2$–CH(OH)–(H)(2,3-O-isopropylidene furanoside, OCH_3) +	302

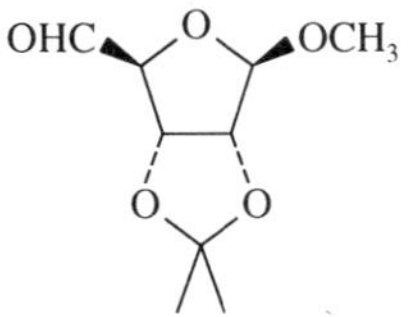

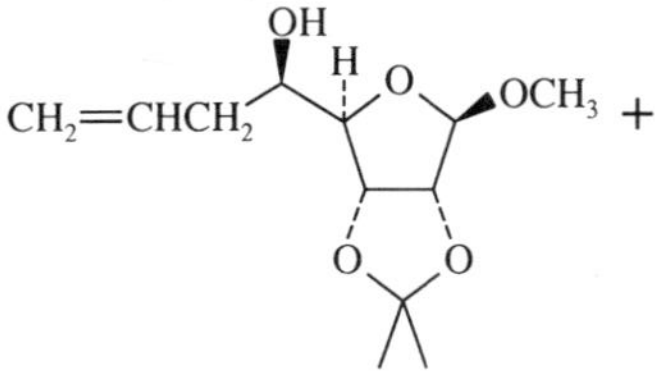

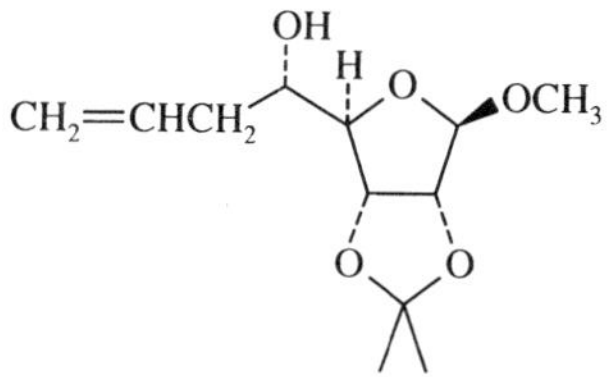

Reactant	Conditions	Product(s) and Yield(s) (%)	Refs.
		$CH_2{=}CHCH_2$ (OH, H) ... OCH_3	
	$BF_3{\cdot}O(C_2H_5)_2$, CH_2Cl_2, −78°, 3 h	>20:1 (80)	
	$TiCl_4$, CH_2Cl_2, −78°	1:20 (89)	
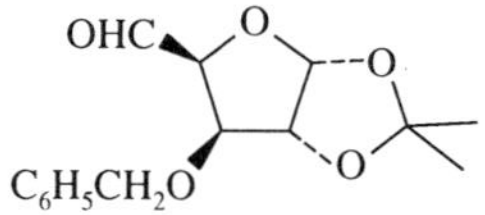 OHC ... $C_6H_5CH_2O$, CH_2Cl_2, −78°		$CH_2{=}CHCH_2$ (OH, H) ... $C_6H_5CH_2O$ +	302
		$CH_2{=}CHCH_2$ (OH, H) ... $C_6H_5CH_2O$	
	$BF_3{\cdot}O(C_2H_5)_2$	>20:1 (69)	
	$TiCl_4$	<1:20 (74)	
	n-$C_5H_{11}CHO$, TBAF, THF, reflux, 15 h	$CH_2{=}CHCH_2CHOHC_5H_{11}$-$n$ (87)[b]	13
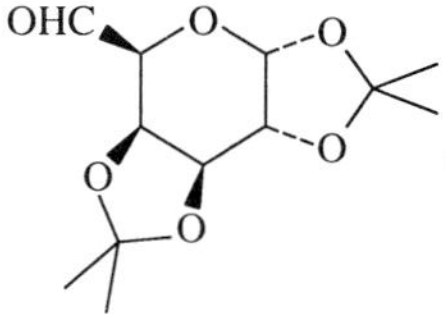 OHC ... , CH_2Cl_2, −78°		$CH_2{=}CHCH_2$ (OH, H) ... +	302
		$CH_2{=}CHCH_2$ (OH, H) ...	

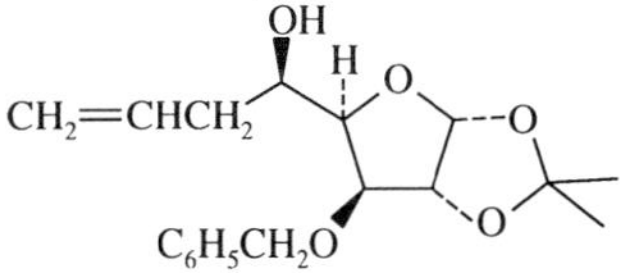

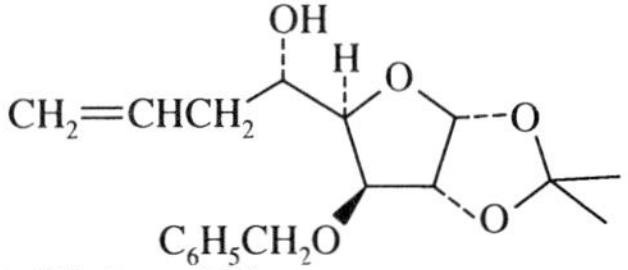

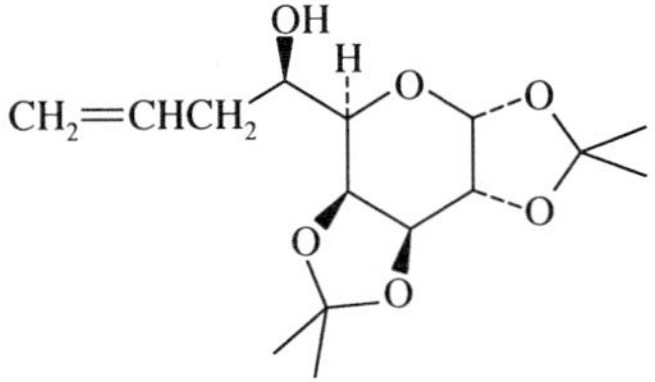

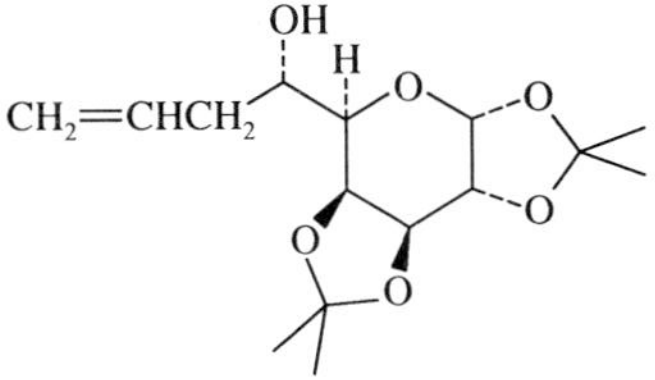

TABLE XI. ALLYLSILANES WITH ALDEHYDES AND KETONES (*Continued*)

Reactant	Conditions	Product(s) and Yield(s)	Refs.
	$BF_3 \cdot O(C_2H_5)_2$ $TiCl_4$	20:1 (78) 1:10 (64)	
	OHC–(pyranose ring, O, ..OCH$_3$, ''OCH$_3$, OCH$_3$)–t-$C_4H_9(CH_3)_2SiO$	CH_2=CHCH$_2$–CH(OH)–(pyranose, H, O, ..OCH$_3$, ''OCH$_3$, OCH$_3$)–t-$C_4H_9(CH_3)_2SiO$ + CH_2=CHCH$_2$–CH(OH)–(pyranose, H, O, ..OCH$_3$, ''OCH$_3$, OCH$_3$)–t-$C_4H_9(CH_3)_2SiO$	302
	$BF_3 \cdot O(C_2H_5)_2$, CH_2Cl_2, −78° $BF_3 \cdot O(C_2H_5)_2$, toluene, −78° $TiCl_4$, CH_2Cl_2, −78°	3:7 (80) 1:20 (—) <1:20 (80)	
	n-$C_3H_7CH(OCH_2C_6H_5)CH_2CHO$, $TiCl_4$, −78°	CH_2=CHCH$_2$CH(OH)CH$_2$CH(OCH$_2$C$_6$H$_5$)C$_3$H$_7$-n + CH_2=CHCH$_2$CH(OH)CH$_2$CH(OCH$_2$C$_6$H$_5$)C$_3$H$_7$-n 95:5 (≥90)	300
	RCHO, AlX_3, CH_2Cl_2, 2 h	2,6-R$_2$-4-X-tetrahydropyran (X, R, O, R)	281

R	X				
C_6H_5	Cl			(72)	
"	Br			(49)	
n-C_6H_{13}	Cl			(51)	
"	Br			(43)	

1. C_6H_5CHO, $TiCl_4$, CH_2Cl_2, 1 min
2. RCHO, 4 h

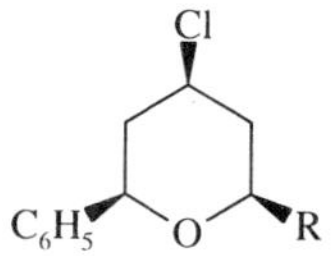

[4-chloro-2-C_6H_5-6-R-tetrahydropyran] R = CH_3 (56); R = n-C_3H_7 (62) — 281

RCHO R		$CH_2{=}CHCH_2CHOHR$	
C_6H_5	$BF_3{\cdot}O(C_2H_5)_2$, CH_2Cl_2, 1 min	(58)	64
"	$(C_6H_5)_2BO_3SCF_3$ (cat.), CH_2Cl_2, 0° 2 h	(83)	356
"	TBAF, THF, reflux, 90 h	(93)[b,c]	13
n-C_6H_{13}	$TiCl_4$, CH_2Cl_2, 1 min	(91)	64
$C_6H_5CH_2$	$AlCl_3$, CH_2Cl_2, −25°, 2 h	(40–45)	626
n-C_7H_{15}	$TiCl_4$, CH_2Cl_2, 1 min	(86)	64
"	Montmorillonite-K10, CH_2Cl_2, −78°, 0.5 h; 0°, 0.5 h; 25°, 0.5 h	(63)	357
$C_6H_5CH{=}CH$	$BF_3{\cdot}O(C_2H_5)_2$	(50)	162
"	TBAF, DMF, HMPA	(86)	162
$C_6H_5(CH_2)_2$	$TiCl_4$, CH_2Cl_2, 1 min	(96)	64
"	TBAF, THF, reflux, 4 h	(86)[b]	13

$C_6H_5CH(CH_3)CHO$, CH_2Cl_2 — $CH_2{=}CHCH_2CH(OH)CH(CH_3)C_6H_5$ + $CH_2{=}CHCH_2CH(OH)CH(CH_3)C_6H_5$

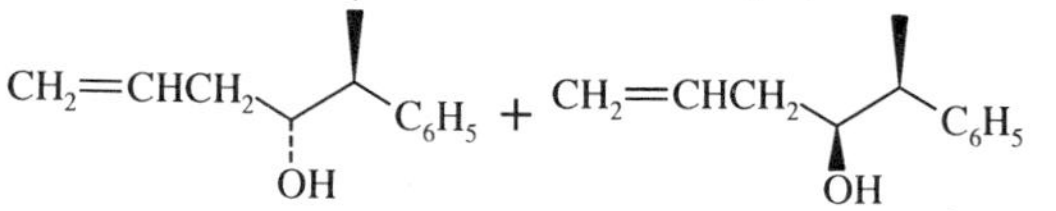

Conditions	Ratio	Yield	Refs.
$TiCl_4$, −78°, 30 min	1:1.6	(86)	299, 629
", 25°, 5 min	1:1.3	(81)	299

TABLE XI. ALLYLSILANES WITH ALDEHYDES AND KETONES (*Continued*)

Reactant	Conditions	Product(s) and Yield(s)	Refs.
	$BF_3 \cdot O(C_2H_5)_2$, −78°, 30 min	1:2 (47)	299, 629
	", 25°, 5 min	1:1.7 (51)	299
	$SnCl_4$, −78°, 30 min	1:2.2 (86)	299, 629
	$AlCl_3$, −78°	26:74 (—)	629
	$C_6H_5CH{=}C(CH_3)CHO$, $BF_3 \cdot O(C_2H_5)_2$	$CH_2{=}CHCH_2CHOHC(CH_3){=}CHC_6H_5$ (45)	162
	", TBAF, DMF, HMPA	" (92)	162
	$(CH_3)_2C{=}CH(CH_2)_3CH(CH_3)CHO$, $AlBr_3$, CH_2Cl_2, 2 h	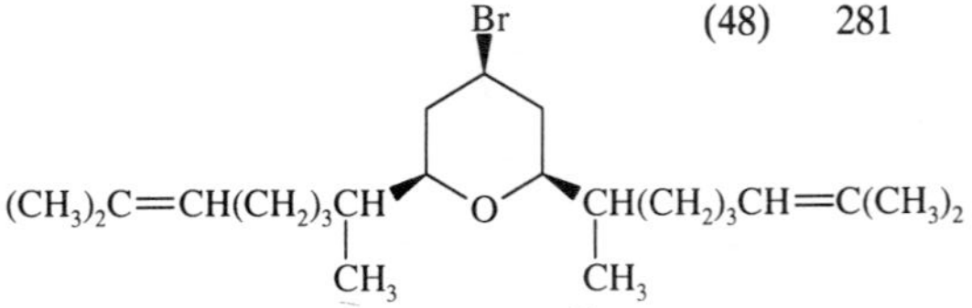 Br; $(CH_3)_2C{=}CH(CH_2)_3CH(CH_3)$–O–$CH(CH_3)(CH_2)_3CH{=}C(CH_3)_2$ (48)	281
	1. $(CH_3)_2C{=}CH(CH_2)_3CH(CH_3)CHO$, $TiCl_4$, CH_2Cl_2, 4 h 2. n-C_3H_7CHO, 4 h	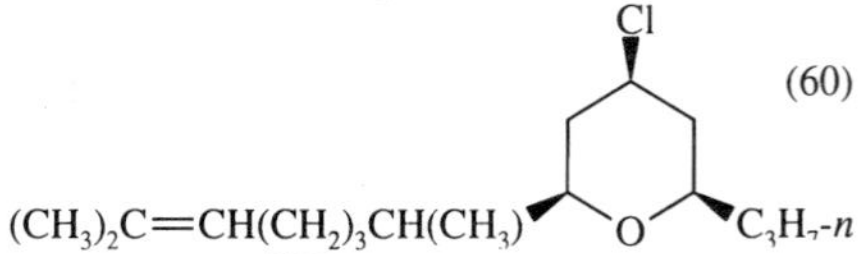 Cl; $(CH_3)_2C{=}CH(CH_2)_3CH(CH_3)$–O–$C_3H_7$-$n$ (60)	281
	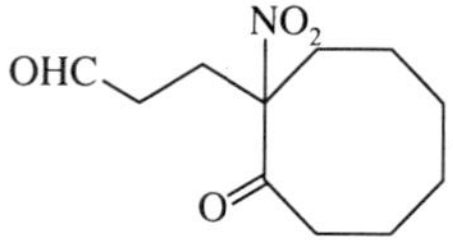 OHC, NO_2, O, $TiCl_4$, CH_2Cl_2, −65°, 20 min	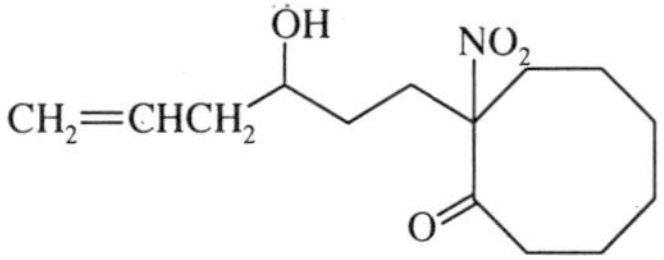 $CH_2{=}CHCH_2$, OH, NO_2, O	289
	$(CH_3)_2CO$, $TiCl_4$, CH_2Cl_2, 1 min	$CH_2{=}CHCH_2COH(CH_3)_2$ (83)	64
	CH_3COCH_2Cl, $AlCl_3$, 100–110°, 24 h	$CH_2{=}CHCH_2COH(CH_3)CH_2Cl$ (60)	313
	$(ClCH_2)_2CO$, "	$CH_2{=}CHCH_2COH(CH_2Cl)_2$ (63)	626

Reactant, Conditions	Product(s) (Yield, %)	Refs.
$CH_3COCH(OCH_3)_2$	$CH_2{=}CHCH_2C(OH)(CH_3)CH(OCH_3)_2$ + $CH_2{=}CHCH_2C(OH)(CH_3)CH(OCH_3)CH_2CH{=}CH_2$	360
$TiCl_4$, CH_2Cl_2, $-78°$, 4 h	6:94 (33)	
", ", 0°, 4 h	0:100 (45)	
$AlCl_3$, 0°, 4 h	100:0 (67)	
CH_3COCO_2R, $TiCl_4$, CH_2Cl_2	$CH_2{=}CHCH_2COH(CH_3)CO_2R$	631
R = CH_3 $-78°$, 30 min	(86)	
C_2H_5 ", "	(85)	
C_4H_9-*n* ", 20 min	(92)	
C_6H_5 ", "	(82)	
$CH_2C_6H_5$ ", "	(82)	
(menthyl) $-75°$, 30 min	55% de (93)	
$-16°$, "	41% de (92)	
2°, "	37% de (82)	
20°, "	34% de (79)	
CH_3COCON(pyrrolidine-2-CO_2CH_3)	$CH_2{=}CHCH_2COH(CH_3)CON$(pyrrolidine-2-$CO_2CH_3$)	632, 633
$TiCl_4$, $-78°$, 3 h	56% de (47)	
$SnCl_4$ ", "	45% de (67)	
$(CF_3)_2CO$, $AlCl_3$, 25°, 24 h	$CH_2{=}CHCH_2C(CF_3)_2OSi(CH_3)_3$ (51)[d] + $(CH_3)_3SiCH_2CH{=}CHC(CF_3)_2OH$ (16) + $(CH_3)_3SiCH{=}CHCH_2C(CF_3)_2OH$ (13)	141
$CH_3COCH_2CH(OCH_3)_2$	See Table XIX	

TABLE XI. ALLYLSILANES WITH ALDEHYDES AND KETONES (*Continued*)

Reactant	Conditions	Product(s) and Yield(s)	Refs.
	$CH_3CH(OCH_2C_6H_5)COCN$, CH_2Cl_2, −78°, 2 h	$CH_2{=}CHCH_2C(CN)(OH)CH(OCH_2C_6H_5)CH_3$ + $CH_2{=}CHCH_2C(CN)(OH)CH(OCH_2C_6H_5)CH_3$ (diastereomer)	634
	$TiCl_4$	96:14 (98)	
	$SnCl_4$	85:15 (—)	
	$CH_3COC_3H_7$-*i*, $TiCl_4$, CH_2Cl_2, 1 min	$CH_2{=}CHCH_2COH(CH_3)C_3H_7$-*i* (44)	64
	Cyclopentanone, $TiCl_4$, CH_2Cl_2, 3 min	1-($CH_2{=}CHCH_2$)-1-OH-cyclopentane (44)	64
	$CH_3CO(CH_2)_2CO_2CH_3$, TBAF, THF, reflux	5-CH_3-5-($CH_2{=}CHCH_2$)-dihydrofuran-2-one (71)[b]	13
	$CH_3CH(OCH_2C_6H_5)CH_2COCN$, $TiCl_4$, CH_2Cl_2, −78°	$CH_2{=}CHCH_2C(CN)(OH)CH_2CH(OCH_2C_6H_5)CH_3$ + $CH_2{=}CHCH_2C(CN)(OH)CH_2CH(OCH_2C_6H_5)CH_3$ (diastereomer) 99:1 (90)	634
	2-R-cyclohexanone	1-($CH_2{=}CHCH_2$)-2-R-cyclohexan-1-ol	

Reactant	Conditions	Product(s) and Yield(s) (%)		Refs.
R = H	$TiCl_4$, CH_2Cl_2, 3 min		(70)	64
"	TBAF, THF, reflux, 16 h		(60)[b]	13
R = OCH_3	$TiCl_4$, 0°		(—)	635
$C_6H_5COCH_3$, TBAF, THF, reflux, 16 h		$CH_2{=}CHCH_2COH(C_6H_5)CH_3$	(60)[b]	13
C_6H_5COCON(pyrrolidine-2-CO_2R), CH_2Cl_2		$CH_2{=}CHCH_2C(OH)(C_6H_5)CON$(pyrrolidine-2-$CO_2R$)		633

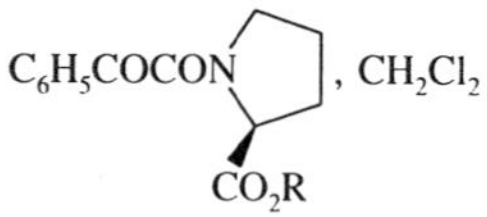

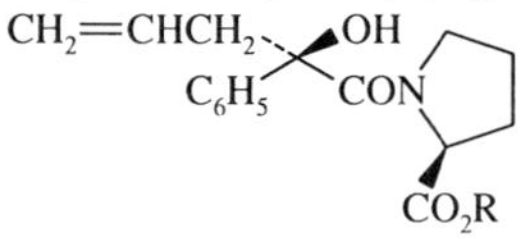

R	Conditions	de	Yield
CH_3	$SnCl_4$, −40°, 3 h	82% de	(44)
"	$SnBr_4$, 44 h	84% de	(59)
"	", 0°, 24 h	83% de	(55)
"	$BF_3{\cdot}O(C_2H_5)_2$, 24 h	66% de	(5)
"	$TiCl_4$, 0°, 1 h	2% de	(84)
"	", −40°, 3 h	24% de	(69)
"	", −78°, 6 h	52% de	(56)
"	", C_6H_{14}, −40°, 3 h	78% de	(72)
"	", ", −78°, 5 h	72% de	(13)
n-C_3H_7	$SnCl_4$, 0°	79% de	(63)
$C_6H_5CH_2$	", −78°, 4.5 h	89% de	(46)
"	$TiCl_4$, ", 5 h	75% de	(58)

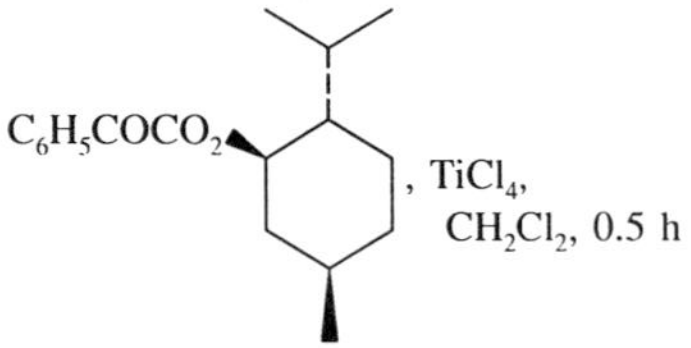

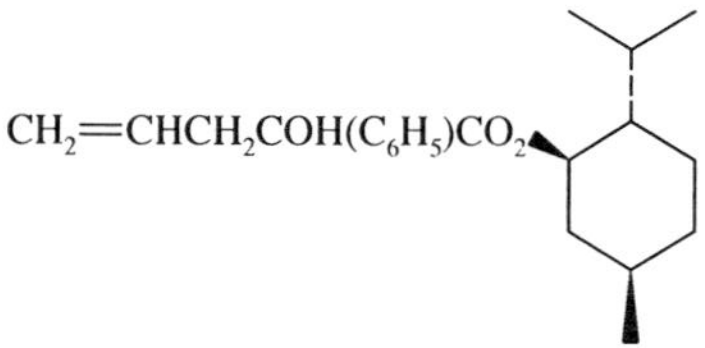

Reactant	Conditions	Product		Refs.
$C_6H_5COCO_2$-(menthyl), $TiCl_4$, CH_2Cl_2, 0.5 h		$CH_2{=}CHCH_2COH(C_6H_5)CO_2$-(menthyl)		631
	20°	16% de	(82)	
	−75°	23% de	(86)	

TABLE XI. ALLYLSILANES WITH ALDEHYDES AND KETONES (*Continued*)

Reactant	Conditions	Product(s) and Yield(s)	Refs.
	4-tert-butylcyclohexanone ($O=C_6H_9$-C_4H_9-t), Al-Montmorillonite, CH_2Cl_2, −30°, 0.1 h; −15°, 3 h	$CH_2{=}CHCH_2$, $OSi(CH_3)_3$ cyclohexane with C_4H_9-t + $CH_2{=}CHCH_2$, $OSi(CH_3)_3$ cyclohexane with C_4H_9-t 13:87 (80)	357
	$(C_6H_5)_2CO$, $TiCl_4$, CH_2Cl_2, reflux, 4 h	$CH_2{=}CHCH_2C(C_6H_5)_2OH$ (25)	246
$RCH(OSi(CH_3)_2CH_2CH{=}CH_2)CH_2CHO$	$TiCl_4$, CH_2Cl_2, −20°, 1–2 h	$RCH(OH)CH_2CH(OH)CH_2CH{=}CH_2$ >95:5 (—)	636
pinane derivative with OCH_3 and $Si(CH_3)_2CH_2CH{=}CH_2$	n-C_3H_7CHO	$CH_2{=}CHCH_2CHC_3H_7$-n (O–$Si(CH_3)_2$ pinane, OCH_3) I + $CH_2{=}CHCH_2CHOHC_3H_7$-n II + 4-Cl-2,6-di-n-C_3H_7-tetrahydropyran III	27
		% ee / I / II / III	
	$TiCl_4$, CH_2Cl_2, −78°, 4 h	46 (65) (4) (11)	
	$BF_3{\cdot}O(C_2H_5)_2$	— (0) (12) (0)	
	$AlCl_3$	— (53) (0) (41)	

Reactant	Conditions	Product(s) and Yield(s) (%)	Refs.
	$C_2H_5AlCl_2$	— (0) (4) (0)	
	ZnI_2	59 (14) (0) (0)	
	BBr_3	31 (33) (39) (26) (Cl=Br)	
	$(C_5H_5)_2TiCl_2$	— (3) (0) (0)	
	120°, 12 h	48 (14) (0) (0)	
	R^1R^2CO, $TiCl_4$, CH_2Cl_2, −78°, 4 h	$CH_2{=}CHCH_2COHR^1R^2$	27
	R^1 / R^2	% ee	
	CH_3 / H	40 (61)	
	n-C_3H_7 / "	46 (58)	
	C_6H_5 / "	21 (45)	
	CH_3 / C_2H_5	40 (39)	
$(CH_3)_3SiCH_2CH{=}CHN$ (pyrrolidine ring)	C_6H_5COR, TBAF, DMF	I (2-pyrrolidinyl-5-R^1,5-R^2-tetrahydrofuran) + II ($CH_2{=}CHCHCR^1R^2OH$ with N-pyrrolidinyl)	115
	R = CH_3	R^1 = C_6H_5, R^2 = CH_3 9:1 (30)	
	R = C_2H_5	R^1 = C_6H_5, R^2 = C_2H_5 9:1 (30)	
$(CH_3)_3SiCHCH{=}CH_2$ (N-pyrrolidinyl on CH)	R^1R^2CO, CsF, DMF	I + II	115
	R^1 / R^2	I:II	
	i-C_3H_7 / H / 30 h	1:1 (53)	
	C_6H_5 / " / "	2:3 (55)	
	CH_3 / CH_3 / "	3:2 (40)	
	C_6H_5 / " / "	3:2 (40)	
	" / C_2H_5 / 24 h	1:1 (40)	
	1. n-C_4H_9Li 2. C_6H_5CHO 3. H_2O	I R^1 = C_6H_5, R^2 = H (58)	115

TABLE XI. ALLYLSILANES WITH ALDEHYDES AND KETONES (*Continued*)

Reactant	Conditions	Product(s) and Yield(s)	Refs.
$(CH_3)_3SiCH_2CH{=}CHCl$	RCHO, $AlCl_3$, CH_2Cl_2	$CH_2{=}CHCHClCH(OCH_3)R^d$	288
	R		
	i-C_4H_9	(60)	
	$CH_3CO(CH_2)_2$	(24)	
	C_6H_5	(51)	
	$C_6H_5CH_2$	(73)	
	n-C_8H_{17}	(63)	
	n-$C_5H_{11}COCH_3$, $AlCl_3$, CH_2Cl_2	No reaction	288
$(CH_3)_3SiCHClCH{=}CH_2$	$C_6H_5(CH_2)_2CHO$, $TiCl_4$, −20°, 8 h	$CHCl{=}CHCH_2CHOH(CH_2)_2C_6H_5$ *E*:*Z* 23:77 (35)	637
$(CH_3)_3SiCH_2CH{=}CHSi(CH_3)_3$	R^1R^2CO, TBAF	$R^1R^2COHCH_2CH{=}CHSi(CH_3)_3$	307
	R^1 R^2		
	CH_3 H 25°, 3 h	(70)	
	C_6H_5 " 40°, 5 h	(90)	
	n-C_6H_{13} " 25°, 3 h	(75)	
	CH_3 CH_3 ", 16 h	(45)	
	—$(CH_2)_4$— ", "	(65)	
	—$(CH_2)_5$— ", "	(62)	
	C_6H_5 CH_3 ", 24 h	(60)	
$(CH_3)_3SiCH_2CBr{=}CH_2$	RCHO, $TiCl_4$, CH_2Cl_2, −78°, 1–2 h	$CH_2{=}CBrCH_2CHOHR$	290
	R		
	H	(96)	
	n-C_4H_9	(96)	
	t-C_4H_9	(80)	
	n-C_5H_{11}	(88)	
	$CH_2{=}CH(CH_2)_8$	(76)	

Reactant	Conditions	Product(s) and Yield(s) (%)	Refs.
	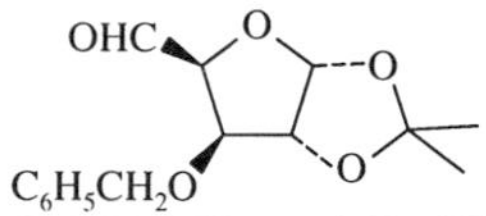 OHC, $C_6H_5CH_2O$ (sugar aldehyde), $TiCl_4$, CH_2Cl_2, $-78°$, 1–2 h	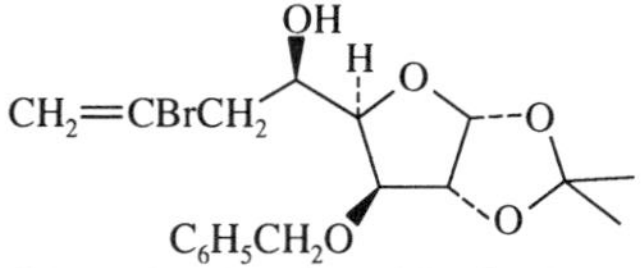 $CH_2{=}CBrCH_2$, OH, H, $C_6H_5CH_2O$ (86)	290
	$CH_3COC_3H_7$-*n*, $TiCl_4$, CH_2Cl_2, $-78°$, 1–2 h	$CH_2{=}CBrCH_2COH(CH_3)C_3H_7$-*n* (86)	290
	O (4-R-cyclohexanone), $TiCl_4$, CH_2Cl_2, $-78°$, 1–2 h	$CH_2{=}CBrCH_2$, OH, R	290
	R = H	(96)	
	R = CH_3	(80)	
	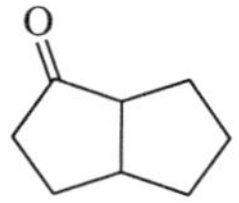 O (bicyclo[3.3.0]octan-2-one), $TiCl_4$, CH_2Cl_2, $-78°$, 1–2 h	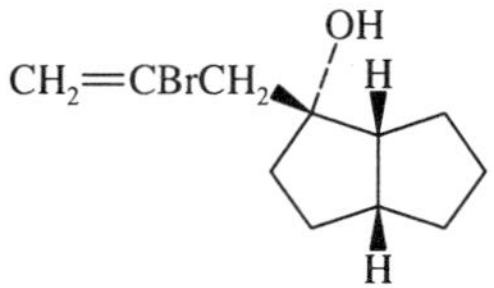 $CH_2{=}CBrCH_2$, OH, H, H (94)	290
$C_6H_5(CH_3)_2SiCH_2CH{=}CF_2$	1. RCHO, conditions **A**, **B**, **C**, or **D** **A** TASF (cat.), HMPA, overnight **B** TASF (cat.), DMPU, overnight **C** TASF (cat.), THF, overnight **D** KOC_4H_9-*t* (cat.), DMPU, overnight 2. HCl, CH_3OH	$CH_2{=}CHCF_2CHOHR$	306

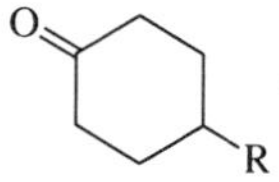

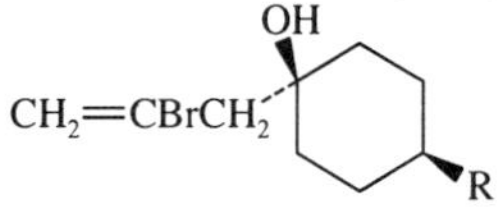

R	Conditions	Yield (%)
t-C_4H_9	**A**	(53)[e]
C_6H_5	**B**	(93)
"	**C**	(100)
"	**D**	(98)[f]
p-ClC_6H_4	**B**	(87)
"	**D**	(100)
p-(*n*-C_4H_9O)C_6H_4	**D**	(100)
$C_6H_5CH{=}CH$	**B**	(52)
n-$C_{10}H_{21}$	**B**	(44)

TABLE XI. ALLYLSILANES WITH ALDEHYDES AND KETONES (*Continued*)

Reactant	Conditions	Product(s) and Yield(s)	Refs.
$C_6H_5(CH_3)_2SiCH_2CH{=}CF_2$	1. $(C_6H_5)_2CO$, TASF (cat.), THF, 12 h 2. HCl, CH_3OH	$CH_2{=}CHCF_2C(C_6H_5)_2OH$ (34)[c]	306
$C_6H_5(CH_3)_2SiCF_2CH{=}CH_2$	C_6H_5CHO, TASF, THF, 12 h	$CH_2{=}CHCF_2CHOHC_6H_5$ (58)	306
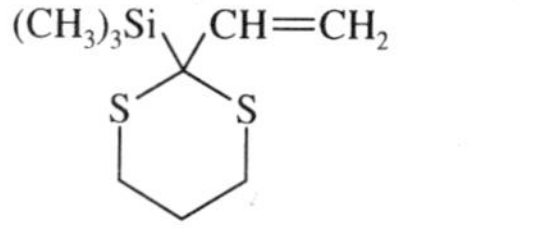 $(CH_3)_3Si$, $CH{=}CH_2$ (1,3-dithiane)	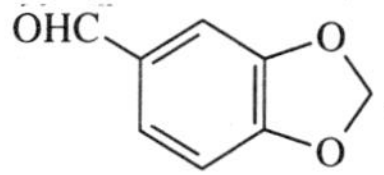 OHC (3,4-methylenedioxyphenyl), TBAF, THF, −20°, 15 min	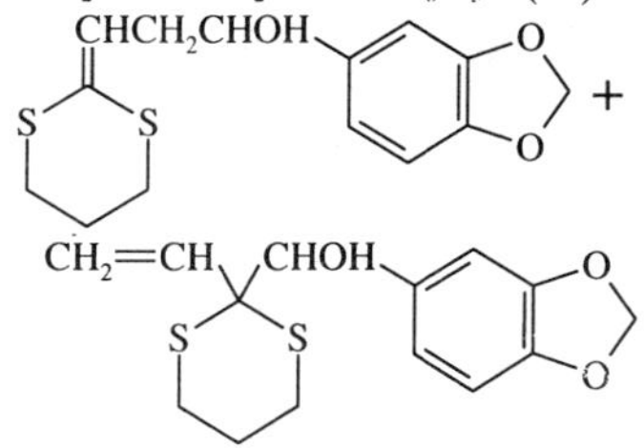 $CHCH_2CHOH$ (dithianylidene, methylenedioxyphenyl) + $CH_2{=}CH$, CHOH (dithiane, methylenedioxyphenyl) 1:3.5 (—)	258
C_4			
(*E*)-$(CH_3)_3SiCH_2CH{=}CHCH_3$	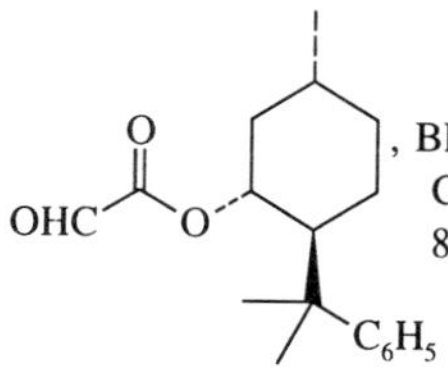 OHC, O (ester), C_6H_5, $BF_3{\cdot}O(C_2H_5)_2$, CH_2Cl_2, −78°, 8 h	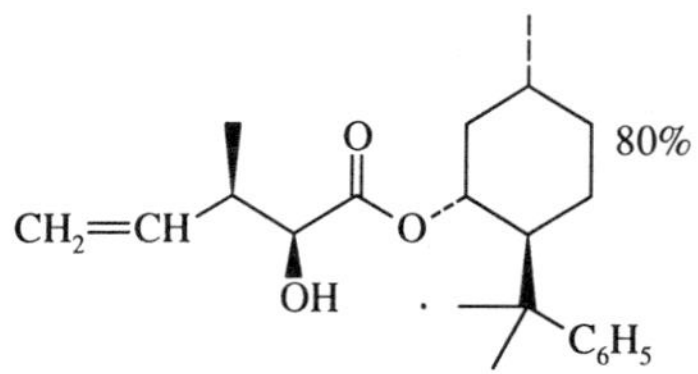 $CH_2{=}CH$, O, OH, O, C_6H_5 80% de (63)	298
	RCHO, $TiCl_4$, CH_2Cl_2	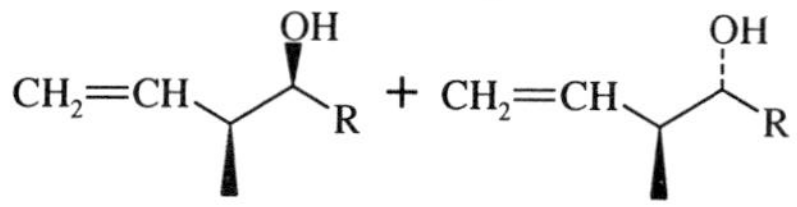 $CH_2{=}CH$, OH, R (I) + $CH_2{=}CH$, OH, R (II)	
	R	I:II	
	C_2H_5 −78°, 1 h	95:5 (91)	293

Allylsilane	Conditions	Product(s) and Yield(s) (%)	Refs.
	n-C_3H_7 0.5 min	— (83)	64
	i-C_3H_7 −78°, 1 h	97:3 (92)	293
	t-C_4H_9 ", "	>99:1 (98)	293
	n-C_3H_7CHO, TBAF, THF, reflux, 24 h	$CH_2{=}CHCH(CH_3)CHOHC_3H_7$-$n$ (23)[b] + $CH_3CH{=}CHCH_2CHOHC_3H_7$-$n$ (37)[b]	13
	$(CH_3)_2CO$, $TiCl_4$, CH_2Cl_2, 0.5 min	$CH_2{=}CHCH(CH_3)C(CH_3)_2OH$ (45)	64
(Z)-$(CH_3)_3SiCH_2CH{=}CHCH_3$	$RCHO$, $TiCl_4$, CH_2Cl_2, −78°	I + II	
	R	I:II	
	C_2H_5 0.5 h	69:31 (98)	293
	n-C_3H_7 0.5 min	— (71)	64
	i-C_3H_7 0.5 h	64:36 (98)	293
	t-C_4H_9 ", "	65:35 (87)	293
	$(CH_3)_2CO$, $TiCl_4$, CH_2Cl_2, 0.5 min	$CH_2{=}CHCH(CH_3)C(CH_3)_2OH$ (51)	64
$(CH_3)_3SiCH(CH_3)CH{=}CH_2$	n-C_3H_7CHO, $TiCl_4$, CH_2Cl_2, 1 min	(E)- + (Z)-$CH_3CH{=}CHCH_2CHOHC_3H_7$-$n$ (89)	64
	", TBAF, THF, reflux, 24 h	(E)-$CH_3CH{=}CHCH_2CHOHC_3H_7$-$n$ (41)[b] + $CH_2{=}CHCH(CH_3)CHOHC_3H_7$-$n$ (30)[b]	13
	$(CH_3)_2CO$, $TiCl_4$, CH_2Cl_2, 1 min	$CH_3CH{=}CHCH_2C(CH_3)_2OH$ E:Z 63:37 (72)	64
(S)-$(C_2H_5)_3SiCH(CH_3)CH{=}CH_2$	t-C_4H_9CHO, $TiCl_4$, CH_2Cl_2, −78°, 10 min	(R)-(E)-$CH_3CH{=}CHCH_2CHOHC_4H_9$-$t$ (22)	638
(S)-$C_6H_5(CH_3)_2SiCH(CH_3)CH{=}CH_2$	", ", ", −78°, 10 min	" (47)	638
(E)-$(CH_3)_3SiCH_2CH{=}CHCH_3$ + $(CH_3)_3SiCH(CH_3)CH{=}CH_2$ 75:25	Cl_3CCHO, $AlCl_3$	$CH_3CH{=}CHCH_2CHOHCCl_3$ + $CH_2{=}CH$-$CH(CH_3)CHOHCCl_3$ 25:75 (—)	313
$(CH_3)_3SiCH_2C(CH_3){=}CH_2$	$RCH_2CH(OCH_2C_6H_5)CHO$	$CH_2{=}C(CH_3)CH_2$–CH(OH)–CH($OCH_2C_6H_5$)–CH_2R (I) + $CH_2{=}C(CH_3)CH_2$–CH(OH)–CH($OCH_2C_6H_5$)–CH_2R (II)	

TABLE XI. ALLYLSILANES WITH ALDEHYDES AND KETONES (*Continued*)

Reactant	Conditions	Product(s) and Yield(s)	Refs.
	R / Conditions	I:II	
	H, $SnCl_4$, −78°, 20 min	45:1 (81)	299
	″, $BF_3 \cdot O(C_2H_5)_2$, −78°, 20 min	1:2.6 (40)	299
	t-$C_4H_9(CH_3)_2SiO$, $SnCl_4$, CH_2Cl_2, −78°	>97:3 (—)	628
	″, $BF_3 \cdot O(C_2H_5)_2$, CH_2Cl_2, −78°	23:77 (—)	628
	$CH_3CH(OCH_2C_6H_5)CH_2CHO$, CH_2Cl_2, −78°	$CH_2{=}C(CH_3)CH_2CH(OH)CH_2CH(OCH_2C_6H_5)CH_3$ + $CH_2{=}C(CH_3)CH_2CH(OH)CH_2CH(OCH_2C_6H_5)CH_3$	
	$SnCl_4$, CH_2Cl_2, −78°, 20 min	7:1 (86)	299
	$TiCl_4$, −78°	95:5 (≥90)	300
	$C_6H_5CH_2OCH_2CH(CH_3)CHO$, $SnCl_4$, CH_2Cl_2, −78°, 20 min	$CH_2{=}C(CH_3)CH_2CH(OH)CH(CH_3)CH_2OCH_2C_6H_5$ + $CH_2{=}C(CH_3)CH_2CH(OH)CH(CH_3)CH_2OCH_2C_6H_5$ 10:1 (83)	299
	n-$C_4H_9CH(OCH_2C_6H_5)CH_2CHO$, $TiCl_4$, −78°	$CH_2{=}C(CH_3)CH_2CH(OH)CH_2CH(OCH_2C_6H_5)C_4H_9$-$n$ + $CH_2{=}C(CH_3)CH_2CH(OH)CH_2CH(OCH_2C_6H_5)C_4H_9$-$n$ 99:1 (≥90)	300

	$C_6H_5CH(CH_3)CHO$, CH_2Cl_2	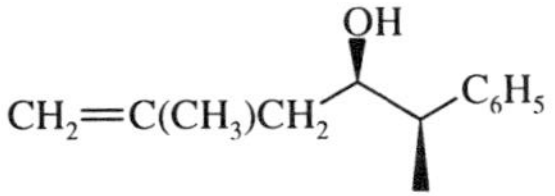 $CH_2{=}C(CH_3)CH_2CH(OH)CH(CH_3)C_6H_5$ + $CH_2{=}C(CH_3)CH_2CH(OH)CH(CH_3)C_6H_5$ (diastereomers)	299
	$TiCl_4$, −78°, 20–30 min	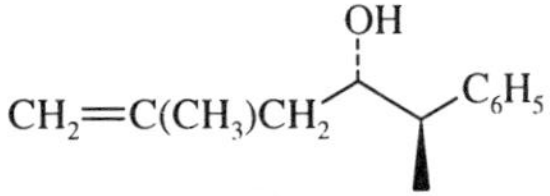 2.8:1 (66)	
	", 25°, 2–5 min	1.3:1 (38)	
	$BF_3{\cdot}O(C_2H_5)_2$, −78°, 20–30 min	7:1 (64)	
	", 25°, 2–5 min	2.8:1 (53)	
	$SnCl_4$, −78°, 20–30 min	3.2:1 (68)	
	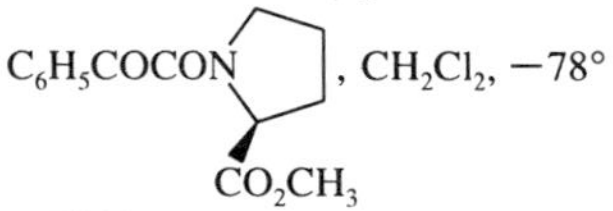 C_6H_5COCON(2-CO_2CH_3-pyrrolidine), CH_2Cl_2, −78°	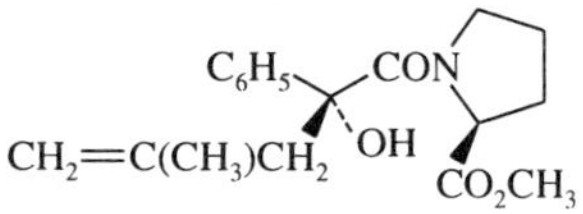 $CH_2{=}C(CH_3)CH_2C(OH)(C_6H_5)CON$(2-$CO_2CH_3$-pyrrolidine)	633
	$TiCl_4$, 1 h	44% de (63)	
	$SnCl_4$, 6 h	12% de (56)	
	$SnBr_4$, 24 h	30% de (90)	
$(CH_3)_3SiCH_2CH{=}CHCH_2Si(CH_3)_3$	C_2H_5CHO, $AlCl_3$, CH_2Cl_2, −25°, 2 h	$CH_2{=}CHCH[CH_2Si(CH_3)_3]CHOHC_2H_5$ (40) + $(CH_3)_3SiCH_2CH(COC_2H_5)(CH_2)_2Si(CH_3)_3$ (15)	626
$(CH_3)_3SiCHClC(CH_3){=}CH_2$	RCHO, $TiCl_4$, CH_2Cl_2, −78°, 3 h	$CHCl{=}C(CH_3)CH_2CHOHR$	637
	R = n-C_5H_{11}	(69)	
	R = $C_6H_5(CH_2)_2$	(93)	
$(CH_3)_3SiCH_2C({=}CH_2)CH_2I$	R^1COCOR^2, SnF_2, THF	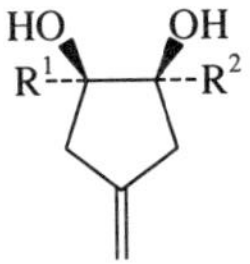 1,2-di-R^1,R^2-4-methylenecyclopentane-1,2-diol (HO, OH, R^1, R^2)	639

TABLE XI. ALLYLSILANES WITH ALDEHYDES AND KETONES (*Continued*)

Reactant	Conditions			Product(s) and Yield(s)		Refs.
	R^1	R^2				
	CH_3	CH_3			(72)	
	”	C_2H_5			(88)	
	”	$i\text{-}C_3H_7$			(24)	
	—$(CH_2)_4$—				(41)	
	$n\text{-}C_3H_7$	$n\text{-}C_3H_7$			(68)	
	CH_3	C_6H_5			(72)	
	$n\text{-}C_4H_9$	$Cl(CH_2)_4$			(73)	
	C_6H_5	C_6H_5			(0)	
	$n\text{-}C_4H_9$	$(R)\text{-}CH_3CH(O_2CCH_3)$			(46)	
$C_6H_5(CH_3)_2SiCH_2C(CH_3)=CF_2$	C_6H_5CHO, TASF, DMPU, 12 h			$CH_2=C(CH_3)CF_2CHOHC_6H_5$ (100)[b]		306
	”, $KOC_4H_9\text{-}t$, DMPU, 12 h			” (98)[b]		306
$(CH_3)_3SiCH_2C(CF_3)=CH_2$	RCHO, CsF, DMF (Method **A**) RCHO, TBAF, THF (Method **B**)			$CH_2=C(CF_3)CH_2CHOHR$		
	R	Method				
	2-furyl	**A**	2 h		(85)	305
	”	**B**	”		(89)	305, 640
	$i\text{-}C_3H_7(CH_2)_2$	”	”		(46)	640
	$n\text{-}C_6H_{13}$	**A**	”		(61)	305
	”	**B**	”		(82)	305, 640
	C_6H_5	**A**	”		(99)	305
	”	**B**	”		(85)	305, 640
	$C_6H_5CH(CH_3)$	**A**	5 h		(77)	305
	”	**B**	”		(59)	305, 640
	$C_6H_5CH=CH$	**A**	1 h		(78)	305
	”	**B**	”		(48)	305
	$(CH_3)_2C=CH(CH_2)_2\text{-}CH(CH_3)CH_2$	**A**	”		(83)	305
	” ”	**B**	”		(80)	305, 640
	R^1R^2CO, CsF, DMF (Method **A**)			$CH_2=C(CF_3)CH_2C[OSi(CH_3)_3]R^1R^2$		

Reactant	Conditions	Product(s) and Yield(s) (%)	Refs.
	R^1R^2CO, TBAF, THF (Method **B**)		
	R^1 / R^2 / Method		
	—$(CH_2)_5$— / / **A** / 5 h	(41)	305
	$C_2H_5CH(CH_3)CH_2$ / C_2H_5 / " / "	(31)[h]	305
	C_6H_5 / CH_3 / " / "	(81)[h]	305
	" / " / **B** / "	(56)[h]	305, 640
	C_6H_5 / C_6H_5 / **A** / "	(94)[h]	305
$C_6H_5(CH_3)_2SiCH(CO_2CH_3)CH{=}CH_2$	R^1R^2CO	$CH_3O_2CCH{=}CHCH_2CXR^1R^2$ I	411
	R^1 / R^2	X = OH / OCH_3 / Cl	
	n-C_4H_9 / H / $TiCl_4$, 12 h	(21) / (22) / (0)	
	" / " / $TiCl_4$,[h] 1.5 h	(60) / (trace) / (0)	
	C_6H_5 / " / $TiCl_4$, 12 h	(0) / (0) / (80)	
	" / " / $AlCl_3$, "	(0) / (24) / (0)	
	CH_3 / CO_2CH_3 / $TiCl_4$, 12 h	(89) / (0) / (0)	
	—$(CH_2)_5$— / ", "	(0) / (0) / (22)	
p-$CH_3C_6H_4(CH_3)_2SiCH(CO_2CH_3)CH{=}CH_2$	R^1R^2CO, $TiCl_4$, 12 h	I	411
	R^1 / R^2	X = OH / OCH_3 / Cl	
	n-C_4H_9 / H	(59) / (trace) / (0)	
	CH_3 / CO_2CH_3	(90) / (0) / (0)	
	—$(CH_2)_5$—	(0) / (0) / (17)	
$(CH_3)_3SiCH_2C(CO_2C_2H_5){=}CH_2$	n-C_3H_7CHO, $TiCl_4$, CH_2Cl_2, 0–25°, 3 h	n-C_3H_7 (O, O) (25)	284
$(CH_3)_3SiCH_2C(CO_2C_4H_9$-$t){=}CH_2$	", ", ", ", 4.5 h	" (23)	284
$(CH_3)_3SiCH_2C(CONHCH(CH_3)C_6H_5){=}CH_2$	n-$C_6H_{13}CHO$, TBAF, THF, 55°, 2 h	$CH_2{=}C(CONHCH(CH_3)C_6H_5)CH_2CHOHC_6H_{13}$-$n$ (58)	641

TABLE XI. ALLYLSILANES WITH ALDEHYDES AND KETONES (*Continued*)

Reactant	Conditions	Product(s) and Yield(s)	Refs.
C_5			
$(CH_3)_3SiCH(CH_3)CH{=}CHCH_3$	t-C_4H_9CHO, $TiCl_4$, CH_2Cl_2, 0°, 3 h	(*E*)-$CH_3CH{=}CH$–CH(CH_3)–CH(OH)–C_4H_9-t I (52)	638
	OHC–CH(CH_3)–CH_2–CH(CH_3)–CO_2CH_3, $TiCl_4$, −60°	(*E*)-$CH_3CH{=}CH$–CH(CH_3)–CH(OH)–CH(CH_3)–CH_2–CH(CH_3)–CO_2CH_3 + (*E*)-$CH_3CH{=}CH$–CH(CH_3)–CH(OH)–CH(CH_3)–CH_2–CH(CH_3)–CO_2CH_3 >90:10 (80)	312
$(C_2H_5)_3SiCH(CH_3)CH{=}CHCH_3$	t-C_4H_9CHO, $TiCl_4$, CH_2Cl_2, 0°, 3 h	I (20)	638
$(CH_3)_3SiCH_2CH{=}C(CH_3)_2$	$(C_6H_5)_2CO$, TBAF, THF, reflux, 48 h	$(CH_3)_2C{=}CHCH_2C(C_6H_5)_2OH$ (87)	13
$(CH_3)_3SiC(CH_3)_2CH{=}CH_2$	RCHO, $TiCl_4$	2,2-dimethyl-5-R-tetrahydrofuran	83
	R		
	s-C_4H_9 5 min	(79)	
	i-C_4H_9 "	(78)	
	$C_6H_5CH_2$ 6 min	(81)	
	" 4 min	$(CH_3)_2CCl(CH_2)_2CHOH(CH_2)_2C_6H_5$ (76) + $(CH_3)_2C{=}CHCH_2CHOH(CH_2)_2C_6H_5$ (13)	83

	$CH_3COCO_2CH_3$, $TiCl_4$, CH_2Cl_2	$(CH_3)_2C{=}CHCH_2COH(CH_3)CO_2CH_3$ + (I) + 2,2,5-trimethyl-5-(methoxycarbonyl)tetrahydrofuran (II)	83
	−78°, 1.5 h; room temp, 15 min	I (—) + II (61)	
	room temp, 4 min	I (61) + II (16)	
	″ ″, 20 min	I (36) + II (30)	
$(CH_3)_3SiCH_2C(CH_2OSi(CH_3)_2C_4H_9\text{-}t){=}CHCH_3$	3,4-$(CH_3O)_2C_6H_3CH_2O$-substituted 1,3-dioxane aldehyde (OHC) bearing 4-glutarimidyl group, $BF_3{\cdot}O(C_2H_5)_2$, CH_2Cl_2, −90°	$CH_2{=}CCH(CH_3)$–CH(OH)– adduct ($CH_2OSi(CH_3)_2C_4H_9\text{-}t$; ArO; 1,3-dioxane; glutarimide NH) (50)[i]	311
(*E*)-$(CH_3)_3SiCH_2CH{=}CHCH{=}CH_2$	RCHO, $TiCl_4$, CH_2Cl_2, −40° to 0°, 15 min	(*E*)-$CH_2{=}CHCH{=}CHCH_2CHOHR$ (I)	282
	R		
	C_2H_5	(79)	
	n-C_3H_7	(70)	
	i-C_3H_7	(80)	
	n-C_4H_9	(80)	
	t-C_4H_9	(60)	
	n-C_5H_{11}	(56)	
	C_6H_5	(51)	
	C_6H_{11}	(58)	
	C_6H_5CHO, $BF_3{\cdot}O(C_2H_5)_2$, −40°, 1 h	I R = C_6H_5 (16)	410
	″, ″, −78°, 2 h	″ ″ (37)	410
	R^1R^2CO, $TiCl_4$, CH_2Cl_2, −40° to 0°, 15 min	(*E*)-$CH_2{=}CHCH{=}CHCOHR^1R^2$	282
	R^1 / R^2		
	CH_3 / CH_3	(48)	

TABLE XI. ALLYLSILANES WITH ALDEHYDES AND KETONES (*Continued*)

Reactant	Conditions	Product(s) and Yield(s)	Refs.
	C_2H_5 C_2H_5	(80)	
	CH_3 i-C_3H_7	(60)	
	" i-C_4H_9	(64)	
	—$(CH_2)_5$—	(64)	
	n-C_3H_7 n-C_3H_7	(51)	
$(CH_3)_3SiCH_2C(=CH_2)CH{=}CH_2$	RCHO, $AlCl_3$, CH_2Cl_2, −78° (Method **A**) ", $TiCl_4$, ", " (Method **B**) ", TBAF, THF (Method **C**)	$CH_2{=}C(CH{=}CH_2)CH_2CHOHR$	
	R — Method		
	n-C_4H_9 — **A** 2 min	(25)	186
	i-C_4H_9 — " "	(30)	186
	" — **B** 5 s	(22)	186, 642
	" — **C** room temp, 30 min	(74)[b]	186, 309
	s-C_4H_9 — **A** 2 min	(44)	186
	" — **B** 1 s	(15)	186
	" — **C** 45°, 1.5 h	(75)[b]	186, 309
	n-C_5H_{11} — " ", 4 h	(38)[b]	186, 309
	n-$C_3H_7CH(CH_3)$ — " ", 35 min	(81)[b]	186, 309
	$(CH_3)_2C{=}CH$ — " 35°, 2 h	(70)[b]	186, 309
	C_6H_5 — " 40°, 1 h	(90)[b]	186, 309
	$C_6H_5(CH_2)_2$ — **A** 2 min	(43)	308
	" — **B** 1 s	(37)	308
	1. R^1R^2CO, TBAF, THF 2. HCl, CH_3OH	$CH_2{=}C(CH{=}CH_2)CH_2COHR^1R^2$	186, 309
	R^1 R^2		
	C_2H_5 C_2H_5 room temp, 4 h	(61)	

Reactant	Conditions	Product(s) and Yield(s) (%)	Refs.
	CH_3 n-C_5H_{11} 45°, 3.5 h	(33)	
	" C_6H_5 room temp, 2.5 h	(87)	
	C_2H_5 C_6H_5 ", 3 h	(89)	
	C_6H_5 " 50°, 2.5 h	(100)	
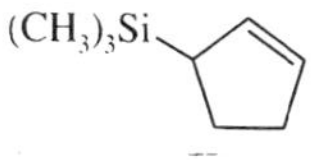	R^1R^2CO, $TiCl_4$, CH_2Cl_2	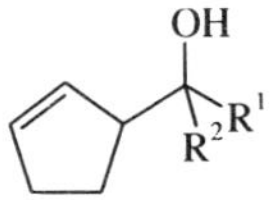 I	
	R^1 R^2		
	n-C_3H_7 H −15°, 1 h	(78)	643
	t-C_4H_9 " −10° → 20°, 1 h	*erythro*:*threo* 83:17 (52)[j]	24
	CH_3 CO_2CH_3 −15°, 2 h	(86)	643
	" n-C_6H_{13} ", 1 h	(40)	643
	C_6H_5 $CO_2C_2H_5$ ", 2 h	(90)	643
	$CH_3COCH(OCH_3)_2$, $AlCl_3$, CH_2Cl_2 −15° → 10°, 3 h	I $R^1 = CH_3$, $R^2 = CH(OCH_3)_2$ (80)	360
	", $TiCl_4$, CH_2Cl_2 −18°, 4 h	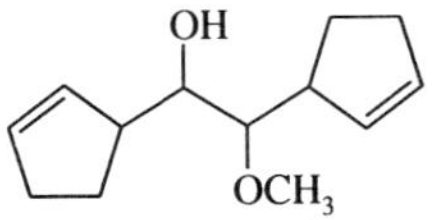 (60)	360
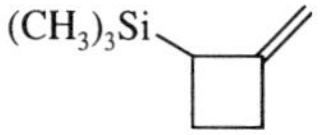	R^1R^2CO, $TiCl_4$, CH_2Cl_2, 0°, 5 min	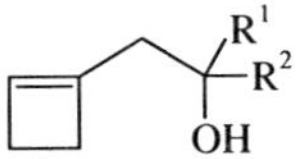	174
	R^1 R^2		
	i-C_4H_9 H	(86)	
	n-C_6H_{13} "	(82)	
	—$(CH_2)_5$—	(86)[k](55)[l]	
$C_6H_5(CH_3)_2SiCH(CO_2CH_3)CH{=}CHCH_3$	RCHO, $TiCl_4$, CH_2Cl_2	(E)-$CH_3O_2CCH{=}CHCH(CH_3)CHOHR$	612
	R = i-C_3H_7 2 h	(48)	
	R = $C_6H_4NO_2$-p, 1 h	(69)[m]	

TABLE XI. ALLYLSILANES WITH ALDEHYDES AND KETONES (*Continued*)

Reactant	Conditions			Product(s) and Yield(s)	Refs.
	1. R^1R^2CO, $TiCl_4$, $-78° \rightarrow$ room temp 2. HCl, H_2O, 90°, 1 h			I (6-R^1-6-R^2-4-methyl-5,6-dihydro-2H-pyran-2-one)	
	R^1	R^2			
	n-C_3H_7	H		(81)	283
	i-C_3H_7	"		(79)	283
	i-C_4H_9	"		(68)	644
	t-C_4H_9	"		(78)	283
	CH_3	CH_3		(80)	283
	—$(CH_2)_5$—			(78)	283
$(CH_3)_3SiCH(CO_2C_2H_5)C(CH_3)=CH_2$	R^1R^2CO, CH_2Cl_2			I + $R^1R^2COHCH_2C(CH_3)=CHCO_2C_2H_5$ (II)	252
	R^1	R^2			
	n-C_5H_{11}	H	TiF_4, $-78°$, 5 h; room temp, 22 h	I (5) + II (26)	
	"	"	$TiCl_4$, $-78°$, 12 h; $-30°$, 3 h	I (2) + II (20)	
	C_6H_5	"	TiF_4, $-78°$, 1.5 h; $-32°$, 1.5 h; room temp, 7 h	II (16) + $C_6H_5CH=CHC(CH_3)=CHCO_2C_2H_5$ (3) + $[C_2H_5O_2CCH=C(CH_3)CH_2CHC_6H_5]_2O$ (64)	
	p-$CH_3OC_6H_4$	"	$TiCl_4$, $-78°$, 16 h	p-$CH_3OC_6H_4CH[CH_2C(CH_3)=CHCO_2C_2H_5]_2$ $E,E:E,Z$ 82:18 (72)	
	p-$O_2NC_6H_4$	"	$TiCl_4$, $-78°$, 16 h	(E)-II (65)	

Reactant	Conditions	Product(s) and Yield(s) (%)	Refs.
	” ” $FeCl_3$, −78°, 4 h; room temp 13.5 h	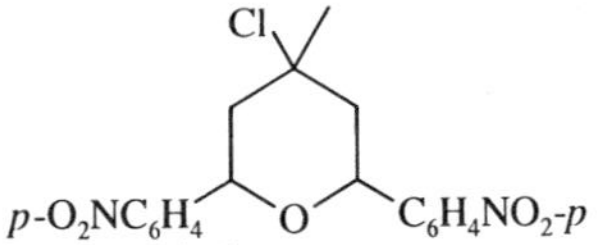 Cl; p-$O_2NC_6H_4$, O, $C_6H_4NO_2$-p (86)	
	CH_3 CH_3 TiF_4	I + II (16)	
	” ” $TiCl_4$, 24 h	I (7) + II (54) + $(CH_3)_2CClCH_2C(CH_3){=}CHCO_2C_2H_5$ (10)	
$C_6H_5(CH_3)_2SiC(CH_3)(CO_2CH_3)CH{=}CH_2$	R^1R^2CO, $TiCl_4$, CH_2Cl_2	(E)-$CH_3O_2CC(CH_3){=}CHCH_2COHR^1R^2$	612
	R^1 R^2		
	n-C_3H_7 H 2 h	(48)	
	p-$O_2NC_6H_4$ ” 1 h	(53)	
	CH_3 CH_3 3 h	(96)	
	RCHO, $TiCl_4$, CH_2Cl_2, −78°, 2 h; room temp, 8–12 h	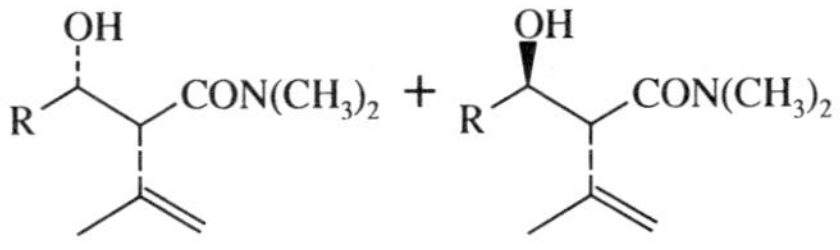 OH, R, $CON(CH_3)_2$ (I) + OH, R, $CON(CH_3)_2$ (II)	287
	R	I:II	
	C_2H_5	97:3 (25)	
	n-C_3H_7	96:4 (65)	
	i-C_3H_7	59:41 (68)	
	C_6H_5	97:3 (62)	
	p-$O_2NC_6H_4$	99:1 (74)	
	3,4-$(CH_3O)_2C_6H_3$	70:30 (47)	
	RCHO, TBAF, THF, −45°, 2 min	I + II + $RCHOHCH_2C(CH_3){=}CHCON(CH_3)_2$ III	287
	R	I:II:(Z)-III	
	4-C_5H_4N	37:46:17 (74)	
	C_6H_5	44:48:8 (69)	
	p-$O_2NC_6H_4$	29:36:35 (85)	

TABLE XI. ALLYLSILANES WITH ALDEHYDES AND KETONES (*Continued*)

Reactant	Conditions	Product(s) and Yield(s)	Refs.
	3,4-$(CH_3O)_2C_6H_3$	43:46:11 (58)	
	RCHO, TBAF (5 mol %), THF, room temp, 15–20 h	I + II + III	287
	R	I:II:(*E*)-III:(*Z*)-III	
	3-C_5H_4N	0:0:19:81 (70)	
	C_6H_5	0:0:20:80 (72)	
	p-$O_2NC_6H_4$	27:33:0:40 (80)	
	3,4-$(CH_3O)_2C_6H_3$	0:0:35:65 (75)	
C_6			
(*Z*,*E*)-$(CH_3)_3SiCH_2(CH{=}CH)_2CH_3$	RCHO, $TiCl_4$, CH_2Cl_2, −78°, 1 h	$CH_2{=}CHCH{=}CH$–CH(CH_3)–CH(OH)R (wedge OH) + $CH_2{=}CHCH{=}CH$–CH(CH_3)–CH(OH)R (hashed OH)	282
	R		
	CH_3	2:1 (61)	
	C_2H_5	74:26 (51)	
	i-C_3H_7	84:16 (37)	
$(CH_3)_3SiCH(CO_2C_2H_5)C(CH_3){=}CHCH_3$	C_6H_5CHO, $TiCl_4$, CH_2Cl_2, −78°, 8 h; room temp, 21 h	$C_6H_5CHClCH(CH_3)C(CH_3){=}CHCO_2C_2H_5$ (44) + $C_6H_5CH{=}C(COCH_3)CO_2C_2H_5$ (21)	252
$(CH_3)_3Si$–CH(CH_2CHO)–CH=CHCH₃ (Z) + $(CH_3)_3Si$–CH(CH_2CHO)–CH=CHCH₃ (E)	$TiCl_4$, CH_2Cl_2, −78°	2-methylcyclopentanol (structure) (100)	296

Reactant	Conditions	Product(s) and Yield(s) (%)	Refs.
C_7			
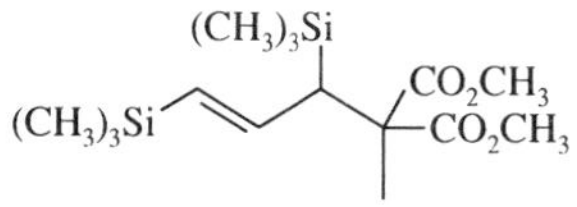 $(CH_3)_3Si$–CH=CH–CH($Si(CH_3)_3$)–C(CH_3)(CO_2CH_3)$_2$	RCHO, TBAF, 1,3-dimethyl-2-imidazolidinone 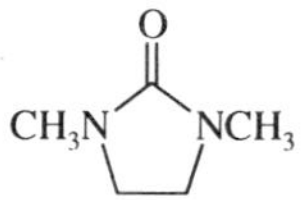	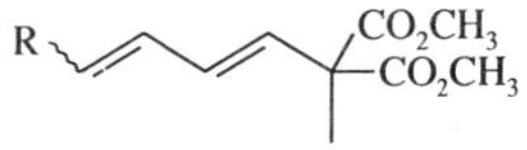 R–CH=CH–CH=CH–C(CH_3)(CO_2CH_3)$_2$	227a
	R	$E,E:Z,E$	
	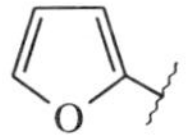 2-furyl	50:50 (86)	
	n-C_5H_{11}	45:55 (29)	
	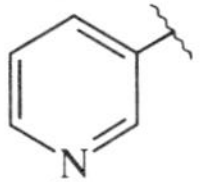 3-pyridyl	47:53 (67)	
	C_6H_{11}	47:53 (39)	
	C_6H_5	48:52 (82)	
	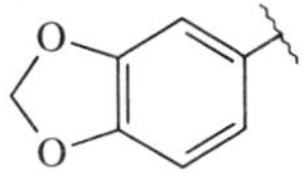 3,4-methylenedioxyphenyl	45:55 (64)	
	1-$C_{10}H_7$	60:40 (78)	
C_8			
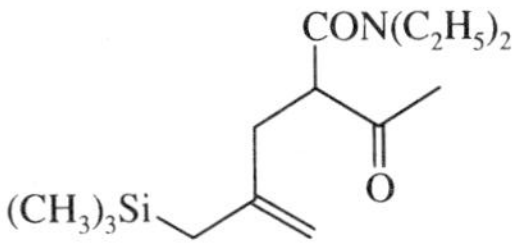 $(CH_3)_3SiCH_2C(=CH_2)CH_2CH(CON(C_2H_5)_2)COCH_3$	$TiCl_4$, CH_2Cl_2, −55°, 6 h	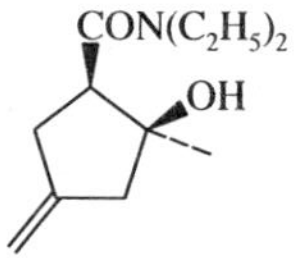 4-methylene-2-$CON(C_2H_5)_2$-1-methylcyclopentan-1-ol (65)	645
$(CH_3)_3SiCH_2C(CO_2C_2H_5)=CH(CH_2)_3CHO$	$TiCl_4$, CH_2Cl_2, −20°	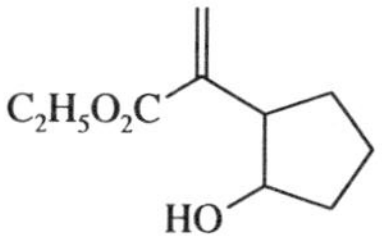 2-[1-($CO_2C_2H_5$)vinyl]cyclopentanol (100)	646

TABLE XI. ALLYLSILANES WITH ALDEHYDES AND KETONES (*Continued*)

Reactant	Conditions	Product(s) and Yield(s)	Refs.
(*E*,*E*)-$(CH_3)_3SiCH_2CH{=}CHCH_2$–$CH_2CH{=}CHCH_2Si(CH_3)_3$ [two units joined at CH_2]	CH_3CHO, $TiCl_4$, CH_2Cl_2, $-78°$, 6 min	$CH_2{=}CHCH(CH_3CHOH)(CH_2)_2CH(HOCHCH_3)CH{=}CH_2$ (65)	310
C_9			
$(CH_3)_3SiCH_2C(CO_2C_2H_5){=}CH(CH_2)_4CHO$	$TiCl_4$, CH_2Cl_2, $-5° \rightarrow 0°$	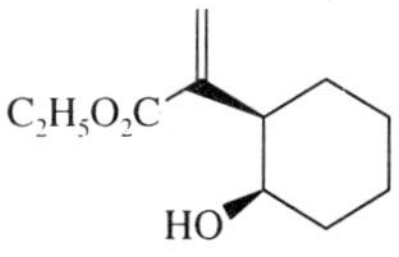 $C_2H_5O_2C$, HO [cyclohexane, cis] (57) +	646
		$C_2H_5O_2C$, HO [cyclohexane, trans] (39)	
(*E*)-$(CH_3)_3SiCH_2CH{=}CHC_6H_5$	RCHO, $TiCl_4$, CH_2Cl_2	$CH_2{=}CH$–$CH(C_6H_5)$–$CH(OH)R$ I + $CH_2{=}CH$–$CH(C_6H_5)$–$CH(OH)R$ II	
	R	I:II	
	CH_3 0°, 1 h	93:7 (76)	293
	C_2H_5 ", 2 h	94:6 (76)	293
	n-C_3H_7 room temp, 0.5 min	— (87)	64
	t-C_4H_9 0°, 3 h	>99:1 (78)	293
	C_6H_5CHO, KF, 18-crown-6, THF, 20°, 12 h	$C_6H_5CH{=}CHCH_2CHOHC_6H_5$ (50)	259

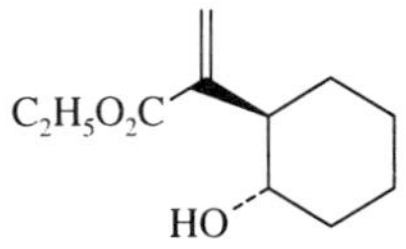

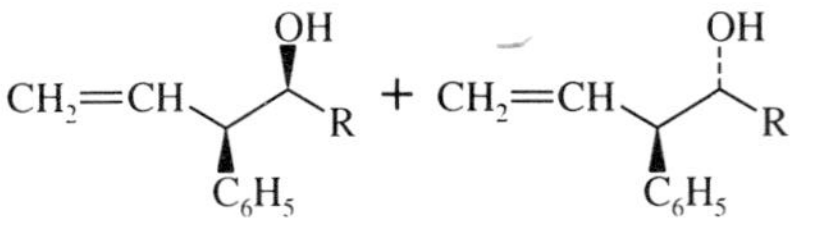

Reactant	Conditions	Product(s) and Yield(s) (%)	Refs.
(E) $(CH_3)_3SiCH_2CH{=}CHC_6H_5$	$RCHO$, $TiCl_4$, CH_2Cl_2, 0°	I + II	293
	R	I:II	
	CH_3 1 h	72:28 (50)	
	C_2H_5 2 h	71:29 (68)	
	t-C_4H_9 3 h	75:25 (74)	
$(CH_3)_3SiCH(C_6H_5)CH{=}CH_2$	n-C_3H_7CHO, $TiCl_4$, CH_2Cl_2, 0.5 min	$C_6H_5CH{=}CHCH_2CHOHC_3H_7$-$n$ (54)	64
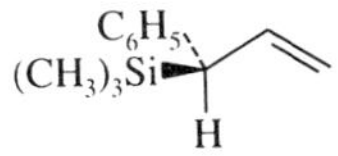 C_6H_5, $(CH_3)_3Si$, H	$RCHO$, $TiCl_4$, CH_2Cl_2, −78°, 2 min	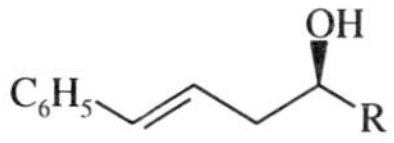 C_6H_5, OH, R	647
% ee	R	% ee	
91	CH_3	64 (83)	
95	i-C_3H_7	91 (66)	
95	t-C_4H_9	91 (71)	
$(CH_3)_3SiCH_2C(C_5H_{11}$-$n){=}C{=}CH_2$	n-$C_8H_{17}CHO$, $TiCl_4$, CH_2Cl_2, −78°	$CH_2{=}CC_5H_{11}$-n / $CH_2{=}CCHOHC_8H_{17}$-n (56)	648
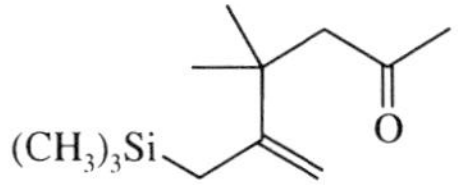 $(CH_3)_3Si$, O	$C_2H_5AlCl_2$, toluene, 0°	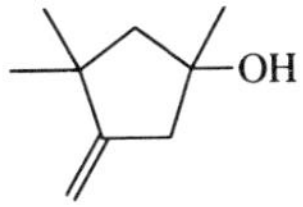 OH (99)	290
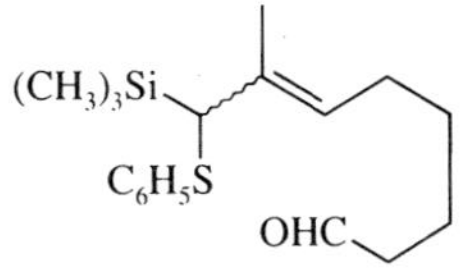 $(CH_3)_3Si$, C_6H_5S, OHC		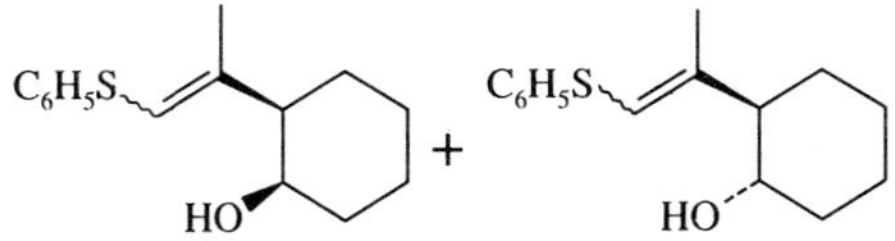 C_6H_5S, HO + C_6H_5S, HO	295
E	$SnCl_4$, C_6H_{14}, −78°, 10 min	65:35 (58)	
”	$TiCl_4$, CH_2Cl_2, ”, ”	70:30 (60)	
”	$BF_3{\cdot}O(C_2H_5)_2$, C_6H_{14}, −78°, 15 min	85:15 (50)	
”	TFA, CH_2Cl_2, 0°, 5 min	>98:2 (54)	
Z	”, CF_3CH_2OH, 0°, 0.5 h	<7:93 (52)	

TABLE XI. ALLYLSILANES WITH ALDEHYDES AND KETONES (*Continued*)

Reactant	Conditions	Product(s) and Yield(s)	Refs.
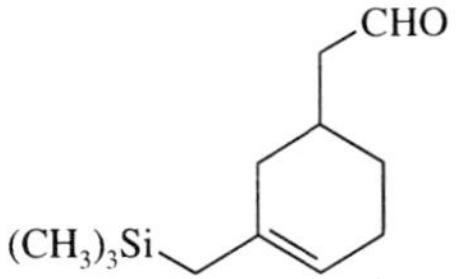 $(CH_3)_3Si$ … CHO		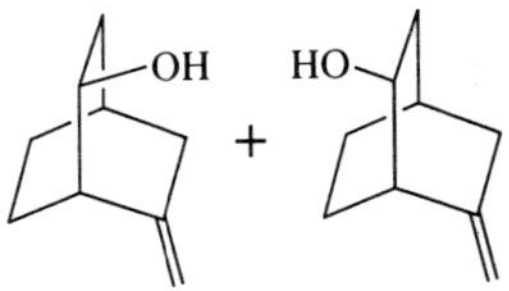 OH + HO	294
	$BF_3{\cdot}O(C_2H_5)_2$, CH_2Cl_2, $-70°$	80:20 (—)	
	$AlCl_3$, ", "	79:21 (—)	
	$FeCl_3$, ", "	70:30 (—)	
	$(C_2H_5)_2AlCl$, ", "	66:34 (—)	
	$SnCl_4$, ", "	49:51 (—)	
	TBAF, THF, reflux	30:70 (—)	
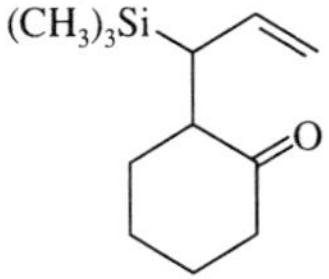 $(CH_3)_3Si$ … O	$TiCl_4$, CH_2Cl_2, $-78°$, a few min	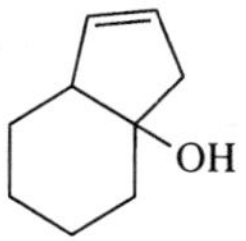 OH (68)	291
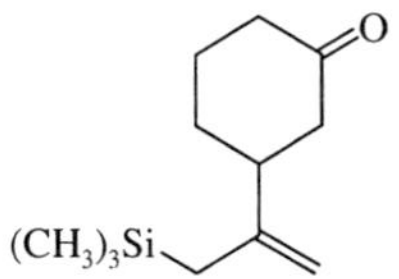 $(CH_3)_3Si$ … O	$C_2H_5AlCl_2$, toluene, 0°	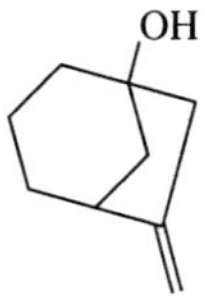 OH (91)	290
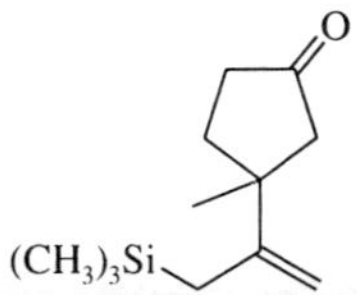 $(CH_3)_3Si$ … O	$C_2H_5AlCl_2$, toluene, 0°	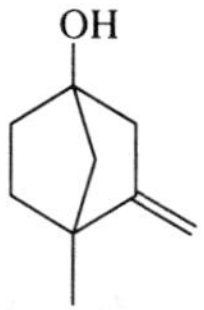 OH (84)	290

Reactant	Conditions	Product(s) and Yield(s) (%)	Refs.
$(CH_3)_3Si$-allyl(=CHOCH_3) cycloalkanone $(CH_2)_n$ + $(CH_3)_3Si$-allyl-CH(OCH_3) cycloalkanone $(CH_2)_n$	$TiCl_4$, $-78°$	CH_3O, OH bicyclic $(CH_2)_n$ + OH, CH_3O bicyclic $(CH_2)_n$	649
$n = 1$ 9:1		(—)"	
$n = 2$ 3.2:1		(—)"	
$n = 3$ 1.27:1		(—)"	
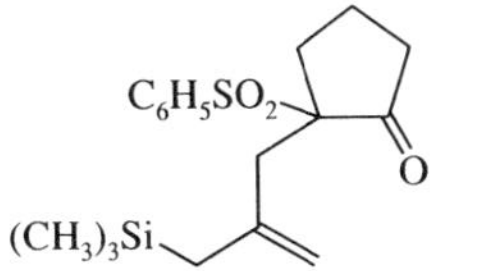 2-($C_6H_5SO_2$)-2-[2-(($CH_3)_3Si$CH_2$)allyl]cyclopentanone	TBAF, THF, 55°, 1 h	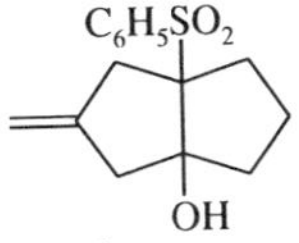 $C_6H_5SO_2$, OH bicyclooctane (94)	650
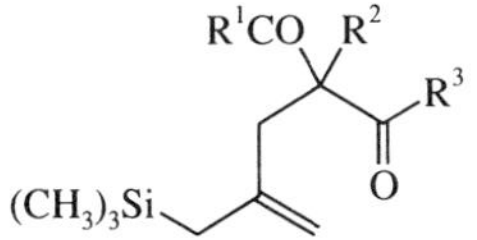 $(CH_3)_3Si$-allyl-C(R^2)(COR^1)-COR^3		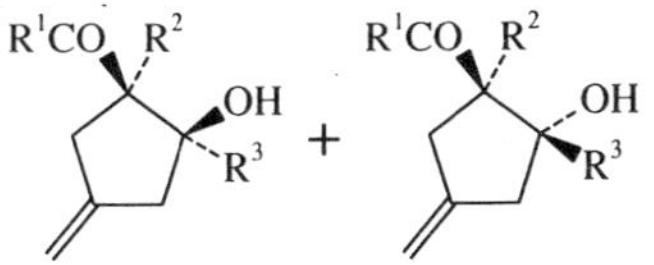 R^1CO, R^2, OH, R^3 cyclopentanol + R^1CO, R^2, OH, R^3 cyclopentanol (diastereomer)	645

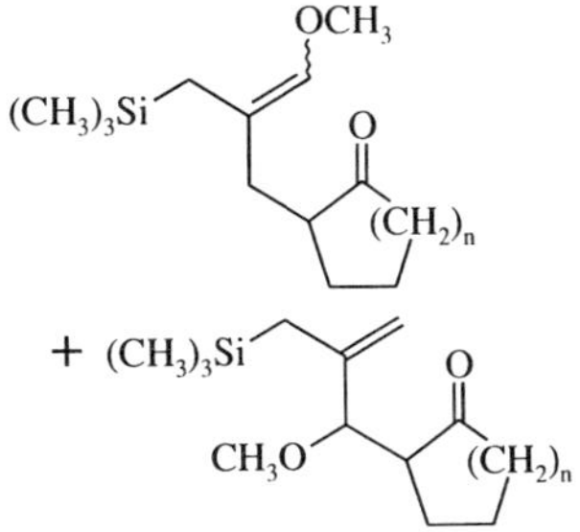

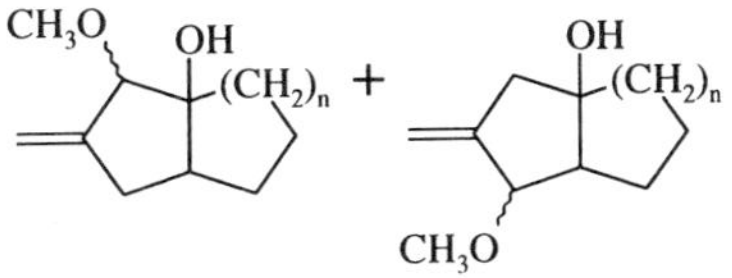

R^1	R^2	R^3	Conditions	Ratio (Yield)
C_2H_5O	CH_3	CH_3	$TiCl_4$, CH_2Cl_2, $-78°$, 2 h	100:0 (88)
"	"	"	$BF_3 \cdot O(C_2H_5)_2$, CH_2Cl_2	1:4.8 (—)
"	"	"	TBAF, THF	2:1 (—)
$(C_2H_5)_2N$	"	"	$TiCl_4$, CH_2Cl_2, $-55°$, 6 h	100:0 (100)
$(CH_3)_2N$	H	C_2H_5	", ", ", "	100:0 (57)

Reactant	Conditions	Product(s) and Yield(s) (%)	Refs.
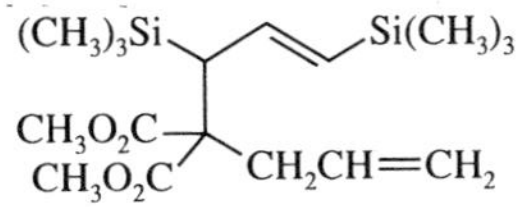 $(CH_3)_3Si$-CH(CH=CH-$Si(CH_3)_3$)-C(CO_2CH_3)$_2$-$CH_2CH{=}CH_2$	C_6H_5CHO, TBAF, 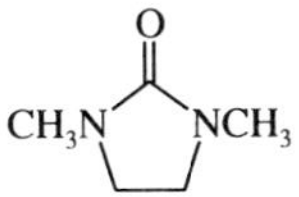 1,3-dimethyl-2-imidazolidinone (CH_3N, NCH_3)	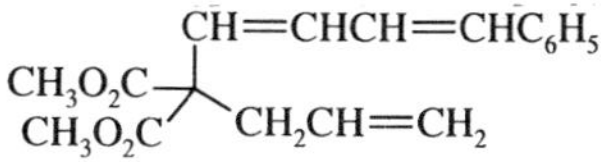 $(CH_3O_2C)_2C(CH{=}CHCH{=}CHC_6H_5)(CH_2CH{=}CH_2)$ (76) E,E:E,Z 48:52	227a

TABLE XI. ALLYLSILANES WITH ALDEHYDES AND KETONES (*Continued*)

Reactant	Conditions	Product(s) and Yield(s)	Refs.
C_{10}			
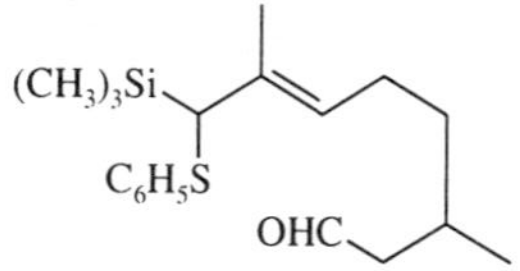 $(CH_3)_3Si$, C_6H_5S, OHC	1. TFA, CH_2Cl_2, 0° 2. Raney Ni	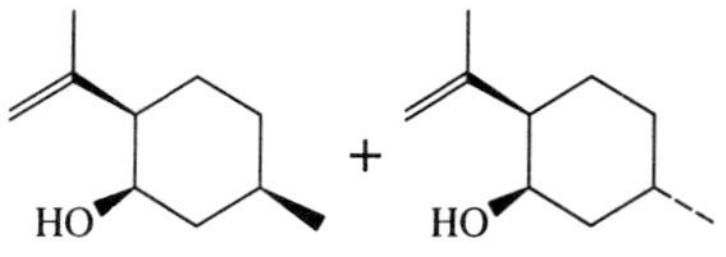 HO + HO (50)	295
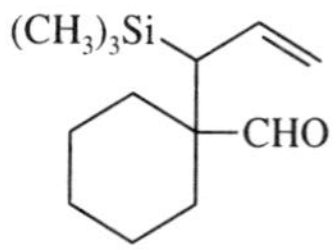 $(CH_3)_3Si$, CHO	$TiCl_4$, CH_2Cl_2, −78°, a few min	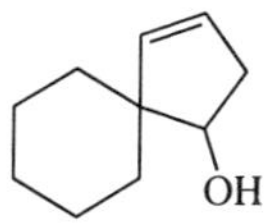 OH (67)	291
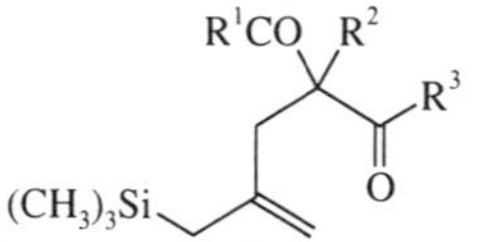 R^1CO R^2, R^3, O, $(CH_3)_3Si$	$TiCl_4$, CH_2Cl_2	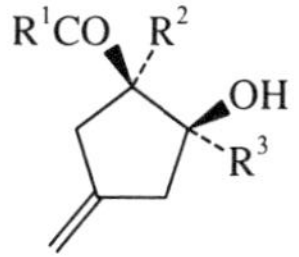 R^1CO R^2, OH, R^3	645
R^1 / R^2 / R^3			
C_2H_5O / CH_3 / C_2H_5	−78°, 2 h	(78)	
" / C_2H_5 / CH_3	", "	(78)	
$(C_2H_5)_2N$ / H / i-C_3H_7	−55°, 6 h	(56)	
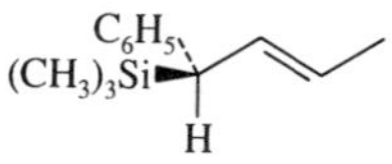 C_6H_5, $(CH_3)_3Si$, H	RCHO, $TiCl_4$, CH_2Cl_2, −78°, 4–5 min	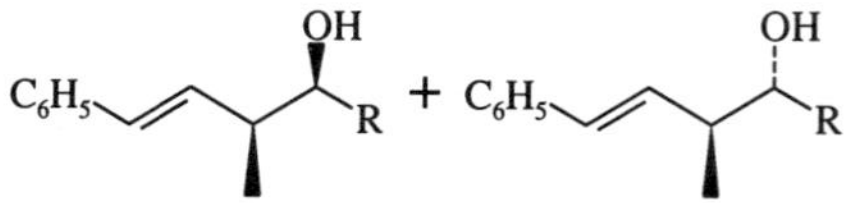 C_6H_5 … OH, R + C_6H_5 … OH, R	
	R		
	H	(—)	21
	CH_3	92:8 (76)	22
	i-C_3H_7	95:5 (67)	22
	t-C_4H_9	>99:1 (47)	22

Reactant	Conditions	Product(s) and Yield(s) (%)	Refs.
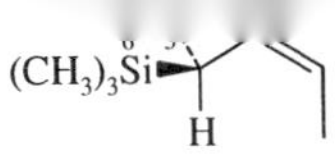 $(CH_3)_3Si$ (structure)	RCHO, $TiCl_4$, CH_2Cl_2, −78°	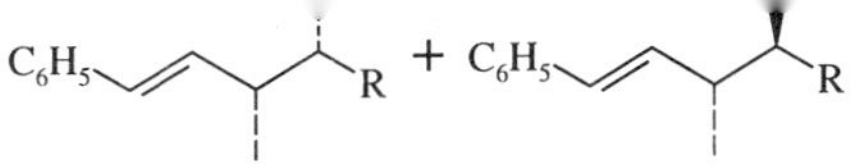 C_6H_5 (structure) R + C_6H_5 (structure) R	22
	R: CH_3	50:50 (82)	
	i-C_3H_7	65:35 (61)	
	t-C_4H_9	>99:1 (27)	
$(CH_3)_3SiCH_2CH{=}C{=}CHC_6H_5$	2RCHO, $TiCl_4$	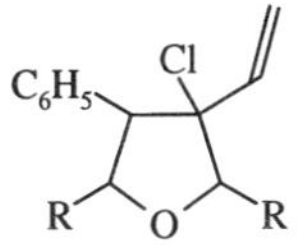 (structure: C_6H_5, Cl, R, O, R) (19)[o]	281
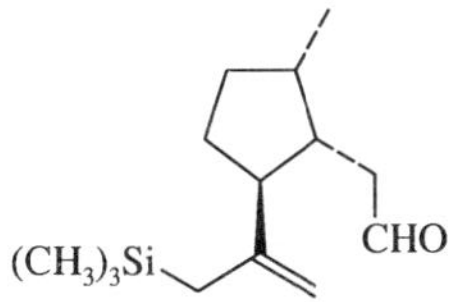 (structure: $(CH_3)_3Si$, CHO)		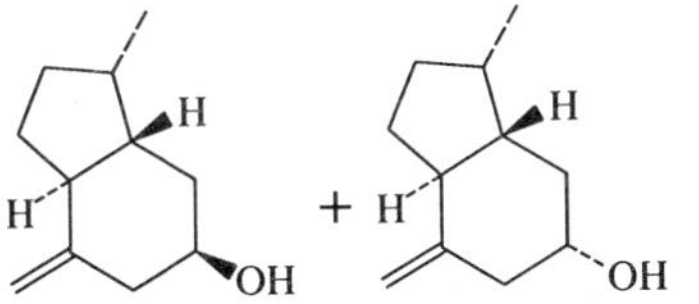 (structure: H, H, OH) + (structure: H, H, OH)	258
	$SnCl_4$, CH_2Cl_2, 0°, 10 min	53:47 (80)[b]	
	$BF_3{\cdot}O(C_2H_5)_2$, CH_2Cl_2, −78°, 15 min	85:15 (—)	
	TBAF, THF, 50°, 1.5 h	18:82 (—)	
C_{11}			
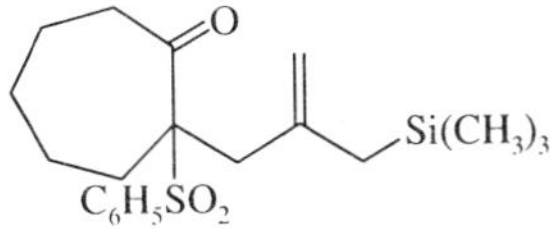 (structure: O, $C_6H_5SO_2$, $Si(CH_3)_3$)	$C_2H_5AlCl_2$, CH_2Cl_2, reflux, 3 h	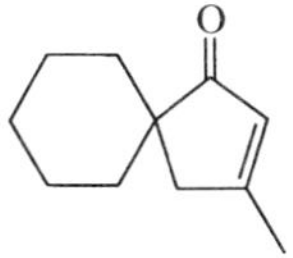 (structure: O) (97)	651
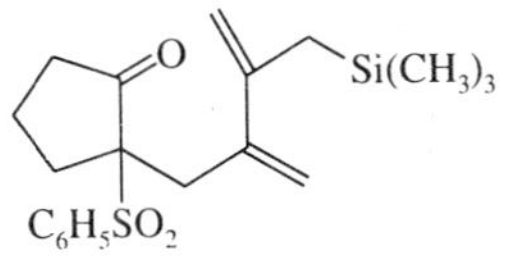 (structure: O, $C_6H_5SO_2$, $Si(CH_3)_3$)	$C_2H_5AlCl_2$, CH_2Cl_2, −78° or TBAF, THF, 55°	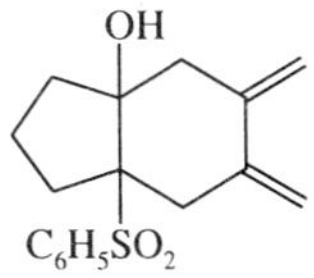 (structure: OH, $C_6H_5SO_2$) (60–78)	652

TABLE XI. ALLYLSILANES WITH ALDEHYDES AND KETONES (*Continued*)

Reactant	Conditions	Product(s) and Yield(s)	Refs.
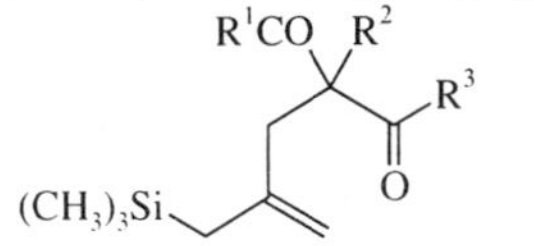 $(CH_3)_3SiCH_2C(=CH_2)CH_2C(COR^1)(R^2)COR^3$		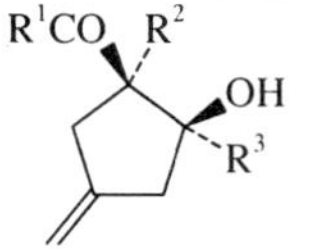 1-R^1CO-1-R^2-2-OH-2-R^3-4-methylenecyclopentane	645
R^1 / R^2 / R^3			
C_2H_5O / CH_3 / i-C_3H_7	$TiCl_4$, CH_2Cl_2, $-78°$, 2 h	(66)	
" / i-C_3H_7 / CH_3	", ", ", "	(78)	
$(CH_3)_2N$ / H / t-C_4H_9	$TiCl_4$, CH_2Cl_2, $-55°$, 6 h	(75)	
$(CH_3)_3SiCH_2C(C_6H_5)=C=CHCF_3$	RCHO, TBAF, 1.5 h	$CH_2=C(C_6H_5)C(CHOHR)=CHCF_3$ + $CH_2=C(C_6H_5)CH=CHCF_3$	653
	R		
	i-C_3H_7 THF	(20) (39)	
	2-furyl "	(51) (25)	
	" DMF	(23) (21)	
	C_6H_5 THF	(30) (16)	
	" DMF	(30) (26)	
C_{12}			
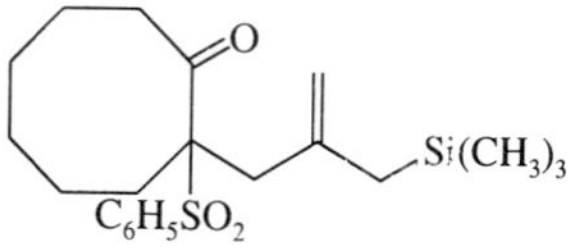 2-(Phenylsulfonyl)-2-[2-(trimethylsilylmethyl)allyl]cyclooctanone ($C_6H_5SO_2$, $Si(CH_3)_3$)	$C_2H_5AlCl_2$, CH_2Cl_2, reflux, 3 h	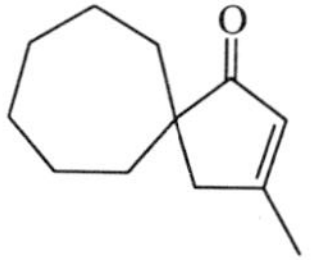 3-methylspiro[4.6]undec-2-en-1-one (95)	651

Reactant	Conditions	Product(s) and Yield(s) (%)	Refs.
$(CH_3)_3Si$ (cyclopropylidene-methyl cyclohexene)	$(CH_3)_2CO$, TBAF, THF, reflux, 10 h	HO (90)	167, 654
CO_2CH_3, $Si(CH_3)_3$	$C_2H_5AlCl_2$, CH_2Cl_2, $-78°$	OH, CO_2CH_3 (60–78)	652
$(CH_3)_3SiCH_2C{=}CHOCH_3$ / $(CH_2)_2COC_6H_5$ + $(CH_3)_3SiCH_2C{=}CH_2$ / $CH(OCH_3)CH_2COC_6H_5$	$TiCl_4$, $-78°$	OCH_3, OH, C_6H_5 + OH, C_6H_5, CH_3O (—)[n]	649
$(CH_3)_3Si$, O, X; X = $OSi(CH_3)_3$	1. $C_2H_5AlCl_2$, toluene, $-78°$, 2 h 2. H_2SO_4, H_2O, THF, 23°	OH, X; X = OH (85)	292
X = SO_2CH_3; $(CH_3)_3Si$, X, O	$C_2H_5AlCl_2$, toluene, room temp	X = SO_2CH_3 (73); HO, X	655

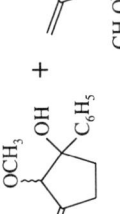

TABLE XI. ALLYLSILANES WITH ALDEHYDES AND KETONES (*Continued*)

Reactant	Conditions	Product(s) and Yield(s)	Refs.
X = $OSi(CH_3)_3$	1. $C_2H_5AlCl_2$, toluene, −78°, 0.5 h 2. H_2SO_4, H_2O, THF, 23°	X = OH (87)	292
X = SO_2CH_3	$C_2H_5AlCl_2$, toluene, room temp	X = SO_2CH_3 (72)	655
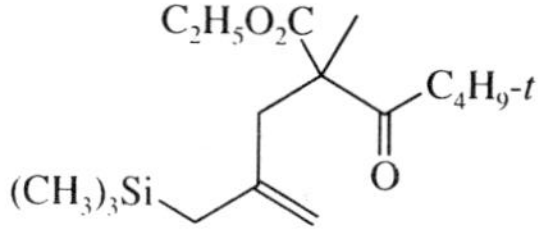 $C_2H_5O_2C$, C_4H_9-*t*, O, $(CH_3)_3Si$	$TiCl_4$, CH_2Cl_2, −78°, 2 h	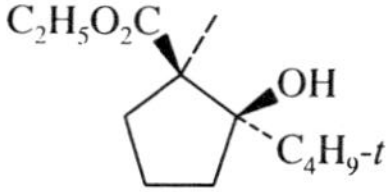 $C_2H_5O_2C$, OH, C_4H_9-*t* (74)	645
C_{13}			
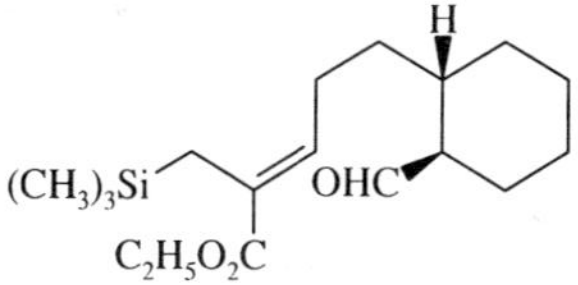 H, $(CH_3)_3Si$, OHC, $C_2H_5O_2C$	TBAF, THF, −5°	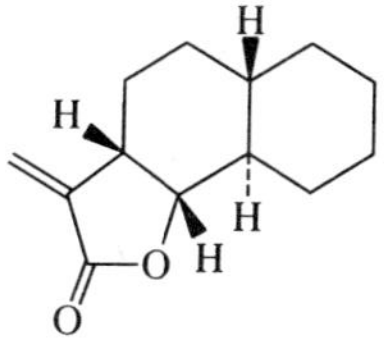 H, H, H, H, O, O (62) +	656
		H, H, $C_2H_5O_2C$, OH (5)	
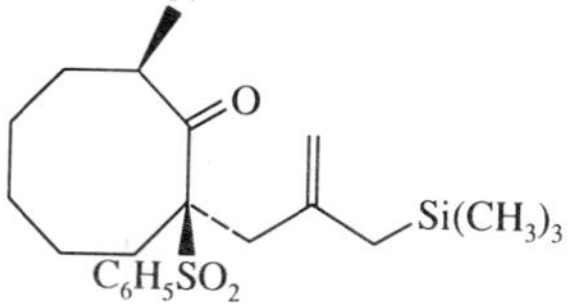 O, $Si(CH_3)_3$, $C_6H_5SO_2$	$C_2H_5AlCl_2$, CH_2Cl_2, reflux, 3 h	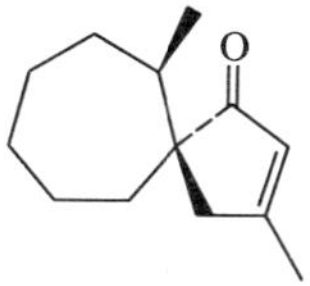 O (86)	651
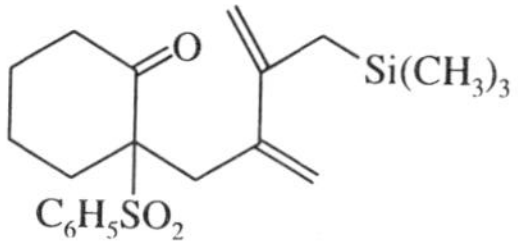 O, $Si(CH_3)_3$, $C_6H_5SO_2$	$C_2H_5AlCl_2$, CH_2Cl_2, −78° or TBAF, THF, 55°	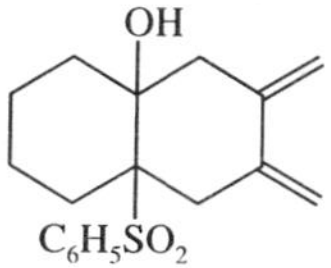 OH, $C_6H_5SO_2$ (60–78)	652

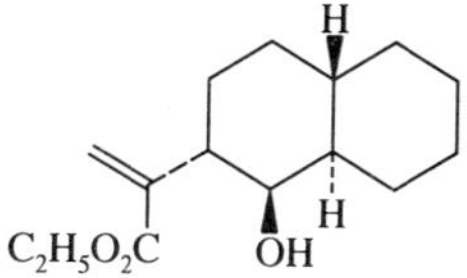

Reactant	Conditions	Product(s) and Yield(s) (%)	Refs.
C_{14}			
$(CH_3)_3SiCH_2C(=CHSi(CH_3)_3)CH_2CH_2CO(CH_2)_2C_6H_5$	TBAF, THF	R = $Si(CH_3)_3$ (54) + R = H (16)	166
$(CH_3)_3SiCH_2C(=CH_2)CH_2C(R^1)(CO_2C_2H_5)COR^2$	$TiCl_4$, CH_2Cl_2, −78°, 2 h		645
R^1 = CH_3, R^2 = C_6H_5		(73)	
R^1 = C_6H_5, R^2 = CH_3		(73)	
C_{15}			
$(CH_3)_3Si(C_6H_5)CH$–CH=CHC_6H_5 (E)	t-C_4H_9CHO, $TiCl_4$, CH_2Cl_2, 0°, 1 h	$C_6H_5CH=CHCH(C_6H_5)CH(OH)C_4H_9$-$t$ + $C_6H_5CH=CHCH(C_6H_5)CH(OH)C_4H_9$-$t$ >99:1 (44)	22
$(CH_3)_3Si(C_6H_5)CH$–CH=CHC_6H_5 (Z)	t-C_4H_9CHO, $TiCl_4$, CH_2Cl_2, 0°, 1 h	$C_6H_5CH=CHCH(C_6H_5)CH(OH)C_4H_9$-$t$ + $C_6H_5CH=CHCH(C_6H_5)CH(OH)C_4H_9$-$t$ >99:1 (10)	22

TABLE XI. ALLYLSILANES WITH ALDEHYDES AND KETONES (*Continued*)

Reactant	Conditions	Product(s) and Yield(s)	Refs.
C_{16}			
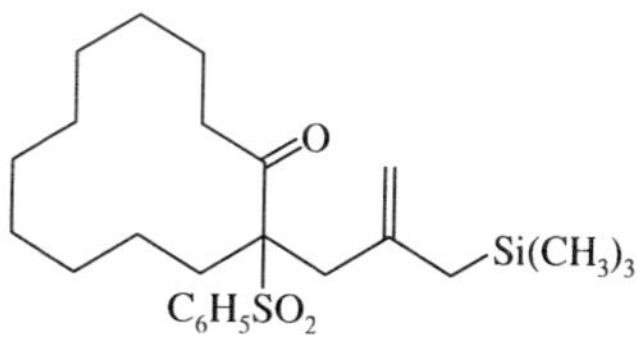 $C_6H_5SO_2$, $Si(CH_3)_3$, O	$C_2H_5AlCl_2$, CH_2Cl_2, reflux, 3 h	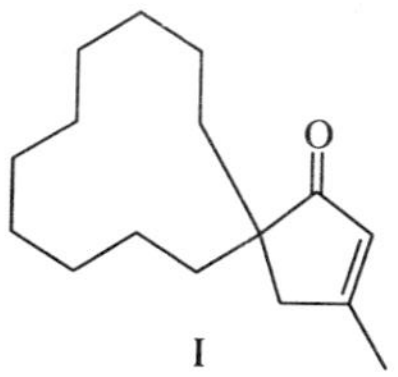 I (97)	651
	$C_2H_5AlCl_2$, CH_2Cl_2, 0° and room temp	I 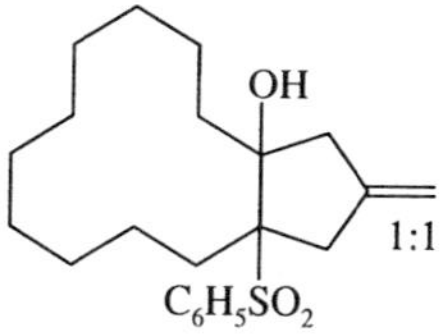 + OH, $C_6H_5SO_2$ 1:1 (—)	651

[a] Debenzylated products are isolated.
[b] This product is isolated after methanolysis of the silyl ether using HCl in CH_3OH.
[c] This alcohol is isolated as a 28:72 mixture with its silyl ether by omitting the methanolysis.
[d] The formation of a methoxy derivative in place of the expected alcohol is commented upon but not explained.
[e] The product isolated in this case is the phenyldimethylsilyl ether, which is not methanolyzed under the reaction conditions.
[f] The yield is reduced to 68% when 1 eq of KOC_4H_9-*t* is used.
[g] Benzophenone is recovered in 50% yield.
[h] 4 Eq of $TiCl_4$ are used in place of the 1.2 eq used in the other reactions recorded in this subtable.
[i] Two diastereoisomers at the C–CH_3 group are produced in a ratio of 1.7:1.
[j] The partly resolved starting material is estimated to have an ee of 22–25%, and each diastereoisomer of the product an ee of 24–26%.
[k] This is the yield given in the experimental section.
[l] This is the yield given in the discussion section.
[m] The diastereoisomers are produced in a ratio of 73:27.
[n] The overall yields are 47–60%.
[o] The R groups are not specified; the structure shown in Ref. 281 is incorrect. The structure shown here is more probable.

TABLE XII. VINYLSILANES WITH ALDEHYDES AND KETONES

Reactant	Conditions	Product(s) and Yield(s)	Refs.
C_2			
(E)-$(CH_3)_3SiCH{=}CHSi(CH_3)_3$	Cl_3CCHO, $AlCl_3$,[a] CH_2Cl_2, 1 h	(E)-$(CH_3)_3SiCH{=}CHCHOHCCl_3$ (80)	313
$(CH_3)_3SiC(SC_6H_5){=}CH_2$	C_6H_5CHO, TBAF	$C_6H_5CHOHC(SC_6H_5){=}CH_2$ (51)	101
$(CH_3)_3SiCH{=}C(OCH_3)OSi(CH_3)_3$	n-$C_8H_{17}CHO$	n-$C_8H_{17}CH{=}CHCO_2CH_3$	657
	$TiCl_4$, CH_2Cl_2, $-95°$, 3 h; room temp, 2 h	E:Z 14:86 (91)	
	$BF_3 \cdot O(C_2H_5)_2$, CH_2Cl_2, $-95°$, 3 h; room temp, 2 h	", 23:77 (90)	
	$AlCl_3$, C_6H_6, reflux, 8 h	", 89:11 (70)	
	", CCl_4 ", "	", 96:4 (75)	
$(C_2H_5)_3SiCF{=}CHF$	n-$C_{10}H_{21}CHO$, TASF (cat.), THF, 24 h	n-$C_{10}H_{21}CH[OSi(C_2H_5)_3]CF{=}CHF$ (47)	658
$C_6H_5(CH_3)_2SiCF{=}CF_2$	1. C_6H_5CHO, TASF (cat.), THF, 24 h 2. H^+, CH_3OH	$C_6H_5CHOHCF{=}CF_2$ (61)	658
$(C_2H_5)_3SiCF{=}CF_2$	RCHO, TASF (cat.), THF	$RCH[OSi(C_2H_5)_3]CF{=}CF_2$	658
	R		
	C_6H_5 24 h	(66)	
	$C_6H_5CH{=}CH$ 12 h	(43)	
	n-$C_{10}H_{21}$ "	(59)	
$(C_2H_5)_3SiCCl{=}CF_2$	C_6H_5CHO, TASF (cat.), THF, 12 h	$C_6H_5CH[OSi(C_2H_5)_3]CCl{=}CF_2$ (38)	658
$(C_2H_5)_3SiCF{=}CFSC_2H_5$	C_6H_5CHO, TASF (cat.), THF, 8 h	$C_6H_5CH[OSi(C_2H_5)_3]CF{=}CFSC_2H_5$ (86)	658
C_3			
(E)-$(CH_3)_3SiC(SC_6H_5){=}CHCF_3$	RCHO, TBAF, DMF, $-60°$	(E)-$RCHOHC(SC_6H_5){=}CHCF_3$	659
	R		
	CH_3	(69)	
	C_2H_5	(83)	
	i-C_3H_7	(72)	
	$CH_3CH{=}CH$	(34)	
	n-C_4H_9	(76)	

TABLE XII. VINYLSILANES WITH ALDEHYDES AND KETONES (*Continued*)

Reactant	Conditions	Product(s) and Yield(s)	Refs.
	t-C_4H_9	(36)	
	C_6H_5	(90)	
	$C_6H_5(CH_2)_2$	(70)	
	i-C_3H_7CHO, TBAF	(E)-i-$C_3H_7CHOHC(SC_6H_5)$=$CHCF_3$ + (E)-C_6H_5SCH=$CHCF_3$	659
	THF, −78°	65:35 (>95)	
	” −18°	52:48 (>95)	
	” room temp	31:69 (>95)	
	$(C_2H_5)_2O$ −78°	41:59 (>95)	
	DMF −60°	83:17 (>95)	
	” −18°	62:38 (>95)	
(Z)-$(CH_3)_3SiC(SC_6H_5)$=$CHCF_3$	RCHO, TBAF, DMF, −60°	(Z)-$RCHOHC(SC_6H_5)$=$CHCF_3$	659
	R		
	i-C_3H_7	(72)	
	C_6H_5	(71)	
C_4			
$(CH_3)_3SiC(CO_2C_4H_9$-$t)$=$CHCF_3$	RCHO, TBAF, THF, −78°, 1.5 h; room temp, 0.5 h	$RCHOHC(CO_2C_4H_9$-$t)$=$CHCF_3$	659
	R	E:Z	
	C_2H_5	34:66 (57)	
	i-C_3H_7	31:69 (69)	
	n-C_4H_9	38:62 (54)	
	t-C_4H_9	4:96 (54)	
	C_6H_{11}	33:67 (62)	
	C_6H_5	44:56 (60)	
	C_6H_5CH=CH	53:47 (34)	
	$C_6H_5(CH_2)_2$	55:45 (65)	

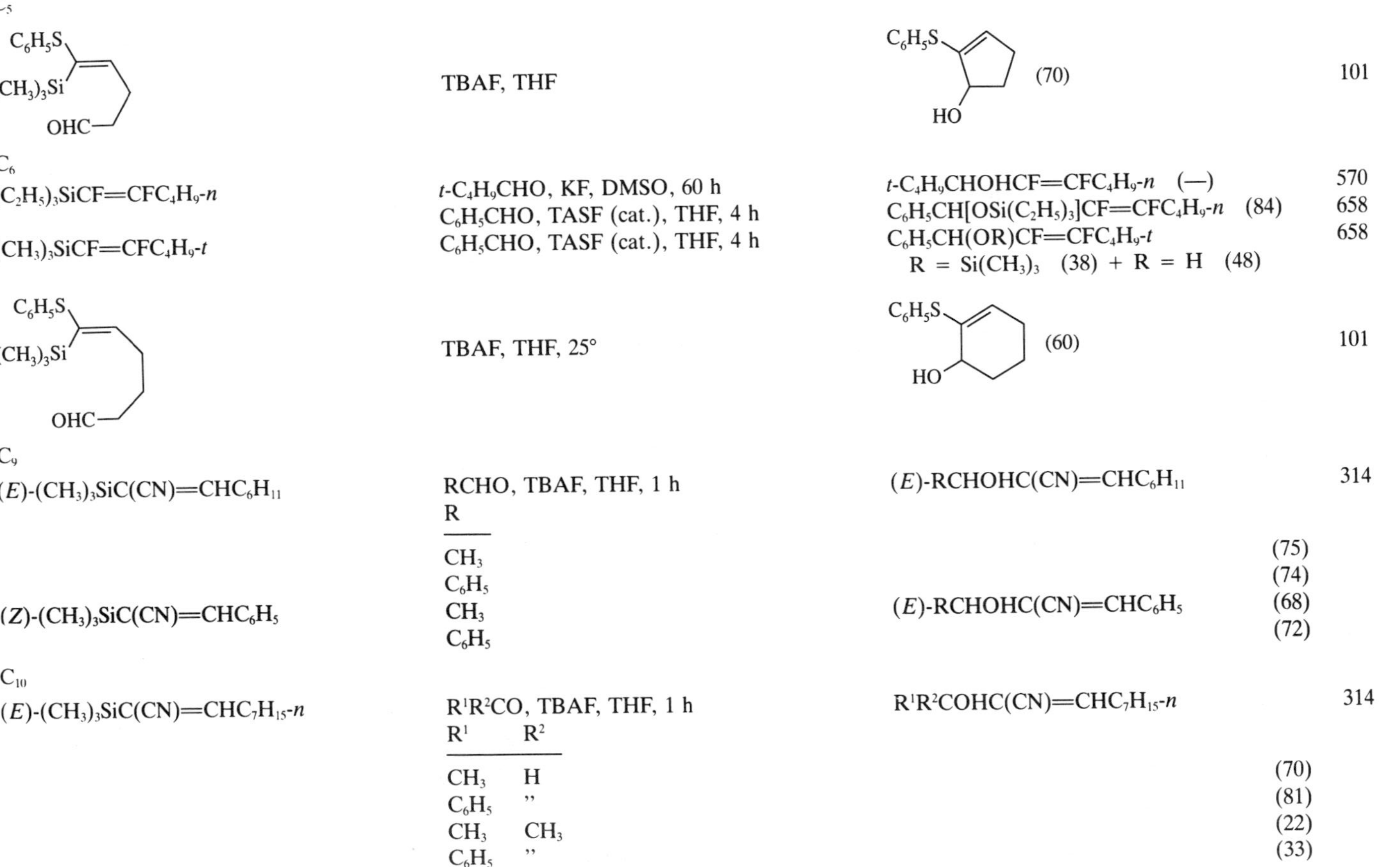

Reactant	Conditions	Product(s) and Yield(s) (%)	Refs.
C_5			
$C_6H_5S(CH_3)_3Si$-alkenyl-CH_2CH_2CHO (structure)	TBAF, THF	2-(C_6H_5S)-cyclopent-2-en-1-ol (structure) (70)	101
C_6			
$(C_2H_5)_3SiCF{=}CFC_4H_9$-*n*	*t*-C_4H_9CHO, KF, DMSO, 60 h	*t*-$C_4H_9CHOHCF{=}CFC_4H_9$-*n* (—)	570
	C_6H_5CHO, TASF (cat.), THF, 4 h	$C_6H_5CH[OSi(C_2H_5)_3]CF{=}CFC_4H_9$-*n* (84)	658
$(CH_3)_3SiCF{=}CFC_4H_9$-*t*	C_6H_5CHO, TASF (cat.), THF, 4 h	$C_6H_5CH(OR)CF{=}CFC_4H_9$-*t* R = $Si(CH_3)_3$ (38) + R = H (48)	658
$C_6H_5S(CH_3)_3Si$-alkenyl-$CH_2CH_2CH_2CHO$ (structure)	TBAF, THF, 25°	2-(C_6H_5S)-cyclohex-2-en-1-ol (structure) (60)	101
C_9			
(*E*)-$(CH_3)_3SiC(CN){=}CHC_6H_{11}$	RCHO, TBAF, THF, 1 h	(*E*)-$RCHOHC(CN){=}CHC_6H_{11}$	314
	R = CH_3	(75)	
	R = C_6H_5	(74)	
(*Z*)-$(CH_3)_3SiC(CN){=}CHC_6H_5$	R = CH_3	(*E*)-$RCHOHC(CN){=}CHC_6H_5$ (68)	
	R = C_6H_5	(72)	
C_{10}			
(*E*)-$(CH_3)_3SiC(CN){=}CHC_7H_{15}$-*n*	R^1R^2CO, TBAF, THF, 1 h	$R^1R^2COHC(CN){=}CHC_7H_{15}$-*n*	314
	R^1 = CH_3, R^2 = H	(70)	
	R^1 = C_6H_5, R^2 = "	(81)	
	R^1 = CH_3, R^2 = CH_3	(22)	
	R^1 = C_6H_5, R^2 = "	(33)	

TABLE XII. VINYLSILANES WITH ALDEHYDES AND KETONES (*Continued*)

Reactant	Conditions	Product(s) and Yield(s)	Refs.
(*E*)-$(CH_3)_3SiC(CN)=CHCH(C_2H_5)C_4H_9$-*n*	CH_3CHO, TBAF, THF, 1 h	$CH_3CHOHC(CN)=CHCH(C_2H_5)C_4H_9$-*n* (76)	314
	C_6H_5CHO ", ", "	$C_6H_5CHOHC(CN)=CHCH(C_2H_5)C_4H_9$-*n* (79)	314

[a] Catalytic quantities of $AlCl_3$, $InCl_3$, or $GaCl_3$ at 100° are also effective.

TABLE XIII. ALLENYLSILANES WITH ALDEHYDES AND KETONES

Reactant	Conditions	Product(s) and Yield(s)	Refs.
C_3			
$(CH_3)_3SiCH{=}C{=}CH_2$	1. R^1R^2CO, $TiCl_4$, CH_2Cl_2, −78° 2. KF, DMSO, 25°[a]	$HC{\equiv}CCH_2COHR^1R^2$	315
	R^1 / R^2		
	$C_6H_5(CH_2)_2$ / H	(84)	
	$-(CH_2)_5-$	(89)	
	$C_6H_5CH_2$ / CH_3	(72)	
C_4			
$(CH_3)_3SiCH{=}C{=}CHCH_3$	1. RCHO, $TiCl_4$, CH_2Cl_2, −78°, 15 min 2. KF, DMSO, 25°, 12 h[a]	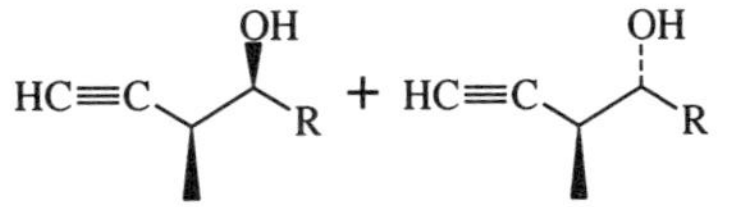 $HC{\equiv}C$–CH(CH_3)–CH(OH)R (syn) + $HC{\equiv}C$–CH(CH_3)–CH(OH)R (anti)	315
	R		
	C_6H_{11}	4.1:1 (81)	
	$C_6H_5(CH_2)_2$	3.1:1 (89)	
	1. 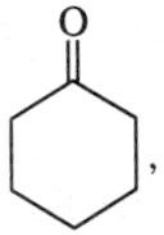 cyclohexanone, $TiCl_4$, CH_2Cl_2, −78°, 1 h; 25°, 1 h 2. KF, DMSO, 25°, 12 h[a]	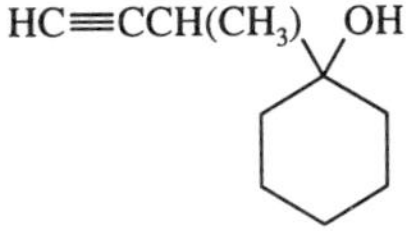 1-($HC{\equiv}CCH(CH_3)$)cyclohexanol (77)	315
$(CH_3)_3SiC(CH_3){=}C{=}CH_2$	R^1R^2CO, $TiCl_4$, CH_2Cl_2	$CH_3C{\equiv}CCH_2COHR^1R^2$	315
	R^1 / R^2		
	C_6H_{11} / H / −78°, 2 h	(68)	
	$C_6H_5(CH_2)_2$ / H / −78°, 1 h	(85)	
	i-C_3H_7 / CH_3 / ", "; 25°, 0.5 h	(89)	

TABLE XIII. ALLENYLSILANES WITH ALDEHYDES AND KETONES (*Continued*)

Reactant	Conditions	Product(s) and Yield(s)	Refs.
C_6			
$(CH_3)_3SiC(C_3H_7\text{-}i){=}C{=}CH_2$	R^1R^2CO, $TiCl_4$, CH_2Cl_2	$i\text{-}C_3H_7C{\equiv}CCH_2COHR^1R^2$	315
	R^1 / R^2		
	$C_6H_5(CH_2)_2$ / H / −78°, 1 h	(89)	
	$i\text{-}C_3H_7$ / CH_3 / ", "; 0°, 1 h	(51)	
	$-(CH_2)_5-$ / ", 2 h; 0°, 1 h	(84)	
$(CH_3)_3SiC(CH_3){=}C{=}C(CH_3)_2$	$C_6H_5(CH_2)_2CHO$, $TiCl_4$	$CH_3C{\equiv}CC(CH_3)_2CHOH(CH_2)_2C_6H_5$ (27) + 2-methyl-3-$(CH_3)_3Si$-4,4-dimethyl-5-$(CH_2)_2C_6H_5$-4,5-dihydrofuran (54)	315
C_{11}			
$(CH_3)_3SiCH{=}C{=}CH(CH_2)_2C_6H_5$	1. $(CH_3)_2CO$, $TiCl_4$, CH_2Cl_2, −78°, 0.5 h; 0°, 1.5 h 2. KF, DMSO, 25°, 14 h[a]	$HC{\equiv}CCH[(CH_2)_2C_6H_5]COH(CH_3)_2$ (38)	315
	1. Cyclohexanone, $TiCl_4$, CH_2Cl_2, −78°; 25°, 2 h 2. KF, DMSO, 25°, 16 h[a]	1-[$HC{\equiv}CCH((CH_2)_2C_6H_5)$]cyclohexanol (49)	315

[a] This treatment is to complete the elimination of the silyl group from the intermediate β-chlorovinylsilane.

TABLE XIV. ALLYLSILANES WITH α,β-UNSATURATED CARBONYL COMPOUNDS AND α,β-UNSATURATED NITRILES

Reactant	Conditions	Product(s) and Yield(s)	Refs.
C_3			
$(CH_3)_3SiCH_2CH{=}CH_2$	$C_6H_5CH{=}CHCHO$, $BF_3{\cdot}O(C_2H_5)_2$	$CH_2{=}CHCH_2CHOHCH{=}CHC_6H_5$ (50)	162
	", TBAF, DMF, HMPA	" (86)	162
	$C_6H_5CH{=}C(CH_3)CHO$, $BF_3{\cdot}O(C_2H_5)_2$	$CH_2{=}CHCH_2CHOHC(CH_3){=}CHC_6H_5$ (45)	162
	", TBAF, DMF, HMPA	" (92)	162
	[structure]CHO, $TiCl_4$, CH_2Cl_2, −78°, 30 min	$CH_2{=}CHCH_2C(CH_3)_2CH{=}CH$–$C(CH_3)_2CH_2CHO$ (OHCCH₂C(CH₃)₂ on CH) (51)	607
	$CH_2{=}CHCO_2CH_2C_6H_5$, TBAF, DMF, HMPA	$CH_2{=}CH(CH_2)_3CO_2CH_2C_6H_5$ (65)	162
	(*E*)-$CH_3CH{=}CHCO_2C_2H_5$, TBAF, DMF, HMPA	$CH_2{=}CHCH_2CH(CH_3)CH_2CO_2C_2H_5$ (27)	162
	(*E*)-$CH_3O_2CCH{=}CHCO_2CH_3$, TBAF, DMF, HMPA	$CH_2{=}CHCH_2CH(CO_2CH_3)CH_2CO_2CH_3$ (80)	162
	$CH_2{=}CHCOCH_3$, $TiCl_4$, CH_2Cl_2, −78°, 1 min	$CH_2{=}CH(CH_2)_3COCH_3$ (59)	317
	", ", ", −78°, 5 h; −30°, 5 h	" (40)	319
	$CH_2{=}CHCOCN$, $TiCl_4$, CH_2Cl_2, −78°, 3 h; −30°, 4 h	$CH_2{=}CH(CH_2)_3COCN$ (30) + [structure: 6-methyl-dihydropyran-2-carbonitrile, O, CN] (35)	331
	(*E*)-$CH_3CH{=}CHCOSi(CH_3)_3$, $TiCl_4$, CH_2Cl_2, −78°, 15 min	$CH_2{=}CHCH_2CH(CH_3)CH_2COSi(CH_3)_3$ (74)	332

TABLE XIV. ALLYLSILANES WITH α,β-UNSATURATED CARBONYL COMPOUNDS AND α,β-UNSATURATED NITRILES (*Continued*)

Reactant	Conditions	Product(s) and Yield(s)	Refs.
	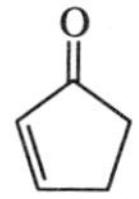 2-cyclopentenone, $TiCl_4$, CH_2Cl_2	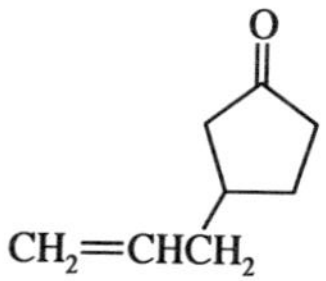 3-(CH_2=$CHCH_2$)cyclopentanone (—)	324
	(*E*)-CH_3CH=$CHCOCN$, $TiCl_4$, CH_2Cl_2, −78°, 3 h; −30°, 4 h	CH_2=$CHCH_2CH(CH_3)CH_2COCN$ (95)[a]	331
	CH_2=$C(CH_3)COCN$, $TiCl_4$, CH_2Cl_2, −78°, 3h; −30°, 4 h	CH_2=$CH(CH_2)_2CH(CH_3)COCN$ (30)[a]	331
	$(CH_3)_2C$=$CHCOSi(CH_3)_3$, $TiCl_4$, CH_2Cl_2, −78°, 15 min	CH_2=$CHCH_2C(CH_3)_2CH_2COSi(CH_3)_3$ (82)	332
	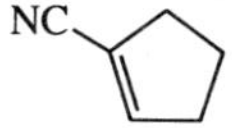 1-cyanocyclopentene (NC), TBAF, DMF, HMPA	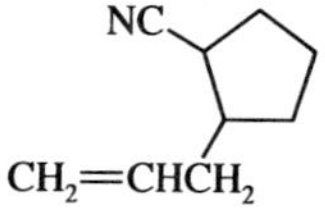 2-(CH_2=$CHCH_2$)cyclopentanecarbonitrile (NC) (44)	162
	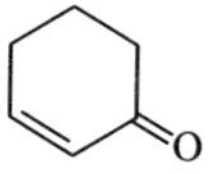 2-cyclohexenone, $TiCl_4$, CH_2Cl_2, −78°, 1 h; −30°, 20 min	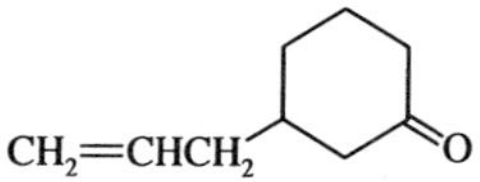 3-(CH_2=$CHCH_2$)cyclohexanone (82)	317, 319
	1. ", $TiCl_4$, −30°, 30 min 2. C_2H_5CHO, −78°, 1 h	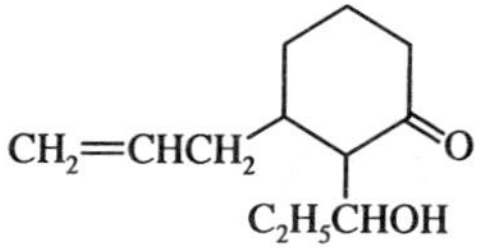 3-(CH_2=$CHCH_2$)-2-(C_2H_5CHOH)cyclohexanone (50)	330
	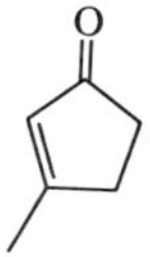 3-methyl-2-cyclopentenone, $TiCl_4$	No reaction	328

Conditions	Product(s) and Yield(s) (%)	Refs.
$(CH_3)_2C{=}CHCOCN$, $TiCl_4$, CH_2Cl_2, −78°, 3 h; −30°, 4 h	$CH_2{=}CHCH_2C(CH_3)_2CH_2COCN$ (95)[a]	331
$(CH_3)_2C{=}CHCOCH_3$, $TiCl_4$, CH_2Cl_2, 5 min	$CH_2{=}CHCH_2C(CH_3)_2CH_2COCH_3$ (87)	317
", ", ", −78°, 5 h; −30°, "	" (50)	319
1. $(CH_3)_2C{=}CHCOCH_3$, $TiCl_4$, CH_2Cl_2, −30°, 30 min 2. C_2H_5CHO, 1 h	$CH_2{=}CHCH_2C(CH_3)_2CH(CH(OH)C_2H_5)COCH_3$ (20)	330
$CH_3(CH{=}CH)_2CO_2C_2H_5$, TBAF, DMF, HMPA	$CH_3CH{=}CHCH(CH_2CH{=}CH_2)CH_2CO_2C_2H_5$ (31)	162
$(CH_3)_2C{=}C(CO_2C_2H_5)_2$, $TiCl_4$, CH_2Cl_2, −78°	$CH_2{=}CHCH_2C(CH_3)_2CH(CO_2C_2H_5)_2$ (35)	162
", TBAF, DMF, HMPA	" (52)	162
1-($C_6H_5CH_2O_2C$)cyclopentene, TBAF, DMF, HMPA	1-($C_6H_5CH_2O_2C$)-2-($CH_2{=}CHCH_2$)cyclopentane (50)	162
(*E*)-$C_6H_5CH_2OCH(CH_3)CH{=}CHCOCH_3$, $TiCl_4$, CH_2Cl_2, −78°, 1.5 h	$C_6H_5CH_2OCH(CH_3)CH(CH_2CH{=}CH_2)CH_2COCH_3$ (syn) + $C_6H_5CH_2OCH(CH_3)CH(CH_2CH{=}CH_2)CH_2COCH_3$ (anti) 7:1 (83)	299
(*Z*)-$C_6H_5CH_2OCH(CH_3)CH{=}CHCOCH_3$, $TiCl_4$, CH_2Cl_2, −78°, 1.5 h	" 1:10 (70)	299
t-$C_4H_9CH{=}CHR$, TBAF, DMF, HMPA	$CH_2{=}CHCH_2CH(C_4H_9\text{-}t)CH_2R$	162
R = CN	" (65)	
R = $CO_2C_2H_5$	" (83)	
R = $CON(C_2H_5)_2$	*t*-$C_4H_9CH{=}CHC(CH_2CH{=}CH_2)_2OH$ (26)	

TABLE XIV. ALLYLSILANES WITH α,β-UNSATURATED CARBONYL COMPOUNDS AND α,β-UNSATURATED NITRILES (*Continued*)

Reactant	Conditions	Product(s) and Yield(s)	Refs.
	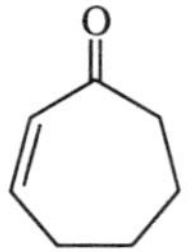 [cycloheptenone], $TiCl_4$, CH_2Cl_2, −78°, 30 min	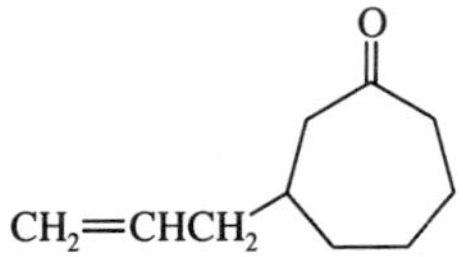 $CH_2{=}CHCH_2$–[cycloheptanone] (85)	325
	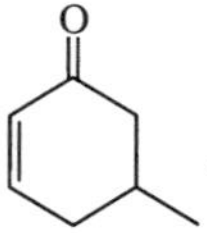 [5-methylcyclohexenone], $TiCl_4$, CH_2Cl_2, −78°, 1.75 h	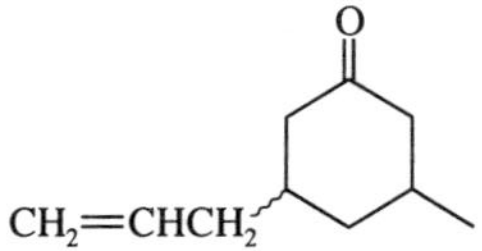 $CH_2{=}CHCH_2$–[methylcyclohexanone] *trans:cis* > 98:2 (83)	326
	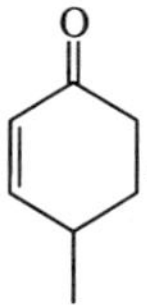 [4-methylcyclohexenone], $TiCl_4$, CH_2Cl_2, −78°, 1.75 h	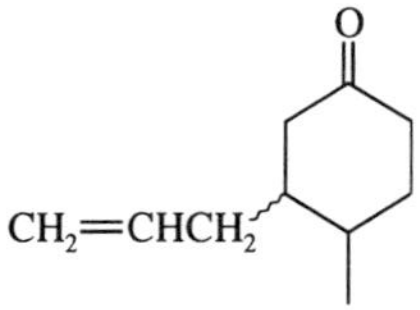 $CH_2{=}CHCH_2$–[methylcyclohexanone] *trans:cis* 32:68 (76)	326
	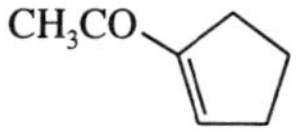 CH_3CO–[cyclopentene], $TiCl_4$, CH_2Cl_2, −78°, 1.5 h	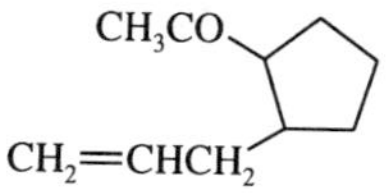 CH_3CO, $CH_2{=}CHCH_2$–[cyclopentane] (83–94) I	660, 321
	", $TiCl_4$, CH_2Cl_2, −78°, 5 h	*trans*-I (78) + *cis*-I (4) 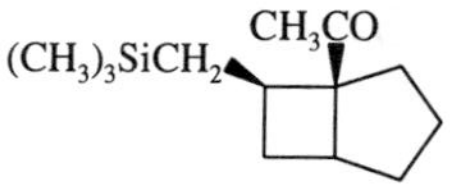 + $(CH_3)_3SiCH_2$, CH_3CO–[bicyclo[3.2.0]heptane] (18)	319
	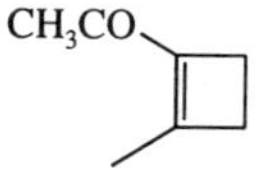 CH_3CO–[methylcyclobutene], $TiCl_4$, CH_2Cl_2, −30°	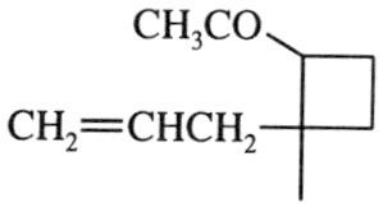 CH_3CO, $CH_2{=}CHCH_2$–[methylcyclobutane] *E:Z* 1:2 (essentially quantitative)	423

$CH_3(CH{=}CH)_2COCN$, $TiCl_4$, CH_2Cl_2, $-78°$, 3 h; $-30°$, 4 h	$CH_3CH(CH_2CH{=}CH_2)CH{=}CHCH_2CO_2H$ (45) + $CH_3CH{=}CHCH(CH_2CH{=}CH_2)CH_2CO_2H$ (30)	331
3-(CH=CHR)furan, TBAF, DMF, HMPA	3-[$CH(CH_2CH{=}CH_2)CH_2R$]furan	162
R = CN	(91)	
R = $CO_2C_2H_5$	(83)	
R = $CON(C_2H_5)_2$	(84)	
1-[$(CH_3)_3SiCO$]cyclohexene, $TiCl_4$, CH_2Cl_2, $-78°$, 15 min	1-[$(CH_3)_3SiCO$]-2-($CH_2{=}CHCH_2$)cyclohexane *trans:cis* 6:1 (65)	332
5-methylcyclohept-2-en-1-one, $TiCl_4$, CH_2Cl_2, $-78°$, 1.75 h	3-($CH_2{=}CHCH_2$)-5-methylcycloheptanone (71) *trans:cis* 11:89	326
6-methylcyclohept-2-en-1-one, $TiCl_4$, CH_2Cl_2, $-78°$, 1.75 h	3-($CH_2{=}CHCH_2$)-6-methylcycloheptanone (76) *trans:cis* >98:2	326

TABLE XIV. ALLYLSILANES WITH α,β-UNSATURATED CARBONYL COMPOUNDS AND α,β-UNSATURATED NITRILES (*Continued*)

Reactant	Conditions	Product(s) and Yield(s)	Refs.
	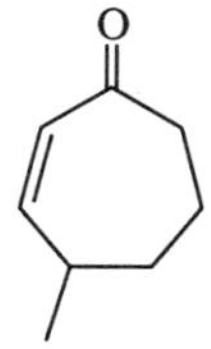 , $TiCl_4$, CH_2Cl_2, −78°, 1.75 h	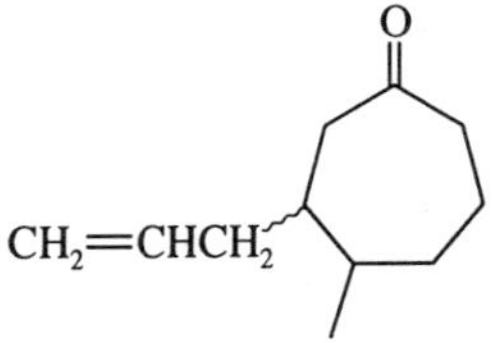 CH_2=$CHCH_2$ (71) *trans*:*cis* 35:65	326
	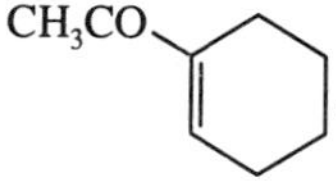 CH_3CO, $TiCl_4$, CH_2Cl_2, −78°, 5 h; −30°, 5 h	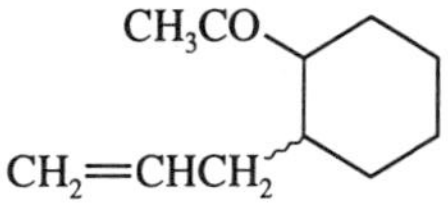 CH_3CO CH_2=$CHCH_2$ *trans* (75) + *cis* (8) 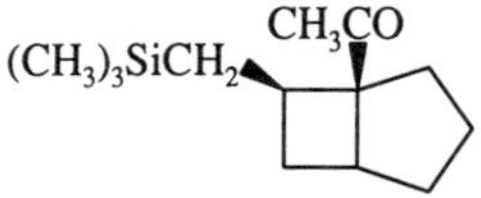 + $(CH_3)_3SiCH_2$ CH_3CO (17)	319
	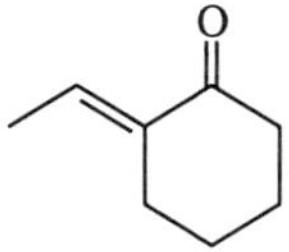 , $TiCl_4$, CH_2Cl_2, −78°, 2 h	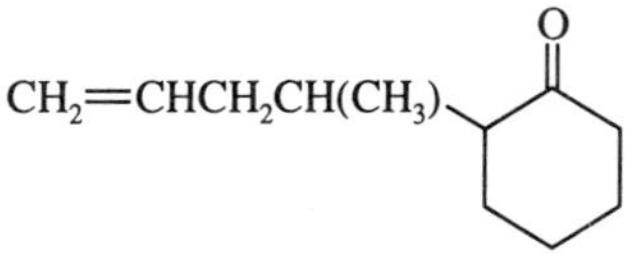 CH_2=$CHCH_2CH(CH_3)$ (45) 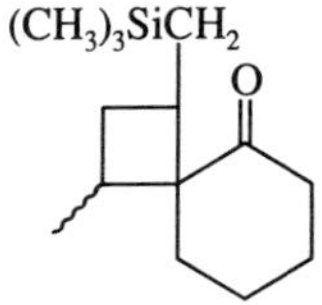 + $(CH_3)_3SiCH_2$ (—)	322
	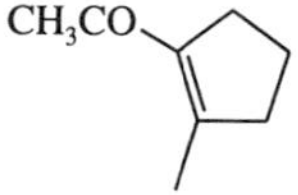 CH_3CO, $TiCl_4$, CH_2Cl_2	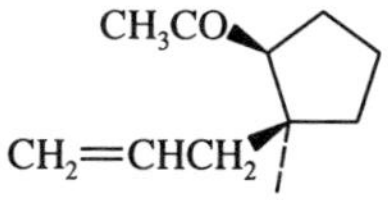 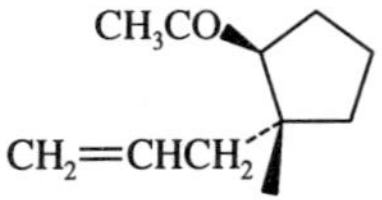 CH_3CO CH_2=$CHCH_2$ + CH_3CO CH_2=$CHCH_2$ 60:40 (80)	319

Reagents and Conditions	Product(s) and Yield(s) (%)	Refs.
$C_6H_5CH{=}CHR$, TBAF, DMF, HMPA	$CH_2{=}CHCH_2CH(C_6H_5)CH_2R$	162
R		
CN	(65)	
CO_2CH_3 10 min	(90)	
$CON(C_2H_5)_2$	(80)	
3,5,5-trimethylcyclohex-2-enone, $TiCl_4$, CH_2Cl_2, $-30°$, 20 min	3-allyl-3,5,5-trimethylcyclohexanone ($CH_2{=}CHCH_2$) (76)	317
$C_6H_5CH{=}CHCOCH_3$, $TiCl_4$, CH_2Cl_2, $-40°$, 30 min	$CH_2{=}CHCH_2CH(C_6H_5)CH_2COCH_3$ (78–80)	318, 162
", TBAF, THF, reflux, 5 h	" (24) + $CH_2{=}CHCH_2C(CH_3)OH$ with $C_6H_5CH{=}CH$ on C (28) + $CH_2{=}CHCH_2CH(C_6H_5)CH{=}C(OSi(CH_3)_3)CH_3$ (22)	13, 162
1. $C_6H_5CH{=}CHCOCH_3$, $TiCl_4$, CH_2Cl_2, $-30°$, 30 min	$CH_2{=}CHCH_2CH(C_6H_5)CH(COCH_3)C(OH)R^1R^2$	330
2. R^1R^2CO		
R^1 / R^2		
C_2H_5 / H, $-30°$, 1 h	(64)	
C_6H_5 / " , ", "	(51)	
$C_6H_5(CH_2)_2$ / " , ", 2 h	(77)	
CH_3 / CH_3, 1 h	(56)	
C_2H_5 / C_2H_5, 4 h	(19)	
1. $C_6H_5CH{=}CHCOCH_3$, $TiCl_4$, CH_2Cl_2, $-30°$, 30 min	$CH_2{=}CHCH_2CH(C_6H_5)CH(COCH_3)CH(OR^2)R^1$	330

TABLE XIV. ALLYLSILANES WITH α,β-UNSATURATED CARBONYL COMPOUNDS AND α,β-UNSATURATED NITRILES (*Continued*)

Reactant	Conditions	Product(s) and Yield(s)	Refs.
	2. $R^1CH(OR^2)_2$, 1 h		
	R^1 — R^2		
	CH_3 — C_2H_5 — −30°	(63)	
	$C_6H_5(CH_2)_2$ — CH_3 — "	(43)	
	CH_3O — CH_3 — −55°	(72)	
	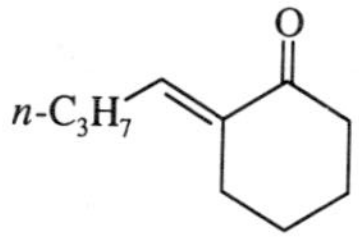 2-(n-butylidene)cyclohexanone (n-C_3H_7), $TiCl_4$, CH_2Cl_2, −78°, 2 h	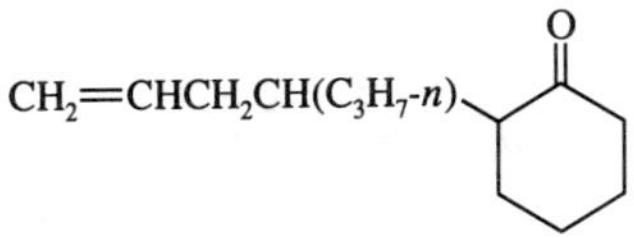 $CH_2{=}CHCH_2CH(C_3H_7\text{-}n)$-cyclohexan-2-one (52) 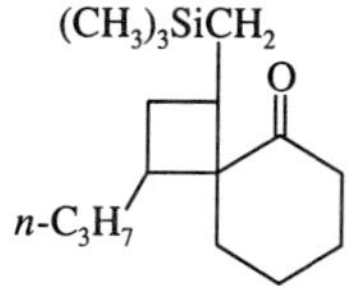 + spiro cyclobutane ketone with $(CH_3)_3SiCH_2$ and n-C_3H_7 (—)	322
	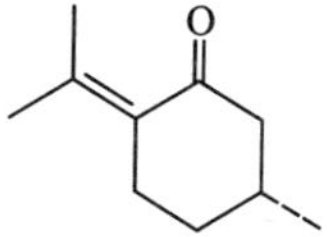 pulegone, $TiCl_4$, CH_2Cl_2, −78°, 5 h; −30°, 5 h	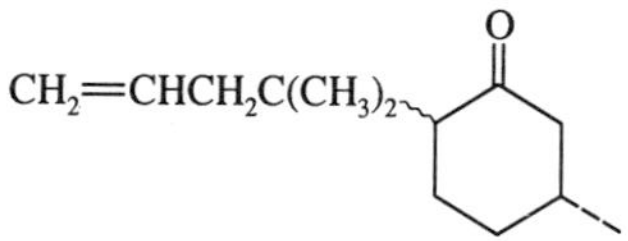 $CH_2{=}CHCH_2C(CH_3)_2$-(5-methylcyclohexanone) (50) *trans*:*cis* 60:40	319
	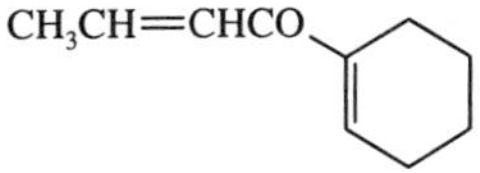 $CH_3CH{=}CHCO$-cyclohexenyl, $TiCl_4$, CH_2Cl_2, −78°, 5 h; −30°, 5 h	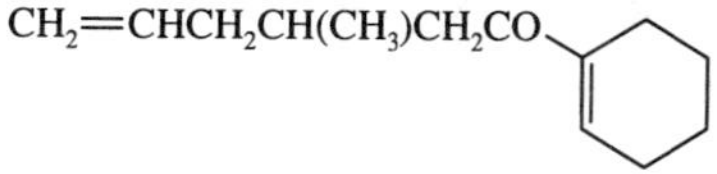 $CH_2{=}CHCH_2CH(CH_3)CH_2CO$-cyclohexenyl (68) 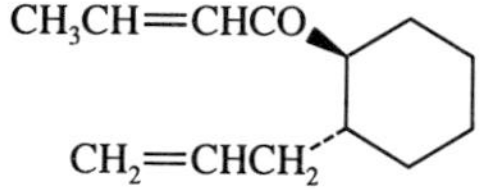 + $CH_3CH{=}CHCO$, $CH_2{=}CHCH_2$-cyclohexane (9) +	319

Reactants and Conditions	Product(s) and Yield(s) (%)	Refs.
	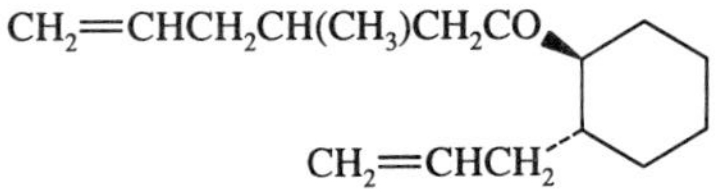 $CH_2{=}CHCH_2CH(CH_3)CH_2CO$; $CH_2{=}CHCH_2$ (9)	
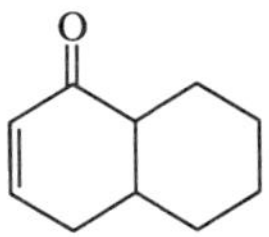 O, , $TiCl_4$, CH_2Cl_2, −78°, 5 h; −30°, 5 h	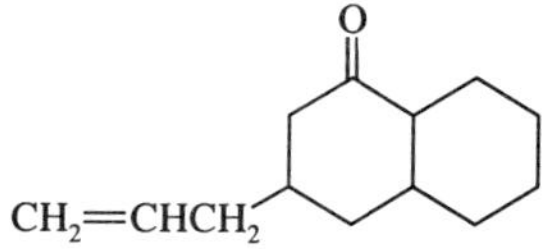 O ; $CH_2{=}CHCH_2$ (100)	319
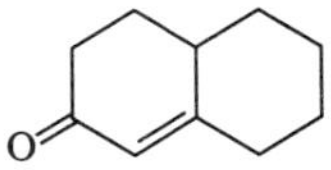 O, , $TiCl_4$, CH_2Cl_2, −78°, 18 h; −30°, 5 h	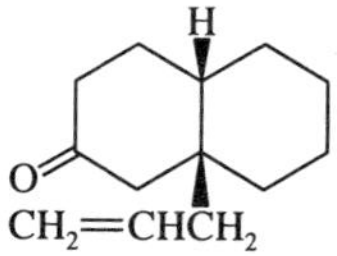 H ; O ; $CH_2{=}CHCH_2$ (85)	317
$C_6H_5CH{=}CHCOCN$, $TiCl_4$, CH_2Cl_2, −78°, 3 h; −30°, 4 h	$CH_2{=}CHCH_2CH(C_6H_5)CH_2COCN$ (80)	331
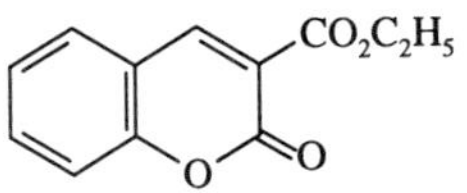 $CO_2C_2H_5$; O ; O	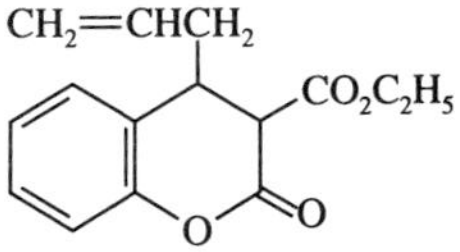 $CH_2{=}CHCH_2$; $CO_2C_2H_5$; O ; O	162
$TiCl_4$, CH_2Cl_2, −78°	(49)	
TBAF, DMF, HMPA	(91)	
$C_6H_5CH{=}C(CH_3)CO_2C_2H_5$, TBAF, DMF, HMPA	$CH_2{=}CHCH_2CH(C_6H_5)CH(CH_3)CO_2C_2H_5$ (80)	162
$C_6H_5C(CH_3){=}CHCO_2C_2H_5$, TBAF, DMF, HMPA	$CH_2{=}CHCH_2C(C_6H_5)(CH_3)CH_2CO_2C_2H_5$ (47)	162
$C_6H_5(CH{=}CH)_2CN$, TBAF, DMF, HMPA	$CH_3CH{=}CHCHCH_2CN$; $CH_2{=}CHCH_2$ (37)	162
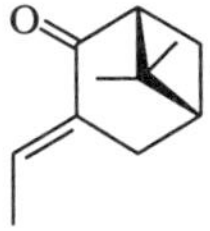 O, , $TiCl_4$, CH_2Cl_2, −78°, 2 h	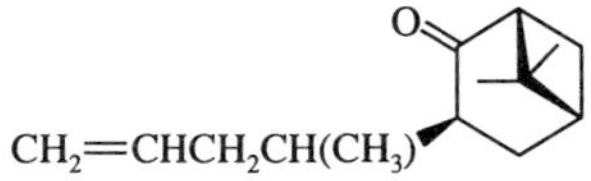 O ; $CH_2{=}CHCH_2CH(CH_3)$ (89) *erythro:threo* 1:4	323

TABLE XIV. ALLYLSILANES WITH α,β-UNSATURATED CARBONYL COMPOUNDS AND α,β-UNSATURATED NITRILES (*Continued*)

Reactant	Conditions	Product(s) and Yield(s)	Refs.
	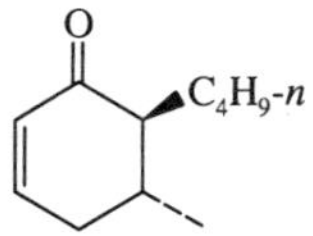 [structure: 6-n-butyl-5-methylcyclohex-2-enone], $TiCl_4$, CH_2Cl_2, −78°, 1.5 h	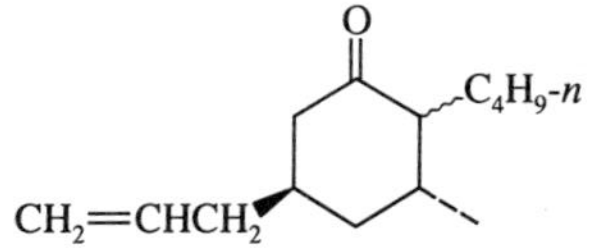 [structure: 3-allyl-5-methyl-6-n-butylcyclohexanone, with $CH_2{=}CHCH_2$ and C_4H_9-*n* labels] (95)[b]	334
	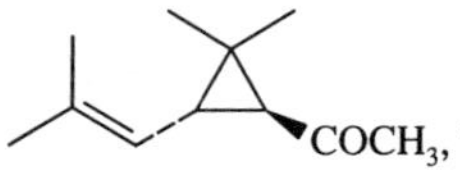 [structure: trans-2,2-dimethyl-3-(2-methylpropenyl)cyclopropyl]$COCH_3$, $TiCl_4$, CH_2Cl_2, −50°, 10 min	$CH_2{=}CHCH_2C(CH_3)_2CH{=}CH$–$C(CH_3)_2CH_2COCH_3$ (53)	607
	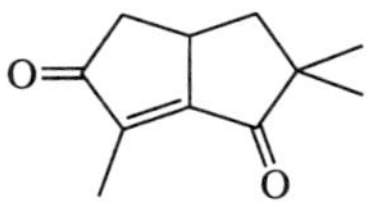 [structure: bicyclic enedione], $TiCl_4$, −78°	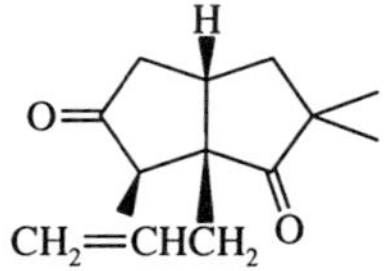 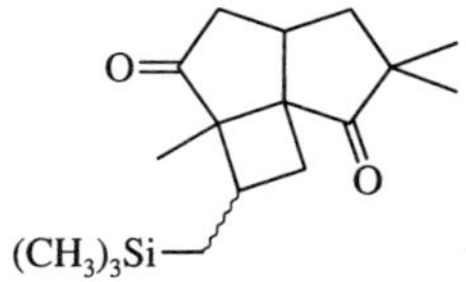 [structure: bicyclic dione with H and $CH_2{=}CHCH_2$ substituents] (12–31) +	327
		[structure: tricyclic dione with $(CH_3)_3Si$– substituent] (—)[c]	
	[structure: 3-acetylcoumarin] $TiCl_4$, CH_2Cl_2, −78°,	[structure: 4-allyl-3-acetyl-3,4-dihydrocoumarin, with $CH_2{=}CHCH_2$ and $COCH_3$ labels] (79)	162
	TBAF, DMF, HMPA	(22)	

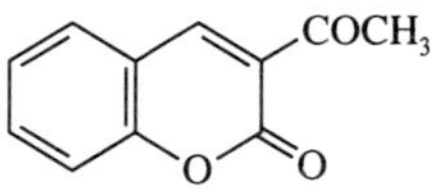

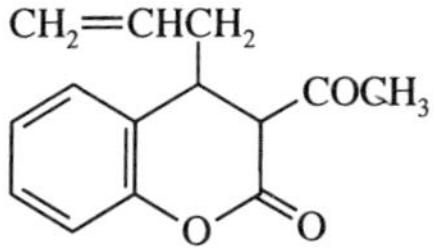

$C_6H_5(CH{=}CH)_2CO_2C_2H_5$, TBAF	$C_6H_5CH{=}CHCH(CH_2CH{=}CH_2)CH_2CO_2C_2H_5$ (63)	162
(*E*)-$CH_3CH(OCH_2C_6H_5)CH{=}CHCOC_6H_5$, $TiCl_4$, CH_2Cl_2, −78°, 1.5 h	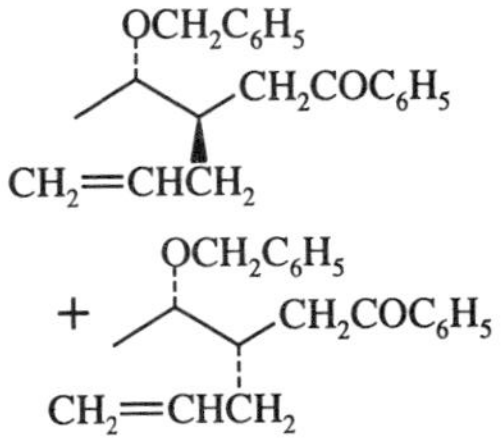 8:1 (82)	299
(*E*)-$C_6H_5CH(CH_3)CH{=}CHCOCH_3$, CH_2Cl_2	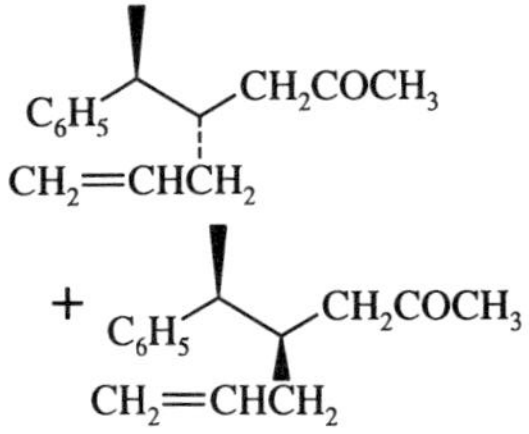	299
$TiCl_4$, −78°, 1 h	4:1 (74)	
$BF_3{\cdot}O(C_2H_5)_2$, −78° to 25°, 24 h	" (<50)	
" −78°, 24 h	" (43)	
$C_6H_5CH{=}CHCOC_4H_9$-*s*, $TiCl_4$, CH_2Cl_2, −78°	$CH_2{=}CHCH_2CH(C_6H_5)CH_2COC_4H_9$-*s* (74) I	162
", TBAF, DMF, HMPA	I (57) + $C_6H_5CH{=}CHC(C_4H_9\text{-}s)(CH_2CH{=}CH_2)OH$ (20)	162
$C_6H_5CH{=}CHCOC_4H_9$-*t*, $TiCl_4$, CH_2Cl_2, −78°	$CH_2{=}CHCH_2CH(C_6H_5)CH_2COC_4H_9$-*t* (92) II	162
", TBAF, DMF, HMPA	II (38) + $C_6H_5CH{=}CHC(C_4H_9\text{-}t)(CH_2CH{=}CH_2)OH$ (26)	162

TABLE XIV. ALLYLSILANES WITH α,β-UNSATURATED CARBONYL COMPOUNDS AND α,β-UNSATURATED NITRILES (*Continued*)

Reactant	Conditions	Product(s) and Yield(s)	Refs.
	$C_6H_5CH{=}CHCOC_6H_5$, $TiCl_4$, CH_2Cl_2, −30°, 5 min	$CH_2{=}CHCH_2CH(C_6H_5)CH_2COC_6H_5$ (95)	322, 317
$(CH_3)_3SiCH_2CH{=}CHSi(CH_3)_3$	1-acetylcyclopentene (CH_3CO-cyclopentene), $TiCl_4$, CH_2Cl_2, −78°, 1 h	1-acetyl-2-[$(CH_3)_3SiCH_2CH{=}CH$]cyclopentane (87)	320
$(CH_3)_3SiCHBrCH{=}CH_2$	2-(n-C_3H_7-methylene)cyclohexanone, $TiCl_4$, CH_2Cl_2, −78°, 4 h	2-[$BrCH{=}CHCH_2CH(C_3H_7$-$n)$]cyclohexanone (63)	637
$(CH_3)_3SiCH_2CBr{=}CH_2$	1-acetylcyclopentene (CH_3CO-cyclopentene), $TiCl_4$, CH_2Cl_2, −78°	1-acetyl-2-($CH_2{=}CBrCH_2$)cyclopentane (75)	290
C_4			
$(CH_3)_3SiCH_2C(CH_3){=}CH_2$	2-cyclopentenone, $TiCl_4$, CH_2Cl_2, −78°, 10 min	3-[$CH_2{=}C(CH_3)CH_2$]cyclopentanone (70)	322
	$CH_3CH{=}CHCOCN$, $TiCl_4$, CH_2Cl_2, −78°, 3 h; −30°, 4 h	$CH_2{=}C(CH_3)CH_2CH(CH_3)CH_2COCN$ (95)	331
	2-cyclohexenone, $TiCl_4$, CH_2Cl_2, −78°, 10 min	3-[$CH_2{=}C(CH_3)CH_2$]cyclohexanone (99)	322
	2-methylenecyclopentanone, $BF_3 \cdot O(C_2H_5)_2$, CH_2Cl_2, −78°, 1 h	2-[$CH_2{=}C(CH_3)(CH_2)_2$]cyclopentanone (45)	322

Reactants and Conditions	Product(s) and Yield(s) (%)	Refs.
$(CH_3)_2C{=}CHCOCN$, $TiCl_4$, CH_2Cl_2, −78°, 3 h; −30°, 4 h	$CH_2{=}C(CH_3)CH_2C(CH_3)_2CH_2COCN$ (60) + [2,2,4,4-tetramethyl-3,4-dihydro-2H-pyran-6-carbonitrile, CN] (35)	331
(*E*)-$CH_3CH(OCH_2C_6H_5)CH{=}CHCOC_6H_5$, $TiCl_4$, CH_2Cl_2, −78°, 1.5 h	$OCH_2C_6H_5$, $CH_2COC_6H_5$, $CH_2{=}C(CH_3)CH_2$ (I) + $OCH_2C_6H_5$, $CH_2COC_6H_5$, $CH_2{=}C(CH_3)CH_2$ (II) 4:1 (80)	299
(*Z*)-$CH_3CH(OCH_2C_6H_5)CH{=}CHCOC_6H_5$, $TiCl_4$, CH_2Cl_2, −78°, 1.5 h	I + II 1:1 (78)	299
CH_3CO-cyclopentenone (O), $TiCl_4$, CH_2Cl_2	CH_3CO, O, $CH_2{=}C(CH_3)CH_2$ [2-acetyl-3-(2-methylallyl)cyclopentanone] (—)	319
$CH_3(CH{=}CH)_2COCN$, $TiCl_4$, CH_2Cl_2, −78°, 3 h; −30°, 4 h	$CH_2{=}C(CH_3)CH_2CH(CH_3)CH{=}CH$–$HO_2CCH_2$ (30) + $CH_3CH{=}CHCHCH_2CO_2H$–$CH_2{=}C(CH_3)CH_2$ (45)	331
2-ethylidenecyclohexanone (O), $TiCl_4$, CH_2Cl_2, −78°, 1 h	$CH_2{=}C(CH_3)CH_2CH(CH_3)$-cyclohexanone (O) (74)	322

TABLE XIV. Allylsilanes with α,β-Unsaturated Carbonyl Compounds and α,β-Unsaturated Nitriles (*Continued*)

Reactant	Conditions	Product(s) and Yield(s)	Refs.
	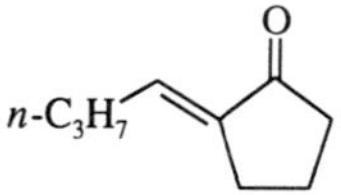 n-C_3H_7 [structure], $TiCl_4$, CH_2Cl_2, −78°, 30 min	$CH_2{=}C(CH_3)CH_2CH(C_3H_7$-$n)$ 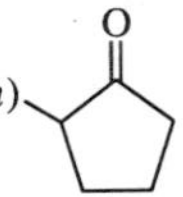 [structure] (82)	322
	$C_6H_5CH{=}CHCOCH_3$, $TiCl_4$, CH_2Cl_2, −78°, 30 s	$CH_2{=}C(CH_3)CH_2CH(C_6H_5)CH_2COCH_3$ (69)	322
	1. ", $TiCl_4$, CH_2Cl_2, −78°, 2 min 2. $HC(OCH_3)_3$, 1 h	$CH_2{=}C(CH_3)CH_2CH(C_6H_5)CH(CH(OCH_3)_2)COCH_3$ (62)	330
	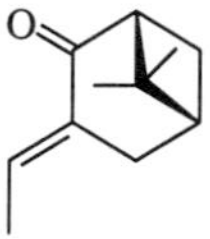 [structure], $TiCl_4$, CH_2Cl_2, −78°, 2 h	$CH_2{=}C(CH_3)CH_2CH(CH_3)$ 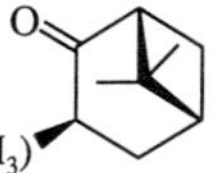 [structure] (81) *erythro*:*threo* 1:3	323
	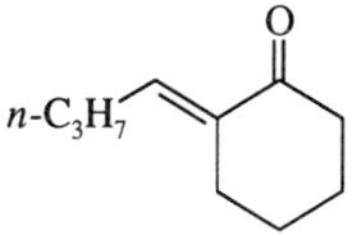 n-C_3H_7 [structure], $TiCl_4$, CH_2Cl_2, −78°, 30 min	$CH_2{=}C(CH_3)CH_2CH(C_3H_7$-$n)$ 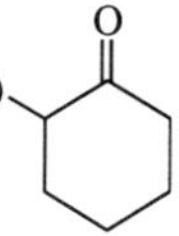 [structure] (67)	322
	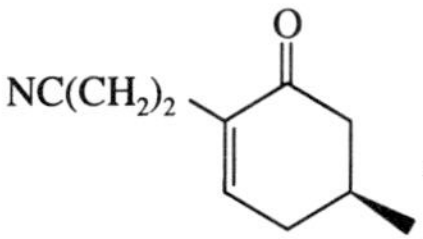 $NC(CH_2)_2$ [structure], $TiCl_4$, CH_2Cl_2, −78°, 1.25 h	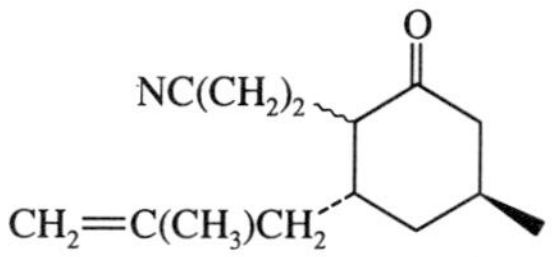 $NC(CH_2)_2$, $CH_2{=}C(CH_3)CH_2$ [structure] (96)	340
$(CH_3)_3SiCH_2C(CH_2OSi(CH_3)_3){=}CH_2$	", $TiCl_4$, CH_2Cl_2	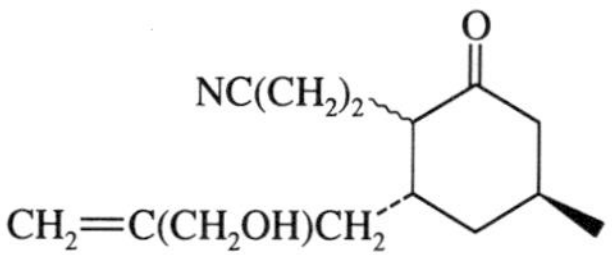 $NC(CH_2)_2$, $CH_2{=}C(CH_2OH)CH_2$ [structure] (94)	342

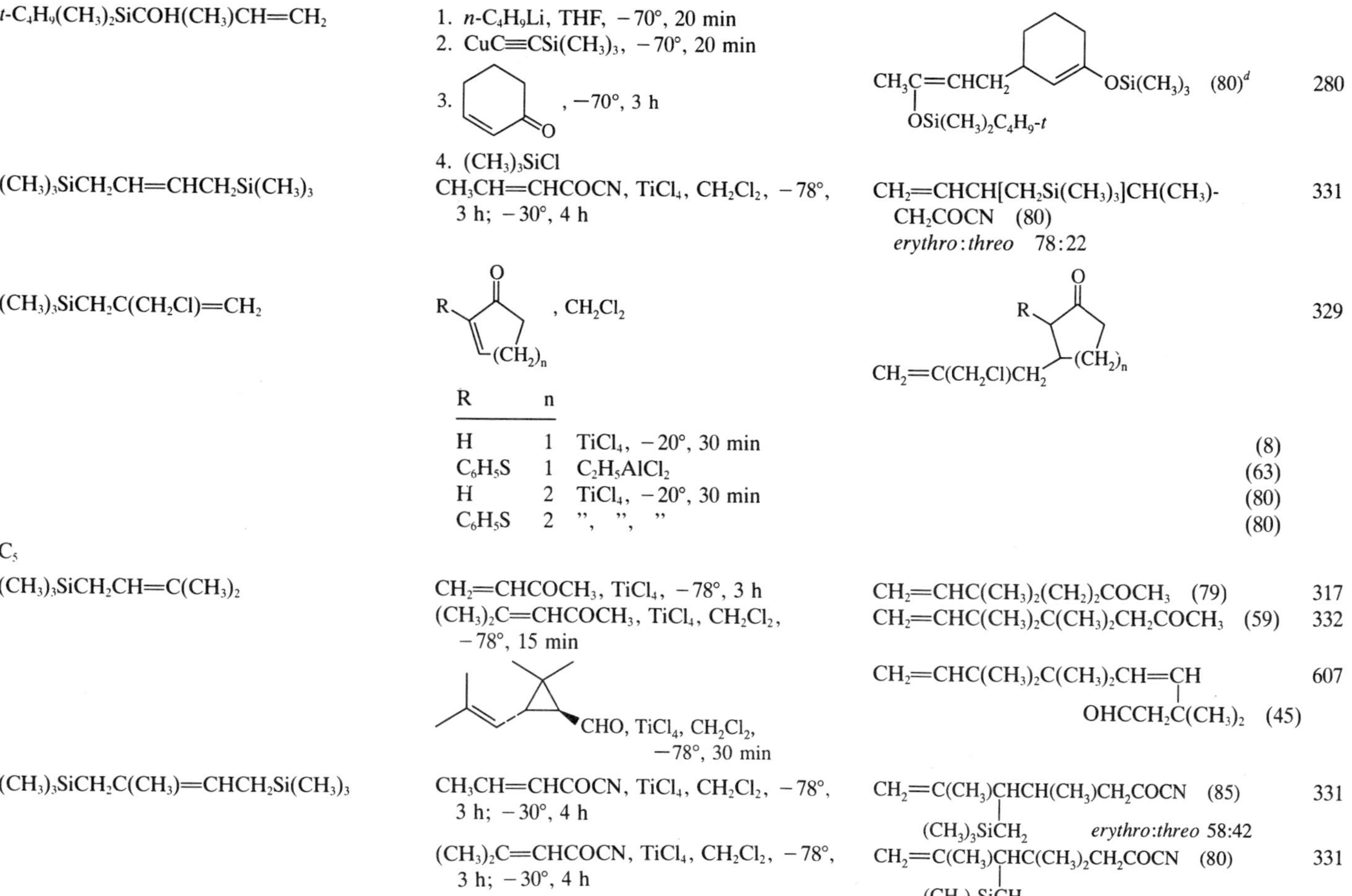

$t\text{-}C_4H_9(CH_3)_2SiCOH(CH_3)CH{=}CH_2$	1. $n\text{-}C_4H_9Li$, THF, −70°, 20 min 2. $CuC{\equiv}CSi(CH_3)_3$, −70°, 20 min 3. [cyclohexenone], −70°, 3 h 4. $(CH_3)_3SiCl$	$CH_3C(OSi(CH_3)_2C_4H_9\text{-}t){=}CHCH_2$–[cyclohexenyl]–$OSi(CH_3)_3$ (80)[d]	280
$(CH_3)_3SiCH_2CH{=}CHCH_2Si(CH_3)_3$	$CH_3CH{=}CHCOCN$, $TiCl_4$, CH_2Cl_2, −78°, 3 h; −30°, 4 h	$CH_2{=}CHCH[CH_2Si(CH_3)_3]CH(CH_3)$-$CH_2COCN$ (80) *erythro*:*threo* 78:22	331
$(CH_3)_3SiCH_2C(CH_2Cl){=}CH_2$	[2-R-cycloalkenone, $(CH_2)_n$], CH_2Cl_2	$CH_2{=}C(CH_2Cl)CH_2$–[2-R-cycloalkanone, $(CH_2)_n$]	329
	R = H, n = 1: $TiCl_4$, −20°, 30 min	(8)	
	R = C_6H_5S, n = 1: $C_2H_5AlCl_2$	(63)	
	R = H, n = 2: $TiCl_4$, −20°, 30 min	(80)	
	R = C_6H_5S, n = 2: ", ", "	(80)	
C_5			
$(CH_3)_3SiCH_2CH{=}C(CH_3)_2$	$CH_2{=}CHCOCH_3$, $TiCl_4$, −78°, 3 h	$CH_2{=}CHC(CH_3)_2(CH_2)_2COCH_3$ (79)	317
	$(CH_3)_2C{=}CHCOCH_3$, $TiCl_4$, CH_2Cl_2, −78°, 15 min	$CH_2{=}CHC(CH_3)_2C(CH_3)_2CH_2COCH_3$ (59)	332
	[2,2-dimethyl-3-(2-methylpropenyl)cyclopropyl]CHO, $TiCl_4$, CH_2Cl_2, −78°, 30 min	$CH_2{=}CHC(CH_3)_2C(CH_3)_2CH{=}CH$–$OHCCH_2C(CH_3)_2$ (45)	607
$(CH_3)_3SiCH_2C(CH_3){=}CHCH_2Si(CH_3)_3$	$CH_3CH{=}CHCOCN$, $TiCl_4$, CH_2Cl_2, −78°, 3 h; −30°, 4 h	$CH_2{=}C(CH_3)CH[CH_2Si(CH_3)_3]CH(CH_3)CH_2COCN$ (85) *erythro*:*threo* 58:42	331
	$(CH_3)_2C{=}CHCOCN$, $TiCl_4$, CH_2Cl_2, −78°, 3 h; −30°, 4 h	$CH_2{=}C(CH_3)CH[CH_2Si(CH_3)_3]C(CH_3)_2CH_2COCN$ (80)	331

TABLE XIV. ALLYLSILANES WITH α,β-UNSATURATED CARBONYL COMPOUNDS AND α,β-UNSATURATED NITRILES (*Continued*)

Reactant	Conditions	Product(s) and Yield(s)	Refs.
$(CH_3)_3SiCH_2CH{=}CHCH{=}CH_2$	$(CH_3)_2C{=}CHCOCH_3$, $TiCl_4$, CH_2Cl_2, $-40°$, 5 min	$CH_2{=}CHCH{=}CHCH_2C(CH_3)_2CH_2COCH_3$ (35) + [cyclohexene with gem-dimethyl, $COCH_3$, $(CH_3)_3SiCH_2$] (24)	282
$(CH_3)_3Si$–[2-cyclopentenyl]	$(CH_3)_2C{=}CHCOCH_3$, $TiCl_4$, CH_2Cl_2, $-15°$, 2 h	[2-cyclopentenyl]–$C(CH_3)_2CH_2COCH_3$ (78)	643
	[2-methylenecyclopentanone], $TiCl_4$, CH_2Cl_2	[2-(2-cyclopentenylmethyl)-4,4-dimethylcyclopentanone] (45)	341
	$C_6H_5CH{=}CHCOCH_3$, $TiCl_4$, CH_2Cl_2, $-15°$, 2 h	[2-cyclopentenyl]–$CH(C_6H_5)CH_2COCH_3$ (77)	643
	$C_6H_5CH{=}CHCOC_6H_5$, $TiCl_4$, CH_2Cl_2, $-15°$, 2 h	[2-cyclopentenyl]–$CH(C_6H_5)CH_2COC_6H_5$ (84)	643
$t\text{-}C_4H_9(CH_3)_2SiCOH(C_2H_5)CH{=}CH_2$	1. $n\text{-}C_4H_9Li$, THF, $-70°$, 20 min 2. $CuC{\equiv}CSi(CH_3)_3$, $-70°$, 20 min 3. $n\text{-}C_3H_7CH{=}CHCOCH_3$ 4. $(CH_3)_3SiCl$	$C_2H_5C(OSi(CH_3)_2C_4H_9\text{-}t){=}CHCH_2CH(C_3H_7\text{-}n)CH{=}C(OSi(CH_3)_3)CH_3$ (51)	280

Reactant	Conditions	Product(s) and Yield(s) (%)	Refs.
	1. n-C_4H_9Li, THF, −70°, 20 min 2. $CuC{\equiv}CSi(CH_3)_3$, −70°, 20 min 3. 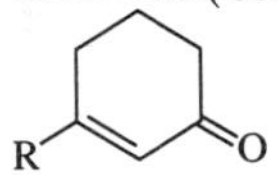 R (cyclohexenone), O 4. $(CH_3)_3SiCl$	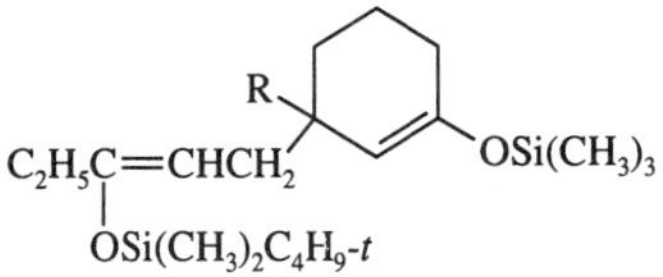 $C_2H_5C{=}CHCH_2$ (R) $OSi(CH_3)_3$ $OSi(CH_3)_2C_4H_9$-t R = H (92) R = CH_3 (77)	280
C_6			
$(CH_3)_3SiCH_2C(CH_3){=}C(CH_3)CH_2Si(CH_3)_3$	$CH_3CH{=}CHCOCN$, $TiCl_4$, CH_2Cl_2, −78°, 3 h; −30°, 4 h	$CH_2{=}C(CH_3)C(CH_3)CH(CH_3)CH_2COCN$ (80) $(CH_3)_3SiCH_2$ *erythro:threo* 52:48	331
	$(CH_3)_2C{=}CHCOCN$, $TiCl_4$, CH_2Cl_2, −78°, 3 h; −30°, 4 h	$CH_2{=}C(CH_3)C(CH_3)C(CH_3)_2CH_2COCN$ (85) $(CH_3)_3SiCH_2$	331
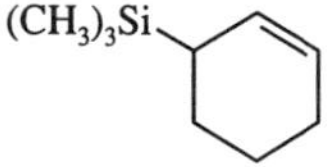 $(CH_3)_3Si$	$CH_2{=}CHCOSi(CH_3)_2R$, $TiCl_4$, CH_2Cl_2, −78°, 15 min R = CH_3 R = C_4H_9-t	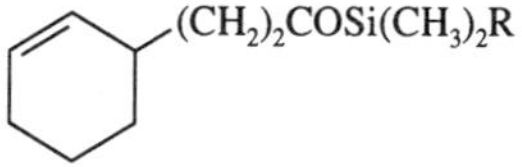 $(CH_2)_2COSi(CH_3)_2R$ (61) (73)	332
	$(CH_3)_2C{=}CHCOCH_3$, $TiCl_4$, CH_2Cl_2, −78°, 15 min	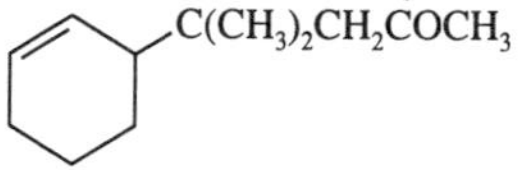 $C(CH_3)_2CH_2COCH_3$ (75)	332
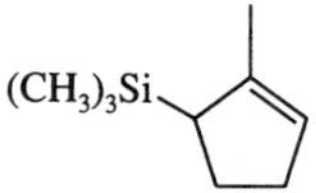 $(CH_3)_3Si$	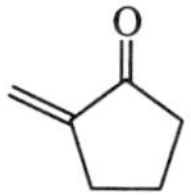 O, $TiCl_4$, CH_2Cl_2, −78°, 30 min	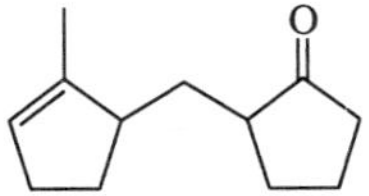 O (71)[e]	341
C_7			
$(CH_3)_3SiCH_2C(C_4H_9$-$t){=}CH_2$	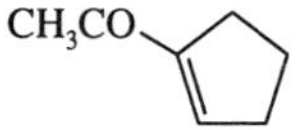 CH_3CO, $TiCl_4$, CH_2Cl_2, −78°, 1 h	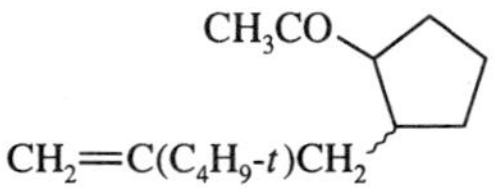 CH_3CO (85) $CH_2{=}C(C_4H_9$-$t)CH_2$	320

TABLE XIV. ALLYLSILANES WITH α,β-UNSATURATED CARBONYL COMPOUNDS AND α,β-UNSATURATED NITRILES (*Continued*)

Reactant	Conditions	Product(s) and Yield(s)	Refs.
C_8			
$(CH_3)_3SiCH_2C(=CH_2)CH_2CR^1_2CH{=}CHR^2$	TBAF, DMF, HMPA, 3–12 h[f]	4-methylene-1-(CH_2R^2)-2,2-(R^1)cyclopentane	163
R^1 / R^2			
H / CN		(59)	
" / $CON(C_2H_5)_2$		(0)	
" / $CO_2CH_2C_6H_5$		(66)	
$-S(CH_2)_3S-$ / CN		(81)	
" / $CO_2C_2H_5$		(52)	
" / $CON(C_2H_5)_2$		(56)	
$(CH_3)_3SiCH_2CH{=}$(cyclohexylidene)	$CH_2{=}CHCOCH_3$, $TiCl_4$, CH_2Cl_2, −78°, 1 h	1-($CH_2{=}CH$)-1-($(CH_2)_2COCH_3$)cyclohexane (55)	170
$[(CH_3)_3SiC(OMgBr)(CH{=}CH_2)CH(C_2H_5)_2]$[g]	1. $CuC{\equiv}CSi(CH_3)_3$, −20°, 20 min 2. 2-cyclohexen-1-one, −20°, 12 h	$(C_2H_5)_2CHC(OSi(CH_3)_3){=}CHCH_2$-(3-oxocyclohexyl) (86)	280
C_9			
$(CH_3)_3SiCH_2C(=CH_2)CH_2CR^1R^2CH{=}CHCOCH_3$	TBAF, DMF, HMPA, 3–12 h[f]	4-methylene-1-(CH_2COCH_3)-2-(R^1)(2-R^2)cyclopentane	163
R^1 / R^2			
H / H		(40)	
$-S(CH_2)_3S-$		(67)	

Reactant	Conditions	Product(s) and Yield(s) (%)	Refs.
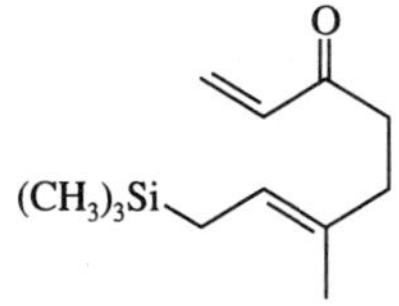 $(CH_3)_3Si$	$BF_3\cdot O(C_2H_5)_2$, $(C_2H_5)_2O$	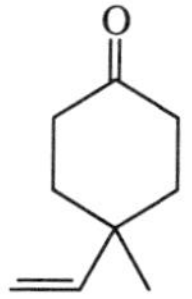 (73)	181
$(CH_3)_3Si$	$TiCl_4$, CH_2Cl_2, $-78°$, 2 h $C_2H_5AlCl_2$, toluene, 0°, 2 h	(50) (84)	163
	TBAF, DMF, HMPA, 3–12 h	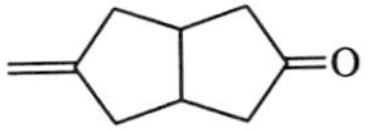 (64)	
$(CH_3)_3SiCH_2CH{=}CHC_6H_5$	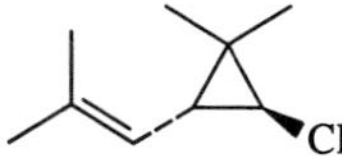 CHO, $TiCl_4$, CH_2Cl_2, $-78°$, 30 min	$CH_2{=}CHCH(C_6H_5)C(CH_3)_2CH{=}CH$ — $OHCCH_2C(CH_3)_2$ (33)	607
C_{10}			
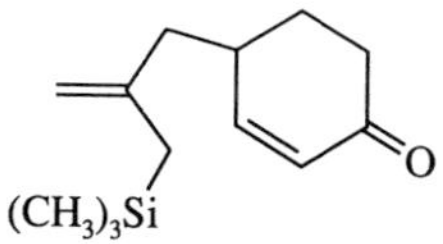 $(CH_3)_3Si$	$TiCl_4$, CH_2Cl_2, $-78°$, 2 h	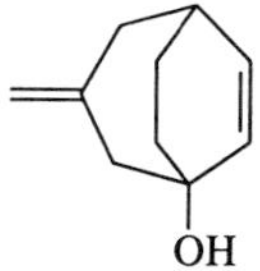 OH (49)	163
	$C_2H_5AlCl_2$, toluene, 0°, 2 h	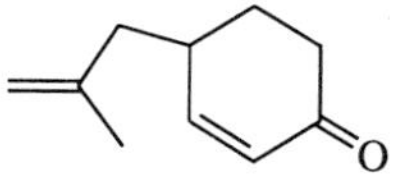 (30)	163

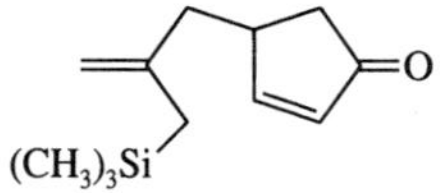

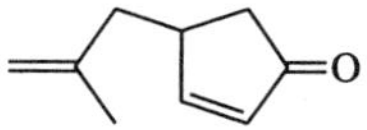

TABLE XIV. ALLYLSILANES WITH α,β-UNSATURATED CARBONYL COMPOUNDS AND α,β-UNSATURATED NITRILES (*Continued*)

Reactant	Conditions	Product(s) and Yield(s)	Refs.
	TBAF, DMF, HMPA, 3–12 h	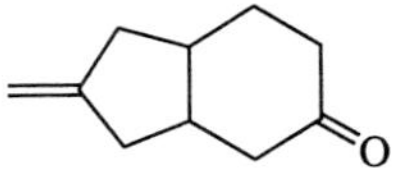 (69)	163
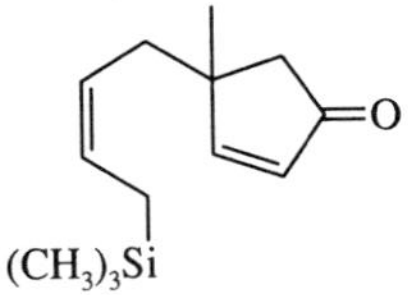 $(CH_3)_3Si$	$TiCl_4$	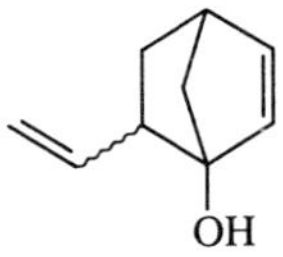 OH (45)	337
	TBAF	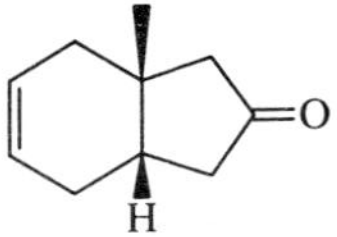 H (85)	337
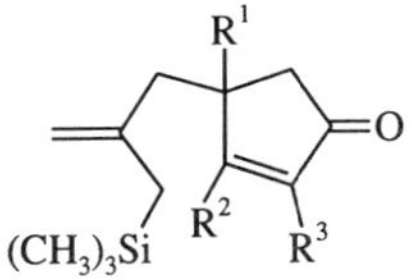 $(CH_3)_3Si$, R^1, R^2, R^3 I	**A** $TiCl_4$, CH_2Cl_2, −78°, 2 h **B** $C_2H_5AlCl_2$, toluene, 0°, 2 h **C** TBAF, DMF, HMPA, 3–12 h	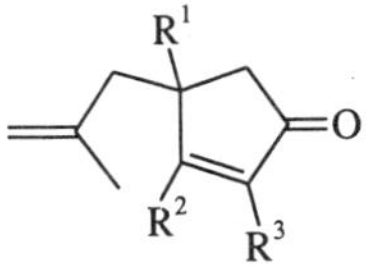 II 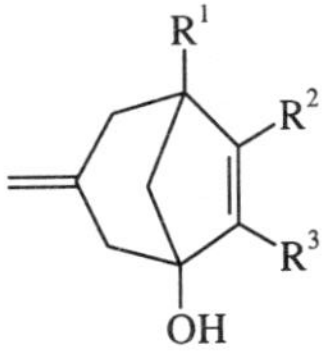 + III 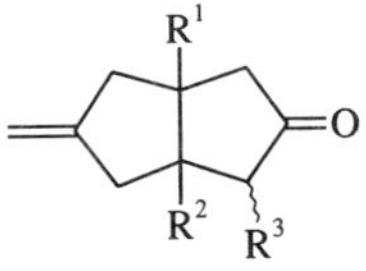 (OH) + IV	163

R^1	R^2	R^3	Conditions	I	II	III	IV
CH_3	H	H	**A**	(0)	(60)	(0)	(0)
"	"	"	**B**	(5)	(75)	(0)	(0)
"	"	"	**C**	(0)	(0)	(0)	(55)
H	CH_3	H	**A**	(0)	(0)	(33)	(0)
"	"	"	**B**	(0)	(92)	(0)	(0)
"	"	"	**C**	(0)	(29)	(0)	(15)
"	H	CH_3	**A**	(0)	(95)	(0)	(0)
"	"	"	**B**	(50)	(40)	(0)	(0)
"	"	"	**C**	(0)	(0)	(4)	(40)

t-$C_4H_9(CH_3)_2SiCOH(C_7H_{15}$-$n)CH{=}CH_2$

1. n-C_4H_9Li, −70°, 20 min
2. $CuC{\equiv}CSi(CH_3)_3$, −70°, 20 min
3. E^+, −70°, 3 h

A 4. H_2O or **B** 4. $(CH_3)_3SiCl$

Product: t-$C_4H_9(CH_3)_2SiO$–C(n-C_7H_{15})$=CHCH_2R$ — 280

E^+	Method	R	
$HC{\equiv}CCO_2CH_3$	**A**	$CH{=}CHCO_2CH_3$	(75)
$CH_2{=}CHCOCH_3$	**A**	$(CH_2)_2COCH_3$	(62)
n-$C_3H_7CH{=}CHCOCH_3$	**B**	$CH(C_3H_7$-$n)CH{=}C(CH_3)OSi(CH_3)_3$	(61)
2-cyclohexenone	**B**	3-[$OSi(CH_3)_3$]-2-cyclohexenyl	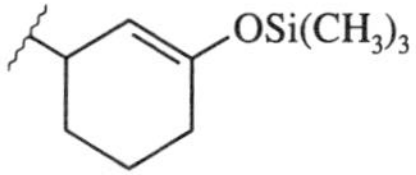 (82)
3-methyl-2-cyclohexenone	**B**	1-methyl-3-[$OSi(CH_3)_3$]-2-cyclohexenyl	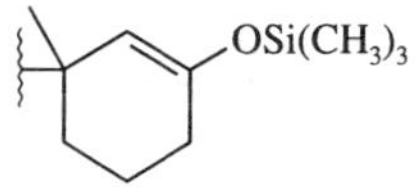 (88)

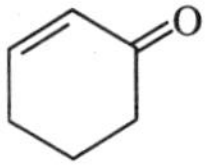

C_{11}

4-[(Z)-6-(trimethylsilyl)-4-hexenyl]-2-cyclohexenone, $(CH_3)_3Si$

Conditions	Product		Ref.
$C_2H_5AlCl_2$, toluene, C_6H_{14}, 0°, 30 min	1-vinyl-hexahydroindan-5-one, α:β 1:1	(84)	661
", toluene, C_6H_{14}, −78°	" " 4:1	(92)	661

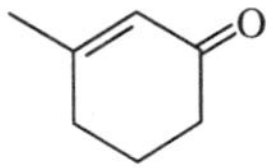

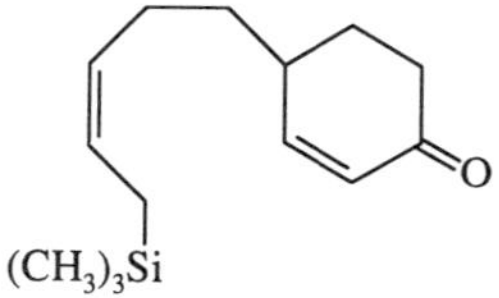

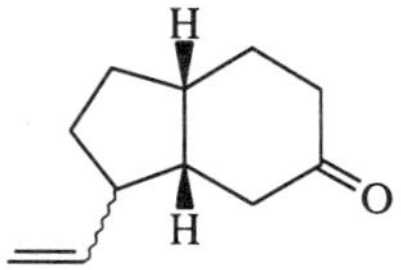

TABLE XIV. ALLYLSILANES WITH α,β-UNSATURATED CARBONYL COMPOUNDS AND α,β-UNSATURATED NITRILES (*Continued*)

Reactant	Conditions	Product(s) and Yield(s)	Refs.
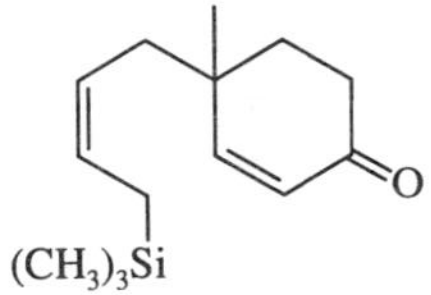 $(CH_3)_3Si$	$TiCl_4$	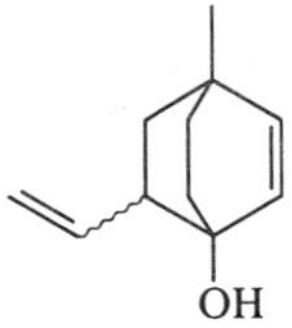 OH (40)	337
	TBAF	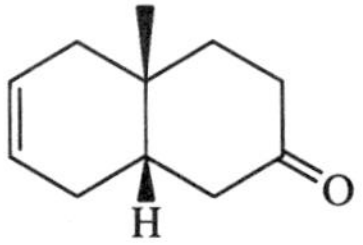 H (76)	337
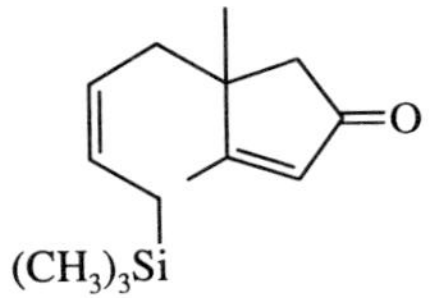 $(CH_3)_3Si$	$TiCl_4$	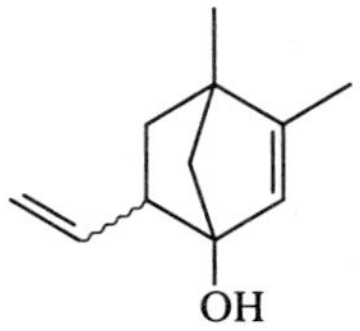 OH (30)	337
	TBAF	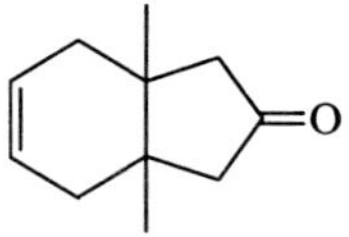 (7)[h]	337

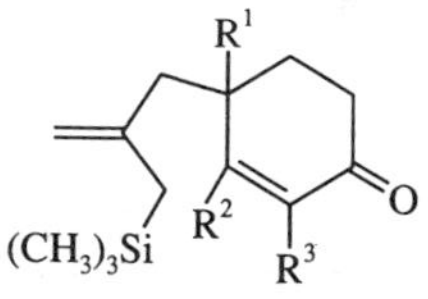

I

A $TiCl_4$, CH_2Cl_2, −78°, 2 h
B $C_2H_5AlCl_2$, toluene, 0°, 2 h
C TBAF, DMF, HMPA, 3–12 h

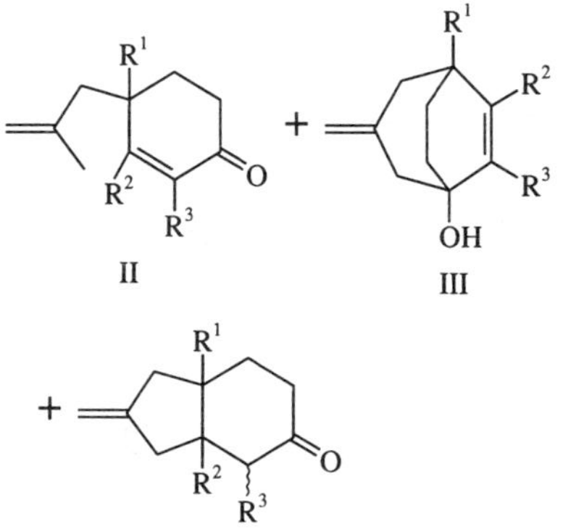

II + III + IV

163

R^1	R^2	R^3	Conditions	I	II	III	IV
CH_3	H	H	**A**	(0)	(0)	(24)	(0)
"	"	"	**B**	(25)	(10)	(0)	(0)
"	"	"	**C**	(0)	(0)	(0)	(63)
H	CH_3	H	**A**	(0)	(92)	(0)	(0)
"	"	"	**B**	(5)	(70)	(0)	(0)
"	"	"	**C**	(0)	(0)	(0)	(46)
"	H	CH_3	**A**	(0)	(60)	(34)	(0)
"	"	"	**B**	(16)	(50)	(0)	(0)
"	"	"	**C**	(0)	(0)	(0)	(57)

TABLE XIV. ALLYLSILANES WITH α,β-UNSATURATED CARBONYL COMPOUNDS AND α,β-UNSATURATED NITRILES (*Continued*)

Reactant	Conditions	Product(s) and Yield(s)	Refs.
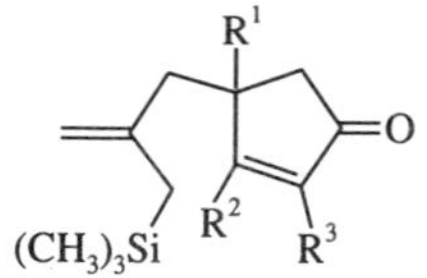 I	**A** $TiCl_4$, CH_2Cl_2, −78°, 2 h **B** $C_2H_5AlCl_2$, toluene, 0°, 2 h **C** TBAF, DMF, HMPA, 3–12 h	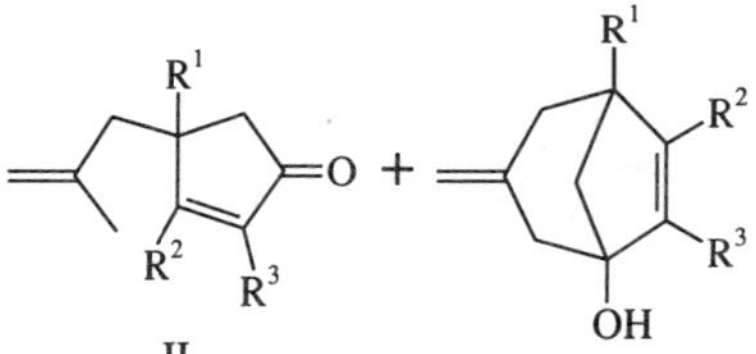 II + III 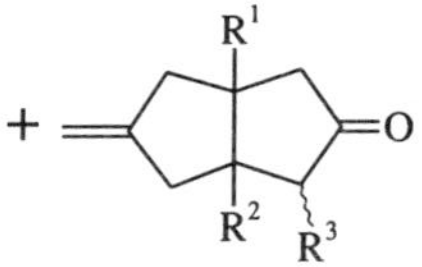 + IV	163
$(CH_3)_3Si$	$C_2H_5AlCl_2$	(41)	339

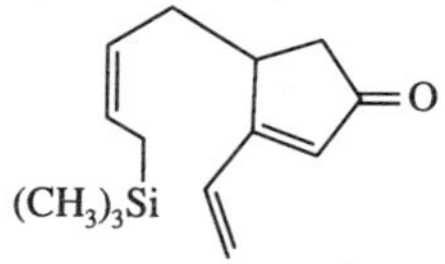

R^1	R^2	R^3	Conditions	I	II	III	IV
CH_3	CH_3	H	**A**	(0)	(0)	(86)	(0)
"	"	"	**B**	(20)	(65)	(0)	(0)
"	"	"	**C**	(0)	(0)	(45)	(10)
"	H	CH_3	**A**	(0)	(0)	(47)[i]	(0)
"	"	"	**B**	(15)	(70)	(0)	(0)
"	"	"	**C**	(0)	(0)	(10)	(40)
H	CH_3	CH_3	**A**	(0)	(86)	(0)	(0)
"	"	"	**B**	(55)	(20)	(0)	(0)
"	"	"	**C**	(0)	(22)	(22)	(0)

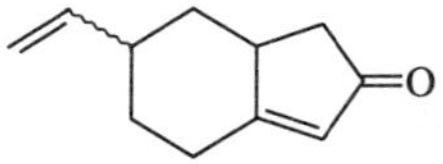

Reactant	Conditions	Product(s) and Yield(s) (%)	Refs.
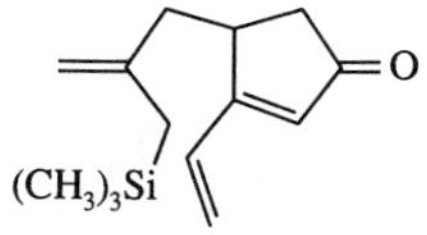	TBAF	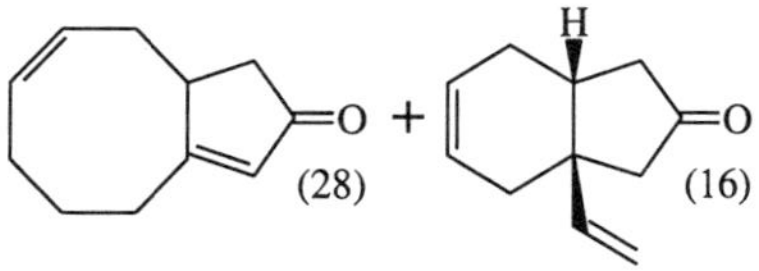 (28) + (16)	339
$(CH_3)_3Si$	$C_2H_5AlCl_2$	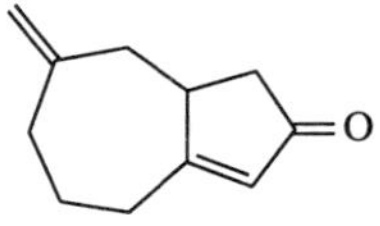 (46) I	338
	TBAF	I (2) + 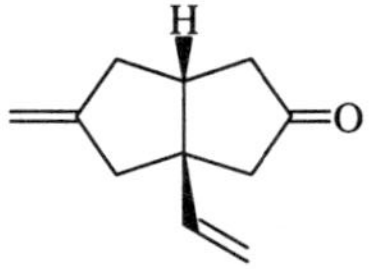 (2)	338
C_{12} $(CH_3)_3Si$	$C_2H_5AlCl_2$	+	226
	CH_2Cl_2, $-78°$	2:1 (85)	
	toluene, $-78°$	7:1 (72)	
	", 0°	3:1 (77)	
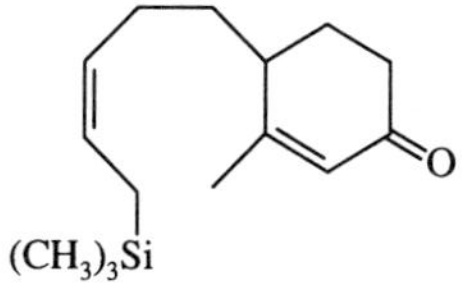 $(CH_3)_3Si$	$C_2H_5AlCl_2$, toluene, C_6H_{14}, 30 min	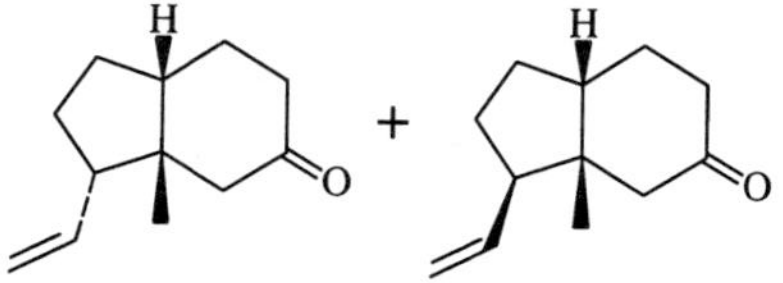 +	661
	0°	2:1 (91)	
	$-30°$	3:1 (77)	

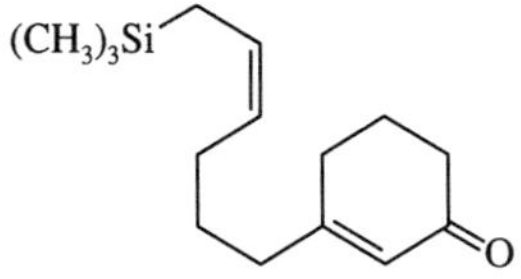

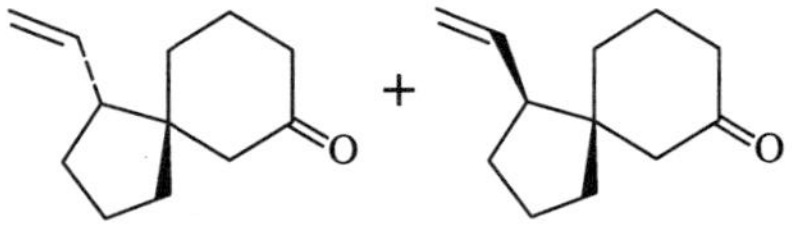

TABLE XIV. ALLYLSILANES WITH α,β-UNSATURATED CARBONYL COMPOUNDS AND α,β-UNSATURATED NITRILES (*Continued*)

Reactant	Conditions	Product(s) and Yield(s)	Refs.
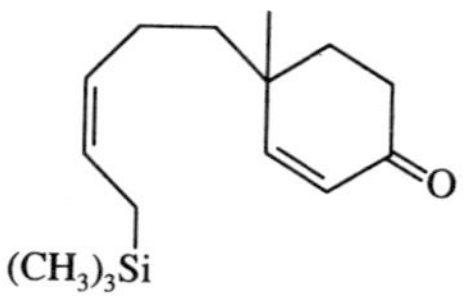 $(CH_3)_3Si$	$C_2H_5AlCl_2$	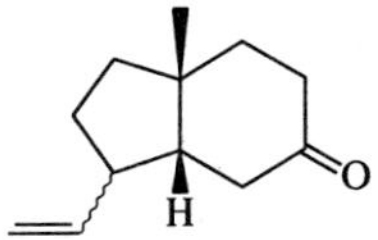 α:β 1:3 (62)	336, 337
	TBAF	" α:β 1:1 (77)	336, 337
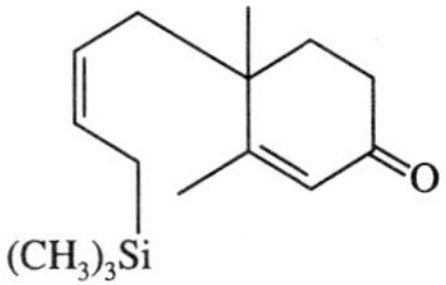 $(CH_3)_3Si$	$TiCl_4$	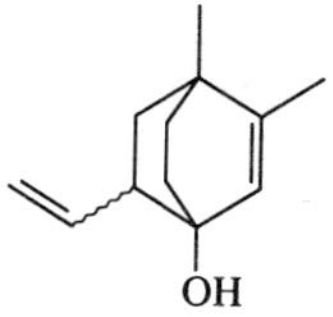 OH (30)[j]	337, 336
	TBAF	" (40) + 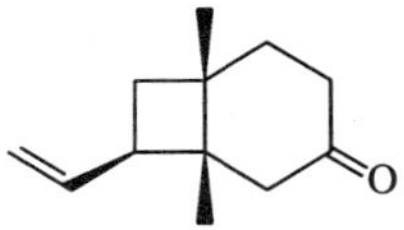 (30)[j]	337, 336
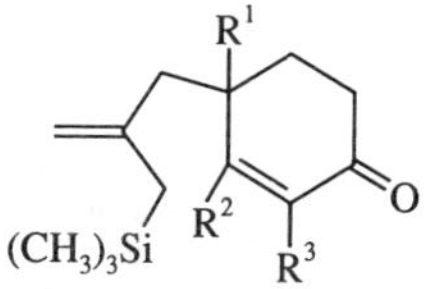 $(CH_3)_3Si$ I	**A** $TiCl_4$, CH_2Cl_2, −78°, 2 h **B** $C_2H_5AlCl_2$, toluene, 0°, 2 h **C** TBAF, DMF, HMPA, 3–12 h	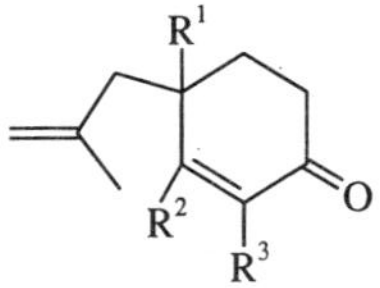 II 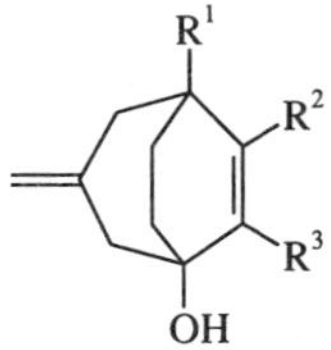 + III 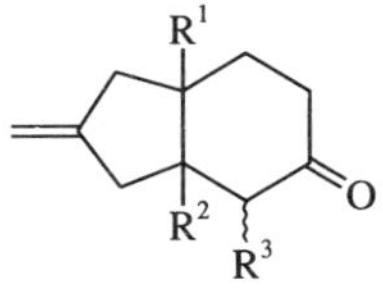 + IV	163

R^1	R^2	R^3	Conditions	I	II	III	IV
CH_3	CH_3	H	**A**	(0)	(10)	(41)	(0)
"	"	"	**B**	(12)	(10)	(40)	(20)
"	"	"	**C**	(0)	(0)	(0)	(65)
"	H	CH_3	**A**	(0)	(60)	(0)	(0)
"	"	"	**B**	(5)	(70)	(0)	(0)
"	"	"	**C**	(0)	(0)	(6)	(55)
H	CH_3	"	**A**	(0)	(20)	(50)	(0)
"	"	"	**B**	(50)	(20)	(0)	(0)
"	"	"	**C**	(0)	(10)	(10)	(43)

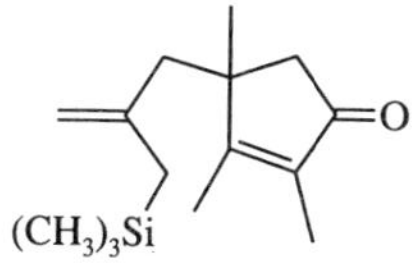

$(CH_3)_3Si$

O

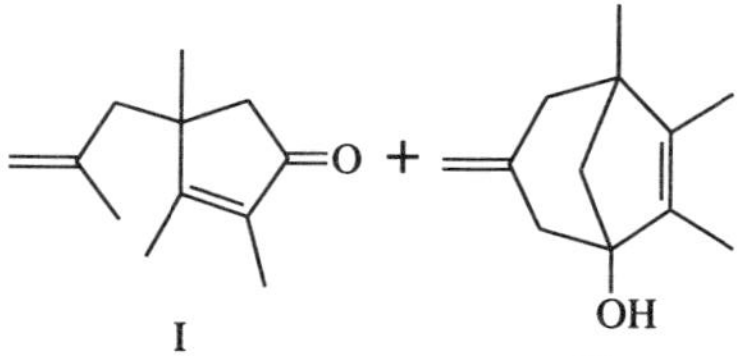

O + OH

I II

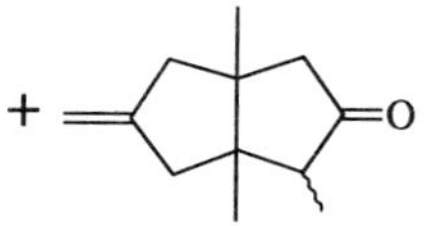

+ O

III

163

Conditions	I	II	III
$TiCl_4$, CH_2Cl_2, −78°, 2 h	(20)	(46)	(0)
$C_2H_5AlCl_2$, toluene, 0°, 2 h	(80)	(0)	(0)
TBAF, DMF, HMPA, 3–12 h	(0)	(54)	(2)

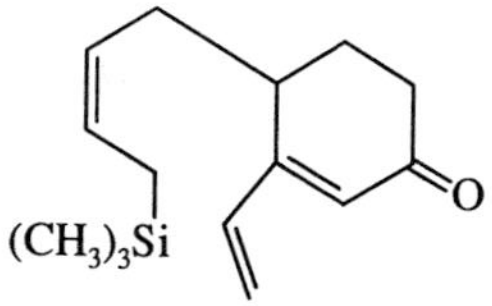

$(CH_3)_3Si$ O

$C_2H_5AlCl_2$

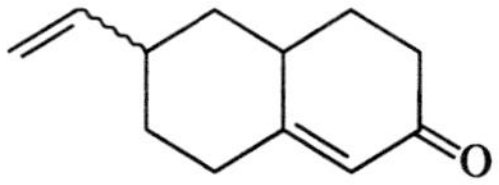

O (68)

339

TABLE XIV. ALLYLSILANES WITH α,β-UNSATURATED CARBONYL COMPOUNDS AND α,β-UNSATURATED NITRILES (*Continued*)

Reactant	Conditions	Product(s) and Yield(s)	Refs.
	TBAF	(12) + (22) OH	339
$(CH_3)_3Si$	$C_2H_5AlCl_2$	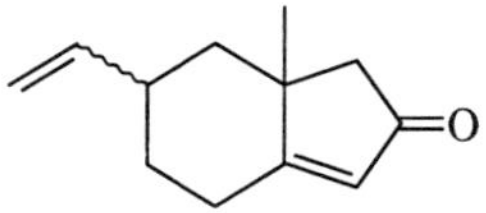 (60)	339
	TBAF	(33) + (23)	339
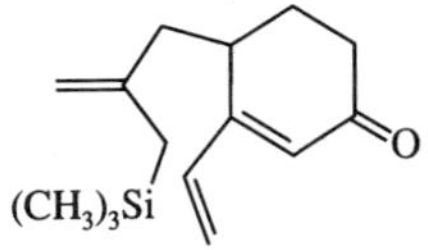 $(CH_3)_3Si$	$C_2H_5AlCl_2$	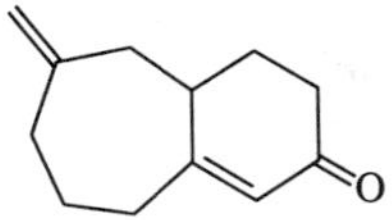 (49)	338

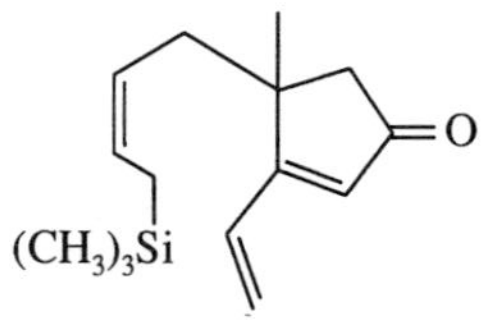

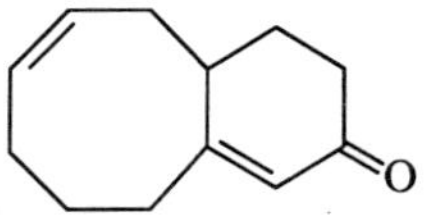

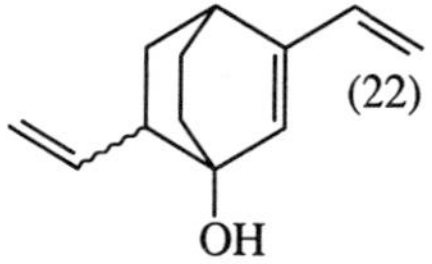

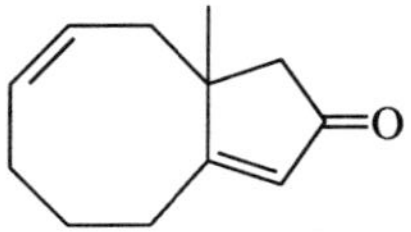

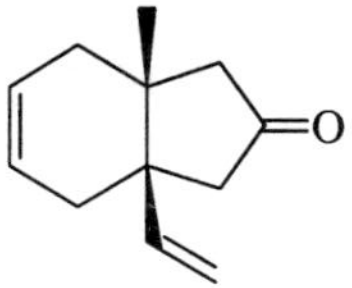

Reactant	Conditions	Product(s) and Yield(s) (%)	Refs.
	TBAF	" (20) + [hydrindanone structure] (11)	338
$(CH_3)_3Si$ [2-vinylcyclopentenone structure with R^1, R^2]		I + II + III	338

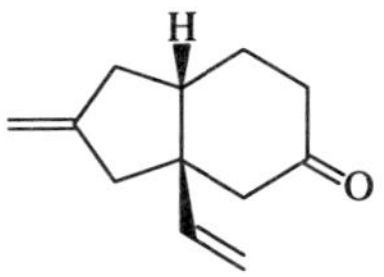

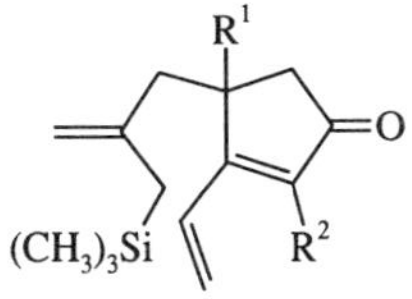

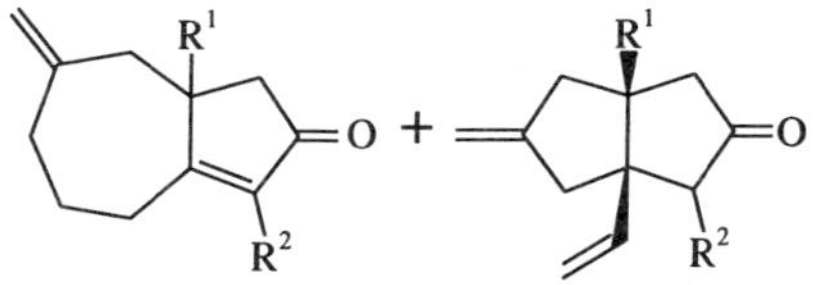

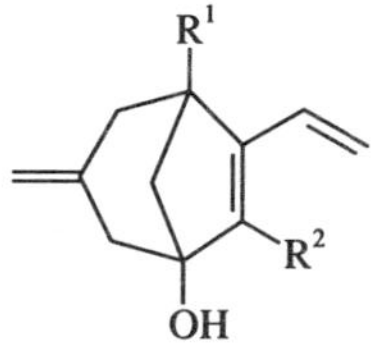

R^1	R^2	Conditions	I	II	III
CH_3	H	$C_2H_5AlCl_2$	(70)	(0)	(0)
"	"	TBAF	(0)	(28)	(33)
H	CH_3	$C_2H_5AlCl_2$	(65)	(0)	(0)
"	"	TBAF	(25)	(0)	(23)

Reactant	Conditions	Product(s) and Yield(s) (%)	Refs.
$[(CH_3)_3SiC(CH{=}CH_2)CH(C_2H_5)C_6H_5$ (OMgBr) $]^g$	1. $CuC{\equiv}CSi(CH_3)_3$, −20°, 20 min 2. E^+, −20°, 2–12 h	$C_6H_5CH(C_2H_5)C(OSi(CH_3)_3){=}CHCH_2R$	280

E^+	R	
n-$C_4H_9CH{=}CHCHO$	$CH(C_4H_9$-$n)CH_2CHO$	(49)
$CH_2{=}CHCO_2C_2H_5$	$(CH_2)_2CO_2C_2H_5$	(58)

TABLE XIV. ALLYLSILANES WITH α,β-UNSATURATED CARBONYL COMPOUNDS AND α,β-UNSATURATED NITRILES (*Continued*)

Reactant	Conditions	Product(s) and Yield(s)	Refs.
	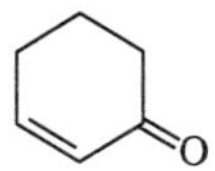	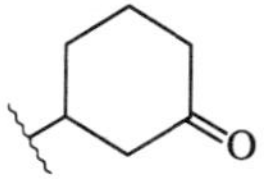 (84)	
	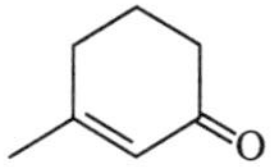	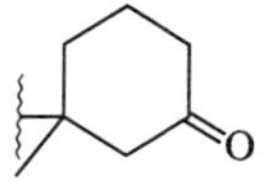 (74)	
	n-$C_3H_7CH{=}CHCOCH_3$	$CH(C_3H_7$-$n)CH_2COCH_3$ (64)	
	n-$C_4H_9CH{=}CHCO_2C_2H_5$	$CH(C_4H_9$-$n)CH_2CO_2C_2H_5$ (49)	
C_{13}			
$(CH_3)_3Si$		+	226
	$C_2H_5AlCl_2$, toluene, 0°	5:1 (75)	
	", ", −78°	7.5:1 (70)	
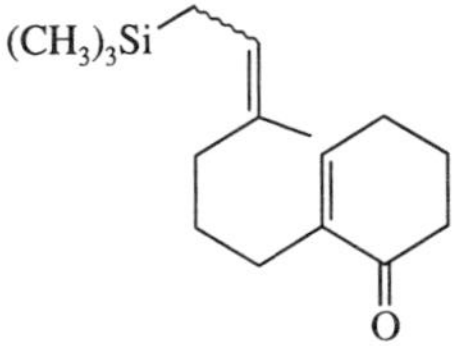 $(CH_3)_3Si$	$TiCl_4$, CH_2Cl_2, −78°	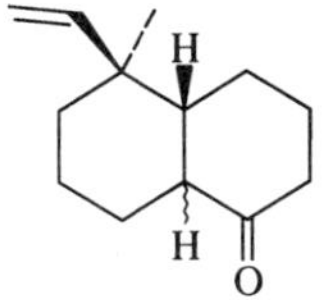 H H *trans:cis* 3:2 (90)	335

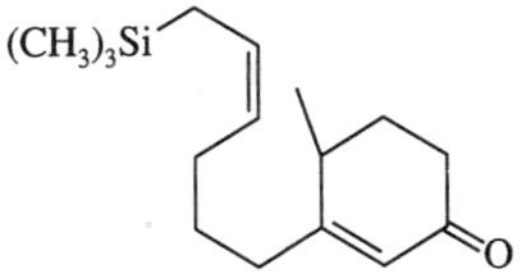

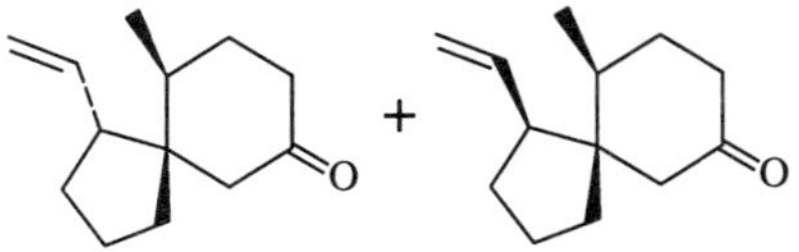

Reactant	Conditions	Product(s) and Yield(s) (%)	Refs.
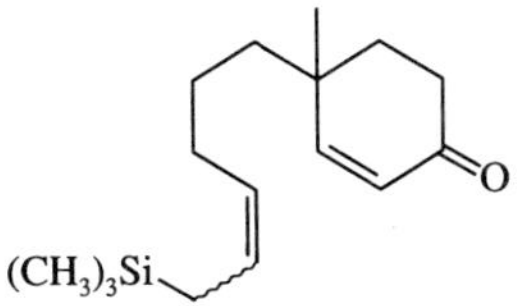 $(CH_3)_3Si$		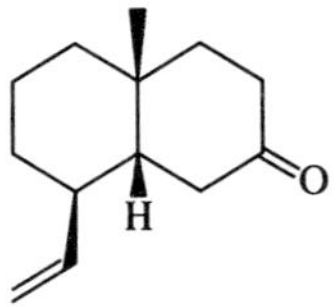	336, 337
E	$TiCl_4$	(77)	
"	TBAF	(0)[k]	
Z	$TiCl_4$	(86)	
"	TBAF	(0)[k]	
$(CH_3)_3Si$; R^1, R^2, R^3		R^1, R^2, R^3 + R^1, R^2, R^3	

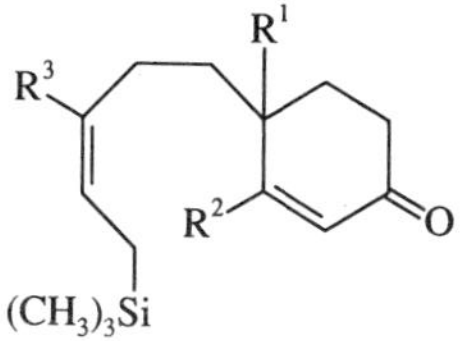

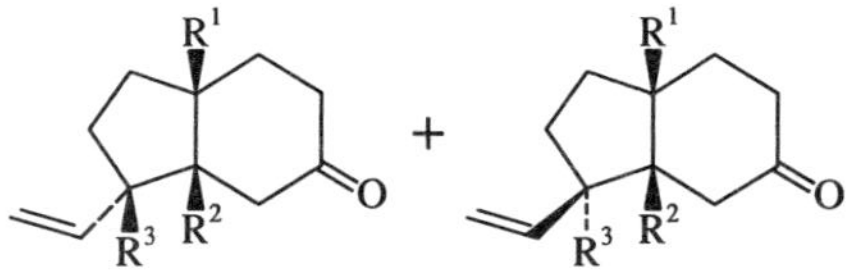

R^1	R^2	R^3	Conditions	Ratio (Yield)	Refs.
H	C_2H_5	H	$C_2H_5AlCl_2$, toluene, C_6H_{14}, 0°, 30 min	2:1 (86)	661
"	"	"	", −30°, 30 min	2.5:1 (90)	661
CH_3	CH_3	"	$TiCl_4$	3–4:1 (77)	336, 337
"	"	"	TBAF	1:6 (82)	336, 337
"	H	CH_3	$C_2H_5AlCl_2$	5:1 (71)	336, 337
"	"	"	TBAF	1:4 (88)	336, 337

Reactant	Conditions	Product and Yield (%)	Refs.
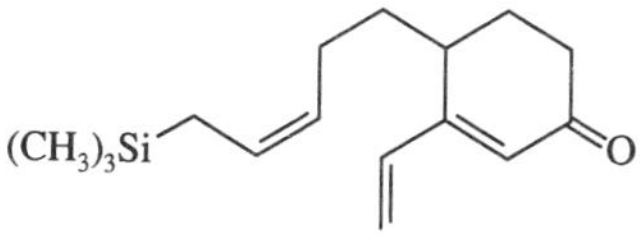 $(CH_3)_3Si$	$C_2H_5AlCl_2$, toluene, −78°	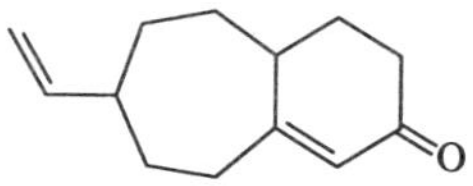 (60)	662

TABLE XIV. ALLYLSILANES WITH α,β-UNSATURATED CARBONYL COMPOUNDS AND α,β-UNSATURATED NITRILES (*Continued*)

Reactant			Conditions	Product(s) and Yield(s)					Refs.
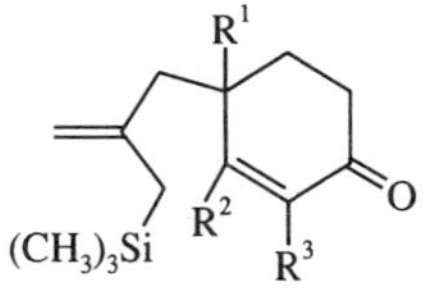 $(CH_3)_3Si$ … I				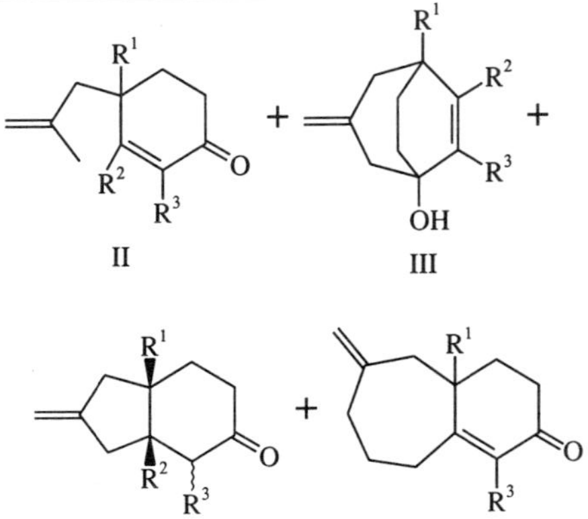 II + III + IV + V					
R^1	R^2	R^3		I	II	III	IV	V	
CH_3	CH_3	CH_3	$TiCl_4$, CH_2Cl_2, $-78°$, 2 h	(0)	(16)	(74)	(0)	(0)	163
"	"	"	$C_2H_5AlCl_2$, toluene, 0°, 2 h	(8)	(60)	(0)	(0)	(0)	163
"	"	"	TBAF, DMF, HMPA, 3–12 h	(0)	(0)	(19)	(32)	(0)	163
"	$CH_2{=}CH$	H	$C_2H_5AlCl_2$	—	—	(0)	(0)	(55)	338
"	"	"	TBAF	—	—	(5)	(45)	(0)	338
H	"	CH_3	$C_2H_5AlCl_2$	—	—	(0)	(0)	(70)	338
"	"	"	TBAF	—	—	(0)	(0)	(72)	338

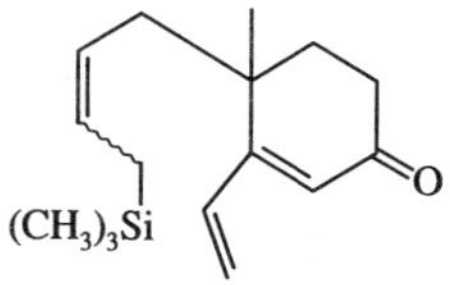

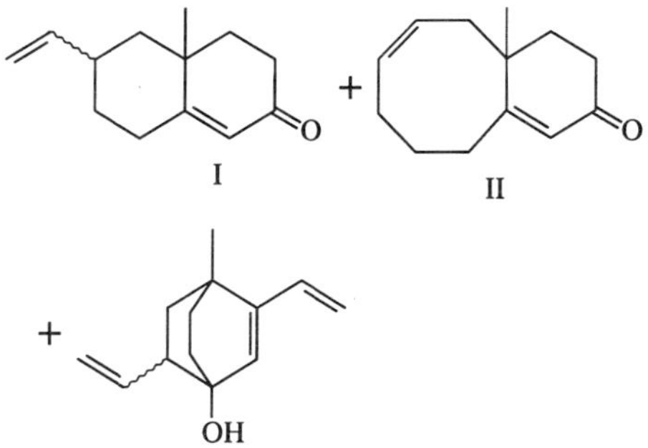

(CH$_3$)$_3$Si

I + II 339

+ OH

III

Substrate	Reagent	I	II	III
Z	$C_2H_5AlCl_2$	(77)	(0)	(0)
E:Z 1:4	”	(75)	(0)	(0)
Z	TBAF	(0)	(35)	(32)
E:Z 1:4	”	(0)	(22)	(32)[l]

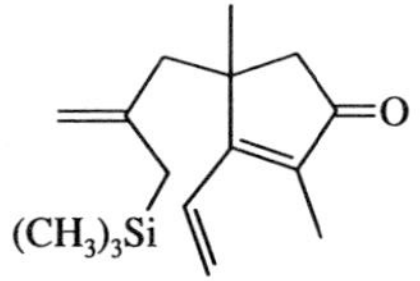

(CH$_3$)$_3$Si — $C_2H_5AlCl_2$ 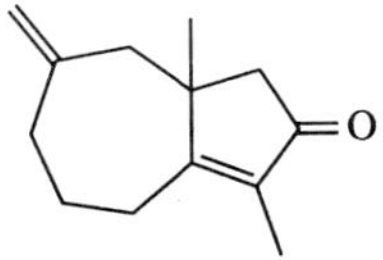 — (57) 338

TBAF — OH (96) 338

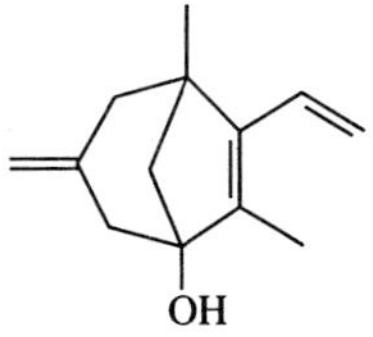

TABLE XIV. ALLYLSILANES WITH α,β-UNSATURATED CARBONYL COMPOUNDS AND α,β-UNSATURATED NITRILES (*Continued*)

Reactant	Conditions	Product(s) and Yield(s)	Refs.
$[(CH_3)_3SiC(OMgBr)(CH=CH_2)CH(C_3H_7\text{-}i)C_6H_5]$ [g]	1. $CuC\equiv CSi(CH_3)_3$, THF, −20°, 20 min 2. E^+, −20°, 2–12 h	$C_6H_5CH(C_3H_7\text{-}i)C(OSi(CH_3)_3)=CHCH_2R$	280
	E^+	R	
	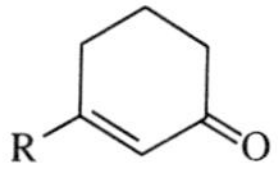 3-R-cyclohex-2-enone (structure)	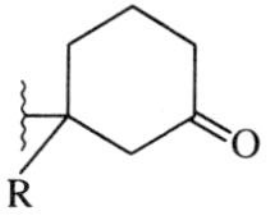 3-R-3-substituted cyclohexanone (structure) R = H (88); R = CH_3 (78)	
	$n\text{-}C_3H_7CH=CHCOCH_3$	$CH(C_3H_7\text{-}n)CH_2COCH_3$ (65)	
C_{14}			
(structure: $(CH_3)_3Si$-containing cyclohexenone)	$TiCl_4$, CH_2Cl_2, −78°	(structure) *trans:cis* 3:1 (80)	335
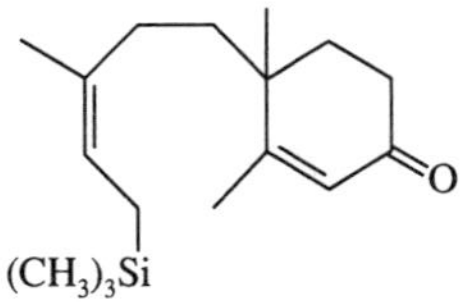 (structure: $(CH_3)_3Si$-containing cyclohexenone)		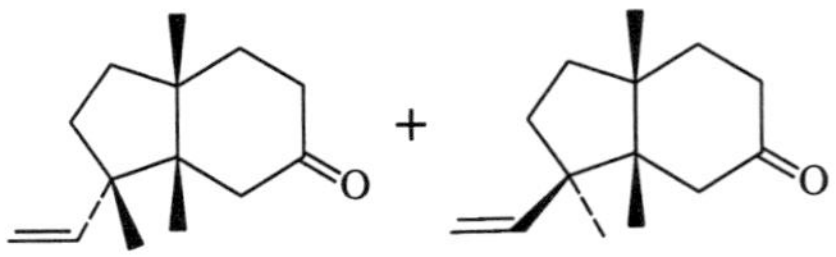 (structure) + (structure)	
	$TiCl_4$	4:1 (78)	336, 337
	TBAF	1:2–3 (70)	336, 337

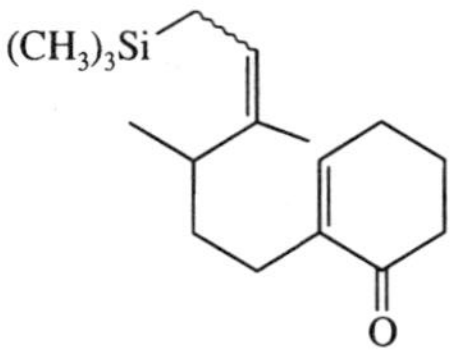

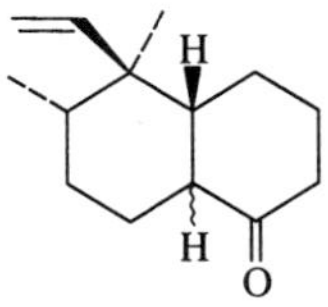

$(CH_3)_3Si$, R^2, R^1, O

I, II, III

R^1	R^2	Reagent	I	II	III	Refs.
CH_3	H	$C_2H_5AlCl_2$	(80)	(0)	(0)	339
,,	,,	TBAF	(0)	(65)	(34)	
H	CH_3	$C_2H_5AlCl_2$	(70)	(0)	(0)	
,,	,,	TBAF	(0)	(42)	(12)[l]	

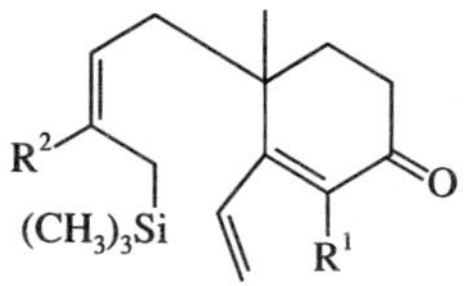

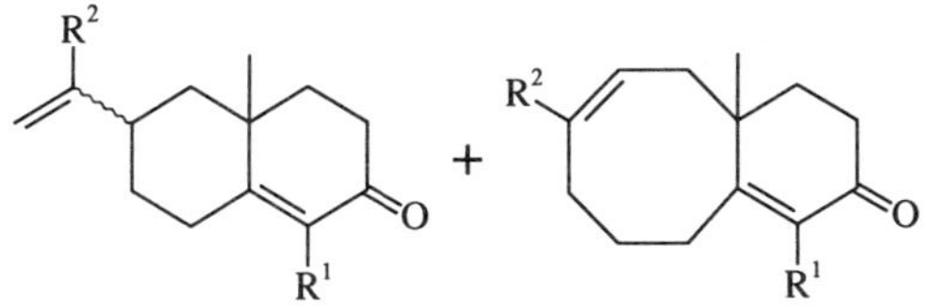

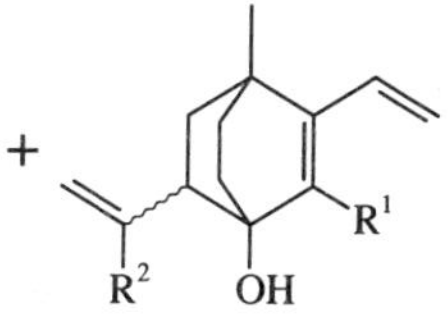

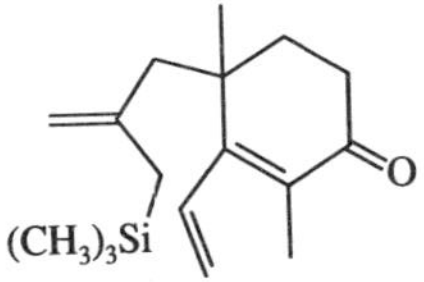

$(CH_3)_3Si$ — $C_2H_5AlCl_2$ — O (90) — 338

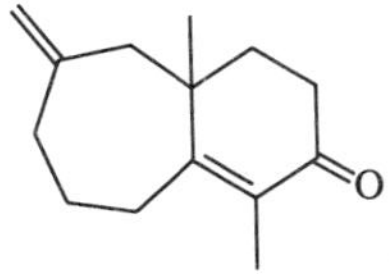

TBAF — O (54) + OH (24) — 338

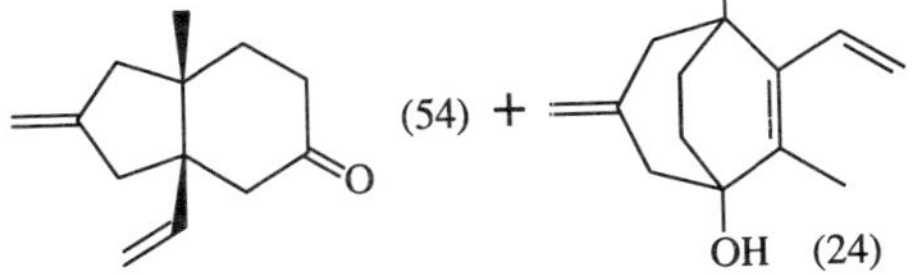

TABLE XIV. Allylsilanes with α,β-Unsaturated Carbonyl Compounds and α,β-Unsaturated Nitriles (*Continued*)

Reactant	Conditions	Product(s) and Yield(s)	Refs.
C_{15}			
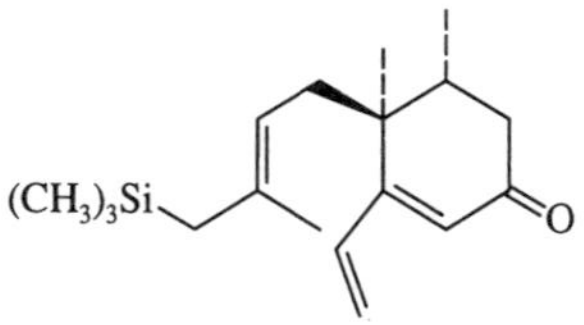 $(CH_3)_3Si$	$C_2H_5AlCl_2$, toluene, 0°, 1 h	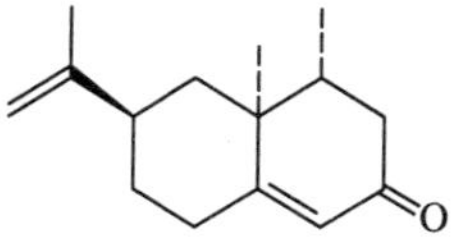 (65)	316

[a] The product can also be obtained as the acid or ester, according to the method of workup.
[b] Mixtures of diastereoisomers at C-2 are obtained as a result of partial or complete equilibration during workup.
[c] The structure of this cycloadduct is only tentatively assigned.
[d] This yield is based on cyclohexenone, and 2 eq of the allylsilane are used; with 1 eq, the yield is 43%.
[e] This product is a 2.16:1 mixture of diastereoisomers.
[f] The use of Lewis acids in these cyclizations is unsuccessful.
[g] This intermediate is prepared in situ by the addition of vinylmagnesium bromide to the corresponding acylsilane.
[h] The major product (81%) is the result of protodesilylation.
[i] The exocyclic double bond is rearranged into the ring.
[j] The product of protodesilylation (10%) is also produced.
[k] The product is the result of protodesilylation.
[l] The product of protodesilylation (5–10%) is also formed.

TABLE XV. VINYLSILANES WITH α,β-UNSATURATED CARBONYL COMPOUNDS

Reactant	Conditions	Product(s) and Yield(s)	Refs.
C_3			
$(CH_3)_3SiCH_2CH{=}CHSi(CH_3)_3$	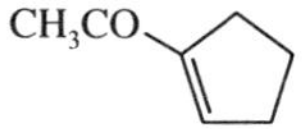 CH_3CO-cyclopentene, $TiCl_4$, CH_2Cl_2, −78°, 1 h	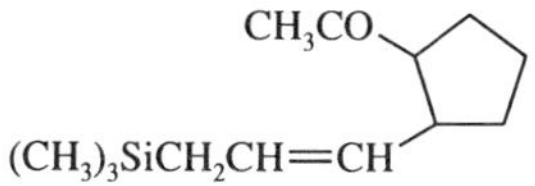 CH_3CO, $(CH_3)_3SiCH_2CH{=}CH$ (cyclopentane) (87)	320
C_6			
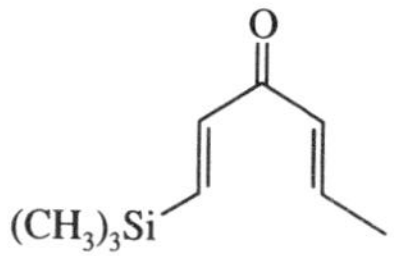 $(CH_3)_3Si$ (dienone)	$FeCl_3$, CH_2Cl_2, 20°, 12 h	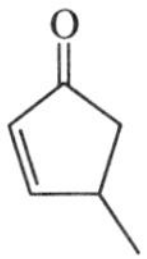 (54)	343
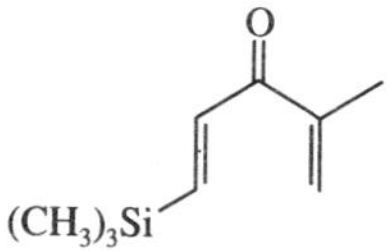 $(CH_3)_3Si$ (dienone)	$FeCl_3$, CH_2Cl_2, 20°, 12 h	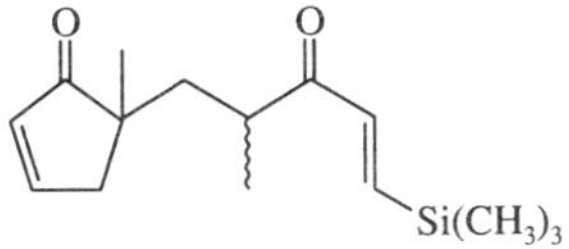 $Si(CH_3)_3$ (42)	343
C_7			
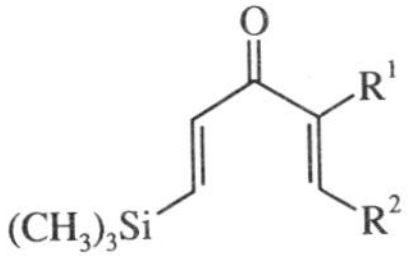 $(CH_3)_3Si$, R^1, R^2 (dienone)	$FeCl_3$, CH_2Cl_2	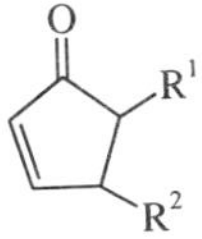 R^1, R^2 (cyclopentenone)	343
R^1 / R^2			
C_2H_5 / H	(0.02 *M*), 0°, 8 h	(33)	
" / "	(0.004 *M*), 0°, 48 h	(51)	
CH_3 / CH_3	−10°, 0.5 h	*trans*:*cis* 41:59 (95)	

TABLE XV. VINYLSILANES WITH α,β-UNSATURATED CARBONYL COMPOUNDS (*Continued*)

Reactant	Conditions	Product(s) and Yield(s)	Refs.
$(CH_3)_3Si$–CH=CH–CO–C(R^1)=CH–R^2	$FeCl_3$, CH_2Cl_2,	2-cyclopentenone, 4-R^2, 5-R^1	343
R^1 = C_2H_5, R^2 = CH_3	0°, 1 h	*trans*:*cis* 57:43 (70)	
R^1, R^2 = —$(CH_2)_3$—	20°, 25 h	*cis* (52)	
[$(CH_3)_3Si$(ArS)C=CH–CO–(1-cyclopentenyl)][a]	$AgBF_4$, CH_2Cl_2, $Cl(CH_2)_2Cl$	bicyclic enone with ArS	119
Ar = 2,4-$(O_2N)_2C_6H_3$		(58)	
Ar = *p*-ClC_6H_4		(15)	
C_9			
$(CH_3)_3Si$–CH=CH–CO–CH=CH–C_4H_9-*t*	$FeCl_3$, CH_2Cl_2, 0°, 1 h	4-C_4H_9-*t*-2-cyclopentenone (97)	343
$(CH_3)_3Si$–CH=CH–CO–(3-R-1-cyclohexenyl)	$FeCl_3$, CH_2Cl_2	cis-fused bicyclic enone (H, H; R)	

Reactant	Conditions	Product(s) and Yield(s) (%)	Refs.
R			
H	20°, 25 h	(84)	343
$OCH_2C_6H_5$	0°, 2 h	R α:β 10:90 (76)	344
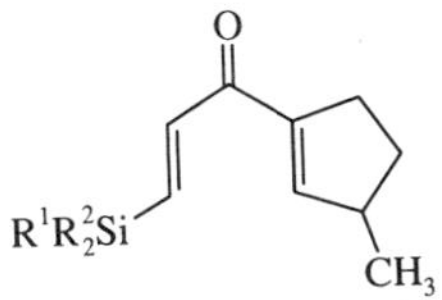 $R^1R^2_2Si$-CH=CH-CO-(3-methylcyclopent-1-enyl)	$FeCl_3$, CH_2Cl_2, 20°	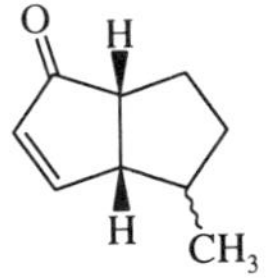 bicyclic enone, CH_3 α:β	344
R^1 / R^2			
CH_3 / CH_3	2.5 h	54:46 (50)	
C_6H_5 / "	3h	59:41 (46)	
CH_3 / C_6H_5	2 h	62:38 (41)	
C_6H_5 / "	4 h	76:24 (13)	
i-C_3H_7 / i-C_3H_7	4 h	79:21 (13)	
C_{10}			
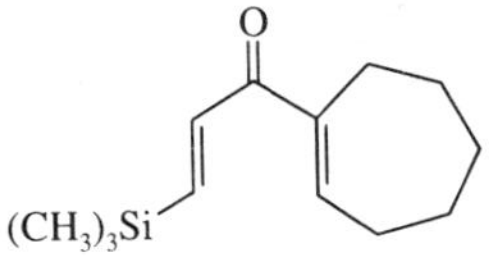 $(CH_3)_3Si$-CH=CH-CO-(cyclohept-1-enyl)	$FeCl_3$, CH_2Cl_2, 20°, 2.5 h	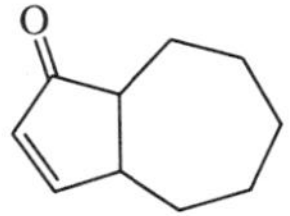 *trans*:*cis* 15:85 (74)	343
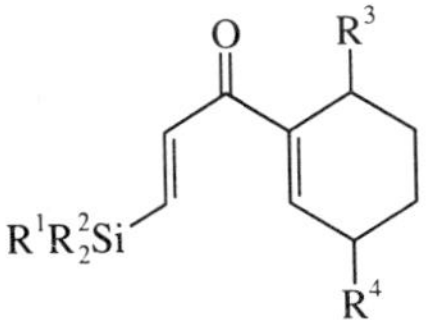 $R^1R^2_2Si$-CH=CH-CO-(cyclohexenyl with R^3, R^4)	$FeCl_3$, CH_2Cl_2, 0°	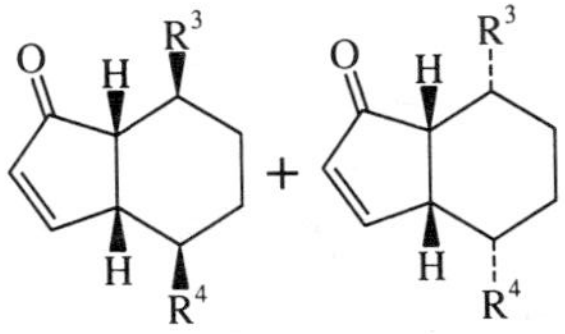 I + II	

TABLE XV. Vinylsilanes with α,β-Unsaturated Carbonyl Compounds (*Continued*)

Reactant				Conditions	Product(s) and Yield(s)		Refs.
R^1	R^2	R^3	R^4		I:II		
CH_3	CH_3	CH_3	H	4 h	72:28	(85)	663
”	”	H	CH_3	”	78:22	(99)	663
”	”	”	$CH_2OCH_2OCH_2C_6H_5$	2 h	93:7	(40)	344
i-C_3H_7	i-C_3H_7	CH_3	H	4 h	78:22	(78)	663
”	”	H	CH_3	”	91:9	(70)	663
C_6H_5	CH_3	”	”	2 h	84:16	(63)	344
CH_3	C_6H_5	”	”	”	86:14	(83)	344
C_6H_5	”	”	”	”	87:13	(15)	344
C_{11}							
$(CH_3)_3Si$ (structure)				$FeCl_3$, $Cl(CH_2)_2Cl$, 20°	(structure) + $(CH_3)_3Si$ (structure) (0)	(40)	345
				$FeCl_3$, toluene, reflux	(20–30)	(—)	
				$BF_3{\cdot}O(C_2H_5)_2$, toluene, reflux, 36 h	(70)	(0)	
$Cl(CH_2)_3$, $Si(CH_3)_2C_4H_9$-n (structure)				$SnCl_4$, −78°, 1 h	$Cl(CH_2)_3$ (structure) (100)		131
$(CH_3)_3Si$ (structure)				$FeCl_3$, CH_2Cl_2, −10°, 2 h	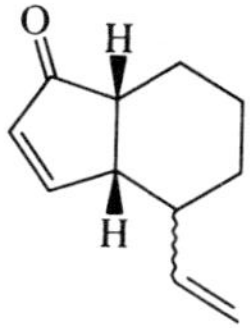 (structure) α:β 30:70 (66)		344

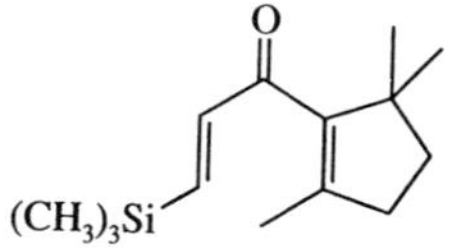

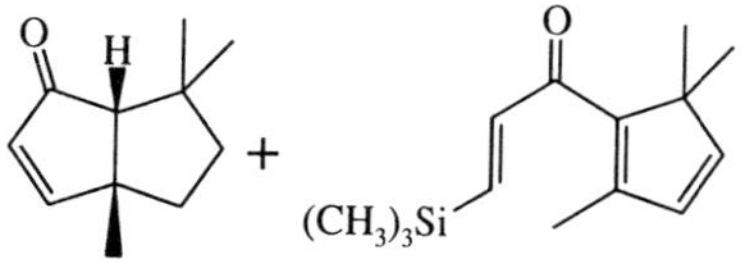

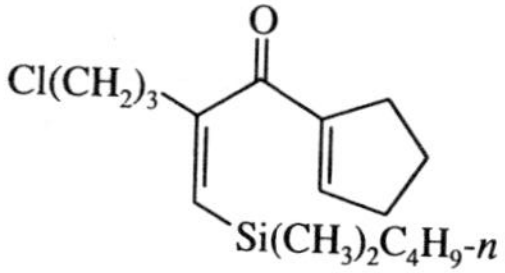

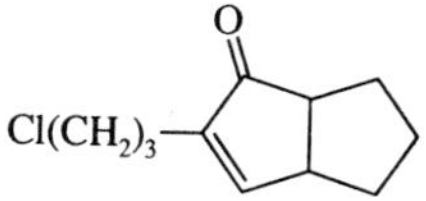

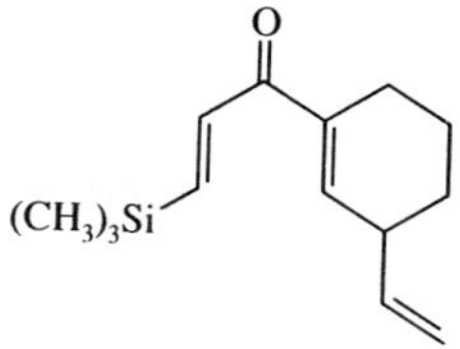

Reactant	Conditions	Product(s) and Yield(s) (%)	Refs.
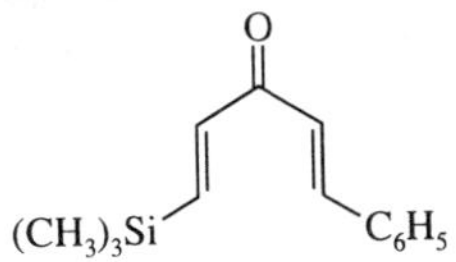 $(CH_3)_3Si$, C_6H_5	$FeCl_3$, toluene, 20°, 12 h	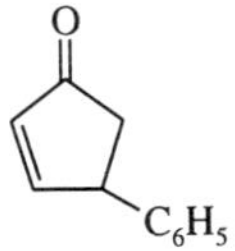 C_6H_5 (27)	343
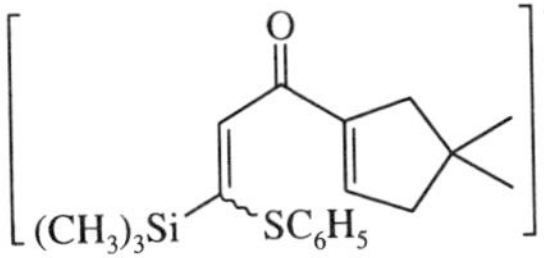 [$(CH_3)_3Si$, SC_6H_5][a]	$AgBF_4$, CH_2Cl_2, 20°	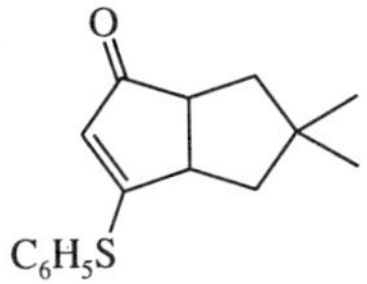 C_6H_5S (38)	119
C_{13}			
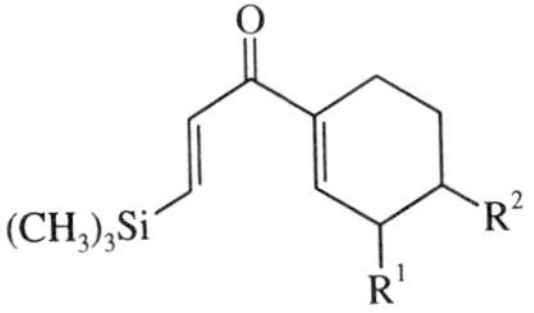 $(CH_3)_3Si$, R^2, R^1	$FeCl_3$, CH_2Cl_2, 0°	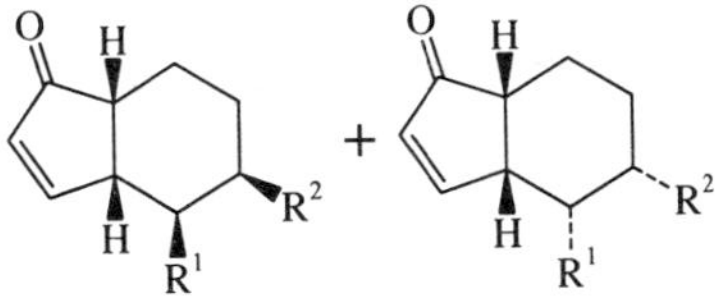 R^2, R^1 I + R^2, R^1 II	
R^1 / R^2		I:II	
H / t-C_4H_9	4 h	76:24 (82)	663
t-C_4H_9 / H	8 h	94:6 (63)	344
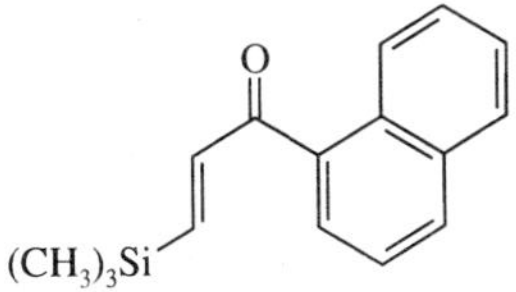 $(CH_3)_3Si$	$FeCl_3$, CH_2Cl_2, 20°, 12 h	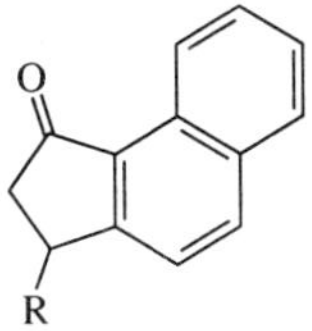 R = $Si(CH_3)_3$ (70)	343
	$FeCl_3$, toluene, 20°, 48 h	R = H (60)	343

TABLE XV. VINYLSILANES WITH α,β-UNSATURATED CARBONYL COMPOUNDS (*Continued*)

Reactant	Conditions	Product(s) and Yield(s)	Refs.
C_{14}			
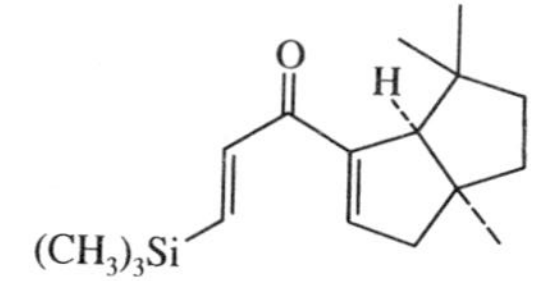 $(CH_3)_3Si$	$BF_3·O(C_2H_5)_2$, toluene, 25°	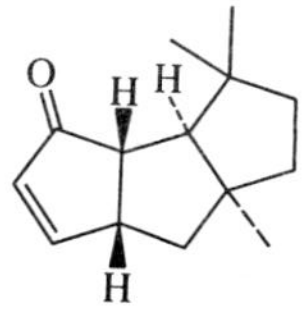 (80)	345
C_{15}			
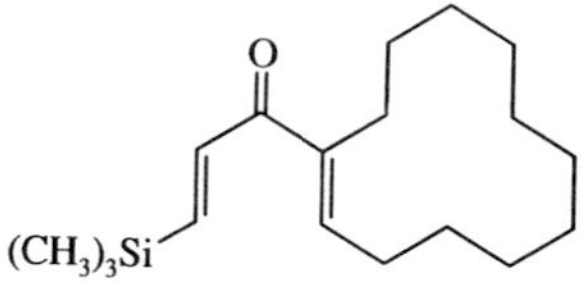 $(CH_3)_3Si$	$FeCl_3$, CH_2Cl_2, 20°, 25 h	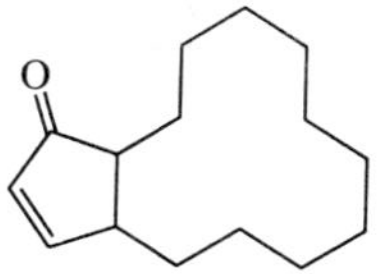 *trans:cis* 54:46 (78)	343
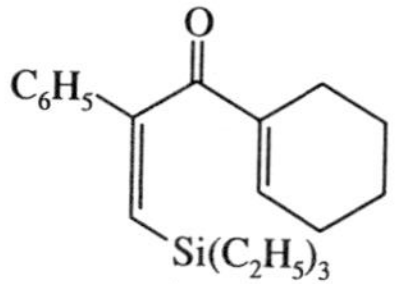 C_6H_5, $Si(C_2H_5)_3$		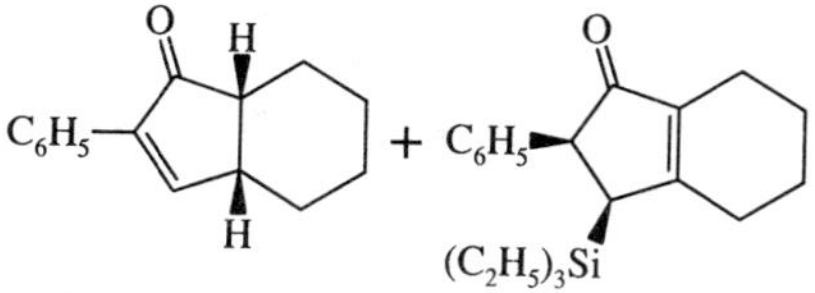 C_6H_5 + C_6H_5, $(C_2H_5)_3Si$	131
	$BF_3·O(C_2H_5)_2$, −20°, 4 h; 0°, 24 h	(30) (37)	
	$SnCl_4$, −78°, 2 h	(18) (18)	
	$FeCl_3$, 0°, 4 h	(18) (24)	
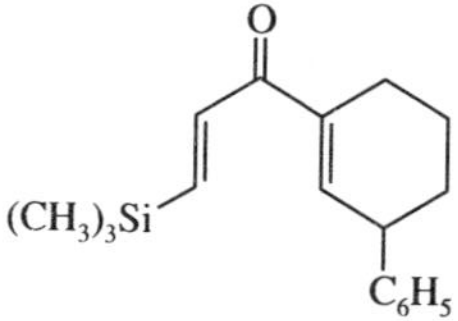 $(CH_3)_3Si$, C_6H_5	$FeCl_3$, CH_2Cl_2, 0°, 4 h	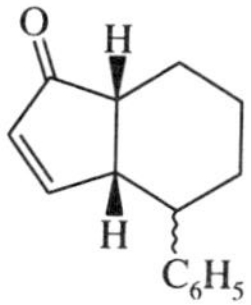 C_6H_5 α:β 6:94 (76)	344

[a] This dienone is a presumed intermediate produced in the acylation of the appropriate 1-arylthiovinyltrimethylsilane.

TABLE XVI. ALLENYLSILANES WITH α,β-UNSATURATED CARBONYL COMPOUNDS

Reactant	Conditions	Product(s) and Yield(s)	Refs.
C_3			
$(CH_3)_3SiCH{=}C{=}CH_2$	$CH_2{=}CHCOCN$, $TiCl_4$, CH_2Cl_2, −78°, 3 h; −30°, 4 h	$HC{\equiv}C(CH_2)_3CO_2H$ (20) + $(CH_3)_3SiC{\equiv}C(CH_2)_3CO_2H$ (30)	331
	$CH_3CH{=}CHCOCN$, $TiCl_4$, CH_2Cl_2, −78°, 3 h; −30°, 4 h	$HC{\equiv}CCH_2CH(CH_3)CH_2COCN$ (80) 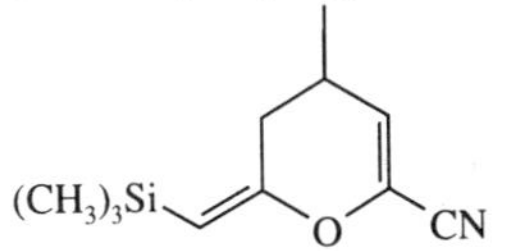 + $(CH_3)_3Si$, O, CN (5)	331
	$CH_2{=}C(CH_3)COCN$, $TiCl_4$, CH_2Cl_2, −78°, 3 h; −30°, 4 h	$HC{\equiv}C(CH_2)_2CH(CH_3)COCN$ (65)	331
	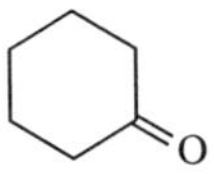 O, $TiCl_4$, CH_2Cl_2, −78°, 1 h	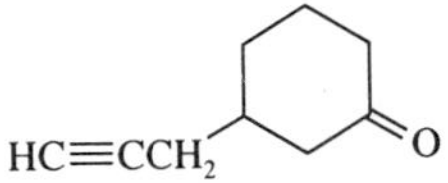 $HC{\equiv}CCH_2$, O (30) 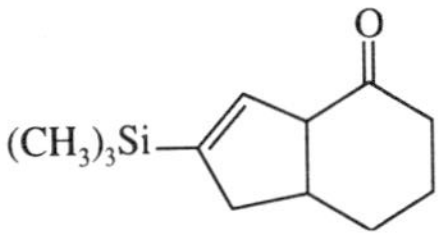 + $(CH_3)_3Si$, O (19)	346
	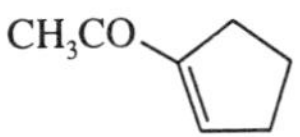 CH_3CO, $TiCl_4$, CH_2Cl_2, −78°, 1 h	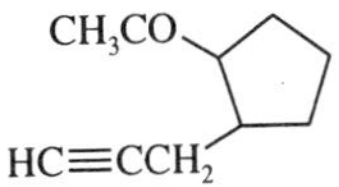 CH_3CO, $HC{\equiv}CCH_2$ (18) 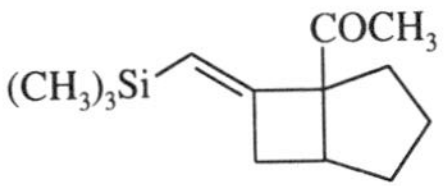 + $(CH_3)_3Si$, $COCH_3$ (12)[a]	319
C_5			
$(CH_3)_3SiCH{=}C{=}C(CH_3)_2$	$CH_2{=}CHCOCN$, $TiCl_4$, CH_2Cl_2, −78°, 3 h; −30°, 4 h	$(CH_3)_3SiC{\equiv}CC(CH_3)_2(CH_2)_2COCN$ (60)	331

TABLE XVI. ALLENYLSILANES WITH α,β-UNSATURATED CARBONYL COMPOUNDS (*Continued*)

Reactant	Conditions	Product(s) and Yield(s)	Refs.
	$CH_3CH{=}CHCOCN$, $TiCl_4$, CH_2Cl_2, −78°, 3 h; −30°, 4 h	$HC{\equiv}CC(CH_3)_2CH(CH_3)CH_2COCN$ (75) + [structure: dihydropyran bearing $(CH_3)_3Si$CH=, CN] (10)	331
	$CH_2{=}C(CH_3)COCN$, $TiCl_4$, CH_2Cl_2, −78°, 3 h; −30°, 4 h	$HC{\equiv}CC(CH_3)_2CH_2CH(CH_3)COCN$ (30) + $(CH_3)_3SiC{\equiv}CC(CH_3)_2CH_2CH(CH_3)COCN$ (20)	331
	$(CH_3)_2C{=}CHCOCN$, $TiCl_4$, CH_2Cl_2, −78°, 3 h; −30°, 4 h	$HC{\equiv}CC(CH_3)_2C(CH_3)_2CH_2COCN$ (55) + $(CH_3)_3SiC{\equiv}CC(CH_3)_2C(CH_3)_2CH_2COCN$ (25) + [structure: dihydropyran bearing $(CH_3)_3Si$CH=, CN] (10)	331
C_6			
$(CH_3)_3SiC(CH_3){=}C{=}C(CH_3)_2$	$(CH_3)_2C{=}CHCOCN$, $TiCl_4$, CH_2Cl_2, 1 h	$CH_3C{\equiv}CC(CH_3)_2C(CH_3)_2CH_2COCN$ (25) + [structure: cyclopentene bearing $(CH_3)_3Si$, $COCH_3$] (7)	346
$(CH_3)_3SiCH{=}C{=}C(CH_3)CH{=}CH_2$	$CH_3CH{=}CHCOCN$, $TiCl_4$, CH_2Cl_2, −78°, 3 h; −30°, 4 h	[structure: cyclohexene bearing COCN, =CHR] R = H (25) + R = $(CH_3)_3Si$ (35)	331

[a] A substantial amount (65%) of starting material is recovered in this experiment.

Reactant	Conditions	Product(s) and Yield(s)	Refs.
C_3			
$(CH_3)_3SiCH_2CH{=}CH_2$	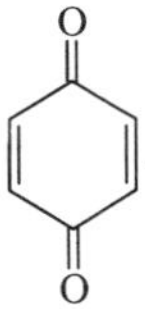 [p-benzoquinone: O, O], $TiCl_4$, CH_2Cl_2, 0° 45 min	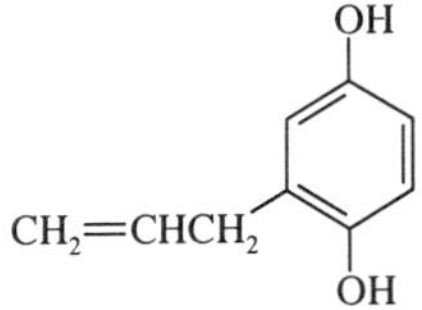 [OH, OH, $CH_2{=}CHCH_2$] (55)	347, 664
	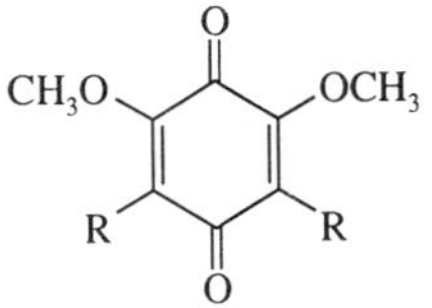 [CH_3O, OCH_3, R, R, O, O], $TiCl_4$, CH_2Cl_2, −78°	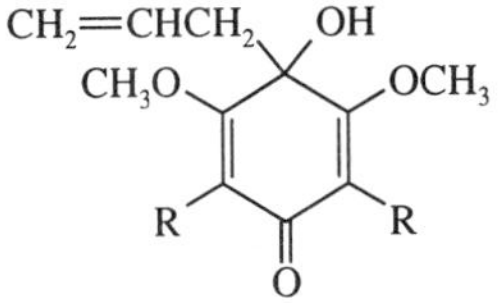 [$CH_2{=}CHCH_2$, OH, CH_3O, OCH_3, R, R, O]	347
	R = H 30 min	(81)	
	R = OCH_3 20 min	(91)	
	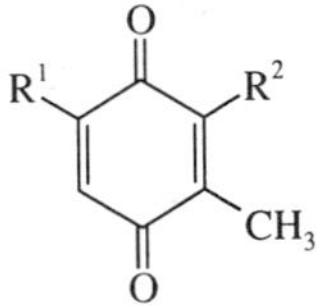 [R^1, R^2, CH_3, O, O], $TiCl_4$, CH_2Cl_2, −78°	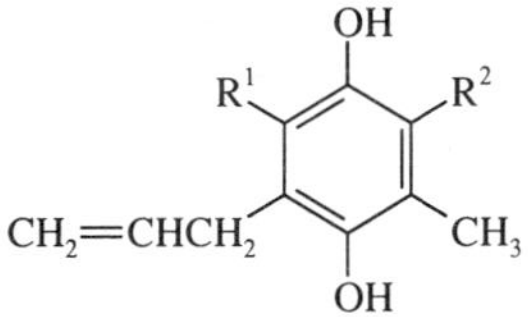 [OH, OH, R^1, R^2, CH_3, $CH_2{=}CHCH_2$]	347
	R^1 R^2		
	CH_3 H 10 min	(47)	
	H CH_3 2 h	(58)	
	1. 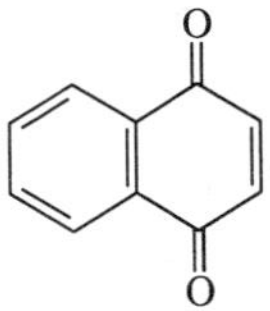 [1,4-naphthoquinone: O, O], $h\nu$, C_6H_6[a]	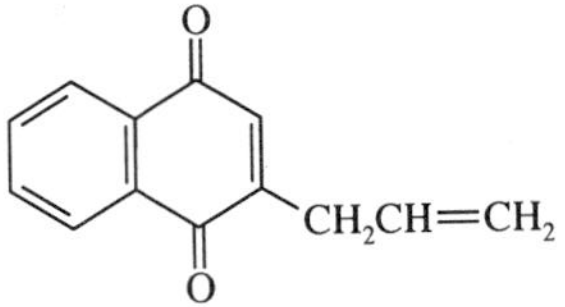 [O, O, $CH_2CH{=}CH_2$] (63)	665
	2. $BF_3{\cdot}O(C_2H_5)_2$, CH_2Cl_2, 0°		
	3. $FeCl_3$, H_2O		

TABLE XVII. ALLYLSILANES WITH QUINONES (*Continued*)

Reactant	Conditions	Product(s) and Yield(s)	Refs.
	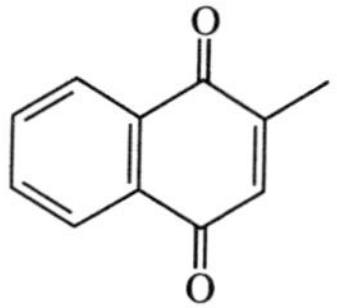 , $TiCl_4$, −45°, 15 min	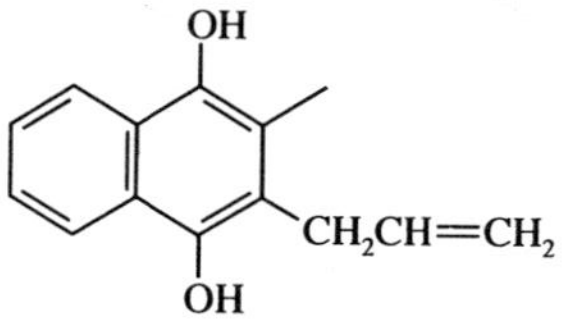 $CH_2CH{=}CH_2$ (7)	347
	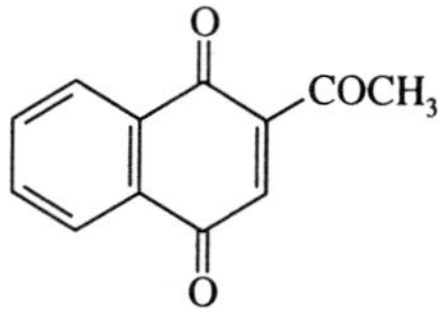 $COCH_3$, $AlCl_3$	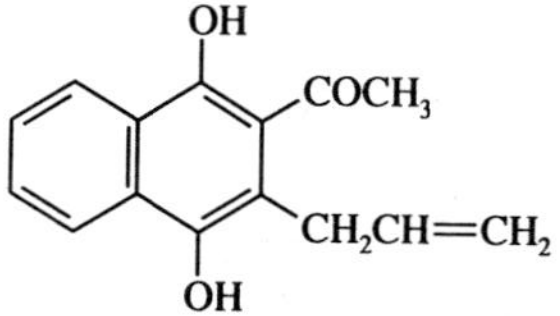 $COCH_3$, $CH_2CH{=}CH_2$ (40)	102
	" $BF_3{\cdot}O(C_2H_5)_2$, −50°, 1 h	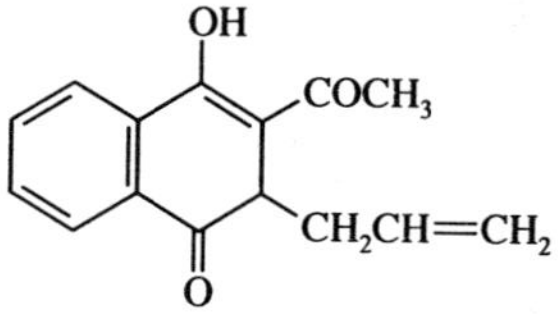 $COCH_3$, $CH_2CH{=}CH_2$ (39) +	348
		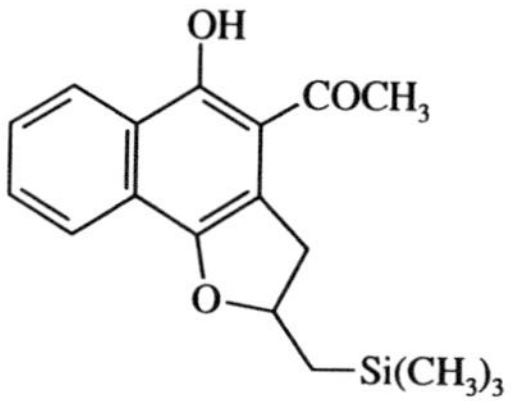 $COCH_3$, $Si(CH_3)_3$ (30)	

Reagent	Substrate and Conditions	Product(s) and Yield(s) (%)	Refs.
	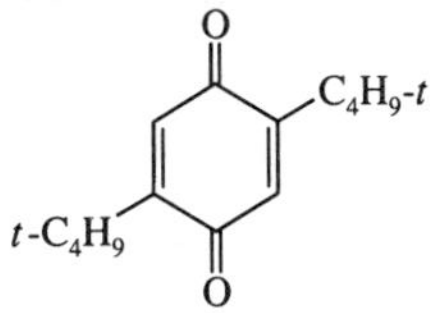 (t-C_4H_9 / C_4H_9-t / O / O), $TiCl_4$, CH_2Cl_2, −78°, 1 h	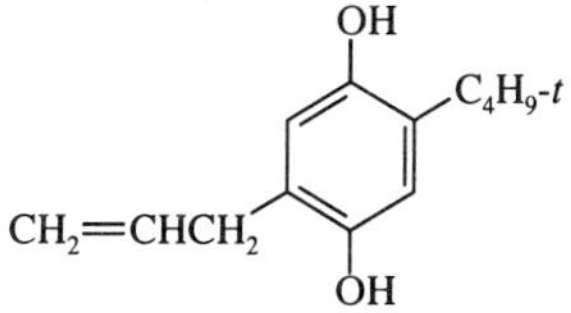 (CH_2=$CHCH_2$ / C_4H_9-t / OH / OH) (59)	347
$(C_6H_5)_3SiCH_2CH$=CH_2	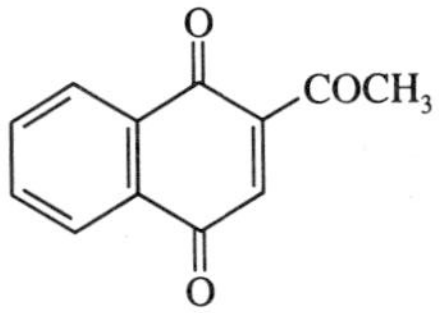 ($COCH_3$ / O / O), $AlCl_3$, CH_2Cl_2, 0°, 0.5 h	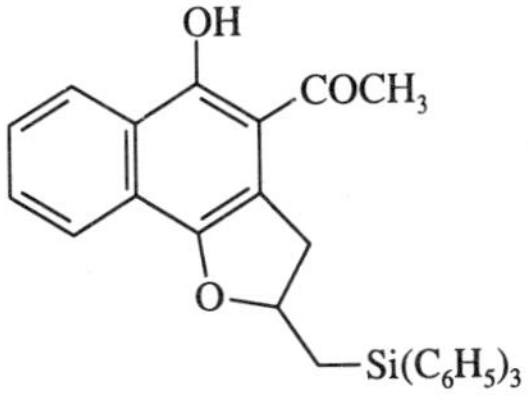 (OH / $COCH_3$ / O / $Si(C_6H_5)_3$) (61)	348
C_4			
$(CH_3)_3SiCH_2C(CH_3)$=CH_2	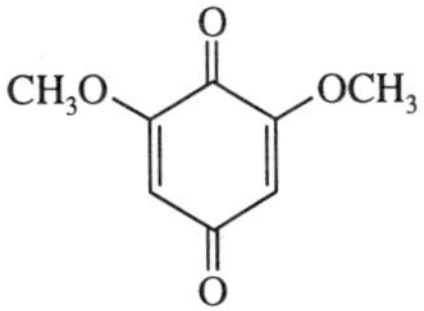 (CH_3O / OCH_3 / O / O), $TiCl_4$, CH_2Cl_2, −78°, 30 min	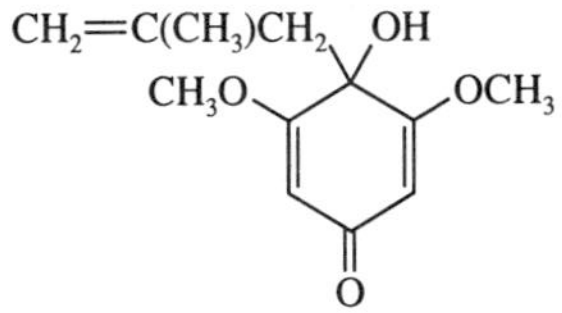 (CH_2=$C(CH_3)CH_2$ / OH / CH_3O / OCH_3 / O) (81)	347, 664
	1. 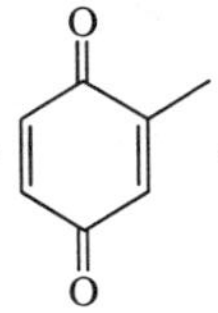 (O / O), $TiCl_4$, CH_2Cl_2, −78°, 20 min 2. I_2 or $FeCl_3$	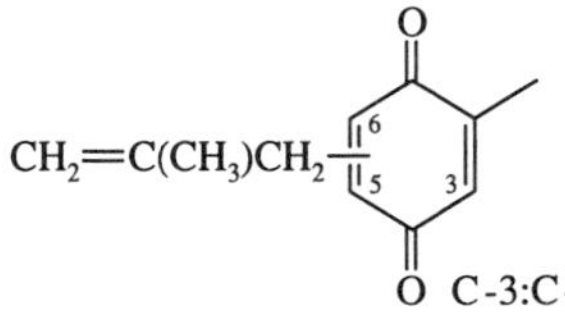 (CH_2=$C(CH_3)CH_2$ / 6 / 5 / 3 / O / O) (30) C-3:C-5:C-6 5:32:65	347

TABLE XVII. ALLYLSILANES WITH QUINONES (*Continued*)

Reactant	Conditions	Product(s) and Yield(s)	Refs.
	1. 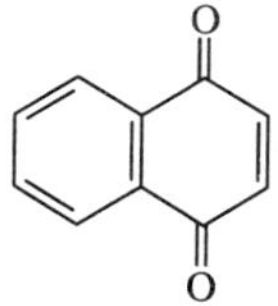 1,4-naphthoquinone, $TiCl_4$, CH_2Cl_2, −78°, 2 h 2. I_2 or $FeCl_3$	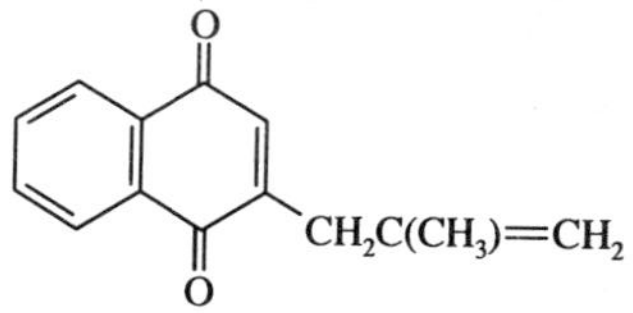 2-[$CH_2C(CH_3)=CH_2$]-1,4-naphthoquinone (38)	347
$C_6H_5(CH_3)_2SiCH(CO_2CH_3)CH=CH_2$	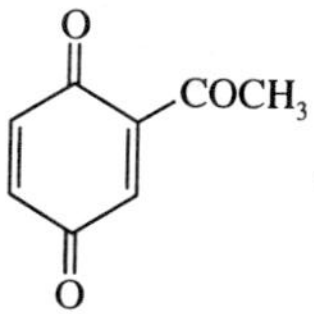 2-($COCH_3$)-1,4-benzoquinone, $SnCl_4$, 1 h	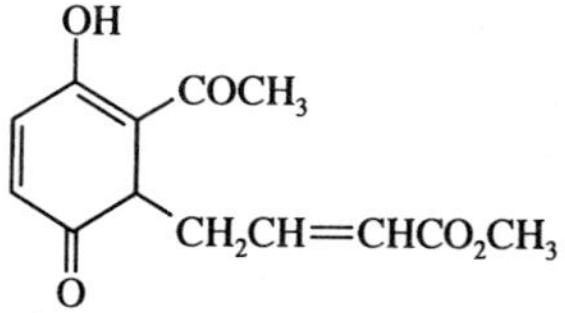 OH, $COCH_3$, $CH_2CH=CHCO_2CH_3$ adduct (72)	411
	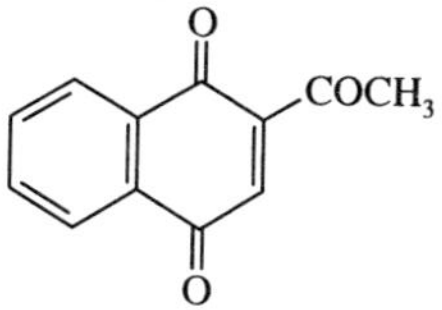 2-($COCH_3$)-1,4-naphthoquinone, $SnCl_4$, CH_2Cl_2, 0°, 2 h	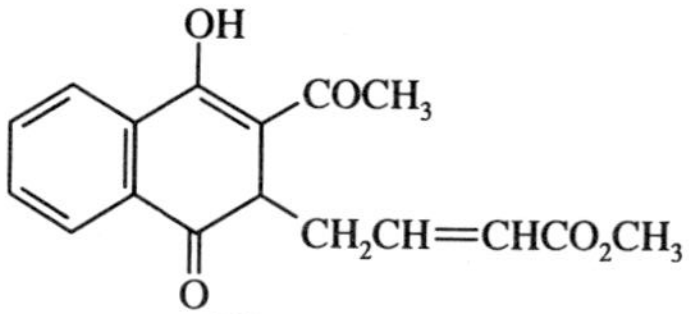 OH, $COCH_3$, $CH_2CH=CHCO_2CH_3$ adduct (100)	348
	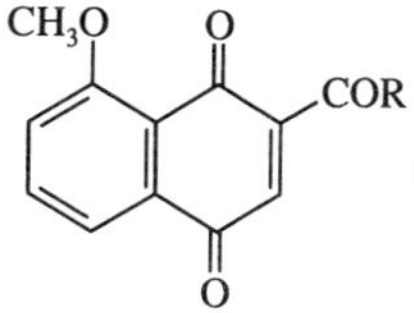 5-CH_3O-2-(COR)-1,4-naphthoquinone, $SnCl_4$, CH_2Cl_2	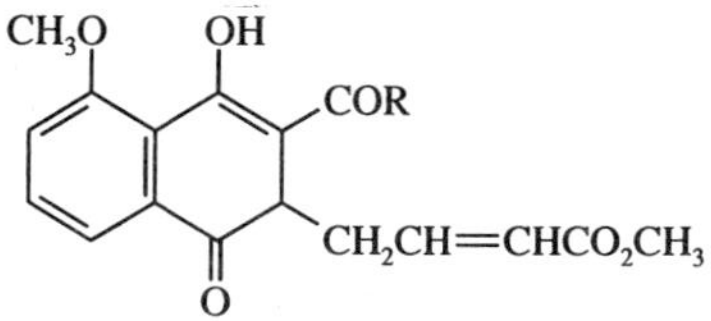 CH_3O, OH, COR, $CH_2CH=CHCO_2CH_3$ adduct	
	R: CH_3, 30 min	(88)	348
	R: n-C_3H_7, 0°, 1 h	(82)	666, 348
	R: n-C_4H_9, ", "	(83)	667
	R: $(CH_2)_3COCH_3$, −20°, 2 h	(39)	668

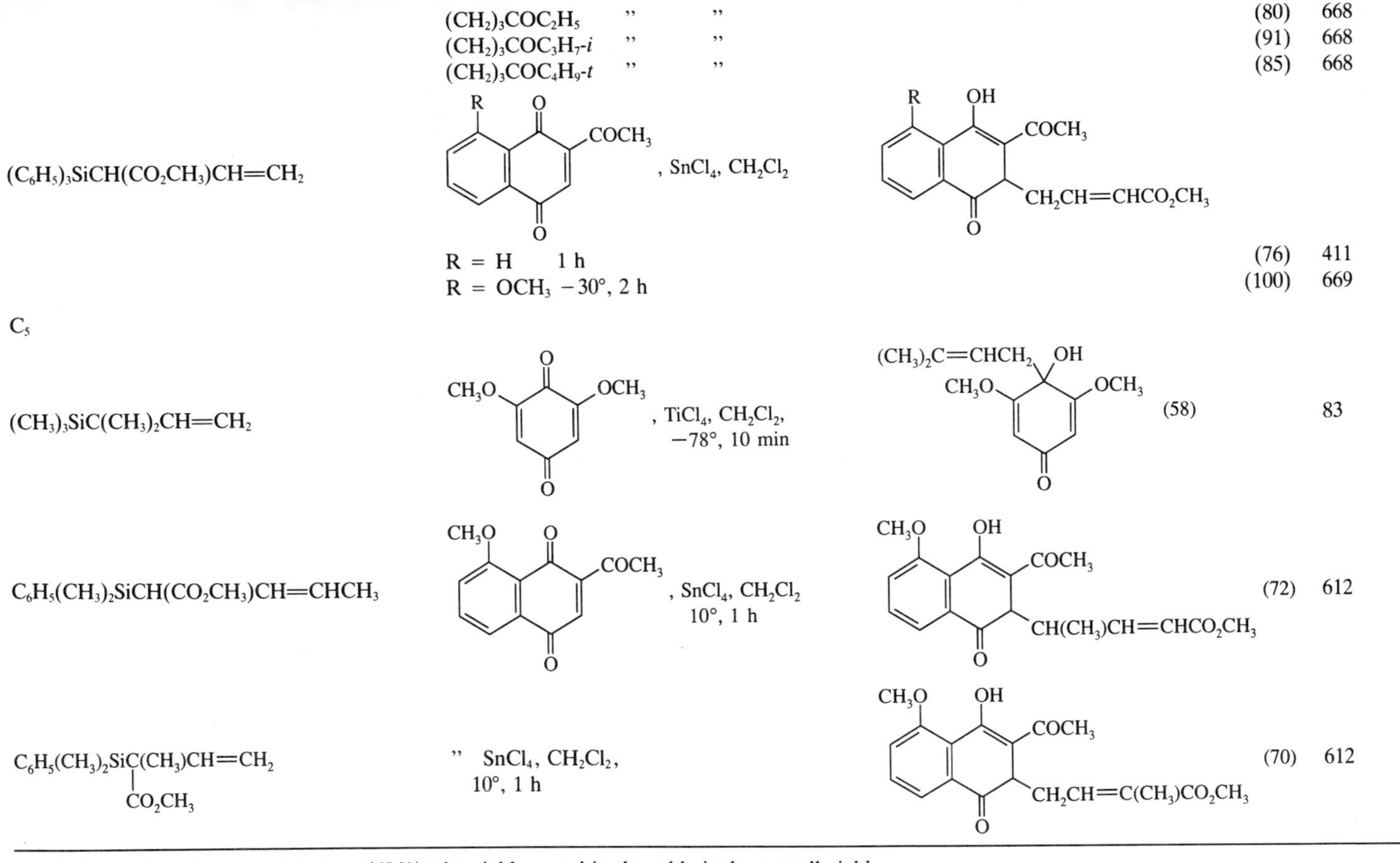

Reactant	Reagent, Conditions	Product(s) and Yield(s) (%)	Refs.
	$(CH_2)_3COC_2H_5$ ” ”	(80)	668
	$(CH_2)_3COC_3H_7$-*i* ” ”	(91)	668
	$(CH_2)_3COC_4H_9$-*t* ” ”	(85)	668
$(C_6H_5)_3SiCH(CO_2CH_3)CH{=}CH_2$	R, $COCH_3$ naphthoquinone, $SnCl_4$, CH_2Cl_2	R, OH, $COCH_3$, $CH_2CH{=}CHCO_2CH_3$	
	R = H 1 h	(76)	411
	R = OCH_3 −30°, 2 h	(100)	669
C_5			
$(CH_3)_3SiC(CH_3)_2CH{=}CH_2$	CH_3O, OCH_3 quinone, $TiCl_4$, CH_2Cl_2, −78°, 10 min	$(CH_3)_2C{=}CHCH_2$, OH, CH_3O, OCH_3 (58)	83
$C_6H_5(CH_3)_2SiCH(CO_2CH_3)CH{=}CHCH_3$	CH_3O, $COCH_3$ naphthoquinone, $SnCl_4$, CH_2Cl_2 10°, 1 h	CH_3O, OH, $COCH_3$, $CH(CH_3)CH{=}CHCO_2CH_3$ (72)	612
$C_6H_5(CH_3)_2SiC(CH_3)(CO_2CH_3)CH{=}CH_2$	” $SnCl_4$, CH_2Cl_2, 10°, 1 h	CH_3O, OH, $COCH_3$, $CH_2CH{=}C(CH_3)CO_2CH_3$ (70)	612

[a] This reaction gives a cyclobutane (69%); the yield quoted in the table is the overall yield.

TABLE XVIII. ALLYLSILANES WITH α,β-UNSATURATED NITRO AND α,β-UNSATURATED NITROSO COMPOUNDS

Reactant	Conditions	Product(s) and Yield(s)	Refs.
C_3			
$(CH_3)_3SiCH_2CH{=}CH_2$	1. $BrCH_2C(=NOH)CO_2C_2H_5$, Na_2CO_3, CH_2Cl_2, 24 h 2. $HClO_4$, CH_3CN, 1–7 h	$CH_2{=}CH(CH_2)_2COCO_2C_2H_5$ (64)	351
	1. $ClCH_2C(=NOH)C_6H_5$, Na_2CO_3, $(C_2H_5)_2O$, 8 d 2. $HClO_4$, CH_3CN, 14 h	$CH_2{=}CH(CH_2)_2COC_6H_5$ (46)	670
	1. $ClCH_2C(=NOH)C_6H_5$, Na_2CO_3, $(C_2H_5)_2O$, 8 d 2. $LiAlH_4$, $(C_2H_5)_2O$, 16 h 3. HCl, H_2O, 20°, 24 h	$CH_2{=}CH(CH_2)_2CH(NH_2)C_6H_5$ (70)	670
	1. $C_6H_5CH{=}CHNO_2$, $TiCl_4$, −15°, 2 h 2. Zn, THF, 1 h	$CH_2{=}CHCH_2CH(C_6H_5)CN$ (55)	350
	1. p-$R^1C_6H_4CH{=}CR^2NO_2$, $AlCl_3$, CH_2Cl_2, −20°, 2 h 2. $TiCl_3$ R^1 — R^2 H — CH_3 CH_3O — " Cl — " H — C_2H_5 Cl — "	$CH_2{=}CHCH_2CH(C_6H_4R^1$-$p)COR^2$ (51) (48–59) (55) (62) (49–55)	349
	1. $C_6H_5(CH_2)_2CH{=}C(C_2H_5)NO_2$, $AlCl_3$, CH_2Cl_2, −20°, 3 h 2. $TiCl_3$	$CH_2{=}CHCH_2CH(CH_2CH_2C_6H_5)COC_2H_5$ (50)	349

	1. $n\text{-}C_{10}H_{21}CH{=}C(C_2H_5)NO_2$, $AlCl_3$, CH_2Cl_2, −20°, 2 h 2. $TiCl_3$	$CH_2{=}CHCH_2CH(C_{10}H_{21}\text{-}n)COC_2H_5$ (74)	349
$C_6H_5(CH_3)_2SiCH_2CH{=}CH_2$	1. $C_6H_5CH{=}CHNO_2$, $TiCl_4$, −15°, 2 h 2. Zn, THF, 1 h	$CH_2{=}CHCH_2CH(C_6H_5)CN$ (61)	350
$(C_6H_5)_3SiCH_2CH{=}CH_2$	1. $C_6H_5CH{=}CHNO_2$, $TiCl_4$, −15°, 2 h 2. Zn, THF, 1 h	$CH_2{=}CHCH_2CH(C_6H_5)CN$ (69)	350
C_4			
$C_6H_5(CH_3)_2SiCH_2CH{=}CHCH_3$	1. $C_6H_5CH{=}CHNO_2$, $TiCl_4$, −15°, 2 h 2. Zn, THF, 1 h	$CH_2{=}CHCH(CH_3)CH(C_6H_5)CN$ (65)[a] I	350
$(CH_3)_3SiCH(CH_3)CH{=}CH_2$	1. $BrCH_2C({=}NOH)CO_2C_2H_5$, Na_2CO_3, CH_2Cl_2, 24 h 2. $HClO_4$, CH_3CN, 1–7 h	$CH_3CH{=}CH(CH_2)_2COCO_2C_2H_5$ (53)	351
$C_6H_5(CH_3)_2SiCH(CH_3)CH{=}CH_2$	1. $C_6H_5CH{=}CHNO_2$, $TiCl_4$, −15°, 2 h 2. Zn, THF, 1 h	$CH_3CH{=}CHCH_2CH(C_6H_5)CN$ (32) + I (13) $E{:}Z$ 1:1	350
$(CH_3)_3SiCH_2C(CH_3){=}CH_2$	1. $BrCH_2C({=}NOH)CO_2C_2H_5$, Na_2CO_3, CH_2Cl_2, 24 h 2. HCl, CH_3CN, 1–7 h	$CH_2{=}C(CH_3)(CH_2)_2COCO_2C_2H_5$ (61)	351
	1. $C_6H_5CH{=}C(C_2H_5)NO_2$, $AlCl_3$, CH_2Cl_2, −20°, 1 h 2. $TiCl_3$	$CH_2{=}C(CH_3)CH_2CH(C_6H_5)COC_2H_5$ (62)	349
	1. $n\text{-}C_{10}H_{21}CH{=}C(C_2H_5)NO_2$, $AlCl_3$, CH_2Cl_2, −20°, 1 h 2. $TiCl_3$	$CH_2{=}C(CH_3)CH_2CH(C_{10}H_{21}\text{-}n)COC_2H_5$ (74)	349
$C_6H_5(CH_3)_2SiCH_2C(CH_3){=}CH_2$	1. $C_6H_5CH{=}CHNO_2$, $TiCl_4$, −15°, 2 h 2. Zn, THF, 1 h	$(CH_3)_2CClCH_2CH(C_6H_5)CN$ (19)	350
C_5			
$C_6H_5(CH_3)_2SiCH_2CH{=}C(CH_3)_2$	1. $C_6H_5CH{=}CHNO_2$, $TiCl_4$, −15°, 2 h 2. Zn, THF, 1 h	$CH_2{=}CHC(CH_3)_2CH(C_6H_5)CN$ (33)	350

TABLE XVIII. ALLYLSILANES WITH α,β-UNSATURATED NITRO AND α,β-UNSATURATED NITROSO COMPOUNDS (*Continued*)

Reactant	Conditions	Product(s) and Yield(s)	Refs.
$C_6H_5(CH_3)_2SiC(CH_3)(CO_2CH_3)CH{=}CH_2$	1. $C_6H_5CH{=}CHNO_2$, $TiCl_4$, −15°, 2 h 2. Zn, THF, 1 h	$CH_3O_2CC(CH_3){=}CHCH_2CH(C_6H_5)CN$ (46)	350
C_9			
$(CH_3)_3SiCH_2CH{=}CHC_6H_5$	1. $BrCH_2C({=}NOH)CO_2C_2H_5$, Na_2CO_3, CH_2Cl_2, 24 h 2. $HClO_4$, CH_3CN, 1–7 h	$CH_2{=}CHCH(C_6H_5)CH_2COCO_2C_2H_5$ (43)	351
$C_6H_5(CH_3)_2SiCH_2CH{=}CHC_6H_5$	1. $C_6H_5CH{=}CHNO_2$, $TiCl_4$, −15°, 2 h 2. Zn, THF, 1 h	$CH_2{=}CHCH(C_6H_5)CH(C_6H_5)CN$ (41)[b]	350
$(CH_3)_3SiCH_2C(C_6H_5){=}CH_2$	1. $BrCH_2C({=}NOH)CO_2C_2H_5$, Na_2CO_3, CH_2Cl_2, 24 h 2. HCl, CH_3CN, 1–7 h	$CH_2{=}C(C_6H_5)(CH_2)_2COCO_2C_2H_5$ (52)	351

[a] The diastereoisomers are produced in a ratio of 75:25.
[b] The diastereoisomers are produced in a ratio of 67:33.

TABLE XIX. ALLYLSILANES WITH ACETALS AND KETALS

Reactant	Conditions	Product(s) and Yield(s)	Refs.
C_3			
$(CH_3)_3SiCH_2CH{=}CH_2$	$XCH_2O(CH_2)_2C_6H_5$, CH_2Cl_2, 30 min	$CH_2{=}CH(CH_2)_2O(CH_2)_2C_6H_5$	
	X		
	CH_3S $SnCl_4$	(70)	365
	$CH_3O(CH_2)_2O$ $TiCl_4$	(93)	364
	C_6H_5S $SnCl_4$	" (26) + $CH_2{=}CH(CH_2)_2SC_6H_5$ (43)	365
	XCH_2O-(4-C_4H_9-*t*-cyclohexyl), CH_2Cl_2, 30 min	$CH_2{=}CH(CH_2)_2O$-(4-C_4H_9-*t*-cyclohexyl)	
	X		
	CH_3S $SnCl_4$	(80)	365
	" $TiCl_4$, $-20°$	(4)	365
	$CH_3O(CH_2)_2O$ "	(95)	364
	$C_6H_5SCH_2OCH_2CH(C_2H_5)C_4H_9$-*n*, $SnCl_4$, CH_2Cl_2, 30 min	$CH_2{=}CH(CH_2)_2SC_6H_5$ (81)	365
	$CH_3O(CH_2)_2OCH(OC_2H_5)CH_3$, $TiCl_4$, CH_2Cl_2, $-20°$, 30 min	$CH_2{=}CHCH_2CH(OC_2H_5)CH_3$ (86)	364
	$(C_2H_5O)_2CHCH_2Br$, $TiCl_4$, CH_2Cl_2, $-78°$, 3 h	$CH_2{=}CHCH_2CH(OC_2H_5)CH_2Br$ (90)	352
	$(CH_3O)_2CHCOCH_3$	See Table XI	
	$(CH_3O)_2CHC_3H_7$-*n*, $(CH_3)_3SiO_3SCF_3$ (1%), CH_2Cl_2, $-60°$ to $-45°$, 15 h	$CH_2{=}CHCH_2CH(OCH_3)C_3H_7$-*n* (87)	353
	" $(CH_3)_3SiO_3SCF_3$ (10%), $-78°$, 1 h	" (95)	353, 354
	$([(S)\text{-}C_6H_5CH(CH_3)O]_2CHC_3H_7\text{-}i)$,[a] $TiCl_4$, CH_2Cl_2	$CH_2{=}CHCH_2CH(C_3H_7\text{-}i)OCH(CH_3)C_6H_5$ (50) 90% de	372

TABLE XIX. ALLYLSILANES WITH ACETALS AND KETALS (*Continued*)

Reactant	Conditions	Product(s) and Yield(s)	Refs.
	$(CH_3O)_2CHCH_2CH(OCH_2C_6H_5)CH_3$, CH_2Cl_2	$CH_2{=}CHCH_2CH(OCH_3)CH_2CH(OCH_2C_6H_5)CH_3$ I + $CH_2{=}CHCH_2CH(OCH_3)CH_2CH(OCH_2C_6H_5)CH_3$ II	367
		I:II	
	$TiCl_4$ (0.05 eq), $-78°$, 2 h	1.55:1 (72)	
	", (0.1 eq), ", "	1.28:1 (81)	
	", (1.5 eq), ", "	1.23:1 (86)	
	", $Ti(OC_3H_7\text{-}i)_4$, ", "	2.8:1 (94)	
	$SnCl_4$, ", "	1.12:1 (53)	
	$BF_3{\cdot}O(C_2H_5)_2$, ", "	1.24:1 (67)	
	$ZnCl_2$, 0°	1:1 (62)	
	$(CH_3)_3SiO_3SCF_3$ (cat.), $-78°$, 6 h	1.1:1 (82)	
	$(CH_3O)_2CHCH_2COCH_3$, $AlCl_3$, CH_2Cl_2, $-15°$ to $-10°$, 3 h	$CH_2{=}CHCH_2CH(OCH_3)CH_2COCH_3$ (34)	360
	", $TiCl_4$, CH_2Cl_2, $-18°$ to $-15°$, 5 h	" (60)	360
	$(CH_3O)_2CHC_4H_9\text{-}n$, $TiCl_4$, CH_2Cl_2, $-78°$, 3 h	$CH_2{=}CHCH_2CH(OCH_3)C_4H_9\text{-}n$ (77)	352
	", $(CH_3)_3SiI$ (cat.), CH_2Cl_2, $-78°$, 1 h; $-20°$, 2 h	" (65)	355
	$(C_2H_5O)_2CHC_4H_9\text{-}n$, $TiCl_4$, CH_2Cl_2, $-78°$, 3 h	$CH_2{=}CHCH_2CH(OC_2H_5)C_4H_9\text{-}n$ (76)	352

Reactant and Conditions	Product(s) and Yield(s) (%)	Refs.
$(CH_3O)_2CHC_4H_9$-*i*, $(CH_3)_3SiI$ (cat.), CH_2Cl_2, −78°, 25 min; −30°, 2 h	$CH_2{=}CHCH_2CH(OCH_3)C_4H_9$-*i* (70)	355
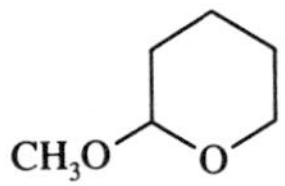 2-methoxytetrahydropyran (CH_3O–), $(CH_3)_3SiO_3SCF_3$ (cat.), −50°, 12 min; −40°, 2 h; 0°, 0.5 h	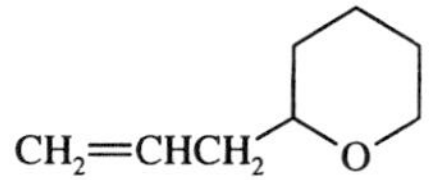 2-allyltetrahydropyran ($CH_2{=}CHCH_2$–) (87)	353, 354
$(C_2H_5O)_2CHCHBrC_3H_7$-*n*, $TiCl_4$, CH_2Cl_2, −78°, 3 h	$CH_2{=}CHCH_2CH(OC_2H_5)CHBrC_3H_7$-*n* (91)	352
4,6-dimethyl-2-[$(CH_2)_2CH{=}CH_2$]-1,3-dioxane	$CH_2{=}CHCH_2CH[(CH_2)_2CH{=}CH_2]$–O–CH($CH_3$)$CH_2CH(OH)CH_3$	
$TiCl_4$, CH_2Cl_2, −78°, 20 min	82% de (92)	671
$TiCl_4$, $Ti(OC_3H_7$-*i*$)_4$, CH_2Cl_2, −78°, 2.5 h	96% de (92)	225
RO-glycoside (cyclohexylidene, $OSi(CH_3)_2C_4H_9$-*t*)	$CH_2{=}CHCH_2$-C-glycoside ($OSi(CH_3)_2C_4H_9$-*t*)	

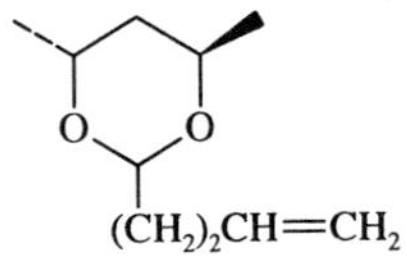

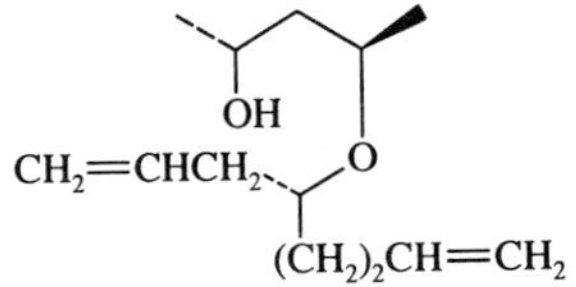

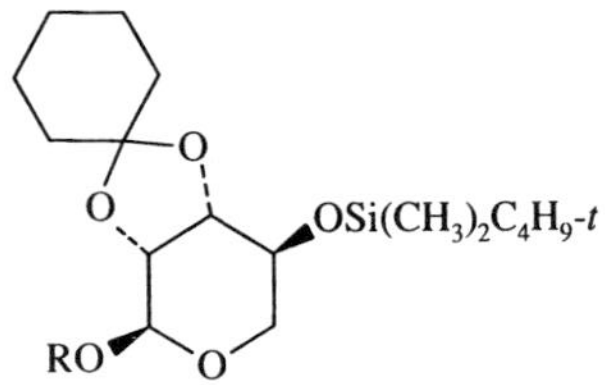

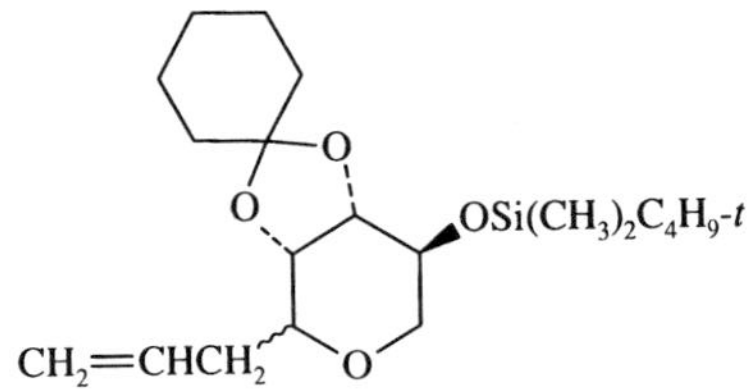

R	Conditions	α:β	Yield	Refs.
CH_3CO	$BF_3{\cdot}O(C_2H_5)_2$, CH_3CN, 3 h	<5:95	(88)[b]	370, 672
"	$ZnBr_2$, 110°, 15 min	1:4	(88)[b]	370, 368
"	$ZnCl_2$, ", 2 h	1:3	(90)[b]	370
"	$(CH_3)_3SiO_3SCF_3$, CH_2Cl_2	0:1	(42)[b]	370
p-$O_2NC_6H_4CO$	$BF_3{\cdot}O(C_2H_5)_2$, CH_3CN, 3 h	1:1	(84)[b]	370
"	$ZnBr_2$, 110°, 15 min	1:5	(88)[b]	370

TABLE XIX. ALLYLSILANES WITH ACETALS AND KETALS (*Continued*)

Reactant	Conditions	Product(s) and Yield(s)	Refs.
	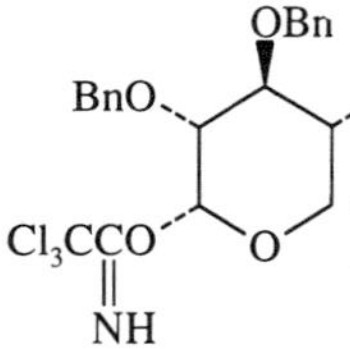 2,3,4-tri-O-benzyl pyranosyl trichloroacetimidate ($Cl_3CCO(=NH)$, BnO, OBn, OBn), $ZnCl_2$, CH_2Cl_2	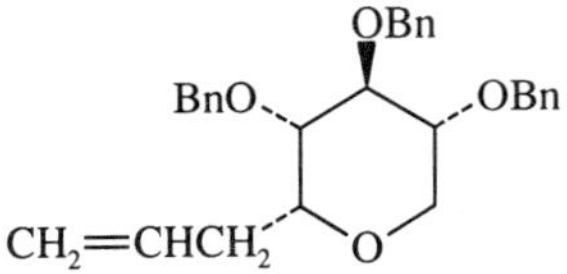 $CH_2=CHCH_2$-pyranoside (BnO, OBn, OBn) α:β 2:1 (59)	673
	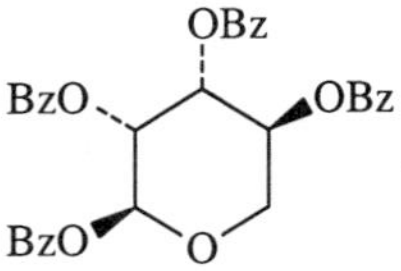 1,2,3,4-tetra-O-benzoyl pyranose (BzO, BzO, OBz, OBz), $ZnBr_2$, 110°, 2 h	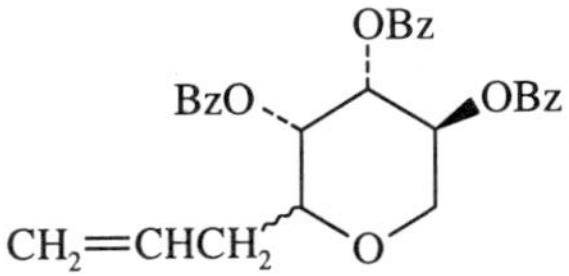 $CH_2=CHCH_2$-pyranoside (BzO, OBz, OBz) α:β 1:1 (100)	370, 368
	" $BF_3{\cdot}O(C_2H_5)_2$, CH_3CN, 10 h	" 5:1 (95)	370
	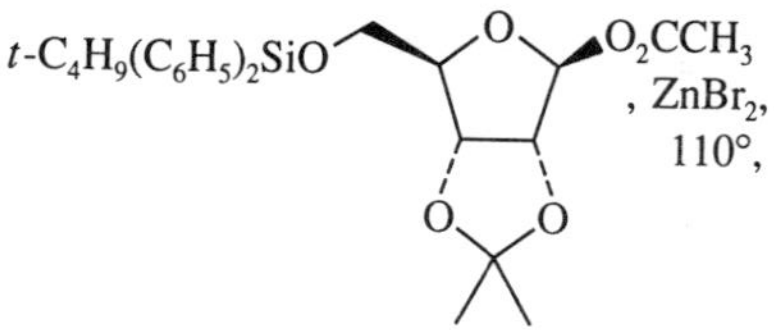 $t\text{-}C_4H_9(C_6H_5)_2SiO$-furanosyl acetate ($O_2CCH_3$), 2,3-O-isopropylidene, $ZnBr_2$, 110°, 3 h 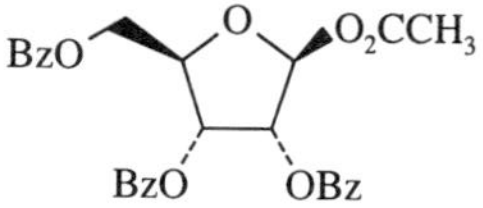	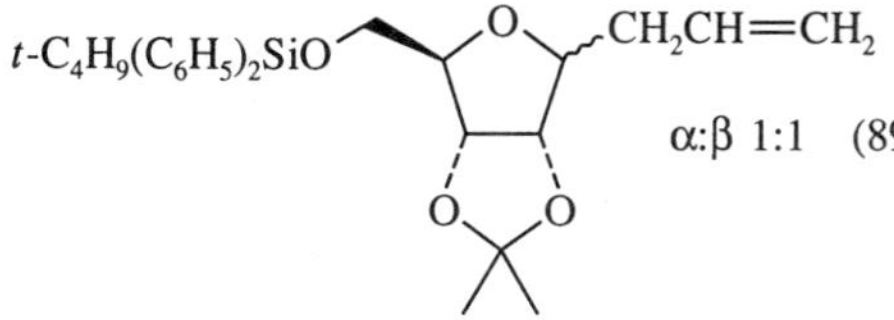 $t\text{-}C_4H_9(C_6H_5)_2SiO$-furanoside, $CH_2CH=CH_2$, 2,3-O-isopropylidene α:β 1:1 (89) 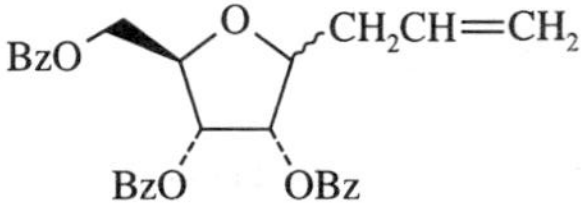	368
	BzO-furanosyl acetate (O_2CCH_3, BzO, OBz)	BzO-furanoside $CH_2CH=CH_2$ (BzO, OBz) α:β	
	$ZnBr_2$, 110°, 2 h	4:1 (85)	370, 368
	$BF_3{\cdot}O(C_2H_5)_2$, CH_3CN, 3 h	7:1 (93)	370
	", CH_2Cl_2, 24 h	2.5:1 (48)	370
	$(CH_3)_3SiO_3SCF_3$, CH_3NO_2, 0°, 5 h	10:1 (83)	674

Reagents and conditions	Product(s) and Yield(s) (%)	Refs.
$(C_6H_5)_3CClO_4$ (cat.), DME, 0°, 1 h	100:0 (90)	675
RCHO, $(CH_3O)_4Si$, $(CH_3)_3SiI$ (cat.), CH_2Cl_2, −78°, 15 min; 0°, 3 h	$CH_2{=}CHCH_2CH(OCH_3)R$	359

R	
n-C_5H_{11}	(94)
n-$C_3H_7CH(CH_3)$	(77)

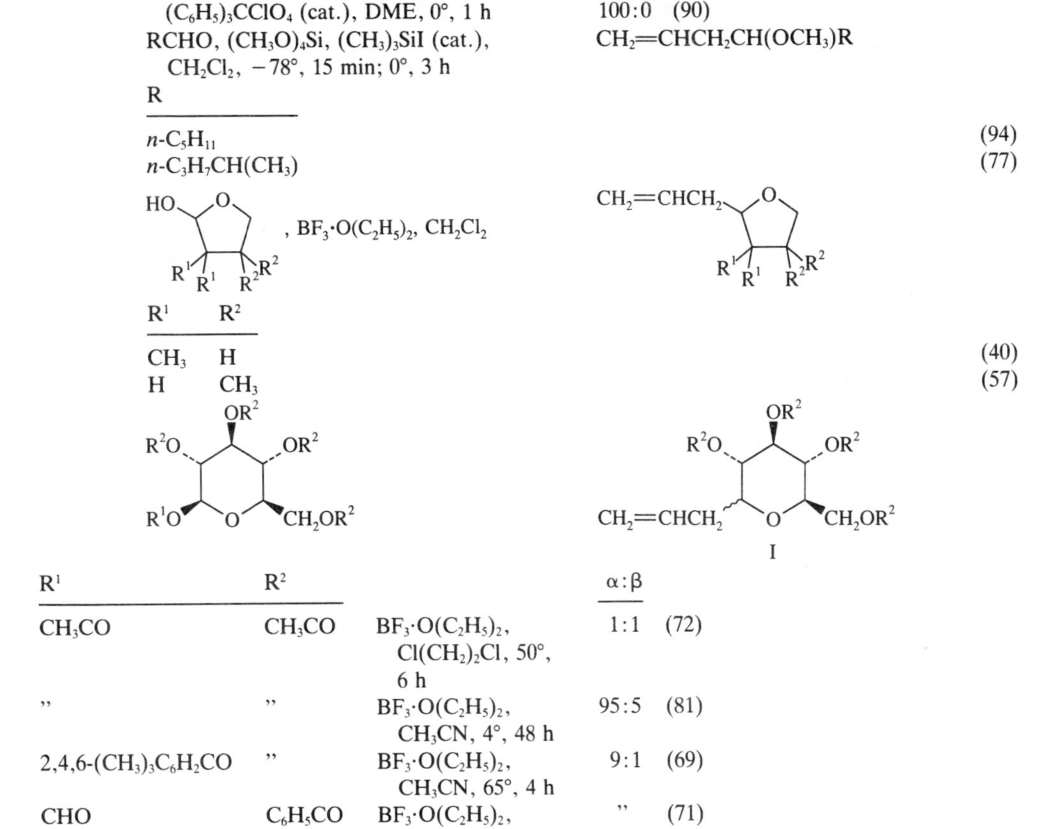

, $BF_3{\cdot}O(C_2H_5)_2$, CH_2Cl_2 — 358

R^1	R^2	
CH_3	H	(40)
H	CH_3	(57)

R^1	R^2	Conditions	α:β	(Yield)	Refs.
CH_3CO	CH_3CO	$BF_3{\cdot}O(C_2H_5)_2$, $Cl(CH_2)_2Cl$, 50°, 6 h	1:1	(72)	673
”	”	$BF_3{\cdot}O(C_2H_5)_2$, CH_3CN, 4°, 48 h	95:5	(81)	673
2,4,6-$(CH_3)_3C_6H_2CO$	”	$BF_3{\cdot}O(C_2H_5)_2$, CH_3CN, 65°, 4 h	9:1	(69)	673
CHO	C_6H_5CO	$BF_3{\cdot}O(C_2H_5)_2$, CH_3CN, 65°, 4 h	”	(71)	676

TABLE XIX. ALLYLSILANES WITH ACETALS AND KETALS (*Continued*)

Reactant	Conditions	Product(s) and Yield(s)	Refs.
	[structure: 2,3,4-tri-O-benzyl-6-O-benzyl pyranoside, RO at anomeric carbon]	I (R^2 = Bn)	
	R	α:β	
	CH_3 — $(CH_3)_3SiO_3SCF_3$, CH_3CN, 16 h	10:1 (86)	254
	$Cl_3CC{=}NH$ — $ZnCl_2$, CH_2Cl_2	" (75)	673
	p-$O_2NC_6H_4$ — $BF_3 \cdot O(C_2H_5)_2$, CH_3CN, 3 h	" (79)	369
	[structure: 1,6-anhydro-2,3,4-tri-O-benzyl sugar], $BF_3 \cdot O(C_2H_5)_2$, CH_3CN, 20 h	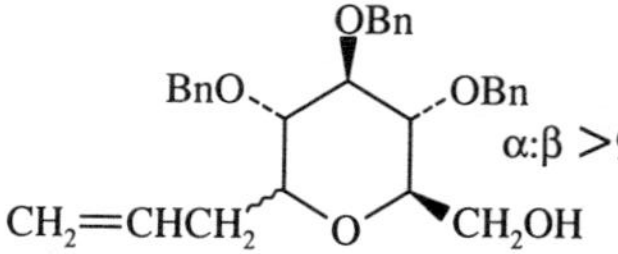 [structure: BnO, OBn, OBn; $CH_2{=}CHCH_2$, CH_2OH] α:β >90:10 (60)	369
	[structure: RO, OAc, OAc, AcO, CH_2OAc pyranose], $BF_3 \cdot O(C_2H_5)_2$,	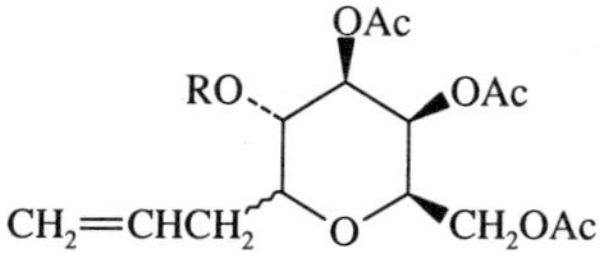 [structure: RO, OAc, OAc; $CH_2{=}CHCH_2$, CH_2OAc]	676
	R	R — α:β	
	H — $Cl(CH_2)_2Cl$, 80°, 24 h	H and CH_3CO[c] — 1:0 (80)	
	" — CH_3CN, 80°, 24 h	H — " (25)	
	CH_3CO — $Cl(CH_2)_2Cl$, 50°, 6 h	CH_3CO — 1:1 (78)	
	" — CH_3CN, 4°, 48 h	" — 95:5 (80)	
	CH_3SO_2 — ", 80°, 24 h	CH_3SO_2 — 99:1 (88)	

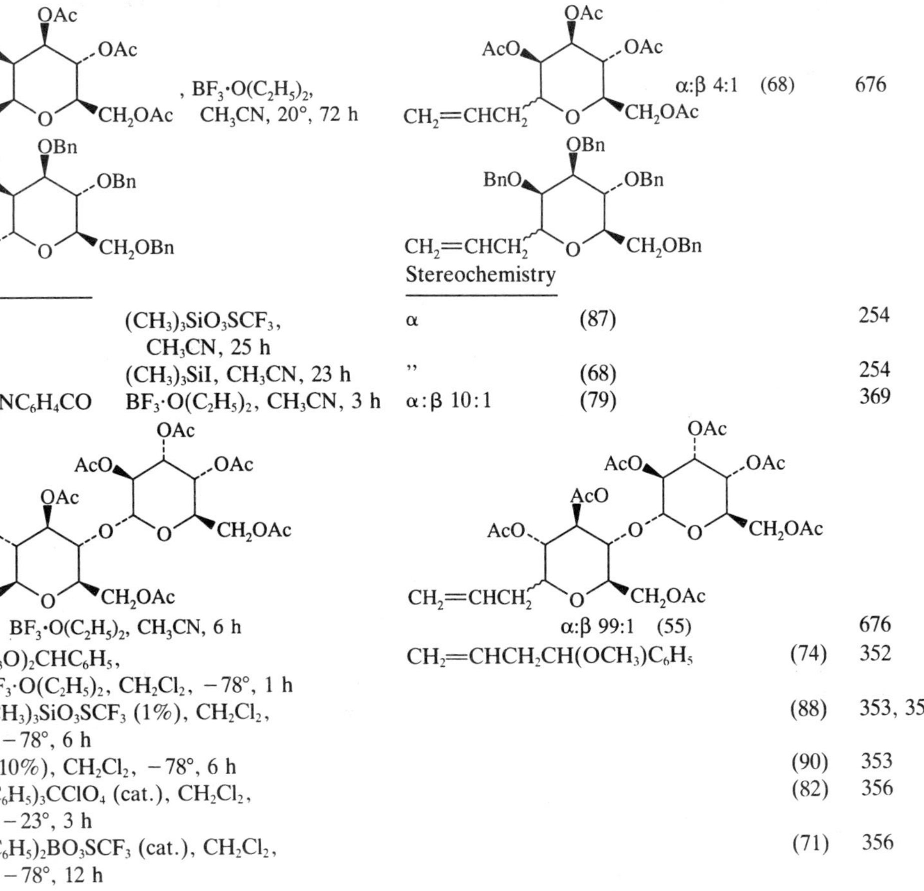

Reactant	Conditions	Product	(%)	Refs.
1,2,3,4,6-penta-O-acetyl pyranose (AcO, AcO, OAc, OAc, CH_2OAc)	, $BF_3 \cdot O(C_2H_5)_2$, CH_3CN, 20°, 72 h	$CH_2{=}CHCH_2$ (tetra-O-acetyl, CH_2OAc) α:β 4:1	(68)	676
RO– tetra-O-benzyl pyranose (BnO, OBn, OBn, CH_2OBn)		$CH_2{=}CHCH_2$ (tetra-O-benzyl, CH_2OBn)		
R		**Stereochemistry**		
CH_3	$(CH_3)_3SiO_3SCF_3$, CH_3CN, 25 h	α	(87)	254
”	$(CH_3)_3SiI$, CH_3CN, 23 h	”	(68)	254
p-$O_2NC_6H_4CO$	$BF_3 \cdot O(C_2H_5)_2$, CH_3CN, 3 h	α:β 10:1	(79)	369
Octa-O-acetyl disaccharide (AcO, AcO, OAc, CH_2OAc; AcO, OAc, OAc, CH_2OAc)	$BF_3 \cdot O(C_2H_5)_2$, CH_3CN, 6 h	$CH_2{=}CHCH_2$ hepta-O-acetyl disaccharide α:β 99:1	(55)	676
$(CH_3O)_2CHC_6H_5$,	$BF_3 \cdot O(C_2H_5)_2$, CH_2Cl_2, −78°, 1 h	$CH_2{=}CHCH_2CH(OCH_3)C_6H_5$	(74)	352
	$(CH_3)_3SiO_3SCF_3$ (1%), CH_2Cl_2, −78°, 6 h		(88)	353, 354
	” (10%), CH_2Cl_2, −78°, 6 h		(90)	353
	$(C_6H_5)_3CClO_4$ (cat.), CH_2Cl_2, −23°, 3 h		(82)	356
	$(C_6H_5)_2BO_3SCF_3$ (cat.), CH_2Cl_2, −78°, 12 h		(71)	356

TABLE XIX. ALLYLSILANES WITH ACETALS AND KETALS (*Continued*)

Reactant	Conditions	Product(s) and Yield(s)	Refs.
	$(CH_3)_3SiI$ (cat.), CH_2Cl_2, −78°, 8 h	(81)	355
	Al-Montmorillonite, CH_2Cl_2, 0°, 0.5 h	(95)	357
	C_6H_5CHO, $(CH_3O)_4Si$, $(CH_3)_3SiI$ (cat.), CH_2Cl_2, −30°, 6 h; room temp, 3 h	(90)	359
	$[(RO)_2CHC_6H_5]$[d], $TiCl_4$, CH_2Cl_2, −75°, 30 min	$CH_2{=}CHCH_2CH(OR)C_6H_5$	372
	R		
	C_2H_5	(95)	
	i-C_3H_7	(90)	
	$CH_2{=}CHCH_2$	(50)	
	[structure: 2-C_6H_{11}-4,6-dimethyl-1,3-dioxane]	[structure: $CH_2{=}CHCH_2CH(C_6H_{11})$–O–CH(CH_3)CH_2CH(OH)CH_3]	
	$TiCl_4$, CH_2Cl_2, −78°, 20 min	82% de (90)	671
	$TiCl_4$, $Ti(OC_3H_7\text{-}i)_4$, CH_2Cl_2, −78°, 2.5 h	90% de (98)	225
	$(CH_3O)_2CHC_7H_{15}\text{-}n$, CH_2Cl_2,	$CH_2{=}CHCH_2CH(OCH_3)C_7H_{15}\text{-}n$	357
	Al-Montmorillonite, 0°, 0.5 h	(95)[e]	
	Silica–alumina, 25°, 2 h	(58)	
	Amberlyst-15, ", "	(47)	
	Nafion-117, ", "	(51)	
	[structure: 1-($CH_3O(CH_2)_2O$)-1,3-dihydroisobenzofuran], $TiCl_4$, CH_2Cl_2, −20°, 30 min	[structure: 1-($CH_2{=}CHCH_2$)-1,3-dihydroisobenzofuran] (90)	364
	$(RO)_2CHCH{=}CHC_6H_5$, CH_2Cl_2	$CH_2{=}CHCH_2CH(OR)CH{=}CHC_6H_5$	

Reactant	Conditions	Product(s) and Yield(s) (%)	Refs.
R			
CH_3	$(CH_3)_3SiI$ (cat.), −30°, 7 h	(84)	355
"	$(C_6H_5)_3CClO_4$ (cat.), −23°, 0.25 h	(80)	356
"	Montmorillonite-K10, −78°, 0.5 h; 0°, 0.5 h; 25°, 0.5 h	(80)	357
C_2H_5	$(CH_3)_3SiI$ (cat.), −78°, 1 h; −60°, 1 h; room temp, 0.5 h	(91)	355
$(RO)_2CH(CH_2)_2C_6H_5$, CH_2Cl_2		$CH_2{=}CHCH_2CH(OR)(CH_2)_2C_6H_5$	
R			
CH_3	$(CH_3)_3SiI$ (cat.), −78°, 6 h	(95)	355
"	$(C_6H_5)_3CClO_4$ (cat.), −23°, 4.5 h	(90)	356
"	$(C_6H_5)_2BO_3SCF_3$ (cat.), −78°, 24 h	(77)	356
$i\text{-}C_3H_7$[e], $TiCl_4$, −75°, 30 min		(65)	372
[cyclic acetal: $(CH_2)_n$, O, O, $C_8H_{17}\text{-}n$], CH_2Cl_2, −78°		[$(CH_2)_n$, HO, O, $CH_2{=}CHCH_2$, $C_8H_{17}\text{-}n$]	
n		de	
0	$TiCl_4$, 20 min	88% (90)	671
1	", "	87% (98)	671
1	$TiCl_4$, $Ti(OC_3H_7\text{-}i)_4$, 2.5 h	96% (98)	225
[dioxanone: O=, O, O, R]	$TiCl_3X$, CH_2Cl_2, −75° → room temp	[HO_2C, O, $CH_2{=}CHCH_2$, R]	371

TABLE XIX. ALLYLSILANES WITH ACETALS AND KETALS (*Continued*)

Reactant	Conditions	Product(s) and Yield(s)	Refs.
	R / X / de	de	
	$C_6H_5(CH_2)_2$, Cl, >98%	87% (95)	
	", OC_3H_7-*i*, "	97%[g] (95)	
	n-C_8H_{17}, ", 90%	96% (99)	
	$(CH_3O)_2CHCH(CH_3)C_6H_5$	CH_2=$CHCH_2$–CH(OCH_3)–CH(CH_3)C_6H_5 + CH_2=$CHCH_2$–CH(OCH_3)–CH(CH_3)C_6H_5 (diastereomer)	367
	BF_3·$O(C_2H_5)_2$, −78°, 1 h	1.7:1 (39)	
	$TiCl_4$, $Ti(OC_3H_7\text{-}i)_4$, −78°, 1 h	1.7:1 (82)	
	$ZnCl_2$, 1 h	1.8:1 (92)	
	$(CH_3)_3SiO_3SCF_3$, −78°, 6 h	2:1 (32)	
	$TiCl_4$, −78°, 1 h	2.8:1 (63)	
	$SnCl_4$ ", "	3.5:1 (76)	
	HO-substituted tetrahydrofuran bearing CO_2CH_3, BF_3·$O(C_2H_5)_2$, CH_2Cl_2	CH_2=$CHCH_2$-substituted tetrahydrofuran bearing CO_2CH_3 (88) *trans*:*cis* 6:1	358
	$([(S)\text{-}C_6H_5CH(CH_3)O]_2CH(CH_2)_2C_6H_5)$[a], $TiCl_4$, CH_2Cl_2	CH_2=$CHCH_2$CH($OCH(CH_3)C_6H_5$)$(CH_2)_2C_6H_5$ (42) 91.5% de	372
	2-X-chroman, CH_2Cl_2,	2-(CH_2=$CHCH_2$)chroman	

Reactant	Conditions	Product(s) and Yield(s) (%)	Refs.
X			
$CH_3O(CH_2)_2O$	$TiCl_4$, −20°, 30 min	(90)	364
C_6H_5S	$SnCl_4$, 30 min	(90)	365
2-X-5-C_6H_{13}-n-tetrahydrofuran	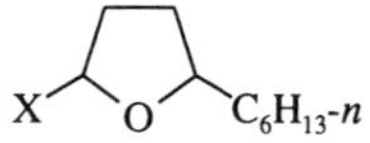 , CH_2Cl_2	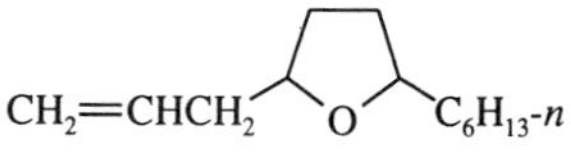 $CH_2{=}CHCH_2$-(tetrahydrofuran)-C_6H_{13}-n I	
X			
$CH_3O(CH_2)_2O$	$TiCl_4$, −20°, 30 min	(96)	364
C_6H_5S	$SnCl_4$, 30 min	(86)	365
"	$TiCl_4$, −20°, 30 min	I (72) + $CH_2{=}CHCH_2CH(SC_6H_5)(CH_2)_2$-$CHOHC_6H_{13}$-n (13)	365
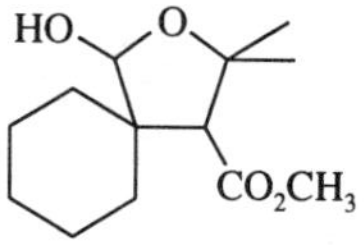 HO-spiro(cyclohexane)tetrahydrofuran, CO_2CH_3	, $BF_3{\cdot}O(C_2H_5)_2$, CH_2Cl_2	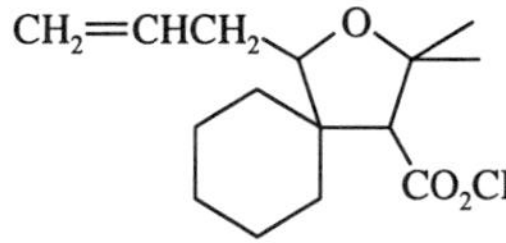 $CH_2{=}CHCH_2$-spiro(cyclohexane)tetrahydrofuran, CO_2CH_3 (67) *trans:cis* 4:1	358
1. 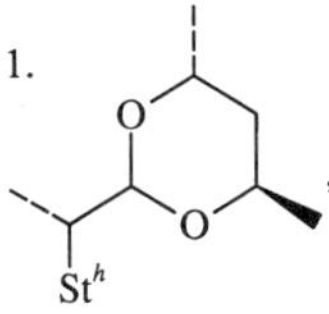 1,3-dioxane acetal (St^h), $TiCl_4$, CH_2Cl_2, −78° 2. PCC, CH_2Cl_2 3. KOH, THF, CH_3OH		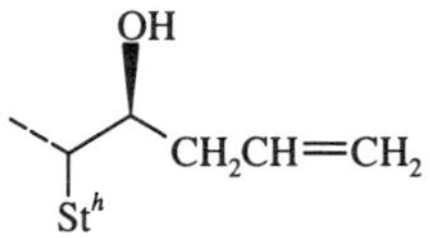 St^h-CH(CH_3)-CH(OH)-$CH_2CH{=}CH_2$ I >99:1 (93)	
1. 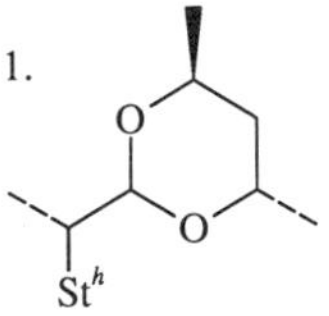 1,3-dioxane acetal (St^h), $TiCl_4$, CH_2Cl_2, −78° 2. PCC, CH_2Cl_2 3. KOH, THF, CH_3OH		I 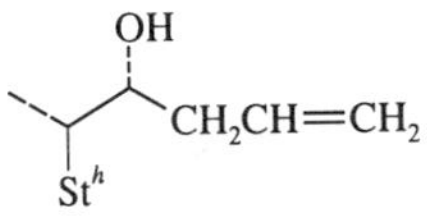 + St^h-CH(CH_3)-CH(OH)-$CH_2CH{=}CH_2$ 90:10 (93)	677

TABLE XIX. ALLYLSILANES WITH ACETALS AND KETALS (*Continued*)

Reactant	Conditions	Product(s) and Yield(s)	Refs.
	$(CH_3O)_2C(CH_3)_2$, $TiCl_4$, CH_2Cl_2, −78°, 3 h	$CH_2{=}CHCH_2C(CH_3)_2OCH_3$ (98)	352
	", $(CH_3)_3SiI$ (cat.), CH_2Cl_2, −78°, 2 h; −50°, 1.5 h	" (83)	355
	$(CH_3O)_2C(C_2H_5)_2$, $(CH_3)_3SiO_3SCF_3$ (cat.) CH_2Cl_2, −78°, 0.2 h; −50°, 1 h; −30°, 1.5 h	$CH_2{=}CHCH_2C(C_2H_5)_2OCH_3$ (89)	353, 354
	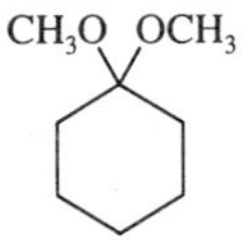 1,1-dimethoxycyclohexane (CH_3O, OCH_3), $TiCl_4$, CH_2Cl_2, −78°, 3 h	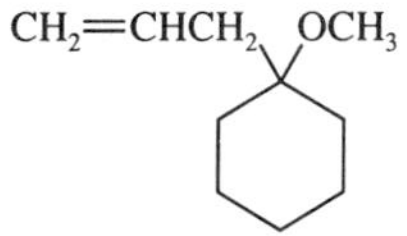 1-allyl-1-methoxycyclohexane ($CH_2{=}CHCH_2$, OCH_3) (71)	352
	", $(CH_3)_3SiO_3SCF_3$ (cat.), CH_2Cl_2, −45°, 18 h	" (81)	353, 354
	", Montmorillonite-K10, CH_2Cl_2, 0°, 0.5 h	" (73)	357
	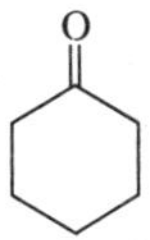 cyclohexanone, $(CH_3O)_4Si$, $(CH_3)_3SiI$ (cat.), CH_2Cl_2, −40°, 30 min; 0°, 4 h	" (90)	359
	$(CH_3O)_2C(CH_3)C_6H_5$, $(CH_3)_3SiO_3SCF_3$, CH_2Cl_2, −78°, 0.2 h; −50°, 2 h	$CH_2{=}CHCH_2C(OCH_3)(CH_3)C_6H_5$ (79)	353, 354
	", $(CH_3)_3SiO_3S(CF_2)_2O$-polymer (cat.), CH_2Cl_2	" (30)	354
	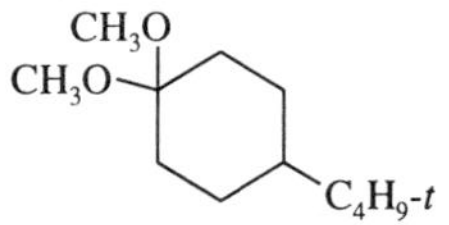 4-C_4H_9-t-1,1-dimethoxycyclohexane (CH_3O, CH_3O, C_4H_9-t), $(CH_3)_3SiO_3SCF_3$, −40°	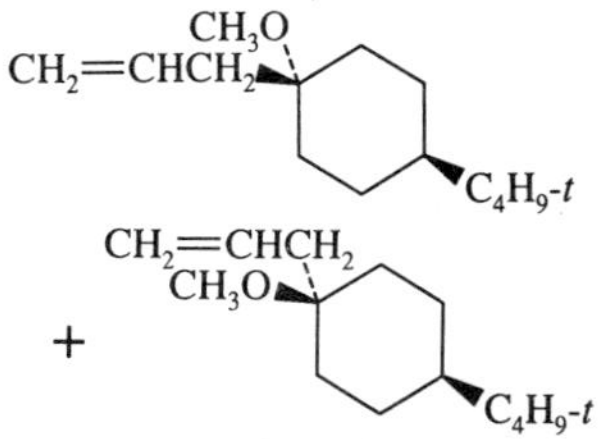 1-allyl-1-methoxy-4-C_4H_9-t-cyclohexane ($CH_2{=}CHCH_2$, CH_3O, C_4H_9-t) + epimer ($CH_2{=}CHCH_2$, CH_3O, C_4H_9-t) 7:93 (89)	353, 354
	Montmorillonite, 0°, 0.5 h	5:95 (88)	357

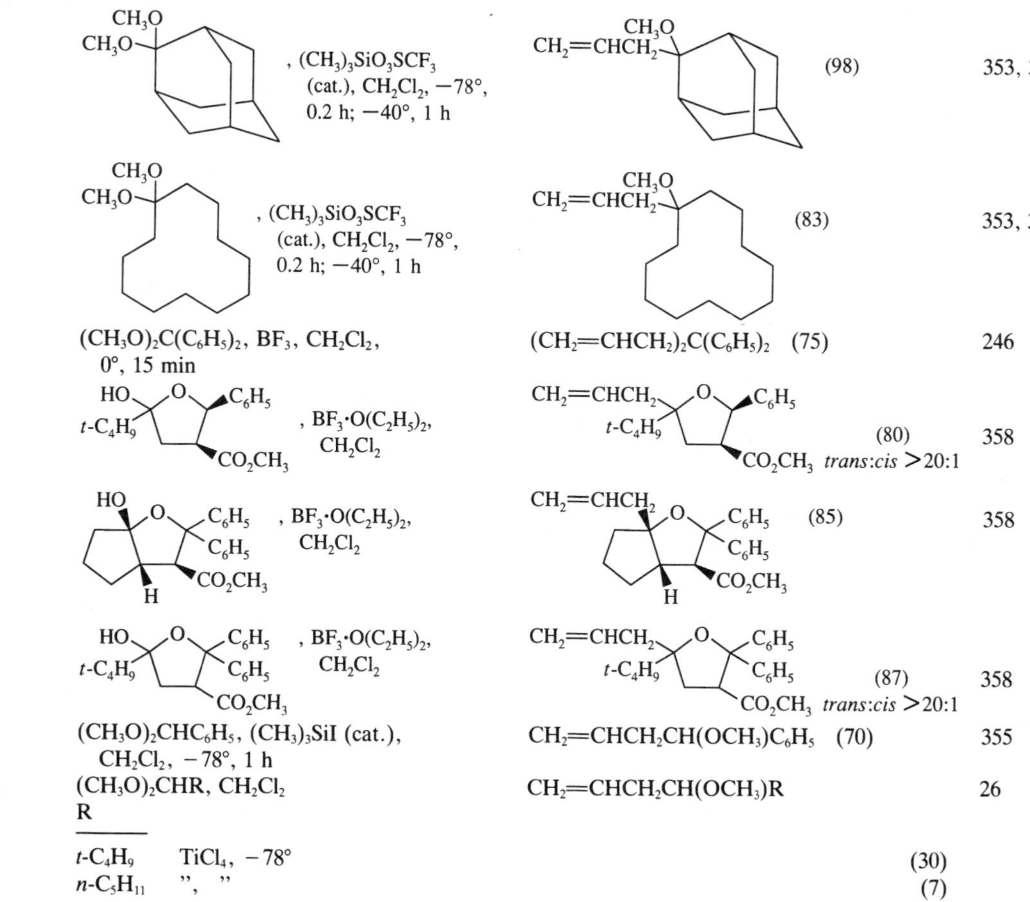

Reagent	Substrate and Conditions	Product(s) and Yield(s) (%)	Refs.
	, $(CH_3)_3SiO_3SCF_3$ (cat.), CH_2Cl_2, −78°, 0.2 h; −40°, 1 h	(98)	353, 354
	, $(CH_3)_3SiO_3SCF_3$ (cat.), CH_2Cl_2, −78°, 0.2 h; −40°, 1 h	(83)	353, 354
	$(CH_3O)_2C(C_6H_5)_2$, BF_3, CH_2Cl_2, 0°, 15 min	$(CH_2{=}CHCH_2)_2C(C_6H_5)_2$ (75)	246
	, $BF_3{\cdot}O(C_2H_5)_2$, CH_2Cl_2	(80) *trans:cis* >20:1	358
	, $BF_3{\cdot}O(C_2H_5)_2$, CH_2Cl_2	(85)	358
	, $BF_3{\cdot}O(C_2H_5)_2$, CH_2Cl_2	(87) *trans:cis* >20:1	358
i-$C_3H_7(CH_3)_2SiCH_2CH{=}CH_2$	$(CH_3O)_2CHC_6H_5$, $(CH_3)_3SiI$ (cat.), CH_2Cl_2, −78°, 1 h	$CH_2{=}CHCH_2CH(OCH_3)C_6H_5$ (70)	355
(S)-1-$C_{10}H_7(C_6H_5)(CH_3)SiCH_2CH{=}CH_2$	$(CH_3O)_2CHR$, CH_2Cl_2	$CH_2{=}CHCH_2CH(OCH_3)R$	26
	R: t-C_4H_9 — $TiCl_4$, −78°	(30)	
	R: n-C_5H_{11} — ", "	(7)	

TABLE XIX. ALLYLSILANES WITH ACETALS AND KETALS (*Continued*)

Reactant	Conditions	Product(s) and Yield(s)	Refs.
	C_6H_5 $BF_3 \cdot O(C_2H_5)_2$, −78°, 1 h	5.5% ee (19)	
	" ", −20°, "	4% ee (29)	
	" ", 20°, "	" (27)	
$(CH_3)_3SiCH_2CH{=}CHCl$	$(CH_3O)_2CHC_6H_5$, $AlCl_3$, CH_2Cl_2, −20°, 2 h	$CH_2{=}CHCHClCH(OCH_3)C_6H_5$ (54)	288
$(CH_3)_3SiCHClCH{=}CH_2$	$(CH_3O)_2CHC_6H_5$, $BF_3 \cdot O(C_2H_5)_2$, CH_2Cl_2, 10 h	$CHCl{=}CHCH_2CH(OCH_3)C_6H_5$ *E*:*Z* 15:85 (85)	637
	$(C_2H_5O)_2CH(CH_2)_2C_6H_5$, CH_2Cl_2, −78°	$CHCl{=}CHCH_2CH(OC_2H_5)(CH_2)_2C_6H_5$ *E*:*Z*	637
	$TiCl_4$, 2 h	10:90 (92)	
	$SnCl_4$, "	12:88 (87)	
	$AlCl_3$, 5 h	13:87 (88)	
	$(C_2H_5O)_2CH(CH_2)_3C_6H_5$, $TiCl_4$, CH_2Cl_2, −78°, 4 h	$CHCl{=}CHCH_2CH(OC_2H_5)(CH_2)_3C_6H_5$ (79) *E*:*Z* 9:91	637
$(CH_3)_3SiCHBrCH{=}CH_2$	$(CH_3O)_2CHC_4H_9$-*n*, $TiCl_4$, CH_2Cl_2, −78°, 5 h	$CHBr{=}CHCH_2CH(OCH_3)C_4H_9$-*n* *E*:*Z* 39:61 (77)	637
	$(CH_3O)_2CHC_6H_5$, $BF_3 \cdot O(C_2H_5)_2$, CH_2Cl_2, 12 h	$CHBr{=}CHCH_2CH(OCH_3)C_6H_5$ *E*:*Z* 21:79 (87)	637
	$(C_2H_5O)_2CH(CH_2)_2C_6H_5$, $TiCl_4$, CH_2Cl_2, −78°, 3 h	$CHBr{=}CHCH_2CH(OC_2H_5)(CH_2)_2C_6H_5$ (83) *E*:*Z* 17:83	637
	$(C_2H_5O)_2CH(CH_2)_3C_6H_5$, $TiCl_4$, CH_2Cl_2, −78°, 2 h	$CHBr{=}CHCH_2CH(OC_2H_5)(CH_2)_3C_6H_5$ (98) *E*:*Z* 29:71	637
$(CH_3)_3SiCH_2CBr{=}CH_2$	[pyranoside: BnO, OBn, OBn, CH_3O, O, CH_2OBn] $(CH_3)_3SiOSO_2CF_3$ (cat.), CH_3CN, 16 h	[pyran: BnO, OBn, OBn, $CH_2{=}CBrCH_2$, O, CH_2OBn] (71)	254

Reactant	Reagent(s) and Conditions	Product(s) and Yield(s) (%)	Refs.
	BnO, OBn, OBn, CH_3O, O, CH_2OBn (pyranoside) $(CH_3)_3SiI$ (cat.), CH_3CN, 12 h	BnO, OBn, OBn, $CH_2{=}CBrCH_2$, O, CH_2OBn (45)	254
C_4			
$(CH_3)_3SiCH_2CH{=}CHCH_3$	$CH_3SCH_2O(CH_2)_2C_6H_5$, $SnCl_4$, CH_2Cl_2, 30 min	$CH_2{=}CHCH(CH_3)CH_2O(CH_2)_2C_6H_5$ (40)	365
	$CH_3O(CH_2)_2OCH_2O(CH_2)_2C_6H_5$, $TiCl_4$, CH_2Cl_2, $-20°$, 30 min	" (94)	364
	CH_3SCH_2O-(cyclohexyl)-C_4H_9-t $SnCl_4$, CH_2Cl_2, 30 min	$CH_2{=}CHCH(CH_3)CH_2O$-(cyclohexyl)-C_4H_9-t (88)	365
	$CH_3O(CH_2)_2OCH_2O$-(cyclohexyl)-C_4H_9-t $TiCl_4$, CH_2Cl_2, $-20°$, 30 min	" (96)	364
	$CH_3O(CH_2)_2OCH(CH_3)OC_2H_5$, $TiCl_4$, CH_2Cl_2, $-20°$, 30 min	$CH_2{=}CHCH(CH_3)CH(OC_2H_5)CH_3$ (90)	364
	$(CH_3O)_2CHC_4H_9$-i, $TiCl_4$, CH_2Cl_2	OCH_3, C_4H_9-i + OCH_3, C_4H_9-i 91:9 (90)[i] 91:9 (99)[j]	366
	$(CH_3O)_2CHC_4H_9$-t, CH_2Cl_2,	OCH_3, C_4H_9-t + OCH_3, C_4H_9-t	366

TABLE XIX. ALLYLSILANES WITH ACETALS AND KETALS (*Continued*)

Reactant	Conditions	Product(s) and Yield(s)	Refs.
	$TiCl_4$	96:4 (63)[i]	
	"	93:7 (83)[j]	
	$(CH_3)_3SiI$ (cat.)	92:8 (71)[j]	
	$(CH_3)_3SiO_3SCF_3$ (cat.)	97:3 (66)[i]	
	"	92:8 (70)[j]	
	2,3,4-Tri-O-Bn-6-O-Bn glycoside (BnO, OBn, OBn, CH_3O, CH_2OBn); $(CH_3)_3SiO_3SCF_3$ (cat.), CH_3CN, 24 h	C-glycoside (BnO, OBn, OBn, CH_2OBn) with $CH_2{=}CHCH(CH_3)$ substituent (68)[k]	254
	$(R^1O)_2CHC_6H_4R^2$-*p*, CH_2Cl_2	syn $CH_2{=}CHCH(CH_3)CH(OR^1)C_6H_4R^2$-*p* + anti $CH_2{=}CHCH(CH_3)CH(OR^1)C_6H_4R^2$-*p*	
	R^1 R^2		
	CH_3 H $BF_3{\cdot}O(C_2H_5)_2$	75:25 (94)[i]	366
	" " "	28:72 (78)[j]	366
	" " $(CH_3)_3SiO_3SCF_3$ (cat.)	79:21 (66)[i]	366
	" " "	20:80 (76)[j]	366
	" " $(CH_3)_3SiI$ (cat.)	69:31 (99)[i]	366
	" " "	28:72 (82)[j]	366
	" " $(C_6H_5)_3CClO_4$ (cat.), −23°, 1 h	71:29 (62)	356
	" " ", −78°, 24 h	64:36 (85)	356
	" " $(C_6H_5)_2BO_3SCF_3$ (cat.), −78°, 24 h	76:24 (69)	356

	C_2H_5	”	$BF_3 \cdot O(C_2H_5)_2$	71:29 (75)[i]	366
	”	”	”	31:69 (82)[j]	366
	CH_3	CH_3O	”	45:55 (56)[i]	366
	”	”	”	46:54 (66)[j]	366
	CH_3	CH_3	”	65:35 (77)[i]	366
	”	”	”	32:68 (87)[j]	366
	CH_3	CN	”	80:20 (84)[i]	366
	”	”	”	20:80 (92)[j]	366
$i\text{-}C_3H_7(CH_3)_2SiCH_2CH{=}CHCH_3$			$(CH_3O)_2CHC_6H_5$, $(CH_3)_3SiI$ (cat.), CH_2Cl_2, −40°, 1.5 h; 0°, 3 h	$CH_2{=}CHCH(CH_3)CH(OCH_3)C_6H_5$ (82)	355
$(CH_3)_3SiCH(CH_3)CH{=}CH_2$			$(CH_3O)_2CHC_6H_5$, $(CH_3)_3SiI$ (cat.), CH_2Cl_2, −40°, 0.5 h	$CH_3CH{=}CHCH_2CH(OCH_3)C_6H_5$ $E + Z$ (85)	355
$(CH_3)_3SiCH_2C(CH_3){=}CH_2$			$(CH_3O)_2CHC_4H_9\text{-}n$, $(CH_3)_3SiI$ (cat.), CH_2Cl_2, −78°, 25 min; −30°, 2.5 h	$CH_2{=}C(CH_3)CH_2CH(OCH_3)C_4H_9\text{-}n$ (69)	355
			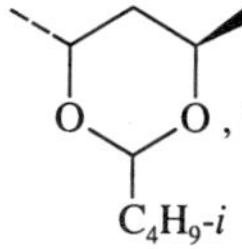 O, O, $C_4H_9\text{-}i$, $TiCl_4$, $Ti(OC_3H_7\text{-}i)_4$, CH_2Cl_2, −78°, 2.5 h	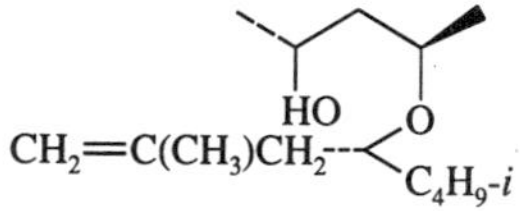 $CH_2{=}C(CH_3)CH_2$, HO, O, $C_4H_9\text{-}i$ >98% de (93)	
			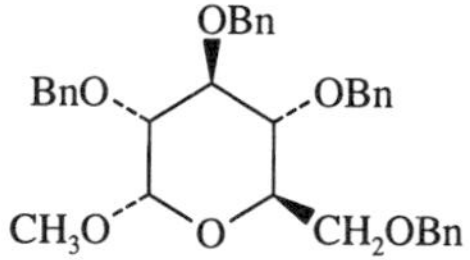 OBn, BnO, OBn, CH_3O, O, CH_2OBn; $(CH_3)_3SiO_3SCF_3$ (cat.), CH_3CN, 10 h	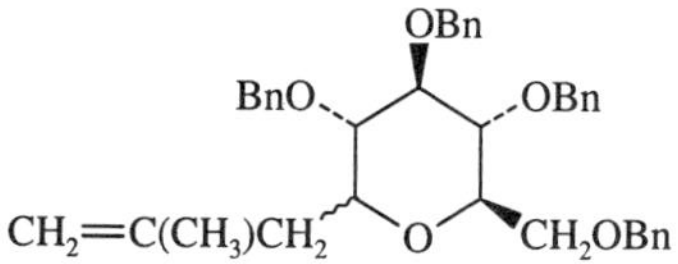 OBn, BnO, OBn, $CH_2{=}C(CH_3)CH_2$, O, CH_2OBn (87); α:β 6:1	254
			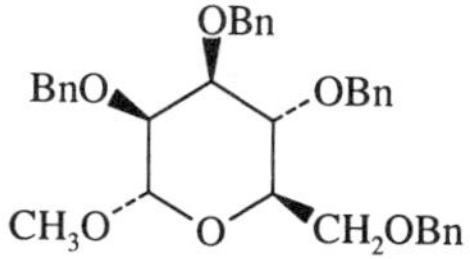 OBn, BnO, OBn, CH_3O, O, CH_2OBn; $(CH_3)_3SiO_3SCF_3$ (cat.), CH_3CN, 8 h	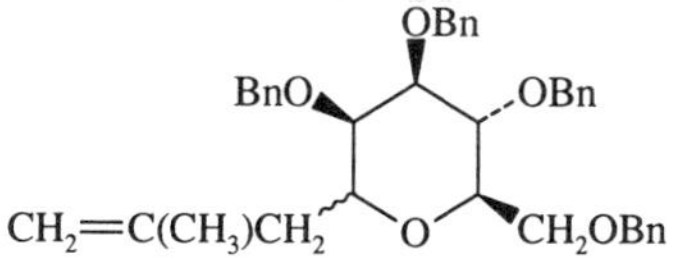 OBn, BnO, OBn, $CH_2{=}C(CH_3)CH_2$, O, CH_2OBn (81); α:β 8:1	254
			$(CH_3O)_2CHC_6H_5$, $(CH_3)_3SiI$ (cat.), CH_2Cl_2, −60°, 1 h; room temp, 0.5 h	$CH_2{=}C(CH_3)CH_2CH(OCH_3)C_6H_5$ (78)	355

TABLE XIX. ALLYLSILANES WITH ACETALS AND KETALS (*Continued*)

Reactant	Conditions	Product(s) and Yield(s)	Refs.
	[structure: 2-R-4,6-dimethyl-1,3-dioxane], $TiCl_4$, $Ti(OC_3H_7\text{-}i)_4$, CH_2Cl_2, −78°, 2.5 h	$CH_2{=}C(CH_3)CH_2CH(R)OCH(CH_3)CH_2CH(OH)CH_3$ [structure]	225
	R		
	C_6H_{11}	>98% de (95)	
	$C_8H_{17}\text{-}n$	" (95)	
	[structure: hydrindane acetal, H, $OCH_2C_6H_5$], 6 $TiCl_4$ + 5 $Ti(OC_3H_7\text{-}i)_4$	[structure: H, $CH_2C(CH_3){=}CH_2$, O, OH, $OCH_2C_6H_5$] (94)	373
$(CH_3)_3SiCHClCH{=}CHCH_3$	$(CH_3O)_2CHC_5H_{11}\text{-}n$, $TiCl_4$, CH_2Cl_2, −78°	$ClCH{=}CHCH(CH_3)CH(OCH_3)C_5H_{11}\text{-}n$ (92)	637
	$(CH_3O)_2CHC_6H_5$, $BF_3{\cdot}O(C_2H_5)_2$, CH_2Cl_2, −78°, 3 h	$ClCH{=}CHCH(CH_3)CH(OCH_3)C_6H_5$ (53) mainly *Z*	637
	$(C_2H_5O)_2CH(CH_2)_2C_6H_5$, $TiCl_4$, CH_2Cl_2, −78°, 3 h	$ClCH{=}CHCH(CH_3)CH(OC_2H_5)(CH_2)_2C_6H_5$ (95) mainly *Z*	637
$(CH_3)_3SiCHClC(CH_3){=}CH_2$	$(C_2H_5O)_2CH(CH_2)_2C_6H_5$, $TiCl_4$, CH_2Cl_2, −78°, 3 h	$ClCH{=}C(CH_3)CH_2CH(OC_2H_5)(CH_2)_2C_6H_5$ (88)	637
$C_6H_5(CH_3)_2SiCH(CO_2CH_3)CH{=}CH_2$	$(CH_3O)_2C(CH_3)_2$, $TiCl_4$ (1.2 eq), 12 h	$CH_3O_2CCH{=}CHCH_2CCl(CH_3)_2$ I (45) + $CH_3O_2CCH{=}CHCH{=}C(CH_3)_2$ II (29)	411
	", ", (4 eq), 1.5 h	I (48) + II (48)	411
$(CH_3)_3SiCH_2C(CO_2C_2H_5){=}CH_2$	$(R^1O)_2CHR^2$, CH_2Cl_2	$CH_2{=}C(CO_2C_2H_5)CH_2CH(OR^1)R^2$	284

Reactant	Conditions	Product(s) and Yield(s) (%)	Refs.
	R^1 R^2		
	C_2H_5 CH_3 — $TiCl_4$, 6 h	(89)	
	CH_3 C_2H_5 — ", 0°, 5 min; 25°, 30 min	(54)	
	" n-C_4H_9 — ", 0°, 11 h	(85)	
	" i-C_4H_9 — ", ", 7 h	(76)	
	" C_6H_5 — $BF_3 \cdot O(C_2H_5)_2$, 0°, 8 h	(89)	
	" $C_6H_5(CH_2)_2$ — $TiCl_4$, 0°, 8 h	(42)	
$(CH_3)_3SiCH_2C(CO_2C_4H_9$-$t)$=$CH_2$	$(CH_3O)_2CHC_6H_5$, $BF_3 \cdot O(C_2H_5)_2$, CH_2Cl_2, 0°, 4 h	CH_2=$C(CO_2C_4H_9$-$t)CH_2$–$CH(OCH_3)C_6H_5$ (67)	284
$(CH_3)_3SiCH_2C[CO_2Si(CH_3)_3]$=$CH_2$	1. $(CH_3O)_2CHC_6H_5$, $BF_3 \cdot O(C_2H_5)_2$, CH_2Cl_2, 0°, 8 h 2. $(CH_3)_3SiI$, CCl_4, 50°, 2 h	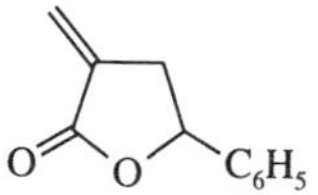 O O C_6H_5 (54)	284
$(CH_3)_3SiCH_2C(CF_3)$=CH_2	$(CH_3O)_2CHC_6H_5$, $TiCl_4$, CH_2Cl_2, 3 h	CH_2=$C(CF_3)CH_2CHClC_6H_5$ (95)	305
	$(CH_3O)_2CHC_6H_5$, $(CH_3)_3SiO_3SCF_3$, CH_2Cl_2, −78°, 3 h; room temp, 24 h	CH_2=$C(CF_3)CH_2CH(OCH_3)C_6H_5$ (38)	305
C_5			
$(CH_3)_3SiCH_2CH$=$C(CH_3)_2$	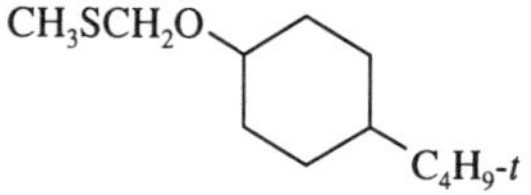 CH_3SCH_2O C_4H_9-t $SnCl_4$, CH_2Cl_2, 30 min	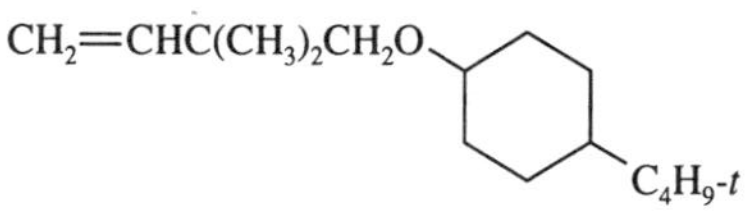 CH_2=$CHC(CH_3)_2CH_2O$ C_4H_9-t (74)	365
	$(CH_3O)_2CHR$, CH_2Cl_2	CH_2=$CHC(CH_3)_2CH(OCH_3)R$	
	R		
	n-C_4H_9 — $TiCl_4$, −78°, 10 min	(83)	83, 352
	C_6H_5 — $BF_3 \cdot O(C_2H_5)_2$, −78°, 30 min; 0°, 5 min	(85)	83
	$C_6H_5(CH_2)_2$ — $TiCl_4$, −78°, 2 min	(93)	83
	" — $(C_6H_5)_3CClO_4$ (cat.), −23°, 20 h	(57)	356

TABLE XIX. Allylsilanes with Acetals and Ketals (*Continued*)

Reactant	Conditions	Product(s) and Yield(s)	Refs.
$(CH_3)_3SiCH_2CH{=}CHCH{=}CH_2$	$(R^1O)_2CHR^2$, CH_2Cl_2; R^1, R^2	(*E*)-$CH_2{=}CHCH{=}CHCH_2CH(OR^1)R^2$	
	R^1 = CH_3, R^2 = *n*-C_4H_9; $TiCl_4$, $-78°$, 10 min	(46)	410
	R^1 = C_2H_5, R^2 = *i*-C_4H_9; ", ", "	(65)	362, 282
	R^1 = CH_3, R^2 = C_6H_5; ", ", "	(79)	362, 282
	R^1 = ", R^2 = "; $BF_3{\cdot}O(C_2H_5)_2$, $-78°$, 1 h	(85)	410
	R^1 = ", R^2 = "; $(CH_3)_3SiI$ (cat.), $-78°$, 10 min	(88)	355
	R^1 = ", R^2 = $C_6H_5(CH_2)_2$; $TiCl_4$, $-78°$, 3 min	(66)	410
	$(CH_3O)_2C(CH_3)_2$, $TiCl_4$, CH_2Cl_2, $-78°$, 10 min	(*E*)-$CH_2{=}CHCH{=}CHCH_2C(CH_3)_2OCH_3$ (46)	282
	[1,1-diethoxycyclohexane], $TiCl_4$, CH_2Cl_2, $-78°$, 10 min	(*E*)-$CH_2{=}CHCH{=}CHCH_2$-[1-ethoxycyclohexyl] (40)	282
$(CH_3)_3SiCH_2C(CH{=}CH_2){=}CH_2$	$(R^1O)_2CHR^2$, CH_2Cl_2; R^1, R^2	$CH_2{=}C(CH{=}CH_2)CH_2CH(OR^1)R^2$	
	R^1 = C_2H_5, R^2 = *n*-C_4H_9; $TiCl_4$, $-78°$, 10 min	(52)	308
	R^1 = CH_3, R^2 = *i*-C_4H_9; ", ", "	(88)	186
	R^1 = ", R^2 = "; $(CH_3)_3SiI$ (cat.), $-78°$, 40 min; $-40°$, 3 h	(90)	355
	R^1 = ", R^2 = $C_6H_5(CH_2)_2$; $TiCl_4$, $-78°$, 5 min	(81)	186
	R^1 = C_2H_5, R^2 = "; ", ", 7 min	(63)	186

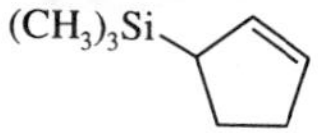 $(CH_3)_3Si$– (cyclopentenyl)	$(CH_3O)_2CHCOCH_3$, $TiCl_4$, −18°, 4 h	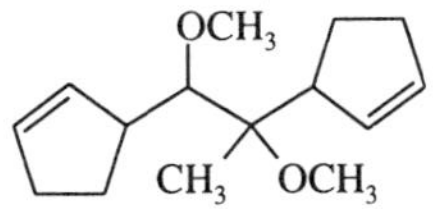 OCH_3, CH_3, OCH_3 (60)	360
	", ", 20°, 3 h	" (83)	360
	$(CH_3O)_2CHCH_2COCH_3$, $TiCl_4$, 0°, 3 h	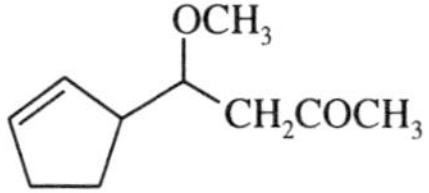 OCH_3, CH_2COCH_3 (55)	360
	", $AlCl_3$, 0°, 4 h	" (59)	360
$(CH_3)_3SiCH(CO_2C_2H_5)C(CH_3){=}CH_2$	$(n\text{-}C_4H_9O)_2CHC_6H_4OCH_3\text{-}p$, $(CH_3)_3SiO_3SCF_3$ (cat.), CH_2Cl_2, −78°, 2 h	$C_2H_5O_2CCH{=}C(CH_3)CH_2CH(OC_4H_9\text{-}n)\text{-}C_6H_4OCH_3\text{-}p$ $E{:}Z$ 97:3 (81)	252
	$(CH_3O)_2C(CH_3)_2$, $TiCl_4$, CH_2Cl_2, −78°, 22 h; 0°, 9 h	$C_2H_5O_2CCH{=}C(CH_3)CH_2C(CH_3)_2OCH_3$ (100) $E{:}Z$ 91:9	252
	", $(CH_3)_3SiO_3SCF_3$, −78°, 26 h	", ", 86:14 (83)	252
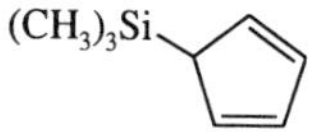 $(CH_3)_3Si$– (cyclopentadienyl)	$(R^1O)_2CR^2R^3$, $(CH_3)_3SiO_3SCF_3$, CH_2Cl_2	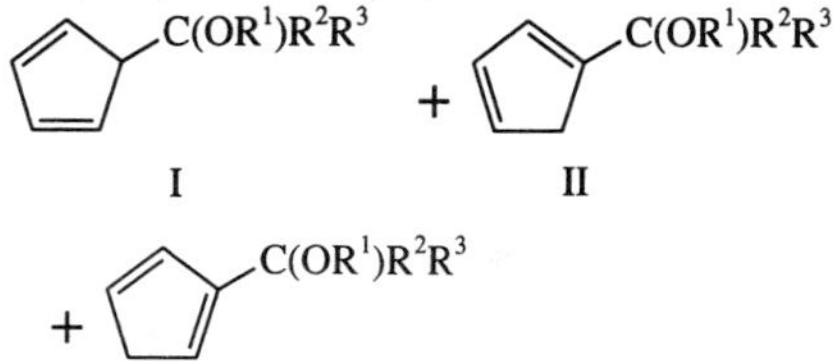 $C(OR^1)R^2R^3$ (I) + $C(OR^1)R^2R^3$ (II) + $C(OR^1)R^2R^3$ (III)	678

R^1	R^2	R^3		I:II:III	
CH_3	$i\text{-}C_3H_7$	H	−40°, 1 h	0:1:0	(78)
C_2H_5	$CH_2{=}CHCH_2C(CH_3)_2$	"	−20°, 0.5 h	0:1:0	(89)
CH_3	CH_3	CH_3	−40°, 0.5 h	8:2:1	(86)

$[(CH_3)_3SiCH_2C(CH_2OC_4H_9\text{-}i){=}C{=}CH_2]$[l]	$(i\text{-}C_4H_9O)_2CH_2$, $TiCl_4$, CH_2Cl_2, −60°, 15 min; 0°, 15 min	$CH_2{=}C(CH_2OC_4H_9\text{-}i)C(CH_2OC_4H_9\text{-}i){=}CH_2$ (70)	679

TABLE XIX. ALLYLSILANES WITH ACETALS AND KETALS (*Continued*)

Reactant	Conditions	Product(s) and Yield(s)	Refs.
C_6			
$(CH_3)_3SiCH_2(CH{=}CH)_2CH_3$	$(CH_3O)_2CH(CH_2)_2C_6H_5$, $TiCl_4$, CH_2Cl_2, $-78°$, 3 min	(E)-$CH_2{=}CHCH{=}CHCH(CH_3)CH(OCH_3)(CH_2)_2C_6H_5$ (66) + (E)-$CH_2{=}CHCH(CH{=}CHCH_3)CH(OCH_3)(CH_2)_2C_6H_5$ (7)	410
$\left[(CH_3)_3SiCH_2C(CH(OC_2H_5)CH_2Br){=}C{=}CH_2\right]^{l}$	$(C_2H_5O)_2CHCH_2Br$, $TiCl_4$, CH_2Cl_2, $-60°$, 15 min; 0°, 15 min	$CH_2{=}C(CH(OC_2H_5)CH_2Br)C(CH(OC_2H_5)CH_2Br){=}CH_2$ (64)	679
C_7			
$(CH_3)_3Si$, OCH_3, OCH_3	$SnCl_4$, CCl_4, 3 min	OCH_3 (72)	363
$(CH_3)_3Si$, OCH_3, OCH_3	$SnCl_4$, CH_2Cl_2, 1°, 5 min	OCH_3 (64)	376
$(CH_3)_3Si$, $Si(CH_3)_3$, CO_2CH_3, CO_2CH_3	$(CH_3O)_2CHC_6H_5$, $(CH_3)_3SiO_3SCF_3$, $(n\text{-}C_4H_9)_2SnCl_2$, CH_2Cl_2	CH_3O_2C, CO_2CH_3, C_6H_5 (44)	227a
	$(CH_3O)_2CHC_{10}H_7$-1, $(CH_3)_3SiO_3SCF_3$, $(n\text{-}C_4H_9)_2SnCl_2$, CH_2Cl_2	CH_3O_2C, CO_2CH_3, $C_{10}H_7$-1 (27)	227a

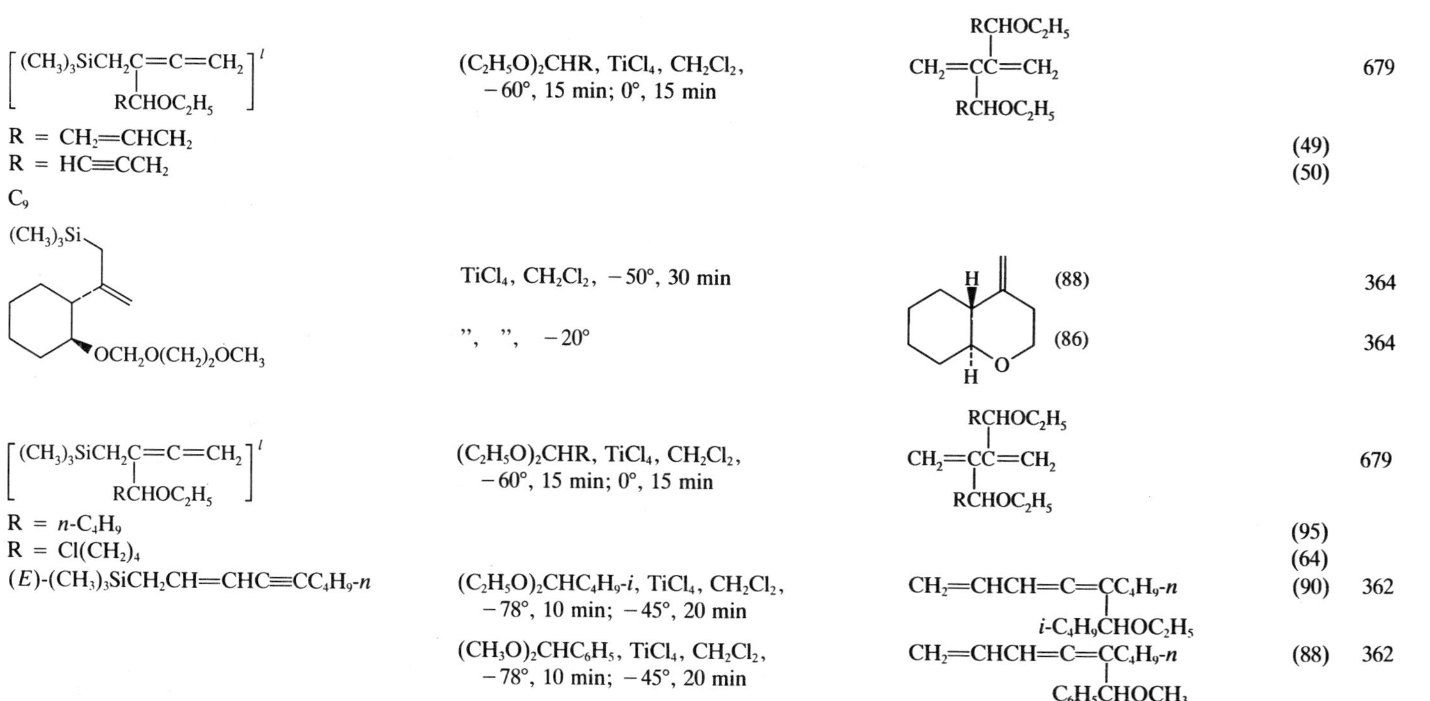

Reactant	Conditions	Product(s) and Yield(s) (%)	Refs.
C_8			
$[(CH_3)_3SiCH_2C(RCHOC_2H_5)=C=CH_2]^{l}$	$(C_2H_5O)_2CHR$, $TiCl_4$, CH_2Cl_2, $-60°$, 15 min; $0°$, 15 min	$CH_2=C(RCHOC_2H_5)C(RCHOC_2H_5)=CH_2$	679
R = $CH_2=CHCH_2$		(49)	
R = $HC\equiv CCH_2$		(50)	
C_9			
$(CH_3)_3Si$ (structure), $OCH_2O(CH_2)_2OCH_3$	$TiCl_4$, CH_2Cl_2, $-50°$, 30 min	(structure) (88)	364
	", ", $-20°$	(86)	364
$[(CH_3)_3SiCH_2C(RCHOC_2H_5)=C=CH_2]^{l}$	$(C_2H_5O)_2CHR$, $TiCl_4$, CH_2Cl_2, $-60°$, 15 min; $0°$, 15 min	$CH_2=C(RCHOC_2H_5)C(RCHOC_2H_5)=CH_2$	679
R = n-C_4H_9		(95)	
R = $Cl(CH_2)_4$		(64)	
(E)-$(CH_3)_3SiCH_2CH=CHC\equiv CC_4H_9$-$n$	$(C_2H_5O)_2CHC_4H_9$-i, $TiCl_4$, CH_2Cl_2, $-78°$, 10 min; $-45°$, 20 min	$CH_2=CHCH=C=C(i\text{-}C_4H_9CHOC_2H_5)C_4H_9$-$n$ (90)	362
	$(CH_3O)_2CHC_6H_5$, $TiCl_4$, CH_2Cl_2, $-78°$, 10 min; $-45°$, 20 min	$CH_2=CHCH=C=C(C_6H_5CHOCH_3)C_4H_9$-$n$ (88)	362

TABLE XIX. ALLYLSILANES WITH ACETALS AND KETALS (*Continued*)

Reactant	Conditions	Product(s) and Yield(s)	Refs.
$(CH_3)_3SiCH_2CH{=}CHC_6H_5$	$(C_2H_5O)_2CHC_4H_9$-*i*, $TiCl_4$, CH_2Cl_2, −78°, 10 min; −45°, 20 min	$CH_2{=}CHCH(C_6H_5)CH(OC_2H_5)C_4H_9$-*i* (86)	362
	$(CH_3O)_2CHC_6H_5$, $TiCl_4$, CH_2Cl_2, −78°, 10 min; −45°, 20 min	$CH_2{=}CHCH(C_6H_5)CH(OCH_3)C_6H_5$ (83)	362
	", $(C_6H_5)_3CClO_4$ (cat.), CH_2Cl_2, −23°, 24 h	" (64)[m]	356
	", $(C_6H_5)_2BO_3SCF_3$ (cat.), CH_2Cl_2, −78°, 48 h	" (73)[n]	356
$(CH_3)_3Si$–CH₂–cyclohexene with CO_2CH_3, CO_2CH_3	$(C_2H_5O)_2CH(CH_2)_2C_6H_5$, $TiCl_4$, CH_2Cl_2	$C_6H_5(CH_2)_2CH(OC_2H_5)$–methylenecyclohexane with CO_2CH_3, CO_2CH_3 (100)	549
$(CH_3)_3Si$–CH₂–cyclohexadiene with CO_2CH_3, CO_2CH_3	$(C_2H_5O)_2CH(CH_2)_2C_6H_5$, $TiCl_4$, CH_2Cl_2	$C_6H_5(CH_2)_2CH(OC_2H_5)$–methylenecyclohexene with CO_2CH_3, CO_2CH_3 (58)	549
C_{10}			
$(CH_3)_3SiCH_2CH{=}CH(CH_2)_4C{-}CCH(OCH_2CH{=}CH_2)_2$ with $(OC)_3Co$ $Co(CO)_3$	$BF_3{\cdot}O(C_2H_5)_2$, CH_2Cl_2	$CH_2{=}CH$, $OCH_2CH{=}CH_2$ cyclooctyne complex with $Co(CO)_3$, $Co(CO)_3$ (75)	614

$(CH_3)_3Si$–CH(CH=CHSi(CH_3)_3)–C(CO_2CH_3)_2(CH_2)_2CH=CH_2	$(CH_3O)_2CHC_6H_5$, $(CH_3)_3SiO_3SCF_3$, $(n\text{-}C_4H_9)_2SnCl_2$, CH_2Cl_2	$CH_2{=}CH(CH_2)_2C(CO_2CH_3)_2CH{=}CHCH{=}CHC_6H_5$ (38)	227a
$(CH_3)_3SiCH(CO_2C_2H_5)CH{=}CHC_6H_5$	$(n\text{-}C_4H_9O)_2CHC_6H_4OCH_3\text{-}p$ $(CH_3)_3SiO_3SCF_3$, CH_2Cl_2, −78°, 4.5 h; room temp, 17 h	$C_2H_5O_2CCH{=}CHCH(C_6H_5)CH(OC_4H_9\text{-}n)\text{-}C_6H_4OCH_3\text{-}p$ $E{:}Z$ 65:35 (60)	252
C_{11}			
$(CH_3)_3SiCH_2CH{=}CHCH_2CH(C_6H_{13}\text{-}n)OCH_2O(CH_2)_2OCH_3$	$TiCl_4$, CH_2Cl_2, −50°, 30 min	3-vinyl-5-(C_6H_{13}-n)tetrahydrofuran: $CH_2{=}CH$, O, C_6H_{13}-n (96)	364
$(CH_3)_3SiCH_2C({=}CH_2)CH(C_6H_5)CH_2OCH_2O(CH_2)_2OCH_3$	$TiCl_4$, CH_2Cl_2, −50°, 30 min	4-methylene-3-C_6H_5-tetrahydropyran: O, C_6H_5 (85)	364
$[(CH_3)_3SiCH_2C(CH(OCH_3)C_6H_5){=}C{=}CH_2]^{l}$	$(CH_3O)_2CHC_6H_5$, $TiCl_4$, CH_2Cl_2, −60°, 15 min; 0°, 15 min	$CH_2{=}C(C_6H_5CHOCH_3)C(C_6H_5CHOCH_3){=}CH_2$ (70)	679
C_{13}			
(E)-$(CH_3)_3SiCH_2C(CO_2C_2H_5){=}CH$–(CH_2)_2–decalin (H) with 1,3-dioxolan-2-yl	TsOH, $(CH_3)_2CO$, reflux, 6 h	α-methylene γ-lactone fused decalin: H, H, H, O, O (78)	656

TABLE XIX. ALLYLSILANES WITH ACETALS AND KETALS (*Continued*)

Reactant	Conditions	Product(s) and Yield(s)	Refs.
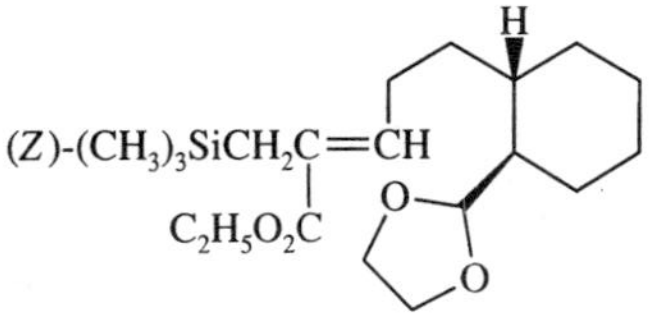	TsOH, $(CH_3)_2CO$, reflux, 7 h	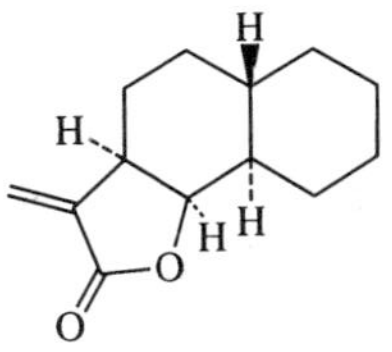 (78)	656

[a] This acetal is prepared in situ from $[(S)\text{-}C_6H_5CH(CH_3)O]_2TiCl_2$ and the aldehyde.
[b] The product is isolated with this yield after desilylation with TBAF.
[c] The product isolated is a 1:1 mixture of the 2-acetate and the free alcohol.
[d] This acetal is prepared in situ from $(RO)_2TiCl_2$ and C_6H_5CHO.
[e] This is the best result from a range of dried clays. The preparative yield is 90%.
[f] This acetal is prepared in situ from $(i\text{-}C_3H_7O)_2TiCl_2$ and $C_6H_5(CH_2)_2CHO$.
[g] Reaction at room temperature gives 95% de.
[h] The steroid skeleton used in this work is

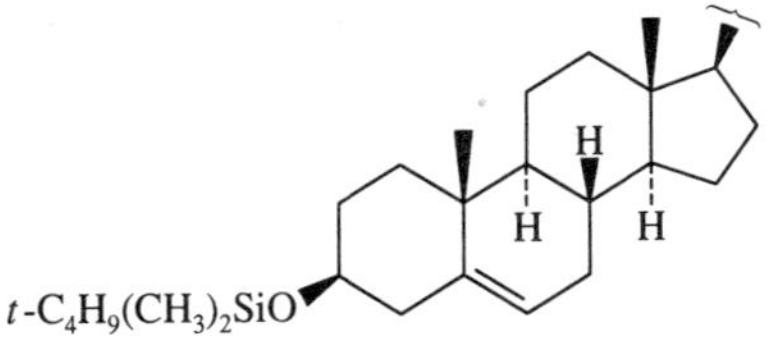

[i] This is the result with the *E* allylsilane, *E*:*Z* 94:6.
[j] This is the result with the *Z* allylsilane, *E*:*Z* 11:89.
[k] All four possible diastereoisomers are detected.
[l] This allylsilane is an intermediate in the reaction of $(CH_3)_3SiCH_2C{\equiv}CCH_2Si(CH_3)_3$ with the appropriate acetal. It is isolated as the product when the acetal is not used in excess. The products shown in the table are the result of using a five-fold excess of the acetal relative to the propargylsilane.
[m] The product is a 67:33 mixture of diastereoisomers.
[n] The product is an 82:18 mixture of diastereoisomers.

TABLE XX. VINYLSILANES WITH ACETALS AND KETALS

Reactant	Conditions	Product(s) and Yield(s)	Refs.
C_4			
$(CH_3)_3SiCX{=}CH(CH_2)_2OCH_2O(CH_2)_2OCH_3$		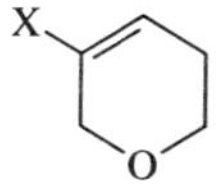 X	377
X			
Z H	$SnCl_4$, CH_2Cl_2, −20°, 1 h	(71)	
E Br	$TiCl_4$, ", −60°, 2 h	(78)	
C_5			
(*Z*)-$(CH_3)_3SiCH{=}C(CH_3)(CH_2)_2OCH_2O(CH_2)_2OCH_3$	$SnCl_4$, CH_2Cl_2, −20°, 1 h	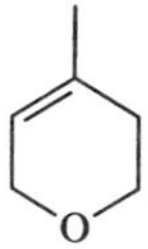 (65)	377
C_6			
$(CH_3)_3SiC({=}CH_2)(CH_2)_4OCH_2O(CH_2)_2OCH_3$	$SnCl_4$, CH_2Cl_2, −15°, 9 h	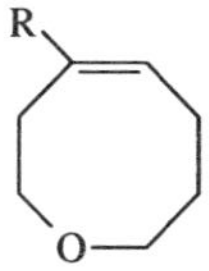 R = H (51) + R = $(CH_3)_3Si$ (10)	379
(*Z*)-$(CH_3)_3SiCH{=}CH(CH_2)_3CH(SCH_3)_2$	$(CH_3)_2\overset{+}{S}SCH_3$ $^-BF_4$, CH_2Cl_2, 20 h	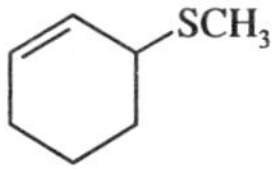 SCH_3 (37)	380
C_7			
(*Z*)-$(CH_3)_3SiCH{=}CH(CH_2)_3C(SCH_3)_2CH_3$	$(CH_3)_2\overset{+}{S}SCH_3$ $^-BF_4$, CH_2Cl_2, 20 h	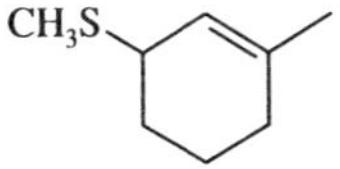 CH_3S (68)	380

TABLE XX. VINYLSILANES WITH ACETALS AND KETALS (*Continued*)

Reactant	Conditions	Product(s) and Yield(s)	Refs.
(*Z*)-$(CH_3)_3SiCH{=}CH(CH_2)_3C(SC_2H_5)_2CH_3$		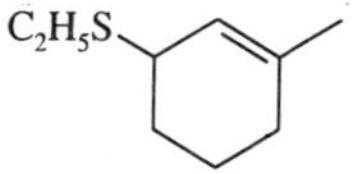 " + [structure: C_2H_5S-substituted methylcyclohexene] (—) 1:1	380
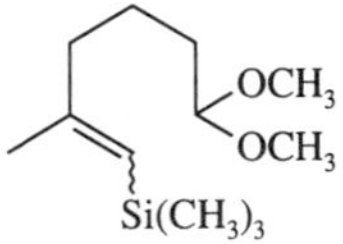 [structure: OCH_3, OCH_3, $Si(CH_3)_3$]	$ZnBr_2$ (cat.), CCl_4, 10°	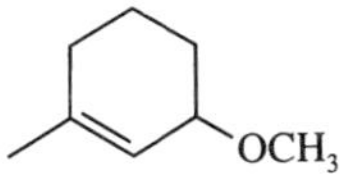 [structure: methylcyclohexene with OCH_3]	376
E	24 h	(55)	
Z	12 h	(70)	
$(CH_3)_3SiC(=CH_2)(CH_2)_5OCH_2O(CH_2)_2OCH_3$	$SnCl_4$, CH_2Cl_2, 25°, 4 d	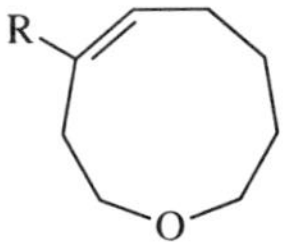 [structure: R-substituted oxacyclic alkene] R = H (50–60) + R = $Si(CH_3)_3$ (15–20)	379
$(CH_3)_3SiC(=CH_2)(CH_2)_3CH(CH_3)OCH(CH_3)OC_2H_5$	$SnCl_4$, CH_2Cl_2, 20°, 1 h	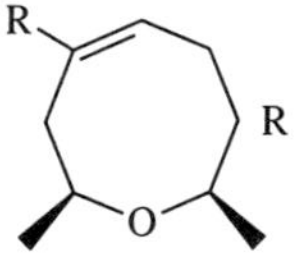 [structure: R-substituted dimethyl oxocene] R = H + R = $(CH_3)_3Si$ (34) *cis*:*trans* 30:1 + $CH_3CHOHCH_2CH[Si(CH_3)_3](CH_2)_3COCH_3$ (51)	379
C_8			
(*E*)-$(CH_3)_3SiC(=CHC_4H_9\text{-}n)(CH_2)_2OCH_2O(CH_2)_2OCH_3$	$SnCl_4$, CH_2Cl_2, −5°, 12 h	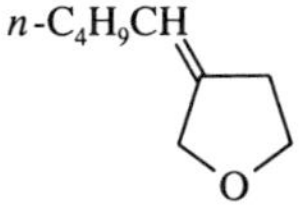 [structure: n-C_4H_9CH= tetrahydrofuran] *E*:*Z* 97:3 (81)	377

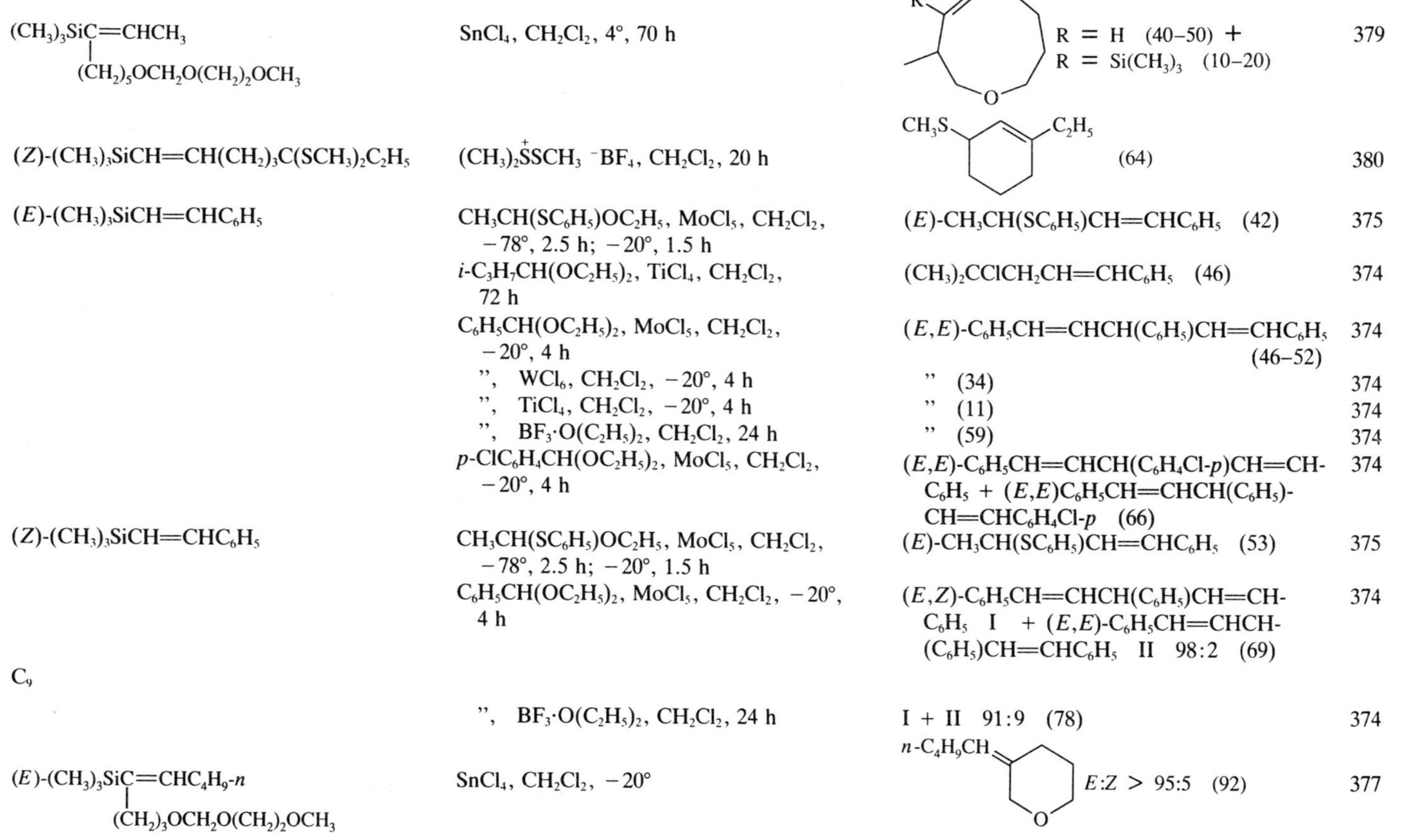

$(CH_3)_3SiC(=CHCH_3)(CH_2)_5OCH_2O(CH_2)_2OCH_3$	$SnCl_4$, CH_2Cl_2, 4°, 70 h	[oxocene structure] R = H (40–50) + R = $Si(CH_3)_3$ (10–20)	379
(Z)-$(CH_3)_3SiCH{=}CH(CH_2)_3C(SCH_3)_2C_2H_5$	$(CH_3)_2\overset{+}{S}SCH_3$ $^-BF_4$, CH_2Cl_2, 20 h	[3-CH_3S-1-C_2H_5-cyclohexene structure] (64)	380
(E)-$(CH_3)_3SiCH{=}CHC_6H_5$	$CH_3CH(SC_6H_5)OC_2H_5$, $MoCl_5$, CH_2Cl_2, −78°, 2.5 h; −20°, 1.5 h	(E)-$CH_3CH(SC_6H_5)CH{=}CHC_6H_5$ (42)	375
	i-$C_3H_7CH(OC_2H_5)_2$, $TiCl_4$, CH_2Cl_2, 72 h	$(CH_3)_2CClCH_2CH{=}CHC_6H_5$ (46)	374
	$C_6H_5CH(OC_2H_5)_2$, $MoCl_5$, CH_2Cl_2, −20°, 4 h	(E,E)-$C_6H_5CH{=}CHCH(C_6H_5)CH{=}CHC_6H_5$ (46–52)	374
	", WCl_6, CH_2Cl_2, −20°, 4 h	" (34)	374
	", $TiCl_4$, CH_2Cl_2, −20°, 4 h	" (11)	374
	", $BF_3{\cdot}O(C_2H_5)_2$, CH_2Cl_2, 24 h	" (59)	374
	p-$ClC_6H_4CH(OC_2H_5)_2$, $MoCl_5$, CH_2Cl_2, −20°, 4 h	(E,E)-$C_6H_5CH{=}CHCH(C_6H_4Cl$-$p)CH{=}CH$-C_6H_5 + (E,E)$C_6H_5CH{=}CHCH(C_6H_5)$-$CH{=}CHC_6H_4Cl$-p (66)	374
(Z)-$(CH_3)_3SiCH{=}CHC_6H_5$	$CH_3CH(SC_6H_5)OC_2H_5$, $MoCl_5$, CH_2Cl_2, −78°, 2.5 h; −20°, 1.5 h	(E)-$CH_3CH(SC_6H_5)CH{=}CHC_6H_5$ (53)	375
	$C_6H_5CH(OC_2H_5)_2$, $MoCl_5$, CH_2Cl_2, −20°, 4 h	(E,Z)-$C_6H_5CH{=}CHCH(C_6H_5)CH{=}CH$-$C_6H_5$ I + (E,E)-$C_6H_5CH{=}CHCH$-$(C_6H_5)CH{=}CHC_6H_5$ II 98:2 (69)	374
C_9			
	", $BF_3{\cdot}O(C_2H_5)_2$, CH_2Cl_2, 24 h	I + II 91:9 (78)	374
(E)-$(CH_3)_3SiC(=CHC_4H_9$-$n)(CH_2)_3OCH_2O(CH_2)_2OCH_3$	$SnCl_4$, CH_2Cl_2, −20°	[3-(n-$C_4H_9CH{=}$)tetrahydropyran structure] E:Z > 95:5 (92)	377

TABLE XX. VINYLSILANES WITH ACETALS AND KETALS (*Continued*)

Reactant	Conditions	Product(s) and Yield(s)	Refs.
(*Z*)-$(CH_3)_3SiC(=CHC_4H_9\text{-}n)(CH_2)_3OCH_2O(CH_2)_2OCH_3$	", ", "	", E:Z > 0.5:99.5 (89)	377
C_{10}			
(*E*)-$(CH_3)_3SiC(=C(C_2H_5)C_4H_9\text{-}n)(CH_2)_2OCH_2O(CH_2)_2OCH_3$	$SnCl_4$, CH_2Cl_2, −10°, 6 h	3-[1-ethylpentylidene]tetrahydrofuran (n-C_4H_9, C_2H_5) E:Z 40:60 (89)	377
(*Z*)-$(CH_3)_3SiC(=C(C_2H_5)C_4H_9\text{-}n)(CH_2)_2OCH_2O(CH_2)_2OCH_3$	", ", ", "	", E:Z 40:60 (86)	377
$(CH_3)_3SiC(=CHC_4H_9\text{-}n)(CH_2)_4OCH_2O(CH_2)_2OCH_3$	$SnCl_4$, CH_2Cl_2, −15°, 19 h	4-pentylideneoxepane (n-C_4H_9) (35) + 4-$(CH_3)_3Si$-3-n-C_4H_9-oxocene (35)	379
(*Z*)-$(CH_3)_3SiCH=CH(CH_2)_3C(SCH_3)_2(CH_2)_2CH(CH_3)OSi(CH_3)_3$	$(CH_3)_2\overset{+}{S}SCH_3$ $^-BF_4$, CH_2Cl_2, 20 h	3-CH_3S-1-[$(CH_2)_2CH(CH_3)OSi(CH_3)_3$]cyclohexene (67)	380

Reactant	Conditions	Product(s) (Yield, %)	Refs.
(Z)-$(CH_3)_3SiCH{=}CH(CH_2)_3C(SCH_3)_2$ / $CH_3CO(CH_2)_3$	$(CH_3)_2\overset{+}{S}SCH_3$ $^{-}BF_4$, CH_2Cl_2	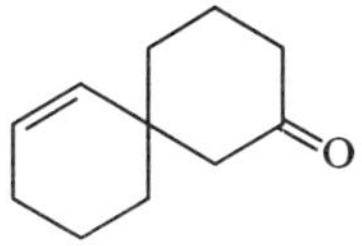 (—)	380
C_{12}			
(*E*)-$(CH_3)_3SiC{=}CHC_4H_9$-*n* / $(CH_2)_3C(CH_3)_2OCH_2O(CH_2)_2OCH_3$	$SnCl_4$, CH_2Cl_2, $-15°$, 24 h	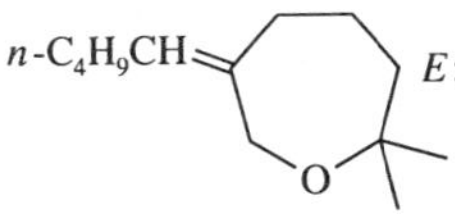 *n*-$C_4H_9CH{=}$ *E*:*Z* > 98:2 (57)	377
(*Z*)-$(CH_3)_3SiC{=}CHC_4H_9$-*n* / $(CH_2)_3C(CH_3)_2OCH_2O(CH_2)_2OCH_3$	", ", ", 12 h	", ", <2:98 (71)	377
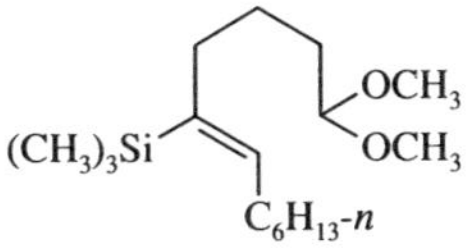 $(CH_3)_3Si$, OCH_3, OCH_3, C_6H_{13}-*n*	$TiCl_4$, CH_2Cl_2, $-78°$, 2 min	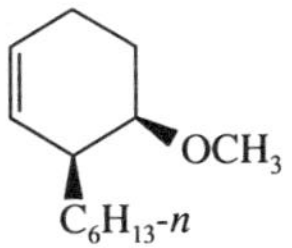 OCH_3, C_6H_{13}-*n* (33)	120
$(CH_3)_3SiCH{=}CHCH_2COHC_6H_4OC_2H_5$-*p* / $(CH_3O)_2CHCH_2$	$TiCl_4$, CH_2Cl_2, $(C_2H_5)_2O$, 0°	$C_6H_5C_6H_4OC_2H_5$-*p* (49)	378
C_{13}			
$(CH_3)_3SiC{=}CH(CH_2)_2OCH_2O(CH_2)_2OCH_3$ / $(CH_2)_3C_6H_5$	$TiCl_3(OC_3H_7$-*i*$)$, CH_2Cl_2, $-20°$, 2 h	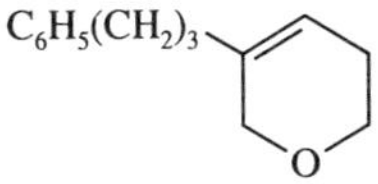 $C_6H_5(CH_2)_3$ (83)	377
$(CH_3)_3SiCH{=}CHCH_2COHC_6H_4CH_3$-*p* / $(CH_3O)_2CHCH_2$	$TiCl_4$, CH_2Cl_2, $(C_2H_5)_2O$, 0°	$C_6H_5C_6H_4CH_3$-*p* (59)	378
C_{14}			
$(CH_3)_3SiCH{=}CHCH_2COHC_6H_4CH_3$-*p* / $(CH_3O)_2CHCHCH_3$	$TiCl_4$, CH_2Cl_2, $(C_2H_5)_2O$, 0°, 4 h	*o*-$CH_3C_6H_4C_6H_4CH_3$-*p* (44)	378

TABLE XX. VINYLSILANES WITH ACETALS AND KETALS (*Continued*)

Reactant	Conditions	Product(s) and Yield(s)	Refs.
1-Hydroxy-1-[$CH_2CH{=}CHSi(CH_3)_3$]-2-[$CH(OCH_3)_2$]-6-CH_3O-1,2,3,4-tetrahydronaphthalene	$TiCl_4$, CH_2Cl_2, $(C_2H_5)_2O$, 0°, 4 h	CH_3O-9,10-dihydrophenanthrene (12)	378
C_{16}			
$(CH_3)_3SiCH{=}CHCH_2C(OH)(C_{10}H_7\text{-}1)CH_2CH(OCH_3)_2$	$TiCl_4$, CH_2Cl_2, $(C_2H_5)_2O$, 0°	$C_6H_5C_{10}H_7$-1 (44)	378

TABLE XXI. ALLENYSILANES WITH ACETALS

Reactant	Conditions	Product(s) and Yield(s)	Refs.
C_3			
$(CH_3)_3SiCH{=}C{=}CH_2$	1. $(CH_3O)_2CH(CH_2)_2C_6H_5$, $TiCl_4$ CH_2Cl_2, −78°, 1 h 2. KF, DMSO, 4 h	$HC{\equiv}CCH_2CH(OCH_3)(CH_2)_2C_6H_5$ (75)	381

TABLE XXII. ALLYLSILANES WITH α,β-UNSATURATED ACETALS AND VINYLOGOUS ACETALS

Reactant	Conditions	Product(s) and Yield(s)	Refs.
C_3			
$(CH_3)_3SiCH_2CH{=}CH_2$	$(CH_3O)_2CHCH{=}CHCH_3$, $TiCl_4$, CH_2Cl_2, −78°, 3 h	$CH_2{=}CHCH_2CH{=}CHCH(CH_3)$-$CH_2CH{=}CH_2$ (61)[a]	382
	$(C_2H_5O)_2CHCH{=}CHCH_3$, $TiCl_4$, CH_2Cl_2, −78°, 3 h	" (33)[a]	382
	$(C_2H_5O)_2CHCH{=}CHC_3H_7$-$n$, $AlCl_3$, CH_2Cl_2, −78°, 3 h	$CH_2{=}CHCH_2CH(OC_2H_5)CH{=}CHC_3H_7$-$n$ (27)	382
	", $TiCl_4$, CH_2Cl_2, −78°, 3 h	$CH_2{=}CHCH_2CH{=}CHCH(C_3H_7$-$n)$-$CH_2CH{=}CH_2$ (21)[a]	382
	3-CH_3CO_2-dihydropyran bearing R^1, R^2 (C-3, R^1, R^2 below), CH_2Cl_2	1-$CH_2{=}CHCH_2$-dihydropyran bearing R^1, R^2; C-1 Configuration	
	C-3: β; R^1: H; R^2: CH_2SCH_3; $TiCl_4$, −78°, 30 min	α (95)	383
	C-3: "; R^1: "; R^2: $CH_2OCH_2C_6H_5$; " −78°, "	α (89)	383
	C-3: "; R^1: α-O_2CCH_3; R^2: $CH_2O_2CCH_3$; " −78°, 20	α:β 16:1 (85)	383, 680
	C-3: "; R^1: "; R^2: "; $BF_3{\cdot}O(C_2H_5)_2$, −50°, 1.5 h	", " (94)	384
	C-3: α; R^1: "; R^2: "; $TiCl_4$, −78°, 20 min	", 6:1 (95)	383
	C-3: β; R^1: β-O_2CCH_3; R^2: "; ", ", "	", 30:1 (93)	383
	C-3: "; R^1: H; R^2: C_3H_7-n; ", ", "	α (82)	383
	C-3: "; R^1: "; R^2: C_6H_5; ", ", "	" (60)	383
	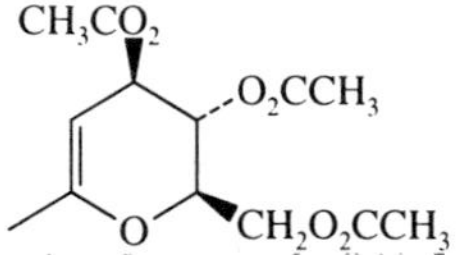 6-methyl-3-CH_3CO_2-4-O_2CCH_3-5-$CH_2O_2CCH_3$-dihydropyran, $TiCl_4$, CH_2Cl_2, −78°	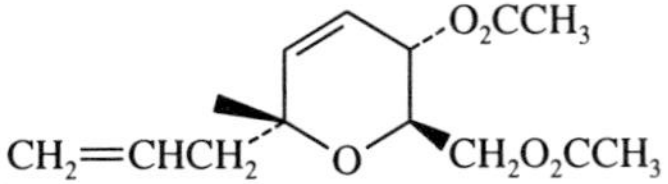 1-methyl-1-($CH_2{=}CHCH_2$)-4-O_2CCH_3-5-$CH_2O_2CCH_3$-dihydropyran (75)	681
	$(CH_3O)_2CHCH{=}CHC_6H_5$, $(CH_3)_3SiO_3SCF_3$ (cat.), CH_2Cl_2, −78°, 20 min	$CH_2{=}CHCH_2CH(OCH_3)CH{=}CHC_6H_5$ (78)	353, 354

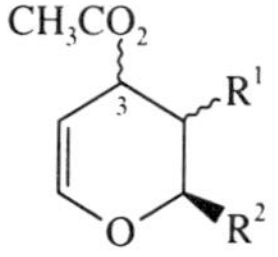

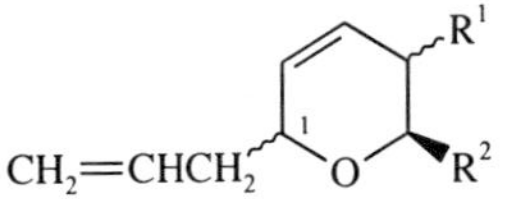

Allylsilane	Reactant and Conditions	Product(s) and Yield(s) (%)	Refs.
	", $TiCl_4$, $Ti(OC_3H_7\text{-}i)_4$, CH_2Cl_2, 0°, 1.5 h	$CH_2{=}CHCH_2CH(OC_3H_7\text{-}i)CH{=}CHC_6H_5$ (81)	382
	", ", CH_2Cl_2, −78°, 3 h	$CH_2{=}CHCH_2CH{=}CHCH(C_6H_5)CH_2$-$CH{=}CH_2$ (48)[a] I + $(CH_2{=}CHCH_2)_2CHCH{=}CHC_6H_5$ (52)[a] II	382
	$(C_2H_5O)_2CHCH{=}CHC_6H_5$, $AlCl_3$, CH_2Cl_2, −78°, 6 h	$CH_2{=}CHCH_2CH(OC_2H_5)CH{=}CHC_6H_5$ (70)	382
	", $BF_3{\cdot}O(C_2H_5)_2$, CH_2Cl_2, −78°, 3 h	" (45)	382
	", $TiCl_4$, CH_2Cl_2, −78°, 3 h	I (41)[a] + II (37)[a]	382
C_4			
(*E*)-$(CH_3)_3SiCH_2CH{=}CHCH_3$	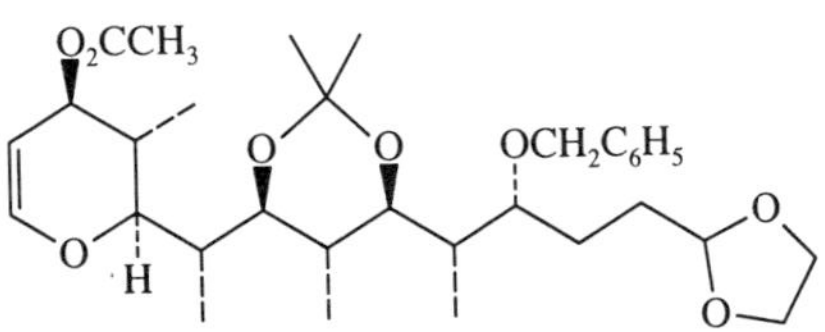	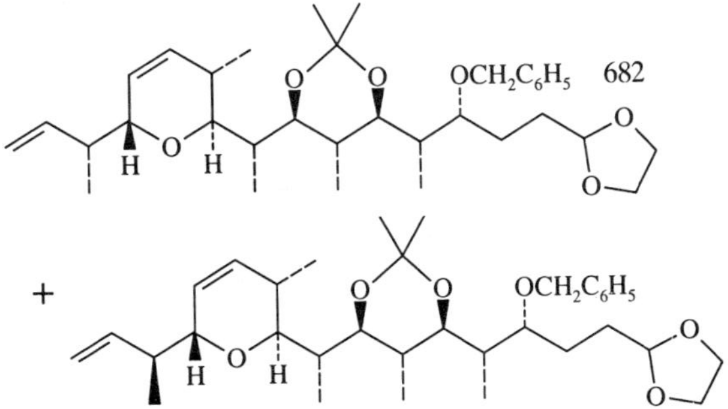 +	682
	$BF_3{\cdot}O(C_2H_5)_2$, C_2H_5CN, −78°	3.5:1 (60)	
	$ZnBr_2$, CH_3NO_2	2.8:1 (77)	
C_5			
$(CH_3)_3SiCH_2CH{=}C(CH_3)_2$	$(C_2H_5O)_2CHCH{=}CHC_3H_7\text{-}n$, $AlCl_3$, CH_2Cl_2, −78°, 9 h	$CH_2{=}CHC(CH_3)_2CH(OC_2H_5)$-$CH{=}CHC_3H_7\text{-}n$ (26)	382
	", $TiCl_4$, CH_2Cl_2, −78°, 16 h	$CH_2{=}CHC(CH_3)_2CH{=}CHCH(C_3H_7\text{-}n)C$-$(CH_3)_2CH{=}CH_2$ (34)[a]	382
	$(C_2H_5O)_2CHCH{=}CHC_6H_5$, $TiCl_4$, CH_2Cl_2, −78°, 16 h	$CH_2{=}CHC(CH_3)_2CH{=}CHCH(C_6H_5)$-$C(CH_3)_2CH{=}CH_2$ (33)[a] + $[CH_2{=}CHC(CH_3)_2]_2CHCH{=}CHC_6H_5$ (4)[a]	382

[a] Two equivalents of allylsilane are used.

TABLE XXIII. ALLYLSILANES WITH IMINIUM IONS[a]

Reactant	Conditions	Product(s) and Yield(s)	Refs.
C_3			
$(CH_3)_3SiCH_2CH{=}CH_2$	$(CH_3OCH_2)_2NCO_2CH_3$, HCO_2H	4-X-piperidine, N-CO_2CH_3; X = O_2CH (73)	390
	", $TiCl_4$, CH_2Cl_2, 12 h	", X = Cl (65)	390
	CH_2O, $C_6H_5CH_2NH_2 \cdot TFA$, LiCl, H_2O, 35°, 24 h	4-X-piperidine, N-$CH_2C_6H_5$; X = OH (81)	391
	", $C_6H_5CH_2NH_2 \cdot HCl$, LiCl, H_2O, 35°, 24 h	", X = Cl (48)	391
	CH_2O, $C_6H_5CH_2NHCH_3 \cdot TFA$, LiCl, H_2O, 50°, 68 h	$CH_2{=}CH(CH_2)_2N(CH_3)CH_2C_6H_5$ (76)	391
	4-CH_3O-3-(CH_3OCH_2)-oxazolidin-2-one, $TiCl_4$, CH_2Cl_2, 12 h	7-Cl-hexahydro-oxazolo[3,4-a]pyridin-3-one (77)	390
	4-CH_3CO_2-2-C_3H_7-i-3-CO_2CH_3-oxazolidine, $TiCl_4$, $Ti(OC_3H_7\text{-}i)_4$	4-($CH_2{=}CHCH_2$)-2-C_3H_7-i-3-CO_2CH_3-oxazolidine (80) trans:cis 1:4	683
	$CH_3O_2CNHCH(OCH_3)CH(OSi(CH_3)_2C_4H_9\text{-}t)CH_3$, $TiCl_4$	$CH_2{=}CHCH_2CH(NHCO_2CH_3)CH(OSi(CH_3)_2C_4H_9\text{-}t)CH_3$ (85) R,R:R,S 4:1	683

Reactants	Conditions	Product(s) and Yield(s) (%)		Refs.
$CH_3OCH(C_3H_7\text{-}i)NHCOCH(CH_3)NHCO_2CH_2C_6H_5$, $TiCl_4$		$CH_2{=}CHCH_2CH(C_3H_7\text{-}i)NHCOCH(CH_3)NHCO_2CH_2C_6H_5$ *S,S*:*S,R* 1:1.1 or 1.1:1	(60)	683
4-X-3-R^1-1-R^2-azetidin-2-one: R^1, R^2, X		4-($CH_2CH{=}CH_2$)-3-R^1-1-R^2-azetidin-2-one: Stereochemistry of $CH_2CH{=}CH_2$ Group		
H, H, O_2CCH_3	$BF_3{\cdot}O(C_2H_5)_2$, CH_2Cl_2	—	(63)	684
β-phthalimido, $C(CO_2CH_3){=}C(CH_3)_2$, α-Cl	$AgBF_4$, CH_2Cl_2, 3 h	α	(51)	685
α-, ", ", α + β-O_2CCH_3	$(CH_3)_3SiO_3SCF_3$, $Cl(CH_2)_2Cl$, 90°, 72 h	β	(76)	686, 687
", ", $CH(CO_2CH_3)C(CH_3){=}CH_2$, "	$(CH_3)_3SiO_3SCF_3$, $Cl(CH_2)_2Cl$, 75°, 29 h	"	(74)	686, 687
1-$C_6H_5CH_2$-5-HO-4-R-imidazolidin-2-one, $(CF_3CO)_2O$, $SnCl_4$, CH_2Cl_2		1-$C_6H_5CH_2$-5-($CH_2{=}CHCH_2$)-4-R-imidazolidin-2-one (trans)		688
R = H	−40°, 2 h; room temp, 10 h		(82)	
R = CH_3	−20°, "; ", 8 h		(67)	
5-C_2H_5O-1-R-pyrrolidin-2-one, CH_2Cl_2		5-($CH_2{=}CHCH_2$)-1-R-pyrrolidin-2-one		

TABLE XXIII. ALLYLSILANES WITH IMINIUM IONS[a] (*Continued*)

Reactant	Conditions	Product(s) and Yield(s)	Refs.
	R: H, $SnCl_4$	(69)	684
	R: n-C_4H_9, $TiCl_4$, 25°, 7 h	(78)	385
	R: $C_6H_5CH_2$, $SnCl_4$	(66)	684
	R: $Br(CH_2)_2$, "	(75)	684
	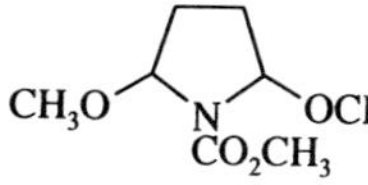 2,5-di(CH_3O)-N-CO_2CH_3-pyrrolidine (CH_3O, OCH_3), $TiCl_4$, CH_2Cl_2, 12 h	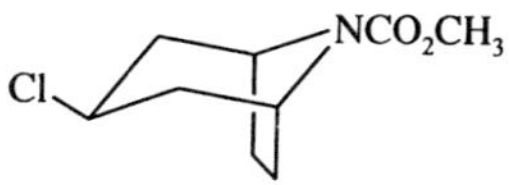 Cl-bicyclic NCO_2CH_3 (85)	390
	n-$C_3H_7CH(OCH_3)N(CO_2CH_3)CH_2OCH_3$, $TiCl_4$, CH_2Cl_2, 12 h	4-Cl-2-n-C_3H_7-N-CO_2CH_3-piperidine (75)	390
	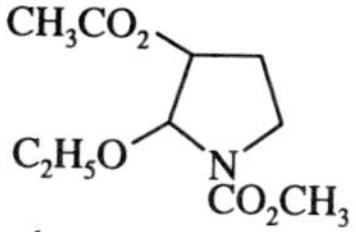 3-CH_3CO_2-2-C_2H_5O-N-CO_2CH_3-pyrrolidine, $SnCl_4$, CH_2Cl_2	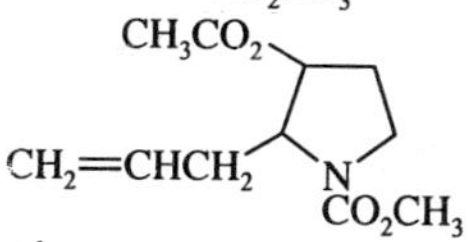 3-CH_3CO_2-2-(CH_2=$CHCH_2$)-N-CO_2CH_3-pyrrolidine (99)	684
	Azetidinone (R^1, X, NR^2)	Azetidinone (R^1, CH_2CH=CH_2, NR^2)	
	R^1 = α-CH_3, R^2 = H, X = α- and β-O_2CCH_3; $TiCl_4$, C_6H_6, 16 h	(—)	689
	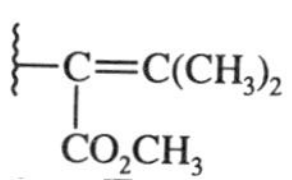 R^1 = ", R^2 = $-C(CO_2CH_3)$=$C(CH_3)_2$, X = β-Cl; $AgBF_4$, CH_2Cl_2, 0°, 1 h	(—)	685
	R^1 = β-CH_3, R^2 = ", X = "; $AgBF_4$, CH_2Cl_2, 0°, 1 h	(—)	685

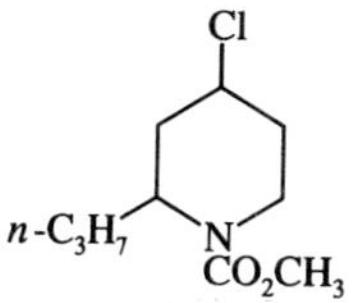

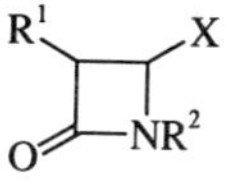

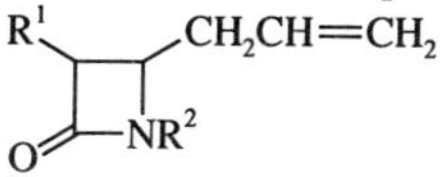

Reactant, Conditions	Product(s) and Yield(s) (%)	Refs.
RN–C(=O)–NH (5-HO, 4,4-dimethyl imidazolidinone), $SnCl_4$, $(CF_3CO)_2O$, CH_2Cl_2	RN–C(=O)–NH (4-$CH_2{=}CHCH_2$ imidazolidinone)	688
R = $C_6H_5CH_2$, 24 h	(94)	
R = p-$CH_3OC_6H_4CH_2$, 84 h	(66)	
CH_3O, R^1, R^2, N-$COCH_3$, $CO_2C_2H_5$ pyrrolidine, $BF_3{\cdot}O(C_2H_5)_2$, CH_2Cl_2, 24 h	$CH_2{=}CHCH_2$, R^1, R^2, N-$COCH_3$, $CO_2C_2H_5$ pyrrolidine	690
R^1 — R^2		
CH_3CO_2 — H	$\alpha:\beta$ 18:82 (71)	
H — CH_3CO_2	" 54:46 (70)	
Pyridine N-oxide (N^+–O^-), TBAF, THF, 5°, 10 min	$CH_3CH{=}CH$-2-pyridine (56)	395
C_2H_5, O_2CCH_3 azetidinone (O=, NH), $TiCl_4$, C_6H_6, 16 h	C_2H_5, $CH_2CH{=}CH_2$ azetidinone (O=, NH) (50)	689
4-OH tetrahydropyridine, N-$CO_2C_2H_5$	$CH_2{=}CHCH_2$ tetrahydropyridine, N-$CO_2C_2H_5$	387
$SnCl_4$, CH_2Cl_2, −78°, 30 min	(90)	
", CH_3CN, 20 min	(64)	
$TiCl_4$, CH_2Cl_2, −78°, 30 min	(89)	
$BF_3{\cdot}O(C_2H_5)_2$, CH_3CN, 30 min	(43)	
$(CH_3)_3SiO_3SCF_3$, CH_2Cl_2, −78°, 1 h	(51)	

TABLE XXIII. ALLYLSILANES WITH IMINIUM IONS[a] (*Continued*)

Reactant	Conditions	Product(s) and Yield(s)	Refs.
	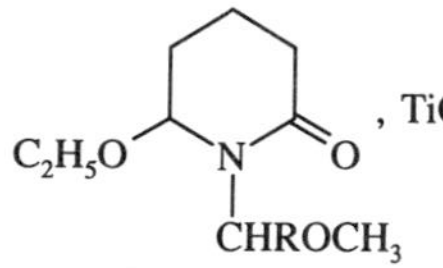 C_2H_5O-piperidinone, N-CHROCH$_3$, $TiCl_4$, CH_2Cl_2, 12 h	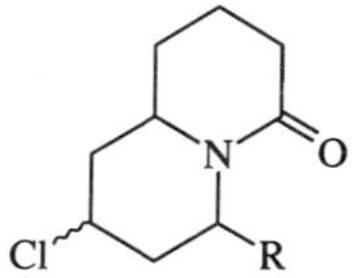 Cl-quinolizidinone, R	390
	R = H	(83)	
	R = CH_3	(50)	
	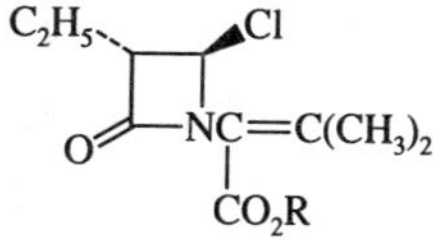 C_2H_5, Cl-azetidinone, $NC{=}C(CH_3)_2$, CO_2R, $AgBF_4$, CH_2Cl_2, 0°, 2 h	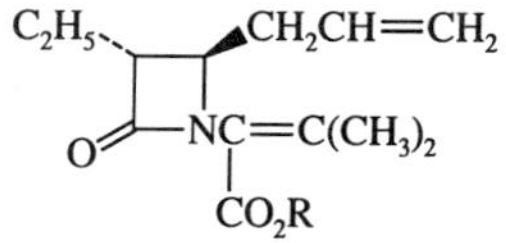 C_2H_5, $CH_2CH{=}CH_2$-azetidinone, $NC{=}C(CH_3)_2$, CO_2R	386
	R = CH_3	(60–70)	
	R = $C_6H_5CH_2$	(50)	
	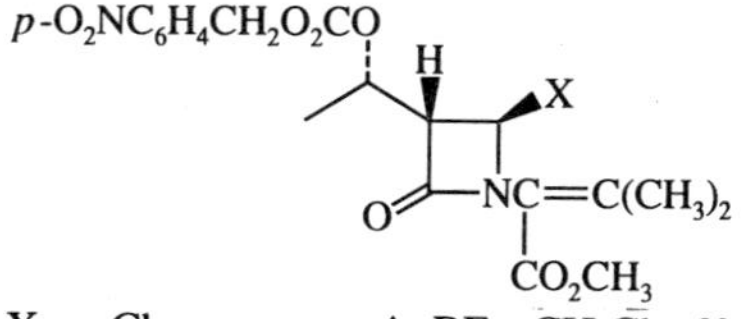 p-$O_2NC_6H_4CH_2O_2CO$, H, X-azetidinone, $NC{=}C(CH_3)_2$, CO_2CH_3	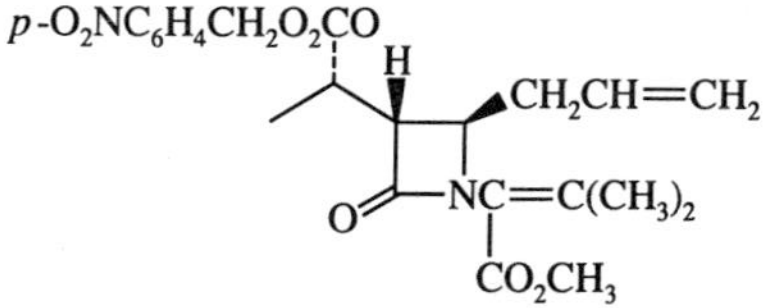 p-$O_2NC_6H_4CH_2O_2CO$, H, $CH_2CH{=}CH_2$-azetidinone, $NC{=}C(CH_3)_2$, CO_2CH_3	
	X = Cl $AgBF_4$, CH_2Cl_2, 0°, 2 h	(78)	386, 685
	X = O_2CCH_3 $(CH_3)_3SiO_3SCF_3$, $Cl(CH_2)_2Cl$, 70°, 29 h	(73)	686
	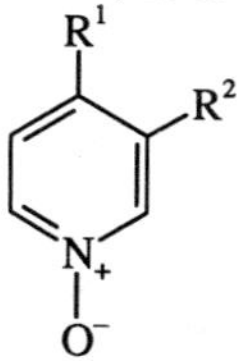 R^1, R^2-pyridine N-oxide (N^+–O^-), TBAF, THF, 5°, 10 min	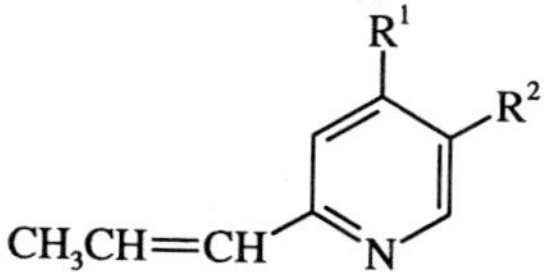 $CH_3CH{=}CH$-pyridine, R^1, R^2	395

Conditions	Product(s) and Yield(s) (%)	Refs.
R^1 = CH_3, R^2 = H	(73)	
R^1 = H, R^2 = CH_3	(69)	
i-C_3H_7-azetidinone-4-O_2CCH_3, $TiCl_4$, C_6H_6, 16 h	i-C_3H_7-azetidinone-4-$CH_2CH{=}CH_2$ (—)	689
7-CH_3O-1-(CH_2OCH_3)azepan-2-one, $TiCl_4$, CH_2Cl_2, 12 h	Cl-substituted pyrido[1,2-a]azepinone (62)	390
$CH_3CO(CH_2)_2$-azetidinone-4-O_2CCH_3, $TiCl_4$, C_6H_6, 16 h	$CH_3CO(CH_2)_2$-azetidinone-4-$CH_2CH{=}CH_2$ (—)	689
1. $C_6H_5CH{=}\overset{+}{N}(O^-)CH_3$ 2. Raney Ni, H_2, NaOH, H_2O, CH_3OH, 12 h 3. H_2SO_4, THF, 16 h	$CH_2{=}CHCH_2CH(C_6H_5)NHCH_3$ (67)	232
1. $C_2H_5O_2CCH{=}\overset{+}{N}(O^-)CH_2C_6H_5$ 2. Raney Ni, H_2, C_2H_5OH, 12 h 3. HCl, C_2H_5OH, 50°, 4 h	$CH_2{=}CHCH_2CH(CO_2C_2H_5)NHCH_2C_6H_5$ (71)	232
1. 2,5-bis($C_6H_5CH_2O$)phenyl-2-C_6H_5-1,3-dioxolane-$CH{=}N^+(O^-)CH_3$	2,5-bis($C_6H_5CH_2O$)phenyl-2-C_6H_5-1,3-dioxolane-$CH_2{=}CHCH_2CHN(CH_3)COCH_3$ (42)	232

TABLE XXIII. ALLYLSILANES WITH IMINIUM IONS[a] (*Continued*)

Reactant	Conditions	Product(s) and Yield(s)	Refs.
	2. Zn, CH_3CO_2H, H_2O, 80°, 18 h 3. $(CH_3CO)_2O$, Py, 6 h		
	2-C_6H_5-1-CH_3-pyrrolinium ClO_4^-, $h\nu$, CH_3OH, 0.5 h	$CH_2{=}CHCH_2$-2-C_6H_5-1-CH_3-pyrrolidine (42)	691
	", ", CH_3CN, 100 min	" (85)	691
$(CH_3)_3SiCH_2CH{=}CHSi(CH_3)_3$	2-C_6H_5-1-CH_3-pyrrolinium ClO_4^-, $h\nu$, CH_3CN, 100 min	(*E*)-$(CH_3)_3SiCH{=}CHCH_2$-2-C_6H_5-1-CH_3-pyrrolidine (85)	691
$(CH_3)_3SiCH_2CBr{=}CH_2$	$(CH_3OCH_2)_2NCO_2CH_3$, $TiCl_4$, CH_2Cl_2, 12 h	4-Br-4-Cl-1-CO_2CH_3-piperidine (48)	390
C_4			
$(CH_3)_3SiCH_2CH{=}CHCH_3$	CH_2O, $C_6H_5CH_2NH_2$, TFA, LiCl, H_2O, 45°, 48 h	4-OH-3-CH_3-1-$CH_2C_6H_5$-piperidine (54)	391
$(CH_3)_3SiCH_2C(=CH_2)CH_2CO_2CH_2C_6H_4NO_2$-*p*	4-Cl-3-[1-($OCO_2CH_2C_6H_4NO_2$-*p*)ethyl]-1-[$C(CO_2CH_3){=}C(CH_3)_2$]azetidin-2-one; $AgBF_4$, CH_2Cl_2, 0°, 2 h	4-[$CH_2C(=CH_2)CH_2CO_2CH_2C_6H_4NO_2$-*p*]-3-[1-($OCO_2CH_2C_6H_4NO_2$-*p*)ethyl]-1-[$C(CO_2CH_3){=}C(CH_3)_2$]azetidin-2-one (69)[b](58)[c]	386
	$(CH_3)_3SiO_3SCF_3$, $Cl(CH_2)_2Cl$, 65°, 63 h	(82)[c]	686

Reactant	Conditions	Product(s) and Yield(s) (%)	Refs.
C_5			
$(CH_3)_3Si$–…–N^+CH_3 ClO_4^-, $(CH_2)_n$, OR		NCH$_3$, $(CH_2)_n$, OR	692
R: CH_3, n: 1	$h\nu$, CH_3CN, 4.5 h	(91)	
R: t-C_4H_9O, n: ”	”, ”, 1.25 h	(87)	
R: CH_3, n: 2	”, ”, 1 h	(98)	
R: t-C_4H_9O, n: ”	”, ”, 1.25 h	(84)	
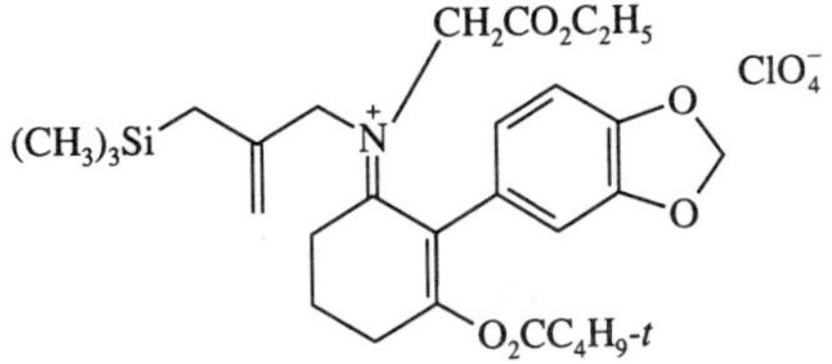 $(CH_3)_3Si$–…–$\overset{+}{N}$($CH_2CO_2C_2H_5$) ClO_4^-, $O_2CC_4H_9$-t	$h\nu$, CH_3CN, 80 min	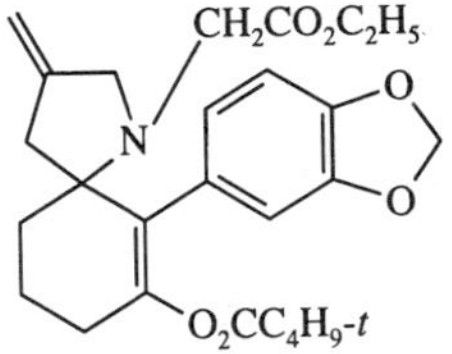 $CH_2CO_2C_2H_5$, N, $O_2CC_4H_9$-t (33)	394
$(CH_3)_3SiCH_2CH{=}C(CH_3)_2$	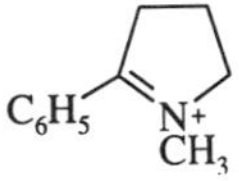 C_6H_5, N^+, CH_3 ClO_4^-, $h\nu$, CH_3CN, 85 min	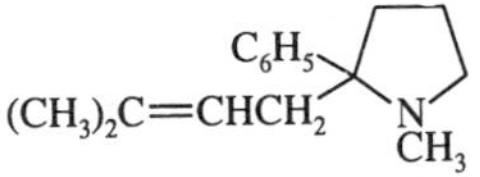 $(CH_3)_2C{=}CHCH_2$, C_6H_5, N, CH_3 (42)[d]	691, 393
$(CH_3)_3SiC(CH_3)_2CH{=}CH_2$	”, ”, ”, ”	” (30)	691
$(CH_3)_3SiCH_2CH{=}CH(CH_2)_2NHCH_2C_6H_5$	CH_2O, TFA, H_2O, THF, 56 h	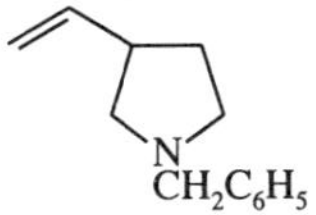 N, $CH_2C_6H_5$ (81)	392
$(CH_3)_3SiCH_2C(=CH_2)(CH_2)_2NH_2$	$C_6H_5CH_2CHO$, TFA, H_2O, THF, 55°, 24 h	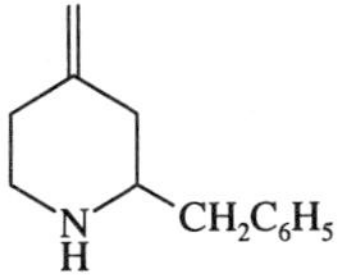 N, H, $CH_2C_6H_5$ (71)	392

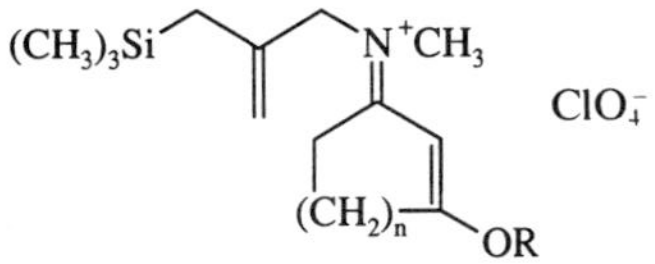

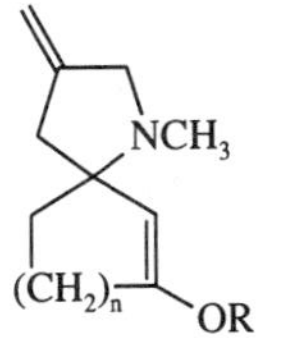

TABLE XXIII. ALLYLSILANES WITH IMINIUM IONS[a] (Continued)

Reactant	Conditions	Product(s) and Yield(s)	Refs.
$(CH_3)_3SiCH_2C(=CH_2)(CH_2)_2NHCH_2C_6H_5$	CH_2O, TFA, H_2O, THF, 19 h	4-methylene-1-($CH_2C_6H_5$)piperidine (73)	392
$(CH_3)_3SiCH_2C(=CH_2)(CH_2)_2OH$	CH_2O, $C_6H_5CH_2NH_2$·TFA, LiCl, H_2O, 25°, 4 h	4-HO-4-$(CH_2)_2OH$-1-($CH_2C_6H_5$)piperidine (100)	391
$(CH_3)_3SiCH_2CH{=}CH(CH_2)_2$–N(5-HO-pyrrolidin-2-one)		$CH_2{=}CH$, H pyrrolizidinone	388
E	HCO_2H, 1 h	(84)	
Z	TFA, CH_2Cl_2, 1 h	(72)	
$(CH_3)_3SiCH_2C(=CH_2)(CH_2)_2$–N(5-$C_2H_5O$ lactam, $(CH_2)_n$)		methylene bicyclic lactam, $(CH_2)_n$	693
n = 1	$TiCl_4$, CH_2Cl_2, 25°, 5 min	(80)	
"	TFA, CH_2Cl_2, 0°, 4 h	(70)	
n = 2	", ", ", "	(94)	
(Z)-$(CH_3)_3SiCH_2CH{=}CH(CH_2)_2N(COR)CHXC_6H_5$		3-($CH_2{=}CH$)-2-C_6H_5-1-COR-pyrrolidine	389

R X			
CH_3 Cl	$(C_2H_5)_2AlCl$, CH_2Cl_2, 2 h	(69)	
OC_2H_5 OC_2H_5	HCO_2H, 17 h	(81)	
" "	$SnCl_4$, CH_2Cl_2, 2 h	(73)	
" "	$(C_2H_5)_2AlCl$, CH_2Cl_2, 2 h	(69)	
$(CH_3)_3SiCH_2CH{=}CHCH{=}CH_2$	CH_2O, $C_6H_5CH_2NHCH_3$·TFA, LiCl, H_2O, 45°, 65 min	$CH_2{=}CHCH{=}CH(CH_2)_2N(CH_3)CH_2C_6H_5$ (95)	391
$(CH_3)_3SiCH_2C(CH{=}CH_2){=}CH_2$	CH_2O, $C_6H_5CH_2NH_2$·TFA, LiCl, H_2O, 25°, 24 h	1-benzyl-4-hydroxy-4-vinylpiperidine (HO, $CH{=}CH_2$; N–$CH_2C_6H_5$) + 1-benzyl-4-[$(CH_3)_3SiCH_2$]-1,2,3,6-tetrahydropyridine (N–$CH_2C_6H_5$) (85) 3.5:1	391
$(CH_3)_3Si$-cyclopent-2-enyl	CH_2O, $C_6H_5CH_2NH_2$·TFA, LiCl, H_2O, 45°, 42 h	cyclopent-2-enyl-$CH_2NHCH_2C_6H_5$ (50)	391
$(CH_3)_3SiCH_2C(CH_2CO_2R^1){=}CH_2$	R^2-azetidinone (4-Cl; $NC{=}C(CH_3)_2$, CO_2CH_3), $AgBF_4$, CH_2Cl_2, 0°	R^2-azetidinone (4-$CH_2C({=}CH_2)CH_2CO_2R^1$; $NC{=}C(CH_3)_2$, CO_2CH_3)	685
R^1	R^2		
CH_3	phthalimido (N–), 1.5 h	(78)	
"	$CH_3\overset{8}{C}H(O_2CCH_2C_6H_4NO_2\text{-}p)$–		
	", 8*R* + 8*S* (1:1) 1.5 h	8*R* + 8*S* (1:1) (58)	
"	", 8*S* 0.5 h	8*S* (73)	
$CH_2C_6H_4NO_2\text{-}p$	", 8*R* 1 h	8*R* (51)	
"	", 8*S* 40 min	8*S* (69)	
"	$(CH_3)_2C(O_2CCH_2C_6H_4NO_2\text{-}p)$– 1 h	(23)	

TABLE XXIII. ALLYLSILANES WITH IMINIUM IONS[a] (*Continued*)

Reactant	Conditions	Product(s) and Yield(s)	Refs.
C_6			
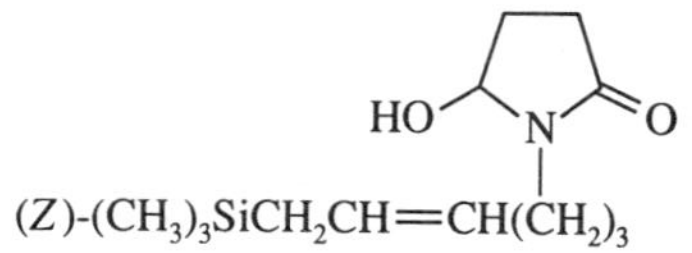 HO, N, O (5-hydroxy-2-pyrrolidinone ring); (Z)-$(CH_3)_3SiCH_2CH{=}CH(CH_2)_3$ on N	TFA, CH_2Cl_2, 1 h	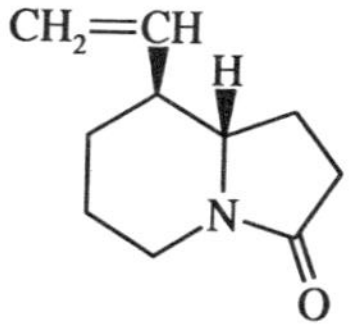 $CH_2{=}CH$, H, N, O (86)	388
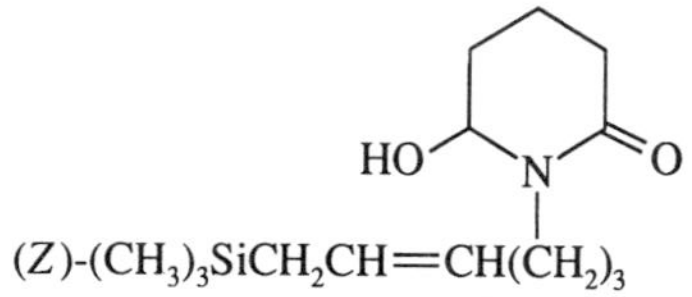 HO, N, O (6-hydroxy-2-piperidinone ring); (Z)-$(CH_3)_3SiCH_2CH{=}CH(CH_2)_3$ on N	TFA, CH_2Cl_2, 1 h	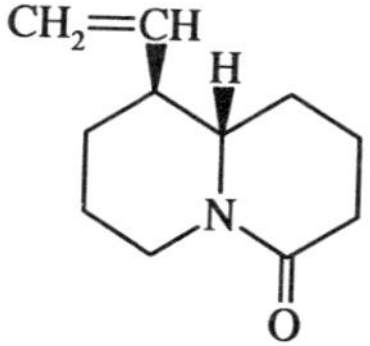 $CH_2{=}CH$, H, N, O (66) 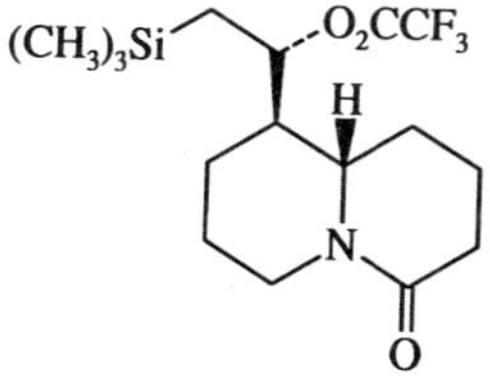 + $(CH_3)_3Si$, O_2CCF_3, H, N, O (5)	388
(Z)-$(CH_3)_3SiCH_2CH{=}CH(CH_2)_3NCOCH_3$ with $C_6H_5CHOC_2H_5$ on N	HCO_2H, 16 h	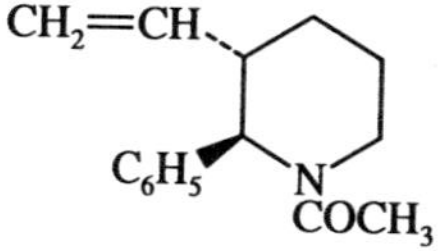 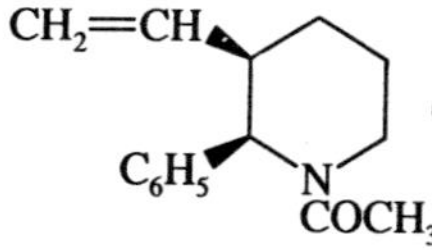 $CH_2{=}CH$, C_6H_5, N, $COCH_3$ + $CH_2{=}CH$, C_6H_5, N, $COCH_3$ (60), 2:1	389

$(CH_3)_3SiCH_2C(=CH_2)(CH_2)_3NHCH_2C_6H_5$	CH_2O, TFA, H_2O, THF, 42 h	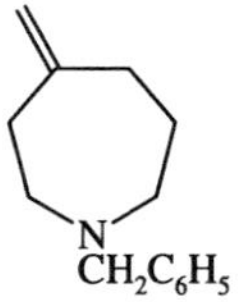 N–$CH_2C_6H_5$ (96)	392
$(CH_3)_3SiCH_2C(=CH_2)(CH_2)_3OH$	CH_2O, $C_6H_5CH_2NH_2$·TFA, LiCl, H_2O, 25°, 6 h 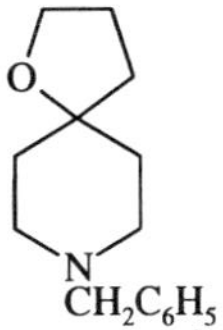	O, N–$CH_2C_6H_5$ (58)	391
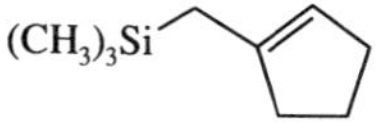 $(CH_3)_3Si$	CH_2O, $C_6H_5CH_2NH_2$·TFA, LiCl, H_2O, 25°, 82 h	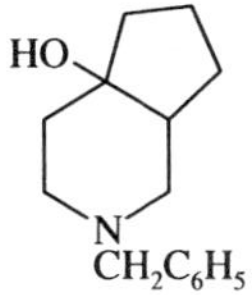 HO, N–$CH_2C_6H_5$ (68)	391
C_7			
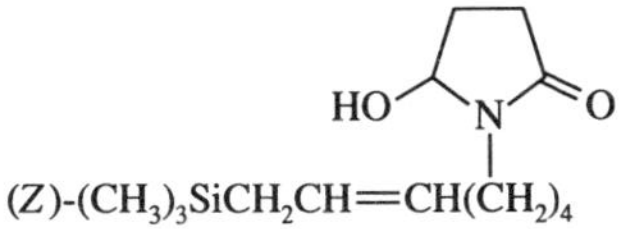 HO, N, O (Z)-$(CH_3)_3SiCH_2CH{=}CH(CH_2)_4$	TFA, CH_2Cl_2, 0°, 2 h	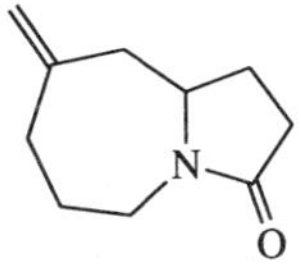 N, O (70–90)	694
$(CH_3)_3SiCH_2C(=CH_2)(CH_2)_4NHCH_2C_6H_5$	CH_2O, TFA, H_2O, THF, 50°, 5 d 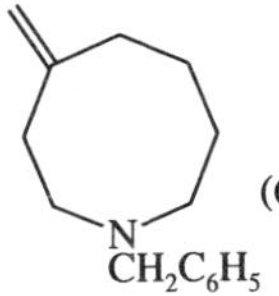	N–$CH_2C_6H_5$ (64)	392

TABLE XXIII. ALLYLSILANES WITH IMINIUM IONS[a] (*Continued*)

Reactant	Conditions	Product(s) and Yield(s)	Refs.
$(CH_3)_3SiCH_2C(=CH_2)(CH_2)_3OH$	CH_2O, $C_6H_5CH_2NH_2$·TFA, LiCl, H_2O, 25°, 6 h	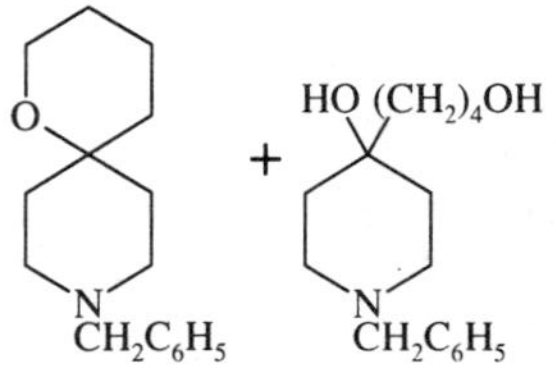 N-$CH_2C_6H_5$ spiro product + HO $(CH_2)_4OH$ N-$CH_2C_6H_5$ piperidine, 1:3.3 (83)	391
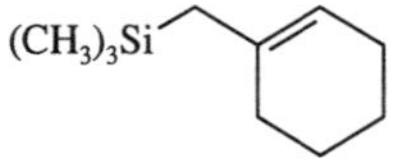 $(CH_3)_3Si$ (cyclohexenylmethyl)	CH_2O, $C_6H_5CH_2NH_2$·TFA, LiCl, H_2O, 35°, 48 h	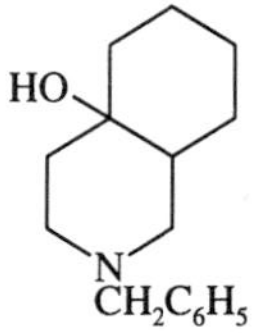 HO, N-$CH_2C_6H_5$ (94)	391
C_8			
$(CH_3)_3SiCH_2C(=CH_2)C_5H_{11}$-*n*	CH_2O, $C_6H_5CH_2NH_2$·TFA, LiCl, H_2O, 30°, 48 h	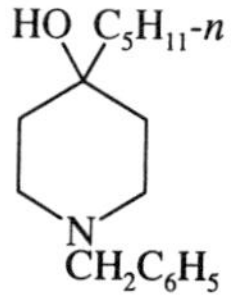 HO C_5H_{11}-*n*, N-$CH_2C_6H_5$ (53)	391
C_9			
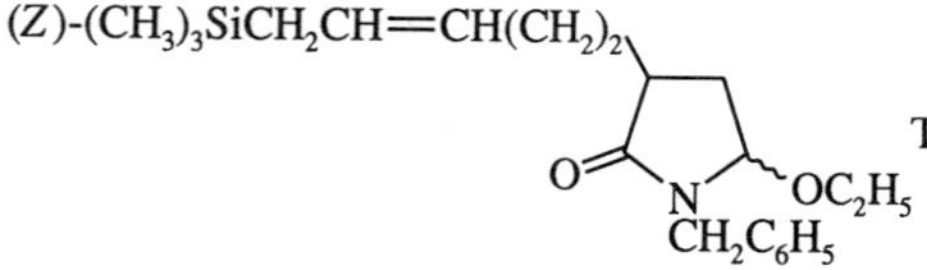 (Z)-$(CH_3)_3SiCH_2CH{=}CH(CH_2)_2$-, O, N-$CH_2C_6H_5$, OC_2H_5	TFA, CH_2Cl_2, 1 h	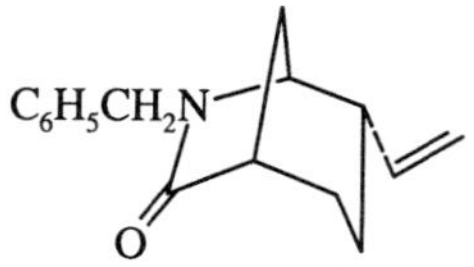 $C_6H_5CH_2N$, O; I (73)	695

Reactant	Conditions	Product(s) and Yield(s) (%)	Refs.
		+ $C_6H_5CH_2N$ (bicyclic lactam, vinyl) II (4)	
		+ $CH_2{=}CH(CH_2)_3$ (pyrrolinone, O, N–$CH_2C_6H_5$) III (10)	
	HCO_2H, 1 h	I (82) + II (5) + III (<2)	695
$(CH_3)_3Si$ (cyclooctenylmethyl)	CH_2O, $C_6H_5CH_2NH_2$·TFA, LiCl, H_2O, 25°, 82 h	(N–$CH_2C_6H_5$ hexahydrocycloocta[c]pyridine) (62)	391
C_{10}			
(Z)-$(CH_3)_3SiCH_2CH{=}CH(CH_2)_2$ (piperidinone, O, N–$CH_2C_6H_5$, ~OC_2H_5)	HCO_2H, 1 h	$C_6H_5CH_2N$ (bicyclic lactam, vinyl) (49)	695
		+ $C_6H_5CH_2N$ (bicyclic lactam, vinyl epimer) (36)	

TABLE XXIII. ALLYLSILANES WITH IMINIUM IONS[a] (*Continued*)

Reactant	Conditions	Product(s) and Yield(s)	Refs.
(Z)-$(CH_3)_3SiCH_2CH{=}CH(CH_2)_3$- (pyrrolidinone, N-$CH_2C_6H_5$, OC_2H_5)	TFA, CH_2Cl_2, 1 h	$C_6H_5CH_2N$ (bicyclic lactam, vinyl) (67) + $C_6H_5CH_2N$ (bicyclic lactam, vinyl) (6) + $CH_2{=}CH(CH_2)_4$- (pyrrolinone, N-$CH_2C_6H_5$) (12)	695
C_{11}			
$(CH_3)_3SiCH_2CH{=}$ (polycyclic, OC_2H_5, OC_2H_5, $C_2H_5O_2CHN$–, O)	$(CH_2O)_3$, HCO_2H, 16 h	$CH_2{=}CH$ (polycyclic, O, $C_2H_5O_2CN$–, O) (85)	696
C_{14}			
(dihydroisoquinolinium, N^+R ClO_4^-, $(CH_3)_3SiCH_2C({=}CH_2)(CH_2)_2$)		(spiro tetrahydroisoquinoline, NR, methylenecyclopentane)	697

R	Conditions	Yield (%)	Refs.
H	$h\nu$, CH_3CN, 1.5 h	(45)	
CH_3	", ", 1 h	(88)	
"	", CH_3OH, 15 min	(87)	
"	CsF, C_2H_5OH, 70°, 40 h	(7)[f]	
$CH_2CO_2C_2H_5$	$h\nu$, CH_3CN, 80 min	(70)	

C_{15}

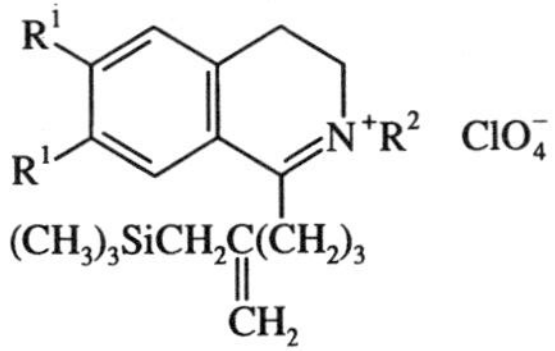

6,7-(R^1)$_2$-1-[$(CH_3)_3SiCH_2C(=CH_2)(CH_2)_3$]-3,4-dihydroisoquinolinium (N^+R^2) ClO_4^- → 6,7-(R^1)$_2$-spiro[tetrahydroisoquinoline-1,1'-(3'-methylenecyclohexane)] (NR^2) — 697

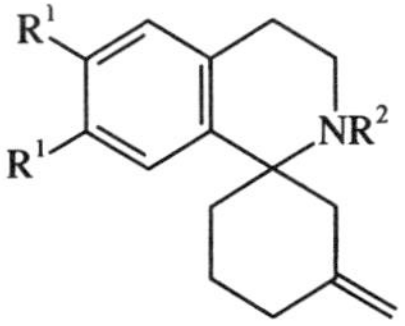

R^1	R^2	Conditions	Yield (%)
H	H	$h\nu$, CH_3CN, 0.5 h	(33)[f]
"	CH_3	", CH_3OH, 22 min	(39)
"	"	CsF, C_2H_5OH, 70°, 3 d	(31)[f]
"	$CH_2CO_2C_2H_5$	$h\nu$, CH_3OH, 25 min	(71)
CH_3O	"	", ", 0.5 h	(60)

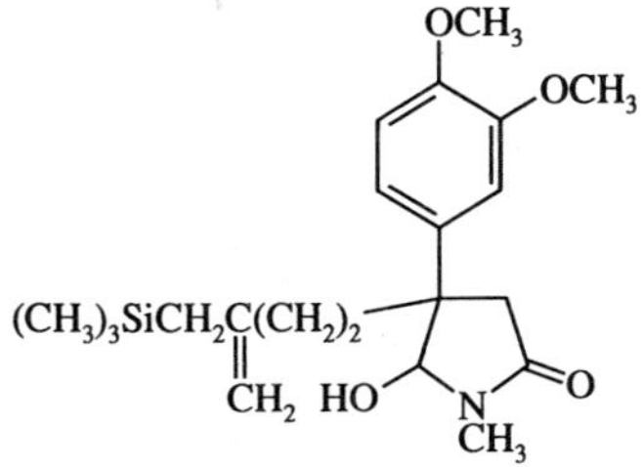

5-Hydroxy-1-methyl-4-(3,4-dimethoxyphenyl)-4-[$(CH_3)_3SiCH_2C(=CH_2)(CH_2)_2$]pyrrolidin-2-one | CH_3SO_2Cl, $(C_2H_5)_3N$, CH_2Cl_2, 12 h | 3a-(3,4-dimethoxyphenyl)-1-methyl-6-methyleneoctahydroindol-2-one (80) | 396

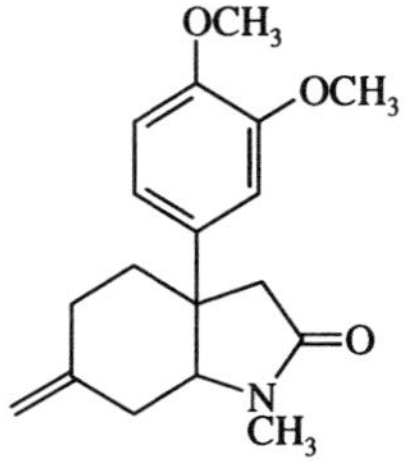

TABLE XXIII. ALLYLSILANES WITH IMINIUM IONS[a] (*Continued*)

Reactant	Conditions	Product(s) and Yield(s)	Refs.
C_{16}			
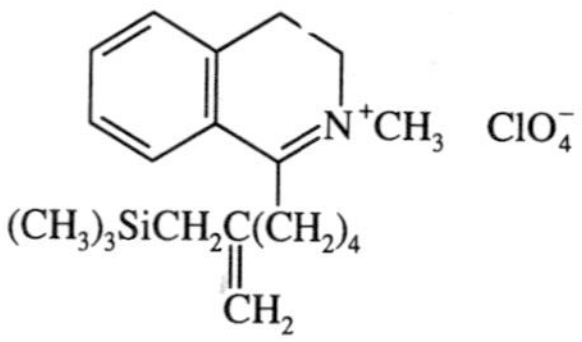 1-[2-((CH$_3$)$_3$SiCH$_2$)-2-propenyl... $(CH_3)_3SiCH_2C(=CH_2)(CH_2)_4$-substituted 2-methyl-3,4-dihydroisoquinolinium (N^+CH_3) ClO_4^-	$h\nu$, CH_3OH, 22 min	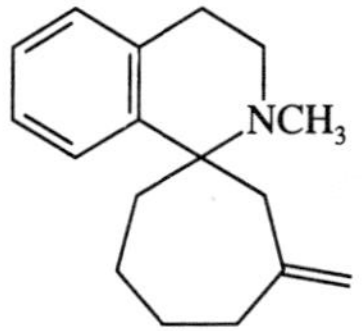 spiro NCH$_3$ product (66)	697
	CsF, C_2H_5OH, 70°, 3 d	", (—)[e]	697
C_{18}			
I (dihydropyridine with R^1O_2C, CO_2R^3, R^2-aryl bearing $CH=CHCH_2Si(CH_3)_3$, N–R^4)	$(CH_3)_3SiO_3SCF_3$, $(C_2H_5)_3N$, CH_2Cl_2, 0°, 30 min	II	698
R^1 = C_2H_5, R^2 = H, R^3 = C_2H_5, R^4 = H		(—)	
R^1 = CH_3, R^2 = CH_3O, R^3 = CH_3, R^4 = CH_3		(—)	
R^1 = ", R^2 = NO_2, R^3 = $CH_2O(CH_2)_2OCH_3$, R^4 = H		(—)	
C_{19}			
I	$(CH_3)_3SiO_3SCF_3$, $(C_2H_5)_3N$, CH_2Cl_2, 0°, 30 min	II	698

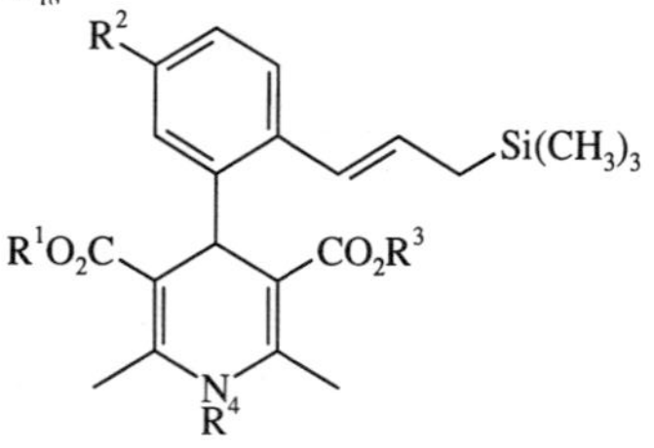

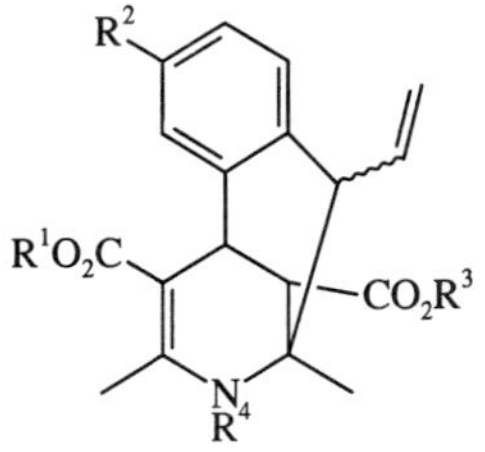

R^1	R^2	R^3	R^4	
$CH_2=CHCH_2$	CN	$CH_2CH=CH_2$	$CH_2C_6H_5$	(—)
CH_3	CF_3	$CH_2N(CH_3)CH_2C_6H_5$	H	(—)

C_{20}

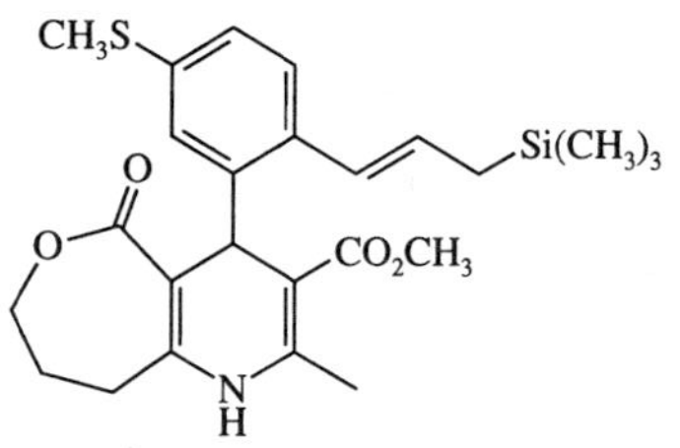

Substrate	Conditions	Product	Yield	Refs.
CH_3S, O, O, $Si(CH_3)_3$, CO_2CH_3, N H	$(CH_3)_3SiO_3SCF_3$, $(C_2H_5)_3N$, CH_2Cl_2, 0°, 30 min	CH_3S, O, O, CO_2CH_3, N H	(—)	698

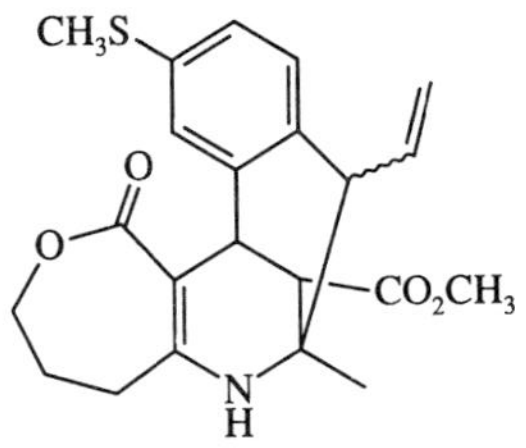

[a] Iminium ions are the likely but presumed intermediates in most of the reactions in this table. The iminium ion itself is the actual reagent in only a few of the entries.

[b] This is the yield from the 8*S* enantiomer.

[c] This is the yield from the 8*R* enantiomer.

[d] Small amounts (5%) of bisallylic products, $[(CH_3)_2C=CHCH_2]_2$ and $(CH_3)_2C=CHCH_2C(CH_3)_2CH=CH_2$, are also produced.

[e] 1-Tetrahydroisoquinolone (0–21%) and the tetrahydroisoquinoline product (21–36%) from reduction of the starting materials are isolated in these reactions.

[f] This product is characterized by hydration of the exocyclic double bond using aqueous acid.

TABLE XXIV. VINYLSILANES WITH IMINIUM IONS[a]

Reactant	Conditions	Product(s) and Yield(s)	Refs.
C_4			
(*E*)-$(CH_3)_3SiCH{=}CH(CH_2)_2NHC_3H_7$-*n*	$(CH_2O)_n$, CSA, CH_3CN, 80°	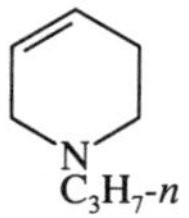 C_3H_7-*n* (73)	398
(*Z*)-$(CH_3)_3SiCH{=}CH(CH_2)_2NHR^1$		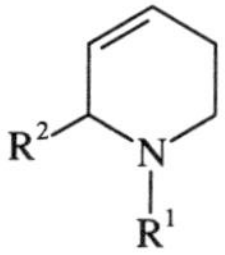 R^2, N, R^1	398
R^1		R^2	
n-C_3H_7	$(CH_2O)_n$, CSA, CH_3CN, 80°	H (90)	
i-C_4H_9	", HCO_2H, ", 1.2 h	" (—)[b]	
C_6H_5	", CSA	" (75)	
"	*n*-$C_6H_{13}CHO$, ", ", 120°	*n*-C_6H_{13} (68)	
C_6H_{11}	$(CH_2O)_n$, ", ", 80°	H (75)	
p-$CH_3OC_6H_4$	", ", ", "	" (92)	
"	*n*-$C_6H_{13}CHO$, ", ", 120°	*n*-C_6H_{13} (45)	
(*Z*)-$(CH_3)_3SiCH{=}CH(CH_2)_2N$-$(CH_2CN)C_6H_{11}$	AgO_2CCF_3, 100°	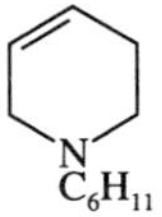 C_6H_{11} (56)	398
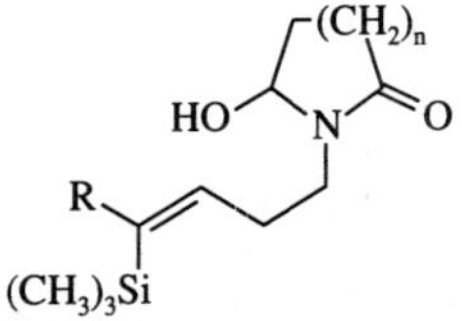 HO, $(CH_2)_n$, N, O, R, $(CH_3)_3Si$	TFA, 25°	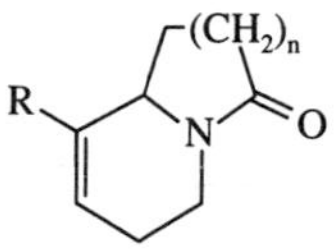 R, $(CH_2)_n$, N, O	398

Reactant	Conditions	Product(s) and Yield(s) (%)	Refs.
R n			
H 1		(92)	
" 2		(91)	
Br 1		(63)	
C_6			
(*Z*)-$(CH_3)_3SiCH{=}CHC_4H_9$-*n*	1. $C_6H_5CH{=}\overset{+}{N}(O^-)CH_3$ 2. Separate regioisomers[c] 3. Zn, HCl, H_2O, THF, 50°, 1 h	(*Z*)-*n*-$C_4H_9CH{=}CHCH(C_6H_5)NHCH_3$ (62)	232
	1. $C_6H_5CH{=}\overset{+}{N}(O^-)CH_3$ 2. Separate regioisomers[c] 3. Raney Ni, H_2, NaOH, H_2O, CH_3OH, 12 h 4. KH, THF, 1 h	(*E*)-*n*-$C_4H_9CH{=}CHCH(C_6H_5)NHCH_3$ (45)	232
C_7			
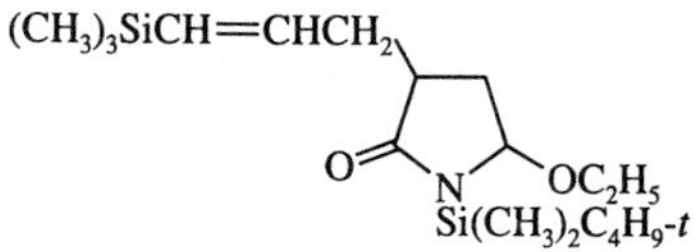 $(CH_3)_3SiCH{=}CHCH_2$-substituted 5-OC_2H_5 pyrrolidin-2-one, N-$Si(CH_3)_2C_4H_9$-*t*	Lewis or protic acid	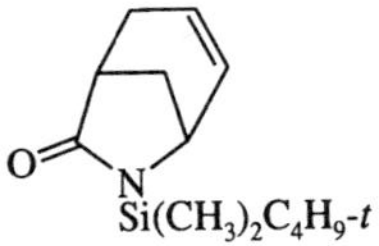 bicyclic lactam, N-$Si(CH_3)_2C_4H_9$-*t* (0)	699
C_8			
(*E*)-$(CH_3)_3SiCH{=}CHC_6H_5$	1. $C_6H_5CH{=}\overset{+}{N}(O^-)CH_3$ 2. Zn, HCl, H_2O, THF, 50°, 1 h	(*E*)-$C_6H_5CH{=}CHCH(C_6H_5)NHCH_3$ (52)	232
	1. $C_6H_5CH{=}\overset{+}{N}(O^-)CH_3$ 2. Raney Ni, H_2, NaOH, CH_3OH, 12 h 3. H_2SO_4, THF, 16 h	" (45)	232

TABLE XXIV. VINYLSILANES WITH IMINIUM IONS[a] (*Continued*)

Reactant	Conditions	Product(s) and Yield(s)	Refs.
	1. $C_6H_5CH{=}\overset{+}{N}(O^-)CH_3$ 2. Raney Ni, H_2, NaOH, CH_3OH, 12 h 3. KH, THF, 1 h	(*Z*)-$C_6H_5CH{=}CHCH(C_6H_5)NHCH_3$ (50)	232
(*Z*)-$(CH_3)_3SiCH{=}CHC_6H_5$	1. $C_6H_5CH{=}\overset{+}{N}(O^-)CH_3$ 2. Zn, HCl, H_2O, THF, 50°, 1 h	" (53)	232
	1. $C_2H_5O_2CCH{=}\overset{+}{N}(O^-)CH_2C_6H_5$ 2. Zn, HCl, H_2O, THF, 50°, 1 h	(*Z*)-$C_6H_5CH{=}CHCH(CO_2C_2H_5)NHCH_2C_6H_5$ (51)	232
C_{13}			
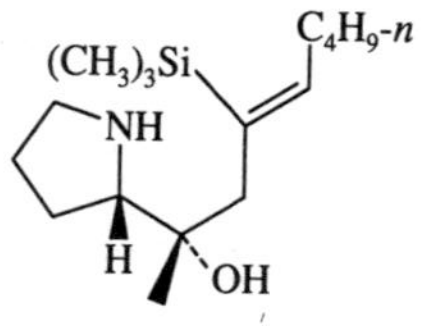 $(CH_3)_3Si$, NH, C_4H_9-*n*, H, OH	$(CH_2O)_n$, CSA, C_2H_5OH, reflux	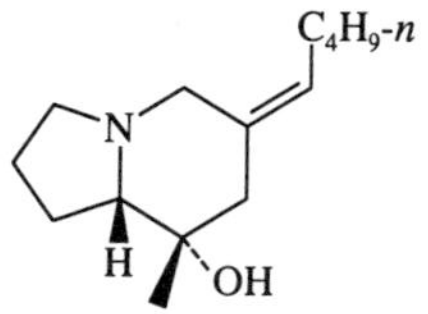 N, C_4H_9-*n*, H, OH (65–80)	400
C_{14}			
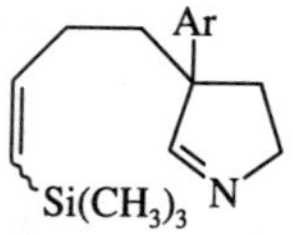 Ar, N, $Si(CH_3)_3$	TFA, CH_3CN, 82°	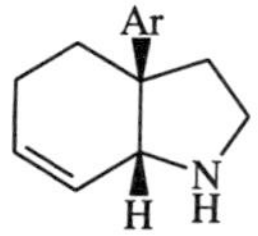 Ar, N, H, H	397

Ar			
E C_6H_5	48 h	(0)	
Z ”	2 h	(90)	
(3,4-methylenedioxyphenyl)	”	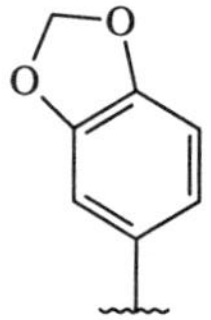 (73)	

C_{16}

Reactant	Conditions	Product (Yield)	Ref.
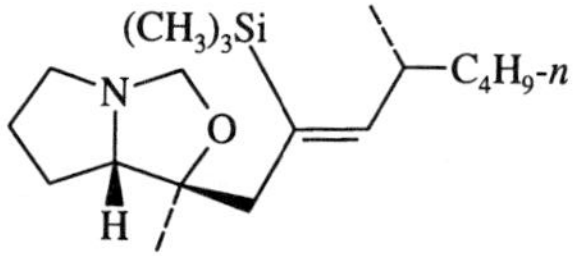 $(CH_3)_3Si$, C_4H_9-*n*, N, O, H	CSA, C_2H_5OH, reflux 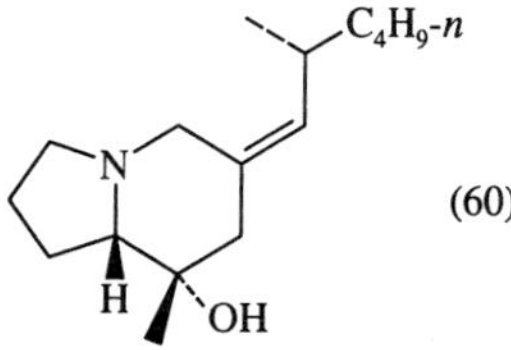	C_4H_9-*n*, N, H, OH (60)	400
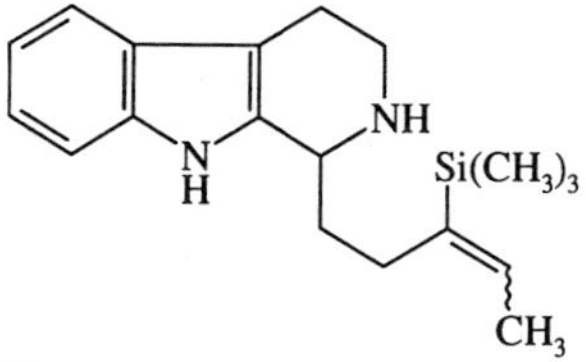 N H, NH, $Si(CH_3)_3$, CH_3 *E* *Z*	$(CH_2O)_n$, CSA, CH_3CN, 82°, 2 h	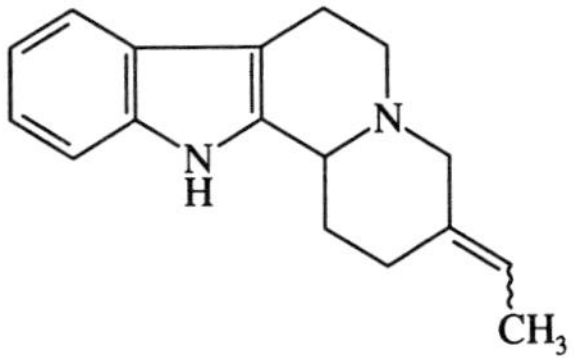 N H, N, CH_3 *E* (83) *Z* (79)	399

TABLE XXIV. VINYLSILANES WITH IMINIUM IONS[a] (*Continued*)

Reactant	Conditions	Product(s) and Yield(s)	Refs.
C_{17}			
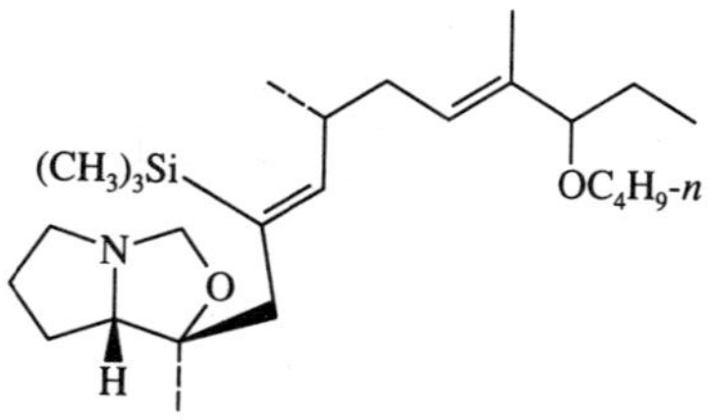 $(CH_3)_3Si$, N, O, H, OC_4H_9-*n*	PyH^+ ^-OTs, Py, CH_3OH, 80°	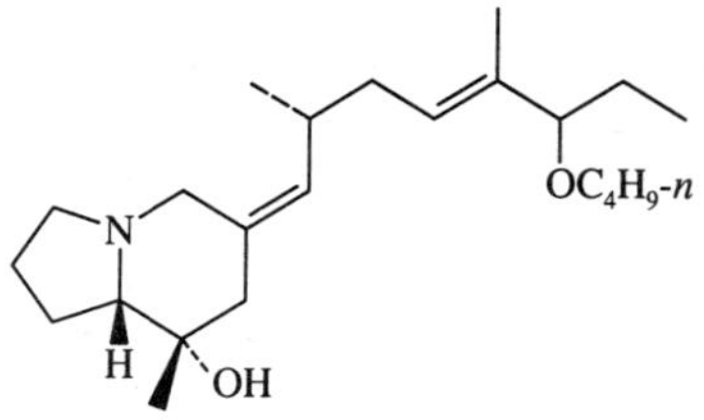 N, H, OH, OC_4H_9-*n* (71)	401

[a] Iminium ions are the likely but presumed intermediates in most of the reactions in this table. The iminium ion itself is the actual reagent in only a few of the entries.

[b] Products of *N*-methylation are also found in the proportions, tetrahydropyridine: *E*-methylamine: *Z*-methylamine, of 6:3:1.

The regioisomers are formed in a ratio of 7:3 in 70% yield; the major product is converted into the final product. The yields given in the table are the overall yields.

TABLE XXV. ALLENYLSILANES WITH IMINIUM IONS

Reactant	Conditions	Product(s) and Yields(s)	Refs.
C_4			
$(CH_3)_3SiC(CH_3){=}C{=}CH_2$	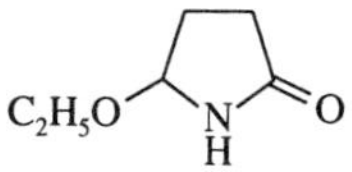 5-ethoxy-2-pyrrolidinone (C_2H_5O, N–H, =O), $TiCl_4$, CH_2Cl_2	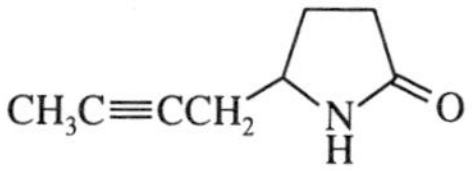 5-($CH_3C{\equiv}CCH_2$)-2-pyrrolidinone (N–H, =O) (59) 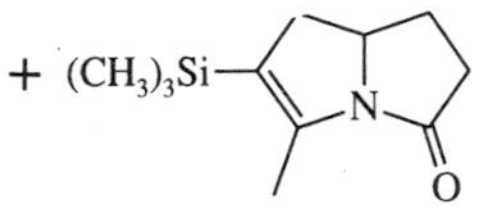 + $(CH_3)_3Si$-substituted methyl dihydropyrrolizinone (28)	129

TABLE XXVI. Allylsilanes with Acid Chlorides and Anhydrides

Reactant	Conditions	Product(s) and Yield(s)	Refs.
C_3			
$(CH_3)_3SiCH_2CH{=}CH_2$	C_2H_5COCl, $ZnCl_2$ or $InCl_3$ or $GaCl_3$	$C_2H_5COCH_2{=}CHCH_2$ + $C_2H_5COCH{=}CHCH_3$ (—)	536
	1-adamantyl-COCl (ClOC–adamantane), $ZnCl_2$, CH_2Cl_2, 14 h	$CH_2{=}CHCH_2CO$–(1-adamantyl) (78)	402
	", $TiCl_4$, CH_2Cl_2, −78°, 5 h	" (70)	402
$(CH_3)_3SiCH[OSi(CH_3)_3]CH{=}CH_2$	i-C_3H_7COCl, $TiCl_4$, CH_2Cl_2, −78°, 3 h	$OHC(CH_2)_2COC_3H_7$-i (20) + $(CH_3)_3SiCH(O_2CC_3H_7$-$i)CH{=}CH_2$ (30)	413, 700
	1. t-C_4H_9COCl, $TiCl_4$, CH_2Cl_2, −78°, 3 h 2. E	$OHCCHRCH_2COC_4H_9$-t	330
	E	**R**	
	$CH_3CH(OC_2H_5)_2$	$CH_3CH(OC_2H_5)$ (53)	
	n-$C_4H_9CH(OCH_3)_2$	n-$C_4H_9CH(OCH_3)$ (59)	
	$C_6H_5CH(OCH_3)_2$	$C_6H_5CH(OCH_3)$ (63)	
	$C_6H_5(CH_2)_2CH(OCH_3)_2$	$C_6H_5(CH_2)_2CH(OCH_3)$ (55)	
$(CH_3)_3SiCH[OSi(C_2H_5)_3]CH{=}CH_2$	i-C_3H_7COCl, $TiCl_4$, CH_2Cl_2, −78°, 3 h	$OHC(CH_2)_2COC_3H_7$-i (43) + $(CH_3)_3SiCH(O_2CC_3H_7$-$i)CH{=}CH_2$ (17)	413
	t-C_4H_9COCl, $TiCl_4$, CH_2Cl_2, −78°, 3 h	$OHC(CH_2)_2COC_4H_9$-t (79)	413
$(CH_3)_3SiCH[OSi(CH_3)_2C_4H_9$-$t]CH{=}CH_2$	$RCOCl$, $TiCl_4$, CH_2Cl_2, −78°	$OHC(CH_2)_2COR$	413
	R		
	n-C_3H_7 3 h	(70)	
	n-C_4H_9 "	(45)	
	s-C_4H_9 "	(67)	
	i-C_4H_9 "	(53)	
	t-C_4H_9 "	(75)[a]	
	$(CH_3)_2C{=}CH$ 4 h	(58)	
	n-C_6H_{13} "	(68)	

Reactant	Conditions	Product(s) and Yield(s) (%)	Refs.
	C_6H_{11} ”	(65)	
	1. $RCOCl$, $TiCl_4$ 2. $RCOCl$, −78°, 5 h	$OHCCH(COR)CH_2COR$	330
	$R = i\text{-}C_3H_7$	(54)	
	$R = t\text{-}C_4H_9$	(59)	
$(CH_3)_3SiCH(SC_6H_5)CH{=}CH_2$	$n\text{-}C_4H_9COCl$, $AlCl_3$, CH_2Cl_2, −78°, 5 h	$(E)\text{-}C_6H_5SCH{=}CHCH_2COC_4H_9\text{-}n$ (57)	412
	$CH_3O_2C(CH_2)_7COCl$, $AlCl_3$, CH_2Cl_2, THF, −78°, 7 h	$C_6H_5SCH{=}CHCH_2CO(CH_2)_7CO_2CH_3$ (75)	701
$(CH_3)_3SiCH_2CH{=}CHCl$	$RCOCl$, $AlCl_3$, CH_2Cl_2, −20°, 2 h	$CH_2{=}CHCHClCOR$	702
	R		
	C_6H_5	(71)	
	C_6H_{11}	(74)	
	$C_6H_5CH_2$	(82)	
	$p\text{-}CH_3C_6H_4$	(50)	
	$n\text{-}C_8H_{17}$	(80)	
	$n\text{-}C_{15}H_{31}$	(77)	
$(CH_3)_3SiCHClCH{=}CH_2$	$n\text{-}C_5H_{11}COCl$, $TiCl_4$, CH_2Cl_2, −78°, 6 h	$ClCH{=}CHCH_2COC_5H_{11}\text{-}n$ $E:Z$ 26:74 (85)	637
$(CH_3)_3SiCH(SeC_6H_5)CH{=}CH_2$	$RCOCl$, $AlCl_3$, CH_2Cl_2	$C_6H_5SeCH{=}CHCH_2COR$ (I) + $(CH_3)_3SiCH{=}CHCH_2SeC_6H_5$ (II) + C_6H_5SeCOR (III)	415

R	Conditions	I	II	III
$n\text{-}C_4H_9$	−78°, 6 h	(26)	(26)	(12)
”	−20°, ”	(36)	(16)	(30)
$n\text{-}C_5H_{11}$	−78°, ”	(22)	(35)	(17)
”	−20°, ”	(35)	(17)	(32)
$n\text{-}C_6H_{13}$	−78°, ”	(27)	(29)	(13)
”	”, 8 h	(38)	(20)	(12)
”	”, 10 h	(32)	(23)	(15)
”	−20°, 6 h	(43)	(15)	(28)
”	”, 8 h	(34)	(11)	(23)
”	0°, 6 h	(42)	(36)	(—)

TABLE XXVI. ALLYLSILANES WITH ACID CHLORIDES AND ANHYDRIDES (*Continued*)

Reactant	Conditions	Product(s) and Yield(s)	Refs.
C_4			
$(CH_3)_3SiCH_2CH{=}CHCH_2Si(CH_3)_3$	RCOCl, $AlCl_3$, CH_2Cl_2, −30°, 55 min	$CH_2{=}CHCH[CH_2Si(CH_3)_3]COR$	403
	R		
	CH_3	(80)	
	C_2H_5	(75)[b]	
	C_6H_5	(75)[c]	
$(CH_3)_3SiCH_2CH{=}CHCH_2Si(C_6H_5)_3$	CH_3COCl, $AlCl_3$, CH_2Cl_2, −30°	$CH_2{=}CHCH[CH_2Si(C_6H_5)_3]COCH_3$ (71)	99
1,1-dimethyl-1-silacyclopent-3-ene (ring structure: Si, $(CH_3)_2$)	CH_3COCl, $AlCl_3$, CH_2Cl_2, −30°, 2 h	$[CH_3CH{=}C(COCH_3)CH_2Si(CH_3)_2]_2O$ (65)	404
	$(CH_3CO)_2O$, BF_3, −15°, 2 h	$CH_3CH{=}C(COCH_3)CH_2Si(CH_3)_2F$ (32)	404
(*E*)-$(CH_3)_3SiCH(SC_6H_5)CH{=}CHCH_3$	RCOCl, $AlCl_3$, CH_2Cl_2, −78°, 8 h	(*E*)-$C_6H_5SCH{=}CHCH(CH_3)COR$	412
	R		
	n-C_3H_7	(75)	
	n-C_4H_9	(67)	
	n-C_6H_{13}	(63)	
$(CH_3)_3SiCH(SC_6H_5)C(CH_3){=}CH_2$	RCOCl, $AlCl_3$, CH_2Cl_2, −78°, 7 h	(*E*)-$C_6H_5SCH{=}C(CH_3)CH_2COR$	412
	R		
	CH_3	(63)	
	n-C_3H_7	(84)	
	n-C_4H_9	(92)	
	n-C_6H_{13}	(84)	
	n-C_7H_{15}	(86)	
$C_6H_5(CH_3)_2SiCH(CO_2CH_3)CH{=}CH_2$	C_2H_5COCl, $TiCl_4$, 0°, 2.5 h	(*E*)-$CH_3O_2CCH{=}CHCH_2COC_2H_5$ (66)	411

C_5

$(CH_3)_3SiCH(CH_3)CH{=}CHCH_3$	[cyclic anhydride structure], $TiCl_4$	$CH_3CH{=}CHCH(CH_3)CO$ ... CO_2H [structure] +	312
		$C_2H_5CH{=}C(CH_3)CO$ ... CO_2H [structure] (75)	
	$COCl$ ... CO_2CH_3 [structure], $TiCl_4$	$CH_3CH{=}CHCH(CH_3)CO$ ... CO_2CH_3 [structure] (—)	312
$(CH_3)_3SiCH_2CH{=}C(CH_3)_2$	$RCOCl$, $AlCl_3$, CH_2Cl_2, $-60°$, 0.75 h	$CH_2{=}CHC(CH_3)_2COR$ I	
	R		
	CH_3	(80)	37
	i-C_4H_9	(70)	37
	$(CH_3)_2C{=}CH$	(90)	37, 406
	$RCOCl$, $TiCl_4$, CH_2Cl_2	I + $(CH_3)_3SiCH_2CH(COR)C(CH_3){=}CH_2$	407
	R		
	CH_3 $-40°$	100:0 (80)	
	2-furyl 0°	50:50 (60)	
	" CS_2, 0°	" (10–20)	
	" CH_3NO_2, 0°	25:75 (65)	
	3-furyl 0°	50:50 (60)	
	2-thiophenyl "	" (55–60)	
	C_6H_5 $-10°$	67:23 (55)	
$F(CH_3)_2SiCH[Si(CH_3)_3]CH{=}C(CH_3)_2$	C_6H_5COCl, $AlCl_3$, CH_2Cl_2, $-70°$	(E)-$(CH_3)_3SiCH{=}CHC(CH_3)_2COC_6H_5$ (76)	586
$(CH_3)_3SiCH_2C(CH_3){=}CHCH_2Si(CH_3)_3$	C_2H_5COCl, $InCl_3$, 25°, 6 h	$CH_2{=}C(CH_3)CH[CH_2Si(CH_3)_3]COC_2H_5$ (55)	403
	C_6H_5COCl, ", ", "	$CH_2{=}C(CH_3)CH[CH_2Si(CH_3)_3]COC_6H_5$ (45)	403

TABLE XXVI. Allylsilanes with Acid Chlorides and Anhydrides (*Continued*)

Reactant	Conditions	Product(s) and Yield(s)	Refs.
	$CH_3CH{=}CHCOCl$, $TiCl_4$, CH_2Cl_2, −80°, 3 h; −30°, 15 h	(70)	703
Si $(CH_3)_2$	CH_3COCl, $AlCl_3$, CH_2Cl_2, −30°, 2 h	$[CH_2{=}C(CH_3)CH(COCH_3)CH_2Si(CH_3)_2]_2O$ + $[(CH_3)_2C{=}C(COCH_3)CH_2Si(CH_3)_2]_2O$ + $CH_2{=}C(CH_3)CH(COCH_3)CH_2Si(CH_3)_2$–O–$(CH_3)_2C{=}C(COCH_3)CH_2Si(CH_3)_2$ (—)[d]	404
	$(CH_3CO)_2O$, BF_3, −15°, 2 h	$CH_2{=}C(CH_3)CH(COCH_3)CH_2Si(CH_3)_2F$ + $(CH_3)_2C{=}C(COCH_3)CH_2Si(CH_3)_2F$ (53)	404
$(CH_3)_3SiCH(CH_2SO_2C_6H_5)C(CH_3){=}CH_2$	i-C_4H_9COCl, $AlCl_3$, CH_2Cl_2, −78°, 30 min	$C_6H_5SO_2CH_2CH{=}C(CH_3)CH_2COC_4H_9$-$i$ (81)	414
$(CH_3)_3SiCH_2CH{=}CHCH{=}CH_2$	t-C_4H_9COCl, $TiCl_4$, CH_2Cl_2, −78°, 1 min	$CH_2{=}CHCH{=}CHCH_2COC_4H_9$-$t$ (34) + $(CH_2{=}CH)_2CHCOC_4H_9$-t (25)	410
$(CH_3)_3SiCH_2C(CH{=}CH_2){=}CH_2$	$RCOCl$, $TiCl_4$, CH_2Cl_2, −78°	$CH_2{=}C(CH{=}CH_2)CH_2COR$	
	R		
	$(CH_3)_2C{=}CH$ 10 min	(71)	186
	i-C_4H_9 1 min	(77)	186, 185
	n-C_5H_{11} "	(66)	186
$(CH_3)_3Si$	**A** $RCOCl$, $AlCl_3$, CH_2Cl_2, −15°, 2 h **B** ", $TiCl_4$, ", ", " **C** ", ", ", −78°, 8 h	COR	

Reactant	Conditions	Product(s) and Yield(s) (%)	Refs.
	R / Method		
	i-C_3H_7 / **A**	(57)	643
	" / **B**	(27)	643
	$(CH_3)_2C{=}CH$ / **A**	(62)	643
	" / **B**	(60)	643
	C_6H_5 / **A**	(68)	643, 704
	" / **B**	(57)	643
	1-adamantyl / **C**	(74)	402
$C_6H_5(CH_3)_2SiCH(CO_2CH_3)CH{=}CHCH_3$	i-C_3H_7COCl, $TiCl_4$, CH_2Cl_2, 1 h	(E)-$CH_3O_2CCH{=}CHCH(CH_3)COC_3H_7$-$i$ (77)	612
$(CH_3)_3SiCH(CO_2C_2H_5)C(CH_3){=}CH_2$	t-C_4H_9COCl, $TiCl_4$, CH_2Cl_2, 24 h	$C_2H_5O_2CCH{=}C(CH_3)CH_2COC_4H_9$-$t$ + 4-methyl-6-(C_4H_9-t)-2-pyranone 82:18 (58)	252
	C_6H_5COCl, $TiCl_4$, CH_2Cl_2, 24 h	$C_2H_5O_2CCH{=}C(CH_3)CH_2COC_6H_5$ + 4-methyl-6-(C_6H_5)-2-pyranone 28:72 (49)	252
$C_6H_5(CH_3)_2SiC(CO_2CH_3)(CH_3)CH{=}CH_2$	i-C_3H_7COCl, $TiCl_4$, CH_2Cl_2, 1 h	$CH_3O_2CC(CH_3){=}CHCH_2COC_3H_7$-$i$ (54)	612
$(CH_3)_3Si$-cyclopentadienyl	1. (p-$CH_3C_6H_4)_2SO$, $(CF_3CO)_2O$, CH_2Cl_2, −10°, 1 h 2. $LiClO_4$	CF_3CO-cyclopentadienide with $(p$-$CH_3C_6H_4)_2S^+$ and $S^+(C_6H_4CH_3$-$p)_2$ substituents, ClO_4^- (46)	705
C_6			
(E)-$(CH_3)_3SiCH_2C(CH_3){=}C(CH_3)$-$CH_2Si(CH_3)_3$	CH_3COCl, $AlCl_3$, CH_2Cl_2, −30°, 55 min	$CH_2{=}C(CH_3)C(CH_3)(CH_2Si(CH_3)_3)COCH_3$ (75)	403

TABLE XXVI. ALLYLSILANES WITH ACID CHLORIDES AND ANHYDRIDES (*Continued*)

Reactant	Conditions	Product(s) and Yield(s)	Refs.
	", $GaCl_3$, 60°, 9 h	" (55)	403
	C_6H_5COCl, $AlCl_3$, CH_2Cl_2, −30°, 55 min	$CH_2{=}C(CH_3)C(CH_3)(CH_2Si(CH_3)_3)COC_6H_5$ (70)	403
	", $InCl_3$, 60°, 9 h	" (25)	403
(*E*)-$(CH_3)_3SiC(CH_3)_2CH{=}C(CH_3)Si(CH_3)_3$	CH_3COCl, $TiCl_4$, CH_2Cl_2, −50°	$(CH_3)_2C{=}CHCH(CH_3)COCH_3$ (65)	547
1,1,3,4-tetramethyl-1-silacyclopent-3-ene [structure: Si$(CH_3)_2$]	CH_3COCl, $AlCl_3$, CH_2Cl_2, −30°, 2 h	$[CH_2{=}C(CH_3)C(CH_3)(COCH_3)CH_2Si(CH_3)_2]_2O$ (30)	404
	$(CH_3CO)_2O$, BF_3, −15°, 2 h	$CH_2{=}C(CH_3)C(CH_3)(COCH_3)CH_2Si(CH_3)_2F$ (28)	404
$(CH_3)_3SiCH_2(CH{=}CH)_2CH_3$	*t*-C_4H_9COCl, $TiCl_4$, CH_2Cl_2, −78°, 1 min	$CH_2{=}CHCH{=}CHCH(CH_3)COC_4H_9$-*t* (12) + $CH_3CH{=}CHCH(CH{=}CH_2)COC_4H_9$-*t* (43)	410
$(CH_3)_3SiC(CO_2C_2H_5)(CH_3)C(CH_3){=}CH_2$	*t*-C_4H_9COCl, $TiCl_4$, CH_2Cl_2, −78°, 8 h; room temp, 40 h	$C_2H_5O_2CC(CH_3){=}C(CH_3)CH_2COC_4H_9$-*t* (26) + 3,4-dimethyl-6-C_4H_9-*t*-2-pyranone [structure] (10)	252
1,2-bis($Si(CH_3)_3$)cyclohex-1-ene [structure: $(CH_3)_3Si$, $Si(CH_3)_3$]	CH_3COCl, $AlCl_3$, CH_2Cl_2, −60°, 1 h; −30°, 1 h	2-$Si(CH_3)_3$-cyclohex-2-enyl $COCH_3$ [structure] (90)	86

Reactant	Conditions	Product(s) and Yield(s) (%)	Refs.
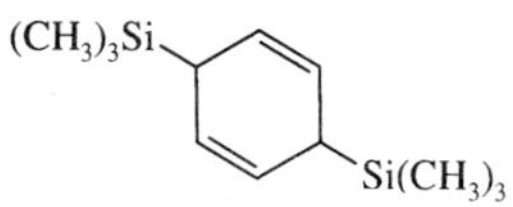 1,4-bis(trimethylsilyl)cyclohexa-2,5-diene: $(CH_3)_3Si$–C_6H_6–$Si(CH_3)_3$	1. CH_3COCl (2 eq), $AlCl_3$, CH_2Cl_2, $-60°$, 1 h; $-40°$, 2 h 2. Air	p-$CH_3COC_6H_4COCH_3$ (70) + $CH_3COC_6H_5$ (10) + p-$CH_3COC_6H_4CHOHCH_3$ (20)	706
C_7			
$(E)(S)$-$F(CH_3)_2SiCH(Si(CH_3)_3)CH{=}C(CH_3)C_3H_7$-$i$	CH_3COCl, $AlCl_3$, CH_2Cl_2, $-70°$	$(E)(S)$-$(CH_3)_3SiCH{=}CHC(CH_3)(COCH_3)C_3H_7$-$i$ (88)	19
$C_6H_5(CH_3)_2SiCH_2CH{=}$(cyclopentylidene)	CH_3COCl, $TiCl_4$, $Ti(OC_3H_7$-$i)_4$, CH_2Cl_2, $-78°$, 4 h	$C_6H_5(CH_3)_2SiCH_2CH(COCH_3)$(cyclopent-1-enyl) (65)	38
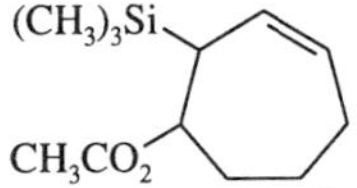 7-(trimethylsilyl)-6-acetoxycycloheptene: $(CH_3)_3Si$, CH_3CO_2	$(CH_3CO)_2O$, $HClO_4$, CH_3CO_2H, $0°$, 0.5 h	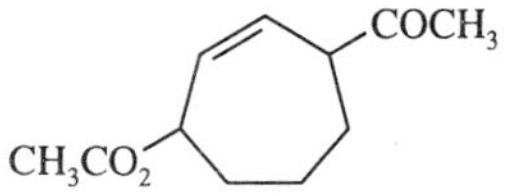 3-acetyl-6-acetoxycycloheptene: $COCH_3$, CH_3CO_2 (—)	405
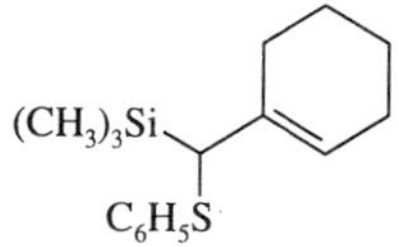 $(CH_3)_3SiCH(SC_6H_5)$(cyclohex-1-enyl)	$RCOCl$, CH_2Cl_2, $-78°$, 6 h	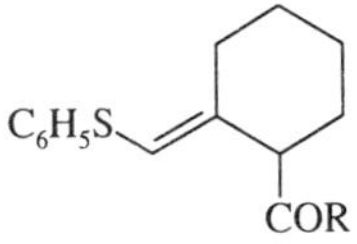 2-($C_6H_5SCH{=}$)cyclohexyl COR	707, 122
	R —		
	CH_3 — $AlCl_3$	(26)	
	" — $TiCl_4$	(44)	
	" — $SnCl_4$	(36)	
	n-C_3H_7 — $AlCl_3$	(65)	
	n-C_4H_9 — "	(80)	
	n-C_6H_{13} — "	(81)	
	n-C_7H_{15} — "	(74)	

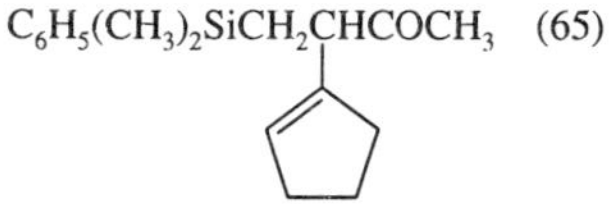

TABLE XXVI. ALLYLSILANES WITH ACID CHLORIDES AND ANHYDRIDES (*Continued*)

Reactant	Conditions	Product(s) and Yield(s)	Refs.
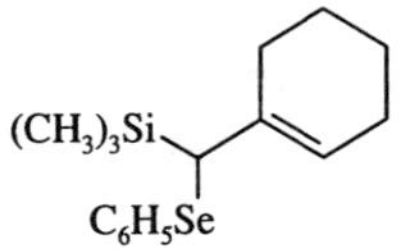 $(CH_3)_3Si$–CH(C_6H_5Se)–(1-cyclohexenyl)	RCOCl, $AlCl_3$, CH_2Cl_2, −78°, 8 h	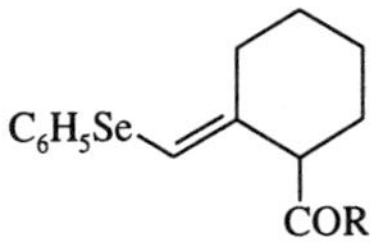 2-(C_6H_5SeCH=)cyclohexyl–COR (I) + C_6H_5SeCOR (II)	415
	R: n-C_3H_7	I $E:Z$ 4:1 (57), II (6)	
	R: n-C_4H_9	I (60), II (13)	
	R: n-C_5H_{11}	I (56), II (12)	
	R: n-C_6H_{13}	I (61), II (10)	
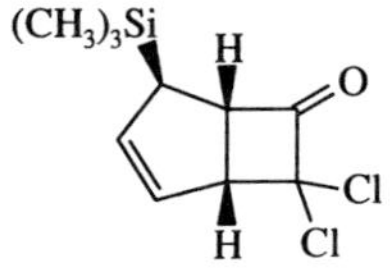 $(CH_3)_3Si$-substituted 7,7-dichlorobicyclo[3.2.0]hept-2-en-6-one	RCOCl, $SnCl_4$, 0°	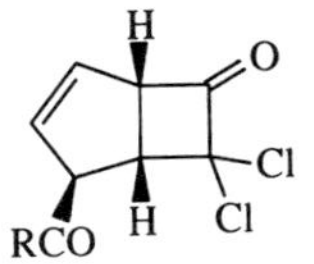 RCO-substituted 7,7-dichlorobicyclo[3.2.0]hept-2-en-6-one	55
	R: CH_3, 1 d	(36)	
	R: C_2H_5, 7 d	(20)	
	R: n-C_7H_{15}, "	(40)	
C_8			
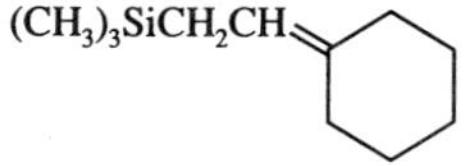 $(CH_3)_3SiCH_2CH$=cyclohexylidene	CH_3COCl, $TiCl_4$, CH_2Cl_2, −78°, 1 h	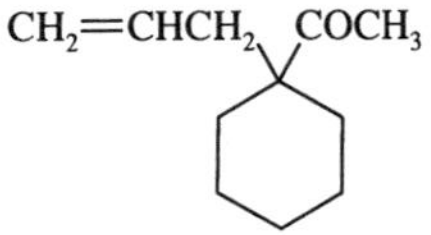 1-(CH_2=$CHCH_2$)-1-($COCH_3$)cyclohexane (82)	170

Reactant	Conditions	Product(s) and Yield(s) (%)	Refs.
$(CH_3)_3Si$ … $Si(CH_3)_3$	CH_3COCl, $AlCl_3$, CH_2Cl_2, $-30°$, 55 min	$COCH_3$, $Si(CH_3)_3$ (70)	403
$(CH_3)_3Si$ … $Si(CH_3)_3$	CH_3COCl (2 eq), $AlCl_3$, CH_2Cl_2	$COCH_3$, $Si(CH_3)_3$, $COCH_3$ (45) + $COCH_3$, $Si(CH_3)_3$ (10) + $COCH_3$ (25) + $COCH_3$, $Si(CH_3)_3$ (10)	706
$(CH_3)_3Si$ … $Si(CH_3)_3$	1. CH_3COCl (2 eq), $AlCl_3$, CH_2Cl_2 2. Air	$COCH_3$, CH_3CO + $COCH_3$, CH_3CO 1:2 (65–70)	706

TABLE XXVI. ALLYLSILANES WITH ACID CHLORIDES AND ANHYDRIDES (*Continued*)

Reactant	Conditions	Product(s) and Yield(s)	Refs.
C_9			
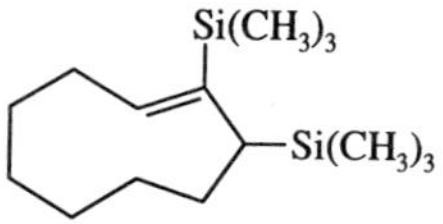 Si(CH$_3$)$_3$, —Si(CH$_3$)$_3$	CH_3COCl, $AlCl_3$	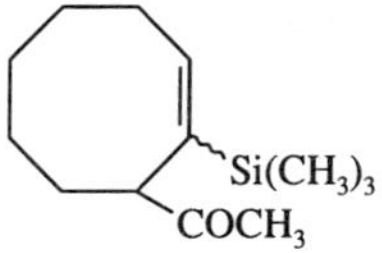 Si(CH$_3$)$_3$, COCH$_3$ *E*:*Z* 60:40 (80)	708
$(CH_3)_3SiCH_2CH{=}CHC_6H_5$	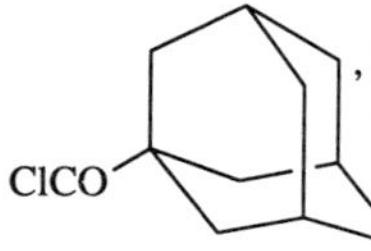 ClCO, $TiCl_4$, CH_2Cl_2, −78°, 3 h	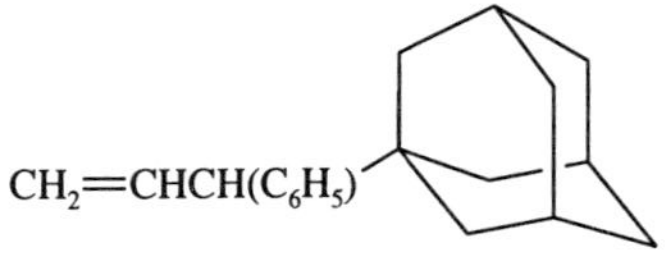 $CH_2{=}CHCH(C_6H_5)$ (34)	402
(*E*)-F$(CH_3)_2$SiCH(Si(CH$_3$)$_3$)CH=CHC$_6$H$_5$	C_6H_5COCl, $AlCl_3$, CH_2Cl_2, −70°	(*E*)-$(CH_3)_3SiCH{=}CHCH(C_6H_5)COC_6H_5$ (61)	586
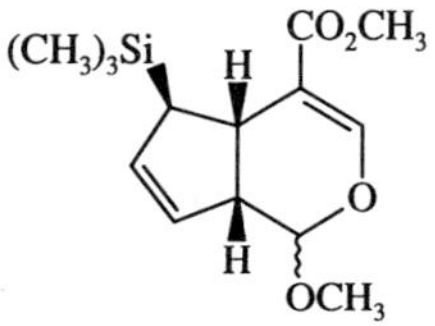 (CH$_3$)$_3$Si, H, CO$_2$CH$_3$, O, H, OCH$_3$	CH_3COCl, $AlCl_3$	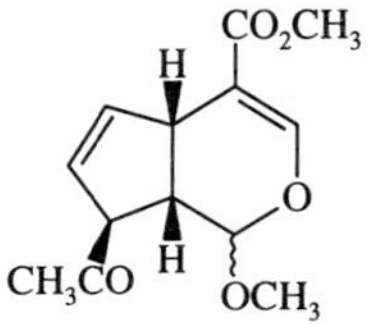 H, CO$_2$CH$_3$, O, CH$_3$CO, H, OCH$_3$ (10)	54
C_{10}			
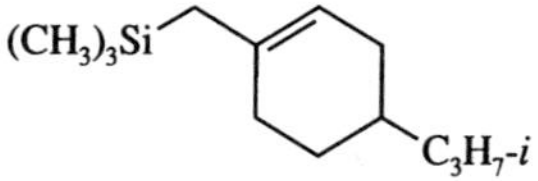 (CH$_3$)$_3$Si, C$_3$H$_7$-*i*	RCOCl, $AlCl_3$, CH_2Cl_2, −90°, 1.25 h	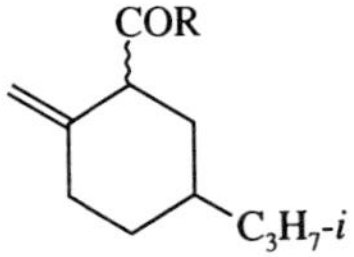 COR, C$_3$H$_7$-*i*	709

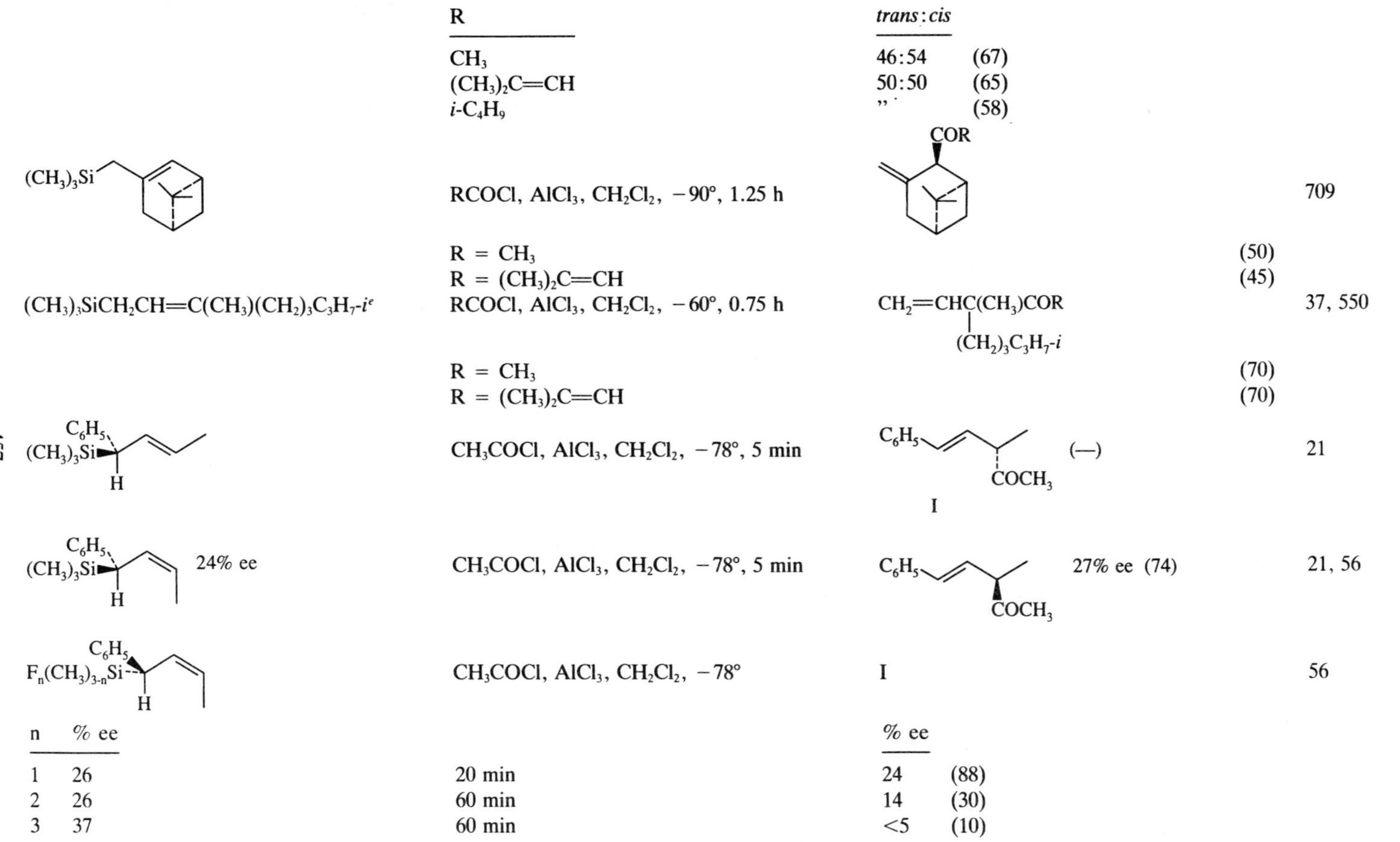

Reactant	Conditions	Product(s) and Yield(s) (%)	Refs.
	R	*trans*:*cis*	
	CH_3	46:54 (67)	
	$(CH_3)_2C{=}CH$	50:50 (65)	
	i-C_4H_9	,, (58)	
$(CH_3)_3Si$ (structure)	$RCOCl$, $AlCl_3$, CH_2Cl_2, $-90°$, 1.25 h	COR (structure)	709
	R = CH_3	(50)	
	R = $(CH_3)_2C{=}CH$	(45)	
$(CH_3)_3SiCH_2CH{=}C(CH_3)(CH_2)_3C_3H_7$-$i$[e]	$RCOCl$, $AlCl_3$, CH_2Cl_2, $-60°$, 0.75 h	$CH_2{=}CHC(CH_3)COR$ with $(CH_2)_3C_3H_7$-i	37, 550
	R = CH_3	(70)	
	R = $(CH_3)_2C{=}CH$	(70)	
C_6H_5, $(CH_3)_3Si$, H (structure)	CH_3COCl, $AlCl_3$, CH_2Cl_2, $-78°$, 5 min	C_6H_5, $COCH_3$ (structure) I (—)	21
C_6H_5, $(CH_3)_3Si$, H (structure) 24% ee	CH_3COCl, $AlCl_3$, CH_2Cl_2, $-78°$, 5 min	C_6H_5, $COCH_3$ (structure) 27% ee (74)	21, 56
C_6H_5, $F_n(CH_3)_{3-n}Si$, H (structure)	CH_3COCl, $AlCl_3$, CH_2Cl_2, $-78°$	I	56
n / % ee		% ee	
1 / 26	20 min	24 (88)	
2 / 26	60 min	14 (30)	
3 / 37	60 min	<5 (10)	

TABLE XXVI. ALLYLSILANES WITH ACID CHLORIDES AND ANHYDRIDES (*Continued*)

Reactant	Conditions	Product(s) and Yield(s)	Refs.
C_{12}			
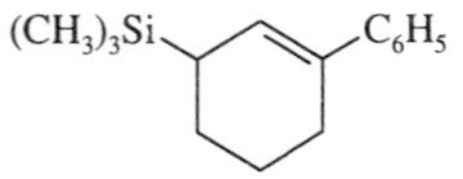 $(CH_3)_3Si$, C_6H_5 (cyclohexene structure)	CH_3COCl, $AlCl_3$, $-25°$, 2 h	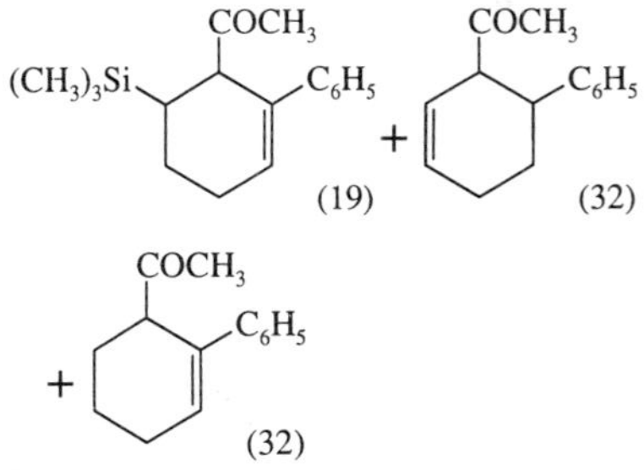 $(CH_3)_3Si$, $COCH_3$, C_6H_5 (19) + $COCH_3$, C_6H_5 (32) + $COCH_3$, C_6H_5 (32)	408
C_{14}			
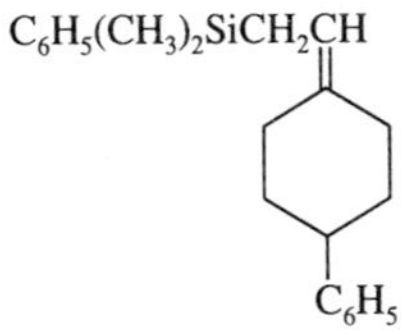 $C_6H_5(CH_3)_2SiCH_2CH$, C_6H_5 (cyclohexylidene structure)	CH_3COCl, $TiCl_4$, CH_2Cl_2, $-78°$, 1 h	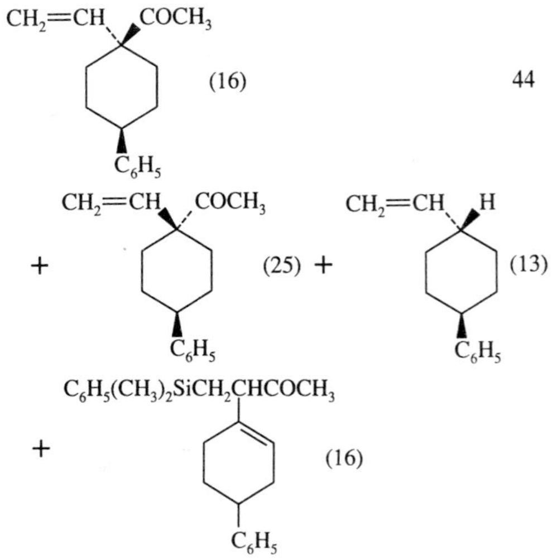 $CH_2=CH$, $COCH_3$, C_6H_5 (16) + $CH_2=CH$, $COCH_3$, C_6H_5 (25) + $CH_2=CH$, H, C_6H_5 (13) + $C_6H_5(CH_3)_2SiCH_2CHCOCH_3$, C_6H_5 (16)	44

C_{15}

Reactant	Conditions	Product(s) and Yield(s) (%)	Refs.
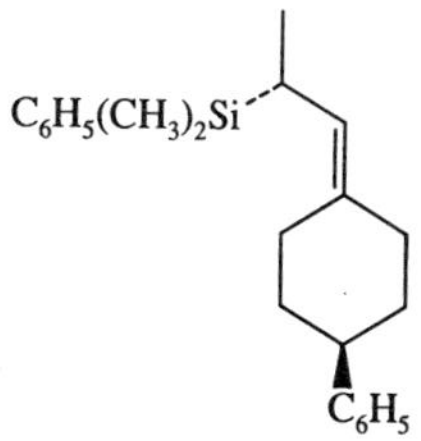 $C_6H_5(CH_3)_2Si$ (structure) ... C_6H_5	CH_3COCl, $TiCl_4$, CH_2Cl_2, $-78°$, 1 h	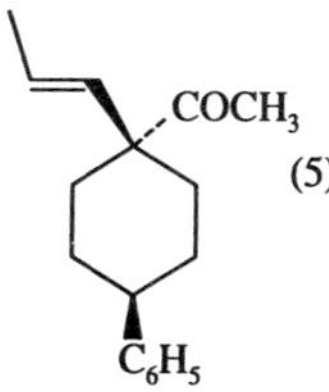 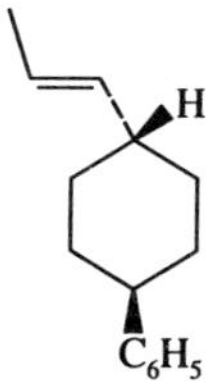 I (COCH$_3$, C_6H_5) (5) + II (H, C_6H_5) (28) 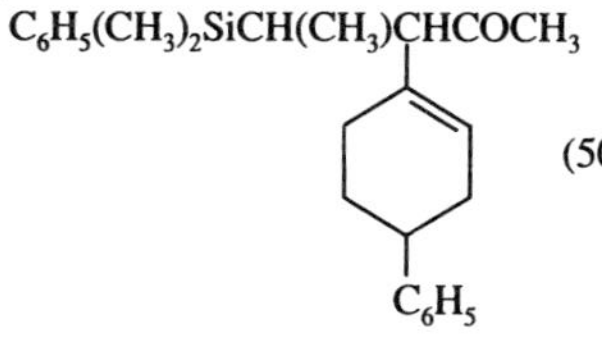 + $C_6H_5(CH_3)_2SiCH(CH_3)CHCOCH_3$ (C_6H_5) III (50)	44
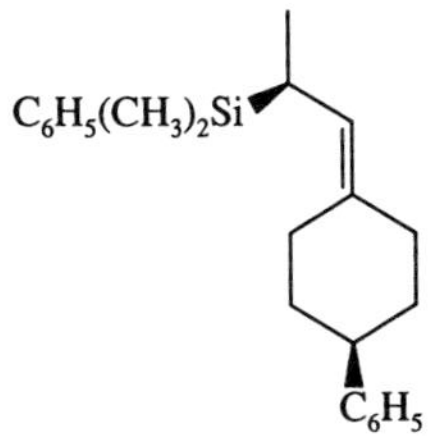 $C_6H_5(CH_3)_2Si$ (structure) ... C_6H_5	CH_3COCl, $TiCl_4$, CH_2Cl_2, $-78°$, 1 h	I (15) + II (16) + III (32)	44

[a] The use of $AlCl_3$ in place of $TiCl_4$ gives a 43% yield.

[b] Gallium chloride at 40° for 6 h gives a 72% yield of the same product.

[c] Gallium chloride at 40° for 6 h gives a 37% yield of the same product.

[d] The major component of this mixture is the first of the three products drawn.

[e] This allylsilane is prepared by selective hydrogenation (H_2, nickel boride) of geranyltrimethylsilane.

TABLE XXVII. VINYLSILANES WITH ACID CHLORIDES

Reactant	Conditions	Product(s) and Yield(s)	Refs.
C_2			
$(CH_3)_3SiCH{=}CH_2$	RCOCl	$RCOCH{=}CH_2$	
	R		
	CH_3 — $AlCl_3$, −20°	(45)	710
	(*E*)-$CH_3CH{=}CH$ — ", $ClCH{=}CHCl$,[a] CH_2Cl_2, −15°, 18 min	(77)	711
	" — $SnCl_4$	(—)	421
	$CH_2{=}C(CH_3)$ — $AlCl_3$, $ClCH{=}CHCl$, CH_2Cl_2	(0)	711
	$(CH_3)_2C{=}CH$ — ", 20°	(53)	710
	" — ", $ClCH{=}CHCl$,[a] 0.5 h	(50–70)	711
	" — $SnCl_4$	(—)	421
	(*E*,*E*)-$CH_3(CH{=}CH)_2$ — $AlCl_3$, $ClCH{=}CHCl$,[a] 0°, 3 min	(66)	711
	(*E*)-$C_6H_5CH{=}CH$ — ", CH_2Cl_2, −10°, 5 min	(55)	711
	$RCH{=}CHCOCl$	4-R-cyclopent-2-en-1-one (structure)	711
	R		
	CH_3 — $AlCl_3$, CCl_4, 50°, 0.5 h	(63)	
	$CH_3CH{=}CH$ — ", $ClCH{=}CHCl$,[a] reflux, 1 h	(18)	
	C_6H_5 — ", CCl_4, 50°, 0.5 h	(46)	

Reactant	Conditions	Product(s) and Yield(s) (%)	Refs.
(*E*)-$CH_3CH{=}C(CH_3)$	", $ClCH{=}CHCl$,[a] 50°, 40 min	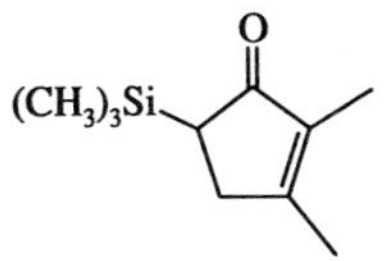 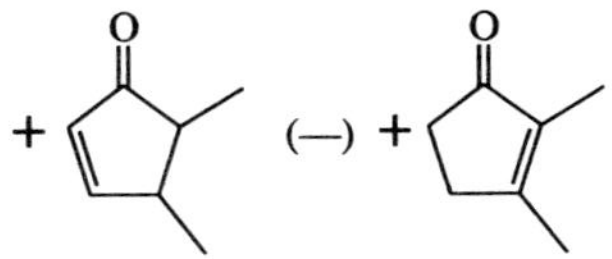 $(CH_3)_3Si$-substituted 2,3-dimethylcyclopentenone (major) + 4,5-dimethylcyclopent-2-enone (—) + 2,3-dimethylcyclopent-2-enone (—)	
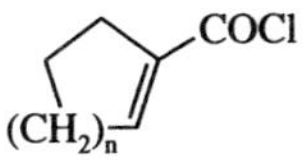	1-(chlorocarbonyl)cycloalkene $(CH_2)_n$, $SnCl_4$, CH_2Cl_2, −30°, 1 h; 20°, 6 h	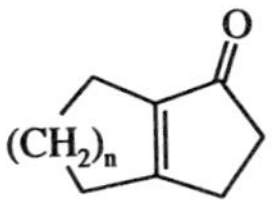 bicyclic cyclopentenone $(CH_2)_n$	421
	n = 1		(53)
	n = 2		(46)
	n = 3		(32)
	R^1,R^2-cyclohexene-1-COCl, $SnCl_4$, CH_2Cl_2, −30°, 1 h; 20°, 6 h	R^1,R^2-substituted tetrahydroindanone	421
	R^1 = CH_3, R^2 = H		(64)
	R^1 = H, R^2 = *i*-C_3H_7		(56)
(*E*)-$(CH_3)_3SiCH{=}CHSi(CH_3)_3$	RCOCl, $AlCl_3$, −20°	(*E*)-$RCOCH{=}CHSi(CH_3)_3$	710
	R = CH_3		(88)
	R = (*E*)-$CH_3CH{=}CH$		(65)
	R = $(CH_3)_2C{=}CH$		(60)

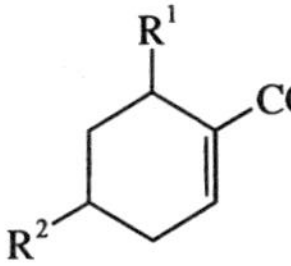

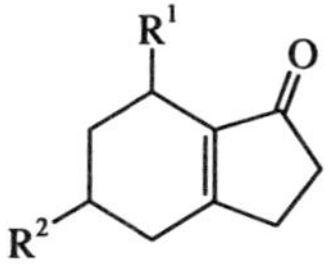

TABLE XXVII. VINYLSILANES WITH ACID CHLORIDES (*Continued*)

Reactant	Conditions	Product(s) and Yield(s)	Refs.
	t-C_4H_9	(65)	
	C_6H_5	(80)	
	$C_6H_5CH{=}CH$	(61)	
$(CH_3)_3SiCH{=}CHSC_6H_5$	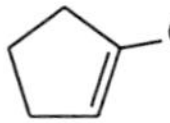 1-cyclopentenyl-COCl, $AlCl_3$, $Cl(CH_2)_2Cl$, 80°, 18 h	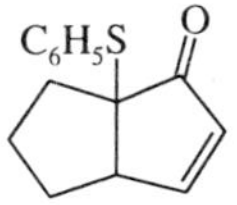 bicyclic enone with C_6H_5S at ring junction (55)	712
$(CH_3)_3SiC(SC_6H_5){=}CH_2$	 4,4-R,R-1-cyclopentenyl-COCl, $AgBF_4$, −20°, 20 h	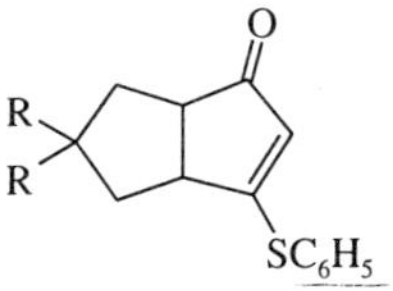 R,R-bicyclic enone with SC_6H_5	
	R = H $Cl(CH_2)_2Cl$, CH_2Cl_2	(35)	712, 421
	R = CH_3 CH_2Cl_2	(38)	712, 119
$(CH_3)_3SiC(SAr){=}CH_2$	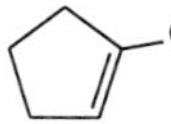 1-cyclopentenyl-COCl, $AlCl_3$, $Cl(CH_2)_2Cl$, CH_2Cl_2, 20°	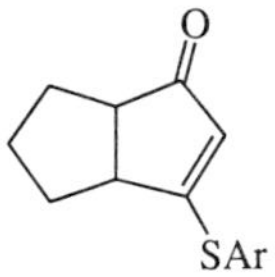 bicyclic enone with SAr	712
Ar = C_6H_4Cl-p	16 h	(15)	
Ar = $C_6H_3(NO_2)_2$-3,4	4 h	(58)	
(Z)-$(CH_3)_3SiCH{=}CHCl$	C_2H_5COCl, $AlCl_3$, CH_2Cl_2, 0°, 0.5 h	(E)-$C_2H_5COCH{=}CHCl$ (30)	416
(Z)-$(CH_3)SiCH{=}CHBr$	CH_3COCl, $AlCl_3$, CH_2Cl_2, 0°, 0.5 h	(E)-$CH_3COCH{=}CHBr$ (60)	416
	C_2H_5COCl, ", ", ", "	(E)-$C_2H_5COCH{=}CHBr$ (45)	416
(E)- + (Z)-$(CH_3)_3SiCH{=}CHI$	CH_3COCl, ", ", ", "	(E)-$CH_3COCH{=}CHI$ (25)	416
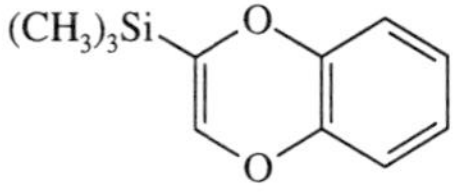 2-$(CH_3)_3Si$-1,4-benzodioxin	RCOCl, $AlCl_3$, CH_2Cl_2, 20 min	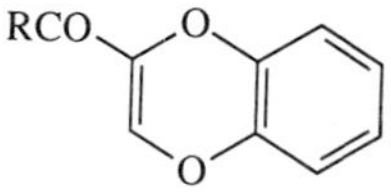 2-RCO-1,4-benzodioxin	713

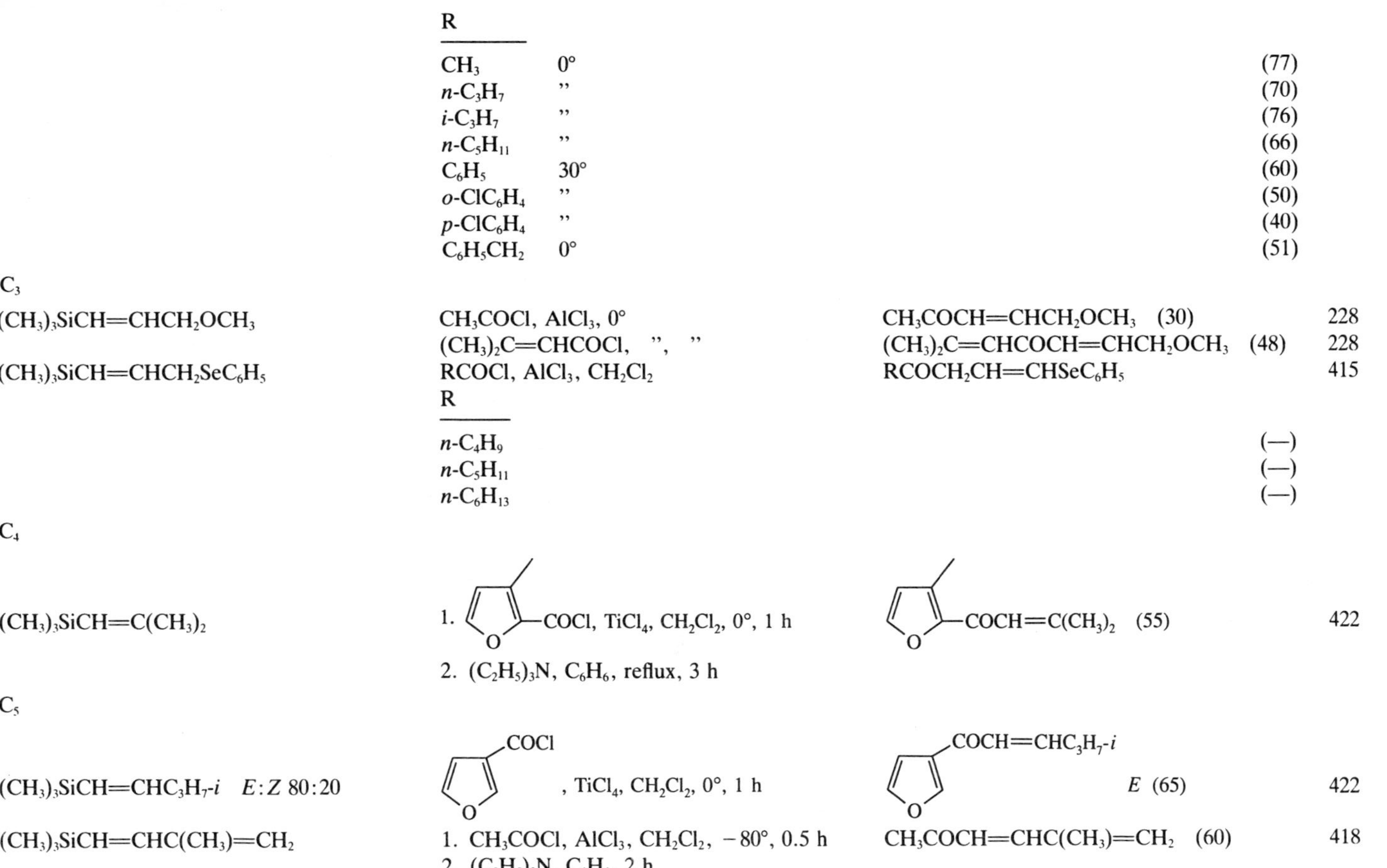

Reactant	Conditions	Product(s) and Yield(s) (%)	Refs.
	R: CH_3, 0°	(77)	
	R: n-C_3H_7, ”	(70)	
	R: i-C_3H_7, ”	(76)	
	R: n-C_5H_{11}, ”	(66)	
	R: C_6H_5, 30°	(60)	
	R: o-ClC_6H_4, ”	(50)	
	R: p-ClC_6H_4, ”	(40)	
	R: $C_6H_5CH_2$, 0°	(51)	
C_3			
$(CH_3)_3SiCH{=}CHCH_2OCH_3$	CH_3COCl, $AlCl_3$, 0°	$CH_3COCH{=}CHCH_2OCH_3$ (30)	228
	$(CH_3)_2C{=}CHCOCl$, ”, ”	$(CH_3)_2C{=}CHCOCH{=}CHCH_2OCH_3$ (48)	228
$(CH_3)_3SiCH{=}CHCH_2SeC_6H_5$	$RCOCl$, $AlCl_3$, CH_2Cl_2	$RCOCH_2CH{=}CHSeC_6H_5$	415
	R: n-C_4H_9	(—)	
	R: n-C_5H_{11}	(—)	
	R: n-C_6H_{13}	(—)	
C_4			
$(CH_3)_3SiCH{=}C(CH_3)_2$	1. 3-methyl-2-furyl–COCl, $TiCl_4$, CH_2Cl_2, 0°, 1 h 2. $(C_2H_5)_3N$, C_6H_6, reflux, 3 h	3-methyl-2-furyl–$COCH{=}C(CH_3)_2$ (55)	422
C_5			
$(CH_3)_3SiCH{=}CHC_3H_7$-i $E{:}Z$ 80:20	3-furyl–COCl, $TiCl_4$, CH_2Cl_2, 0°, 1 h	3-furyl–$COCH{=}CHC_3H_7$-i E (65)	422
$(CH_3)_3SiCH{=}CHC(CH_3){=}CH_2$	1. CH_3COCl, $AlCl_3$, CH_2Cl_2, −80°, 0.5 h 2. $(C_2H_5)_3N$, C_6H_6, 2 h	$CH_3COCH{=}CHC(CH_3){=}CH_2$ (60)	418

TABLE XXVII. Vinylsilanes with Acid Chlorides (*Continued*)

Reactant	Conditions	Product(s) and Yield(s)	Refs.
$(CH_3)_3Si$ 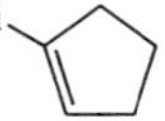	1. $CH_3CH{=}CHCOCl$, $AlCl_3$, CH_2Cl_2, $-78°$, 15 min[b] 2. $BF_3{\cdot}O(C_2H_5)_2$, C_6H_6, reflux, 3 d	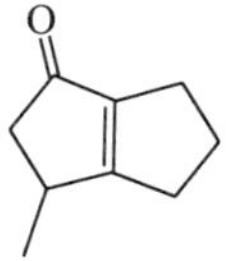 (58)	76
	1. $(CH_3)_2C{=}CHCOCl$, $AlCl_3$, CH_2Cl_2, $-78°$, 15 min[b] 2. $SnCl_4$, CH_2Cl_2, reflux, 8–12 h 3. $RhCl_3$, C_2H_5OH[c]	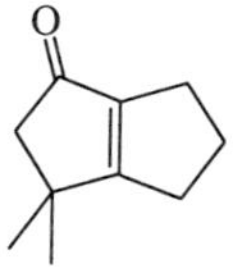 (62)	76
$(CH_3)_3Si$ 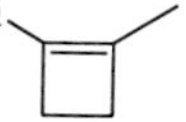	1. CH_3COCl, $AlCl_3$, CH_2Cl_2, 0° 2. NaOH, H_2O	CH_3CO 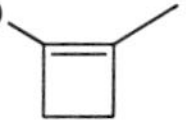 (88)	423
C_6			
(E)-$(CH_3)_3SiCH{=}CHC_4H_9$-n	CH_3COCl, $AlCl_3$, CH_2Cl_2, 0°, 30 min	(E)-$CH_3COCH{=}CHC_4H_9$-n (69)	567
$(CH_3)_3SiCH{=}CHC(C_2H_5){=}CH_2$	1. CH_3COCl, $AlCl_3$, CH_2Cl_2, $-80°$, 0.5 h 2. $(C_2H_5)_3N$, C_6H_6, 2 h	$CH_3COCH{=}CHC(C_2H_5){=}CH_2$ (55)	418
(Z)-$(CH_3)_3SiCF{=}CFC_4H_9$-n	CH_3COCl, $AlCl_3$, CH_2Cl_2, $-30°$, 1 h	$CH_3COCF{=}CFC_4H_9$-n $E{:}Z$ 98:2 (92)	570
$(CH_3)_3Si$ 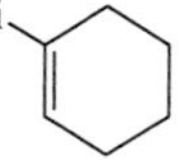	$CH_2{=}CHCOCl$, $AlCl_3$, CH_2Cl_2, 2 h	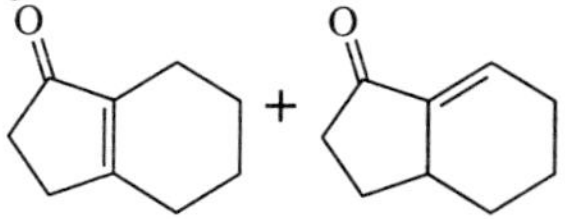 + (27) 1:1.25	76
	1. $CH_3CH{=}CHCOCl$, $AlCl_3$, CH_2Cl_2, 1 h 2. $BF_3{\cdot}O(C_2H_5)_2$, C_6H_6, reflux	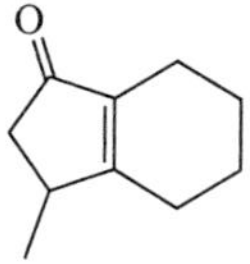 (44)	76

Reactant	Conditions	Product(s) and Yield(s) (%)	Refs.
	1. $(CH_3)_2C{=}CHCOCl$, $AlCl_3$, CH_2Cl_2, $-78°$, 15 min[b] 2. $SnCl_4$, CH_2Cl_2, reflux, 8–12 h 3. $RhCl_3$, C_2H_5OH[c]	O (70)	76
	ArCOCl, $AlCl_3$	ArCO	714
	Ar		
	o-IC_6H_4	(—)	
	2-I-5-$CH_3C_6H_3$	(—)	
$(CH_3)_3Si$	$(CH_3)_2C{=}CHCOCl$, $AlCl_3$, CH_2Cl_2, $-78°$, 5 min	$(CH_3)_2C{=}CHCO$ (>55)	76
(*E*)- + (*Z*)-$(CH_3)_3SiCH{=}CHCH_2$ $(CH_3)_3SiCH{=}CHCH_2$	$(CH_3)_2C{=}CHCOCl$, $AlCl_3$, 4 h	(*E*)- + (*Z*)-$(CH_3)_2C{=}CHCOCH{=}CHCH_2$ (45) $(CH_3)_2C{=}CHCOCH{=}CHCH_2$	505
$CH_3CH{=}C(CH_2)_2COCl$ $(CH_3)_3Si$ *E*:*Z* 52:48	$AlCl_3$, CH_2Cl_2, 13 h	O (62) + O (38) *E*:*Z* > 95:5	116
$CH_2{=}CCH(CH_3)CH_2COCl$ $(CH_3)_3Si$	$AlCl_3$, CH_2Cl_2, 13 h	O (>95)	116

TABLE XXVII. VINYLSILANES WITH ACID CHLORIDES (*Continued*)

Reactant	Conditions	Product(s) and Yield(s)	Refs.
C_7			
(*E*)-$(CH_3)_3SiC(C_3H_7\text{-}n){=}CHC_2H_5$	$CH_2{=}CHCOCl$, $AlCl_3$, CH_2Cl_2, 2 h	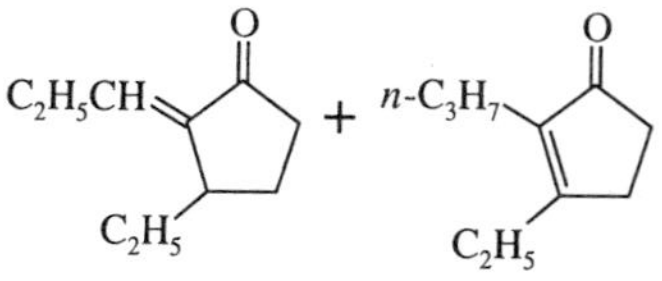 2-propylidene-3-ethylcyclopentanone ($C_2H_5CH{=}$, C_2H_5) + 2-(n-C_3H_7)-3-(C_2H_5)cyclopent-2-enone (24); 3.75:1	76
	1. $(CH_3)_2C{=}CHCOCl$, $AlCl_3$, CH_2Cl_2, −78°, 15 min[b] 2. $BF_3{\cdot}O(C_2H_5)_2$, C_6H_6, reflux, 1–3 d	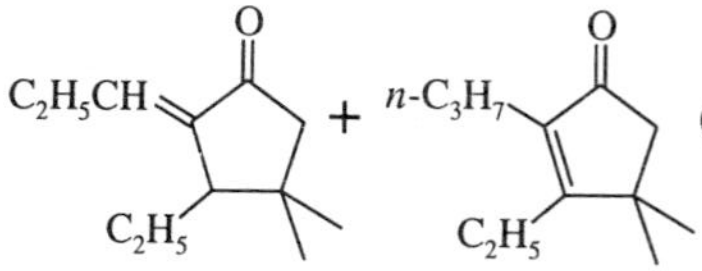 2-propylidene-3-ethyl-4,4-dimethylcyclopentanone ($C_2H_5CH{=}$, C_2H_5) + 2-(n-C_3H_7)-3-(C_2H_5)-4,4-dimethylcyclopent-2-enone (62); 2:1	76
$(CH_3)_3SiCH{=}CHC(C_3H_7\text{-}i){=}CH_2$	1. CH_3COCl, $AlCl_3$, CH_2Cl_2, −80°, 0.5 h 2. $(C_2H_5)_3N$, C_6H_6, 2 h	$CH_3COCH{=}CHC(C_3H_7\text{-}n){=}CH_2$ (67)	418
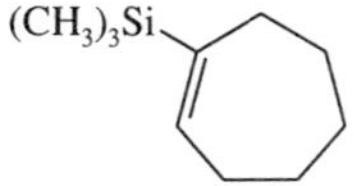 1-$(CH_3)_3Si$-cycloheptene	1. $CH_2{=}CHCOCl$, $AlCl_3$, NaO_2CCH_3, CH_2Cl_2, −45°, 15 min[b] 2. TFA, 3 h	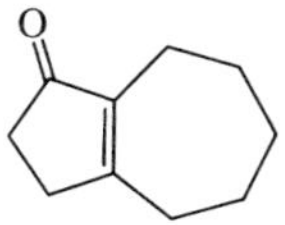 bicyclic cyclopentenone fused to cycloheptane (10)	76
	1. $(CH_3)_2C{=}CHCOCl$, $AlCl_3$, CH_2Cl_2, −78°, 15 min[b] 2. $SnCl_4$, CH_2Cl_2, reflux, 8–12 h	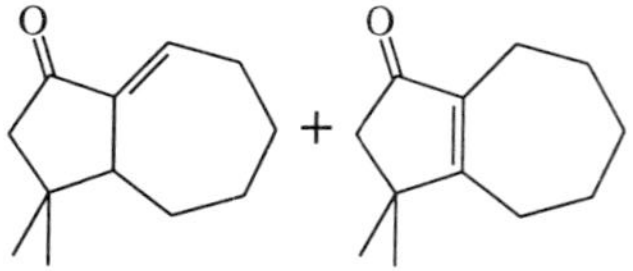 gem-dimethyl bicyclic enone (exocyclic C=C in seven-membered ring) + gem-dimethyl bicyclic cyclopentenone (57); 3.5:1	76

Reactant	Conditions	Product(s) and Yield(s) (%)	Refs.
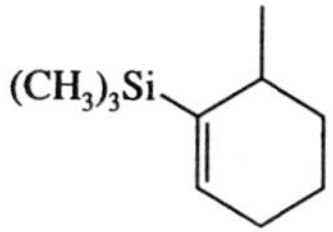 $(CH_3)_3Si$– (6-methylcyclohexenyl)	1. $(CH_3)_2C{=}CHCOCl$, $AlCl_3$, CH_2Cl_2, $-78°$, 15 min[b] 2. $BF_3{\cdot}O(C_2H_5)_2$, C_6H_6, reflux, 1–3 d 3. $RhCl_3$, C_2H_5OH[c]	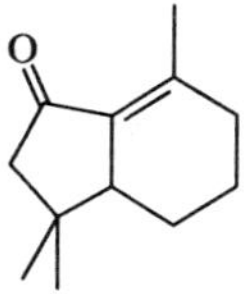 (60)	76
$CH_3CH{=}C(CH_2)_3COCl$ with $(CH_3)_3Si$ on C, E:Z 80:20	$AlCl_3$, CH_2Cl_2, 0°, 3 h; room temp, 10 h	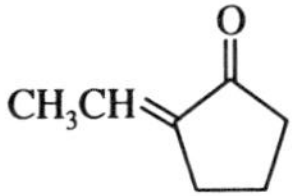 $CH_3CH{=}$ (cyclopentanone) E:Z > 95:5 (>95)	116
$CH_2{=}CCH(CH_3)(CH_2)_2COCl$ with $(CH_3)_3Si$ on C	$AlCl_3$, CH_2Cl_2, 13 h	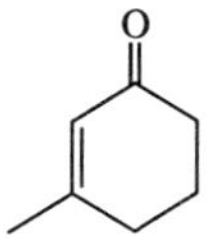 (76)	116
C_8			
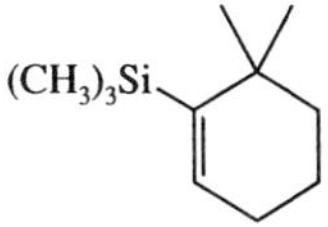 $(CH_3)_3Si$– (6,6-dimethylcyclohexenyl)	CH_3COCl, $AlCl_3$, CH_2Cl_2, 0°, 6 h	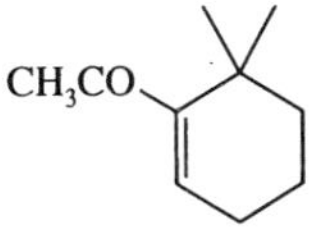 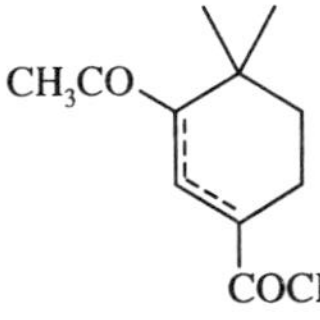 CH_3CO– (11) + CH_3CO–, $COCH_3$ (34)	117
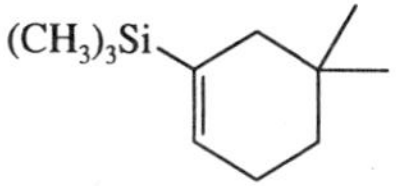 $(CH_3)_3Si$– (5,5-dimethylcyclohexenyl)	", ", ", ", "	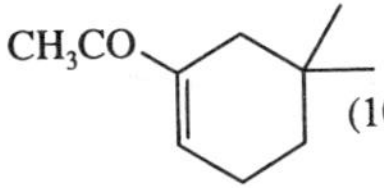 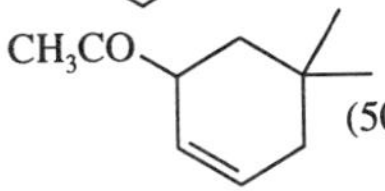 CH_3CO– (10) + CH_3CO– (50)	117

TABLE XXVII. VINYLSILANES WITH ACID CHLORIDES (*Continued*)

Reactant	Conditions	Product(s) and Yield(s)	Refs.
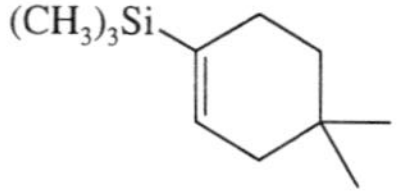 $(CH_3)_3Si$	", ", ", ", "	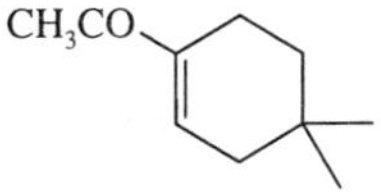 CH_3CO (77)	117
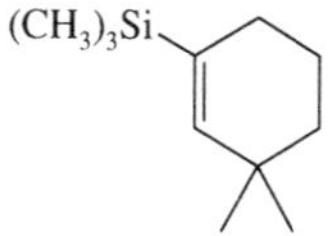 $(CH_3)_3Si$	", ", ", ", "	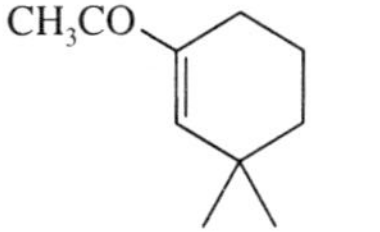 CH_3CO (49)	117
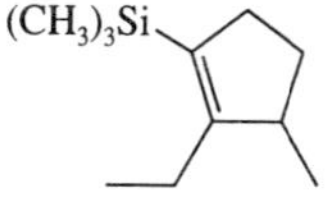 $(CH_3)_3Si$	CH_3COCl, $AlCl_3$, CH_2Cl_2, −78°, 1.5 h	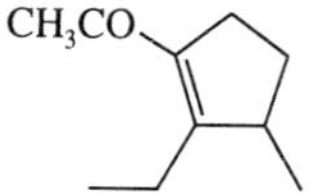 CH_3CO (61)	486
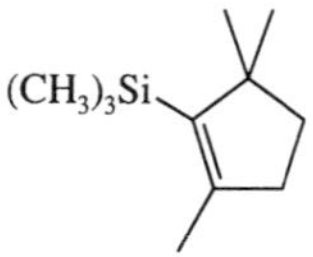 $(CH_3)_3Si$	1. $Cl(CH_2)_2COCl$, $AlCl_3$, CH_2Cl_2, −78°, 15 min 2. DBU, THF, 1 h	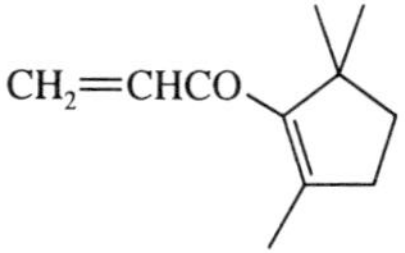 $CH_2{=}CHCO$ (33)	76
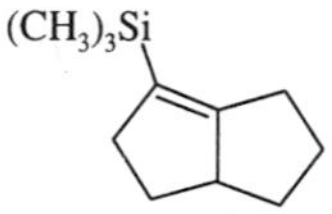 $(CH_3)_3Si$	CH_3COCl, $AlCl_3$, CH_2Cl_2, −78°, 1.5 h	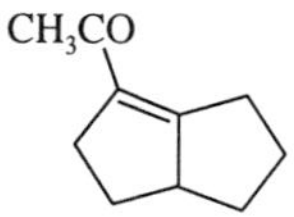 CH_3CO (30)	486
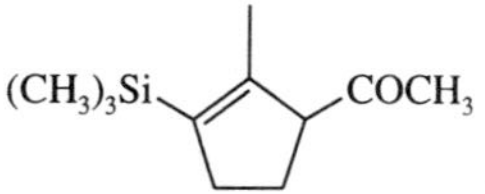 $(CH_3)_3Si$ $COCH_3$	CH_3COCl, $AlCl_3$, CH_2Cl_2, 0°, 1 h	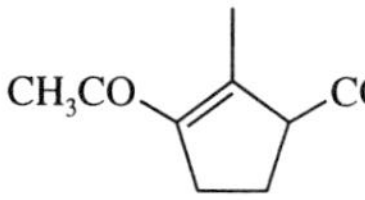 CH_3CO $COCH_3$ (95)	346
(*E*)-$(CH_3)_3SiCH{=}CHC_6H_5$	1. C_6H_5COCl, $AlCl_3$, CH_2Cl_2, 0°, 30 min 2. $(C_2H_5)_3N$, $(C_2H_5)_2O$, reflux, 18 h	(*E*)-$C_6H_5COCH{=}CHC_6H_5$ (73) I	117

Reactant	Conditions	Product(s) and Yield(s) (%)	Refs.
	1. C_6H_5COCl, $AlCl_3$, CH_2Cl_2, 0°, 30 min 2. HCl, $(C_2H_5)_2O$, 0°, 1.5 h	$C_6H_5COCH_2CHClC_6H_5$ (80) II	117
	C_6H_5COCl, $AlCl_3$, CH_2Cl_2, 0°, 30 min	I (27) + II (37)	117
	$C_6H_5CH_2COCl$, $AlCl_3$, CH_2Cl_2, 0°, 30 min	(*E*)-$C_6H_5CH_2COCH{=}CHC_6H_5$ (86)	117
(*Z*)-$(CH_3)_3SiCH{=}C(C_6H_5)Sn(C_4H_9\text{-}n)_3$	CH_3COCl, $AlCl_3$, CH_2Cl_2, 1 h	(*E*)$CH_3COCH{=}C(C_6H_5)Sn(C_4H_9\text{-}n)_3$ (45)	715
C_{10}			
(*E*)-$(CH_3)_3SiC(C_2H_5){=}CHC_6H_{11}$	CH_3COCl, $AlCl_3$, CH_2Cl_2, 0°, 20 min	(*E*)-$CH_3COC(C_2H_5){=}CHC_6H_{11}$ (70)	271, 1
(*Z*)-$(CH_3)_3SiC(C_2H_5){=}CHC_6H_{11}$	CH_3COCl, $AlCl_3$, CH_2Cl_2, 0°, 20 min	(*Z*)-$CH_3COC(C_2H_5){=}CHC_6H_{11}$ (70)	271, 1
(*E*)-$(CH_3)_3SiC(C_4H_9\text{-}n){=}CHC_4H_9\text{-}n$	*o*-IC_6H_4COCl, $AlCl_3$	*o*-$IC_6H_4COC(C_4H_9\text{-}n){=}CHC_4H_9\text{-}n$ (—)	714
$n\text{-}C_5H_{11}CH{=}C(CH_2)_2COCl$ $(CH_3)_3Si$	$AlCl_3$, CH_2Cl_2, 13 h	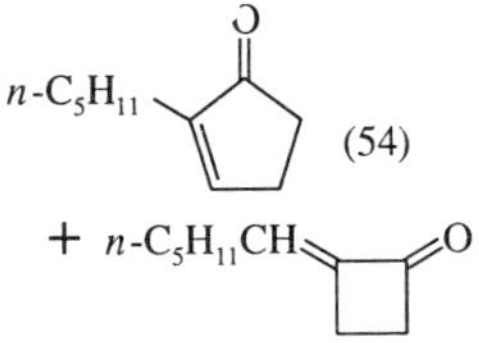 $n\text{-}C_5H_{11}$, O (54) + $n\text{-}C_5H_{11}CH$, O (38) *E*:*Z* > 95:5	116
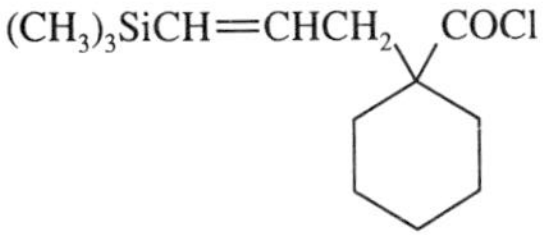 $(CH_3)_3SiCH{=}CHCH_2$, COCl	$TiCl_4$, CH_2Cl_2	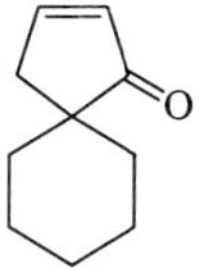 O (65–70)	420
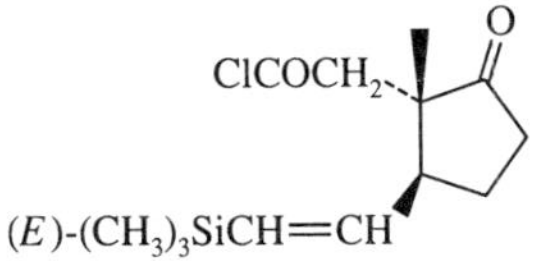 $ClCOCH_2$, O (*E*)-$(CH_3)_3SiCH{=}CH$	$AlCl_3$, CH_2Cl_2, 2 h	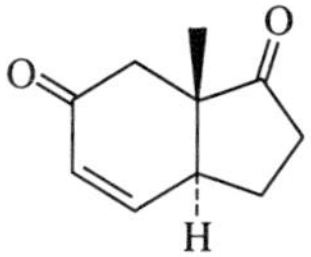 O, O, H (54)	716
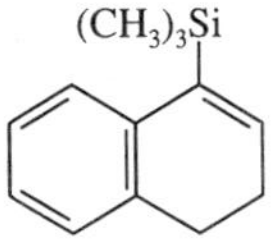 $(CH_3)_3Si$	CH_3COCl, $AlCl_3$, CH_2Cl_2, 40 min	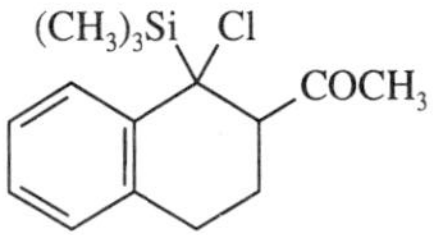 $(CH_3)_3Si$, Cl, $COCH_3$ (>61)	117

TABLE XXVII. VINYLSILANES WITH ACID CHLORIDES (*Continued*)

Reactant	Conditions	Product(s) and Yield(s)	Refs.
C_{11}			
(*E*)-$(CH_3)_3SiCH{=}CHCH_2CH(C_6H_{13}\text{-}n)COCl$	$TiCl_4$, CH_2Cl_2, 0–25°	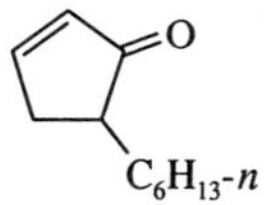 O, C_6H_{13}-*n* (79)	420
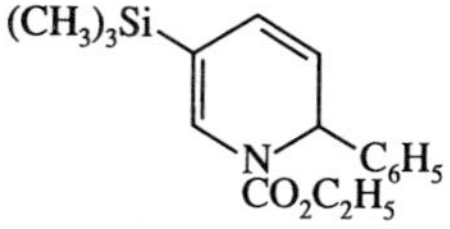 $(CH_3)_3Si$, N, C_6H_5, $CO_2C_2H_5$	CH_3COCl, $SnCl_4$, CH_2Cl_2, −78°	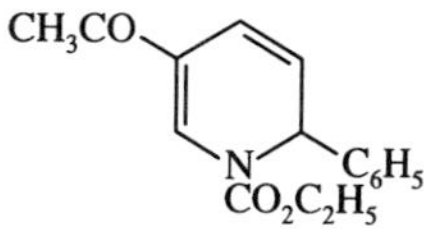 CH_3CO, N, C_6H_5, $CO_2C_2H_5$ (35)	717
C_{12}			
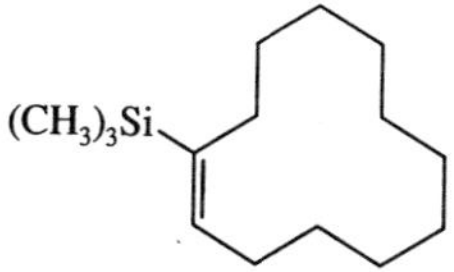 $(CH_3)_3Si$	1. $CH_3CH{=}CHCOCl$, $AlCl_3$, CH_2Cl_2, 1 h[b] 2. $BF_3{\cdot}O(C_2H_5)_2$, C_6H_6, reflux, 3 d 3. $RhCl_3$, C_2H_5OH, reflux, 2 d[c]	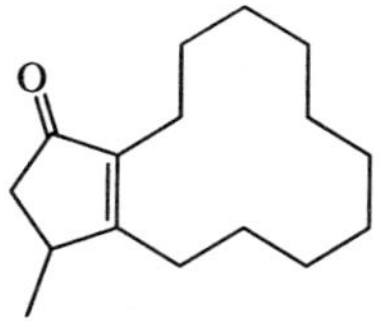 O (66)	76
	1. $(CH_3)_2C{=}CHCOCl$, $AlCl_3$, CH_2Cl_2, −78°, 15 min[b] 2. $BF_3{\cdot}O(C_2H_5)_2$, C_6H_6, reflux, 3 d	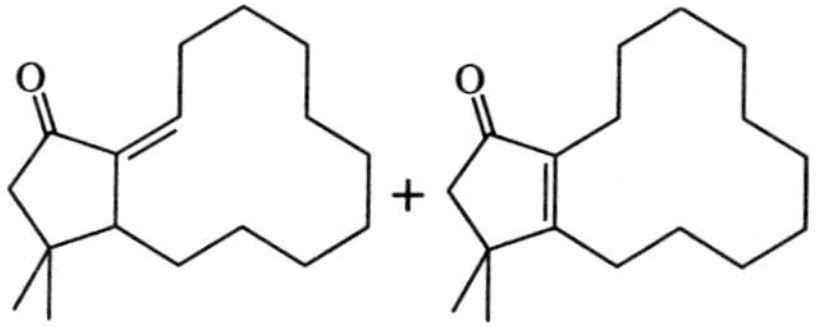 O + O 3:2 (43)	76
(*E*)-$(CH_3)_3SiCH{=}CH$ 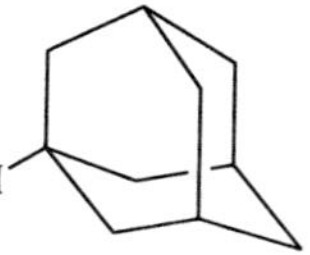	CH_3COCl, $TiCl_4$, CH_2Cl_2, 0°, 5 h	(*E*)-$CH_3COCH{=}CH$ 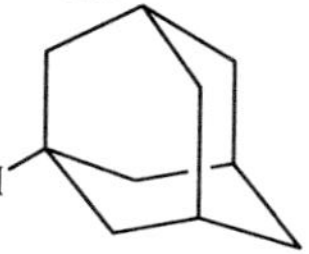 (54)	402

Reactant	Conditions	Product(s) and Yield(s) (%)	Refs.
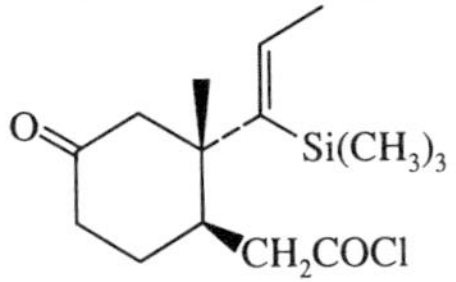 [structure: 3-methyl-3-(1-trimethylsilylpropenyl)-4-(chlorocarbonylmethyl)cyclohexanone; labels $Si(CH_3)_3$, CH_2COCl]	$AgBF_4$, CH_3NO_2	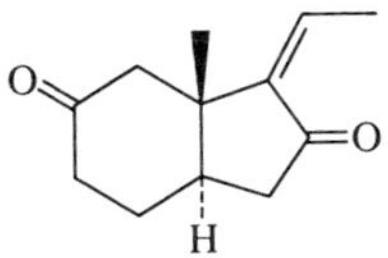 [structure: bicyclic dione, H] (71)	417
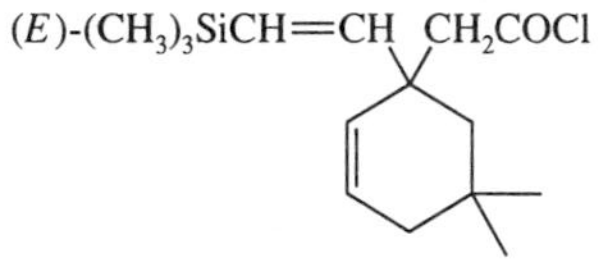 (E)-$(CH_3)_3SiCH{=}CH$ CH_2COCl [structure: 5,5-dimethylcyclohexenyl]	$TiCl_4$, CH_2Cl_2, $-30°$, 2 h; 25°, 15 h	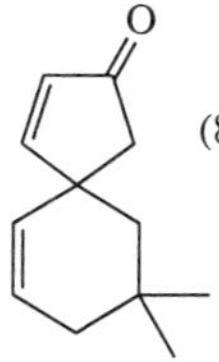 [structure: spiro cyclopentenone] (82)	419
$(CH_3)_3SiCH{=}CHCH_2CH(C_6H_5)$-$CH_2CO_2C_2H_5$	$RCOCl$, $AlCl_3$, 4 h	$RCOCH{=}CHCH_2CH(C_6H_5)CH_2CO_2C_2H_5$	505
	R = CH_3	(75–80)	
	R = $CH_3CH{=}CH$	(75–80)	
	R = $(CH_3)_2C{=}CH$	(75–80)	
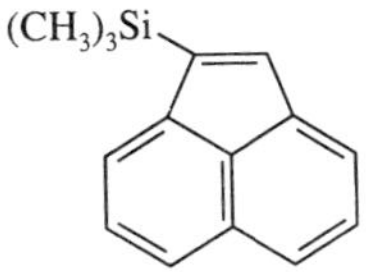 $(CH_3)_3Si$ [structure: 1-acenaphthylenyl]	$RCOX$, CH_2Cl_2,	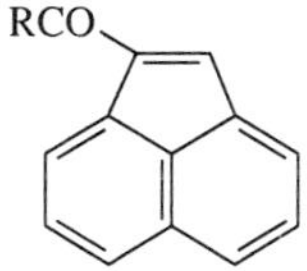 RCO [structure: 1-acenaphthylenyl]	718
	R X		
	CH_3 Cl $AlCl_3$, $-80°$	(30–35)	
	" " $AgBF_4$, 0°	(30–35)	
	" F BF_3, $-10°$	(30–35)	
	C_2H_5 " "	(86)	
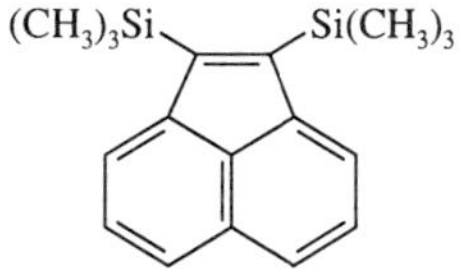 $(CH_3)_3Si$ $Si(CH_3)_3$ [structure: 1,2-acenaphthylene]	CH_3COCl, $AlCl_3$	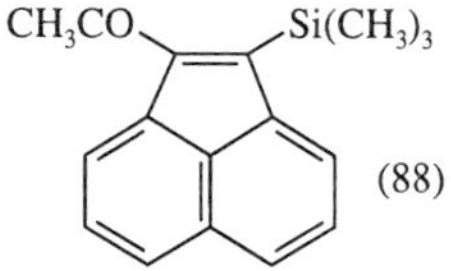 CH_3CO $Si(CH_3)_3$ [structure: 1,2-acenaphthylene] (88)	718

TABLE XXVII. VINYLSILANES WITH ACID CHLORIDES (*Continued*)

Reactant	Conditions	Product(s) and Yield(s)	Refs.
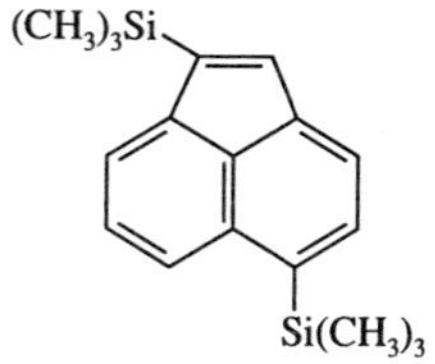 $(CH_3)_3Si$ … $Si(CH_3)_3$	CH_3COCl, $AlCl_3$, CH_2Cl_2, 0–5°	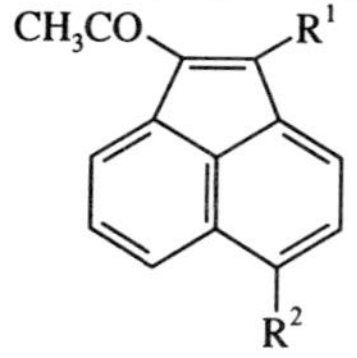 CH_3CO … R^1 … R^2; R^1 = H, R^2 = H (10); + R^1 = ", R^2 = $Si(CH_3)_3$ (10); + R^1 = $(CH_3)_3Si$, R^2 = " (10); + R^1 = ", R^2 = H (30)	718
C_{13}			
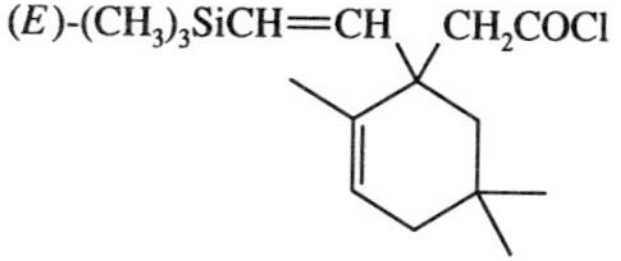 (E)-$(CH_3)_3SiCH{=}CH$ … CH_2COCl	$TiCl_4$, CH_2Cl_2, −30°, 2 h; 25°, 14 h	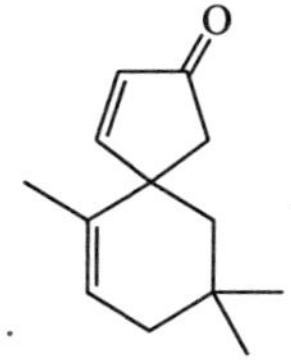 O (84)	419
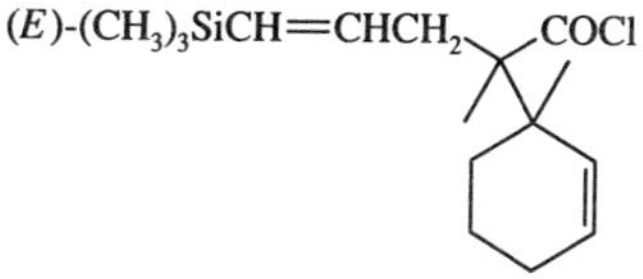 (E)-$(CH_3)_3SiCH{=}CHCH_2$ … COCl	$TiCl_4$ or $SnCl_4$, CH_2Cl_2	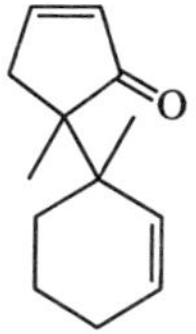 O (60–70)	420
C_{14}			
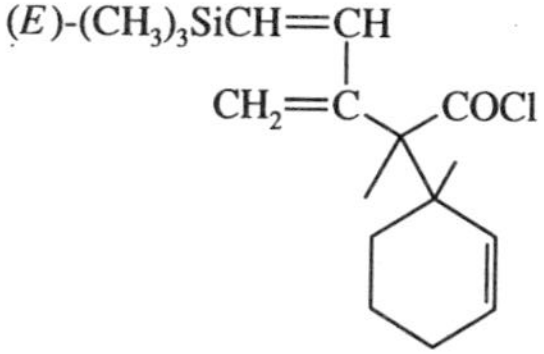 (E)-$(CH_3)_3SiCH{=}CH$ / $CH_2{=}C$ … COCl	$SnCl_4$, CH_2Cl_2	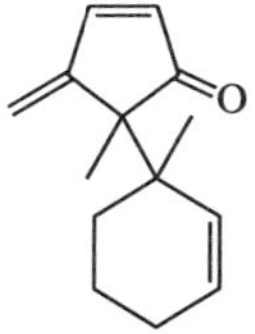 O (80)	420

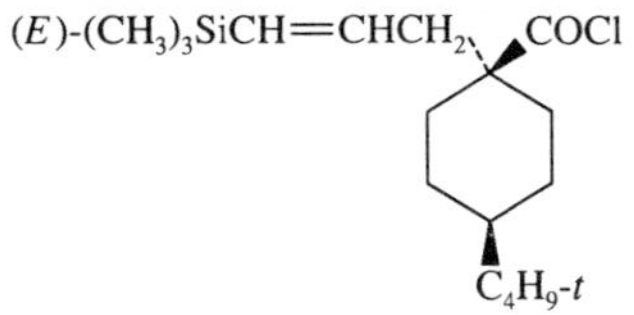 (*E*)-$(CH_3)_3SiCH{=}CHCH_2$, COCl, C_4H_9-*t*	$TiCl_4$ or $SnCl_4$, CH_2Cl_2	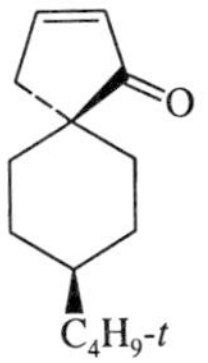 O, C_4H_9-*t* (69)	420
(*E*)-$(CH_3)_3SiCH{=}CHCH_2CHCOCl$ / $(CH_2)_7CH{=}CH_2$	$TiCl_4$ or $SnCl_4$, CH_2Cl_2	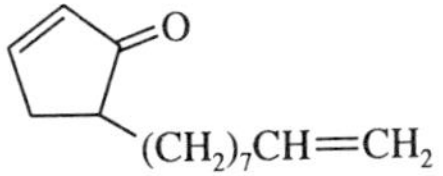 O, $(CH_2)_7CH{=}CH_2$ (60)	420
(*E*)-$(CH_3)_3SiCH{=}CHCH_2CCOCl$ / $CH_2{=}CHCH_2$, C_6H_{13}-*n*	$TiCl_4$ or $SnCl_4$, CH_2Cl_2	O, C_6H_{13}-*n*, $CH_2CH{=}CH_2$ (69)	420
C_{15}			
(*E*)-$(CH_3)_3SiCH{=}CHCH_2CCOCl$ / $CH_3CH{=}CHCH_2$, C_6H_{13}-*n*	$TiCl_4$ or $SnCl_4$, CH_2Cl_2	O, C_6H_{13}-*n*, $CH_2CH{=}CHCH_3$ (70)	420
(*E*)-$(CH_3)_3SiCH{=}CHCH_2CCOCl$ / CH_3CCH_2 ($=CH_2$), C_6H_{13}-*n*	$TiCl_4$ or $SnCl_4$, CH_2Cl_2	O, C_6H_{13}-*n*, $CH_2CH{=}CH_2$ (50)	420
C_{17}			
(*E*)-$(CH_3)_3SiC(C_5H_{11}$-*n*$){=}CHC_{10}H_{21}$-*n*	CH_3COCl, $AlCl_3$, CH_2Cl_2, 0°, 20 min	(*E*)-$CH_3COC(C_5H_{11}$-*n*$){=}CHC_{10}H_{21}$-*n* (—)	271
(*Z*)-$(CH_3)_3SiC(C_5H_{11}$-*n*$){=}CHC_{10}H_{21}$-*n*	", ", ", ", "	(*Z*)-$CH_3COC(C_5H_{11}$-*n*$){=}CHC_{10}H_{21}$-*n* (—)	271

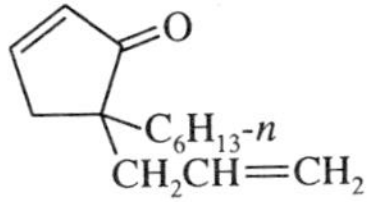

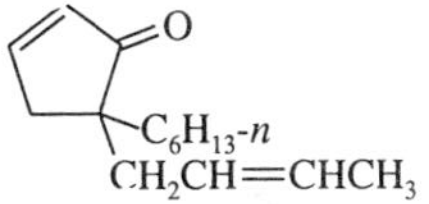

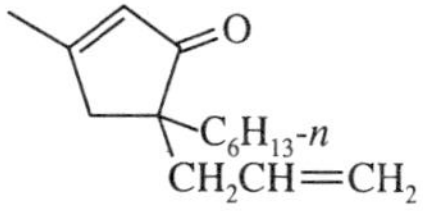

[a] The use of this cosolvent suppresses the addition of HCl across the double bond of the product.
[b] This step gives the dienone, which is not isolated.
[c] This step equilibrates the regioisomeric enones.

TABLE XXVIII. ALLENYLSILANES WITH ACID CHLORIDES

Reactant	Conditions	Product(s) and Yield(s)	Refs.
C_3			
$[(CH_3)_3Si]_2C{=}C{=}C[Si(CH_3)_3]_2$	CH_3COCl, $AlCl_3$	$[(CH_3)_3Si]_2C{=}C{=}C[Si(CH_3)_3]COCH_3$ (—)	274
C_5			
$(CH_3)_3SiCH{=}C{=}C(CH_3)_2$	CH_3COCl, $AlCl_3$, CH_2Cl_2, $-60°$	$(CH_3)_3SiCH{=}C(COCH_3)C(CH_3){=}CH_2$ (30)	719

TABLE XXIX. ALLYLSILANES WITH NITRILES AND AMIDES

Reactant	Conditions	Product(s) and Yield(s)	Refs.
C_3			
$(CH_3)_3SiCH_2CH{=}CH_2$	RCN, BCl_3, CH_2Cl_2, 3 h	$CH_2{=}CHCH_2COR$	425
	R		
	CH_3	(68)	
	$ClCH_2$	(86)	
	Cl_3C	(76)	
	C_2H_5	(69)	
	$Cl(CH_2)_2$	(70)	
	$CH_3O_2CCH_2$	(72)	
	$Br(CH_2)_4$	(62)	
	$C_6H_5CH_2$	(38)	
	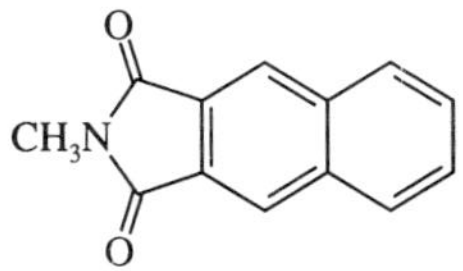 N-methylnaphthalene-2,3-dicarboximide, $h\nu$, CH_3CN	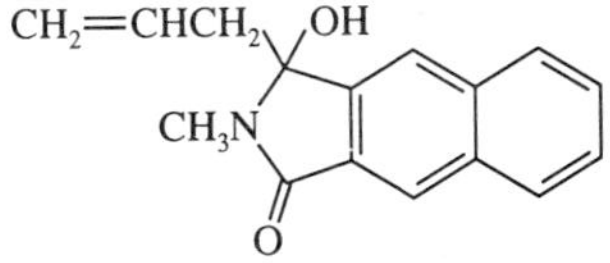 1-allyl-1-hydroxy-2-methylbenz[f]isoindolin-3-one ($CH_2{=}CHCH_2$, OH, CH_3N, O) (73)	720
	N-methylnaphthalene-1,2-dicarboximide (I), $h\nu$, CH_3CN	II: $CH_2{=}CHCH_2$, OH, CH_3N, O (27) +	720

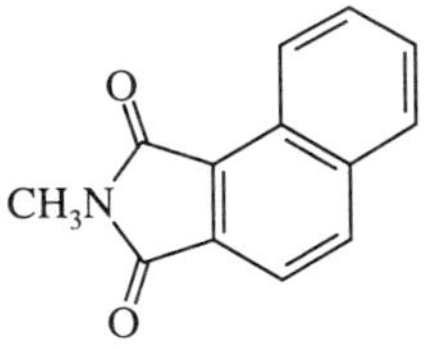

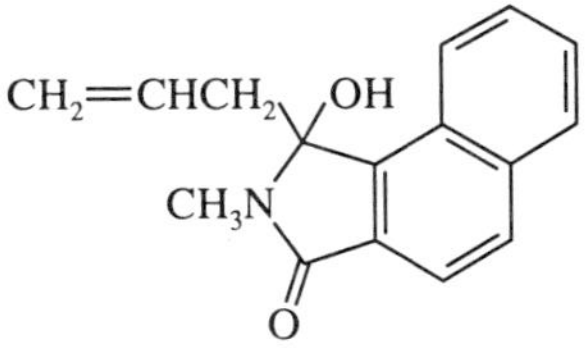

TABLE XXIX. ALLYLSILANES WITH NITRILES AND AMIDES (*Continued*)

Reactant	Conditions	Product(s) and Yield(s)	Refs.
		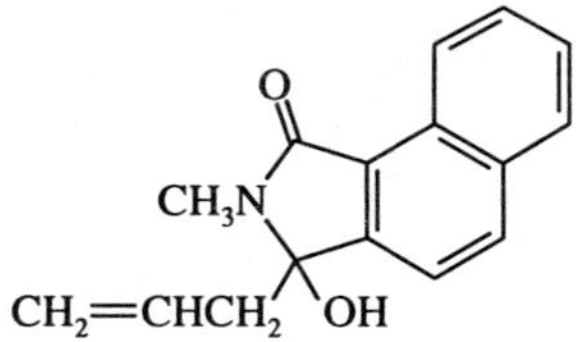 [structure, CH_3N, O, $CH_2{=}CHCH_2$, OH] (17) +	
		III	
		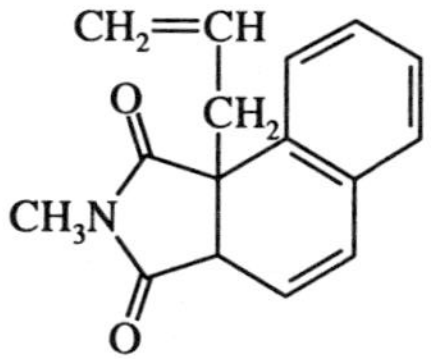 [structure, $CH_2{=}CH$, CH_2, CH_3N, O, O] (16) +	
		IV	
		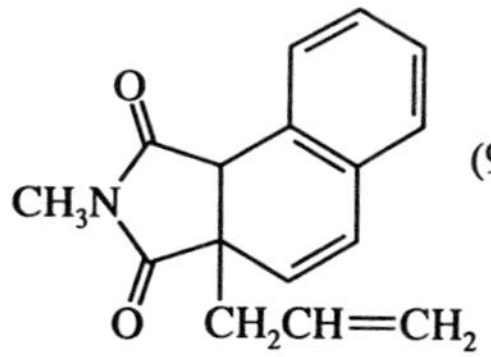 [structure, CH_3N, O, O, $CH_2CH{=}CH_2$] (9)	
		V	
	I, $h\nu$, CH_3OH	II (35) + III (26) + IV (6) + V (trace)	720
	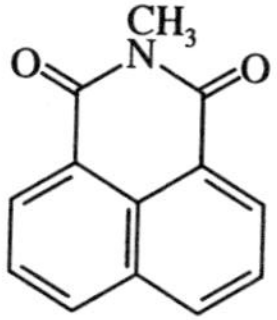 [structure, CH_3, N, O, O], $h\nu$, CH_3CN	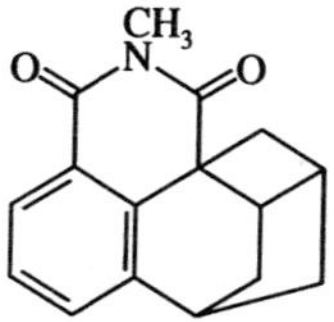 [structure, CH_3, N, O, O] (45)	720

	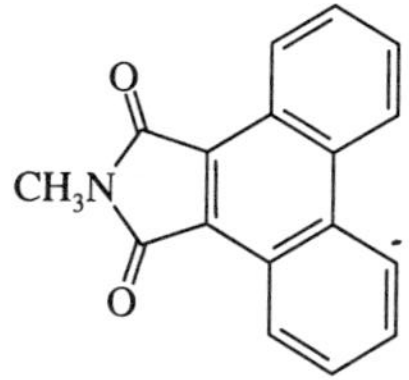 CH_3N, O, O, $h\nu$, CH_3CN	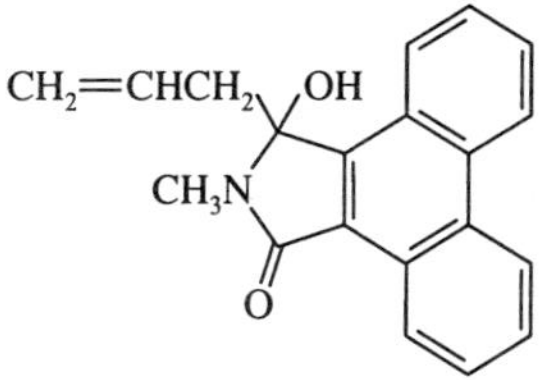 $CH_2{=}CHCH_2$, OH, CH_3N, O (36) +	720
		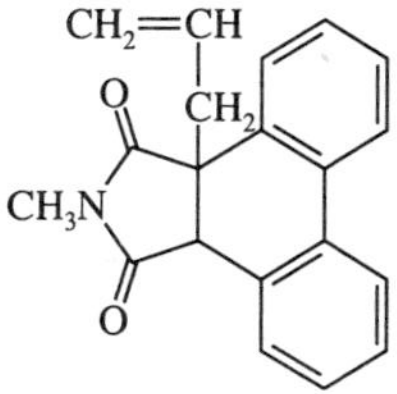 $CH_2{=}CH$, CH_2, CH_3N, O, O (12)	
C_4			
$(CH_3)_3SiCH_2CH{=}CHCH_3$	Cl_3CCN, BCl_3, CH_2Cl_2	$CH_2{=}CHCH(CH_3)COCCl_3$ (79)	425
$(CH_3)_3SiCH_2C(CH_3){=}CH_2$	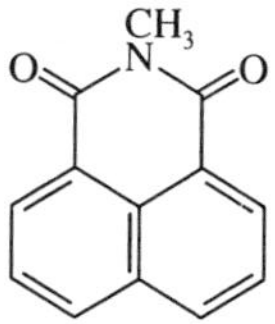 CH_3, N, O, O, $h\nu$, CH_3CN	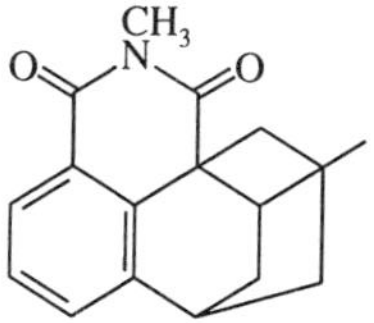 CH_3, N, O, O (20) +	720
		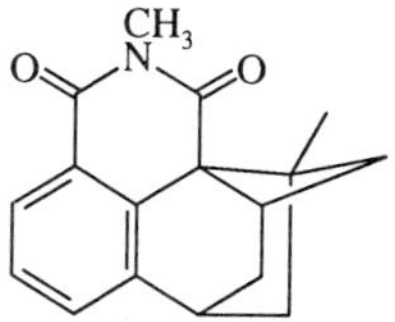 CH_3, N, O, O (21)	

TABLE XXIX. ALLYLSILANES WITH NITRILES AND AMIDES (*Continued*)

Reactant	Conditions	Product(s) and Yield(s)	Refs.
C_5			
(*E*)-$(CH_3)_3SiCH_2C(CH_3){=}CHCO_2CH_3$	Cl_3CCN, BCl_3, CH_2Cl_2	$(CH_3)_2C{=}CHCOCCl_3$ (40)	425
(*Z*)-$(CH_3)_3SiCH_2C(CH_3){=}CHCO_2CH_3$	", ", "	" (51)	425
$(CH_3)_3SiCH_2C(CH_2CO_2CH_3){=}CH_2$	RCN, BCl_3, CH_2Cl_2	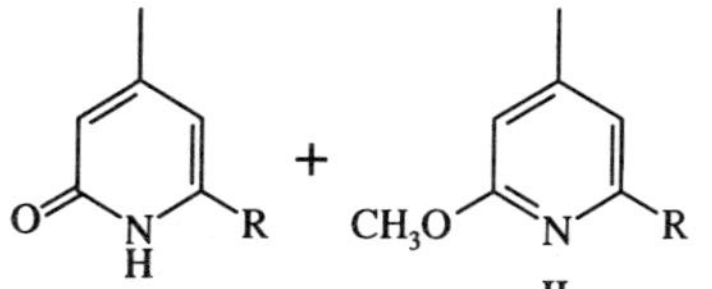 I (O, N–H, R) + II (CH_3O, N, R) + $CH_2{=}C(CH_2CO_2CH_3)CH_2COR$ III	425
	R: $ClCH_2$, 16 h	I (0), II (14), III (58)	
	R: Cl_3C, −78°, 2 h	I (64), II (16), III (0)	
	R: C_2H_5, 16 h	I (0), II (0), III (53)	
C_6			
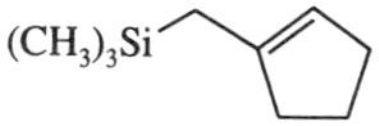 $(CH_3)_3Si$ (cyclopentenylmethyl)	Cl_3CCN, BCl_3, CH_2Cl_2	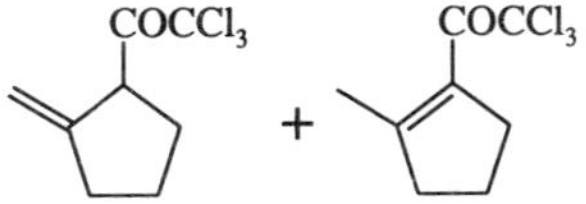 $COCCl_3$ + $COCCl_3$ (53)	425
C_7			
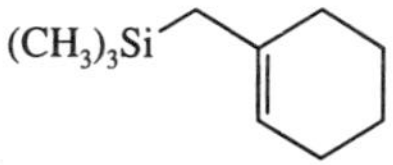 $(CH_3)_3Si$ (cyclohexenylmethyl)	Cl_3CCN, BCl_3, CH_2Cl_2	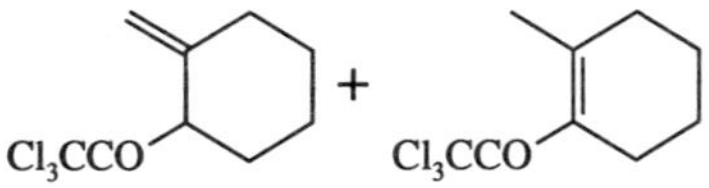 Cl_3CCO + Cl_3CCO (83)	425

$(CH_3)_3SiCH_2C(=CH_2)CH_2CH(CN)CO_2CH_3$	BCl_3, CH_2Cl_2	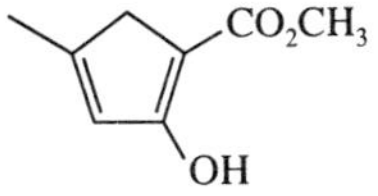 CO_2CH_3, OH (70)	425
C_{11}			
$[(CH_3)_3SiCH_2C(=CH_2)CH_2]_2C(CN)CO_2CH_3$	BCl_3, CH_2Cl_2, 0° 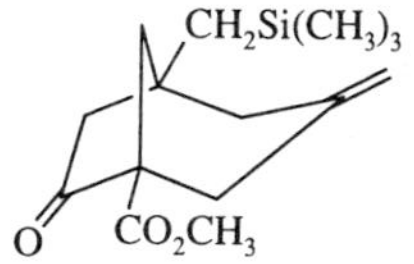	$CH_2Si(CH_3)_3$, O, CO_2CH_3 (75)	425

TABLE XXX. ALLYLSILANES WITH ORTHOESTERS

Reactant	Conditions	Product(s) and Yield(s)	Refs.
C_3			
$(CH_3)_3SiCH_2CH{=}CH_2$	$HC(OC_2H_5)_3$, $TiCl_4$, CH_2Cl_2, 3 h	$(CH_2{=}CHCH_2)_2CHOC_2H_5$ (24)	352
C_4			
$(CH_3)_3SiCH_2C(N\text{-morpholino}){=}CHCO_2CH_3$	$HC(OCH_3)_3$, $TiCl_4$, CH_2Cl_2, $-78°$, 3 h	$CH_3C(N\text{-morpholino}){=}C(CHO)CO_2CH_3$ (50)	285
C_5			
$(CH_3)_3Si$-(cyclopentadien-5-yl)	$HC(OCH_3)_3$, $(CH_3)_3SiO_3SCF_3$ (cat.), CH_2Cl_2, $-50°$, 0.5 h	(cyclopentadienyl)-$CH(OCH_3)_2$ (70)[a]	678
	$HC(OC_2H_5)_3$, $(CH_3)_3SiO_3SCF_3$ (cat.), CH_2Cl_2, $-50°$, 1 h	(cyclopentadienyl)-$CH(OC_2H_5)_2$ (76)	678

[a] After longer times, the other double bond isomers in the cyclopentadiene ring are formed; at equilibrium, only the 1- and 2-substituted isomers are present in the ratio 4.8:1.

TABLE XXXI. VINYLSILANES WITH ORTHOESTERS

Reactant	Conditions	Product(s) and Yield(s)	Refs.
C_8			
(E)- or (Z)-$(CH_3)_3SiCH{=}CHC_6H_5$	2-ethoxy-1,3-dithiolane (C_2H_5O–), $BF_3 \cdot O(C_2H_5)_2$, CH_2Cl_2, 0°, 30 min; room temp, 5 h	(E)-2-(1,3-dithiolanyl)–$CH{=}CHC_6H_5$ (20–22)	375

TABLE XXXII. ALLYLSILANES WITH CARBON ELECTROPHILES AT THE OXIDATION STATE OF CARBON DIOXIDE

Reactant	Conditions	Product(s) and Yield(s)	Refs.
C_7			
$(CH_3)_3Si$-allyl structure with OCOCl	$SnCl_4$, CH_2Cl_2, 0°, 3 h	lactone structure (57)	426
C_{12}			
$(CH_3)_3Si$-cyclopropylidene structure	C_6H_5OCN, TBAF, THF, reflux	R = CN (69) + R = H (20)	167

TABLE XXXIII. ALLYLSILANES WITH NITROGEN ELECTROPHILES

Reactant	Conditions	Product(s) and Yield(s)	Refs.
C_3			
$(CH_3)_3SiCH_2CH{=}CH_2$	NO_2BF_4, CH_2Cl_2, $-78°$, 30 min	$CH_2{=}CHCH_2NO_2$ (80)	431
	4-R-1,2,4-triazoline-3,5-dione (N=N), 25°, 1 h	$CH_2{=}CHCH_2N{-}NH$ (4-R-urazole, 1-allyl) (I) + $(CH_3)_3SiCH{=}CHCH_2N{-}NH$ (4-R-urazole) (II)	427
	R — Solvent	(I) — (II)	
	CH_3 — C_6H_6	(3) — (89)[a]	
	" — $Cl(CH_2)_2Cl$	(11) — (83)[a]	
	" — CH_3CN	(44) — (52)[a]	
	C_6H_5 — C_6H_6	(9) — (Major)[a]	
	" — $Cl(CH_2)_2Cl$	(17) — (Major)[a]	
	" — CH_2Cl_2	(18) — (Major)[a]	
	" — $CH_3CO_2C_2H_5$	(20) — (Major)[a]	
	" — $(CH_3)_2CO$	(48) — (—)[a]	
	" — $(CH_3)_2CO$, H_2O	(73) — (—)[a]	
	" — CH_3CN	(67) — (—)[a]	
	" — CH_3CN, H_2O	(84) — (—)[a]	
	" — CH_3CN, $-25°$	(85) — (—)[a]	
	3-chloro-5,5-dimethyloxazolidin-2-one (ClN), AIBN	$CH_2{=}CHCH_2N$ (5,5-dimethyloxazolidin-2-one) (14)	158
$(CH_3)_2Si(CH_2CH{=}CH_2)_2$	4-methyl-1,2,4-triazoline-3,5-dione (N=N, N-CH_3), C_6H_6, 30 min	$CH_2{=}CHCH_2Si(CH_3)_2CH{=}CHCH_2N{-}NH$ (4-methylurazole) (—)	427

Reactant	Conditions	Product(s) and Yield(s) (%)	Refs.
$(CH_3)_2Si(CH_2CH{=}CH_2)CH{=}CHCH_2N$–NH (4-methyl-1,2,4-triazoline-3,5-dione ring, N-CH_3)	4-phenyl-1,2,4-triazoline-3,5-dione (N=N, N-C_6H_5), CH_3CN, 2 h	$CH_2{=}CHCH_2N$–N(4-phenyl-triazolidinedione)–$Si(CH_3)_2CH{=}CHCH_2N$–NH (4-methyl-triazolidinedione) (46) + HN–$NCH_2CH{=}CHSi(CH_3)_2CH{=}CHCH_2N$–NH (N-$C_6H_5$ and N-CH_3 triazolidinediones) (—)	427
C_4			
$(CH_3)_3SiCH_2CH{=}CHCH_3$	NO_2BF_4, CH_2Cl_2, $-78°$, 30 min	$CH_2{=}CHCH(CH_3)NO_2$ (75)	431
	3-chloro-5,5-dimethyloxazolidin-2-one (ClN), AIBN	$CH_2{=}CHCH(CH_3)N$ (5,5-dimethyloxazolidin-2-one) (6)	158
$(CH_3)_3SiCH_2C(CH_3){=}CH_2$	NO_2BF_4, CH_2Cl_2, $-78°$, 30 min	$CH_2{=}C(CH_3)CH_2NO_2$ (65)	431
C_9			
$(CH_3)_3SiCH_2CH{=}CHC_6H_5$	1. NOCl, $CHCl_3$, $-60°$ 2. H_2, Raney Ni	$C_2H_5CH(NH_2)C_6H_5$ (11)	430

[a] A cyclization product (see Eq. 271) is also formed in this reaction in unspecified yield.

TABLE XXXIV. VINYLSILANES WITH NITROGEN ELECTROPHILES

Reactant	Conditions	Product(s) and Yield(s)	Refs.
C_2			
$(CH_3)_3SiCH{=}CH_2$	$(CH_3)_3SiN_3$, 140°, 3 d	$[(CH_3)_3Si]_2NCH{=}CH_2$ (65)	433
$(C_2H_5)_3SiCH{=}CH_2$	$(CH_3)_3SiN_3$, 140°, 3 d	$(C_2H_5)_3SiN(Si(CH_3)_3)CH{=}CH_2$ (—)	433
C_6			
$(CH_3)_3SiC(CO_2CH_3){=}C(CO_2CH_3)C{\equiv}CN(C_2H_5)_2$	p-$CH_3C_6H_4SO_2N_3$, $(C_2H_5)_2O$, 4 h	3,4-bis(CO_2CH_3)-5-[$C(N(C_2H_5)_2){=}NSO_2C_6H_4CH_3$-$p$]-1H-pyrazole (46)	573
$(C_6H_5)_3SiC(CO_2CH_3){=}C(CO_2CH_3)C{\equiv}CN(C_2H_5)_2$	p-$CH_3C_6H_4SO_2N_3$, $(C_2H_5)_2O$, 4 h	" (61)	573
C_8			
$(CH_3)_3SiCH{=}CHC_6H_5$	$(CH_3)_3SiN_3$, reflux, 5 d	$[(CH_3)_3Si]_2NCH{=}CHC_6H_5$ (55)	433
	1. N_2O_3 2. KOH, CH_3OH	$O_2NCH{=}CHC_6H_5$ (—)	429

TABLE XXXV. Allylsilanes with Oxygen Electrophiles

Reactant	Conditions	Product(s) and Yield(s)	Refs.
C_3			
(E)-$(CH_3)_3SiCH_2CH{=}CHSi(CH_3)_3$	1. MCPBA, CH_2Cl_2, 0° (68%) 2. RMgX, $(C_2H_5)_2O$, 30 min 3. $BF_3{\cdot}O(C_2H_5)_2$, $(C_2H_5)_2O$, 30 min	$RCH{=}CHCH_2Si(CH_3)_3$	721
	R — X	E:Z	
	CH_3 — I	92:8 (56)	
	n-C_3H_7 — Br	>99:1 (55)	
	n-C_4H_9 — "	94:6 (52)	
	t-C_4H_9 — "	>99:1 (10)	
	C_6H_{11} — Cl	>99:1 (56)	
	C_6H_5 — Br	>99:1 (55)	
	1. MCPBA, CH_2Cl_2, 0° (68%) 2. RMgX, $(C_2H_5)_2O$, 30 min 3. NaH, DME, reflux, 3–5 h	$RCH{=}CHCH_2Si(CH_3)$	721
	R — X	E:Z	
	CH_3 — I	11:89 (35)	
	n-C_3H_7 — Br	<1:99 (27)	
	n-C_4H_9 — "	2:98 (42)	
	t-C_4H_9 — "	<1:99 (6)	
	C_6H_{11} — Cl	<1:99 (32)	
	C_6H_5 — Br	6:94 (56)	
	1. MCPBA, CH_2Cl_2, 0° (68%) 2. $(n$-$C_4H_9)_2CuLi$, $(C_2H_5)_2O$, −40°, 24 h 3. $BF_3{\cdot}O(C_2H_5)_2$, $(C_2H_5)_2O$, 30 min	n-$C_4H_9CH{=}CHCH_2Si(CH_3)_3$ (45) E:Z 48:52	721
	1. MCPBA, CH_2Cl_2, 0° (68%) 2. $(n$-$C_4H_9)_2CuLi$, $(C_2H_5)_2O$, −40°, 24 h 3. NaH, DME, reflux, 3–5 h	n-$C_4H_9CH{=}CHCH_2Si(CH_3)_3$ (40) E:Z > 99:1	721

TABLE XXXV. ALLYLSILANES WITH OXYGEN ELECTROPHILES (*Continued*)

Reactant	Conditions	Product(s) and Yield(s)	Refs.
C_4			
Si $(CH_3)_2$	1. MCPBA (almost quantitative) 2. CH_3Li, $(C_2H_5)_2O$, reflux, 5 h	$(CH_3)_3SiCH_2CHOHCH{=}CH_2$ (77)	722
Si $(C_6H_5)_2$	1. MCPBA (95%) 2. HX X = Cl X = Br	$(C_6H_5)_2SiXCH_2CHOHCH{=}CH_2$ (72) (80)	436
C_5			
Si $(CH_3)_2$	1. *p*-$O_2NC_6H_4CO_3H$, $(C_2H_5)_2O$, 30° 2. HCl, H_2O, 20°	$CH_2{=}C(CH_3)CH{=}CH_2$ (70)	436
$C_6H_5(CH_3)_2SiCH(CH_3)CH{=}CHCH_3$ *E*:*Z* 9:1	1. MCPBA, $NaHCO_3$, CD_2Cl_2, 0° 2. CsF, CD_3CN	$CH_3CH{=}CHCHOHCH_3$ *E*:*Z* 61:39 (—)	53
	1. OsO_4, NMMO, $(CH_3)_2CO$ 2. KOC_4H_9-*t*	", *E*:*Z* 66:34 (—)	53
", *E*:*Z* 5:95	1. MCPBA, $NaHCO_3$, CD_2Cl_2, 0° 2. CsF, CD_3CN	", *E*:*Z* > 95:5 (—)	53
	1. OsO_4, NMMO, $(CH_3)_2CO$ 2. KOC_4H_9-*t*	", *E*:*Z* 22:78 (—)	53
$(CH_3)_3Si$ 22–25% ee	1. MCPBA, $NaHCO_3$, CH_2Cl_2, 0°, 1 h 2. CH_3CO_2H, CH_3OH 3. TBAF, THF	OH 19% ee (—)	437
C_6			
(*Z*)-$(CH_3)_3SiCH_2CH{=}CHCH_2CHOHCH_3$	1. $VO(acac)_2$, *t*-$C_4H_9O_2H$, $Cl(CH_2)_2Cl$, or	$CH_2{=}CHCHOHCH_2CHOHCH_3$ (35)	723

Reactant	Conditions	Product(s) and Yield(s) (%)	Refs.
	toluene, CH_2Cl_2, 0°, 8 h; room temp, 24 h 2. SiO_2	*erythro*:*threo* 97:3	
(*E*)-$(CH_3)_3SiCH(CH_2CO_2H)CH{=}CHCH_3$	1. MCPBA, CH_2Cl_2, −20°[b] 2. TBAF	(*E*)-$(n\text{-}C_4H_9)_4NO_2CCH_2CH{=}CHCHOHCH_3$ (—)	724
(*E*)-$(CH_3)_3SiCH(CH_2CO_2C_2H_5)CH{=}CHCH_3$	1. MCPBA, CH_2Cl_2, −20°[c] 2. TBAF	" (—)	724
(*S*)(*E*)-$(CH_3)_3SiCH(CH_2CON(CH_3)_2)CH{=}CHCH_3$	1. MCPBA, CH_2Cl_2, −20° (75%)[d] 2. Separate major product (81%) 3. TBAF	(*S*)(*Z*)-$(n\text{-}C_4H_9)_4NO_2CCH_2CH{=}CHCHOHCH_3$ (—)	725
	1. MCPBA, CH_2Cl_2, −20° (75%)[d] 2. Separate minor product (16%) 3. TBAF	(*R*)(*E*)-$(n\text{-}C_4H_9)_4NO_2CCH_2CH{=}CHCHOHCH_3$ (—)	725
C_7			
$C_6H_5(CH_3)_2SiCH(C_3H_7\text{-}i)CH{=}CHCH_3$ *E*:*Z* 99:1	1. MCPBA, Na_2HPO_4, CH_2Cl_2, 5 h 2. TBAF, THF, CH_2Cl_2, 12 h	$i\text{-}C_3H_7CH{=}CHCHOHCH_3$ *E*:*Z* > 95:5 (90)	51
	1. OsO_4, Py, 18 h 2. $(CH_3CO)_2O$, $(C_2H_5)_3N$, DMAP, CH_2Cl_2, 20 h 3. CsF, DMSO, 80°, 3 h	", *E*:*Z* > 95:5 (54)[a]	51
E:*Z* 11:89	1. MCPBA, Na_2HPO_4, CH_2Cl_2, 5 h 2. TBAF, THF, CH_2Cl_2, 12 h	", *E*:*Z* > 95:5 (54)	51
	OsO_4, Py, 18 h	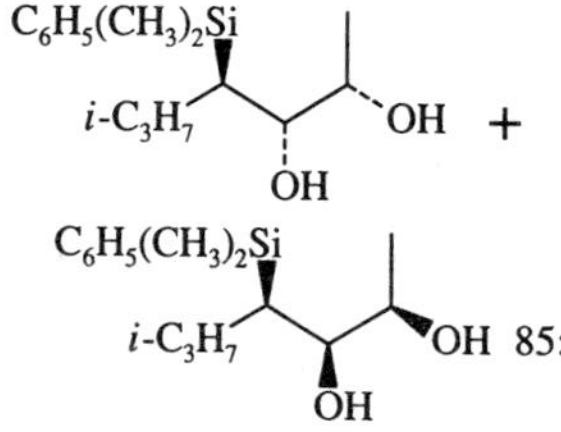 $C_6H_5(CH_3)_2Si$, $i\text{-}C_3H_7$, OH, OH (structure) + $C_6H_5(CH_3)_2Si$, $i\text{-}C_3H_7$, OH, OH (structure) 85:15 (70)	51

TABLE XXXV. ALLYLSILANES WITH OXYGEN ELECTROPHILES (*Continued*)

Reactant	Conditions	Product(s) and Yield(s)	Refs.
(*Z*)-$(CH_3)_3SiCH_2CH{=}CHCH_2CHOHCH_2R$	1. $VO(acac)_2$, t-$C_4H_9O_2H$, $Cl(CH_2)_2Cl$, or toluene, CH_2Cl_2, 0°, 8 h; room temp, 24 h 2. SiO_2	$CH_2{=}CHCHOHCH_2CHOHCH_2R$	723
	R	*erythro*:*threo*	
	$CH_2OCH_2O_2CCH_3$	95:5 (25)	
	$CH_2OCH_2C_6H_5$	83:17 (15)	
	$CO_2CH_2C_6H_5$	95:5 (25)	
$(CH_3)_2NCO$–, $(CH_3)_3Si$– [structure]	1. MCPBA 2. Separate major product 3. $BF_3{\cdot}O(C_2H_5)_2$	[lactone structure] (48)	725
$C_2H_5O_2C$–, $(CH_3)_3Si$– [structure]	1. MCPBA, CH_2Cl_2, −20°[f] 2. TFA	$C_2H_5O_2C$–, OH [structure] (—)[g]	724
$(CH_3)_3Si$–, CH_3CO_2– [cycloheptene structure]	CH_3CO_3H, CH_3CO_2H, 20°, 7.5 h	CH_3CO_2–, OH [cycloheptene structure] (—)	405
$(CH_3)_3Si$–, OCH_3 [furanoside structure]	1. OsO_4, NMMO, THF, 24 h 2. $(CH_3CO)_2O$, Py, DMAP, 4 h 3. TBAF, CH_3CN, 3 h	CH_3CO_2, H, OCH_3 [furanoside structure]	302

Reactant	Conditions	Product(s) and Yield(s) (%)	Refs.
		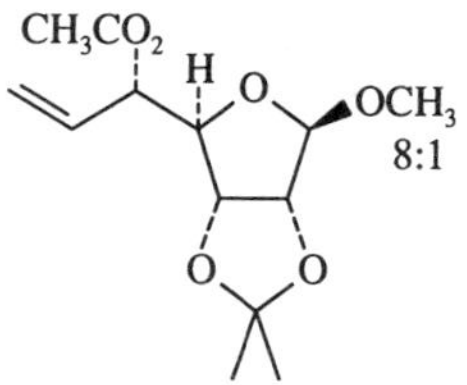 + CH_3CO_2, H, O, OCH_3, O, O; 8:1 (96)	
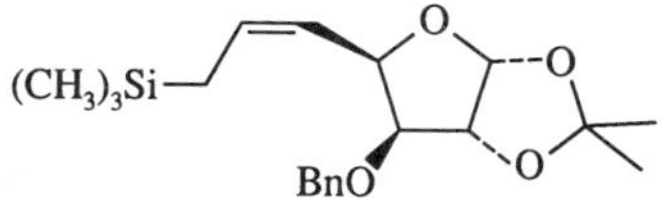 $(CH_3)_3Si$—, O, O, O, BnO	1. OsO_4, NMMO, THF, 24 h 2. $BF_3\cdot O(C_2H_5)_2$, CH_2Cl_2, $-5°$, 2.5 h	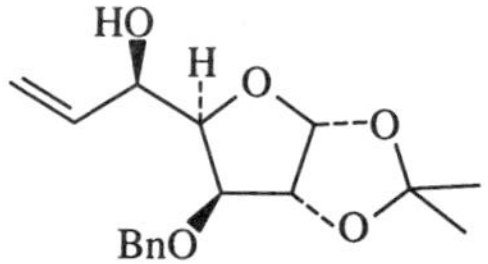 HO, H, O, O, O, BnO (85)	302
$(CH_3)_3SiCH_2C(=CH_2)CH_2CHOHC_2H_5$	1. MCPBA, $NaHCO_3$, CH_2Cl_2, 0°, 1.5 h 2. H_2SO_4, THF	$CH_2{=}C(CH_2OH)CH_2CHOHC_2H_5$ (80)	726
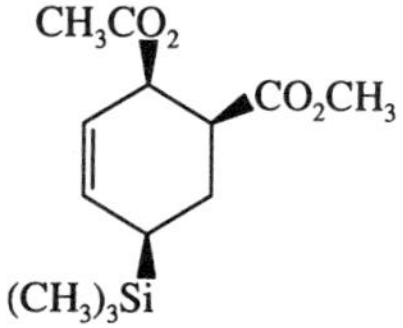 CH_3CO_2, CO_2CH_3, $(CH_3)_3Si$	1. OsO_4, NMMO, t-C_4H_9OH, $(CH_3)_2CO$, H_2O, 10 h 2. TsOH, C_6H_6, reflux, 20 min	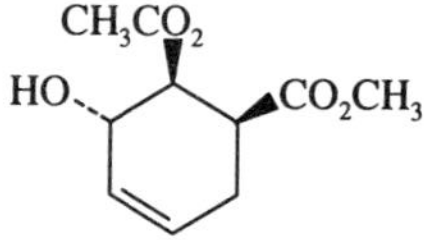 CH_3CO_2, HO, CO_2CH_3 (94)	441
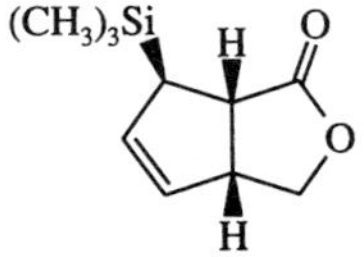 $(CH_3)_3Si$, H, O, O, H	1. MCPBA, $NaHCO_3$, CH_2Cl_2, 28°, 48 h 2. HCl, H_2O, CH_3OH, 28°, 12 h	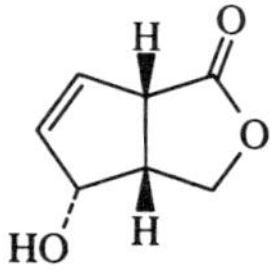 H, O, O, HO, H (72)	440
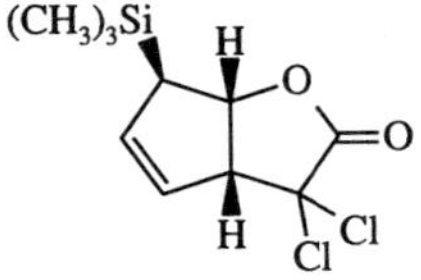 $(CH_3)_3Si$, H, O, O, H, Cl, Cl	1. MCPBA, $NaHCO_3$, CH_2Cl_2, 3.5 d 2. H_3O^+	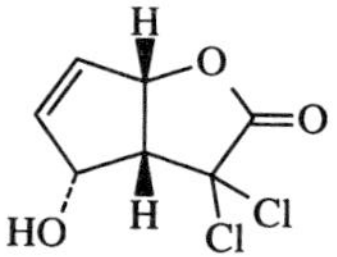 H, O, O, HO, H, Cl, Cl (73)	55

TABLE XXXV. ALLYLSILANES WITH OXYGEN ELECTROPHILES (*Continued*)

Reactant	Conditions	Product(s) and Yield(s)	Refs.
C_8			
[structure]	CH_3CO_3H, $(C_2H_5)_2O$, 24 h	[structure] (66)	134
[structure]	1. OsO_4, NMMO, THF, 24 h 2. $BF_3{\cdot}O(C_2H_5)_2$, CH_2Cl_2, −5°, 2.5 h	[structure] + [structure] 20:1 (80)	302
C_9			
$(CH_3)_3SiCH_2CH{=}CHC_6H_5$	Electrolysis, CH_3OH, CH_3CN, $(C_2H_5)_4NOTs$	$CH_2{=}CHCH(OCH_3)C_6H_5$ + $CH_3OCH_2CH{=}CHC_6H_5$ 63:37 (76)	444
C_{10}			
$(CH_3)_3SiCH_2CH{=}C(CH_3)R^1$	Electrolysis, R^2OH, CH_3CN, $(C_2H_5)_4NOTs$	$CH_2{=}CHC(CH_3)(OR^2)R^1$ + $R^2OCH_2CH{=}C(CH_3)R^1$	444

Reactant	Conditions	Product(s) and Yield(s) (%)	Refs.
R^1	R^2		
$n\text{-}C_6H_{13}$	CH_3	61:39 (100)	
$(CH_3)_2C{=}CH(CH_2)_2$	"	68:32 (69)	
	H	60:40 (62)	
	C_2H_5	63:37 (56)	
	CH_3CO	63:37 (26)[h]	
$(R)(E)$-$(CH_3)_3SiCH(C_6H_5)CH{=}CHCH_3$ 81% ee	1. MCPBA, $NaHCO_3$, CH_2Cl_2, 0°, 1 h 2. CH_3CO_2H, CH_3OH	$C_6H_5CH{=}CHCHOHCH_3$ *E*:*Z* 81:19 (98) *S,E* 78% ee *R,Z* 72% ee	437
$(R)(Z)$-$(CH_3)_3SiCH(C_6H_5)CH{=}CHCH_3$ 24% ee	1. MCPBA, $NaHCO_3$, CH_2Cl_2, 0°, 1 h 2. CH_3CO_2H, CH_3OH	$(R)(E)$-$C_6H_5CH{=}CHCHOHCH_3$ 19% ee (68)	437
$C_6H_5(CH_3)_2SiCH(C_6H_5)CH{=}CHCH_3$ *E*:*Z* > 99:1	1. MCPBA, Na_2HPO_4, CH_2Cl_2, 5 h 2. TBAF, THF, CH_2Cl_2, 12 h	$C_6H_5CH{=}CHCHOHCH_3$ *E*:*Z* 89:11 (90)	51
	1. OsO_4, Py, 18 h 2. NaH, THF, 18 h	", ", 8:92 (88)	51
E:*Z* < 5:95	1. MCPBA, Na_2HPO_4, CH_2Cl_2, 5 h 2. TBAF, THF, CH_2Cl_2, 12 h	", ", >95:5 (60)	51
	OsO_4, Py, 18 h	$C_6H_5(CH_3)_2Si$, C_6H_5, OH, OH (diol) + $C_6H_5(CH_3)_2Si$, C_6H_5, OH, OH (diol) 83:17 (93)	51
$(S)(Z)$-$(CH_3)_3SiCH(CH_3)CH{=}CHC_6H_5$ 44% ee	1. MCPBA, $NaHCO_3$, CH_2Cl_2, 0°, 1 h 2. CH_3CO_2H, CH_3OH	$(S)(E)$-$CH_3CH{=}CHCHOHC_6H_5$ 35% ee (—)	437
C_{11}			
$(CH_3)_3SiCH_2C({=}CH_2)CH_2CHOHC_6H_{13}\text{-}n$	1. MCPBA, $NaHCO_3$, CH_2Cl_2, 0°, 1.5 h 2. H_2SO_4, THF	$CH_2{=}C(CH_2OH)CH_2CHOHC_6H_{13}\text{-}n$ (66)	726

TABLE XXXV. ALLYLSILANES WITH OXYGEN ELECTROPHILES (*Continued*)

Reactant	Conditions	Product(s) and Yield(s)	Refs.
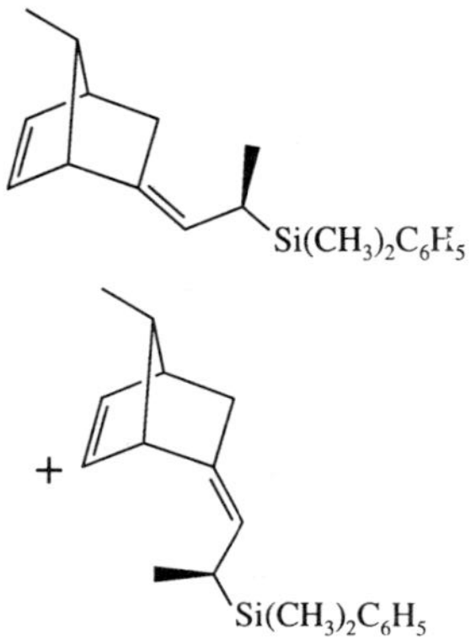 [structure] $Si(CH_3)_2C_6H_5$ + [structure] $Si(CH_3)_2C_6H_5$	1. MCPBA, NaH_2PO_4, CH_2Cl_2, 0°, 1 h 2. TBAF, THF, 15 min	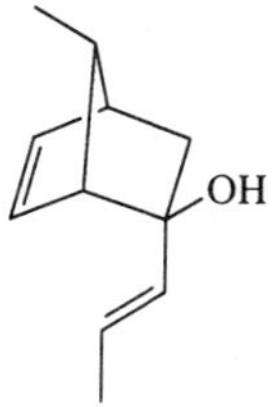 [structure] OH (90)	439
$C_6H_5(CH_3)_2SiCH_2CH{=}CHCH(CH_3)C_6H_5$	1. MCPBA, Na_2HPO_4, CH_2Cl_2, 0°, 1 h 2. TBAF, THF, 4 h	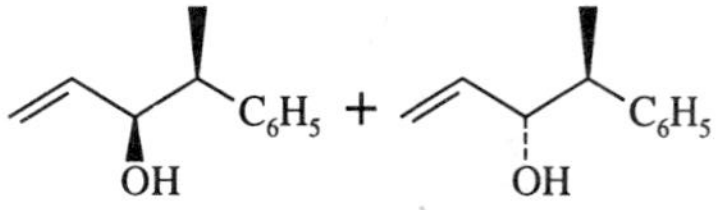 [structure] OH C_6H_5 + [structure] OH C_6H_5 (76) 53:47	409
$(CH_3)_3SiCH_2C({=}CH_2)$ with $C_6H_5CHCH_2OH$	1. MCPBA, $NaHCO_3$, CH_2Cl_2, 0°, 1.5 h 2. H_2SO_4, THF	$CH_2{=}CCH_2OH$ with $C_6H_5CHCH_2OH$ (51)	726

Reactant	Conditions	Product(s) and Yield(s) (%)	Refs.
C_{12}			
$C_6H_5(CH_3)_2SiCH_2CH{=}C(CH_3)CH(CH_3)C_6H_5$ $E{:}Z$ 64:36	1. MCPBA, Na_2HPO_4, CH_2Cl_2, 0°, 1 h 2. TBAF, THF, 4 h	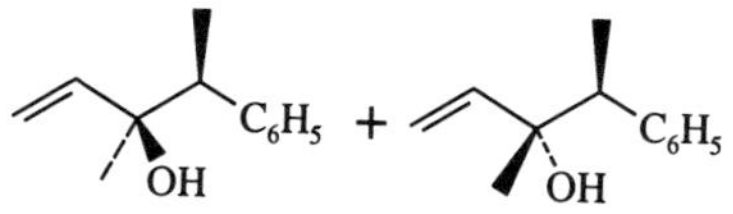 [structure] C_6H_5 + [structure] C_6H_5 (76) 57:43	409
$C_6H_5(CH_3)_2SiCH_2C(CH_3){=}CHCH(CH_3)C_6H_5$	1. MCPBA, Na_2HPO_4, CH_2Cl_2, 0°, 1 h 2. TBAF, THF, 4 h	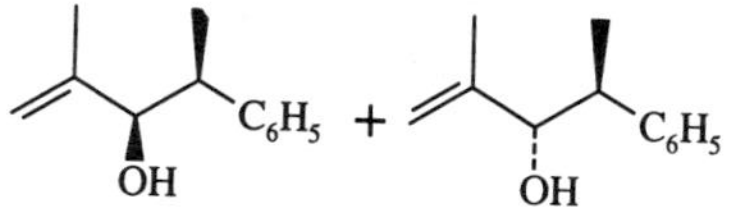 [structure] C_6H_5 + [structure] C_6H_5 (85) 92:8	409
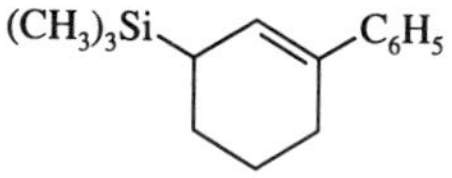 $(CH_3)_3Si$-[cyclohexenyl]-C_6H_5	1. CH_3CO_3H, NaO_2CCH_3, $(C_2H_5)_2O$, 7 h 2. H_2SO_4, H_2O, 2 d	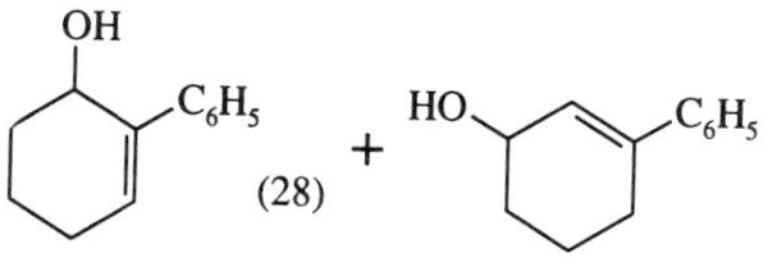 OH, C_6H_5 [cyclohexene] (28) + HO, C_6H_5 [cyclohexene] (28)	408
C_{13}			
$(CH_3)_3SiCH_2C(=CH_2)CH_2COH(C_2H_5)C_6H_{13}$-*n*	1. MCPBA, $NaHCO_3$, CH_2Cl_2, 0°, 1.5 h 2. H_2SO_4, THF	$CH_2{=}C(CH_2OH)CH_2COH(C_2H_5)C_6H_{13}$-*n* (53)	726
$C_6H_5(CH_3)_2SiCH_2C(CH_3){=}C(CH_3)CH(CH_3)$-$C_6H_5$ $E{:}Z$ 74:26	1. MCPBA, Na_2HPO_4, CH_2Cl_2, 0°, 1 h 2. TBAF, THF, 4 h	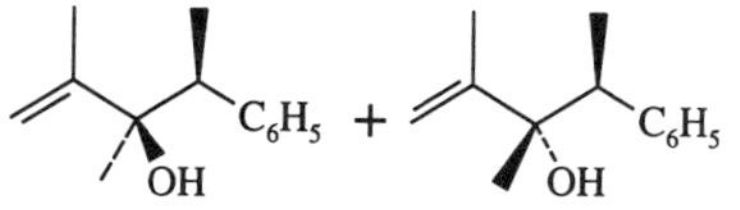 [structure] C_6H_5 + [structure] C_6H_5 (81) 73:27	409

TABLE XXXV. ALLYLSILANES WITH OXYGEN ELECTROPHILES (*Continued*)

Reactant	Conditions	Product(s) and Yield(s)	Refs.
C_{14}			
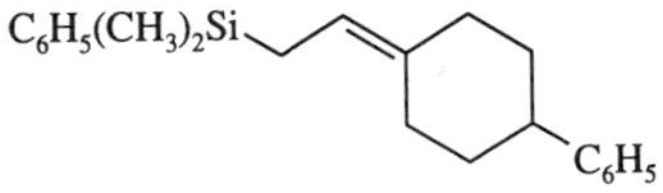 $C_6H_5(CH_3)_2Si$ [structure], C_6H_5	1. MCPBA, Na_2HPO_4, CH_2Cl_2, 0°, 1 h 2. TBAF, THF, 18 h	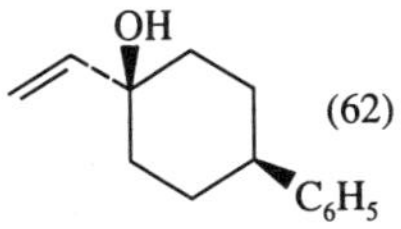 OH, C_6H_5 [structure] (62) 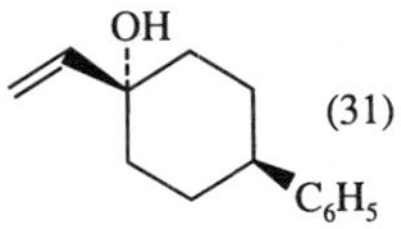 + OH, C_6H_5 [structure] (31)	44
C_{15}			
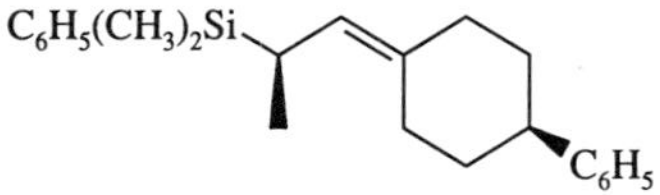 $C_6H_5(CH_3)_2Si$ [structure], C_6H_5	1. MCPBA, Na_2HPO_4, CH_2Cl_2, 0°, 1 h 2. TBAF, THF, 18 h	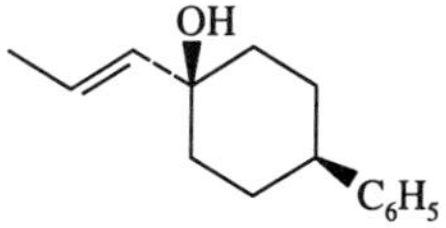 OH, C_6H_5 [structure] I (91)	44
	1. OsO_4, Py, 15 h 2. $(CH_3CO)_2O$, CH_2Cl_2, $(C_2H_5)_3N$, DMAP 3. TBAF, THF, 3 h	I (75)	44
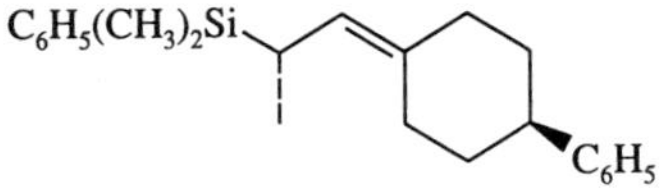 $C_6H_5(CH_3)_2Si$ [structure], C_6H_5	1. MCPBA, Na_2HPO_4, CH_2Cl_2, 0°, 1 h 2. TBAF, THF, 18 h	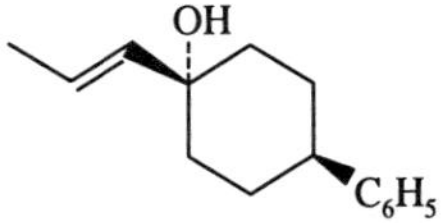 OH, C_6H_5 [structure] II (96) + I (3)	44
	1. OsO_4, Py, 15 h 2. Separate diastereoisomers 3. $(CH_3CO)_2O$, CH_2Cl_2, $(C_2H_5)_3N$, DMAP 4. TBAF, THF, 3 h	II (from major diastereoisomer) (53) + *Z*-I (from minor diastereoisomer) (26)	44
	or 3. KH, THF, 30 min	I (from minor diastereoisomer) (27)	44

C_{18}

Reactant	Conditions	Product(s) and Yield(s) (%)	Refs.
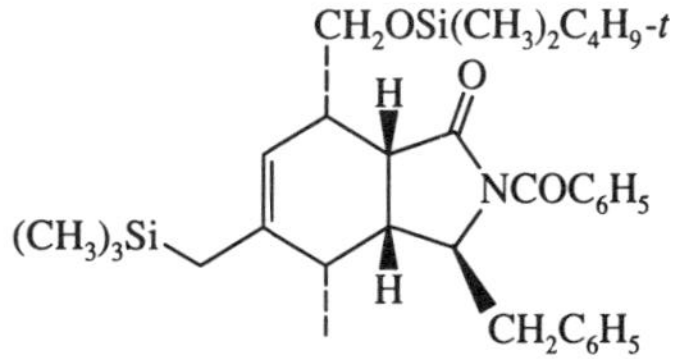 (CH$_3$)$_3$SiCH$_2$-substituted bicyclic lactam: CH$_2$OSi(CH$_3$)$_2$C$_4$H$_9$-*t*, NCOC$_6$H$_5$, CH$_2$C$_6$H$_5$	MCPBA, CH_2Cl_2, 0°, 20 min; 20°, 15 min	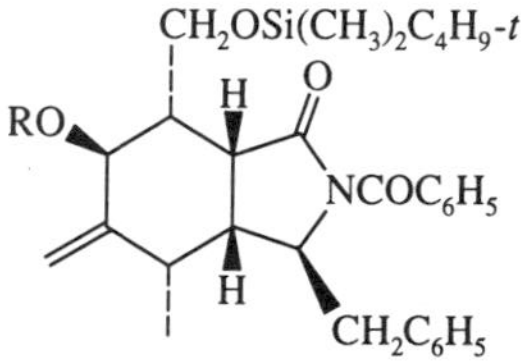 RO-substituted exo-methylene bicyclic lactam: CH$_2$OSi(CH$_3$)$_2$C$_4$H$_9$-*t*, NCOC$_6$H$_5$, CH$_2$C$_6$H$_5$; R=H (60)	442
	1. OsO_4, Py, 1 h 2. $(CH_3CO)_2O$, Py, 14 h 3. $SOCl_2$, Py, DMAP, 0°, 1.5 h	R=COCH$_3$ (66)	442

[a] Of the diastereoisomeric diols produced, only the major one undergoes elimination. The ratio of the diols is 67:33.
[b] Three diastereoisomeric γ lactones are produced in a ratio of 4:25:71.
[c] Two diastereoisomeric γ lactones are produced in a ratio of 22:78.
[d] Three diastereoisomeric γ lactones are produced in a ratio of 81:16:3.
[e] Three diastereoisomeric γ lactones are produced in a ratio of 53:15:15.
[f] Two diastereoisomeric epoxides are produced in a ratio of 75:25.
[g] Both diastereoisomers are produced, reflecting the proportion of the epoxides.
[h] Linalool and geraniol are also produced (31%).

TABLE XXXVI. VINYLSILANES WITH OXYGEN ELECTROPHILES

Reactant	Conditions	Product(s) and Yield(s)	Refs.
C_2			
$(CH_3)_3SiCH{=}CH_2$	1. Epoxidation[a] 2. 310°, 100 min	$(CH_3)_3SiOCH{=}CH_2$ (100)[b]	456
	1. Epoxidation[a] 2. 600°	" (68)[b]	455
	1. Epoxidation[a] 2. $(n\text{-}C_4H_9)_2CuLi$, $(C_2H_5)_2O$, −25°, 5 h 3. KH, THF, 1 h	$n\text{-}C_4H_9CH{=}CH_2$ (95)[b]	727
	1. MCPBA, CH_2Cl_2, 2 h 2. $HC{\equiv}C(CH_2)_nCO_2H$, CH_2Cl_2, 2 h 3. DCC, CH_3CO_2H, DMAP, $(C_2H_5)_2O$, 18 h 4. TBAF, THF, 1 h	$HC{\equiv}C(CH_2)_nCO_2CH{=}CH_2$ $n = 0$ (23) $n = 2$ (35) $n = 8$ (39)	728
	1. MCPBA, CH_2Cl_2, 2 h 2. $HC{\equiv}C(CH_2)_nOH$, $BF_3{\cdot}O(C_2H_5)_2$, CH_2Cl_2, 1 h 3. KH, THF, 1 h	$HC{\equiv}C(CH_2)_nOCH{=}CH_2$ $n = 1$ (26) $n = 2$ (28) $n = 3$ (37)	728
$(CH_3)_3SiCD{=}CH_2$	1. Epoxidation[a] 2. 310°, 100 min	$(CH_3)_3SiOCH{=}CHD$ *E* (4)[b] + *Z* (95)[b]	456
(*E*)-$(CH_3)_3SiCH{=}CHD$	1. Epoxidation[a] 2. 310°, 100 min	$(CH_3)_3SiOCD{=}CH_2$ (20)[b] + $(CH_3)_3SiOCH{=}CHD$ *E* (70)[b] + *Z* (10)[b]	456
$C_6H_5(CH_3)_2SiCH{=}CH_2$	1. CF_3CO_3H, Na_2CO_3, CH_2Cl_2, 4 h (32%) 2. 187°	$C_6H_5(CH_3)_2SiOCH{=}CH_2$ (—)	729
	1. CF_3CO_3H, Na_2CO_3, CH_2Cl_2, 4 h (32%) 2. 2,4-$(O_2N)_2C_6H_3N_2H_3$, H_2SO_4, C_2H_5OH	2,4-$(O_2N)_2C_6H_3NHN{=}CHCH_3$ (—)	729
$(C_6H_5)_3SiCH{=}CH_2$	1. CF_3CO_3H, Na_2CO_3, CH_2Cl_2, 4 h (84%) 2. $MgBr_2$, $(n\text{-}C_4H_9)_2O$, 60°, 5 h	$(C_6H_5)_3SiCH_2CHO$ (73)[c]	450
(*E*)-$(CH_3)_3SiCH{=}CHSi(CH_3)_3$	1. Epoxidation[a] 2. 258°, 4 h	$(CH_3)_3SiOCH{=}CHSi(CH_3)_3$ *E*:*Z* 80:20 (—) I	456

Reactant	Conditions	Product(s) and Yield(s) (%)	Refs.
	1. MCPBA, CH_2Cl_2, 20 h (84%) 2. 600°	*E*-I + *Z*-I + $(CH_3)_3SiCH{=}C{=}O$ (83)[c] 58:20:22	455
	1. MCPBA, CH_2Cl_2, 20 h (84%) 2. $MgBr_2$, $(C_2H_5)_2O$, 0°, 1.5 h	$[(CH_3)_3Si]_2CHCHO$ (90)[c]	454
	1. MCPBA, CH_2Cl_2, 20 h (84%) 2. $BF_3{\cdot}O(C_2H_5)_2$, $(CH_3CO)_2O$, CH_3CO_2H	(*Z*)-$CH_3CO_2CH{=}CHSi(CH_3)_3$ (—)	451
$[(CH_3)_3Si]_2C{=}CH_2$	1. Epoxidation[a] 2. 256°, 2 h	$(CH_3)_3SiOCH{=}CHSi(CH_3)_3$ *E*:*Z* 1:1 (100)[b]	456
	1. MCPBA, CH_2Cl_2, 20 h (70%) 2. 600°	", ", 29:71 (67)[c]	455
	1. MCPBA, CH_2Cl_2, 20 h (70%) 2. $MgBr_2$, THF, 4 h	$[(CH_3)_3Si]_2CHCHO$ (88)[c]	454
	1. MCPBA, CH_2Cl_2, 20 h (70%) 2. $BF_3{\cdot}O(C_2H_5)_2$, $(CH_3CO)_2O$, CH_3CO_2H	$(CH_3)_3SiC(O_2CCH_3){=}CH_2$ (—)	451
C_3			
(*E*)-$(CH_3)_3SiCH{=}CHCH_3$	1. Epoxidation[a] 2. 256°, 315 min	$(CH_3)_3SiOC(CH_3){=}CH_2$ + (*E*)-$(CH_3)_3SiOCH{=}CHCH_3$ 1:1 (100)[b]	456
(*Z*)-$(CH_3)_3SiCH{=}CHCH_3$	1. Epoxidation[a] 2. 256°, 315 min	$(CH_3)_3SiOC(CH_3){=}CH_2$ (100)[b]	456
$(CH_3)_3SiC(CH_3){=}CH_2$	1. Epoxidation[a] 2. 256°, 1 h	$(CH_3)_3SiOC(CH_3){=}CH_2$ + (*E*)-$(CH_3)_3SiOCH{=}CHCH_3$ 1:1 (100)[b]	456
C_4			
$(CH_3)_3SiCH{=}C(CH_3)_2$	1. Epoxidation[a] 2. 233°, 2 h	$(CH_3)_3SiOCH{=}C(CH_3)_2$ (70)[b,d]	456
	1. MCPBA, CH_2Cl_2 (79%) 2. 600°	(*E* and *Z*)-$(CH_3)_3SiOC(CH_3){=}CHCH_3$ + $(CH_3)_3SiOCH{=}C(CH_3)_2$ 77:23 (77)[c]	455
	1. MCPBA, CH_2Cl_2 (79%) 2. $(n\text{-}C_4H_9)_2CuLi$, $(C_2H_5)_2O$, −25°, 5 h 3. NaO_2CCH_3, CH_3CO_2H, 1 h	*n*-$C_4H_9CH{=}C(CH_3)_2$ (61)[c]	727
	1. MCPBA, CH_2Cl_2 (79%) 2. $MgBr_2$, $(C_2H_5)_2O$, 12 h or HBr, $(C_2H_5)_2O$, −78°, 1 h 3. $BF_3{\cdot}O(C_2H_5)_2$, CCl_4, 0°, 1 h	$BrCH{=}C(CH_3)_2$ (90)[c]	451

TABLE XXXVI. VINYLSILANES WITH OXYGEN ELECTROPHILES (*Continued*)

Reactant	Conditions	Product(s) and Yield(s)	Refs.
C_5			
(*E*)-$(CH_3)_3SiCH{=}CHC_3H_7$-*n*	1. MCPBA, CH_2Cl_2 (87%) 2. 600°	(*E* and *Z*)-$(CH_3)_3SiOCH{=}CHC_3H_7$-*n* + $(CH_3)_3SiOC(C_3H_7$-*n*$){=}CH_2$ 41:59 (66)[c]	455
	1. MCPBA, CH_2Cl_2 (87%) 2. HBr, $(C_2H_5)_2O$, −25°, 30 min 3. $BF_3{\cdot}O(C_2H_5)_2$, CH_2Cl_2, 0°, 15 min	$BrCH{=}CHC_3H_7$-*n* $E:Z < 1:99$ (80)[c]	451
	1. MCPBA, CH_2Cl_2 (87%) 2. (*n*-$C_3H_7)_2CuLi$, −5°, 4 h 3. $BF_3{\cdot}O(C_2H_5)_2$, CH_2Cl_2, 0°, 1 h	*n*-$C_3H_7CH{=}CHC_3H_7$-*n* $E:Z$ 0.5:99.5 (80)[c]	727
	1. MCPBA, CH_2Cl_2 (87%) 2. (*n*-$C_3H_7)_2CuLi$, −5°, 4 h 3. H_2SO_4, H_2O, THF, 18 h	", $E:Z$ 0.5:99.5 (77)[c]	727
	1. MCPBA, CH_2Cl_2 (87%) 2. (*n*-$C_3H_7)_2CuLi$, −5°, 4 h 3. KH, THF, 1 h	", $E:Z$ >99.5:0.5 (76)[c]	727
(*Z*)-$(CH_3)_3SiCH{=}CHC_3H_7$-*n*	1. MCPBA, CH_2Cl_2 (65%) 2. 600°	(*E* and *Z*)-$(CH_3)_3SiOCH{=}CHC_3H_7$-*n* + $(CH_3)_3SiOC(C_3H_7$-*n*$){=}CH_2$ 21:79 (72)[c]	455
	1. MCPBA, CH_2Cl_2 (65%) 2. HBr, $(C_2H_5)_2O$, −25°, 30 min 3. $BF_3{\cdot}O(C_2H_5)_2$, CH_2Cl_2, 0°, 10 h	$BrCH{=}CHC_3H_7$-*n* $E:Z$ >99:1 (95)[c]	451
	1. MCPBA, CH_2Cl_2 (65%) 2. $MgBr_2$, $(C_2H_5)_2O$, 6 h	$(CH_3)_3SiCH_2COC_3H_7$-*n* (93)[c]	454
	1. MCPBA, CH_2Cl_2 (65%) 2. (*n*-$C_3H_7)_2CuLi$, −5°, 4 h 3. $BF_3{\cdot}O(C_2H_5)_2$, CH_2Cl_2, 0°, 1 h	*n*-$C_3H_7CH{=}CHC_3H_7$-*n* $E:Z$ 98:2 (70)[c]	727
	1. MCPBA, CH_2Cl_2 (65%) 2. (*n*-$C_3H_7)_2CuLi$, −5°, 4 h 3. H_2SO_4, H_2O, THF, 18 h	", $E:Z$ 99:1 (67)[c]	727
	1. MCPBA, CH_2Cl_2 (65%)	", $E:Z$ 2:98 (69)[c]	727

Reactant	Conditions	Product(s) (Yield, %)	Refs.
	2. $(n\text{-}C_3H_7)_2CuLi$, −5°, 4 h 3. KH, THF, 1 h		
$(CH_3)_3SiC(C_3H_7\text{-}n){=}CH_2$	1. MCPBA, CH_2Cl_2, 6 h (84%) 2. 600°	(E and Z)-$(CH_3)_3SiOCH_2CH{=}CHC_2H_5$ (26)[c] + (E)-$(CH_3)_3SiOCH{=}CHC_3H_7\text{-}n$ (50)[c]	455
	1. MCPBA, CH_2Cl_2, 6 h (84%) 2. $MgBr_2$, $(C_2H_5)_2O$, 2 h	$(CH_3)_3SiOCH{=}CHC_3H_7\text{-}n$ (82)[c]	451
	1. MCPBA, CH_2Cl_2, 6 h (84%) 2. $n\text{-}C_3H_7MgBr$, $(C_2H_5)_2O$, 20 h 3. $BF_3{\cdot}O(C_2H_5)_2$, CH_2Cl_2, 1 h	(E)-$n\text{-}C_3H_7CH{=}CHC_3H_7\text{-}n$ (—)[c]	453
	1. MCPBA, CH_2Cl_2, 6 h (84%) 2. $n\text{-}C_3H_7MgBr$, $(C_2H_5)_2O$, 20 h 3. KH, THF, 1 h	(Z)-$n\text{-}C_3H_7CH{=}CHC_3H_7\text{-}n$ (—)[c]	453
$(CH_3)_3SiC(CH_3){=}CHC_2H_5$ $E{:}Z$ 1:24	1. CH_3CO_3H, NaO_2CCH_3, CH_2Cl_2 (—) 2. H_2SO_4, CH_3OH, 0°, 1.5 h 3. NaH, DMF, 2 h	$CH_3OC(CH_3){=}CHC_2H_5$ $E{:}Z$ 5:95 (—)	73
(E)-$C_6H_5(CH_3)_2SiCH{=}C(CH_3)C_2H_5$	1. MCPBA, CH_2Cl_2, 0°, 1 h (77%) 2. $BF_3{\cdot}O(C_2H_5)_2$, CH_2Cl_2, −78°, 5 min	$C_6H_5(CH_3)_2SiOCH{=}C(CH_3)C_2H_5$ $E{:}Z$ 95:5 (69)[c]	457
(Z)-$C_6H_5(CH_3)_2SiCH{=}C(CH_3)C_2H_5$	1. MCPBA, CH_2Cl_2, 0°, 1 h (89%) 2. $BF_3{\cdot}O(C_2H_5)_2$, CH_2Cl_2, −78°, 5 min	$C_6H_5(CH_3)_2SiOCH{=}C(CH_3)C_2H_5$ $E{:}Z$ 4:96 (68)[c]	457
C_6			
(Z)-$(CH_3)_3SiCH{=}CHC_4H_9\text{-}t$	1. OsO_4, $(CH_3)_3NO$, $t\text{-}C_4H_9OH$, Py, reflux 2. $CDCl_3$	$t\text{-}C_4H_9CH_2CHO$ (—) + $t\text{-}C_4H_9CH[Si(CH_3)_3]CHO$ (—)	448
$(CH_3)_3Si$-(cyclohex-1-enyl)	1. MCPBA, CH_2Cl_2, 0°, 3 h (86%) 2. 600°	$(CH_3)_3SiO$-(cyclohex-1-enyl) (19) + I $(CH_3)_3SiO$-(cyclohex-2-enyl) (58)[c]	455
	1. MCPBA, CH_2Cl_2, 0°, 3 h (86%) 2. $MgBr_2$, $(C_2H_5)_2O$, 7 h	$(CH_3)_3SiOCH{=}$(cyclopentylidene) (84)[c]	454

TABLE XXXVI. VINYLSILANES WITH OXYGEN ELECTROPHILES (*Continued*)

Reactant	Conditions	Product(s) and Yield(s)	Refs.
	1. MCPBA, CH_2Cl_2, 0°, 3 h (86%) 2. H_2SO_4, H_2O, THF 3. KH, $(C_2H_5)_2O$, 30 min 4. $(CH_3)_3SiCl$	I 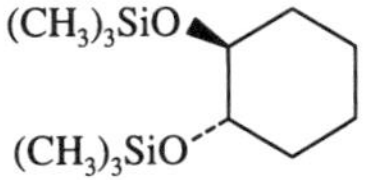 + $(CH_3)_3SiO$, $(CH_3)_3SiO$ (cyclohexane) II 7:86 (74)[c]	452
	1. MCPBA, CH_2Cl_2, 0°, 3 h (86%) 2. H_2SO_4, H_2O, THF 3. KH, THF, 0°, 2 h 4. $(CH_3)_3SiCl$	I + II 8:91 (65)[c]	452
	1. CH_3CO_3H, NaO_2CCH_3, CH_2Cl_2, 30 min (83%) 2. H_2SO_4, ROH, 0° (R = CH_3), room temp (R = $CH_2{=}CHCH_2$), 5 min 3. NaH, DMF, 15 min	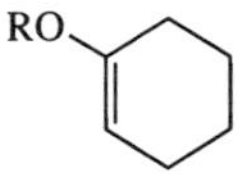 RO (cyclohexene) R = CH_3 (—) R = $CH_2{=}CHCH_2$ (31)[c]	73
	1. CH_3CO_3H, NaO_2CCH_3, CH_2Cl_2, 30 min (83%) 2. H_2SO_4, H_2O, $(CH_3)_2CO$ (83%) 3. $(CH_3CO)_2O$, Py, 40°, 40 h (96%) 4. DMSO, CH_3CO_2H, $(CH_3CO)_2O$, 15°, 68 h (87%) 5. KOH, CH_3OH, 50°, 18 h (80%) 6. KH, THF, 10 min (90%)	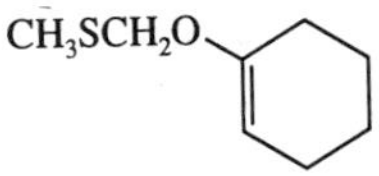 CH_3SCH_2O (cyclohexene) I (41)	730
	1. OsO_4, NMMO, t-C_4H_9OH, H_2O, $(CH_3)_2CO$, 50°, 36 h (70%) 2. $(CH_3CO)_2O$, Py, 40°, 40 h (97%) 3. DMSO, CH_3CO_2H, $(CH_3CO)_2O$, 15°, 68 h (75%) 4. KOH, CH_3OH, 50°, 18 h (72%) 5. CH_3SO_2Cl, Py, 18°, 18 h (94%)	I (34)	730

Reactant	Conditions	Product(s) (Yield, %)	Refs.
	1. OsO_4, $(CH_3)_3NO$, t-C_4H_9OH, Py, reflux, 24 h 2. NaH, $(C_2H_5)_2O$, 5 h	$(CH_3)_3SiO$-cyclohexenyl (50)	731
$(CH_3)_3SiCH{=}CHCH_2C(OH)(CH_3)CH(OCH_3)_2$	1. MCPBA, CH_2Cl_2 (94%) 2. $BF_3{\cdot}O(C_2H_5)_2$, CH_3OH	2-CH_3O-5-methyl-5-$CH(OCH_3)_2$-tetrahydrofuran (10)[c]	132
$(CH_3)_3Si$-cyclohexa-1,3-dienyl	1. MCPBA, CH_2Cl_2 (60–70%) 2. Br_2, CH_2Cl_2 (84–100%) 3. DBN, THF (—) 4. TFA, $CDCl_3$	C_6H_5OH (100)[c]	74
C_7			
$(CH_3)_3SiC(CH_3){=}CHC_4H_9$-$n$	1. MCPBA, CH_2Cl_2, 0°, 15 min; room temp, 6 h 2. H_2SO_4, CH_3OH, 1 h	$CH_3COC_5H_{11}$-n (70)	732
$(CH_3)_3SiC(CH_3){=}CHC_4H_9$-$t$	1. MCPBA, CH_2Cl_2, 0°, 15 min; room temp, 6 h 2. H_2SO_4, CH_3OH, 1 h	$CH_3COCH_2C_4H_9$-t (67)	732
$(CH_3)_3SiC(=CH_2)CH_2C_4H_9$-$t$	1. MCPBA, CH_2Cl_2, 2.5 h 2. 2,4-$(O_2N)_2C_6H_3N_2H_3$, H_2SO_4, C_2H_5OH	$CH_3C(=NNHC_6H_3(NO_2)_2$-2,4$)CH_2C_4H_9$-t (76)	733
$(CH_3)_3SiCH_2C(=CHSi(CH_3)_3)(CH_2)_2COCH_2SO_2C_6H_5$	MCPBA, CH_2Cl_2, −78°, 30 min; −20°, 1 h	$(CH_3)_3SiCH_2C[Si(CH_3)_3]CHO$ with $(CH_2)_2COCH_2SO_2C_6H_5$ (100)	166
$(CH_3)_3Si$∼CH= pyrrolizidinone (H, O_2CCH_3) *E* *Z*	1. MCPBA, 12 h 2. HCO_2H, reflux, 0.5 h	OHC-pyrrolizidinone (H, O_2CCH_3), 4:1 mixture (72) (91)	578

TABLE XXXVI. VINYLSILANES WITH OXYGEN ELECTROPHILES (*Continued*)

Reactant	Conditions	Product(s) and Yield(s)	Refs.
$(CH_3)_3Si$-cyclohexadiene bearing R^1 and R^2 (structure)	1. MCPBA, CH_2Cl_2 (60–70%) 2. Br_2, CH_2Cl_2 (84–100%) 3. DBN, THF (—) 4. TFA, $CDCl_3$		74
R^1 = CH_3, R^2 = H		o-$CH_3C_6H_4OH$ (100)[e]	
R^1 = H, R^2 = CH_3		m- + p-$CH_3C_6H_4OH$ 1:1 (100)[e]	
C_8			
(E)-$(CH_3)_3SiCH{=}CHC_6H_{13}$-$n$	1. MCPBA, CH_2Cl_2, 12 h 2. H_2SO_4, CH_3OH, 90°, 10 min	$(CH_3O)_2CHC_7H_{15}$-n (60)	445
	1. MCPBA, CH_2Cl_2, 12 h 2. 2,4-$(O_2N)_2C_6H_3N_2H_3$, H_2SO_4, C_2H_5OH	2,4-$(O_2N)_2C_6H_3NHN{=}CHC_7H_{15}$-$n$ (65)	445
	1. Epoxidation[a] 2. H_2SO_4, H_2O, THF 3. KH, $(C_2H_5)_2O$ 4. $(CH_3)_3SiCl$	$(CH_3)_3SiOCH{=}CHC_6H_{13}$-$n$ E:Z 1:2 (—)	447
	1. OsO_4, $(CH_3)_3NO$, t-C_4H_9OH, reflux, 24 h 2. NaH, $(C_2H_5)_2O$, 18 h	$(CH_3)_3SiOCH{=}CHC_6H_{13}$-$n$ E:Z 99:1 (31)	731
	1. Epoxidation[a] 2. $BF_3{\cdot}O(C_2H_5)_2$, CH_3OH, 0°, 30 h 3. KH, THF, 0°, 60 min	$CH_3OCH{=}CHC_6H_{13}$-n E:Z 97:3 (81)[b]	451
	1. Epoxidation[a] 2. $(CH_3CO)_2O$, CH_3CO_2H, $BF_3{\cdot}O(C_2H_5)_2$, 18 h	$CH_3CO_2CH{=}CHC_6H_{13}$-n E:Z 3:97 (81)[b]	451
	1. Epoxidation[a] 2. $BF_3{\cdot}O(C_2H_5)_2$, CH_3CN, −25°, 20 min 3. H_2SO_4, H_2O, THF, 12 h 4. KH, THF, 50 min	$CH_3CONHCH{=}CHC_6H_{13}$-n E:Z >99:1 (62)[b]	451

	1. Epoxidation[a] 2. HBr, $(C_2H_5)_2O$, −25°, 30 min 3. $BF_3{\cdot}O(C_2H_5)_2$, CH_2Cl_2, 0°, 15 min	$BrCH{=}CHC_6H_{13}$-*n* $E{:}Z$ <1:99 (85)[b]	451
(*Z*)-$(CH_3)_3SiCH{=}CHC_6H_{13}$-*n*	1. Epoxidation[a] 2. H_2SO_4, H_2O, THF, 2 h 3. KH, $(C_2H_5)_2O$, 0°, 1 h 4. $(CH_3)_3SiCl$	$(CH_3)_3SiOCH{=}CHC_6H_{13}$-*n* $E{:}Z$ 5:1 (81)[b]	447
	1. Epoxidation[a] 2. H_2SO_4, H_2O, THF, 2 h 3. KH, $(C_2H_5)_2O$, 0°, 1 h 4. $NaHCO_3$, H_2O	$(CH_3)_3SiOCH{=}CHC_6H_{13}$-*n* $E{:}Z$ 96:4 (52)[b]	447
	1. OsO_4, $(CH_3)_3NO$, *t*-C_4H_9OH, reflux, 24 h 2. NaH, $(C_2H_5)_2O$, 18 h	$(CH_3)_3SiOCH{=}CHC_6H_{13}$-*n* $E{:}Z$ <1:99 (51)	731
	1. Epoxidation[a] 2. TFA, CH_3OH, 0°, 5 h 3. KH, THF, 0°, 60 min	$CH_3OCH{=}CHC_6H_{13}$-*n* $E{:}Z$ 14:86 (85)[b]	451
	1. Epoxidation[a] 2. $(CH_3CO)_2O$, CH_3CO_2H, $BF_3{\cdot}O(C_2H_5)_2$, 2 h	$CH_3CO_2CH{=}CHC_6H_{13}$-*n* $E{:}Z$ 97:3 (84)[b]	451
	1. Epoxidation[a] 2. $BF_3{\cdot}O(C_2H_5)_2$, CH_3CN, −25°, 20 min 3. H_2SO_4, H_2O, THF, 10 h 4. KH, THF, 50 min	$CH_3CONHCH{=}CHC_6H_{13}$-*n* $E{:}Z$ <1:99 (80)[b]	451
	1. Epoxidation[a] 2. HBr, $(C_2H_5)_2O$, −25°, 30 min 3. $BF_3{\cdot}O(C_2H_5)_2$, CH_2Cl_2, 0°, 10 h	$BrCH{=}CHC_6H_{13}$-*n* $E{:}Z$ 98:2 (90)[b]	451
(*E*)-$(CH_3)_3SiC(C_3H_7$-*n*$){=}CHC_3H_7$-*n*	1. OsO_4, $(CH_3)_3NO$, *t*-C_4H_9OH, reflux, 24 h 2. NaH, $(C_2H_5)_2O$, 5 h	(*E*)-$(CH_3)_3SiOC(C_3H_7$-*n*$){=}CHC_3H_7$-*n* (52)	731
(*Z*)-$(CH_3)_3SiC(C_3H_7$-*n*$){=}CHC_3H_7$-*n*	1. OsO_4, $(CH_3)_3NO$, *t*-C_4H_9OH, reflux, 24 h 2. NaH, $(C_2H_5)_2O$, 5 h	(*Z*)-$(CH_3)_3SiOC(C_3H_7$-*n*$){=}CHC_3H_7$-*n* (47)	731
$(CH_3)_3SiC(C_6H_{13}$-*n*$){=}CH_2$	1. Epoxidation[a] 2. H_2SO_4, H_2O, THF, 30 h 3. Stronger acid or longer times	$CH_3COC_6H_{13}$-*n* (—)	446

TABLE XXXVI. VINYLSILANES WITH OXYGEN ELECTROPHILES (*Continued*)

Reactant	Conditions	Product(s) and Yield(s)	Refs.
	1. Epoxidation[a] 2. H_2SO_4, H_2O, THF, 3 h 3. KH, $(C_2H_5)_2O$, 0°, 1 h 4. $(CH_3)_3SiCl$	$(CH_3)_3SiOC(C_6H_{13}\text{-}n){=}CH_2$ (52–89)[b]	447
	1. OsO_4, $(CH_3)_3NO$, $t\text{-}C_4H_9OH$, reflux, 24 h 2. NaH, $(C_2H_5)_2O$, 4 h	$(CH_3)_3SiOC(C_6H_{13}\text{-}n){=}CH_2$ (52)	731
	1. Epoxidation[a] 2. TFA, ROH, 3 h 3. KH, THF, 1 h	$ROC(C_6H_{13}\text{-}n){=}CH_2$ R = CH_3 (77)[b] R = $CH_2{=}CHCH_2$ (—)	451
	1. Epoxidation[a] 2. HBr, $(C_2H_5)_2O$, −25°, 30 min 3. $BF_3{\cdot}O(C_2H_5)_2$, CH_2Cl_2, 0°, 10 h	$n\text{-}C_6H_{13}CBr{=}CH_2$ (82)[b]	451
	1. Epoxidation[a] 2. C_2H_5MgBr, $(C_2H_5)_2O$, 2 h 3. $BF_3{\cdot}O(C_2H_5)_2$, CH_2Cl_2, 1 h	$(E)\text{-}n\text{-}C_6H_{13}CH{=}CHC_2H_5$ (76)[b]	453
	1. Epoxidation[a] 2. C_2H_5MgBr, $(C_2H_5)_2O$, 2 h 3. KH, THF, 1 h	$(Z)\text{-}n\text{-}C_6H_{13}CH{=}CHC_2H_5$ (76)[b]	453
$(C_2H_5)_3SiC(C_3H_7\text{-}n){=}CHC_3H_7\text{-}n$	1. MCPBA, CH_2Cl_2 (—) 2. H_2SO_4, CH_3OH, reflux	$n\text{-}C_3H_7COC_4H_9\text{-}n$ (—)	449
$(CH_3)_3Si$ (1-cyclooctenyl)	1. OsO_4, NMMO, $t\text{-}C_4H_9OH$, H_2O, $(CH_3)_2CO$, 50°, 10 d (74%) 2. $(CH_3CO)_2O$, Py, DMAP, CH_2Cl_2, −20°, 75 min (97%) 3. DMSO, CH_3CO_2H, $(CH_3CO)_2O$, 18°, 40 h (72%) 4. KOH, CH_3OH, 50°, 18 h (73%) 5. KH, THF, 0°, 20 min (—)	CH_3SCH_2O (1-cyclooctenyl) (—)	730

Reactant	Conditions	Product(s) and Yield(s) (%)	Refs.
(1st row: structure continued from previous page)	1. OsO_4, NMMO, t-C_4H_9OH, H_2O, $(CH_3)_2CO$, 50°, 10 d (74%) 2. $(CH_3CO)_2O$, Py, DMAP, CH_2Cl_2, −20°, 75 min (97%) 3. DMSO, CH_3CO_2H, $(CH_3CO)_2O$, 18°, 40 h (72%) 4. Raney Ni, C_6H_6, 1 h (94%) 5. KOH, CH_3OH, 45°, 6 h (79%) 6. KH, THF, 0°, 30 min (—)	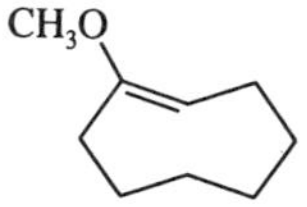 1-methoxycyclononene (CH_3O) (—)	730
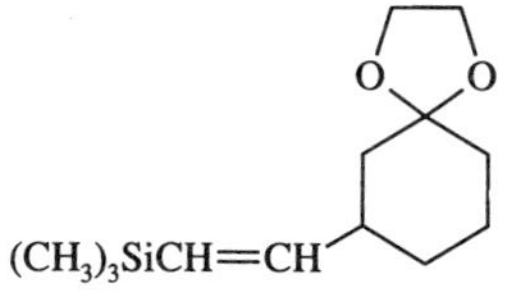 $(CH_3)_3SiCH{=}CH$–(cyclohexane ethylene ketal, 3-substituted)	1. MCPBA, CH_2Cl_2, 18 h (73%) 2. $BF_3{\cdot}O(C_2H_5)_2$, CH_3OH, H_2O, 2 h	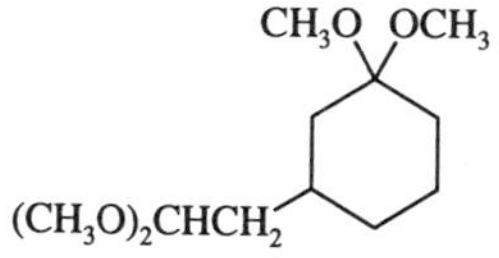 $(CH_3O)_2CHCH_2$–(1,1-dimethoxycyclohexane, 3-substituted; CH_3O OCH_3) (53)[c]	734
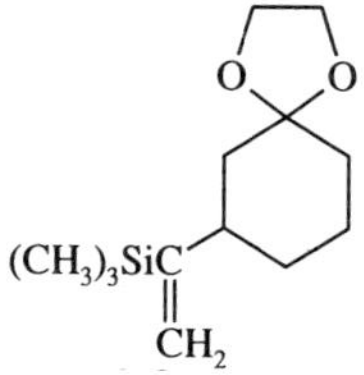 $(CH_3)_3SiC({=}CH_2)$–(cyclohexane ethylene ketal, 3-substituted)	1. MCPBA, CH_2Cl_2, 18 h (79%) 2. $HClO_4$, CH_3OH, H_2O, 60°, 0.5 h	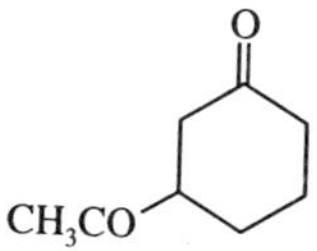 3-CH_3CO-cyclohexanone (41)[c]	734
	1. MCPBA, CH_2Cl_2, 18 h (79%) 2. $BF_3{\cdot}O(C_2H_5)_2$, CH_3OH, H_2O, 2 h	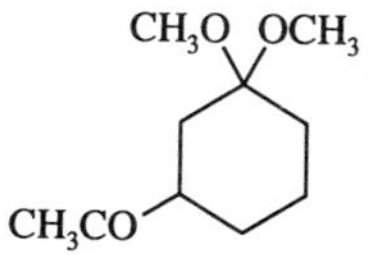 CH_3CO–(1,1-dimethoxycyclohexane, 3-substituted; CH_3O OCH_3) (77)[c]	734
(E)-$(CH_3)_3SiCH{=}CHC_6H_5$	1. Epoxidation[a] 2. 200°, 150 min	(E)-$(CH_3)_3SiOCH{=}CHC_6H_5$ (60)[b]	456
	1. Epoxidation[a] 2. 250°, 134 min	$(CH_3)_3SiOCH{=}CHC_6H_5$ $E{:}Z$ 55:45 (100)[b]	456
(Z)-$(CH_3)_3SiCH{=}CHC_6H_5$	1. Epoxidation[a] 2. 160°, 20 h	(E)-$(CH_3)_3SiOCH{=}CHC_6H_5$ (50)[b]	456

TABLE XXXVI. VINYLSILANES WITH OXYGEN ELECTROPHILES (*Continued*)

Reactant	Conditions	Product(s) and Yield(s)	Refs.
	1. Epoxidation[a] 2. 320°, 1 h	$(CH_3)_3SiOCH{=}CHC_6H_5$ $E{:}Z$ 85:15 (100)[b]	456
(E)-$(CH_3)_3SiCD{=}CHC_6H_5$	1. Epoxidation[a] 2. 170°, 22 h	(E)-$(CH_3)_3SiOCD{=}CHC_6H_5$ (100)[b]	456
	1. Epoxidation[a] 2. 310°, 30 min	$(CH_3)_3SiOCD{=}CHC_6H_5$ $E{:}Z$ 85:15 (100)[b]	456
$(CH_3)_3SiC(C_6H_5){=}CH_2$	1. Epoxidation[a] 2. 250°, 15 min	$(CH_3)_3SiOCH{=}CHC_6H_5$ $E{:}Z$ 55:45 (100)[b]	456
C_9			
$(CH_3)_3SiC(CH_3){=}CHC_6H_{11}$	1. MCPBA, CH_2Cl_2, 0°, 15 min 2. H_2SO_4, H_2O, CH_3OH, 1 h	$CH_3COCH_2C_6H_{11}$ (65)	732
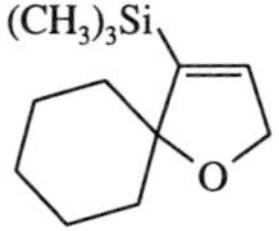 $(CH_3)_3Si$ (spiro, O)	1. MCPBA or CH_3CO_3H (82–90%) 2. H_2SO_4, CH_3OH, 50°	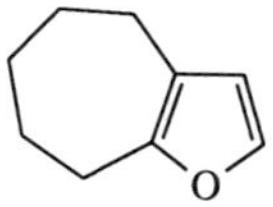 (O) (92)[c]	460
$(CH_3)_3SiCH{=}CHCHOHC_6H_5$	1. MCPBA, CH_2Cl_2, 12 h (—) 2. H_2SO_4, CH_3OH, 90°, 10 min	$C_6H_5CH{=}CHCHO$ (70)[b]	445
$(C_6H_5)_3SiC(CH_2Cl){=}CHC_6H_5$	1. MCPBA, $CHCl_3$, reflux, 5 d (75%) 2. KF, CH_3CN,  (X)	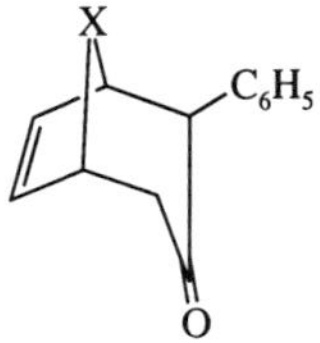 (X, C_6H_5, O) X = CH_2 (66) X = $NHCO_2CH_3$ (49) X = O (—)	735 735 735
C_{10}			
$(CH_3)_3SiC(C_4H_9\text{-}n){=}CHC_4H_9\text{-}n$	1. MCPBA, CH_2Cl_2, 0°, 15 min; room temp, 6 h	n-$C_4H_9COC_5H_{11}$-n (40)	732

Reactant	Conditions	Product(s) and Yield(s) (%)	Refs.
	2. H_2SO_4, CH_3OH, H_2O, 1 h		
$(CH_3)_3SiC(C_4H_9\text{-}s){=}CHC_4H_9\text{-}n$	1. MCPBA, CH_2Cl_2, 0°, 15 min; room temp, 6 h 2. H_2SO_4, CH_3OH, H_2O, 1 h	$s\text{-}C_4H_9COC_5H_{11}\text{-}n$ (69)	732
$(CH_3)_3SiC(=CHCH_2CH{=}CH_2)(CH_2)_2C_3H_7\text{-}i$	1. MCPBA, CH_2Cl_2, 5 h 2. TFA, CH_3OH, H_2O, reflux, 3 h	$i\text{-}C_3H_7(CH_2)_2CO(CH_2)_2CH{=}CH_2$ (75)	736
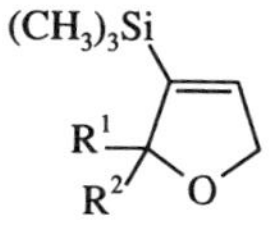 3-(trimethylsilyl)-2,5-dihydrofuran, 2-R^1,2-R^2 R^1 / R^2: H / $n\text{-}C_6H_{13}$ $n\text{-}C_3H_7$ / $n\text{-}C_3H_7$ —$(CH_2)_6$—	1. MCPBA or CH_3CO_3H (82–90%) 2. H_2SO_4, CH_3OH, 50°	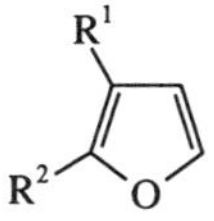 3-R^1-2-R^2-furan R^1 / R^2: H / $n\text{-}C_6H_{13}$ (20)[c] + $n\text{-}C_6H_{13}$ / H (60)[c] $n\text{-}C_3H_7$ / $n\text{-}C_3H_7$ (95)[c] —$(CH_2)_6$— (95)[c]	460
$(CH_3)_3SiC(=CHC_5H_{11}\text{-}n)(CH_2)_2CO_2H$ *E* *Z*	1. MCPBA, CH_2Cl_2, 0–25°, 2 h[f] 2. CH_3COCl, Py, THF, 8 h 3. TBAF, HMPA, THF, 25°, 0.5 h	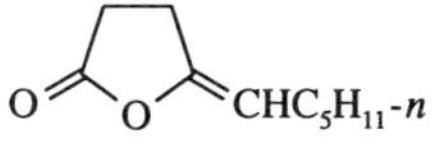 5-(n-hexylidene)-dihydrofuran-2-one ($O{=}C$ lactone $=CHC_5H_{11}\text{-}n$) *E*:*Z* 1:99 (78) *E*:*Z* 99:1 (56)	737
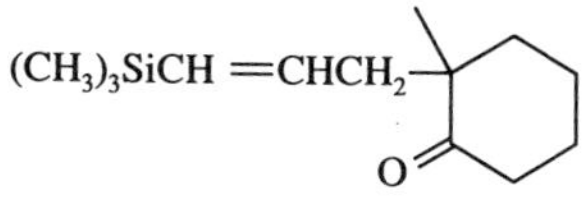 2-methyl-2-[$(CH_3)_3SiCH{=}CHCH_2$]cyclohexanone	MCPBA, CH_2Cl_2, 17 h	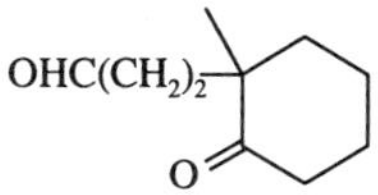 2-methyl-2-[$OHC(CH_2)_2$]cyclohexanone (60)	449
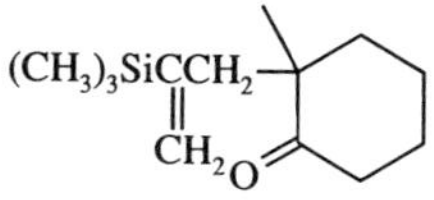 2-methyl-2-[$(CH_3)_3SiC(=CH_2)CH_2$]cyclohexanone	1. MCPBA (100%)[g] 2. H_2SO_4, CH_3OH, reflux	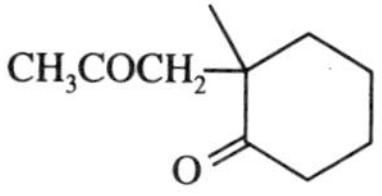 2-methyl-2-[CH_3COCH_2]cyclohexanone (100)	449

TABLE XXXVI. VINYLSILANES WITH OXYGEN ELECTROPHILES (*Continued*)

Reactant	Conditions	Product(s) and Yield(s)	Refs.
C_{11}			
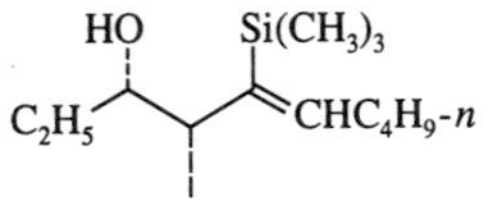 [structure: C_2H_5, HO, $Si(CH_3)_3$, $=CHC_4H_9$-*n*]	1. $VO(acac)_2$, *t*-C_4H_9OH 2. CH_3COCl, $(C_2H_5)_3N$ 3. H_2SO_4, CH_3OH, 90°	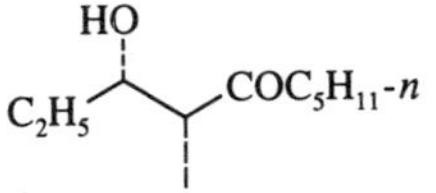 [structure: C_2H_5, HO, COC_5H_{11}-*n*] (90)	210
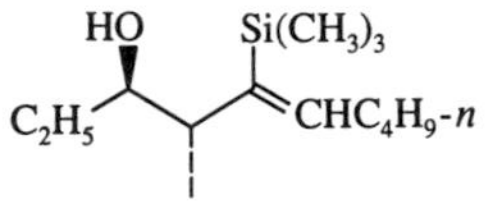 [structure: C_2H_5, HO, $Si(CH_3)_3$, $=CHC_4H_9$-*n*]	1. $VO(acac)_2$, *t*-C_4H_9OH 2. CH_3COCl, $(C_2H_5)_3N$ 3. H_2SO_4, CH_3OH, 90°	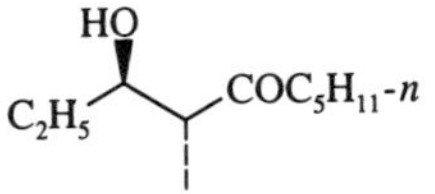 [structure: C_2H_5, HO, COC_5H_{11}-*n*] (81)	210
$(CH_3)_3SiC(C_6H_{13}\text{-}n)=CHCH_2CH=CH_2$	1. MCPBA, CH_2Cl_2, 5 h 2. TFA, CH_3OH, H_2O, reflux, 3 h	*n*-$C_6H_{13}CO(CH_2)_2CH=CH_2$ (75)	736
$(CH_3)_3SiC((CH_2)_2COCH_3)=CHC_5H_{11}$-*n*	1. MCPBA, CH_2Cl_2 2. H_2SO_4, CH_3OH, H_2O, reflux, 13 h	$CH_3CO(CH_2)_2COC_6H_{13}$-*n* (~100)	461
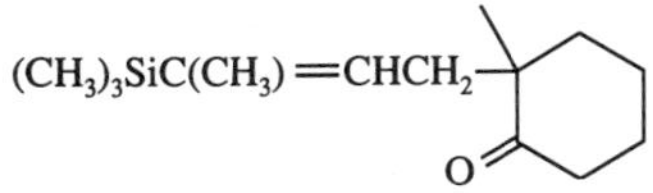 $(CH_3)_3SiC(CH_3)=CHCH_2$– [2-methylcyclohexanone structure]	MCPBA, CH_2Cl_2, 4 h	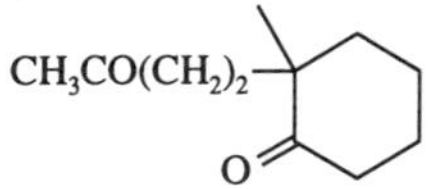 $CH_3CO(CH_2)_2$– [2-methylcyclohexanone structure] (90)	449
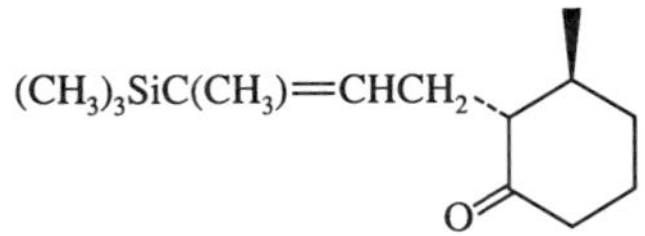 $(CH_3)_3SiC(CH_3)=CHCH_2$– [3-methylcyclohexanone structure]	1. MCPBA, CH_2Cl_2 2. HCO_2H	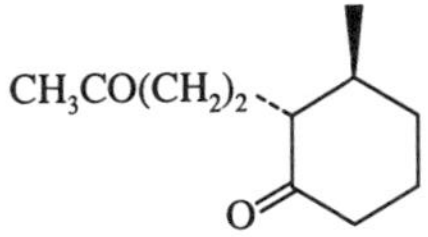 $CH_3CO(CH_2)_2$– [3-methylcyclohexanone structure] (83)	449
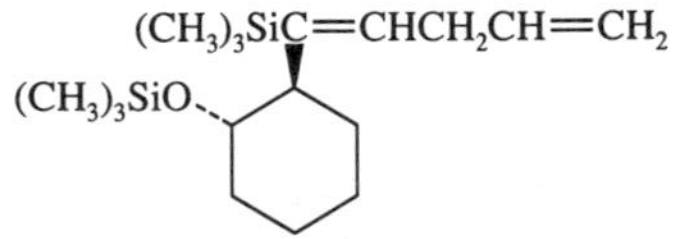 $(CH_3)_3SiC(=CHCH_2CH=CH_2)$– [2-$(CH_3)_3SiO$-cyclohexyl structure]	1. MCPBA, CH_2Cl_2, 3 h 2. TFA, CH_3OH, H_2O, reflux, 6.5 h	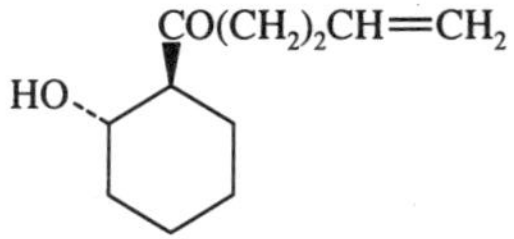 $CO(CH_2)_2CH=CH_2$– [2-HO-cyclohexyl structure] (48)	736

Reactant	Conditions	Product(s) and Yield(s) (%)	Refs.
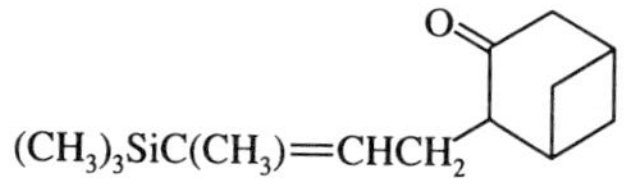 $(CH_3)_3SiC(CH_3){=}CHCH_2$– (bicyclic ketone)	1. MCPBA, CH_2Cl_2, 0°, 3 h 2. HCO_2H, CH_2Cl_2, 25°, 30 min	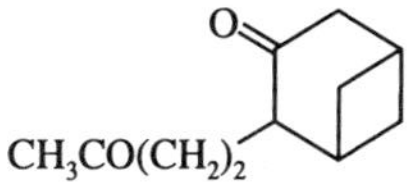 $CH_3CO(CH_2)_2$– (bicyclic ketone) (87)	738
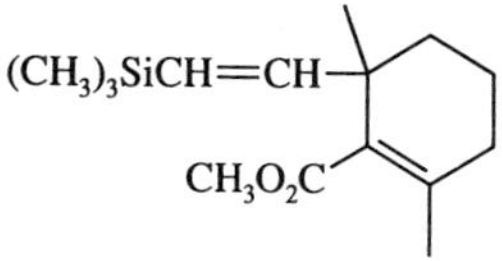 $(CH_3)_3SiCH{=}CH$– (cyclohexene, CH_3O_2C)	1. MCPBA, $CDCl_3$ 2. $BF_3{\cdot}O(C_2H_5)_2$, CH_3OH, 25°, 2 h	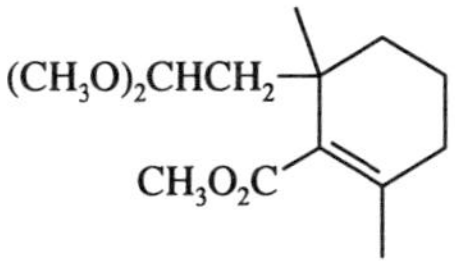 $(CH_3O)_2CHCH_2$– (cyclohexene, CH_3O_2C) (20)	739
C_{12}			
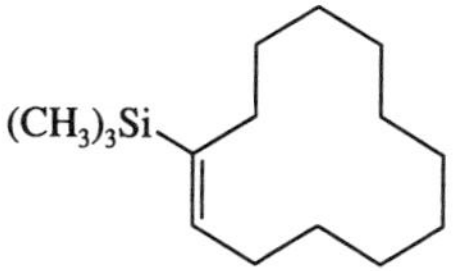 $(CH_3)_3Si$– (cyclododecene)	1. MCPBA (90%) 2. H_2SO_4, H_2O, dioxane, 7 d	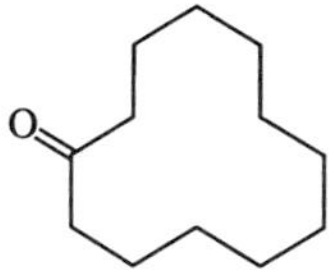 cyclododecanone (80)[c]	458
	1. MCPBA (90%) 2. H_2SO_4, H_2O, dioxane, 100°, 5 h	" (79)[c]	458
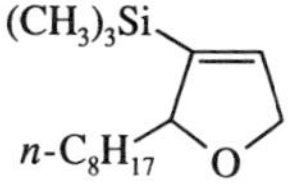 $(CH_3)_3Si$–, n-C_8H_{17} (dihydrofuran)	1. MCPBA or CH_3CO_3H (82–90%) 2. H_2SO_4, CH_3OH, 50°	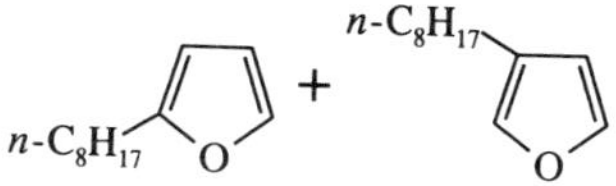 2-(n-C_8H_{17})furan + 3-(n-C_8H_{17})furan (49) 13:1	460
$(CH_3)_3SiC(CH_2C_6H_5){=}CHCH_2CH{=}CH_2$	1. MCPBA, CH_2Cl_2, 5 h 2. TFA, CH_3OH, H_2O, reflux, 3 h	$C_6H_5CH_2CO(CH_2)_2CH{=}CH_2$ (62)	736
C_{14}			
(Z)-$(CH_3)_3SiC(C_6H_{13}$-$n){=}CHC_6H_{13}$-n	1. MCPBA, CH_2Cl_2, 0°, 12 h (68%) 2. HI, $(C_2H_5)_2O$, 0° 3. CH_3Li, $(C_2H_5)_2O$, 1 h 4. CH_3Li, THF, 1 h 5. CH_3CO_2H, NaO_2CCH_3, −20°, 0.5 h; room temp, 12 h	n-$C_6H_{13}C(CH_3){=}CHC_6H_{13}$-$n$ (74)[c] E:Z 9:91	740

TABLE XXXVI. VINYLSILANES WITH OXYGEN ELECTROPHILES (*Continued*)

Reactant	Conditions	Product(s) and Yield(s)	Refs.
	1. MCPBA, CH_2Cl_2, 0°, 12 h (68%) 2. HI, $(C_2H_5)_2O$, 0° 3. CH_3Li, $(C_2H_5)_2O$, 1 h 4. CH_3Li, THF, 1 h 5. KOC_4H_9-*t*, THF, 1 h	*n*-$C_6H_{13}C(CH_3)$=CHC_6H_{13}-*n* (79)[c] *E*:*Z* 91:9	740
$(CH_3)_3SiCH$=$CHCH_2$– (bicyclic lactone)	1. MCPBA, CH_2Cl_2, 0°, 30 min; room temp, 35 h 2. H_2SO_4, CH_3OH, reflux, 24 h	$R(CH_2)_2$– (bicyclic lactone) (74) R = CHO : R = $CH(OCH_3)_2$ 45:55	462
C_{15}			
$(CH_3)_3SiC(=CH_2)(CH_2)_2$– (bicyclic ketone)	1. MCPBA, $NaHCO_3$, CH_2Cl_2, 0°, 2 h; room temp, 20 h (90%)[f] 2. H_2SO_4, CH_3OH, H_2O, 12 h	$CH_3CO(CH_2)_2$– (bicyclic ketone) (97)	462
$(CH_3)_3SiC(=CHCH_3)$– (bicyclic ketone)	1. MCPBA, CH_2Cl_2, $NaHCO_3$, 16 h 2. H_2SO_4, CH_3OH, H_2O, reflux, 16 h	$C_2H_5COCH_2$– (bicyclic ketone) (65)	463

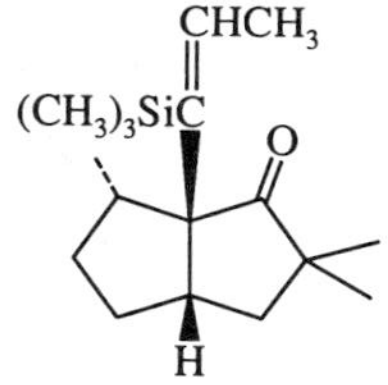 $(CH_3)_3SiC(=CHCH_3)$	1. MCPBA, CH_2Cl_2, $NaHCO_3$, 15 h 2. H_2SO_4, CH_3OH, H_2O, reflux, 2 h	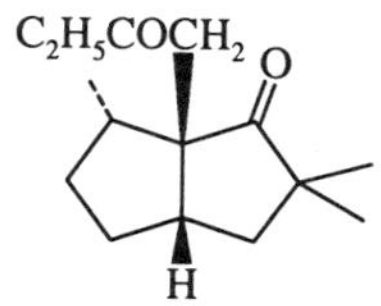 $C_2H_5COCH_2$ (93)	463
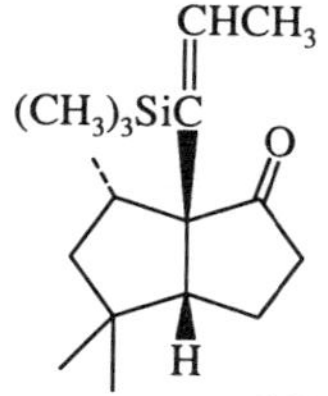 $(CH_3)_3SiC(=CHCH_3)$	1. MCPBA, CH_2Cl_2, $NaHCO_3$, 30 h 2. H_2SO_4, CH_3OH, H_2O, reflux, 15 h	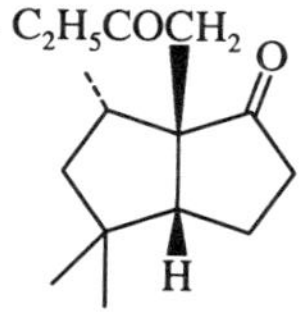 $C_2H_5COCH_2$ (32)	463
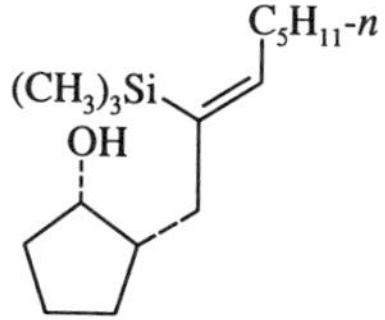 $(CH_3)_3Si$, OH, C_5H_{11}-n	1. MCPBA, CH_2Cl_2, 0–25° 2. KH, 1–2 h	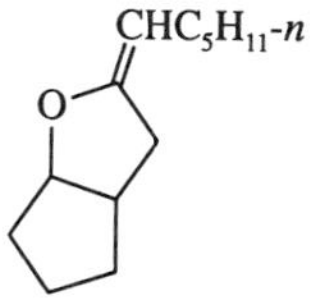 CHC_5H_{11}-n *E*:*Z* 3:97 (75)	737
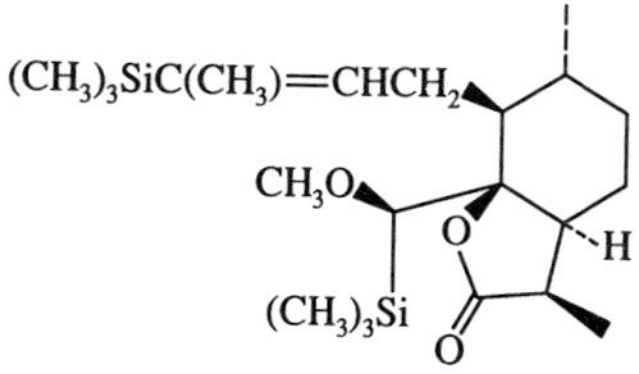 $(CH_3)_3SiC(CH_3){=}CHCH_2$, CH_3O, $(CH_3)_3Si$	1. MCPBA, CH_2Cl_2 2. TFA, CH_2Cl_2, 0°, 3 min	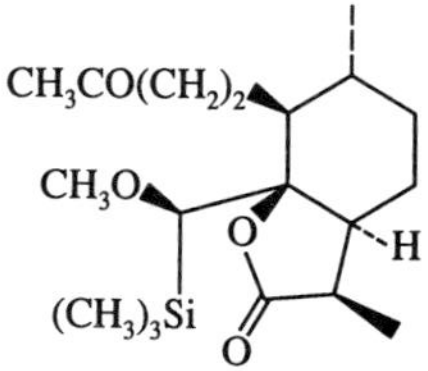 $CH_3CO(CH_2)_2$, CH_3O, $(CH_3)_3Si$ (72)	464

TABLE XXXVI. VINYLSILANES WITH OXYGEN ELECTROPHILES (*Continued*)

Reactant	Conditions	Product(s) and Yield(s)	Refs.
C_{18}			
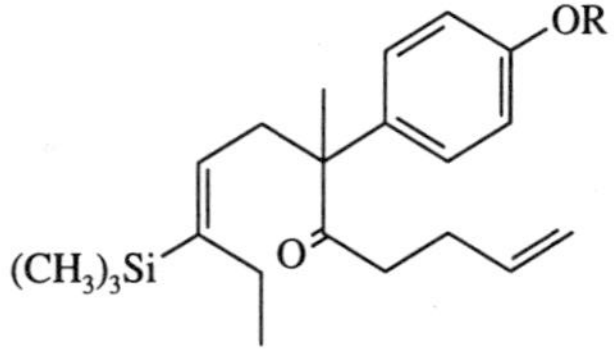 $(CH_3)_3Si$ … OR (structure)	1. MCPBA, CH_2Cl_2, Na_2CO_3, 2 h 2. $BF_3{\cdot}O(C_2H_5)_2$, $(C_2H_5)_2O$, 0°, 0.5 h	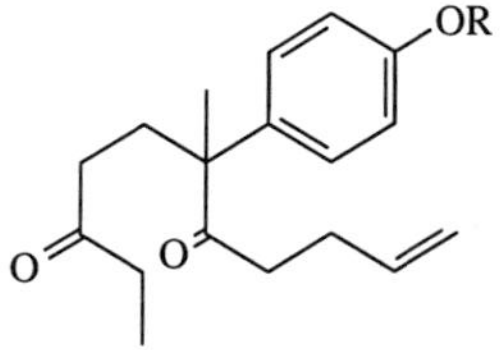 OR (structure)	
R = CH_3		(90)	
R = $CH_2C_6H_5$		(71)	
C_{20}			
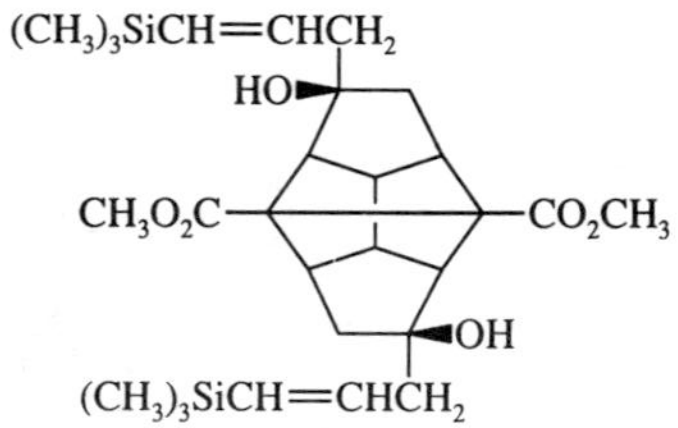 $(CH_3)_3SiCH{=}CHCH_2$, HO, CH_3O_2C, CO_2CH_3, OH, $(CH_3)_3SiCH{=}CHCH_2$ (structure)	CH_3CO_3H, H_2SO_4, CH_3CO_2H, 4 h	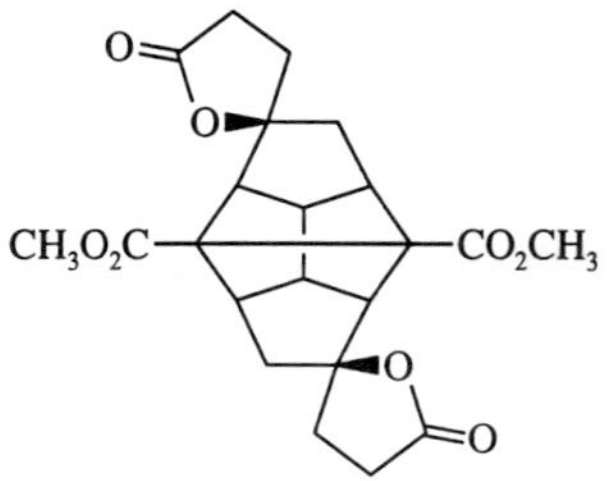 CH_3O_2C, CO_2CH_3 (structure) (24)	742

Reactant	Conditions	Product(s) and Yield(s) (%)	Refs.
C_{25}			
$(CH_3)_3SiC(CH_3){=}CHCH_2$ (steroid: O, H, H, H, H, OH)	1. MCPBA, CH_2Cl_2, 4 h 2. HCO_2H	$CH_3CO(CH_2)_2$ (steroid: O, H, H, H, H, OH) (71)	449

[a] Details of the epoxidation are not given; for general methods, see refs. 435 and 743.

[b] This yield is based on the epoxide, since the yield of the epoxide from the vinylsilane is not given.

[c] This yield is based on the epoxide, the yield of which is given in the table.

[d] Vinylsilane epoxide (30%) is recovered in this reaction.

[e] This yield is based on the 1-silylareneoxide-oxepin used, the yield of which is not given.

[f] The product of this step is the γ lactone derived from the epoxide.

[g] The product of this step is the internal ketal of the diol corresponding to the epoxide.

TABLE XXXVII. ALLYLSILANES WITH PHOSPHORUS ELECTROPHILES

Reactant	Conditions	Product(s) and Yield(s)	Refs.
C_4			
$(CH_3)_3SiCH_2C(CH_3)=CH_2$	1. PCl_5, −20° 2. SO_2	$CH_2=C(CH_3)CH_2POCl_2$ I + $(CH_3)_2C=CHPOCl_2$ II + $(CH_3)_2CClCH_2POCl_2$ III (—)	469
$(C_2H_5)_3SiCH_2C(CH_3)=CH_2$	1. PCl_5, −20° 2. SO_2	I + II + III (—)	469

TABLE XXXVIII. ALLYLSILANES WITH SULFUR ELECTROPHILES

Reactant	Conditions	Product(s) and Yield(s)	Refs.
C_3			
$(CH_3)_3SiCH_2CH{=}CH_2$	CH_3SO_2Cl, CH_2Cl_2, CH_3CN, 115°, 10 h	$CH_2{=}CHCH_2SO_2CH_3$ (35)	471
	$C_6H_5SO_2Cl$, n-C_3H_7CN, CuCl, reflux, 15 h	$CH_2{=}CHCH_2SO_2C_6H_5$ (50)	471
$C_6H_5(CH_3)_2SiCH_2CH{=}CH_2$	$(CH_3)_3SiSO_3Cl$, 0°, 2 h	$CH_2{=}CHCH_2SO_3Si(CH_3)_2C_6H_5$ (85)	8
$C_6H_5(1\text{-}C_{10}H_7)(C_2H_5)SiCH_2CH{=}CH_2$	", C_6H_{12}	$CH_2{=}CHCH_2SO_3Si(C_2H_5)(1\text{-}C_{10}H_7)C_6H_5$ (77)	8
2-R-2-($SiCH_2CH{=}CH_2$)-tetrahydro-2-silanaphthalene	", "	2-R-2-($SiO_3SCH_2CH{=}CH_2$)-tetrahydro-2-silanaphthalene	8
R = C_6H_5		(80)	
R = 1-$C_{10}H_7$		(62)	
$(CH_3)_3SiCH_2CH{=}CHSi(CH_3)_3$	", 0°, 0.25 h	$CH_2{=}CHCH[Si(CH_3)_3]SO_3Si(CH_3)_3$ (70)	8
$(CH_3)_3SiCH[SO_3Si(CH_3)_3]CH{=}CH_2$	", CH_2Cl_2, reflux, a few h	$(CH_3)_3SiO_3SCH_2CH{=}CHSO_3Si(CH_3)_3$ (50)	8
C_4			
$(CH_3)_3SiCH_2CH{=}CHCH_3$	1. C_6H_5SCl, CH_2Cl_2, −78° 2. SiO_2	$CH_2{=}CHCH(CH_3)SC_6H_5$ (94)	472
1,1-dimethyl-1-silacyclopent-3-ene ($Si(CH_3)_2$ ring)	$(CH_3)_3SiSO_3Cl$, CCl_4, 20°, 4 h	3-vinyl-cyclic sulfonate ($CH_2{=}CH$, SO_2–O–$Si(CH_3)_2$ ring) (90)	135
C_5			
$(CH_3)_3Si$-cyclopentadiene	R^1R^2SO	$R^1R^2S^+$–cyclopentadienide$^-$	705

TABLE XXXVIII. ALLYLSILANES WITH SULFUR ELECTROPHILES (*Continued*)

Reactant	Conditions	Product(s) and Yield(s)	Refs.
	R^1 R^2		
	CH_3 CH_3 3 h	(35)	
	$-(CH_2)_4-$	(59)	
	$-(CH_2)_5-$	(30)	
	$-(CH_2)_2O(CH_2)_2-$	(43)	
	R^1R^2SO, $(CF_3CO)_2O$	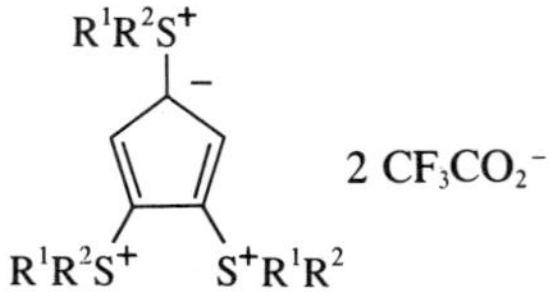 1,2,4-tris($R^1R^2S^+$)cyclopentadienide $2\ CF_3CO_2^-$	705
	R^1 R^2		
	CH_3 CH_3	(45)	
	C_2H_5 C_2H_5	(30)	
	$-(CH_2)_4-$	(55)	
	$-(CH_2)_5-$	(48)	
	$-(CH_2)_2O(CH_2)_2-$	(49)	
	1. $(p\text{-}CH_3C_6H_4)_2SO$, $(CF_3CO)_2O$, CH_2Cl_2, $-10°$, 1 h 2. $LiClO_4$	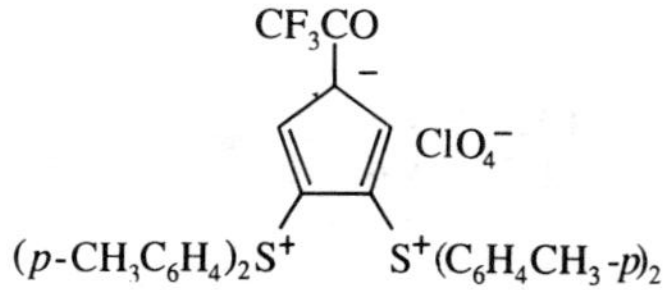 CF_3CO-cyclopentadienide with $(p\text{-}CH_3C_6H_4)_2S^+$ and $S^+(C_6H_4CH_3\text{-}p)_2$ ClO_4^- (46)	705
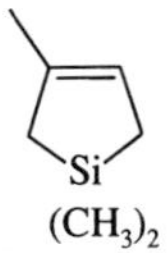 3-methyl-1,1-dimethyl-1-silacyclopent-3-ene [$Si(CH_3)_2$]	$(CH_3)_3SiSO_3Cl$, CCl_4, 20°, 4 h	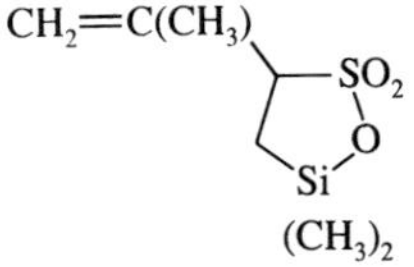 $CH_2{=}C(CH_3)$-substituted cyclic sulfonate (SO_2, O, $Si(CH_3)_2$) (57) +	135

Reactant	Conditions	Product(s) and Yield(s) (%)	Refs.
		3-methyl-1,1-dimethyl-4-($SO_3Si(CH_3)_3$)-1-silacyclopent-2-ene (31)	
C_7			
$[i\text{-}C_3H_7CH{=}CHC(OLi)(CH_3)Si(CH_3)_3]$[a]	CH_3SSCH_3, THF	$i\text{-}C_3H_7CH(SCH_3)CH{=}C(OSi(CH_3)_3)CH_3$ + $i\text{-}C_3H_7CH{=}CHC(OSi(CH_3)_3)(CH_3)SCH_3$ 55:35 (85)	260
$(CH_3)_3Si$-substituted 7,7-dichlorobicyclo[3.2.0]hept-2-en-6-one	1. C_6H_5SCl, CH_2Cl_2, 0°, 20 h 2. NaF, CH_3OH, H_2O, 22 h	C_6H_5S-substituted 7,7-dichlorobicyclo[3.2.0]hept-2-en-6-one (80)	55
$(CH_3)_3Si$-substituted dichloro bicyclic lactone	1. C_6H_5SCl, CH_2Cl_2, 0°, 18 h 2. NaF, CH_3OH, H_2O, 20 h	C_6H_5S-substituted dichloro bicyclic lactone (65)	55
C_8			
$(CH_3)_3SiCH_2CH{=}C_6H_{10}$ (cyclohexylidene)	1. C_6H_5SCl, CH_2Cl_2, −78° 2. SiO_2	1-(C_6H_5S)-1-($CH_2{=}CH$)cyclohexane + $C_6H_5SCH_2CH{=}C_6H_{10}$ (cyclohexylidene) 78:22 (86)	472

TABLE XXXVIII. ALLYLSILANES WITH SULFUR ELECTROPHILES (*Continued*)

Reactant	Conditions	Product(s) and Yield(s)	Refs.
[structure: cyclohexene with 4,5-bis(CO_2CH_3) and 3-$(CH_3)_3Si$]	$C_6H_5SBF_4$, CH_3NO_2	[structure: cyclohexene with C_6H_5S and 4,5-bis(CO_2CH_3)] (78)	134
C_9			
$(CH_3)_3SiCH_2CH{=}CHC_6H_5$	1. C_6H_5SCl, CH_2Cl_2, −78° 2. SiO_2	$CH_2{=}CHCH(C_6H_5)SC_6H_5$ + $C_6H_5SCH_2CH{=}CHC_6H_5$ 72:28 (83)	472
C_{11}			
$(CH_3)_3SiCH_2CH{=}CH(CH_2)_2C_6H_5$	1. C_6H_5SCl, CH_2Cl_2, −78° 2. SiO_2	$CH_2{=}CHCH(SC_6H_5)(CH_2)_2C_6H_5$ (91)	472
$[C_6H_5(CH_2)_2C(OLi)[Si(CH_3)_3]CH{=}CH_2]$[b]	CH_3SSCH_3, THF	$C_6H_5(CH_2)_2C(OSi(CH_3)_3){=}CHCH_2SCH_3$ (89)	260
$(CH_3)_3SiCH_2C(CH_3){=}C{=}C(C_6H_5)Si(CH_3)_3$	$(CH_3)_3SiSO_3Cl$, CH_2Cl_2, 1 h	$CH_2{=}C(CH_3)C(SO_3Si(CH_3)_3){=}C(C_6H_5)Si(CH_3)_3$ (65)	223

[a] This intermediate is prepared by addition of methyllithium to 5-methyl-1-trimethylsilyl-2-pentenone.
[b] This intermediate is prepared by addition of vinyllithium to 3-phenyl-1-trimethylsilylpropanone.

TABLE XXXIX. VINYLSILANES WITH SULFUR ELECTROPHILES

Reactant	Conditions	Product(s) and Yield(s)	Refs.
C_2			
$(CH_3)_3SiCH{=}CH_2$	1. ArSCl, CH_2Cl_2, −20° 2. KF, THF, 12 h	$ArSCH{=}CH_2$	421
	Ar = C_6H_5	(87)	
	Ar = o-$O_2NC_6H_4$	(93)	
	Ar = 2,4-$(O_2N)_2C_6H_3$	(95)	
	1. ArSCl, CH_2Cl_2, −20° 2. KF, DMSO, 70°, 25 h	"	421
	Ar = p-ClC_6H_4	(86)	
	Ar = p-$CH_3C_6H_4$	(85)	
	$(CH_3)_3SiSO_3Cl$, 0.5 h	$(CH_3)_3SiO_3SCH{=}CH_2$ (88)	9
$(C_2H_5)_3SiCH{=}CH_2$	" "	$(C_2H_5)_3SiO_3SCH{=}CH_2$ (50)	9
$C_6H_5(CH_3)_2SiCH{=}CH_2$	", 2 h	$C_6H_5(CH_3)_2SiO_3SCH{=}CH_2$ (15) + $(CH_3)_2Si(O_3SC_6H_5)CH{=}CH_2$ (65)	8
$(C_6H_5)_3SiCH{=}CH_2$	", cyclohexane	$(C_6H_5)_2Si(O_3SC_6H_5)CH{=}CH_2$ (60)	8
(E)-$(CH_3)_3SiCH{=}CHSi(CH_3)_3$	", "	(E)-$(CH_3)_3SiO_3SCH{=}CHSi(CH_3)_3$ (—)	8
R $SiCH{=}CH_2$	$(CH_3)_3SiSO_3Cl$, cyclohexane	SiXY X / Y	8
R = C_6H_5		C_6H_5 / $O_3SCH{=}CH_2$ (6) + $O_3SC_6H_5$ / $CH{=}CH_2$ (74)	
R = 1-$C_{10}H_7$		$O_3SC_{10}H_7$-1 / " (76)	
C_4			
$(CH_3)_3SiCH{=}C(CH_3)_2$	$(CH_3)_3SiSO_3Cl$, 0.5 h	$(CH_3)_3SiO_3SCH{=}(CH_3)_2$ (81)	9

TABLE XXXIX. VINYLSILANES WITH SULFUR ELECTROPHILES (*Continued*)

Reactant	Conditions	Product(s) and Yield(s)	Refs.
C_8			
(*E*)-$(CH_3)_3SiCH{=}CHC_6H_5$	SO_3, CCl_4, reflux, 0.5 h	(*E*)-$(CH_3)_3SiO_3SCH{=}CHC_6H_5$ (60)	744
(*E*)- + (*Z*)-$(CH_3)_3SiCH{=}CHC_6H_5$	CH_3SO_2Cl, CuCl, CH_3CN, 130°, 4 h	(*E*)-$CH_3SO_2CH{=}CHC_6H_5$ (30)	471
C_{12}			
1-$(CH_3)_3Si$-acenaphthylene	$(CH_3)_3SiSO_3Cl$, CCl_4, reflux, 2 h	1-$(CH_3)_3SiO_3S$-acenaphthylene (85)	718
1,2-bis[$(CH_3)_3Si$]-acenaphthylene	$(CH_3)_3SiSO_3Cl$ (1 eq)	1-$(CH_3)_3SiO_3S$-2-R-acenaphthylene, R = $Si(CH_3)_3$ (—)	718
"	", (2 eq)	", R = $SO_3Si(CH_3)_3$ (—)	718
1-$(CH_3)_3Si$-5-$Si(CH_3)_3$-acenaphthylene	", "	1-$(CH_3)_3SiO_3S$-5-$Si(CH_3)_3$-acenaphthylene (87)	718
C_n			
$[-CH_2C(Si(CH_3)_3){=}CHCH_2-]_n$	C_6H_5SCl, CH_2Cl_2	$[-CH_2C(SC_6H_5){=}CHCH_2-]_n$ (—)	470

TABLE XL. ALLENYLSILANES WITH SULFUR ELECTROPHILES

Reactant	Conditions	Product(s) and Yield(s)	Refs.
C_3			
$(CH_3)_3SiCH{=}C{=}CH_2$	SO_3, dioxane, CH_2Cl_2, 1 h	$HC{\equiv}CCH_2SO_3Si(CH_3)_3$ (80)	223
	$(CH_3)_3SiSO_3Cl$, CH_2Cl_2, 1 h	" (80)	223
C_5			
$[(CH_3)_3Si]_2C{=}C{=}C(CH_3)CH_2Si(CH_3)_3$	$(CH_3)_3SiSO_3Cl$, 1 h	$(CH_3)_3SiC(SO_3Si(CH_3)_3){=}C{=}C(CH_3)CH_2Si(CH_3)_3$ (20) + $(CH_3)_3SiC{\equiv}CC(SO_3Si(CH_3)_3)(CH_3)CH_2Si(CH_3)_3$ (60)	223
C_{11}			
$(CH_3)_3SiC(C_6H_5){=}C{=}C(CH_3)CH_2Si(CH_3)_3$	see Table XXXVIII		

TABLE XLI. ALLYLSILANES WITH SELENIUM ELECTROPHILES

Reactant	Conditions	Product(s) and Yield(s)	Refs.
C_3			
$(CH_3)_3SiCH_2CH{=}CH_2$	1. C_6H_5SeCl, CH_2Cl_2, $-78°$, 10 min 2. $SnCl_2$ or Florisil	$CH_2{=}CHCH_2SeC_6H_5$ (90)	472
$(CH_3)_3SiCH_2CD{=}CD_2$	1. C_6H_5SeCl, CH_2Cl_2, $-78°$, 10 min 2. $SnCl_2$ or Florisil	$CD_2{=}CDCH_2SeC_6H_5$ + $CH_2{=}CDCD_2SeC_6H_5$ (82) 46:54	472
C_4			
$(CH_3)_3SiCH_2CH{=}CHCH_3$	1. C_6H_5SeCl, CH_2Cl_2, $-78°$, 10 min 2. $SnCl_2$ or Florisil	$C_6H_5SeCH_2CH{=}CHCH_3$ (91)	472
$(CH_3)_3SiCH_2C(CH_3){=}CH_2$	1. C_6H_5SeCl, CH_2Cl_2, $-78°$, 10 min 2. $SnCl_2$ or Florisil	$C_6H_5SeCH_2C(CH_3){=}CH_2$ (87)	472
C_6			
$(CH_3)_3SiCH_2CH{=}CHCH_2CH(O_2CCH_3)CH_3$	1. C_6H_5SeCl, CH_2Cl_2, $-78°$, 10 min 2. $SnCl_2$ or Florisil	$C_6H_5SeCH_2CH{=}CHCH_2CH(O_2CCH_3)CH_3$ (91)	473
$(CH_3)_3SiCH(CH_2CH(O_2CCH_3)CH_3)CH{=}CH_2$	1. C_6H_5SeCl, CH_2Cl_2, $-78°$, 10 min 2. $SnCl_2$ or Florisil	" (80)	472
C_7			
$(CH_3)_3SiCH(CH_2CH(O_2CCH_3)CH_3)CH{=}CHCH_3$	1. C_6H_5SeCl, CH_2Cl_2, $-78°$, 10 min 2. $SnCl_2$ or Florisil	$C_6H_5SeCH(CH_3)CH{=}CHCH_2CH(O_2CCH_3)CH_3$ + $C_6H_5SeCH(CH_2CH(O_2CCH_3)CH_3)CH{=}CHCH_3$ 55:45 (65)	472
$(CH_3)_3Si$, CH_3CO_2 (cycloheptene structure)	$(C_6H_5SeO)_2O$, $BF_3\cdot O(C_2H_5)_2$, CH_2Cl_2	HO, CH_3CO_2 (cycloheptene structure) (—)	405

Reactant	Conditions	Product(s) and Yield(s) (%)	Refs.
$(CH_3)_3Si$, H, O, =O, H, Cl, Cl	1. C_6H_5SeCl, CH_2Cl_2 2. Florisil 3. H_2O_2, Py, CH_2Cl_2, 0°, 20 min	H, O, =O + X = OH (36); X = Cl (39); X, H, Cl, Cl	472
C_8			
$(CH_3)_3SiCH_2CH=$ (cyclohexylidene)	1. C_6H_5SeCl, CH_2Cl_2, −78°, 10 min 2. $SnCl_2$ or Florisil	$C_6H_5SeCH_2CH=$ (cyclohexylidene) (80)	472
$(CH_3)_3SiCH_2CH{=}CH$, CH_3CO_2 (cyclopentyl)	1. C_6H_5SeCl, CH_2Cl_2, −78°, 10 min 2. $SnCl_2$ or Florisil	$C_6H_5SeCH_2CH{=}CH$, CH_3CO_2 (cyclopentyl) (81)	472
C_9			
$(CH_3)_3SiCH_2CH{=}CH$, CH_3CO_2 (cyclohexyl)	SeO_2, THF, reflux, 5 h	$RCOCH{=}CH$, CH_3CO_2 (cyclohexyl) R = H (16) + R = $Si(CH_3)_3$ (34)	474
	1. C_6H_5SeCl, CH_2Cl_2, −78°, 10 min 2. $SnCl_2$	$C_6H_5SeCH_2CH{=}CH$, CH_3CO_2 (cyclohexyl) (78)	473
CH_2, $(CH_3)_3SiCH_2C$, HO (cyclohexyl)	1. C_6H_5SeCl, $(C_2H_5)_3N$, CH_2Cl_2, −78°, 15 min 2. H_2O_2, 0°, 30 min	CH_2, $HOCH_2C$, HO (cyclohexyl) (53)	726
$(CH_3)_3SiCH_2CH{=}CHC_6H_5$	1. C_6H_5SeCl, CH_2Cl_2, −78°, 10 min 2. $SnCl_2$ or Florisil	$C_6H_5SeCH_2CH{=}CHC_6H_5$ (89)	472

TABLE XLI. ALLYLSILANES WITH SELENIUM ELECTROPHILES (*Continued*)

Reactant	Conditions	Product(s) and Yield(s)	Refs.
C_{11}			
$(CH_3)_3SiCH_2CH{=}CHCH_2CH(O_2CCH_3)C_6H_{13}$-*n*	1. C_6H_5SeCl, CH_2Cl_2, −78°, 10 min 2. $SnCl_2$	$C_6H_5SeCH_2CH{=}CHCH_2CH(O_2CCH_3)C_6H_{13}$-*n* (88)	473
$(CH_3)_3SiCH_2C({=}CH_2)CH_2CHOHC_6H_{13}$-*n*	1. C_6H_5SeCl, $(C_2H_5)_3N$, CH_2Cl_2, −78°, 15 min 2. H_2O_2, 0°, 30 min	$HOCH_2C({=}CH_2)CH_2CHOHC_6H_{13}$-*n* (69)	726
$(CH_3)_3SiCH_2C({=}CH_2)CH_2COH(C_2H_5)C_4H_9$-*n*	1. C_6H_5SeCl, $(C_2H_5)_3N$, CH_2Cl_2, −78°, 15 min 2. H_2O_2, 0°, 30 min	$HOCH_2C({=}CH_2)CH_2COH(C_2H_5)C_4H_9$-*n* (64)	726
$(CH_3)_3SiCH_2CH{=}CH(CH_2)_2C_6H_5$	SeO_2, THF, reflux, 3 h	$RCOCH{=}CH(CH_2)_2C_6H_5$ R = H (27) + R = $(CH_3)_3Si$ (13)	474
	1. C_6H_5SeCl, CH_2Cl_2, −78°, 10 min 2. $SnCl_2$ or Florisil	$C_6H_5SeCH_2CH{=}CH(CH_2)_2C_6H_5$ (97)	472
$(CH_3)_3SiCH_2C({=}CH_2)(CH_2)_2C_6H_5$	1. C_6H_5SeCl, CH_2Cl_2, −78°, 10 min 2. $SnCl_2$ or Florisil	$C_6H_5SeCH_2C({=}CH_2)(CH_2)_2C_6H_5$ (87)	472
$(CH_3)_3SiCH_2CH{=}CHCH(C_6H_5)CH_2O_2CCH_3$	1. C_6H_5SeCl, CH_2Cl_2, −78°, 10 min 2. $SnCl_2$	$C_6H_5SeCH_2CH{=}CHCH(C_6H_5)CH_2O_2CCH_3$ (86)	473

TABLE XLII. ALLYLSILANES WITH HALOGEN ELECTROPHILES

Reactant	Conditions	Product(s) and Yield(s)	Refs.
C_3			
$(CH_3)_3SiCH_2CH{=}CH_2$	1. Cl_2, $-70°$ 2. $90°$	$CH_2{=}CHCH_2Cl$ (87)	5
	Br_2, $(C_2H_5)_2O$, $-70°$	$CH_2{=}CHCH_2Br$ (10) + $BrCH_2CHBrCH_2Br$ (7)	5
	I_2, 0°, 6 h	$CH_2{=}CHCH_2I$ (100)[a]	745, 748
	C_6H_5IO, $BF_3{\cdot}O(C_2H_5)_2$, C_2H_5OH, 7 h	$CH_2{=}CHCH_2OC_2H_5$ (97)	746
	", ", C_6H_6, $-20°$, 1 h	$CH_2{=}CHCH_2C_6H_5$ (73)	746
	", ", $C_6H_5OCH_3$, ", "	$CH_2{=}CHCH_2C_6H_4OCH_3$ *o*:*p* 1:3.4 (71)	746
	", ", *p*-$C_6H_4(OCH_3)_2$, $-78°$, 2 h; $-30°$, 1 h	$CH_2{=}CHCH_2C_6H_3(OCH_3)_2$-2,5 (44)	746
	", ", *p*-$C_6H_4(CH_3)_2$, $-20°$, 1 h	$CH_2{=}CHCH_2C_6H_3(CH_3)_2$ (42)	746
C_4			
$[(CH_3)_3Si]_2CHCH{=}CHCH[Si(CH_3)_3]_2$	1. ICl, 0°, 3 h; 25°, 2 h 2. $Na_2S_2O_3$, H_2O	$[(CH_3)_3Si]_2CHCH_2CH{=}CH_2$ (75)	747
C_5			
$(CH_3)_3SiCH_2C({=}CH_2)(CH_2)_2O_2CCH_3$	C_6H_5IO, $BF_3{\cdot}O(C_2H_5)_2$, dioxane, 12 h	$CH_2{=}C(CHO)(CH_2)_2O_2CCH_3$ (63)	480
	", ", ROH	$CH_2{=}C(CH_2OR)(CH_2)_2O_2CCH_3$	746
	R: CH_3, 0.5 h	(87)	
	R: C_2H_5, 1 h	(80)	
	R: $CH_3O(CH_2)_2$, 0°, 2 h	(84)	
	R: $C_6H_5CH_2$, 0.5 h	(72)	

TABLE XLII. ALLYLSILANES WITH HALOGEN ELECTROPHILES (*Continued*)

Reactant	Conditions	Product(s) and Yield(s)	Refs.
C_6			
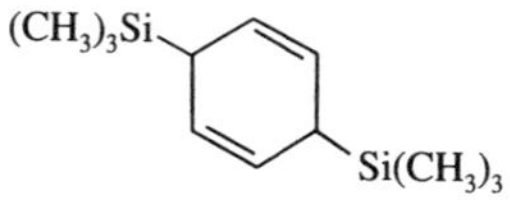 $(CH_3)_3Si$ … $Si(CH_3)_3$	I_2, aprotic solvents, 0–25°	C_6H_6 + $2(CH_3)_3SiI$ (100)[a]	478
C_7			
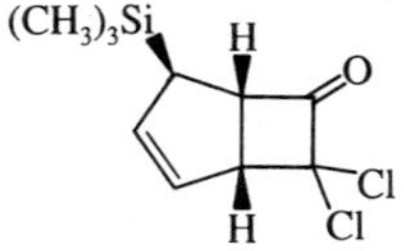 $(CH_3)_3Si$, H, O, Cl, Cl, H	Br_2, hexane, −70°, 1 h	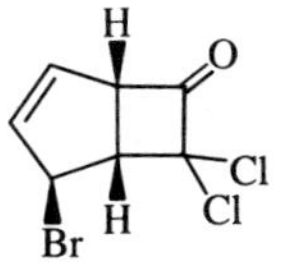 H, O, Cl, Cl, Br, H (63)	55
C_9			
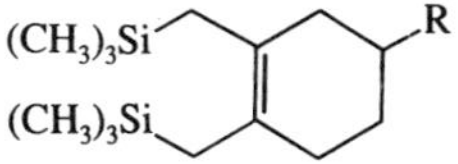 $(CH_3)_3Si$, $(CH_3)_3Si$, R		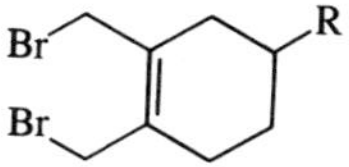 Br, Br, R	
R = $CH_2OSi(CH_3)_2C_4H_9$-*t*	1. Br_2, propylene oxide, CuBr, THF, CCl_4, −78°, 15 min; 0°, 15 min 2. Br_2, −78°, 15 min; 0°, 5 min	(86)	476
R = CO_2CH_3	1. Br_2, propylene oxide, CuBr, THF, CCl_4, −78°, 15 min; 0°, 15 min 2. Br_2, −78°, 15 min; 0°, 15 min	(35)	476
	1. NBS, propylene oxide, THF, −100°, 10 min; −78°, 1 h 2. CH_2=$CHCOCH_3$	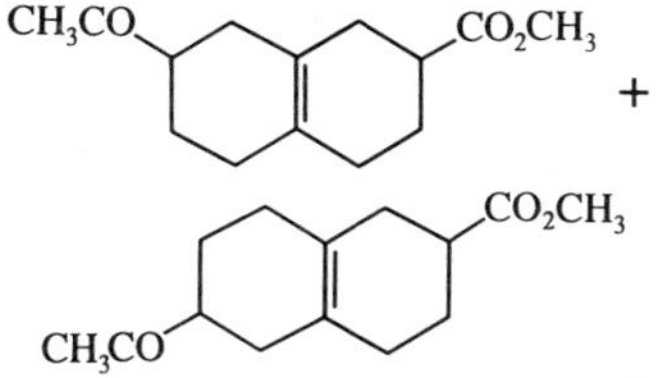 CH_3CO … CO_2CH_3 + CH_3CO … CO_2CH_3 (66)	477

Reactant	Conditions	Product(s) and Yield(s) (%)	Refs.
$(CH_3)_3SiCH_2C(=C_6H_{10})CON(C_2H_5)_2$ 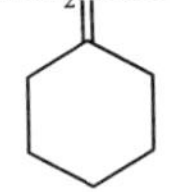 (cyclohexylidene)	1. Br_2, CH_2Cl_2, 10 min 2. C_6H_5SeH	$C_6H_5SeCH_2C(=C_6H_{10})CON(C_2H_5)_2$ 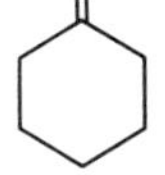 (cyclohexylidene) (89)	179
$(CH_3)_3SiCH_2CH{=}CHC_6H_5$	C_6H_5IO, $BF_3{\cdot}O(C_2H_5)_2$, ROH R = CH_3 0.75 h R = C_2H_5 3.5 h	$CH_2{=}CHCH(C_6H_5)OR$ + $ROCH_2CH{=}CHC_6H_5$ 64:36 (93) 63:37 (67)	746
1-$(CH_3)_3Si$-indene	Br_2, dioxane, THF	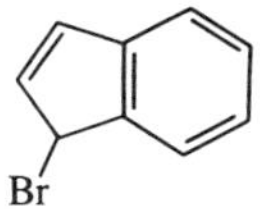 1-Br-indene (66)	748
C_{10}			
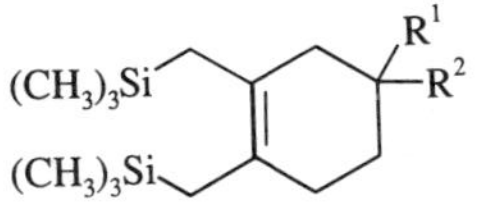 1,2-bis[$(CH_3)_3SiCH_2$]-4-R^1-4-R^2-cyclohexene	1. Br_2, propylene oxide, CuBr, THF, CCl_4, $-78°$, 15 min; 0°, 15 min 2. Br_2, $-78°$, 15 min; 0°, 15 min	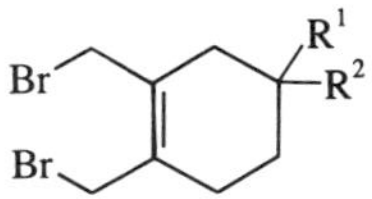 1,2-bis($BrCH_2$)-4-R^1-4-R^2-cyclohexene	
R^1 = CH_3, R^2 = $CH_2OSi(CH_3)_2C_4H_9$-t		(66)	476
R^1 = ", R^2 = CO_2CH_3		(66)	476
R^1 = H, R^2 = $COCH_3$		(41)	476
R^1 = CO_2CH_3, R^2 = CO_2CH_3		(83)	476
" "	1. NBS, propylene oxide, THF, $-100°$, 10 min; $-78°$, 1 h 2. $CH_2{=}CHCOCH_3$	2,2-bis(CO_2CH_3)-6-CH_3CO-octahydronaphthalene + 2,2-bis(CO_2CH_3)-7-CH_3CO-octahydronaphthalene (83)	477

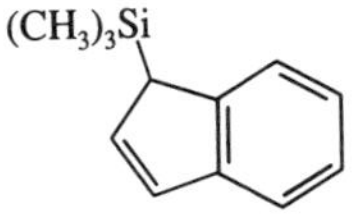

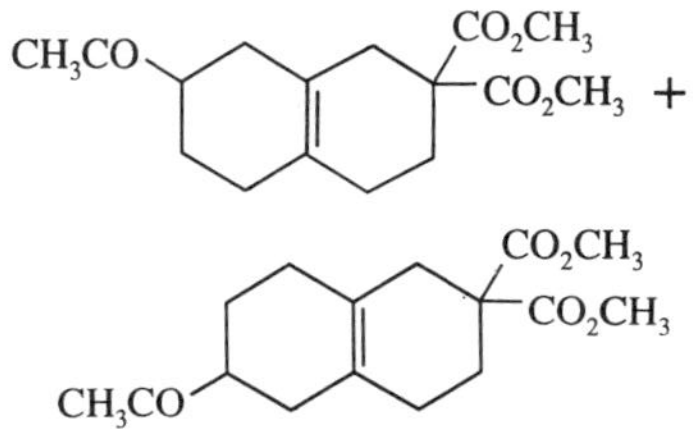

TABLE XLII. Allylsilanes with Halogen Electrophiles (*Continued*)

Reactant		Conditions	Product(s) and Yield(s)	Refs.
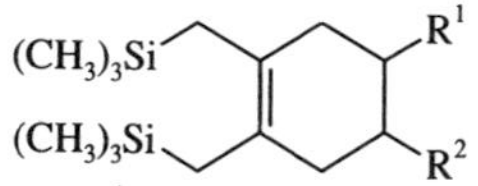 $(CH_3)_3Si$, $(CH_3)_3Si$ (cyclohexene with R^1, R^2)		1. Br_2, propylene oxide, CuBr, THF, CCl_4, −78°, 15 min; 0°, 15 min 2. Br_2, −78°, 15 min; 0°, 15 min	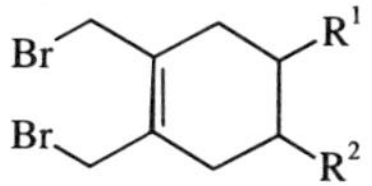 Br, Br (cyclohexene with R^1, R^2)	
R^1	R^2			
$C_6H_5CH_2OCH_2$	$C_6H_5CH_2OCH_2$		(82)	476
t-$C_4H_9(CH_3)_2SiOCH_2$	t-$C_4H_9(CH_3)_2SiOCH_2$		(80)	476
—$CON(C_6H_5)CO$—			(81)	476
CH_3CO	H	1. NBS, propylene oxide, THF, −100°, 10 min; −78°, 1 h 2. *N*-Phenylmaleimide	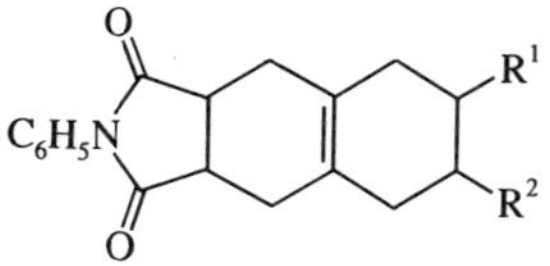 C_6H_5N (O, O; R^1, R^2) (60)	477
CO_2CH_3	CO_2CH_3	1. NBS, propylene oxide, THF, −100°, 10 min; −78°, 1 h 2. *N*-Phenylmaleimide	" (45)	477
—$CON(C_6H_5)CO$—		1. NBS, propylene oxide, THF, −100°, 10 min; −78°, 1 h 2. CH_2=$CHCOCH_3$	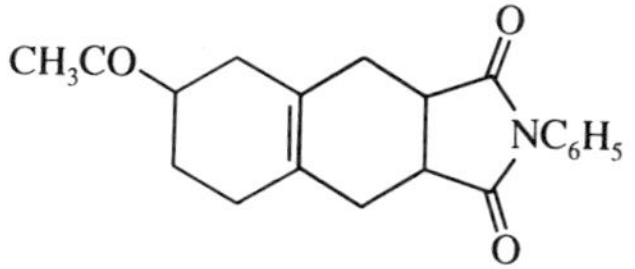 CH_3CO (O, O) NC_6H_5 (59)	477
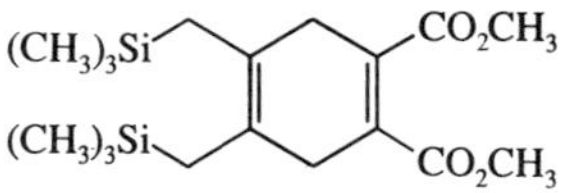 $(CH_3)_3Si$, $(CH_3)_3Si$ (cyclohexadiene with CO_2CH_3, CO_2CH_3)		1. Br_2, propylene oxide, CuBr, THF, CCl_4, −78°, 15 min; 0°, 15 min 2. Br_2, −78°, 15 min; 0°, 15 min	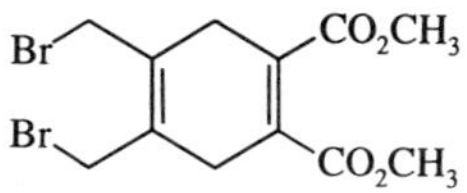 Br, Br (CO_2CH_3, CO_2CH_3) (73)	476

Substrate	Conditions	Product(s) and Yield(s) (%)	Refs.
	1. NBS, propylene oxide, THF, $-100°$, 10 min; $-78°$, 1 h 2. *N*-Phenylmaleimide	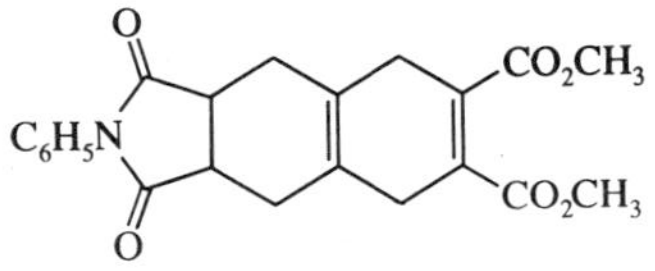 [structure: C_6H_5N imide with CO_2CH_3, CO_2CH_3] (43)	477
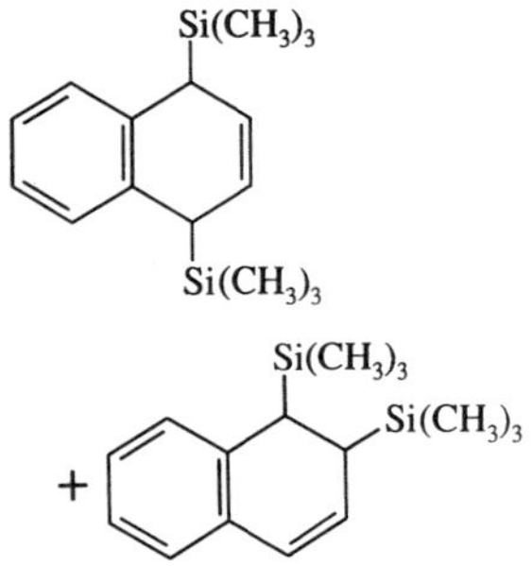 [structure: 1,4-bis($Si(CH_3)_3$)-1,4-dihydronaphthalene] + [structure: 1,2-bis($Si(CH_3)_3$)-1,2-dihydronaphthalene]	Br_2, 15–25°	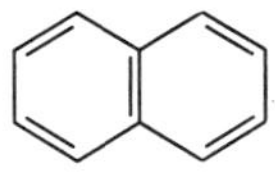 [structure: naphthalene] I (—) + $(CH_3)_3SiBr$ (94)	749, 750
	I_2, 10–27°	I (—) + $(CH_3)_3SiI$ (92)	749, 750
C_{12}			
$(CH_3)_3SiCH_2C(=CH_2)(CH_2)_2CHOHC_6H_{13}$-*n*	C_6H_5IO, $BF_3{\cdot}O(C_2H_5)_2$, dioxane, 3 h	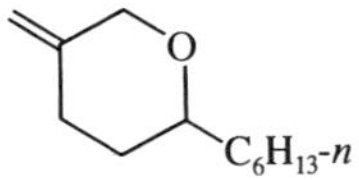 [structure: methylenetetrahydropyran with C_6H_{13}-*n*] (65)	751
$(CH_3)_3SiCH_2C(=CH_2)(CH_2)_2CH(O_2CCH_3)C_6H_{13}$-*n*	", ", ", 12 h	$CH_2{=}C(CHO)(CH_2)_2CH(O_2CCH_3)C_6H_{13}$-*n* (72)	480
[structure: $(CH_3)_3Si$-cyclohexene-C_6H_5]	Br_2, hexane	$C_6H_5C_6H_5$ + $(CH_3)_3SiC_6H_4C_6H_5$-*m* 72:28 (—)	408
	NBS, CCl_4, *hν*, reflux, 1 h	58:42 (—)	
C_{13}			
$(CH_3)_3SiCH_2C(=CH_2)(CH_2)_nCHOHR$	C_6H_5IO, $BF_3{\cdot}O(C_2H_5)_2$	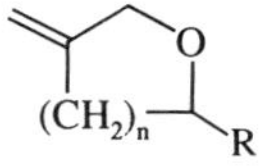 [structure: methylene cyclic ether with $(CH_2)_n$ and R]	751

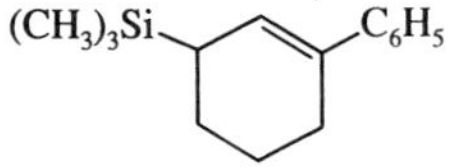

TABLE XLII. ALLYLSILANES WITH HALOGEN ELECTROPHILES (*Continued*)

Reactant	Conditions	Product(s) and Yield(s)	Refs.
n R			
2 n-C_7H_{15}	CH_2Cl_2, −20°, 1 h	(24)	
" "	$(C_2H_5)_2O$, −20°, 2 h; 0°, 2 h	(58)	
" "	THF, −70°, 4 h; −20°, 2 h	(53)	
" "	Dioxane, 3 h	(63)	
1 $(CH_2)_2C_6H_5$	THF, 0°, 5 h	(19)	
" "	DME, 0°, 1.5 h	(52)	
" "	Dioxane, 2 h	(68)	
$(CH_3)_3SiCH_2C(=CH_2)(CH_2)_nCH(O_2CCH_3)R$	C_6H_5IO, $BF_3 \cdot O(C_2H_5)_2$, dioxane, 12 h	$CH_2=C(CHO)(CH_2)_nCH(O_2CCH_3)R$	480
n R			
2 n-C_7H_{15}		(71)	
1 $(CH_2)_2C_6H_5$		(65)	
$(CH_3)_3SiCHDC(=CH_2)CH_2CHOH(CH_2)_2C_6H_5$	C_6H_5IO, $BF_3 \cdot O(C_2H_5)_2$, dioxane, 5 h	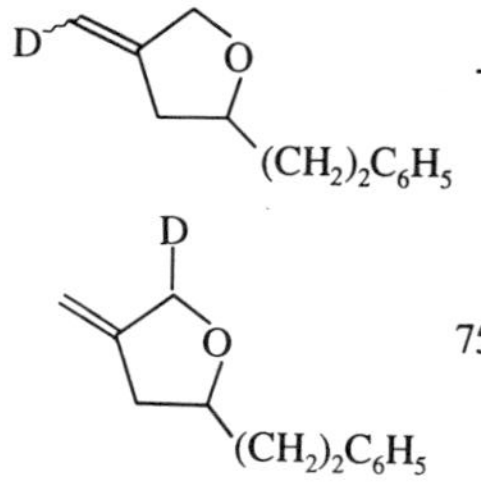 D, O, $(CH_2)_2C_6H_5$ + D, O, $(CH_2)_2C_6H_5$ 75:25 (52)	751

$(CH_3)_3SiCH_2C(=CH_2)CH_2CH(CO_2H)CH_2C_6H_5$	C_6H_5IO, $BF_3{\cdot}O(C_2H_5)_2$, dioxane, 48 h	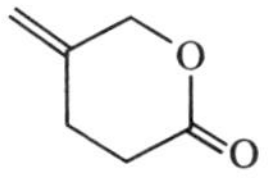 O, O (20) + $CH_2C_6H_5$ $(CH_3)_3Si(CH_2)_2COCH_2CH(CO_2H)CH_2C_6H_5$ (35)	751
C_{15}			
$(CH_3)_3SiCH_2C(=CH_2)CH_2CHOH(CH_2)_8CH{=}CH_2$	C_6H_5IO, $BF_3{\cdot}O(C_2H_5)_2$, dioxane, 4 h	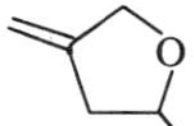 O, $(CH_2)_8CH{=}CH_2$ (40)	751
$(CH_3)_3SiCH_2C(=CH_2)CH_2CH(O_2CCH_3)(CH_2)_8CH{=}CH_2$	C_6H_5IO, $BF_3{\cdot}O(C_2H_5)_2$, dioxane, 12 h	$CH_2{=}C(CHO)CH_2CH(O_2CCH_3)(CH_2)_8CH{=}CH_2$ (63)	480
C_{17}			
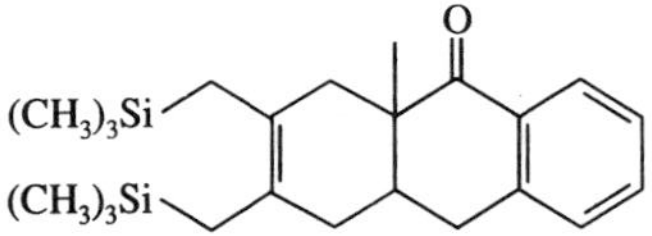 $(CH_3)_3Si$, $(CH_3)_3Si$, O	1. NBS, propylene oxide, THF, −100°, 10 min; −78°, 1 h 2. *N*-Phenylmaleimide	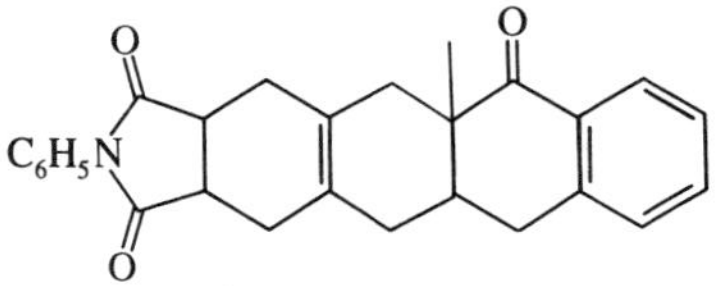 O, C_6H_5N, O, O (41)	477

[a] This product is mixed with hexamethyldisiloxane, which codistils with it.

TABLE XLIII. VINYLSILANES WITH HALOGEN ELECTROPHILES

Reactant	Conditions	Product(s) and Yield(s)	Refs.
C_2			
$(CH_3)_3SiCH{=}CH_2$	1. Cl_2, −78° 2. KOH, CH_3OH, H_2O, 10 min	$ClCH{=}CH_2$ (35)	6
	1. Br_2, −78° 2. KOH, CH_3OH, H_2O, 10 min	$BrCH{=}CH_2$ (61)	6
	1. Br_2, −78°, 1 h 2. $(C_2H_5)_2NH$, reflux, 12 h	$(CH_3)_3SiCBr{=}CH_2$ (65–68)	752
$(C_2H_5)_3SiCH{=}CH_2$	1. Br_2, gentle heat, 1 h 2. $(C_2H_5)_2NH$, reflux, 5 h	$(C_2H_5)_3SiCBr{=}CH_2$ (31)	482
(*E*)-$(CH_3)_3SiCH{=}CHSi(CH_3)_3$	1. Cl_2, CH_2Cl_2, −80° 2. KF, DMSO, 60°, 24 h	(*Z*)-$ClCH{=}CHSi(CH_3)_3$ (51)	416
	1. Br_2, −80°, 1 h 2. KF, DMSO, 1 h	(*Z*)-$BrCH{=}CHSi(CH_3)_3$ (64)	416
	I_2, CCl_4, 90°, 48 h	$ICH{=}CHSi(CH_3)_3$ *E*:*Z* 1:1 (80) + $ICH{=}CHI$ (8)	416
$[(CH_3)_3Si]_2C{=}CH_2$	1. Br_2, CCl_4, −10° 2. $NaHCO_3$, CH_3OH, reflux, 2 h	$(CH_3)_3SiCBr{=}CH_2$ (57)	753
$(CH_3)_3SiC(SC_6H_5){=}CH_2$	1. Br_2, CCl_4, 0° 2. $NaOCH_3$, CH_3OH	$BrC(SC_6H_5){=}CH_2$ (89)	118
(*E*)-$(CH_3)_3SiCCl{=}CClSi(CH_3)_3$	1. Cl_2, $ZnCl_2$, $(C_2H_5)_2O$ 2. distil	$Cl_2C{=}CClSi(CH_3)_3$ (9)	754
C_3			
(*E*)-$(CH_3)_3SiCH{=}CHCH_3$	1. Br_2, 20° 2. C_2H_5OH	$BrCH{=}CHCH_3$ *E*:*Z* 0.3:99.7[a] (—)	14
	1. Br_2, 20° 2. HCO_2H	", *E*:*Z* 15:85[a] (—)	14
(*E*)-$(CH_3)_3SiCH{=}CHCH_2OC_6H_4Br$-*p*	C_6H_5IO, $BF_3 \cdot O(C_2H_5)_2$, CH_2Cl_2, 0°, 7 h	(*E*)-$C_6H_5\overset{+}{I}CH{=}CHCH_2OC_6H_4Br$-*p* $^{-}BF_4$ (72)	755

Reactant	Conditions	Product(s) and Yield(s) (%)	Refs.
C_5			
$(CH_3)_3SiC(=CH_2)CH_2CHOHCH_3$	1. Br_2, CH_2Cl_2, −78° 2. $NaOCH_3$, CH_3OH	$BrC(=CH_2)CH_2CHOHCH_3$ (68)	756
C_6			
(E)-$(CH_3)_3SiCH{=}CHC_4H_9$-n	1. Cl_2, CH_2Cl_2, −78° 2. $KF{\cdot}2H_2O$, DMSO, 9 h	$ClCH{=}CHC_4H_9$-n E:Z 14:86 (63)	57
	1. Cl_2, CH_2Cl_2, −78° 2. $NaOCH_3$, CH_3OH	", E:Z 1:99 (63)	757
	1. Cl_2, CH_2Cl_2, −78° 2. Al_2O_3, C_5H_{12}, 5 h	", E:Z 1:99 (59)	757
	1. Br_2, CH_2Cl_2, −78° 2. $KF{\cdot}2H_2O$, DMSO, 9 h	$BrCH{=}CHC_4H_9$-n E:Z 7:93 (69)	57
	1. Br_2, CH_2Cl_2, −78° 2. $NaOCH_3$, CH_3OH	", E:Z 2:98 (81)	757
	1. Br_2, CH_2Cl_2, −78° 2. Al_2O_3, C_5H_{12}, 2 h	", E:Z 1:99 (64)	757
	1. ICl, CCl_4, 0°, 30 min 2. $KF{\cdot}2H_2O$, DMSO, 4 h	$ClCH{=}CHC_4H_9$-n E:Z 8:92 (5) + $ICH{=}CHC_4H_9$-n E:Z 5:95 (66)	489
(Z)-$(CH_3)_3SiCH{=}CHC_4H_9$-n	1. Cl_2, CH_2Cl_2, −78° 2. $KF{\cdot}2H_2O$, DMSO, 9 h	$ClCH{=}CHC_4H_9$-n E:Z 98:2 (82)	57
	1. Cl_2, CH_2Cl_2, −78° 2. $NaOCH_3$, CH_3OH	", E:Z 98:2 (66)	757
	1. Cl_2, CH_2Cl_2, −78° 2. Al_2O_3, C_5H_{12}, 5 h	", E:Z 77:23 (63)	757
	1. Br_2, CH_2Cl_2, −23° 2. $NaOCH_3$, CH_3OH, 2 h	$BrCH{=}CHC_4H_9$-n, E:Z 95:5 (80)	757, 758
	1. Br_2, CH_2Cl_2, −23° 2. Al_2O_3, C_5H_{12}, 2 h	", E:Z 85:15 (75)	757
	1. ICl, CCl_4, 0°, 30 min 2. $KF{\cdot}2H_2O$, DMSO, 4 h	$ClCH{=}CHC_4H_9$-n E:Z 88:12 (8) + $ICH{=}CHC_4H_9$-n E:Z 77:23 (66)	489
	1. ICl 2. $NaOCH_3$, CH_3OH, 1 h	(E)-$ICH{=}CHC_4H_9$-n + (Z)-$(CH_3)_3SiCI{=}CHC_4H_9$-n 3:1 (—)	490

TABLE XLIII. VINYLSILANES WITH HALOGEN ELECTROPHILES (*Continued*)

Reactant	Conditions	Product(s) and Yield(s)	Refs.
	I_2, CH_2Cl_2	(*Z*)-ICH=CHC$_4$H$_9$-*n* (60)	491, 490
	1. I_2, AgO_2CCF_3, CH_2Cl_2 2. KF·2H$_2$O, DMSO	(*E*)-ICH=CHC$_4$H$_9$-*n* (38)	491
(*E*)-(CH$_3$)$_3$SiCH=CHC$_4$H$_9$-*t*	1. Cl_2, CCl_4, 0° 2. NaOCH$_3$, CH$_3$OH, 2 h	ClCH=CHC$_4$H$_9$-*t* *E*:*Z* 1:99 (82)	57
	1. Br_2, CCl_4, 0° 2. NaOCH$_3$, CH$_3$OH, 2 h	BrCH=CHC$_4$H$_9$-*t* *E*:*Z* 13:87 (31)	57
	1. ICl, CCl_4, 0°, 30 min 2. KF·2H$_2$O, DMSO, 4 h	ClCH=CHC$_4$H$_9$-*t* *E*:*Z* 9:91 (9) + ICH=CHC$_4$H$_9$-*t* *E*:*Z* 5:95 (42)	489
(*Z*)-(CH$_3$)$_3$SiCH=CHC$_4$H$_9$-*t*	1. Cl_2, CCl_4, −78° 2. KF·2H$_2$O, DMSO, 9 h	ClCH=CHC$_4$H$_9$-*t* *E*:*Z* 8:92 (91)	57
	1. Cl_2, CCl_4, −78° 2. NaOCH$_3$, CH$_3$OH, reflux	(CH$_3$)$_3$SiCCl=CHC$_4$H$_9$-*t* mostly *Z* (—)	57
	1. Br_2, CCl_4, −78° 2. KF·2H$_2$O, DMSO, 9 h	BrCH=CHC$_4$H$_9$-*t* *E*:*Z* 1:99 (70)	57
	1. Br_2, CCl_4, −78° 2. NaOCH$_3$, CH$_3$OH, reflux	(CH$_3$)$_3$SiCBr=CHC$_4$H$_9$-*t* mostly *Z* (—)	57
	1. ICl, CCl_4, 0°, 30 min 2. KF·2H$_2$O, DMSO, 4 h	ClCH=CHC$_4$H$_9$-*t* *E*:*Z* 25:75 (6) + ICH=CHC$_4$H$_9$-*t* *E*:*Z* 5:95 (63)	489
(CH$_3$)$_3$SiC(=CH$_2$)CH$_2$CHOHC$_2$H$_5$	1. Br_2, CH_2Cl_2, −78° 2. NaOCH$_3$, CH$_3$OH	BrC(=CH$_2$)CH$_2$CHOHC$_2$H$_5$ (—)	756
(CH$_3$)$_3$SiCX=CHR X R		BrClC=CHR *E*:*Z*	759
E Cl *n*-C$_4$H$_9$	1. Br_2, CH_2Cl_2, −78°, 30 min 2. NaOCH$_3$, CH$_3$OH, 0°, 30 min; 25°, 2 h	99:1 (80)	
" " *t*-C$_4$H$_9$	1. Br_2, CH_2Cl_2, −78°, 30 min 2. NaOCH$_3$, CH$_3$OH, 0°, 30 min; 25°, 2 h	57:43 (82)	
Z Br *n*-C$_4$H$_9$	1. Cl_2, $CHCl_3$, −60°, 30 min 2. NaOCH$_3$, CH$_3$OH, 0°, 30 min; 25°, 2 h	98:2 (83)	

" " t-C_4H_9	1. Cl_2, $CHCl_3$, −60°, 30 min 2. $NaOCH_3$, CH_3OH, 0°, 30 min; 25°, 2 h	73:27 (—)	
$(CH_3)_3SiCH{=}CH(CH_2)_2CH{=}CHSi(CH_3)_3$	Br_2, CH_2Cl_2, −8°, 4 h	$BrCH{=}CH(CH_2)_2CH{=}CHBr$ (50)	505
C_7			
$(CH_3)_3SiC(CH_3){=}CHC_4H_9$-$n$	1. Br_2, CH_2Cl_2, −78° 2. $NaOCH_3$, CH_3OH, 0°, 1 h; room temp, 2 h	$BrC(CH_3){=}CHC_4H_9$-n	481
		$E:Z$	
E		<1:99 (87)	
Z		99:1 (91)	
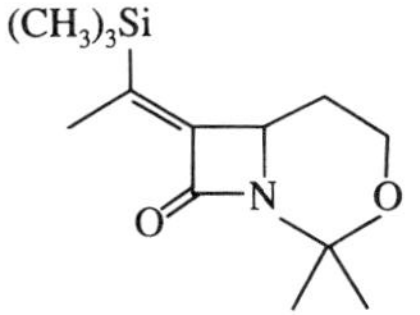 $(CH_3)_3Si$	1. NBS, DMF, H_2O (82%) 2. KF, CH_3CN	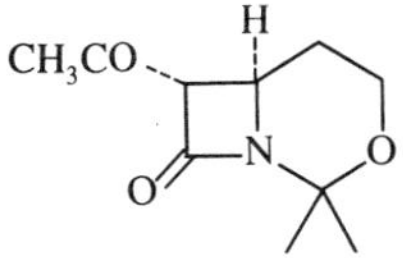 CH_3CO (−)	760
C_8			
$(CH_3)_3SiC(C_2H_5){=}CHC_4H_9$-$n$	1. Br_2, CH_2Cl_2, −78° 2. $NaOCH_3$, CH_3OH, 0°, 1 h; room temp, 2 h	$BrC(C_2H_5){=}CHC_4H_9$-n	481
		$E:Z$	
E		2:98 (84)	
Z		97:3 (93)	
(E)-$(CH_3)_3SiC(C_2H_5){=}C(CH_3)C_3H_7$-$i$	BrCN, $AlCl_3$	(E)-$BrC(C_2H_5){=}C(CH_3)C_3H_7$-$i$ (—)	492
	I_2	(E)-$IC(C_2H_5){=}C(CH_3)C_3H_7$-$i$ (—)	492
(E)-$(CH_3)_3SiCH{=}CH(CH_2)_6OTHP$	1. ICl, CCl_4, 0°, 15 min 2. KF·DMSO, 4 h	(Z)-$ICH{=}CH(CH_2)_6OTHP$ (84)	500
(E)-$(CH_3)_3SiCH{=}CHC_6H_{11}$	1. Cl_2, CH_2Cl_2, −78° 2. $NaOCH_3$, CH_3OH, 2 h	$ClCH{=}CHC_6H_{11}$ $E:Z$ 5:95 (88)	57
	1. Br_2, CH_2Cl_2, −78° 2. $NaOCH_3$, CH_3OH, 2 h	$BrCH{=}CHC_6H_{11}$ $E:Z$ 1:99 (93)	57

TABLE XLIII. VINYLSILANES WITH HALOGEN ELECTROPHILES (*Continued*)

Reactant	Conditions	Product(s) and Yield(s)	Refs.
	1. ICl, CCl_4, 0°, 30 min 2. $KF\cdot 2H_2O$, DMSO, 4 h	$ClCH{=}CHC_6H_{11}$ *E*:*Z* 2:98 (15) + $ICH{=}CHC_6H_{11}$ *E*:*Z* 9:91 (72)	489
(*Z*)-$(CH_3)_3SiCH{=}CHC_6H_{11}$	1. Cl_2, CH_2Cl_2, −78° 2. $NaOCH_3$, CH_3OH, 2 h	$ClCH{=}CHC_6H_{11}$ *E*:*Z* 99:1 (97)	57
	1. Br_2, CH_2Cl_2, −78° 2. $NaOCH_3$, CH_3OH, 2 h	$BrCH{=}CHC_6H_{11}$ *E*:*Z* 99:1 (98)	57
	1. ICl, CCl_4, 0°, 30 min 2. $KF\cdot 2H_2O$, DMSO, 4 h	$ClCH{=}CHC_6H_{11}$ *E*:*Z* 96:4 (40) + $ICH{=}CHC_6H_{11}$ *E*:*Z* 66:34 (38)	489
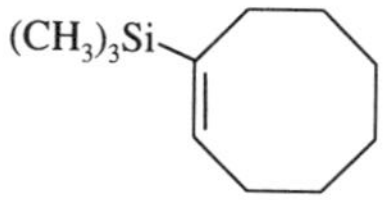 $(CH_3)_3Si$	Br_2, CH_2Cl_2, −78°	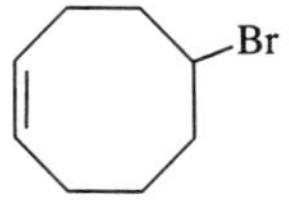 Br (70)	488
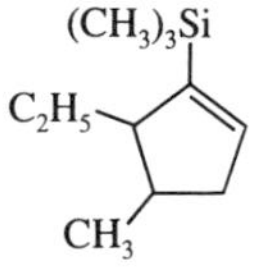 $(CH_3)_3Si$, C_2H_5, CH_3	1. Br_2, CH_2Cl_2 2. TBAF, THF, 0°, 3 h	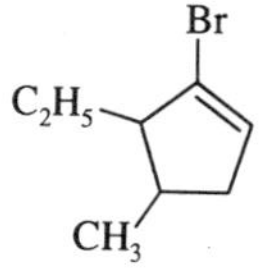 Br, C_2H_5, CH_3	487
trans		(45)	
cis		(43)	
$(CH_3)_3SiCX{=}CHR$	1. Br_2, CH_2Cl_2, −78°, 30 min 2. $NaOCH_3$, CH_3OH, 0°, 30 min; 25°, 2 h	$BrCX{=}CHR$	759
X R		*E*:*Z*	
E Cl *n*-C_6H_{13}		99:1 (69)	
" Br "		— (89)	
" Cl C_6H_{11}		91:9 (96)	
" Br "		— (87)	
" " C_6H_5		— (76)	
Z " C_6H_{11}	1. Cl_2, $CHCl_3$, −60°, 30 min 2. $NaOCH_3$, CH_3OH, 0°, 30 min; 25°, 2 h	$BrClC{=}CHC_6H_{11}$ *E*:*Z* 95:5 (84)	759

Reactant	Conditions	Product(s) and Yield(s) (%)	Refs.
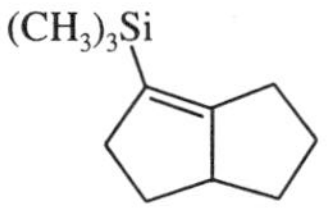 $(CH_3)_3Si$ (bicyclic structure)	Br_2, CH_2Cl_2, $-78°$	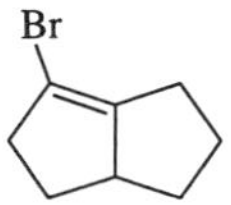 Br (bicyclic structure) (30)	486
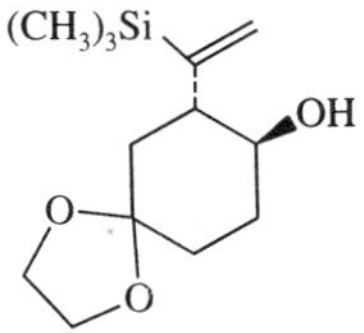 $(CH_3)_3Si$, OH, O, O (structure)	1. Br_2, CH_2Cl_2, 0°, 3 min 2. $NaOCH_3$, CH_3OH, 3 h	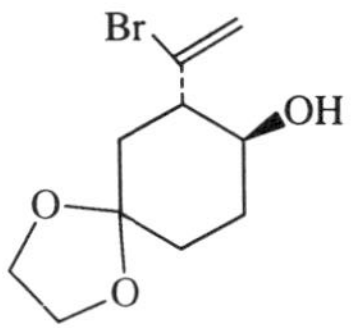 Br, OH, O, O (structure) (45)	761
(*E*)-$(CH_3)_3SiCH{=}CHC_6H_5$	1. Cl_2, CH_2Cl_2, $-78°$ 2. $NaOCH_3$, CH_3OH, 3 h	$ClCH{=}CHC_6H_5$ $E:Z$ 92:8 (66)	57
	1. Br_2, CH_2Cl_2, $-78°$ 2. $NaOCH_3$, CH_3OH, 3 h	$BrCH{=}CHC_6H_5$ $E:Z$ 99:1 (99)	57
	1. Br_2, CS_2, $-100°$ 2. CH_3CN	(*E*)-$BrCH{=}CHC_6H_5$ (—)	484
	C_6H_5IO, $BF_3{\cdot}O(C_2H_5)_2$, CH_2Cl_2, 18 h	$C_6H_5\overset{+}{I}CH{=}CHC_6H_5$ $^{-}BF_4$ (61)	497
(*Z*)-$(CH_3)_3SiCH{=}CHC_6H_5$	1. Cl_2, CH_2Cl_2, $-78°$ 2. $NaOCH_3$, CH_3OH, 3 h	$ClCH{=}CHC_6H_5$ $E:Z$ 10:90 (58)	57
	1. Br_2, CH_2Cl_2, $-78°$ 2. $NaOCH_3$, CH_3OH, 3 h	$BrCH{=}CHC_6H_5$ $E:Z$ 4:96 (94)	57
	1. Br_2, CS_2, $-100°$ 2. CH_3CN	", $E:Z$ 15:85 (—)	484
(*E*)-$(C_6H_5)_3SiCH{=}CHC_6H_5$	1. Br_2, CH_2Cl_2, $-78°$ 2. DMSO, 25°, 20 h, or CH_3CN, 25°, 12 d	(*E*)-$BrCH{=}CHC_6H_5$ (80)	762
(*Z*)-$(C_6H_5)_3SiCH{=}CHC_6H_5$	1. Br_2, CH_2Cl_2, $-78°$ 2. DMSO, 25°, 20 h, or CH_3CN, 25°, 12 d	(*Z*)-$BrCH{=}CHC_6H_5$ (—)	762
C_9			
(*E*)-$(CH_3)_3SiCH{=}CH(CH_2)_4C_3H_7$-*i*	1. ICl, CCl_4, 0°, 15 min 2. KF, DMSO, 4 h	(*Z*)-$ICH{=}CH(CH_2)_4C_3H_7$-*i* (85)	500
(*Z*)-$(CH_3)_3SiCH{=}C(CH_3)C_6H_{13}$-*n*	I_2	(*Z*)-$ICH{=}C(CH_3)C_6H_{13}$-*n* (>60)	493
(*E*)-$(CH_3)_3SiCH{=}CH(CH_2)_7OTHP$	1. ICl, CCl_4, 0°, 15 min 2. KF, DMSO, 4 h	(*Z*)-$ICH{=}CH(CH_2)_7OTHP$ (86)	500

TABLE XLIII. Vinylsilanes with Halogen Electrophiles (*Continued*)

Reactant	Conditions	Product(s) and Yield(s)	Refs.
(*E*)-$(CH_3)_3SiCH{=}CHCH_2C_6H_5$	C_6H_5IO, $BF_3{\cdot}O(C_2H_5)_2$, 25°, 20 h	(*E*)-$C_6H_5\overset{+}{I}CH{=}CHCH_2C_6H_5$ $^-BF_4$ (69)	497
C_{10}			
(*E*)-$(CH_3)_3SiCH{=}CHC_8H_{17}$-*n*	1. ICl 2. $NaOCH_3$, CH_3OH, 1 h	$ICH{=}CHC_8H_{17}$-*n* *E*:*Z* 2:98 (77)	490
	I_2, CH_2Cl_2, 2 h	(*Z*)-$ICH{=}CHC_8H_{17}$-*n* (65)	490
	C_6H_5IO, $BF_3{\cdot}O(C_2H_5)_2$, 25°, 4 h	(*E*)-$C_6H_5\overset{+}{I}CH{=}CHC_8H_{17}$-*n* $^-BF_4$ (72)	497
(*Z*)-$(CH_3)_3SiCH{=}CHC_8H_{17}$-*n*	C_6H_5IO, $BF_3{\cdot}O(C_2H_5)_2$, 25°, 4 h	$HC{\equiv}CC_8H_{17}$-*n* (90)	497
$(CH_3)_3SiCR^1{=}CHR^2$	**A** 1. Br_2, CH_2Cl_2, −78° 2. $NaOCH_3$, CH_3OH, 0°, 1 h; room temp, 1 h **B** BrCN, $AlCl_3$, CH_2Cl_2, 0° **C** I_2, CH_2Cl_2, 1 h	$XR^1C{=}CHR^2$	
R¹ R²	Conditions	X	
E *n*-C_4H_9 *n*-C_4H_9	**A**	Br *E*:*Z* 3:97 (91)	481
Z " "	"	" " 99:1 (92)	481
E *n*-C_5H_{11} *i*-C_3H_7	"	" *Z* (—)	271
" " "	**C**	I *E* (—)	271
Z " "	**A**	Br " (—)	271
" " "	**C**	I *Z* (—)	271
E C_2H_5 C_6H_{11}	**A**	Br " (87)	271, 1
" " "	**B**	" *E* (73)	271, 1
" " "	**C**	I " (—)	271
Z " "	**A**	Br " (—)	271
" " "	**B**	" *Z* (—)	271
" " "	**C**	I " (—)	271
" *s*-C_4H_9 *n*-C_4H_9	**A**	Br *E*:*Z* 98:2 (75)	481
$(CH_3)_3SiCH{=}CH(CH_2)_8OTHP$	1. Br_2, CH_2Cl_2, −78° 2. TBAF, CH_2Cl_2, THF, −20°	$BrCH{=}CH(CH_2)_8OTHP$	763
E		*Z* (60)	
Z		*E* (58)	

Reactant	Conditions	Product(s) and Yield(s) (%)	Refs.
$(CH_3)_3Si$–cyclohexenyl–C_4H_9-t	C_6H_5IO, $(C_2H_5)_3OBF_4$, CH_2Cl_2, 0°, 1 h; 25°, 2.5 h	$C_6H_5I^+$–cyclohexenyl–C_4H_9-t $^-BF_4$ (74)	497
(E)-$(CH_3)_3SiCH{=}CH(CH_2)_2C_6H_4R$-$p$	C_6H_5IO, $(C_2H_5)_3OBF_4$, CH_2Cl_2, 0°, 7 h	(E)-$C_6H_5\overset{+}{I}CH{=}CH(CH_2)_2C_6H_4R$-$p$ $^-BF_4$	763
R = H		(77)	
R = Cl		(75)	
C_{11}			
$(CH_3)_3SiCH{=}C(CH_3)(CH_2)_2C_6H_5$ $E:Z$ 90:10	C_6H_5IO, $(C_2H_5)_3OBF_4$, CH_2Cl_2, 25°, 3.5 h	$C_6H_5\overset{+}{I}CH{=}C(CH_3)(CH_2)_2C_6H_5$ $^-BF_4$ (89) $E:Z$ 90:10	497
$(CH_3)_3Si$–CH= lactone (C_6H_5)	1. Br_2, THF, 0°, 30 min 2. TBAF, THF, 0°, 4 h	Br–CH= lactone (C_6H_5) E:Z 85:15 (72)	764
(E)-$(CH_3)_3SiC(CH(CH_3)C_6H_5){=}CHD$	1. Br_2, CH_2Cl_2, −78°, 20 min 2. $NaOCH_3$, CH_3OH, CH_2Cl_2, −10° → 25°, 3 h	(Z)-$BrC(CH(CH_3)C_6H_5){=}CHD$ (99)	584
C_{12}			
(E)-$(CH_3)_3SiCH{=}CH(CH_2)_{10}O_2CCH_3$	I_2, $AlCl_3$, CH_2Cl_2, 0°	$ICH{=}CH(CH_2)_{10}O_2CCH_3$ $E:Z$ 80:20 (95)	494
$(CH_3)_3SiCH{=}CHCH_2CH(C_6H_5)$-$CH_2CO_2C_2H_5$	Br_2, CH_2Cl_2, −80°, 4 h	$BrCH{=}CHCH_2CH(C_6H_5)CH_2CO_2C_2H_5$ (60)	505
$(CH_3)_3Si$–bicyclic–$CON(C_2H_5)_2$	I_2, THF, 50°	I–bicyclic–$CON(C_2H_5)_2$ (79)	765
$(CH_3)_3Si$, R-acenaphthylene	ICl, CCl_4, 0°, 3 h; room temp, 1–2 h	X, R-acenaphthylene	718

TABLE XLIII. VINYLSILANES WITH HALOGEN ELECTROPHILES (*Continued*)

Reactant	Conditions	Product(s) and Yield(s)	Refs.
R		X R	
H		Cl H (10–30)	
		+ I " (50–60)	
$Si(CH_3)_3$		I I (73)	
C_{13}			
(*E*)-$(CH_3)_3SiCH{=}CH(CH_2)_{11}O_2CCH_3$	I_2, CH_2Cl_2, −78°, 1 h,	$ICH{=}CH(CH_2)_{11}O_2CCH_3$ I	494
	$SnCl_4$ (1.1 eq)	I *E*:*Z* 8:1 (90)	
	$SnCl_4$ (2 eq)	I *E*:*Z* 13:1 (92)	
	$SbCl_5$ (1.1 eq)	I *E*:*Z* 3:1 (82) + $ClCH{=}CH(CH_2)_{11}O_2CCH_3$ II (10)	
	$SbCl_5$ (2 eq)	I *E*:*Z* 7:1 (60) + II (26)	
(*E*)-$(CH_3)_3SiCH{=}CHCH_2$–(2-hydroxy-2-adamantyl) 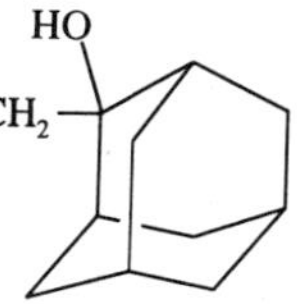	1. NBS, THF, 0°, 3 h 2. KF·$2H_2O$, DMSO, 45°, 6 h	(*Z*)-$BrCH{=}CHCH_2$–(2-hydroxy-2-adamantyl) (70) 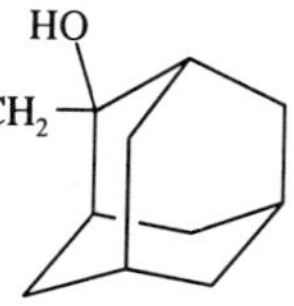	132
C_{14}			
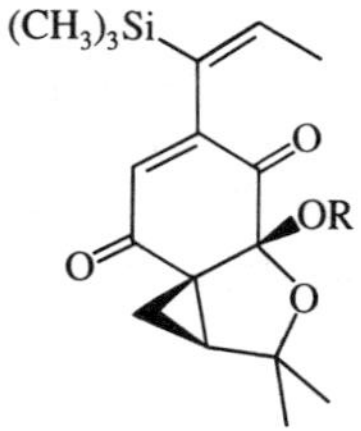 [structure: $(CH_3)_3Si$ vinyl dione, OR, O]	1. Cl_2, −78° 2. KF, DMSO, 25°	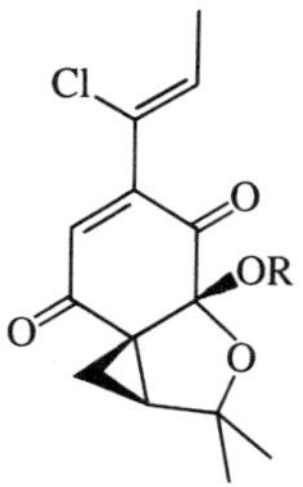 [structure: Cl vinyl dione, OR, O]	499

Reactant	Conditions	Product(s) and Yield(s) (%)	Refs.
R = H		(73)	
R = CH_3		(60)	
C_{15}			
(*E*)-$(CH_3)_3SiCH{=}CHC_{13}H_{27}$-*n*	1. ICl, CCl_4, 0°, 15 min 2. KF, DMSO, 4 h	(*Z*)-$ICH{=}CHC_{13}H_{27}$-*n* (83)	500
	I_2, $SnCl_4$ (1.1 eq), CH_2Cl_2, −78°, 1 h	$ICH{=}CHC_{13}H_{27}$-*n* *E*:*Z* 14:1 (95)	494
	I_2, $SnCl_4$ (2 eq), CH_2Cl_2, −78°, 1 h	", *E*:*Z* 18:1 (95)	494
	I_2, $AlCl_3$, CH_2Cl_2, 0°	", *E*:*Z* 4:1 (82)	494
C_{17}			
(*E*)-$(CH_3)_3SiC(C_5H_{11}$-*n*$){=}CHC_{10}H_{21}$-*n*	1. Br_2, CH_2Cl_2, −78° 2. CH_3CN	(*Z*)-$BrC(C_5H_{11}$-*n*$){=}CHC_{10}H_{21}$-*n* (—)	271
	BrCN, $AlCl_3$, CH_2Cl_2, 0°	(*E*)-$BrC(C_5H_{11}$-*n*$){=}CHC_{10}H_{21}$-*n* (—)	271
	I_2, CH_2Cl_2, 1 h	(*E*)-$IC(C_5H_{11}$-*n*$){=}CHC_{10}H_{21}$-*n* (—)	271
(*Z*)-$(CH_3)_3SiC(C_5H_{11}$-*n*$){=}CHC_{10}H_{21}$-*n*	1. Br_2, CH_2Cl_2, −78° 2. CH_3CN	(*E*)-$BrC(C_5H_{11}$-*n*$){=}CHC_{10}H_{21}$-*n* (—)	271
	BrCN, $AlCl_3$, CH_2Cl_2, 0°	(*Z*)-$BrC(C_5H_{11}$-*n*$){=}CHC_{10}H_{21}$-*n* (—)	271
	I_2, CH_2Cl_2, 1 h	(*Z*)-$IC(C_5H_{11}$-*n*$){=}CHC_{10}H_{21}$-*n* (—)	271
C_{20}			
$(CH_3)_3SiC(C_6H_5){=}C(C_6H_5)_2$	1. Br_2, CH_2Cl_2, −78° 2. $NaOCH_3$, CH_3OH, −78°, 5 min; room temp, 2 h	$BrC(C_6H_5){=}C(C_6H_5)_2$ (85)	766

[a] These values are the extremes of seven measurements using solvents of *Y* values ranging from −2.03 to +2.05.

TABLE XLIV. ALLYLSILANES WITH METAL ION ELECTROPHILES

Reactant	Conditions	Product(s) and Yield(s)	Refs.
C_3			
$(CH_3)_3SiCH_2CH{=}CH_2$	Li_2PdCl_4, C_2H_5OH, reflux, 2 h	$[(\eta^3\text{-allyl})PdCl]_2$ (65)	513
	$PdCl_2(C_6H_5CN)_2$, C_6H_6, 3 d	" (87)	767
	$Pd(O_2CCH_3)_2$, $(o\text{-}CH_3C_6H_4)_3P$, C_6H_5I, 120°, 7 h	$C_6H_5CH_2CH{=}CH_2$ (I[a]) + $C_6H_5C(CH_3){=}CH_2$ (II[a])	768
		I / II	
	CH_3CN	(9) / (15)	
	DMSO	(13) / (14)	
	", $AgNO_3$	(20) / (6)	
	Toluene	(20) / (12)	
	", $AgNO_3$	(10) / (12)	
	$Pd(O_2CCF_3)_2$, O_2, $(CH_3)_2CO$, $h\nu$, 160 h	$CH_2{=}CHCHO$ (32)	514
	$HgCl_2$, CH_3CN, 12 h	$CH_2{=}CHCH_2HgCl$ (51)	769
	$Hg(O_2CCH_3)_2$, CH_3CO_2H, 1 h	$CH_2{=}CHCH_2HgO_2CCH_3$ (40)	521
	$Tl(O_2CCH_3)_3$, CH_3CO_2H, 1 h	$CH_2{=}CHCH_2O_2CCH_3$ (81)	525
	$Tl(NO_3)_3$, dioxane, 15 min	$CH_2{=}CHCH_2ONO_2$ (81)	498
	$Tl(O_2CCF_3)_3$, C_2H_5OH, 0°, 30 min	$CH_2{=}CHCH_2OC_2H_5$ (86)	525
	$Tl(O_2CCF_3)_3$, RCN,	$CH_2{=}CHCH_2NHCOR$	524
	R		
	CH_3 −20°, 1 h; 0°, 1 h	(48)	
	$CH_2{=}CH$ ", "; ", "	(37)	
	C_2H_5 ", "; ", "	(29)	
	C_6H_5 −15°, "; ", "	(21)	
	A $Tl(O_2CCF_3)_3$, ArH	$CH_2{=}CHCH_2Ar$	
	B $C_6H_5Tl(O_2CCF_3)_2$, ArH, 2.5 h		

Reactant					Ref.
	ArH	Conditions	Ar		
	Furan	**A**, 0°, 1 h	2-Furyl	(89)	770
	”	**B**	”	(65)	526
	Thiophene	**A**, 0°, 1 h	2-Thiophenyl + 3-thiophenyl 1:0.9	(41)	770
	C_6H_6	**A**, CH_2Cl_2, 0°, 5 h; room temp, 30 min	C_6H_5	(56)	770
	”	**B**	”	(64)	526
	$C_6H_5OCH_3$	**A**, CH_2Cl_2, 30 min	*o*- and *p*-$CH_3OC_6H_4$ 1:1.7	(53)	770
	”	**B**	”, 1:1.4	(88)	526
	C_6H_5Cl	**A**, CH_2Cl_2, 1 h	*o*- and *p*-ClC_6H_4 1:0.7	(19)	770
	p-$(CH_3O)_2C_6H_4$	**A**, ”, 45 min	2,5-$(CH_3O)_2C_6H_3$	(46)	770
	”	**B**	”	(77)	526
	1,3,5-$(CH_3O)_3C_6H_3$	**A**, CH_2Cl_2, 1.5 h	2,4,6-$(CH_3O)_3C_6H_2$	(53)	770
	p-$(CH_3)_2C_6H_4$	**A**, ”, 30	2,5-$(CH_3)_2C_6H_3$	(84)	770
	”	**B**	”	(80)	526
$(CH_3)_3SiCH_2CH{=}CHSi(CH_3)_3$		$Pd(O_2CCH_3)_2$, $(C_6H_5)_3P$, C_6H_5I, $AgNO_3$, $(C_2H_5)_3N$, CH_3CN, 120°, 2 h	$CH_2{=}C(C_6H_5)CH_2Si(CH_3)_3$	(53)	768
		Li_2PdCl_4, CH_3OH, 10 h	[π-allyl–Si(CH3)3 / PdCl]$_2$	(—)	771
$(CH_3)_3SiCH(SO_2R)CH{=}CH_2$		Cat, O_2, $(CH_3)_2CO$, $h\nu$	$RSO_2CH{=}CHCHO$		514
R	Cat[b]		Conversion		
C_6H_5	$Pd(O_2CCF_3)_2$	38 h	100%	(95)	
$C_6H_4CH_3$-*p*	”	”	100%	(95)	
”	$PdCl_2(CH_3CN)_2$	40 h	95%	(90)	
”	$Pd(O_2CCH_3)_2$	”	90%	(60)	
”	$Pd[P(C_6H_5)_3]_4$	”	38%	(32)	
C_4					
$(CH_3)_3SiCH_2CH{=}CHCH_3$		$Tl(O_2CCF_3)_3$, $C_6H_4(OCH_3)_2$-*p*, CH_2Cl_2, 2.5 h	$CH_2{=}CHCH(CH_3)C_6H_3(OCH_3)_2$-2,5 + $CH_3CH{=}CHCH_2C_6H_3(OCH_3)_2$-2,5 1:3.3	(19)	770

TABLE XLIV. ALLYLSILANES WITH METAL ION ELECTROPHILES (*Continued*)

Reactant	Conditions	Product(s) and Yield(s)	Refs.
$(CH_3)_3SiCH_2C(CH_3){=}CH_2$	$Tl(NO_3)_3$, dioxane, 10 min	$CH_2{=}C(CH_3)CH_2ONO_2$ (71)	498
	", DME, "	" (92)	498
	$Tl(O_2CCF_3)_3$, RCN, $-20°$, 1 h; 0°, 1 h	$CH_2{=}C(CH_3)CH_2NHCOR$	524
	R = CH_3	(37)	
	R = $CH_2{=}CH$	(26)	
	$TlX(O_2CCF_3)_2$, ArH, CH_2Cl_2	$CH_2{=}C(CH_3)CH_2Ar$	
	X — ArH	Ar	
	O_2CCF_3 — C_6H_6	C_6H_5 (28)	770
	" — p-$(CH_3O)_2C_6H_4$	$C_6H_3(OCH_3)_2$-2,5 (59)	770
	" — 1,3,5-$(CH_3O)_3C_6H_3$	$C_6H_2(OCH_3)_3$-2,4,6 (54)	770
	C_6H_5 — p-$(CH_3O)_2C_6H_4$, 2.5 h	$C_6H_3(OCH_3)_2$-2,5 (63)	526
$[(CH_3)_3SiCH_2]_2C{=}CH_2$	$Tl(NO_3)_3$, dioxane, 15 min	$CH_2{=}C(CH_2ONO_2)_2$ (85)	498
$(CH_3)_3SiCH(SO_2C_6H_4CH_3$-$p)CH{=}CHCH_3$	PdX_2 (cat.), O_2, $(CH_3)_2CO$, $h\nu$	p-$CH_3C_6H_4SO_2CH{=}CHCOCH_3$	514
	X[b]	Conversion	
	O_2CCF_3 — 64 h	83% (61)	
	"[c] — 24 h	36% (20)	
	Cl — 64 h	63% (47)	
$(CH_3)_3SiCH(SO_2C_6H_4CH_3$-$p)C(CH_3){=}CH_2$	PdX_2 (cat.), O_2, $(CH_3)_2CO$, $h\nu$, 40 h	p-$CH_3C_6H_4SO_2CH{=}C(CH_3)CHO$	514
	X	Conversion	
	O_2CCF_3	85% (83)	
	Cl	64% (90)	
C_5			
$(CH_3)_3SiCH_2C(=CH_2)(CH_2)_2O_2CCH_3$	$Tl(O_2CCH_3)_3$, CH_3CO_2H, 1 h	$CH_2{=}C(CH_2O_2CCH_3)(CH_2)_2O_2CCH_3$ (70)	525
	$Tl(O_2CCF_3)_3$, CH_3OH, $-20°$, 30 min	$CH_2{=}C(CH_2OCH_3)(CH_2)_2O_2CCH_3$ (49)	525

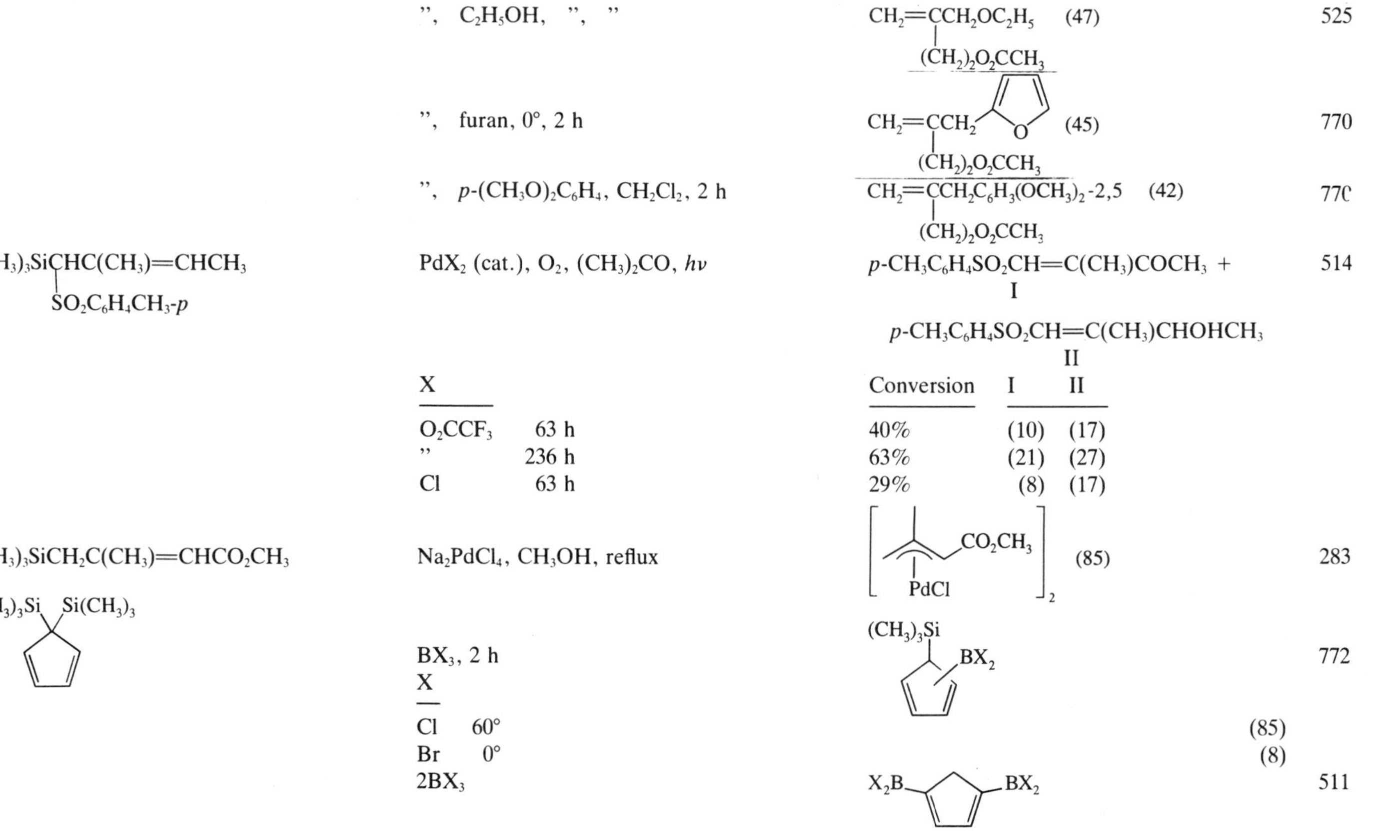

Reactant	Conditions	Product(s) and Yield(s) (%)	Refs.
	", C_2H_5OH, ", "	$CH_2{=}C(CH_2OC_2H_5)(CH_2)_2O_2CCH_3$ (47)	525
	", furan, 0°, 2 h	$CH_2{=}C((CH_2)_2O_2CCH_3)CH_2$-2-furyl (45)	770
	", p-$(CH_3O)_2C_6H_4$, CH_2Cl_2, 2 h	$CH_2{=}C((CH_2)_2O_2CCH_3)CH_2C_6H_3(OCH_3)_2$-2,5 (42)	770
$(CH_3)_3SiCH(SO_2C_6H_4CH_3$-$p)C(CH_3){=}CHCH_3$	PdX_2 (cat.), O_2, $(CH_3)_2CO$, $h\nu$	p-$CH_3C_6H_4SO_2CH{=}C(CH_3)COCH_3$ (I) + p-$CH_3C_6H_4SO_2CH{=}C(CH_3)CHOHCH_3$ (II)	514
	X: O_2CCF_3, 63 h	Conversion 40%; I (10); II (17)	
	X: ", 236 h	Conversion 63%; I (21); II (27)	
	X: Cl, 63 h	Conversion 29%; I (8); II (17)	
$(CH_3)_3SiCH_2C(CH_3){=}CHCO_2CH_3$	Na_2PdCl_4, CH_3OH, reflux	$[\pi$-allyl-PdCl (with CO_2CH_3)$]_2$ (85)	283
5,5-bis(trimethylsilyl)cyclopentadiene, $(CH_3)_3Si$, $Si(CH_3)_3$	BX_3, 2 h	$(CH_3)_3Si$-cyclopentadienyl-BX_2	772
	X: Cl, 60°	(85)	
	X: Br, 0°	(8)	
	$2BX_3$	X_2B-cyclopentadiene-BX_2	511

TABLE XLIV. ALLYLSILANES WITH METAL ION ELECTROPHILES (*Continued*)

Reactant	Conditions	Product(s) and Yield(s)	Refs.
1,5-bis[$(CH_3)_3Si$]-$(CH_3)_3Si$-cyclopentadiene (structure)	BX_3, 2 h	5,5-bis[$(CH_3)_3Si$]-BX_2-cyclopentadiene (structure)	772
	X: Cl, 80°, 7 h	(58)	
	X: Br, pentane, −30°, 1 h	(52)	
	X: I, ", 0°, 1 h	(48)	
	X: Cl, 60°	(65)	
	X: Br, 0°	(75)	
C_6			
$(CH_3)_3SiCH_2C(=CH_2)CH_2CH(O_2CCH_3)CH_3$	$Tl(NO_3)_3$, dioxane, 10 min	$CH_2=C(CH_2ONO_2)CH_2CH(O_2CCH_3)CH_3$ (85)	498
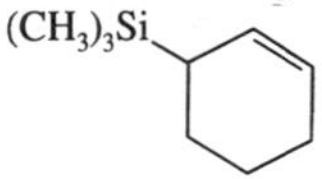 3-$(CH_3)_3Si$-cyclohexene (structure)	", ", "	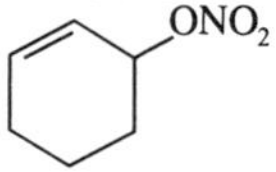 3-ONO_2-cyclohexene (structure) (91)	498
	$Tl(O_2CCF_3)_3$, $C_6H_5OCH_3$, CH_2Cl_2, 0°, 2 h	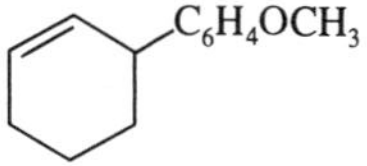 3-$C_6H_4OCH_3$-cyclohexene (structure) *o:p* 1:3 (23)	770
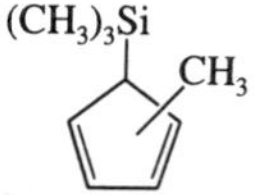 5-$(CH_3)_3Si$-CH_3-cyclopentadiene (structure)	BCl_3, −45°	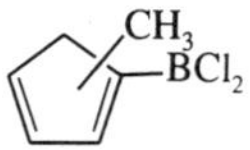 CH_3-BCl_2-cyclopentadiene (structure) (10)	772

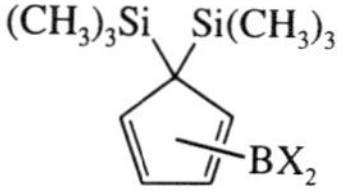

Reactant	Conditions	Product(s) and Yield(s) (%)	Refs.
C_7			
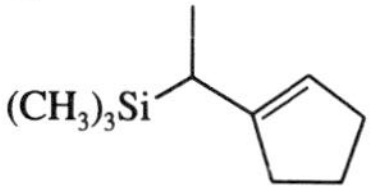 $(CH_3)_3Si$–CH(CH_3)–(1-cyclopentenyl)	$PdCl_2(CH_3CN)_2$, CH_3OH, 8 h	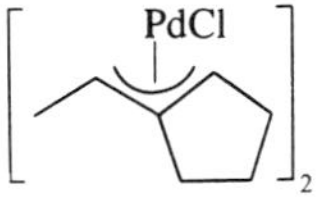 [(π-allyl)PdCl]$_2$ (39)	638
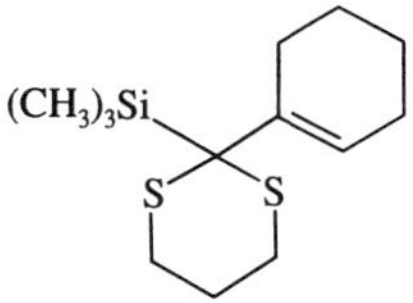 2-(1-cyclohexenyl)-2-(trimethylsilyl)-1,3-dithiane	$SnCl_4$, −30°	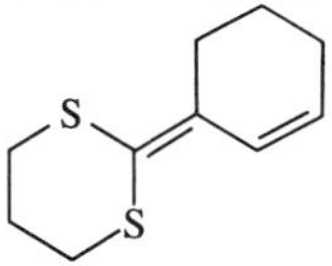 2-(cyclohex-2-enylidene)-1,3-dithiane (—)	258
C_9			
$(CH_3)_3SiCH_2CH{=}CHC_6H_5$	$Tl(NO_3)_3$, dioxane, 20 min	$C_6H_5CH{=}CHCH_2ONO_2$ (74)	498
	TlX_3, ROH	$CH_2{=}CHCH(OR)C_6H_5$ + $ROCH_2CH{=}CHC_6H_5$	
	X / R		
	NO_3 / CH_3, 0°, 30 min	73:27 (70)[d]	525
	O_2CCF_3 / " , " , "	75:25 (88)	525
	" / C_2H_5, 1.5 h	70:30 (43)	525
	" / $C_6H_5CH_2$	1.4:1 (38)	773
C_{10}			
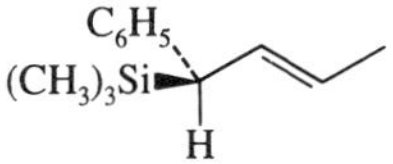 $(CH_3)_3Si$C(H)(C_6H_5)CH=CHCH$_3$ (E)	Li_2PdCl_4, CH_3OH	 [C_6H_5(π-allyl)PdCl]$_2$ **I** (84)	23
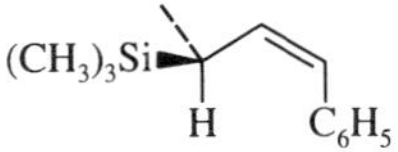 $(CH_3)_3Si$C(H)(CH_3)CH=CHC_6H_5 (Z)	" , "	" (76)	23

TABLE XLIV. ALLYLSILANES WITH METAL ION ELECTROPHILES (*Continued*)

Reactant	Conditions	Product(s) and Yield(s)	Refs.
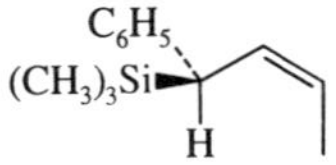 $(CH_3)_3Si(C_6H_5)CH$–CH=CH–CH_3 (Z)	", "	 $[C_6H_5$–allyl–PdCl$]_2$ **II** (100)	23
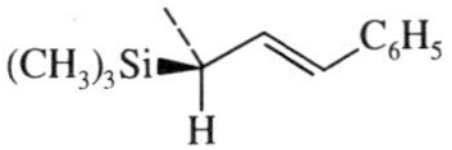 $(CH_3)_3Si(CH_3)CH$–CH=CH–C_6H_5	", "	" (79)	23
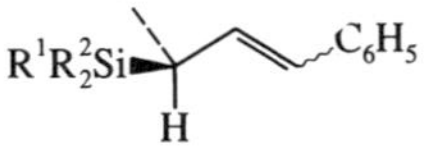 $R^1R^2_2Si(CH_3)CH$–CH=CH–C_6H_5	Li_2PdCl_4, CH_3OH, 0°, 20 h		638
R[1] R[2]			
E C_6H_5 CH_3		II (68)	
" C_2H_5 C_2H_5		II (90)	
Z " "		I (—)	
C_{11}			
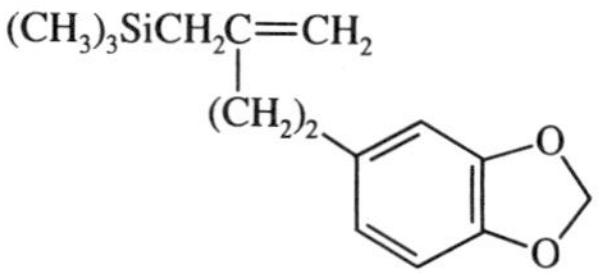 $(CH_3)_3SiCH_2C(=CH_2)(CH_2)_2$–(3,4-methylenedioxyphenyl)	TlX_3	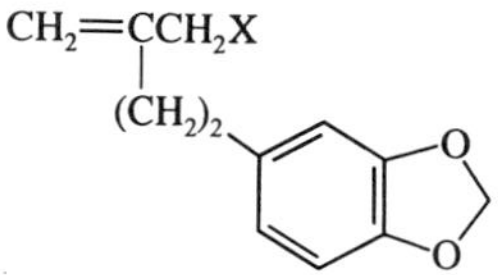 $CH_2=C(CH_2X)(CH_2)_2$–(3,4-methylenedioxyphenyl)	
	X	X	
	O_2CCH_3 CH_3CO_2H	O_2CCH_3 (72)	773
	NO_3 DME, 0°, 10 min	ONO_2 (58)	498
	O_2CCF_3 CH_3CN, −20°, 40 min; 0°, 2 h	$NHCOCH_3$ (9)	524

Reactant	Conditions	Product(s) and Yield(s) (%)	Refs.
C_{12}			
$(CH_3)_3SiCH_2CH{=}CHC_9H_{19}$-*n* + $(CH_3)_3SiCH(C_9H_{19}$-*n*$)CH{=}CH_2$ 3:1	$Pd(O_2CCF_3)_2$ (cat.), O_2, $(CH_3)_2CO$, *hν*, 30 h	*n*-$C_9H_{19}CH{=}CHCH_2OH$ I (23) + *n*-$C_9H_{19}CHOHCH{=}CH_2$ II (25)	514
	$PdCl_2(CH_3CN)_2$ (cat.), O_2, $(CH_3)_2CO$, *hν*, 30 h	I (21) + II (20)	514
$(CH_3)_3SiCH_2C(=CH_2)(CH_2)_2CH(O_2CCH_3)C_6H_{13}$-*n*	$Tl(NO_3)_3$, dioxane, 15 min	$CH_2{=}C(CH_2ONO_2)(CH_2)_2CH(O_2CCH_3)C_6H_{13}$-*n* (75)	498
	", hexane, 0°, 0.5 h; room temp, 0.5 h	" (52)	498
C_{13}			
$(CH_3)_3SiCH_2C(=CH_2)(CH_2)_2CHOHC_7H_{15}$-*n*	$Tl(O_2CCF_3)_3$, THF, 0°, 30 min	3-methylene-6-(C_7H_{15}-*n*)tetrahydropyran (71)	525
$(CH_3)_3SiCH_2C(=CH_2)CH_2CH(O_2CCH_3)(CH_2)_2C_6H_5$	$Tl(NO_3)_3$, dioxane, 15 min	$CH_2{=}C(CH_2ONO_2)CH_2CH(O_2CCH_3)(CH_2)_2C_6H_5$ (79)	498

[a] These are minor products. The major products are the silicon-containing equivalents of these products. The list given in the table is a selection from a larger list with different solvents, additives, temperatures, and times.

[b] This list is a selection of the results reported with varying amounts of catalyst.

[c] This reaction also had $Cu(O_2CCF_3)_2$ (cat.) added.

[d] The nitrate, $C_6H_5CH{=}CHCH_2ONO_2$, is a minor product (13%).

TABLE XLV. VINYLSILANES WITH METAL ION ELECTROPHILES

Reactant	Conditions	Product(s) and Yield(s)	Refs.
C_2			
$(CH_3)_3SiCH{=}CH_2$	$PdCl_2$, CH_3OH, DME, 20 h	$[\pi\text{-}(CH_2CHCH_2Si(CH_3)_3)PdCl]_2$ (92)	774, 767
	A $Pd(O_2CCH_3)_2$, $(C_6H_5)_3P$, ArX, $(C_2H_5)_3N$, DMF, 125°, 30 min **B** $(\pi\text{-}C_3H_5PdCl)_2$, ArX, DMF, 100°, 5 h	$ArCH{=}CH_2$	
	ArX — Conditions		
	$2\text{-}BrC_5H_4N$ — **B**	(30)	775
	$2\text{-}IC_5H_4N$ — **B**	(23)	775
	C_6H_5I — **A**	(60)	519
	" — **B**[a]	(53)	775
	$p\text{-}CH_3OC_6H_4I$ — **A**	(60)	519
	$p\text{-}C_2H_5OC_6H_4I$ — **B**	(14)	775
	$p\text{-}O_2NC_6H_4I$ — **A**	(58)	519
	" — **B**	(10)	775
	$p\text{-}C_6H_4I_2$ — "	(5)[b]	775
	$p\text{-}CH_3C_6H_4I$ — **A**	(51)	519
	$1\text{-}C_{10}H_7Br$ — **B**	(2)	775
	$1\text{-}C_{10}H_7I$ — "	(55)	775
	$Pd(C_6H_5CH{=}CHCOCH{=}CHC_6H_5)_2$, ArN_2BF_4, CH_3CN, 25°, 10–60 min	$ArCH{=}CH_2$ (I) + $(E)\text{-}ArCH{=}CHSi(CH_3)_3$ (II) + $ArC[Si(CH_3)_3]{=}CH_2$ (III)	776
	Ar — $[ArN_2BF_4]:[vinylsilane]$	I:II:III	
	C_6H_5 — 1:1[c]	82:14:4 (80)	
	" — 10:1[c]	24:71:5 (100)	
	$p\text{-}O_2NC_6H_4$ — 3:1	23:70:7 (79)	
	" — 10:1	11:84:5 (100)	

Reactant	Conditions	Product(s) (%)	Refs.
	p-BrC_6H_4 3:1	37:58:5 (100)	
	" 10:1	7:88:5 (97)	
	p-IC_6H_4 3:1	60:35:5 (100)	
	" 10:1	41:54:5 (100)	
	p-$CH_3C_6H_4$ 3:1	91:6:3 (100)	
	" 10:1	65:32:3 (100)	
	C_6H_5HgX, PdX_2, $(C_2H_5)_3N$, 50°, 2 h	$CH_2{=}CHC_6H_5$ + (E)-$(CH_3)_3SiCH{=}CHC_6H_5$	520
	X = Cl CH_3CN	(40) (35)	
	", DMSO	(26) (49)	
	X = O_2CCH_3 CH_3CN	(<1) (94)	
	$[(CH_2{=}CH_2)PtCl_2]_2$, $CDCl_3$, H_2O, 20°, 20 min	$CH_2{=}CH_2$ (—)	165
	1. $Hg(O_2CCH_3)_2$, H_2O, NaOH 2. NaCl	$(CH_3)_3SiCH(HgCl)CH_2OH$ (88)	523
	1. $Hg(O_2CCH_3)_2$, H_2O, THF 2. $NaBH_4$	$(CH_3)_3Si(CH_2)_2OH$ (90)	521
C_3			
$(CH_3)_3SiCH{=}CHCH_3$	$PdCl_2(C_6H_5CN)_2$, C_6H_6	[$Si(CH_3)_3$, ClPd, C_2H_5]$_2$ (21)	767
	1. $Hg(O_2CCH_3)_2$, H_2O, THF 2. $NaBH_4$	$(CH_3)_3SiCH_2CHOHCH_3$ (44) + n-C_3H_7OH (7) + i-C_3H_7OH (19)	521
$(CH_3)_3SiC(CH_3){=}CH_2$	1. $Hg(O_2CCH_3)_2$, H_2O, THF 2. $NaBH_4$	n-C_3H_7OH (9) + i-C_3H_7OH (39)	521
(E)-$(CH_3)_3SiCH{=}CHCH_2OCH_3$	$Pd(C_6H_5CH{=}CHCOCH{=}CHC_6H_5)_2$, $C_6H_5N_2BF_4$, CH_3CN, 25°	$CH_2{=}C(C_6H_5)CH_2OCH_3$ + $CH_3C(C_6H_5){=}CHOCH_3$ 66:16 (23)	588
(Z)-$(CH_3)_3SiCH{=}CHCH_2OCH_3$	", ", ", "	" 56:41 (13)	588
C_4			
$(CH_3)_3SiC(CH_3){=}CHCH_3$	$[(CH_2{=}CH_2)PtCl_2]_2$, $CDCl_3$, H_2O, 35°, 20 min	$CH_3CH{=}CHCH_3$	165
E:Z 95:5		E:Z 95:5 (—)	
E:Z 11:89		E:Z 13:87 (—)	

TABLE XLV. VINYLSILANES WITH METAL ION ELECTROPHILES (*Continued*)

Reactant	Conditions	Product(s) and Yield(s)	Refs.
C_8			
(*E*)-$(CH_3)_3SiCH{=}CHC_6H_{13}$-*n*	$Pd(C_6H_5CH{=}CHCOCH{=}CHC_6H_5)_2$, $C_6H_5N_2BF_4$, CH_3CN, 25°	$C_6H_5CH{=}CHC_6H_{13}$-*n* I + $C_6H_5CH_2CH{=}CHC_5H_{11}$-*n* II + $C_6H_5C(C_6H_{13}$-*n*$){=}CH_2$ III + $C_6H_5C(CH_3){=}CHC_5H_{11}$-*n* IV + $C_6H_5C[Si(CH_3)_3]{=}CHC_6H_{13}$-*n* V + $C_6H_5C(C_6H_{13}$-*n*$){=}CHSi(CH_3)_3$ VI I:II:III:IV:V:VI 18:11:39:21:6:2 (76)	588
(*Z*)-$(CH_3)_3SiCH{=}CHC_6H_{13}$-*n*	", ", ", "	", 17:17:36:17:9:5 (71)	588
(*E*)-$(CH_3)_3SiCH{=}CHC_6H_5$	$PdCl_2$, $(C_6H_{11})_2NC_2H_5$	(*E*,*E*)-$C_6H_5(CH{=}CH)_2C_6H_5$ (79)	517
	$PdCl_2$ (cat.), CH_3OH, LiCl, $CuCl_2$, 22 h	" (52)	517
	$PdCl_2$ (cat.), CH_3OH, LiCl, $CuCl_2$, $CH_2{=}CHCO_2CH_3$	(*E*,*E*)-$C_6H_5(CH{=}CH)_2CO_2CH_3$ (35)	517
	$PdCl_2$ (cat.), CH_3CN, LiCl, $CuCl_2$, $CH_2{=}CH_2$, 18 h	(*E*)-$C_6H_5CH{=}CH(CH_2)_2Cl$ (30)	517
	$Pd(C_6H_5CH{=}CHCOCH{=}CHC_6H_5)_2$, ArN_2X, CH_3CN, 25°	(*E*)-$ArCH{=}CHC_6H_5$ (I) + $C_6H_5C(Ar){=}CH_2$ (II)	588
	Ar[d] X[d]	I:II	
	C_6H_5 BF_4	67:33 (98)	
	" PF_6	85:15 (86)	
	p-$O_2NC_6H_4$ BF_4	86:14 (99)	
	p-BrC_6H_4 "	65:35 (100)	
	" PF_6	86:14 (100)	
	p-$CH_3C_6H_4$ BF_4	58:42 (97)	
(*E*)-$(CH_3)_3SiCH{=}CHAr$	$Pd(C_6H_5CH{=}CHCOCH{=}CHC_6H_5)_2$, ArN_2BF_4, CH_3CN, 25°	I + II	588

Reactant	Conditions	Product(s) and Yield(s) (%)	Refs.
Ar		I:II	
p-$O_2NC_6H_4$		65:35 (67)	
p-$CH_3C_6H_4$		74:26 (100)	
(Z)-$(CH_3)_3SiCH{=}CHC_6H_5$	$Pd(C_6H_5CH{=}CHCOCH{=}CHC_6H_5)_2$, ArN_2X, CH_3CN, 25°	I + II	588
	Ar[d] X[d]	I:II	
	C_6H_5 BF_4	80:20 (97)	
	p-$O_2NC_6H_4$ "	76:24 (84)	
	p-BrC_6H_4 "	76:24 (100)	
	" PF_6	86:14 (100)	
	p-$CH_3C_6H_4$ BF_4	70:30 (68)	
(E)-$(CH_3)_3SiCD{=}CHC_6H_5$	$Pd(C_6H_5CH{=}CHCOCH{=}CHC_6H_5)_2$, p-$O_2NC_6H_4N_2BF_4$, CH_3CN, 25°	(E)-p-$O_2NC_6H_4CD{=}CHC_6H_5$ + (Z)-$C_6H_5C(C_6H_4NO_2$-$p){=}CHD$ 89:11 (86)	518
(Z)-$(CH_3)_3SiCD{=}CHC_6H_5$	$Pd(C_6H_5CH{=}CHCOCH{=}CHC_6H_5)_2$, p-$O_2NC_6H_4N_2BF_4$, CH_3CN, 25°	(E)-p-$O_2NC_6H_4CD{=}CHC_6H_5$ + (E)-$C_6H_5C(C_6H_4NO_2$-$p){=}CHD$ 83:17 (91)	518
C_{15}			
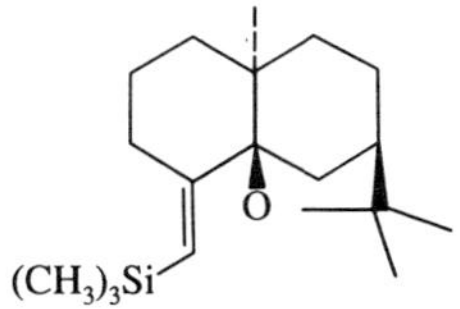 $(CH_3)_3Si$	1. $Hg(O_2CCH_3)_2$, CH_3CO_2H, H_2O, 20 h 2. $LiAlH_4$	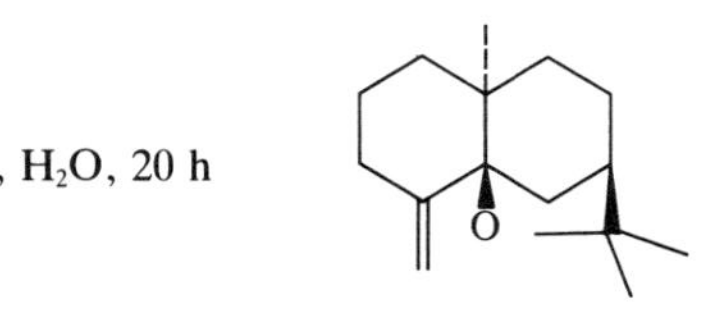 (—)	214

[a] This reaction is carried out in the presence of $(C_2H_5)_3N$ for 15 h.
[b] The product is either p-vinylstyrene or p-iodostyrene, but the report is ambiguous.
[c] These are the extreme values of a range.
[d] These results are a selection from several different concentrations of catalyst and counterion.

ADDENDA TO THE TABLES

Listed below are references that have appeared in some of the major journals since the tables were prepared. The cut-off date is December 1988, but a few references from 1989 are also included.

Table I.	Refs. 777–789
Table II.	Refs. 790–801
Table IV.	Ref. 802
Table V.	Refs. 803, 804
Table VI.	Refs. 805–818
Table IX.	Refs. 819–822
Table XI.	Refs. 823–851
Table XII.	Ref. 852
Table XIV.	Refs. 853–872
Table XV.	Refs. 873–876
Table XIX.	Refs. 877–895
Table XX.	Refs. 896–897
Table XXII.	Refs. 898–915
Table XXIV.	Refs. 916–922
Table XXVI.	Refs. 923–927
Table XXVII.	Refs. 928, 929
Table XXX.	Refs. 930, 931
Table XXXII.	Refs. 932, 933
Table XXXV.	Refs. 934–936
Table XXXVI.	Refs. 937–939
Table XXXVII.	Ref. 940
Table XXXVIII.	Refs. 941–945
Table XLII.	Refs. 946–948
Table XLIII.	Refs. 949–959
Table XLIV.	Refs. 960–962
Table XLV.	Refs. 963–965

REFERENCES

[1] T. H. Chan and I. Fleming, *Synthesis*, **1979**, 761.
[2] I. Fleming, in *Comprehensive Organic Chemistry*, D. H. R. Barton and W. D. Ollis, Eds., Vol. 3, Pergamon, Oxford, 1979, p. 541.
[3] E. W. Colvin, *Silicon in Organic Synthesis*, Butterworths, London, 1981.
[4] W. P. Weber, *Silicon Reagents for Organic Synthesis*, Springer Verlag, Berlin, 1983.
[5] L. H. Sommer, L. J. Tyler, and F. C. Whitmore, *J. Am. Chem. Soc.*, **70**, 2872 (1948).
[6] L. H. Sommer, D. L. Bailey, G. M. Goldberg, C. E. Buck, T. S. Bye, F. J. Evans, and F. C. Whitmore, *J. Am. Chem. Soc.*, **76**, 1613 (1954).
[7] P. Bourgeois, Thèse d'Etat, Bordeaux, 1970.
[8] M. Grignon–Dubois, J.-P. Pillot, J. Dunoguès, N. Duffaut, R. Calas, and B. Henner, *J. Organomet. Chem.*, **124**, 135 (1977).

[9] R. Calas, P. Bourgeois, and N. Duffaut, *C.R. Hebd. Séances Acad. Sci., Sér. C*, **263**, 243 (1966).
[10] A. Z. Shikhamedbekova and R. A. Sultanov, *Zh. Obshch. Khim.*, **40**, 77 (1970); *J. Gen. Chem. USSR* (*Engl. Transl.*), **40**, 72 (1970).
[11] E. Frainnet and R. Calas, *C.R. Hebd. Séances Acad. Sci.*, **240**, 203 (1955).
[12] I. Fleming and A. Pearce, *J. Chem. Soc., Chem. Commun.*, **1975**, 633; full paper, ref. 117.
[13] A. Hosomi, A. Shirata, and H. Sakurai, *Tetrahedron Lett.*, **1978**, 3043.
[14] A. W. P. Jarvie, A. Holt, and J. Thompson, *J. Chem. Soc.* (*B*), **1969**, 852.
[15] E. Rosenburg and J. J. Zuckerman, *J. Organomet. Chem.*, **33**, 321 (1971).
[16] K. E. Koenig and W. P. Weber, *J. Am. Chem. Soc.*, **95**, 3416 (1973).
[17] M. J. Carter and I. Fleming, *J. Chem. Soc.*, *Chem. Commun.*, **1976**, 182; full paper ref. 134.
[18] B.-W. Au-Yeung and I. Fleming, *J. Chem. Soc.*, *Chem. Commun.*, **1977**, 79; full paper ref. 55.
[19] H. Wetter, P. Scherer, and W. B. Schweizer, *Helv. Chim. Acta*, **62,** 1985 (1979).
[20] A. Eschenmoser, P. R. Jenkins, and V. Matassa, ETH, Zurich, personal communication.
[21] T. Hayashi, M. Konishi, H. Ito, and M. Kumada, *J. Am. Chem. Soc.*, **104**, 4962 (1982).
[22] T. Hayashi, M. Konishi, and M. Kumada, *J. Am. Chem. Soc.*, **104**, 4963 (1982).
[23] T. Hayashi, M. Konishi, and M. Kumada, *J. Chem. Soc.*, *Chem. Commun.*, **1983**, 736.
[24] T. Hayashi, K. Kabeta, T. Yamamoto, K. Tamao, and M. Kumada, *Tetrahedron Lett.*, **24**, 5661 (1983).
[25] R. G. Daniels and L. A. Paquette, *Organometallics*, **1**, 1449 (1982).
[26] S. J. Hathaway and L. A. Paquette, *J. Org. Chem.*, **48**, 3351 (1983).
[27] L. Coppi, A. Mordini, and M. Taddei, *Tetrahedron Lett.*, **28**, 969 (1987).
[28] C. Eaborn and R. W. Bott in *Organometallic Compounds of the Group IV Elements*, A. G. MacDiarmid, Ed., Vol. 1, Part 1, Dekker, New York, 1968, pp. 359–437.
[29] I. Fleming and J. A. Langley, *J. Chem. Soc.*, *Perkin Trans. 1*, **1981**, 1421.
[30] G. D. Hartman and T. G. Traylor, *Tetrahedron Lett.*, **1975**, 939.
[31] W. Hanstein, H. J. Berwin, and T. G. Traylor, *J. Am. Chem. Soc.*, **92**, 829 (1970).
[32] C. Eaborn, *J. Chem. Soc.*, *Chem. Commun.*, **1972**, 1255.
[33] I. Fleming, *Frontier Orbitals and Organic Chemical Reactions*, Wiley, London, 1976, p. 81.
[34] S. G. Wierschke, J. Chandrasekhar, and W. L. Jorgensen, *J. Am. Chem. Soc.*, **107**, 1496 (1985).
[35] C. Eaborn, T. A. Emokpae, V. I. Sidorov, and R. Taylor, *J. Chem. Soc.*, *Perkin Trans. 2*, **1974**, 1454.
[36] P. R. Wells, H. H. Feldman, and D. W. Hawker, *J. Chem. Res.* (*S*), **1984**, 60.
[37] G. Déléris, J.-P. Pillot, and J.-C. Rayez, *Tetrahedron*, **36**, 2215 (1980).
[38] I. Fleming, D. Marchi, and S. K. Patel, *J. Chem. Soc.*, *Perkin Trans. 1*, **1981**, 2518.
[39] I. Fleming and S. K. Patel, *Tetrahedron Lett.*, **1981**, 2321.
[40] T. Hayashi, H. Ito, and M. Kumada, *Tetrahedron Lett.*, **23**, 4605 (1982).
[41] G. Wickham and W. Kitching, *Organometallics*, **2**, 541 (1983).
[42] G. Wickham and W. Kitching, *J. Org. Chem.*, **48**, 612 (1983).
[43] H. Wetter and P. Scherer, *Helv. Chim. Acta*, **66**, 118 (1983).
[44] I. Fleming and N. K. Terrett, *J. Organomet. Chem.*, **264**, 99 (1984).
[45] A. Eschenmoser, lecture at Sixth International Symposium on Synthesis in Organic Chemistry, Cambridge, 1979.
[46] M. N. Paddon–Row, N. G. Rondan, and K. N. Houk, *J. Am. Chem. Soc.*, **104**, 7162 (1982).
[47] S. Profeta, R. J. Unwalla, and F. K. Cartledge, *J. Org. Chem.*, **51**, 1884 (1986).
[48] S. D. Kahn, C. F. Pau, A. R. Chamberlin, and W. J. Hehre, *J. Am. Chem. Soc.*, **109**, 650 (1987).
[49] B. Beagley, A. Foord, R. Montran, and B. Roszandai, *J. Mol. Struct.*, **42**, 117 (1977).
[50] M. Hayashi, M. Imachi, and M. Saito, *Chem. Lett.* (*Jpn.*), **1977**, 221.
[51] I. Fleming, A. Sarkar, and A. P. Thomas, *J. Chem. Soc.*, *Chem. Commun.*, **1987**, 157.
[52] P. R. Jenkins, R. Gut, H. Wetter, and A. Eschenmoser, *Helv. Chim. Acta*, **62**, 1922 (1979).

[53] E. Vedejs and C. K. McClure, *J. Am. Chem. Soc.*, **108**, 1094 (1986).
[54] R. K. Chaudhuri, T. Ikeda, and C. R. Hutchinson, *J. Am. Chem. Soc.*, **106**, 6004 (1984).
[55] I. Fleming and B.-W. Au-Yeung, *Tetrahedron*, **37**, Supplement No. 1, 13 (1981).
[56] T. Hayashi, Y. Matsumoto, and Y. Ito, *Organometallics*, **6**, 884 (1987).
[57] R. B. Miller and G. McGarvey, *J. Org. Chem.*, **43**, 4424 (1978).
[58] F. Franke, M. J. Cuthbertson, and P. R. Wells, *J. Org. Chem.*, **49**, 1258 (1984).
[59] T. Yamazaki, K. Takita, and N. Ishikawa, *J. Fluorine Chem.*, **30**, 357 (1985).
[60] C. H. DePuy, V. M. Bierbaum, L. A. Flippin, J. J. Grabowski, G. K. King, R. J. Schmitt, and S. A. Sullivan, *J. Am. Chem. Soc.*, **102**, 5012 (1980); G. Klass, V. C. Trenerry, J. C. Sheldon, and J. H. Bowie, *Aust. J. Chem.*, **34**, 519 (1981).
[61] R. Noyori, I. Nishida, J. Sakata, and M. Nishizawa, *J. Am. Chem. Soc.*, **102**, 1223 (1980); R. Noyori, I. Nishida, and J. Sakata, *ibid.*, **103**, 2106 (1981).
[62] O. W. Webster, W. R. Hertler, D. Y. Sogah, W. B. Farnham, and T. V. RajanBabu, *J. Am. Chem. Soc.*, **105**, 5706 (1983).
[63] R. J. P. Corriu, R. Perz, and C. Reye, *Tetrahedron*, **39**, 999 (1983).
[64] A. Hosomi and H. Sakurai, *Tetrahedron Lett.*, **1976**, 1295.
[65] A. Hosomi, A. Shirata, and H. Sakurai, *Chem. Lett. (Jpn.)*, **1978**, 901.
[66] J. Yoshida, K. Tamao, M. Takahashi, and M. Kumada, *Tetrahedron Lett.*, **1978**, 2161.
[67] K. Tamao, T. Kakui, and M. Kumada, *Tetrahedron Lett.*, **1979**, 619.
[68] K. Tamao, T. Kakui, and M. Kumada, *Tetrahedron Lett.*, **21**, 4105 (1980).
[69] W. F. Reynolds, G. K. Hamer, and A. R. Bassindale, *J. Chem. Soc., Perkin Trans. 2*, **1977**, 971.
[70] J. C. Giordan, *J. Am. Chem. Soc.*, **105**, 6544 (1983).
[71] C. Eaborn and R. W. Bott in *Organometallic Compounds of the Group IV Elements*, A.G. MacDiarmid, Ed., Vol. 1, Part 1, Dekker, New York, 1968, p. 351.
[72] K. Utimoto, M. Kitai, and H. Nozaki, *Tetrahedron Lett.*, **1975**, 2825.
[73] A. P. Davies, G. J. Hughes, P. R. Lowndes, C. M. Robbins, E. J. Thomas, and G. H. Whitham, *J. Chem. Soc., Perkin Trans. 1*, **1981**, 1934.
[74] D. R. Boyd and G. A. Berchtold, *J. Org. Chem.*, **44**, 468 (1979).
[75] P. E. Peterson, D. J. Nelson, and R. Risener, *J. Org. Chem.*, **51**, 2381 (1986).
[76] L. A. Paquette, W. E. Fristad, D. S. Dime, and T. R. Bailey, *J. Org. Chem.*, **45**, 3017 (1980).
[77] C. G. Pitt, *J. Organomet. Chem.*, **61**, 49 (1973).
[78] U. Weidner and A. Schweig, *Angew. Chem. Int. Ed. Engl.*, **11**, 146 (1972).
[79] H. Bock and W. Kaim, *Chem. Ber.*, **111**, 3552 (1978).
[80] H. Bock and W. Kaim, *J. Am. Chem. Soc.*, **102**, 4429 (1980).
[81] W. Oppolzer, R. L. Snowden, and D. P. Simmons, *Helv. Chim. Acta*, **64**, 2002 (1981), footnote 2.
[82] H. Mayr and R. Pock, *Tetrahedron*, **42**, 4211 (1986).
[83] A. Hosomi and H. Sakurai, *Tetrahedron Lett.*, **1978**, 2589.
[84] W. T. Brady and T. C. Cheng, *J. Org. Chem.*, **42**, 732 (1977).
[85] R. L. Danheiser and H. Sard, *Tetrahedron Lett.*, **24**, 23 (1983).
[86] M. Laguerre, M. Grignon–Dubois, and J. Dunoguès, *Tetrahedron*, **37**, 1161 (1981).
[87] H. Gilman, A. G. Brook, and L. S. Miller, *J. Am. Chem. Soc.*, **75**, 4531 (1953).
[88] R. M. G. Roberts and F. El Kaissi, *J. Organomet. Chem.*, **12**, 79 (1968).
[89] C. L. Agre and W. Hilling, *J. Am. Chem. Soc.*, **74**, 3895 (1952).
[90] G. H. Wagner and A. N. Pines, *J. Am. Chem. Soc.*, **71**, 3567 (1949).
[91] R. N. Sterlin, I. L. Knunyants, L. N. Pinkina, and R. D. Yatsenko, *Izv. Akad. Nauk SSSR, Otdel Khim. Nauk*, **1959**, 1492; *Bull. Acad. Sci. USSR, Div. Chem. Sci.* (*Engl. Transl.*), **1959**, 1442.
[92] J. Slutsky and H. Kwart, *J. Am. Chem. Soc.*, **95**, 8678 (1973).
[93] R. B. Larrabee and B. F. Dowden, *Tetrahedron Lett.*, **1970**, 915.
[94] A. J. Ashe, *J. Am. Chem. Soc.*, **92**, 1233 (1970).
[95] W. Adcock, G. L. Aldous, and W. Kitching, *Tetrahedron Lett.*, **1978**, 3387.

[96] T. A. Dixon, K. P. Steele, and W. P. Weber, *J. Organomet. Chem.*, **231**, 299 (1982).
[97] H. Mayr and G. Hagen, *J. Chem. Soc. Chem. Commun.*, **1989**, 91.
[98] K. Tamao, J. Yoshida, M. Akita, Y. Sugihara, T. Iwahara, and M. Kumada, *Bull. Chem. Soc. Jpn.*, **55**, 255 (1982).
[99] T. Hiyama, M. Obayashi, I. Mori, and H. Nozaki, *J. Org. Chem.*, **48**, 912 (1983).
[100] H. Oda, M. Sato, Y. Morizawa, K. Oshima, and H. Nozaki, *Tetrahedron*, **41**, 3257 (1985).
[101] H. Oda, M. Sato, Y. Morizawa, K. Oshima, and H. Nozaki, *Tetrahedron Lett.*, **24**, 2877 (1983).
[102] Y. Naruta, H. Uno, and K. Maruyama, *Tetrahedron Lett.*, **22**, 5221 (1981).
[103] D. J. Nelson and P. J. Cooper, *Tetrahedron Lett.*, **27**, 4693 (1986).
[104] J. A. Soderquist and H. C. Brown, *J. Org. Chem.*, **45**, 3571 (1980).
[105] I. Fleming and N. J. Lawrence, *Tetrahedron Lett.*, **29**, 2073 (1988).
[106] D. Seyferth, *J. Inorg. Nucl. Chem.*, **7**, 152 (1952).
[107] A. G. Brook and J. B. Pierce, *J. Org. Chem.*, **30**, 2566 (1965).
[108] W. K. Musker and G. L. Larson, *Tetrahedron Lett.*, **1968**, 3481.
[109] J.-C. Richer, M.-A. Poirier, Y. Maroni, and G. Manuel, *Can. J. Chem.*, **58**, 39 (1980).
[110] T. F. O. Lim, J. K. Myers, G. T. Rogers, and P. R. Jones, *J. Organomet. Chem.*, **135**, 249 (1977).
[111] G. Manuel, P. Mazerolles, and J. Gril, *J. Organomet. Chem.*, **122**, 335 (1976).
[112] S. Niwayama, S. Dan, Y. Inouye, and K. Kakisawa, *Chem. Lett. (Jpn.)*, **1985**, 957.
[113] A. A. Panasenko, L. M. Khalilov, I. M. Salimgareeva, and V. P. Yur'ev, *Izv. Akad. Nauk SSSR, Ser. Khim.*, **1978**, 938; *Bull. Acad. Sci. USSR, Div. Chem. Sci. (Engl. Transl.)*, **1978**, 812.
[114] M. Wada, T. Shigehisa, H. Kitani, and K. Akiba, *Tetrahedron Lett.*, **24**, 1715 (1983).
[115] R. J. P. Corriu, V. Huynh, and J. J. E. Moreau, *J. Organomet. Chem.*, **259**, 283 (1983).
[116] K. Mikami, N. Kishi, and T. Nakai, *Tetrahedron Lett.*, **24**, 795 (1983).
[117] I. Fleming and A. Pearce, *J. Chem. Soc., Perkin Trans. 1*, **1980**, 2485.
[118] D. J. Ager, *Tetrahedron Lett.*, **23**, 1945 (1982).
[119] P. Magnus and D. Quagliato, *J. Org. Chem.*, **50**, 1621 (1985).
[120] H. Oda, K. Oshima, and H. Nozaki, *Chem. Lett. (Jpn.)*, **1985**, 53.
[121] H. Oda, Y. Morizawa, K. Oshima, and H. Nozaki, *Tetrahedron Lett.*, **25**, 3221 (1984).
[122] L.-C. Chen, L.-M. Chen, and L.-H. Lin, *J. Chin. Chem. Soc. (Taipei)*, **31**, 409 (1984) [*C.A.*, **104**, 109067u (1986)].
[123] Y. Yamamoto, H. Yatagai, and K. Maruyama, *J. Am. Chem. Soc.*, **103**, 3229 (1981).
[124] M. Obayashi, K. Utimoto, and H. Nozaki, *J. Organomet. Chem.*, **177**, 145 (1979).
[125] G. Zweifel and W. Lewis, *J. Org. Chem.*, **43**, 2739 (1978).
[126] J. E. Wrobel and B. Ganem, *J. Org. Chem.*, **48**, 3761 (1983).
[127] R. B. Miller and M. I. Al-Hassan, *J. Org. Chem.*, **48**, 4113 (1983).
[128] R. K. Boeckman, Jr., *J. Am. Chem. Soc.*, **95**, 6867 (1973).
[129] R. L. Danheiser, C. A. Kwasigroch, and Y.-M. Tsai, *J. Am. Chem. Soc.*, **107**, 7233 (1985).
[130] J. M. Muchowski, R. Naef, and M. L. Maddox, *Tetrahedron Lett.*, **26**, 5375 (1985).
[131] B. L. Chenard, C. M. Van Zyl, and D. R. Sanderson, *Tetrahedron Lett.*, **27**, 2801 (1986).
[132] E. Ehlinger and P. Magnus, *J. Am. Chem. Soc.*, **102**, 5004 (1980).
[133] R. Calas, J. Dunoguès, J.-P. Pillot, and N. Ardoin, *J. Organomet. Chem.*, **73**, 211 (1974).
[134] M. J. Carter, I. Fleming, and A. Percival, *J. Chem. Soc., Perkin Trans. 1*, **1981**, 2415.
[135] M. Grignon–Dubois, J. Dunoguès, N. Duffaut, and R. Calas, *J. Organomet. Chem.*, **188**, 311 (1980).
[136] C. M. Robbins and G. H. Whitham, *J. Chem. Soc., Chem. Commun.*, **1976**, 697.
[137] B. B. Snider and M. Karras, *J. Org. Chem.*, **47**, 4588 (1982).
[138] T. Hayama, S. Tomoda, Y. Takeuchi, and Y. Nomura, *J. Org. Chem.*, **49**, 3235 (1984).
[139] A. Laporterie, J. Dubac, and M. Lesbre, *J. Organomet. Chem.*, **101**, 187 (1975).
[140] M. F. Salomon, S. N. Pardo, and R. G. Salomon, *J. Org. Chem.*, **49**, 2446 (1984) and *J. Am. Chem. Soc.*, **106**, 3797 (1984).

[141] E. W. Abel and R. J. Rowley, *J. Organomet. Chem.*, **84**, 199 (1975).
[142] A. Laporterie, J. Dubac, P. Mazerolles, and H. Iloughmane, *J. Organomet. Chem.*, **216**, 321 (1981); J. Dubac, A. Laporterie, H. Iloughmane, J.-P. Pillot, G. Déléris, and J. Dunoguès, *J. Organomet. Chem.*, **281**, 149 (1985).
[143] W. E. Fristad, T. R. Bailey, and L. A. Paquette, *J. Org. Chem.*, **45**, 3028 (1980).
[144] A. Gopalan, R. Moerck, and P. Magnus, *J. Chem. Soc., Chem. Commun.*, **1979**, 548.
[145] A. Laporterie, J. Dubac, G. Manuel, G. Déléris, J. Kowalski, J. Dunoguès, and R. Calas, *Tetrahedron*, **34**, 2669 (1978).
[146] I. Fleming, J. Goldhill, and D. A. Perry, *J. Chem. Soc., Perkin Trans. 1*, **1982**, 1563.
[147] T. J. Barton, S. A. Burns, I. M. T. Davidson, S. Ijadi–Maghsoodi, and I. T. Wood, *J. Am. Chem. Soc.*, **106**, 6367 (1984).
[148] B. M. Trost and S. Mignani, *J. Org. Chem.*, **51**, 3435 (1986).
[149] J. A. Butcher and R. M. Pagni, *J. Am. Chem. Soc.*, **101**, 3997 (1979).
[150] F. G. Yusupova, *Khim. Vysl. Neft.*, **1975**, 58 [*C.A.*, **85**, 94440y (1976)].
[151] G. Zweifel and H. P. On, *Synthesis*, **1980**, 803.
[152] E. Krochmal, D. H. O'Brien, and P. S. Mariano, *J. Org. Chem.*, **40**, 1137 (1975).
[153] H. Sakurai, A. Hosomi, and M. Kumada, *J. Org. Chem.*, **34**, 1764 (1969).
[154] C.-N. Hsiao and H. Shechter, *Tetrahedron Lett.*, **23**, 1963 (1982).
[155] M. Kosugi, K. Kurino, K. Takayama, and T. Migita, *J. Organomet. Chem.*, **56**, C11 (1973); J. Grignon, C. Servens, and M. Pereyre, *ibid.*, **96**, 225 (1975).
[156] G. T. Burns and T. J. Barton, *J. Organomet. Chem.*, **209**, C25 (1981).
[157] G. Stork, M. E. Jung, E. Colvin, and Y. Noel, *J. Am. Chem. Soc.*, **96**, 3684 (1974).
[158] M. Kosugi, K. Yano, M. Chiba, and T. Migita, *Chem. Lett. (Jpn.)*, **1977**, 801.
[159] I. Saito, H. Ikehira, and T. Matsuura, *J. Org. Chem.*, **51**, 5148 (1986).
[160] H. Sakurai, A. Hosomi, and M. Kumada, *J. Org. Chem.*, **34**, 1764 (1969).
[161] B. M. Trost and C. R. Self, *J. Am. Chem. Soc.*, **105**, 5942 (1983).
[162] G. Majetich, A. M. Casares, D. Chapman, and M. Behnke, *J. Org. Chem.*, **51**, 1745 (1986).
[163] G. Majetich, R. W. Desmond, and J. J. Soria, *J. Org. Chem.*, **51**, 1753 (1986).
[164] F. G. Yusupova, G. A. Gailiunas, I. I. Furley, A. A. Panasenko, V. D. Sheludyakov, G. A. Tolstikov, and V. P. Yur'ev, *J. Organomet. Chem.*, **155**, 15 (1978).
[165] D. Mansuy, J. Pusset, and J. C. Chottard, *J. Organomet. Chem.*, **110**, 139 (1976).
[166] M. Ochiai, K. Sumi, and E. Fujita, *Tetrahedron Lett.*, **23**, 5419 (1982).
[167] L. A. Paquette, T.-H. Yan, and G. J. Wells, *J. Org. Chem.*, **49**, 3610 (1984).
[168] G. Majetich, R. Desmond, and A. M. Casares, *Tetrahedron Lett.*, **24**, 1913 (1983).
[169] D. J. Coughlin and R. G. Salomon, *J. Org. Chem.*, **44**, 3784 (1979).
[170] I. Fleming and I. Paterson, *Synthesis*, **1979**, 446.
[171] M. Grignon–Dubois, J. Dunoguès, and R. Calas, *J. Organomet. Chem.*, **181**, 285 (1979).
[172] I. Fleming and D. Marchi, *Synthesis*, **1981**, 560.
[173] R. Calas, P. Bourgeois, J. Dunoguès, F. Pisciotti, and B. Arréguy, *Bull. Soc. Chim. Fr.*, **1974**, 2556.
[174] S. R. Wilson, L. R. Phillips, and K. J. Natalie, *J. Am. Chem. Soc.*, **101**, 3340 (1979).
[175] R. E. Ireland and M. D. Varney, *J. Am. Chem. Soc.*, **106**, 3668 (1984).
[176] M. Shibasaki, E. Fukasawa, and S. Ikegami, *Tetrahedron Lett.*, **24**, 3497 (1983).
[177] T. Morita, Y. Okamoto, and H. Sakurai, *Synthesis*, **1981**, 745.
[178] D. J. Ager, I. Fleming, and S. K. Patel, *J. Chem. Soc., Perkin Trans. 1*, **1981**, 2520.
[179] S. I. Pennanen, *Synth. Commun.*, **10**, 373 (1980).
[180] P. R. Jones, R. A. Pierce, and A. H. B. Cheng, *Organometallics*, **2**, 12 (1983).
[181] S. R. Wilson and M. F. Price, *J. Am. Chem. Soc.*, **104**, 1124 (1982).
[182] S. R. Wilson and M. F. Price, *Tetrahedron Lett.*, **24**, 569 (1983).
[183] C. Eaborn, I. D. Jenkins, and G. Seconi, *J. Organomet. Chem.*, **131**, 387 (1977).
[184] I. M. Salimgareeva, O. Zh. Zhebarov, N. G. Bogatova, and V. P. Yur'ev, *Zh. Obshch. Khim.*, **51**, 420 (1981) [*C.A.*, **95**, 81087v (1981)].
[185] A. Hosomi, H. Iguchi, J. Sasaki, and H. Sakurai, *Tetrahedron Lett.*, **23**, 551 (1982).

[186] H. Sakurai, A. Hosomi, M. Saito, K. Sasaki, H. Iguchi, J. Sasaki, and Y. Araki, *Tetrahedron*, **39**, 883 (1983).
[187] E. Vedejs and J. G. Reid, *J. Am. Chem. Soc.*, **106**, 4617 (1984).
[188] K. S. Kyler and D. S. Watt, *J. Org. Chem.*, **46**, 5182 (1981).
[189] T. K. Sarkar and N. H. Andersen, *Tetrahedron Lett.*, **1978**, 3513.
[190] H. Urabe and I. Kuwajima, *Tetrahedron Lett.*, **24**, 4241 (1983).
[191] H.-F. Chow and I. Fleming, *Tetrahedron Lett.*, **26**, 397 (1985).
[192] T. Morita, Y. Okamoto, and H. Sakurai, *Tetrahedron Lett.*, **21**, 835 (1980).
[193] A. Hosomi and H. Sakurai, *Chem. Lett. (Jpn.)*, **1981**, 85.
[194] M. Yalpani and G. Wilke, *Chem. Ber.*, **118**, 661 (1985).
[195] G. A. Olah, A. Husain, B. G. B. Gupta, G. F. Salem, and S. C. Naraing, *J. Org. Chem.*, **46**, 5212 (1981); G. A. Olah, A. Husain, and B. P. Singh, *Synthesis*, **1983**, 892.
[196] K. Tamao and N. Ishida, *Tetrahedron Lett.*, **25**, 4249 (1984).
[197] G. Büchi and H. Wuest, *Tetrahedron Lett.*, **1977**, 4305.
[198] S. Fujikura, M. Inoue, K. Utimoto, and H. Nozaki, *Tetrahedron Lett.*, **25**, 1999 (1984).
[199] T. H. Chan, W. Mychajlowskij, B. S. Ong, and D. N. Harpp, *J. Organomet. Chem.*, **107**, C1 (1976).
[200] V. D. Sheludyakov, V. I. Zhun, S. V. Loginov, V. S. Vershinin, A. P. Tsilyurik, and N. I. Peryatinskaya, U.S.S.R. Pat., 1193153 (1984) [*C.A.*, **105**, 153296n (1986)].
[201] P. F. Hudrlik and A. K. Kulkarni, *Tetrahedron Lett.*, **26**, 1389 (1985).
[202] L. H. Sommer and F. J. Evans, *J. Am. Chem. Soc.*, **76**, 1186 (1954).
[203] J. Dunoguès, P. Bourgeois, J.-P. Pillot, G. Mérault, and R. Calas, *J. Organomet. Chem.*, **87**, 169 (1975).
[204] P. F. Hudrlik, R. H. Schwartz, and J. C. Hogan, *J. Org. Chem.*, **44**, 155 (1979).
[205] G. Srivastava, *J. Organomet. Chem.*, **152**, 39 (1978).
[206] K. Tamao, N. Miyake, Y. Kiso, and M. Kumada, *J. Am. Chem. Soc.*, **97**, 5603 (1975).
[207] H. P. On, W. Lewis, and G. Zweifel, *Synthesis*, **1981**, 999.
[208] Y. Sato and Y. Niinomi, *J. Chem. Soc., Chem. Commun.*, **1982**, 56.
[209] T. H. Chan and W. Mychajlowskij, *Tetrahedron Lett.*, **1974**, 3479.
[210] F. Sato, M. Kusakabe, and Y. Kobayashi, *J. Chem. Soc., Chem. Commun.*, **1984**, 1130.
[211] A. K. Sarkar, Ph.D. Thesis, University of Cambridge, 1988.
[212] W. Mychajlowskij and T. H. Chan, *Tetrahedron Lett.*, **1976**, 4439.
[213] K. Uchida, K. Utimoto, and H. Nozaki, *Tetrahedron*, **33**, 2987 (1977).
[214] G. Büchi and H. Wuest, *J. Org. Chem.*, **44**, 546 (1979).
[215] V. N. Odinokov, G. G. Balezina, G. Y. Ishmuratov, I. M. Salimgareeva, N. G. Bogatova, L. M. Zelenova, R. R. Muslukhov, and G. A. Tolstikov, *Khim. Prir. Soedin.*, **1984**, 514; *Chem. Nat. Compd. (Engl. Transl.)*, **1984**, 486.
[216] K. Suzuki, T. Ohkuma, and G. Tsuchihashi, *Tetrahedron Lett.*, **26**, 861 (1985).
[217] K. Suzuki, K. Tomooka, E. Katayama, T. Matsumoto, and G. Tsuchihashi, *J. Am. Chem. Soc.*, **108**, 5221 (1986).
[218] H. Uchiyama, Y. Kobayashi, and F. Sato, *Chem. Lett. (Jpn.)*, **1985**, 467.
[219] Y. Kobayashi, Y. Kitano, Y. Takeda, and F. Sato, *Tetrahedron*, **42**, 2937 (1986).
[220] F. Sato, Y. Kobayashi, O. Takahashi, and T. Kato, *Tennen Yuki Kagobutsu Toronkai Koen*, **27**, 40 (1985) [*C.A.*, **105**, 133554s (1986)].
[221] K. Suzuki, M. Shimazaki, and G. Tsuchihashi, *Tetrahedron Lett.*, **27**, 6237 (1986).
[222] M. Ochiai, T. Masahito, S. Tuda, S. Kenzo, and E. Fujita, *J. Chem. Soc., Chem. Commun.*, **1982**, 281.
[223] P. Bourgeois, R. Calas, and G. Mérault, *J. Organomet. Chem.*, **141**, 23 (1977).
[224] A. A. Petrov, V. A. Kormer, and M. D. Stadnichuk, *Zh. Obshch. Khim.*, **30**, 2243 (1960) [*C.A.*, **55**, 13301f (1961)].
[225] W. S. Johnson, P. H. Crackett, and J. D. Elliott, *Tetrahedron Lett.*, **25**, 3951 (1984).
[226] D. Schinzer, *Angew. Chem. Int. Ed. Engl.*, **23**, 308 (1984).
[227a] B. M. Trost and A. Brandi, *J. Org. Chem.*, **49**, 4811 (1984).

[227b] R. K. Sharma and J. L. Fry, *J. Org. Chem.*, **43**, 2112 (1983).
[228] J.-P. Pillot, J. Dunoguès, and R. Calas, *Bull. Soc. Chim. Fr.*, **1975**, 2143.
[229] G. Déléris, J. Dunoguès, and R. Calas, *J. Organomet. Chem.*, **116,** C45 (1976).
[230] R. E. Ireland, J. D. Godfrey, and S. Thaisrivongs, *J. Org. Chem.*, **49**, 1001 (1984).
[231] A. Hosomi, H. Shoji, and H. Sakurai, *Chem. Lett. (Jpn.)*, **1985**, 1049.
[232] P. DeShong, J. M. Leginus, and S. W. Lander, *J. Org. Chem.*, **51**, 574 (1986).
[233] C. Westerlund, *Tetrahedron Lett.*, **23**, 4835 (1982).
[234] J. E. O'Boyle and K. M. Nicholas, *Tetrahedron Lett.*, **21**, 1595 (1980).
[235] L. F. Kelly, A. S. Narula, and A. J. Birch, *Tetrahedron Lett.*, **21**, 871 (1980).
[236] A. J. Birch, L. F. Kelly, and D. J. Thompson, *J. Chem. Soc., Perkin Trans. 1*, **1981**, 1006.
[237] A. J. Birch, L. F. Kelly, and A. S. Narula, *Tetrahedron*, **38**, 1813 (1982).
[238] T. J. Barton and R. J. Rogido, *J. Org. Chem.*, **40**, 582 (1975).
[239] P. DeShong and J. M. Leginus, *J. Org. Chem.*, **49**, 3421 (1984).
[240] Y. Morizawa, S. Kanemoto, K. Oshima, and H. Nozaki, *Tetrahedron Lett.*, **23**, 2953 (1982).
[241] A. Hosomi, T. Imai, M. Endo, and H. Sakurai, *J. Organomet. Chem.*, **285**, 95 (1985).
[242] T. Fujisawa, M. Kawashima, and S. Ando, *Tetrahedron Lett.*, **25**, 3213 (1984).
[243] B. M. Trost and E. Keinan, *Tetrahedron Lett.*, **21**, 2595 (1980).
[244] T. Fuchikama and I. Ojima, *Tetrahedron Lett.*, **25**, 307 (1984).
[245] N. Ono, T. Yanai, A. Kamimura, and A. Kagi, *J. Chem. Soc., Chem. Commun.*, **1986**, 1285.
[246] J. A. Cella, *J. Org. Chem.*, **47**, 2125 (1982).
[247] G. Manuel, G. Bertrand, P. Mazerolles, and J. Ancelle, *J. Organomet. Chem.*, **212**, 311 (1981).
[248] R. J. Armstrong and L. Weiler, *Can. J. Chem.*, **61**, 2530 (1983).
[249] H. M. R. Hoffmann, A. Weber, and R. J. Giguere, *Chem. Ber.*, **117**, 3325 (1984).
[250] L. R. Hughes, R. Schmidt, and W. S. Johnson, *Bioorg. Chem.*, **8**, 513 (1979).
[251] H. Sakurai, Y. Sakata, and A. Hosomi, *Chem. Lett. (Jpn.)*, **1983**, 409.
[252] P. Albaugh–Robertson and J. A. Katzenellenbogen, *J. Org. Chem.*, **48**, 5288 (1983).
[253] M. Wada, T. Shigehisa, and K. Akiba, *Tetrahedron Lett.*, **26**, 5191 (1985).
[254] A. Hosomi, Y. Sakata, and H. Sakurai, *Tetrahedron Lett.*, **25**, 2383 (1984).
[255] K. C. Nicolaou, R. E. Dolle, A. Chucholowski, and J. L. Randall, *J. Chem. Soc., Chem. Commun.*, **1984**, 1153.
[256] K. Yamamoto, J. Yoshitake, N. T. Qui, and J. Tsuji, *Chem. Lett. (Jpn.)*, **1978**, 859.
[257] H. Nishiyama, T. Naritomi, K. Sakuta, and K. Itoh, *J. Org. Chem.*, **48**, 1557 (1983).
[258] N. H. Andersen, D. A. McCrae, D. B. Grotjahn, S. Y. Gabhe, L. J. Theodore, R. M. Ippolito, and T. K. Sarkar, *Tetrahedron*, **37**, 4069 (1981).
[259] A. Ricci, A. Degl'lnnocenti, M. Fiorenza, M. Taddei, M. A. Spartera, and D. R. M. Walton, *Tetrahedron Lett.*, **23**, 577 (1982).
[260] H. J. Reich, R. E. Olson, and M. C. Clark, *J. Am. Chem. Soc.*, **102,** 1423 (1980).
[261] K. Mizuno, M. Ikeda, and Y. Otsuji, *Tetrahedron Lett.*, **26**, 461 (1985).
[262] W. S. Johnson, Y.-Q. Chen, and M. S. Kellogg, *J. Am. Chem. Soc.*, **105**, 6653 (1983).
[263] W. S. Johnson, C. Newton, and S. D. Lindell, *Tetrahedron Lett.*, **27**, 6027 (1986).
[264] A. Itoh, T. Saito, K. Oshima, and H. Nozaki, *Bull. Chem. Soc. Jpn.*, **54**, 1456 (1981).
[265] R. J. Armstrong and L. Weiler, *Can. J. Chem.*, **64**, 584 (1986).
[266] R. J. Armstrong, F. L. Harris, and L. Weiler, *Can. J. Chem.*, **64**, 1002 (1986).
[267] E. J. Corey and W. L. Seibel, *Tetrahedron Lett.*, **27**, 905 (1986).
[268] L. D. Boardman, V. Bagheri, H. Sawada, and E. Negishi, *J. Am. Chem. Soc.*, **106**, 6105 (1984).
[269] L. R. Robinson, G. T. Burns, and T. J. Barton, *J. Am. Chem. Soc.*, **107**, 3935 (1985).
[270] S. D. Burke, J. O. Saunders, J. A. Oplinger, and C. W. Murtiashaw, *Tetrahedron Lett.*, **26**, 1131 (1985).
[271] T. H. Chan, P. W. K. Lau, and W. Mychajlowskij, *Tetrahedron Lett.*, **1977**, 3317.
[272] K. E. Stevens and L. A. Paquette, *Tetrahedron Lett.*, **22**, 4393 (1981).
[273] K. Yamamoto, M. Ohta, and J. Tsuji, *Chem. Lett. (Jpn.)*, **1979**, 713.
[274] M. Ali, *Dacca Univ. Stud. Pat. B*, **28**, 33 (1980) [*C.A.*, **94**, 65749 (1981)].

[275] S. A. Carr and W. P. Weber, *J. Org. Chem.*, **50**, 2782 (1985).
[276] I. Paterson, Ph.D. Thesis, Cambridge University, 1979.
[277] T. S. Tan, A. N. Mather, G. Procter, and A. H. Davidson, *J. Chem. Soc., Chem. Commun.*, **1984**, 585.
[278] D. Wang and T. H. Chan, *J. Chem. Soc., Chem. Commun.*, **1984**, 1273.
[279] R. P. Alexander and I. Paterson, *Tetrahedron Lett.*, **24**, 5911 (1983).
[280] J. Enda and I. Kuwajima, *J. Am. Chem. Soc.*, **107**, 5495 (1985).
[281] L. Coppi, A. Ricci, and M. Taddei, *Tetrahedron Lett.*, **28**, 973 (1987).
[282] D. Seyferth, J. Pornet, and R. M. Weinstein, *Organometallics*, **1**, 1651 (1982).
[283] K. Itoh, M. Fukui, and Y. Kurachi, *J. Chem. Soc., Chem. Commun.*, **1977**, 500.
[284] A. Hosomi, H. Hashimoto, and H. Sakurai, *Tetrahedron Lett.*, **21**, 951 (1980).
[285] T. H. Chan and G. J. Kang, *Tetrahedron Lett.*, **23**, 3011 (1982).
[286] T. H. Chan and G. J. Kang, *Can. J. Chem.*, **63**, 3102 (1985).
[287] J. R. Green, M. Majewski, B. I. Alo, and V. Snieckus, *Tetrahedron Lett.*, **27**, 535 (1986).
[288] M. Ochiai and E. Fujita, *J. Chem. Soc., Chem. Commun.*, **1980**, 1118.
[289] T. Aono and M. Hesse, *Helv. Chim. Acta*, **67**, 1448 (1984).
[290] B. M. Trost and B. M. Coppola, *J. Am. Chem. Soc.*, **104**, 6879 (1982).
[291] I. Kuwajima, T. Tanaka, and K. Atsumi, *Chem. Lett. (Jpn.)*, **1979**, 779.
[292] B. M. Trost and M. J. Fray, *Tetrahedron Lett.*, **25**, 4605 (1984).
[293] T. Hayashi, K. Kabeta, I. Hamachi, and M. Kumada, *Tetrahedron Lett.*, **24**, 2865 (1983).
[294] S. E. Denmark and E. J. Weber, *Helv. Chim. Acta*, **66**, 1655 (1983).
[295] A. Itoh, K. Oshima, and H. Nozaki, *Tetrahedron Lett.*, **1979**, 1783.
[296] K. Mikami, T. Maeda, N. Kishi, and T. Nakai, *Tetrahedron Lett.*, **25**, 5151 (1984).
[297] I. Ojima, Jpn. Kokai Tokkyo Koho 78 34718 (1976) [*C.A.*, **89**, 108154e (1978)].
[298] P. Grossen, P. Herold, P. Mohr, and C. Tamm, *Helv. Chim. Acta*, **67**, 1625 (1984).
[299] C. H. Heathcock, S. Kiyooka, and T. A. Blumenkopf, *J. Org. Chem.*, **49**, 4214 (1984).
[300] M. T. Reetz and A. Jung, *J. Am. Chem. Soc.*, **105**, 4833 (1983).
[301] M. T. Reetz, K. Kesseler, and A. Jung, *Tetrahedron Lett.*, **25**, 729 (1984).
[302] S. J. Danishefsky, M. P. DeNinno, G. B. Phillips, R. L. Zelle, and P. A. Lartey, *Tetrahedron*, **42**, 2809 (1986).
[303] I. Kuwajima, E. Nakamura, and M. Shimizu, *J. Am. Chem. Soc.*, **104**, 1025 (1982).
[304] L. A. Paquette, Ohio State University, Columbus, personal communication.
[305] T. Yamazaki and N. Ishikawa, *Chem. Lett. (Jpn.)*, **1984**, 521; T. Yamazaki, K. Takita, and N. Ishikawa, *Nippon Kagaku Kaishi*, **1985**, 2131 [*C.A.*, **105**, 190436d (1986)].
[306] T. Hiyama, M. Obayashi, and M. Sawahata, *Tetrahedron Lett.*, **24**, 4113 (1983).
[307] R. Corriu, N. Escudié, and C. Guérin, *J. Organomet. Chem.*, **271**, C7 (1984).
[308] A. Hosomi, M. Saito, and H. Sakurai, *Tetrahedron Lett.*, **1979**, 429.
[309] A. Hosomi, Y. Araki, and H. Sakurai, *J. Org. Chem.*, **48**, 3122 (1983).
[310] H. Sakurai, Y. Eriyama, Y. Kamiyama, and Y. Nakadaira, *J. Organomet. Chem.*, **264**, 229 (1984).
[311] M. J. Wanner, N. P. Willard, G.-J. Koomen, and U. K. Pandit, *J. Chem. Soc., Chem. Commun.*, **1986**, 396.
[312] C. Santelli–Bouvier, *Tetrahedron Lett.*, **25**, 4371 (1984).
[313] G. Déléris, J.Dunoguès, and R. Calas, *J. Organomet. Chem.*, **93**, 43 (1975).
[314] Y. Sato and K. Hitomi, *J. Chem. Soc., Chem. Commun.*, **1983**, 170.
[315] R. L. Danheiser, D. J. Carini, and C. A. Kwasigroch, *J. Org. Chem.*, **51**, 3870 (1986).
[316] G. Majetich, M. Behnke, and K. Hull, *J. Org. Chem.*, **50**, 3615 (1985).
[317] A. Hosomi and H. Sakurai, *J. Am. Chem. Soc.*, **99**, 1673 (1977).
[318] H. Sakurai, A. Hosomi, and J. Hayashi, *Org. Synth.*, **62**, 86 (1984).
[319] R. Pardo, J.-P. Zahra, and M. Santelli, *Tetrahedron Lett.*, **1979**, 4557.
[320] H. O. House, P. C. Gaa, J. H. C. Lee, and D. VanDerveer, *J. Org. Chem.*, **48**, 1670 (1983).
[321] H. O. House, T. S. B. Sayer, and C.-C. Yau, *J. Org. Chem.*, **43**, 2153 (1978).
[322] A. Hosomi, H. Kobayashi, and H. Sakurai, *Tetrahedron Lett.*, **21**, 955 (1980).
[323] T. Yanami, M. Miyashita, and A. Yoshikoshi, *J. Org. Chem.*, **45**, 607 (1980).

[324] K. B. Becker and R. W. Pfluger, *Tetrahedron Lett.*, **1979**, 3713.
[325] H. Rudolph, H. Schon, and R. Keese, *Chimia*, **35**, 12 (1981).
[326] T. A. Blumenkopf and C. H. Heathcock, *J. Am. Chem. Soc.*, **105**, 2354 (1983).
[327] S. Danishefsky and M. Kahn, *Tetrahedron Lett.*, **22**, 485 (1981).
[328] A. A. Ponoras, *Tetrahedron Lett.*, **21**, 4803 (1980).
[329] S. Knapp, U. O'Connor, and D. Mobilio, *Tetrahedron Lett.*, **21**, 4557 (1980).
[330] A. Hosomi, H. Hashimoto, H. Kobayashi, and H. Sakurai, *Chem. Lett. (Jpn.)*, **1979**, 245.
[331] M. Santelli, D. El-Abed, and A. Jellal, *J. Org. Chem.*, **51**, 1199 (1986).
[332] R. L. Danheiser and D. M. Fink, *Tetrahedron Lett.*, **26**, 2509 (1985).
[333] D. El-Abed, A. Jellal, and M. Santelli, *Tetrahedron Lett.*, **25**, 1463 (1984).
[334] W. R. Roush and A. E. Watts, *J. Am. Chem. Soc.*, **106**, 721 (1984).
[335] T. Tokoroyama, M. Tsukamoto, and H. Iio, *Tetrahedron Lett.*, **25**, 5067 (1984).
[336] G. Majetich, J. Defauw, K. Hull, and T. Shawe, *Tetrahedron Lett.*, **26**, 4711 (1985).
[337] G. Majetich, K. Hull, J. Defauw, and T. Shawe, *Tetrahedron Lett.*, **26**, 2755 (1985).
[338] G. Majetich, K. Hull, J. Defauw, and R. Desmond, *Tetrahedron Lett.*, **26**, 2747 (1985).
[339] G. Majetich, K. Hull, and R. Desmond, *Tetrahedron Lett.*, **26**, 2751 (1985).
[340] C. H. Heathcock, E. F. Kleinman, and E. S. Binkley, *J. Am. Chem. Soc.*, **104**, 1054 (1982).
[341] R. L. Funk, G. L. Bolton, J. U. Daggett, M. M. Hansen, and L. H. M. Horcher, *Tetrahedron*, **41**, 3479 (1985).
[342] C. H. Heathcock, K. M. Smith, and T. A. Blumenkopf, *J. Am. Chem. Soc.*, **108**, 5022 (1986).
[343] T. K. Jones and S. E. Denmark, *Helv. Chim. Acta*, **66**, 2377 (1983).
[344] S. E. Denmark, K. L. Habermas, G. A. Hite, and T. K. Jones, *Tetrahedron*, **42**, 2821 (1986).
[345] G. T. Crisp, W. J. Scott, and J. K. Stille, *J. Am. Chem. Soc.*, **106**, 7500 (1984).
[346] R. L. Danheiser, D. J. Carini, D. M. Fink, and A. Basak, *Tetrahedron*, **39**, 935 (1983).
[347] A. Hosomi and H. Sakurai, *Tetrahedron Lett.*, **1977**, 4041.
[348] H. Uno, *J. Org. Chem.*, **51**, 350 (1986).
[349] M. Ochiai, M. Arimoto, and E. Fujita, *Tetrahedron Lett.*, **22**, 1115 (1981).
[350] H. Uno, S. Fujika, and H. Suzuki, *Bull. Chem. Soc. Jpn.*, **59**, 1267 (1986).
[351] S. Nakanishi, M. Higuchi, and T. C. Flood, *J. Chem. Soc., Chem. Commun.*, **1986**, 30.
[352] A. Hosomi, M. Endo, and H. Sakurai, *Chem. Lett. (Jpn.)*, **1976**, 941.
[353] T. Tsunoda, M. Suzuki, and R. Noyori, *Tetrahedron Lett.*, **21**, 71 (1980).
[354] Mitsui Petrochemical Industries Ltd., Jpn. Kokai Tokkyo Koho 56 118032 (1981) [*C.A.*, **96**, 84742m (1982)].
[355] H. Sakurai, K. Sasaki, and A. Hosomi, *Tetrahedron Lett.*, **22**, 745 (1981).
[356] T. Mukaiyama, H. Nagaoka, M. Murakami, and M. Ohshima, *Chem. Lett. (Jpn.)*, **1985**, 977.
[357] M. Kawai, M. Onaka, and Y. Izumi, *Chem. Lett. (Jpn.)*, **1986**, 381.
[358] C. Brückner, H. Lorey, and H.-U. Reissig, *Angew. Chem. Int. Ed. Engl.*, **25**, 556 (1986).
[359] H. Sakurai, K. Sasaki, J. Hayashi, and A. Hosomi, *J. Org. Chem.*, **49**, 2808 (1984).
[360] I. Ojima and M. Kumagai, *Chem. Lett. (Jpn.)*, **1978**, 575.
[361] M. Wada, T. Shigehisa, and K. Akiba, *Tetrahedron Lett.*, **24**, 1711 (1983).
[362] J. Pornet, *Tetrahedron Lett.*, **21**, 2049 (1980).
[363] I. Fleming and A. Pearce, *J. Chem. Soc., Perkin Trans. 1*, **1981**, 251.
[364] H. Nishiyama and K. Itoh, *J. Org. Chem.*, **47**, 2496 (1982).
[365] H. Nishiyama, S. Narimatsu, K. Sakuta, and K. Itoh, *J. Chem. Soc., Chem. Commun.*, **1982**, 459.
[366] A. Hosomi, M. Ando, and H. Sakurai, *Chem. Lett. (Jpn.)*, **1986**, 365.
[367] S. Kiyooka, H. Sasaoka, R. Fujiyama, and C. H. Heathcock, *Tetrahedron Lett.*, **25**, 5331 (1984).
[368] A. P. Kozikowski and K. L. Sorgi, *Tetrahedron Lett.*, **23**, 2281 (1982).
[369] M. D. Lewis, J. K. Cha, and Y. Kishi, *J. Am. Chem. Soc.*, **104**, 4976 (1982).
[370] A. P. Kozikowski and K. L. Sorgi, *Tetrahedron Lett.*, **24**, 1563 (1983).
[371] D. Seebach, R. Imwinkelried, and G. Stucky, *Angew. Chem. Int. Ed. Engl.*, **25**, 178 (1986).
[372] R. Imwinkelried and D. Seebach, *Angew. Chem. Int. Ed. Engl.*, **24**, 765 (1985).
[373] W. S. Johnson and M. F. Chan, *J. Org. Chem.*, **50**, 2598 (1985).

[374] T. Hirao, S. Kohno, J. Enda, Y. Ohshiro, and T. Agawa, *Tetrahedron Lett.*, **22**, 3633 (1981).
[375] T. Hirao, S. Kohno, Y. Ohshiro, and T. Agawa, *Bull. Chem. Soc. Jpn.*, **56**, 1569 (1983).
[376] H.-F. Chow and I. Fleming, *J. Chem. Soc., Perkin Trans. 1*, **1984**, 1815.
[377] L. E. Overman, A. Castaneda, and T. A. Blumenkopf, *J. Am. Chem. Soc.*, **108**, 1303 (1986).
[378] M. A. Tius, *Tetrahedron Lett.*, **22**, 3335 (1981).
[379] L. E. Overman, T. A. Blumenkopf, A. Castaneda, and A. S. Thompson, *J. Am. Chem. Soc.*, **108**, 3516 (1986).
[380] B. M. Trost and E. Murayama, *J. Am. Chem. Soc.*, **103**, 6529 (1981).
[381] R. L. Danheiser and D. J. Carini, *J. Org. Chem.*, **45**, 3925 (1980).
[382] A. Hosomi, M. Endo, and H. Sakurai, *Chem. Lett. (Jpn.)*, **1978**, 499.
[383] S. Danishefsky and J. F. Kerwin, *J. Org. Chem.*, **47**, 3803 (1982).
[384] Y. Ichikawa, M. Isobe, and T. Goto, *Tetrahedron Lett.*, **25**, 5049 (1984).
[385] D. J. Hart and Y.-M. Tsai, *Tetrahedron Lett.*, **22**, 1567 (1981).
[386] M. Aratani, K. Sawada, and M. Hashimoto, *Tetrahedron Lett.*, **23**, 3921 (1982).
[387] A. P. Kozikowski and P. Park, *J. Org. Chem.*, **49**, 1674 (1984).
[388] H. Hiemstra, M. H. A. M. Sno, R. J. Vijn, and W. N. Speckamp, *J. Org. Chem.*, **50**, 4014 (1985).
[389] H. Hiemstra, H. P. Fortgens, and W. N. Speckamp, *Tetrahedron Lett.*, **26**, 3155 (1985).
[390] T. Shono, Y. Matsumura, K. Uchida, and H. Kobayashi, *J. Org. Chem.*, **50**, 3243 (1985).
[391] S. D. Larsen, P. A. Grieco, and W. F. Fobare, *J. Am. Chem. Soc.*, **108**, 3512 (1986).
[392] P. A. Grieco and W. F. Fobare, *Tetrahedron Lett.*, **27**, 5067 (1986).
[393] K. Ohga and P. S. Mariano, *J. Am. Chem. Soc.*, **104**, 617 (1982).
[394] F.-T. Chiu, J. W. Ullrich, and P. S. Mariano, *J. Org. Chem.*, **49**, 228 (1984).
[395] H. Vorbrüggen and K. Krolikiewicz, *Tetrahedron Lett.*, **24**, 889 (1983).
[396] J.-C. Gramain and R. Remuson, *Tetrahedron Lett.*, **26**, 4083 (1985).
[397] L. E. Overman and R. M. Burk, *Tetrahedron Lett.*, **25**, 5739 (1984).
[398] L. E. Overman, T. C. Malone, and G. P. Meier, *J. Am. Chem. Soc.*, **105**, 6993 (1983).
[399] L. E. Overman and T. C. Malone, *J. Org. Chem.*, **47**, 5297 (1982).
[400] L. E. Overman, K. L. Bell, and F. Ito, *J. Am. Chem. Soc.*, **106**, 4192 (1984).
[401] L. E. Overman and N.-H. Lin, *J. Org. Chem.*, **50**, 3669 (1985).
[402] T. Sasaki, A. Nakanishi, and M. Ohno, *J. Org. Chem.*, **47**, 3219 (1982).
[403] R. Calas, J. Dunoguès, J.-P. Pillot, C.Biran, F. Pisciotti, and B. Arréguy, *J. Organomet. Chem.*, **85**, 149 (1975).
[404] M. Grignon–Dubois, J. Dunoguès, and R. Calas, *J. Organomet. Chem.*, **97**, 31 (1975).
[405] P. Magnus, F. Cooke, and T. Sarkar, *Organometallics*, **1**, 562 (1982).
[406] J.-P. Pillot, J. Dunoguès, and R. Calas, *Tetrahedron Lett.*, **1976**, 1871.
[407] B. Bennetau, J.-P. Pillot, and J. Dunoguès, unpublished results. B. Bennetau, Thèse d'Etat, Bordeaux, 1985.
[408] W. E. Fristad, Y.-K. Han, and L. A. Paquette, *J. Organomet. Chem.*, **174**, 27 (1979).
[409] J. J. Lewis, Ph.D. Thesis, Cambridge University, 1986.
[410] A. Hosomi, M. Saito, and H. Sakurai, *Tetrahedron Lett.*, **21**, 3783 (1980).
[411] Y. Naruta, H. Uno, and K. Maruyama, *Chem. Lett. (Jpn.)*, **1982**, 961.
[412] K. Hiroi and L.-M. Chen, *J. Chem. Soc., Chem. Commun.*, **1981**, 377.
[413] A. Hosomi, H. Hashimoto, and H. Sakurai, *J. Org. Chem.*, **43**, 2551 (1978).
[414] M. Ochiai, K. Sumi, E. Fujita, and S. Tada, *Chem. Pharm. Bull.*, **31**, 3346 (1983).
[415] K. Hiroi and H. Sato, *Chem. Lett. (Jpn.)*, **1986**, 1723.
[416] J.-P. Pillot, J. Dunoguès, and R. Calas, *Synth. Commun.*, **9**, 395 (1979).
[417] K. Fukuzaki, E. Nakamura, and I. Kuwajima, *Tetrahedron Lett.*, **25**, 3591 (1984).
[418] J.-P. Pillot, J. Dunoguès, and R. Calas, *J. Chem. Res. (S)*, **1977**, 268.
[419] S. D. Burke, C. W. Murtiashaw, M. S. Dike, S. M. Strickland, and J. O. Saunders, *J. Org. Chem.*, **46**, 2400 (1981).
[420] E. Nakamura, K. Fukuzaki, and I. Kuwajima, *J. Chem. Soc., Chem. Commun.*, **1983**, 499.
[421] F. Cooke, R. Moerck, J. Schwindemann, and P. Magnus, *J. Org. Chem.*, **45**, 1046 (1980).
[422] J.-P. Pillot, B. Bennetau, J. Dunoguès, and R. Calas, *Tetrahedron Lett.*, **21**, 4717 (1980).

[423] E. Negishi, L. D. Boardman, J. M. Tour, H. Sawada, and C. L. Rand, *J. Am. Chem. Soc.*, **105**, 6344 (1983).
[424] P. Bourgeois, G. Mérault, N. Duffaut, and R. Calas, *J. Organomet. Chem.*, **59**, 145 (1973).
[425] H. Hamana and T. Sugasawa, *Chem. Lett. (Jpn.)*, **1985**, 921.
[426] K. Isaac, P. Kocienski, and S. Campbell, *J. Chem. Soc., Chem. Commun.*, **1983**, 249.
[427] S. Ohashi, W. E. Ruch, and G. B. Butler, *J. Org. Chem.*, **46**, 614 (1981).
[428] H. Jolibois, A. Doucet, and R. Perrot, *Helv. Chim. Acta*, **59**, 1352 (1976).
[429] H. Jolibois, A. Doucet, and R. Perrot, *Helv. Chim. Acta*, **58**, 1801 (1975).
[430] A. Doucet and R. Perrot, *Ann. Sci. Univ. Besançon, Chim.*, **1979**, 15 and 17 [*C.A.*, **95**, 150746y (1981)].
[431] G. A. Olah and C. Rochin, *J. Org. Chem.*, **52**, 701 (1987).
[432] E. J. Corey and H. Estreicher, *Tetrahedron Lett.*, **21**, 1113 (1980).
[433] A. R. Bassindale, A. G. Brook, P. F. Jones, and J. A. G. Stewart, *J. Organomet. Chem.*, **152**, C25 (1978).
[434] E. P. Plueddemann and G. Fanger, *J. Am. Chem. Soc.*, **81**, 2632 (1959).
[435] V. Bažant and V. Matousek, *Coll. Czech. Chem. Commun.*, **24**, 3758 (1959).
[436] G. Manuel, P. Mazerolles, M. Lesbre, and J.-P. Pradel, *J. Organomet. Chem.*, **61**, 147 (1973).
[437] T. Hayashi, Y. Okamoto, K. Kabeta, T. Hagihara, and M. Kumada, *J. Org. Chem.*, **49**, 4224 (1984).
[438] P. F. Hudrlik and G. P. Withers, *Tetrahedron Lett.*, **1976**, 29.
[439] I. Fleming and N. K. Terrett, *Tetrahedron Lett.*, **25**, 5103 (1984).
[440] B.-W. Au-Yeung, J.-W. Xu, and J. S. Qui, *Huaxue Xuebao*, **44**, 479 (1986) [*C.A.*, **105**, 226137y (1986)].
[441] M. Koreeda and M. A. Ciufolini, *J. Am. Chem. Soc.*, **104**, 2308 (1982).
[442] E. Vedejs, J. B. Campbell, R. C. Gadwood, J. D. Rogers, K. L. Spear, and Y. Watanabe, *J. Org. Chem.*, **47**, 1534 (1982).
[443] J. M. Reuter, A. Sinha, and R. G. Salomon, *J. Org. Chem.*, **43**, 2438 (1978).
[444] J. Yoshida, T. Murata, and S. Isoe, *Tetrahedron Lett.*, **27**, 3373 (1986).
[445] G. Stork and E. Colvin, *J. Am. Chem. Soc.*, **93**, 2080 (1971).
[446] P. F. Hudrlik, J. P. Arcoleo, R. H. Schwartz, R. N. Misra, and R. J. Rona, *Tetrahedron Lett.*, **1977**, 591.
[447] P. F. Hudrlik, R. H. Schwartz, and A. K. Kulkarni, *Tetrahedron Lett.*, **1979**, 2233.
[448] R. F. Cunico, *Tetrahedron Lett.*, **27**, 4269 (1986).
[449] G. Stork and M. E. Jung, *J. Am. Chem. Soc.*, **96**, 3682 (1974).
[450] J. J. Eisch and J. T. Trainor, *J. Org. Chem.*, **28**, 2870 (1963).
[451] P. F. Hudrlik, A. M. Hudrlik, R. J. Rona, R. N. Misra, and G. P. Withers, *J. Am. Chem. Soc.*, **99**, 1993 (1977).
[452] P. F. Hudrlik, G. Nagendrappa, A. K. Kulkarni, and A. M. Hudrlik, *Tetrahedron Lett.*, **1979**, 2237.
[453] P. F. Hudrlik, A. M. Hudrlik, R. N. Misra, D. Peterson, G. P. Withers, and A. K. Kulkarni, *J. Org. Chem.*, **45**, 4444 (1980).
[454] P. F. Hudrlik, R. N. Misra, A. M. Hudrlik, R. J. Rona, and J. P. Arcoleo, *Tetrahedron Lett.*, **1976**, 1453.
[455] P. F. Hudrlik, C.-N. Wan, and G. P. Withers, *Tetrahedron Lett.*, **1976**, 1449.
[456] A. R. Bassindale, A. G. Brook, P. Chen, and J. Lennon, *J. Organomet. Chem.*, **94**, C21 (1975).
[457] I. Fleming and T. W. Newton, *J. Chem. Soc., Perkin Trans. 1*, **1984**, 119.
[458] G. Nagendrappa and T. J. Vidyapati, *J. Organomet. Chem.*, **280**, 31 (1985).
[459] P. F. Hudrlik, A. M. Hudrlik, and A. K. Kulkarni, *J. Am. Chem. Soc.*, **107**, 4260 (1985).
[460] F. Sato, H. Kanbara, and Y. Tanaka, *Tetrahedron Lett.*, **25**, 5063 (1984).
[461] F. Sato, H. Watanabe, Y. Tanaka, T. Yamaji, and M. Sato, *Tetrahedron Lett.*, **24**, 1041 (1983).
[462] L. A. Paquette and Y.-K. Han, *J. Am. Chem. Soc.*, **103**, 1831 (1981).
[463] L. A. Paquette, R. A. Galemmo, J.-C. Caille, and R. S. Valpey, *J. Org. Chem.*, **51**, 686 (1986).

[464] G. Schmid and W. Hofheinz, *J. Am. Chem. Soc.*, **105**, 624 (1983).
[465] G. Büchi and H. Wuest, *J. Am. Chem. Soc.*, **100**, 294 (1978).
[466] B.-W. Au-Yeung and Y. Wang, *J. Chem. Soc., Chem. Commun.*, **1985**, 825.
[467] M. Bertrand, J.-P. Dulcere, and G. Gil, *Tetrahedron Lett.*, **21**, 1945 (1980).
[468] T. H. Chan, B. S. Ong, and W. Mychajlowskij, *Tetrahedron Lett.*, **1976**, 3253; B. S. Ong and T. H. Chan, *ibid.*, **1976**, 3257.
[469] V. V. Kormachov, Y. N. Mitrasov, and V.A. Kukhtin, *Zh. Obshch. Khim.*, **50**, 1884 (1980) [*C.A.*, **94**, 4068 (1981)].
[470] T. J. Katz, S. J. Lee, and M. A. Shippey, *J. Mol. Catal.*, **8**, 219 (1980).
[471] J.-P. Pillot, J. Dunoguès, and R. Calas, *Synthesis*, **1977**, 469.
[472] H. Nishiyama, S. Narimatsu, and K. Itoh, *Tetrahedron Lett.*, **22**, 5285 (1981).
[473] H. Nishiyama, S. Narimatsu, and K. Itoh, *Tetrahedron Lett.*, **22**, 5289 (1981).
[474] Unpublished results, quoted in H. Nishiyama and K. Itoh, *Yuki Gosei Kagaku Kyokaishi*, **40**, 518 (1982) [*C.A.*, **97**, 110062g (1982)].
[475] E. A. Chernyshev, N. G. Komalenkova, O. B. Afanasova, A. V. Kisin, V. M. Nosova, N. N. Silkina, and T. F. Slyusarenko, *Zh. Obshch. Khim.*, **50**, 1040 (1980) [*C.A.*, **93**, 186446u (1980)].
[476] B. M. Trost and M. Shimizu, *J. Am. Chem. Soc.*, **105**, 6757 (1983).
[477] B. M. Trost and M. Shimizu, *J. Am. Chem. Soc.*, **104**, 4299 (1982).
[478] M. E. Jung and T. A. Blumenkopf, *Tetrahedron Lett.*, **1978**, 3657.
[479] M. Ochiai, E. Fujita, M. Arimoto, and H. Yamaguchi, *J. Chem. Soc., Chem. Commun.*, **1982**, 1108.
[480] M. Ochiai and E. Fujita, *Tetrahedron Lett.*, **24**, 777 (1983).
[481] R. B. Miller and G. McGarvey, *J. Org. Chem.*, **44**, 4623 (1979).
[482] R. Nagel and H. W. Post, *J. Org. Chem.*, **17**, 1379 (1952).
[483] A. G. Brook, J. M. Duff, P. Hitchcock, and R. Mason, *J. Organomet. Chem.*, **113**, C11 (1976).
[484] K. E. Koenig and W. P. Weber, *Tetrahedron Lett.*, **1973**, 2533.
[485] H. C. Brown, D. H. Bowman, S. Misumi, and M. K. Unni, *J. Am. Chem. Soc.*, **89**, 4531 (1967).
[486] L. A. Paquette, G. J. Wells, K. A. Horn, and T. H. Yan, *Tetrahedron*, **39**, 913 (1983).
[487] W. Barth and L. A. Paquette, *J. Org. Chem.*, **50**, 2438 (1985).
[488] D. Dhanak, C. B. Reese, and D. E. Williams, *J. Chem. Soc., Chem. Commun.*, **1984**, 988.
[489] R. B. Miller and G. McGarvey, *Synth. Commun.*, **8**, 291 (1978).
[490] C. Huynh and G. Linstrumelle, *Tetrahedron Lett.*, **1979**, 1073.
[491] R. B. Miller and T. Reichenbach, *Tetrahedron Lett.*, **1974**, 543.
[492] R. Amouroux and T. H. Chan, *Tetrahedron Lett.*, **1978**, 4453.
[493] B. B. Snider and M. Karras, *J. Organomet. Chem.*, **179**, C37 (1979).
[494] T. H. Chan and K. Koumaglo, *Tetrahedron Lett.*, **27**, 883 (1986).
[495] E. J. Thomas and G. H. Whitham, *J. Chem. Soc., Chem. Commun.*, **1979**, 212.
[496] F. Duboudin, *J. Organomet. Chem.*, **174**, C18 (1979).
[497] M. Ochiai, K. Sumi, Y. Nagao, and E. Fujita, *Tetrahedron Lett.*, **26**, 2351 (1985).
[498] M. Ochiai, K. Sumi, Y. Nagao, E. Fujita, M. Arimoto, and H. Yamaguchi, *J. Chem. Soc., Chem. Commun.*, **1985**, 697.
[499] E. R. Koft and A. B. Smith, *J. Am. Chem. Soc.*, **104**, 2659 (1982).
[500] T. H. Chan and K. Koumaglo, *J. Organomet. Chem.*, **285**, 109 (1985).
[501] T. C. Wu, D. Wittenberg, and H. Gilman, *J. Org. Chem.*, **25**, 596 (1960).
[502] P. W. K. Lau and T. H. Chan, *Tetrahedron Lett.*, **1978**, 2383.
[503] K. Koumaglo and T. H. Chan, *Tetrahedron Lett.*, **25**, 717 (1984).
[504] R. J. P. Corriu, J. Massé, and D. Samaté, *J. Organomet. Chem.*, **93**, 71 (1975).
[505] R. J. P. Corriu, C. Guérin, and J. M'Boula, *Tetrahedron Lett.*, **22**, 2985 (1981).
[506] L. F. Cason and H. G. Brooks, *J. Am. Chem. Soc.*, **74**, 4582 (1952) and *J. Org. Chem.*, **19**, 1278 (1954).
[507] T. H. Chan, E. Chang, and E. Vinokur, *Tetrahedron Lett.*, **1970**, 1137.

[508] G. R. Buell, R. Corriu, C. Guérin, and L. Spialter, *J. Am. Chem. Soc.*, **92**, 7424 (1970).
[509] D. Seyferth, T. Wada, and G. Raab, *Tetrahedron Lett.*, **1960**, 20.
[510] R. J. P. Corriu, J. M. Fernandez, and C. Guérin, *J. Organomet. Chem.*, **192**, 347 (1980).
[511] P. Jutzi and A. Seufert, *J. Organomet. Chem.*, **169**, 357 (1979).
[512] M. Berglund, C. Andersson, and R. Larsson, *J. Organomet. Chem.*, **314**, 61 (1986).
[513] J. M. Kliegman, *J. Organomet. Chem.*, **29**, 73 (1971).
[514] A. Riahi, J. Cossy, J. Muzart, and J.-P. Pete, *Tetrahedron Lett.*, **26**, 839 (1985).
[515] B. M. Trost and D. M. T. Chan, *J. Am. Chem. Soc.*, **101**, 6429 and 6432 (1979).
[516] B. M. Trost, *Angew. Chem. Int. Ed. Engl.*, **25**, 1 (1986).
[517] W. P. Weber, R. A. Felix, A. K. Willard, and K. E. Koenig, *Tetrahedron Lett.*, **1971**, 4701.
[518] K. Kikukawa, K. Ikenaga, F. Wada, and T. Matsuda, *Chem. Lett. (Jpn.)*, **1983**, 1337.
[519] A. Hallberg and C. Westerlund, *Chem. Lett. (Jpn.)*, **1982**, 1993.
[520] K. Karabelas and A. Hallberg, *J. Org. Chem.*, **51**, 5286 (1986).
[521] J. A. Soderquist and K. L. Thompson, *J. Organomet. Chem.*, **159**, 237 (1978).
[522] M. G. Voronkov, N. F. Chernov, and I. D. Kalikhman, *Dokl. Akad. Nauk SSSR*, **233**, 361 (1977); *Dokl. Chem. (Engl. Transl.)*, **233**, 138 (1977) [*C.A.*, **87**, 135706b (1977)].
[523] D. Seyferth and N. Kahlen, *Zeit. Naturforsh.*, **14b**, 137 (1959).
[524] M. Ochiai, S. Tada, M. Arimoto, and E. Fujita, *Chem. Pharm. Bull.*, **30**, 2836 (1982).
[525] M. Ochiai, E. Fujita, M. Arimoto, and H. Yamaguchi, *Chem. Pharm. Bull.*, **32**, 5027 (1984).
[526] M. Ochiai, E. Fujita, M. Arimoto, and H. Yamaguchi, *Chem. Pharm. Bull.*, **30**, 3994 (1982).
[527] P. Brownbridge, *Synthesis*, **1983**, 1 and 85.
[528] K. Utimoto, Department of Industrial Chemistry, Faculty of Engineering, Kyoto University, personal communication.
[529] R. P. Alexander, Ph.D. Thesis, University of London, 1985.
[530] Y. Sato, Faculty of Pharmaceutical Sciences, Nagoya City University, personal communication.
[531] S. E. Denmark, University of Illinois, Urbana, personal communication.
[532] V. M. Vdovin, K. S. Pushcheyeva, and A. D. Petrov, *Izv. Akad. Nauk SSSR, Otd. Khim. Nauk*, **1961**, 1275; *Bull. Acad. Sci. USSR, Div. Chem. Sci. (Engl. Transl.)*, **1961**, 1185.
[533] C. S. Kraihanzel and J. E. Poist, *J. Organomet. Chem.*, **8**, 239 (1967).
[534] S. F. Chen and P. S. Mariano, *Tetrahedron Lett.*, **26**, 47 (1985).
[535] L. Birkofer and W. Weniger, *Chem. Ber.*, **106**, 3595 (1973).
[536] R. Calas and J. Dunoguès, *J. Organomet. Chem.*, **27**, C21 (1971).
[537] M. Laguerre, J. Dunoguès, and R. Calas, *Tetrahedron*, **34**, 1823 (1978).
[538] M. Keil and F. Effenberger, *Chem. Ber.*, **115**, 1103 (1982).
[539] J. Dunoguès, D. N'Gabe, M. Laguerre, N. Duffaut, and R. Calas, *Organometallics*, **1**, 1525 (1982).
[540] I. Fleming and R. V. Williams, *J. Chem. Soc., Perkin Trans. 1*, **1981**, 684.
[541] W. Kitching, B. Laycock, I. Maynard, and K. Pennan, *J. Chem. Soc., Chem. Commun.*, **1986**, 954.
[542] I. Fleming and D. A. Perry, *Tetrahedron*, **37**, 4027 (1981).
[543] S. R. Wilson and M. S. Haque, *J. Org. Chem.*, **47**, 5413 (1982).
[544] L. A. Paquette, K. A. Horn, and G. J. Wells, *Tetrahedron Lett.*, **23**, 259 (1982).
[545] R. G. Salomon, M. F. Salomon, M. G. Zagorski, J. M. Reuter, and D. J. Coughlin, *J. Am. Chem. Soc.*, **104**, 1008 (1982).
[546] B. Psaume, M. Montury, and J. Goré, *Synth. Commun.*, **12**, 409 (1982).
[547] D. Pandy–Szekeres, G. Déléris, J.-P. Picard, J.-P. Pillot, and R. Calas, *Tetrahedron Lett.*, **21**, 4267 (1980).
[548] L. H. Sommer and N. S. Marans, *J. Am. Chem. Soc.*, **73**, 5135 (1951).
[549] A. Hosomi, M. Saito, and H. Sakurai, *Tetrahedron Lett.*, **21**, 355 (1980).
[550] R. Calas, J.-P. Pillot, and J. Dunoguès, *C.R. Hebd. Séances Acad. Sci., Sér. C*, **292**, 669 (1981).
[551] D. Young, W. Kitching, and G. Wickham, *Tetrahedron Lett.*, **24**, 5789 (1983).
[552] M. Laguerre, G. Félix, J. Dunoguès, and R. Calas, *J. Org. Chem.*, **44**, 4275 (1979).
[553] M. Kanazashi, *Bull. Chem. Soc. Jpn.*, **38**, 44 (1955).

[554] E. Larsson, *Trans. Chalmers Univ. Technol., Gothenburg Sweden*, **115**, 25 (1951) [*C.A.*, **47**, 10470 (1953)].
[555] R. I. Damja, C. Eaborn, and W.-C. Sham, *J. Organomet. Chem.*, **291**, 25 (1985).
[556] M. G. Steinmetz, M. A. Langston, R. T. Mayes, and B. S. Udayakumar, *J. Org. Chem.*, **51**, 5051 (1986).
[557] H. Sakurai, *Yuki Gosei Kagaku Kyokaishi* **40**, 472 (1982) [*C.A.*, **97**, 198235z (1982)].
[558] L. L. Shchukovskaya and R. Pal'chik, *Izv. Akad. Nauk SSSR, Ser. Khim.*, **1964**, 2228; *Bull. Acad. Sci. USSR Div. Chem. Sci.* (*Engl. Transl.*), **1964**, 2129.
[559] K. Wakamatsu, T. Nonaka, Y. Okuda, W. Tuckmantel, K. Oshima, K. Utimoto, and H. Nozaki, *Tetrahedron*, **42**, 4427 (1986).
[560] R. F. Cunico and Y.-K. Han, *J. Organomet. Chem.*, **105**, C29 (1976).
[561] R. K. Boeckman, Jr. and D. M. Blum, *J. Org. Chem.*, **39**, 3307 (1974).
[562] M. Kusakabe and F. Sato, *Chem. Lett.* (*Jpn.*), **1986**, 1473.
[563] A. Padwa and M. W. Wannamaker, *Tetrahedron Lett.*, **27**, 2555 (1986).
[564] W. A. Donaldson and R. P. Hughes, *Synth. Commun.*, **11**, 999 (1981).
[565] J. D. Buynak, M. N. Rao, R. Y. Chandrasekaran, and E. Haley, *Tetrahedron Lett.*, **26**, 5001 (1985).
[566] E. R. F. Gesing, J. P. Tane, and K. P. C. Vollhardt, *Angew. Chem. Int. Ed. Engl.*, **19**, 1023 (1980).
[567] I. Fleming, T. W. Newton, and F. Roessler, *J. Chem. Soc., Perkin Trans. 1*, **1981**, 2527.
[568] K. Suzuki, E. Katayama, T. Matsumoto, and G. Tsuchihashi, *Tetrahedron Lett.*, **25**, 3715 (1984).
[569] S. Martin, R. Sauvêtre, and J.-F. Normant, *Bull. Soc. Chim. Fr.*, **1986**, 900.
[570] S. Martin, R. Sauvêtre, and J.-F. Normant, *J. Organomet. Chem.*, **264**, 155 (1984).
[571] S. R. Wilson, R. N. Misra, and G. M. Georgiadis, *J. Org. Chem.*, **45**, 2460 (1980).
[572] W. E. Fristad, T. R. Bailey, L. A. Paquette, R. Gleiter, and M. C. Bohm, *J. Am. Chem. Soc.*, **101**, 4420 (1979).
[573] G. Himbert, *J. Chem. Res.* (*S*), **1978**, 104.
[574] T. Ohnuma, N. Hata, H. Fujiwara, and Y. Ban, *J. Org. Chem.*, **47**, 4713 (1982).
[575] Y. Kobayashi, Y. Kitano, and F. Sato, *J. Chem Soc., Chem. Commun.*, **1984**, 1329.
[576] G. L. Larson, E. Torres, C. B. Morales, and G. J. McGarvey, *Organometallics*, **5**, 2274 (1986).
[577] J.-K. Choi, D. J. Hart, and Y. M. Tsai, *Tetrahedron Lett.*, **23**, 4765 (1982).
[578] J.-K. Choi and D. J. Hart, *Tetrahedron*, **41**, 3959 (1985).
[579] E. Negishi, D. R. Swanson, F. E. Cederbaum, and T. Takahashi, *Tetrahedron Lett.*, **28**, 917 (1987).
[580] J. Buck, N. G. Clemo, and G. Pattenden, *J. Chem. Soc., Perkin Trans. 1*, **1985**, 2399.
[581] B. B. Snider, M. Karras, and R. S. E. Conn, *J. Am. Chem. Soc.*, **100**, 4624 (1978).
[582] N. Chatani, T. Takeyasu, and T. Hanafusa, *Tetrahedron Lett.*, **27**, 1841 (1986).
[583] K. Uchida, K. Utimoto, and H. Nozaki, *J. Org. Chem.*, **41**, 2215 (1976).
[584] B. Eaton, J. A. King, and K. P. C. Vollhardt, *J. Am. Chem. Soc.*, **108**, 1359 (1986).
[585] S. R. Wilson and M. F. Price, *J. Org. Chem.*, **48**, 4143 (1983).
[586] H. Wetter, *Helv. Chim. Acta*, **61**, 3072 (1978).
[587] R. K. Boeckman, Jr. and T. E. Barta, *J. Org. Chem.*, **50**, 3421 (1985).
[588] K. Ikenaga, K. Kikukawa, and T. Matsuda, *J. Chem. Soc., Perkin Trans. 1*, **1986**, 1959.
[589] K. Utimoto, M. Kitai, M. Naruse, and H. Nozaki, *Tetrahedron Lett.*, **1975**, 4233.
[590] M. Ishikawa, H. Sugisawa, O. Harata, and M. Kumada, *J. Organomet. Chem.*, **217**, 43 (1981).
[591] P. Bourgeois and G. Mérault, *C.R. Hebd. Séances Acad. Sci., Sér. C*, **273**, 714 (1971).
[592] B. Bennetau, D. Y. N'Gabe, and J. Dunoguès, *Tetrahedron Lett.*, **26**, 3813 (1985).
[593] B. J. M. Bennetau and J. P. Dunoguès, *Organomet. Synth.*, **3**, 494 (1986).
[594] A. A. Petrov, V. A. Kormer, and M. D. Stadnichuk, *Zh. Obshch. Khim.*, **31**, 1135 (1961); *J. Gen. Chem. USSR* (*Engl. Transl.*), **31**, 1049 (1961).
[595] M. D. Stadnichuk and A. A. Petrov, *Zh. Obshch. Khim.*, **31**, 411 (1961); *J. Gen. Chem. USSR* (*Engl. Transl.*), **31**, 373 (1961).

[596] G. Mérault, P. Bourgeois, J. Dunoguès, and N. Duffaut, *J. Organomet. Chem.*, **76**, 17 (1974).
[597] E. Wenkert, M. H. Leftin, and E. L. Michelotti, *J. Org. Chem.*, **50**, 1122 (1985).
[598] D. P. Curran and B. H. Kim, *Synthesis*, **1986**, 312.
[599] S. Padmanabhan and K. M. Nicholas, *Tetrahedron Lett.*, **23**, 2555 (1982).
[600] Y. Hashimoto and T. Mukaiyama, *Chem. Lett.* (*Jpn.*), **1986**, 755.
[601] S. F. Karaev and S. O. Guseinov, *Zh. Obshch. Khim.*, **54**, 1926 (1984); *J. Gen. Chem. USSR* (*Engl. Transl.*), **54**, 1717 (1984).
[602] A. S. Arakelyan and A. A. Gevorkyan, *Arm. Khim. Zh.*, **37**, 663 (1984) [*C.A.*, **102**, 220697x (1985)].
[603] I. Ojima and T. Fuchikami, Eur. Pat. 115943 (1984) [*C.A.*, **101**, 191119g (1984)].
[604] G. A. Artamkina, S. V. Kovalenko, I. P. Beletskaya, and O. A. Reutov, *Izv. Akad. Nauk SSSR, Ser. Khim.*, **1985**, 2411; *Bull. Acad. Sci. USSR Div. Chem. Sci.* (*Engl. Transl.*), **1985**, 2234.
[605] T. Fujisawa, Jpn. Kokai Tokkyo Koho 60 260537 (1984) [*C.A.*, **105**, 97750s (1986)].
[606] M. Uemura, T. Kobayashi, and Y. Hayashi, *Synthesis*, **1986**, 385.
[607] M. Ohno, S. Matsuoka, and S. Eguchi, *J. Org. Chem.*, **51**, 4553 (1986).
[608] T. Sasaki, A. Usuki, and M. Ohno, *J. Org. Chem.*, **45**, 3559 (1980).
[609] Idemitsu Kosan Co., Jpn. Kokai Tokkyo Koho 80 59115 [*C.A.*, **93**, 220433d (1980)].
[610] G. A. Kraus and Y.-S. Hon, *J. Am. Chem. Soc.*, **107**, 4341 (1985).
[611] G. A. Kraus and Y.-S. Hon, *J. Org. Chem.*, **50**, 4605 (1985).
[612] H. Uno, *Bull. Chem. Soc. Jpn.*, **59**, 2471 (1986).
[613] R. Henning and H. M. R. Hoffmann, *Tetrahedron Lett.*, **23**, 2305 (1982).
[614] S. L. Schreiber, T. Sammakia, and W. E. Crowe, *J. Am. Chem. Soc.*, **108**, 3128 (1986).
[615] R. Mohan and J. A. Katzenellenbogen, *J. Org. Chem.*, **49**, 1238 (1984).
[616] E. Keinan and Z. Roth, *J. Org. Chem.*, **48**, 1769 (1983).
[617] J. Ipaktschi and G. Lauterbach, *Angew. Chem. Int. Ed. Engl.*, **25**, 354 (1986).
[618] R. J. Armstrong, F. L. Harris, and L. Weiler, *Can. J. Chem.*, **60**, 673 (1982).
[619] H. M. R. Hoffmann and R. Henning, *Helv. Chim. Acta*, **66**, 828 (1983).
[620] W. S. Johnson, Y.-Q. Chen, and M. S. Kellogg, *Biologically Active Principles of Natural Products*, W. Voelter and D. G. Davies, Eds., Georg Thieme, Stuttgart, 1984, p. 55.
[621] K. Yamamoto, O. Nunokawa, and J. Tsuji, *Synthesis*, **1977**, 721.
[622] K. Tamao, T. Nakajima, and M. Kumada, *Organometallics*, **3**, 1655 (1984).
[623] S. Martin, R. Sauvêtre, and J.-F. Normant, *Tetrahedron Lett.*, **27**, 1027 (1986).
[624] V. A. Smit and I. P. Smolyakova, *Izv. Akad. Nauk SSSR, Ser. Khim.*, **1985**, 485; *Bull. Acad. Sci. USSR Div. Chem. Sci.* (*Engl. Transl.*), **1985**, 443.
[625] G. A. Molander and D. C. Shubert, *J. Am. Chem. Soc.*, **109**, 576 (1987).
[626] G. Déléris, J. Dunoguès, and R. Calas, *Tetrahedron Lett.*, **1976**, 2449.
[627] S. Kiyooka and C. H. Heathcock, *Tetrahedron Lett.*, **24**, 4765 (1983).
[628] M. T. Reetz and K. Kesseler, *J. Org. Chem.*, **50**, 5436 (1985).
[629] Y. Yamamoto, T. Komatsu, and K. Maruyama, *J. Organomet. Chem.*, **285**, 31 (1985).
[630] D. R. Williams and F. D. Klinger, *Tetrahedron Lett.*, **28**, 869 (1987).
[631] I. Ojima, Y. Miyazawa, and M. Kumagai, *J. Chem. Soc., Chem. Commun.*, **1976**, 927.
[632] K. Soai and M. Ishizaki, *J. Chem. Soc., Chem. Commun.*, **1984**, 1016.
[633] K. Soai and M. Ishizaki, *J. Org. Chem.*, **51**, 3290 (1986).
[634] M. T. Reetz, K. Kesseler, and A. Jung, *Angew. Chem. Int. Ed. Engl.*, **24**, 989 (1985).
[635] M. T. Reetz, K. Kesseler, S. Schmidtberger, B. Wenderoth, and R. Steinbach, *Angew. Chem. Int. Ed. Engl.*, **22**, 989 (1983).
[636] M. T. Reetz, *Pure Appl. Chem.*, **57**, 1781 (1985).
[637] A. Hosomi, M. Ando, and H. Sakurai, *Chem. Lett.* (*Jpn.*), **1984**, 1385.
[638] T. Hayashi, M. Konishi, Y. Okamoto, K. Kabeta, and M. Kumada, *J. Org. Chem.*, **51**, 3772 (1986).
[639] G. A. Molander and D. C. Shubert, *J. Am. Chem. Soc.*, **108**, 4683 (1986).
[640] N. Ishikawa and T. Kitazume, Ger. Pat., DE 3427821 (1985) [*C.A.*, **103**, 70940a (1985)].
[641] K. Tanako, H. Yoda, Y. Isobe, and A. Kaji, *J. Org. Chem.*, **51**, 1856 (1986).
[642] H. Sakurai and A. Hosomi, Jpn. Kokai Tokkyo Koho 80 27162 [*C.A.*, **93**, 185744g (1980)].

643 I. Ojima, M. Kumagai, and Y. Miyazawa, *Tetrahedron Lett.*, **1977**, 1385.
644 K. Ito and Y. Ishii, Jpn. Kokai Tokkyo Koho 79 22369 [*C.A.*, **91**, 39317n (1979)].
645 G. A. Molander and S. W. Andrews, *Tetrahedron Lett.*, **27**, 3115 (1986).
646 K. Nishitani and K. Yamakawa, *Tetrahedron Lett.*, **28**, 655 (1987).
647 T. Hayashi, M. Konishi, and M. Kumada, *J. Org. Chem.*, **48**, 281 (1983).
648 M. Montury, B. Psaume, and J. Goré, *Tetrahedron Lett.*, **21**, 163 (1980).
649 T. V. Lee, K. A. Richardson, and D. A. Taylor, *Tetrahedron Lett.*, **27**, 5021 (1986).
650 B. M. Trost and J. E. Vincent, *J. Am. Chem. Soc.*, **102**, 5680 (1980).
651 B. M. Trost and B. R. Adams, *J. Am. Chem. Soc.*, **105**, 4849 (1983).
652 B. M. Trost and R. Remuson, *Tetrahedron Lett.*, **24**, 1129 (1983).
653 N. Ishikawa and T. Kitatsume, Jpn. Kokai Tokkyo Koho 60 181039 (1985) [*C.A.*, **104**, 88107q (1986)].
654 T. H. Yan and L. A. Paquette, *Tetrahedron Lett.*, **23**, 3227 (1982).
655 B. M. Trost and H. Hiemstra, *J. Am. Chem. Soc.*, **104**, 886 (1982).
656 C. Kuroda, S. Shimizu, and J. Y. Satoh, *J. Chem. Soc., Chem. Commun.*, **1987**, 286.
657 I. Matsuda and Y. Izumi, *Tetrahedron Lett.*, **22**, 1805 (1981).
658 M. Fujita and T. Hiyama, *J. Am. Chem. Soc.*, **107**, 4085 (1985).
659 T. Yamazaki, K. Takita, and N. Ishikawa, *J. Fluorine Chem.*, **30**, 357 (1985).
660 H. O. House, P. C. Gaa, and D. VanDerveer, *J. Org. Chem.*, **48**, 1661 (1983).
661 D. Schinzer, S. Solyom, and M. Becker, *Tetrahedron Lett.*, **26**, 1831 (1985).
662 D. Schinzer, J. Steffen, and S. Solyom, *J. Chem. Soc., Chem. Commun.*, **1986**, 829.
663 T. K. Jones and S. Denmark, *Helv. Chim. Acta*, **66**, 2397 (1983).
664 H. Sakurai and A. Hosomi, Jpn. Kokai Tokkyo Koho 78 98933 [*C.A.*, **90**, 54667z (1979)].
665 M. Ochiai, M. Arimoto, and E. Fujita, *J. Chem. Soc., Chem. Commun.*, **1981**, 460.
666 Y. Naruta, H. Uno, and K. Maruyama, *Chem. Lett. (Jpn.)*, **1982**, 609.
667 Yoshitomi Pharm. Ind. Ltd., Jpn. Kokai Tokkyo Koho 60 84280 (1983) [*C.A.*, **103**, 215171n (1986)].
668 H. Uno, Y. Naruta, and K. Maruyama, *Tetrahedron*, **40**, 4725 (1984).
669 K. Maruyama, Jpn. Kokai Tokkyo Koho 58 49336 (1981) [*C.A.*, **99**, 5384g (1983)].
670 C. Hippeli and H.-U. Reissig, *Synthesis*, **1987**, 77.
671 P. A. Bartlett, W. S. Johnson, and J. D. Elliott, *J. Am. Chem. Soc.*, **105**, 2088 (1983).
672 A. P. Kozikowski and K. L. Sorgi, *Tetrahedron Lett.*, **25**, 2085 (1984).
673 M. G. Hoffmann and R. R. Schmidt, *Justus Liebigs Ann. Chem.*, **1985**, 2403.
674 T. L. Cupps, D. S. Wise, and L. B. Townsend, *J. Org. Chem.*, **47**, 5115 (1982).
675 T. Mukaiyama, S. Kobayashi, and S. Shoda, *Chem. Lett. (Jpn.)*, **1984**, 1529.
676 A. Giannis and K. Sandhoff, *Tetrahedron Lett.*, **26**, 1479 (1985).
677 Y. Yamamoto, S. Nishii, and J. Yamada, *J. Am. Chem. Soc.*, **108**, 7116 (1986).
678 D. D. Sternbach and S. H. Hobbs, *Synth. Commun.*, **14**, 1305 (1984).
679 J. Pornet and M. Kolani, *Tetrahedron Lett.*, **22**, 3609 (1981).
680 K. C. Nicolaou, M. E. Duggan, C.-K. Hwang, and P. K. Somers, *J. Chem. Soc., Chem. Commun.*, **1985**, 1359.
681 K. C. Nicolaou, C.-K. Hwang, and M. E. Duggan, *J. Chem. Soc., Chem. Commun.*, **1986**, 925.
682 S. J. Danishefsky, H. G. Selnick, M. P. DeNinno, and R. E. Zelle, *J. Am. Chem. Soc.*, **109**, 1572 (1987).
683 P. Renaud and D. Seebach, *Angew. Chem. Int. Ed. Engl.*, **25**, 843 (1986).
684 G. A. Kraus and K. Neuenschwander, *J. Chem. Soc., Chem. Commun.*, **1982**, 134.
685 M. Hashimoto, M. Aratani, and K. Sawada, U.S. Pat., 4,383,945 (1981); M. Hashimoto, M. Aratani, and K. Sawata, Eur. Pat. 35689 (1981) [*C.A.*, **96**, 68708z (1983)].
686 M. Aratani, H. Hirai, K. Sawada, and M. Hashimoto, *Heterocycles*, **23**, 1889 (1985).
687 Fujisawa Pharm. Co. Ltd., Jpn. Kokai Tokkyo Koho 59 93045 (1984) [*C.A.*, **102**, 6052q (1985)].
688 Z.-K. Liao and H. Kohn, *J. Org. Chem.*, **49**, 4745 (1984).
689 K. Prasad, K. Adlgasser, R. Sharma, and P. Stütz, *Heterocycles*, **19**, 2099 (1982).

[690] M. Thaning and L.-G. Wistrand, *Helv. Chim. Acta*, **69**, 1711 (1986).
[691] K. Ohga, U. C. Yuon, and P. S. Mariano, *J. Org. Chem.*, **49**, 213 (1984).
[692] J. W. Ullrich, R.-T. Chu, T. Tiner–Harding, and P. S. Mariano, *J. Org. Chem.*, **49**, 220 (1984).
[693] J.-C. Gramain and R. Remuson, *Tetrahedron Lett.*, **26**, 327 (1985).
[694] H. Hiemstra and W. N. Speckamp, *Tetrahedron Lett.*, **24**, 1407 (1983).
[695] H. Hiemstra, W. J. Klaver, and W. N. Speckamp, *J. Org. Chem.*, **49**, 1149 (1984).
[696] C. Clarke, I. Fleming, J. M. D. Fortunak, P. T. Gallagher, M. C. Honan, A. Mann, C. O. Nübling, P. R. Raithby, and J. J. Wolff, *Tetrahedron*, **43**, 3931 (1988).
[697] R. Ahmed–Schofield and P. S. Mariano, *J. Org. Chem.*, **50**, 5667 (1985).
[698] D. A.Claremon, Merck & Co. Inc., U.S. Pat. 4552881 (1984) [*C.A.*, **104**, 186312x (1986)].
[699] H. Hiemstra, W. J. Klaver, and W. N. Speckamp, *Tetrahedron Lett.*, **27**, 1411 (1986).
[700] H. Sakurai, A. Hosomi, and H. Hashimoto, Jpn. Kokai Tokkyo Koho, 79 48707 [*C.A.*, **91**, 74207y (1979)].
[701] L.-C. Chen and S.-S. Wu, *J. Chin. Chem. Soc.* (*Taipei*), **32**, 481 (1985) [*C.A.*, **105**, 225828n (1986)].
[702] M. Ochiai and E. Fujita, *Tetrahedron Lett.*, **21**, 4369 (1980).
[703] D. El-Abed, Thèse d'Etat, Marseille, 1986.
[704] I. Ojima and M. Tajima, Jpn. Kokkai Tokkyo Koho 78 105460 [*C.A.*, **90**, 22425c (1979)].
[705] G. Hartke and W. Morick, *Chem. Ber.*, **118**, 4821 (1986).
[706] M. Laguerre, J. Dunoguès, and R. Calas, *Tetrahedron Lett.*, **21**, 831 (1980).
[707] K. Hiroi, H. Sato, and K. Kotsuji, *Chem. Lett.* (*Jpn.*), **1986**, 743.
[708] M. Laguerre, J. Dunoguès, and R. Calas, *Tetrahedron Lett.*, **1978**, 57.
[709] J.-P. Pillot, G. Déléris, J. Dunoguès, and R. Calas, *J. Org. Chem.*, **44**, 3397 (1979).
[710] J.-P. Pillot, J. Dunoguès, and R. Calas, *C.R. Hebd. Séances Acad Sci., Sér. C*, **278**, 789 (1974).
[711] G. Kjeldsen, J. S. Knudsen, L. S. Ravn–Petersen, and K. B. G. Torsell, *Tetrahedron*, **39**, 2237 (1983).
[712] P. Magnus, D. A. Quagliato, and J. C. Huffman, *Organometallics*, **1**, 1240 (1982).
[713] G. Guillaumet, M. Trumtel, G. Coudert, and C. Zeggaf, *Synthesis*, **1986**, 337.
[714] E. Negishi and J. M. Tour, *Tetrahedron Lett.*, **27**, 4869 (1986).
[715] B. L. Chenard and C. M. Van Zyl, *J. Org. Chem.*, **51**, 3561 (1986).
[716] S. E. Denmark and J. P. Germanas, *Tetrahedron Lett.*, **25**, 1231 (1984).
[717] D. L. Comins and N. B. Mantlo, *Tetrahedron Lett.*, **24**, 3683 (1983).
[718] G. Félix, M. Laguerre, J. Dunoguès, and R. Calas, *J. Org. Chem.*, **47**, 1423 (1982).
[719] J.-P. Pillot, B. Bennetau, J. Dunoguès, and R. Calas, *Tetrahedron Lett.*, **22**, 3401 (1981).
[720] Y. Kubo, T. Imaoka, T. Shiragami, and T. Araki, *Chem. Lett.* (*Jpn.*), **1986**, 1749.
[721] N. Shimizu, F. Shibata, and Y. Tsuno, *Chem. Lett.* (*Jpn.*), **1985**, 1593.
[722] G. Manuel, G. Bertrand, and F. El Anba, *Organometallics*, **2**, 391 (1983).
[723] P. Mohr and C. Tamm, *Tetrahedron Lett.*, **28**, 391 (1987).
[724] A. T. Russell and G. Procter, *Tetrahedron Lett.*, **28**, 2045 (1987).
[725] A. T. Russell and G. Procter, *Tetrahedron Lett.*, **28**, 2041 (1987).
[726] H. Nishiyama, H. Yokoyama, S. Narimatsu, and K. Itoh, *Tetrahedron Lett.*, **23**, 1267 (1982).
[727] P. F. Hudrlik, D. Peterson, and R. J. Rona, *J. Org. Chem.*, **40**, 2263 (1975).
[728] M. C. Croudace and N. E. Schore, *J. Org. Chem.*, **46**, 5357 (1981).
[729] J. W. Wilt, O. Kolewe, and J. F. Kramer, *J. Am. Chem. Soc.*, **91**, 2624 (1969).
[730] M. J. Prior and G. H. Whitham, *J. Chem. Soc., Perkin Trans. 1*, **1986**, 683.
[731] P. F. Hudrlik, A. M. Hudrlik, and A. K. Kulkarni, *J. Am. Chem. Soc.*, **107**, 4260 (1985).
[732] R. B. Miller, M. I. Al-Hassan; and G. McGarvey, *Synth. Commun.*, **13**, 969 (1983).
[733] D. Seebach, R. Burstinghaus, B.-T. Gröbel, and M. Kolb, *Justus Liebigs Ann. Chem.*, **1977**, 830.
[734] R. K. Boeckman, Jr. and K. J. Bruza, *J. Org. Chem.*, **44**, 4781 (1979).
[735] T. H. Chan, P. L. Ming, W. Mychajlowskij, and D. N. Harpp, *Tetrahedron Lett.*, **1974**, 3511.
[736] R. Yamaguchi, H. Kawasaki, and M. Kawanisi, *Synth. Commun.*, **12**, 1027 (1982).
[737] F.-T. Luo and E. Negishi, *J. Org. Chem.*, **48**, 5144 (1983).
[738] S. Ohuchida, N. Hamanaka, and M. Hayashi, *Tetrahedron Lett.*, **1979**, 3661.

[739] D. J. Belmont and L. A. Paquette, *J. Org. Chem.*, **50**, 4102 (1985).
[740] M. Obayashi, K. Utimoto, and H. Nozaki, *Bull. Chem. Soc. Jpn.*, **52**, 1760 (1979).
[741] V. L. Bell, P. J. Giddings, A. B. Holmes, G. A. Mock, and R. A. Raphael, *J. Chem. Soc., Perkin Trans. 1*, **1986**, 1515.
[742] L. A. Paquette, M. J. Wyvratt, O. Schallner, J. L. Muthard, W. J. Begley, R. M. Blankenship, and D. Balogh, *J. Org. Chem.*, **44**, 3616 (1979).
[743] J. J. Eisch and J. T. Trainor, *J. Org. Chem.*, **28**, 487 (1963).
[744] M. Ali, *J. Bangladesh Acad. Sci.*, **31**, 97 (1979) [*C.A.*, **94**, 30834h (1981)].
[745] D. Grafstein, *J. Am. Chem. Soc.*, **77**, 6650 (1955).
[746] M. Ochiai, E. Fujita, M. Arimoto, and H. Yamaguchi, *Chem. Pharm. Bull.*, **33**, 41 (1985).
[747] M. Laguerre, J. Dunoguès, N. Duffaut, and R. Calas, *J. Organomet. Chem.*, **193**, C17 (1980).
[748] J. B. Woell and P. Boudjouk, *J. Org. Chem.*, **45**, 5213 (1980).
[749] A. H. Schmidt and M. Russ, *Synthesis*, **1981**, 67.
[750] L. Birkofer and E. Kramer, *Chem. Ber.*, **100**, 2776 (1967).
[751] M. Ochiai, E. Fujita, M. Arimoto, and H. Yamaguchi, *Chem. Pharm. Bull.*, **33**, 989 (1985).
[752] R. K. Boeckman, Jr., D. M. Blum, B. Ganem, and N. Halvey, *Org. Synth.*, **58**, 152 (1978).
[753] B.-T. Gröbel and D. Seebach, *Chem. Ber.*, **110**, 867 (1977).
[754] L. Birkofer and T. Kühn, *Chem. Ber.*, **111**, 3119 (1978).
[755] M. Ochiai, Y. Takaoka, K. Sumi, and Y. Nagao, *J. Chem. Soc., Chem. Commun.*, **1986**, 1382.
[756] L. D. Martin and J. K. Stille, *J. Org. Chem.*, **47**, 3630 (1982).
[757] R. B. Miller and G. McGarvey, *Synth. Commun.*, **7**, 475 (1977).
[758] H. Neumann and D. Seebach, *Chem. Ber.*, **111**, 2785 (1978).
[759] R. P. Fisher, H. P. On, J. T. Snow, and G. Zweifel, *Synthesis*, **1982**, 127.
[760] J. D. Buynak, J. Mathew, and M. N. Rao, *J. Chem. Soc., Chem. Commun.*, **1986**, 941.
[761] R. K. Boeckman, Jr. and M. Ramaiah, *J. Org. Chem.*, **42**, 1581 (1977).
[762] A. G. Brook, J. M. Duff, and W. E. Reynolds, *J. Organomet. Chem.*, **121**, 293 (1976).
[763] F. Björkling, T. Norin, C. R. Unelius, and R. B. Miller, *J. Org. Chem.*, **52**, 292 (1987).
[764] G. A. Krafft and J. A. Katzenellenbogen, *J. Am. Chem. Soc.*, **103**, 5459 (1981).
[765] R. Neidlein and W. Wirth, *Helv. Chim. Acta*, **69**, 1851 (1986).
[766] R. B. Miller and M. I. Al-Hassan, *J. Org. Chem.*, **50**, 2121 (1985).
[767] K. Yamamoto, K. Shinohara, T. Ohuchi, and M. Kumada, *Tetrahedron Lett.*, **1974**, 1153.
[768] K. Karabelas, C. Westerlund, and A. Hallberg, *J. Org. Chem.*, **50**, 3896 (1985).
[769] R. M. G. Roberts, *J. Organomet. Chem.*, **12**, 89 (1968).
[770] M. Ochiai, E. Fujita, M. Arimoto, and H. Yamaguchi, *Chem. Pharm. Bull.*, **31**, 86 (1983).
[771] R. J. P. Corriu, N. Escudié, and C. Guérin, *J. Organomet. Chem.*, **271**, C7 (1984).
[772] P. Jutzi and A. Seufert, *Angew. Chem. Int. Ed. Engl.*, **15**, 295 (1976).
[773] M. Ochiai and E. Fujita, *Yuki Gosei Kagaku Kyokaishi*, **40**, 508 (1982) [*C.A.*, **97**, 110059m (1982)].
[774] T. Ohta, T. Hosokawa, S.-I. Murahashi, K. Miki, and N. Kasai, *Organometallics*, **4**, 2080 (1985).
[775] N. M. Chistovalova, I. S. Akhren, E. V. Reshetova, and M. E. Vol'pin, *Izv. Akad. Nauk SSSR, Ser. Khim.*, **1984**, 2342; *Bull. Acad. Sci. USSR Div. Chem. Sci. (Engl. Transl.)*, **1984**, 2139.
[776] K. Kikukawa, K. Ikenaga, K. Kono, K. Toritani, F. Wada, and T. Matsuda, *J. Organomet. Chem.*, **270**, 277 (1984).
[777] M. Suzuki, H. Koyano, and R. Noyori, *J. Org. Chem.*, **52**, 5583 (1987).
[778] A. Takahashi and M. Shibasaki, *Tetrahedron Lett.*, **28**, 1893 (1987).
[779] P. W. Rabideau and G. L. Karrick, *Tetrahedron Lett.*, **28**, 2481 (1987).
[780] C. Nativi, A. Ricci, and M. Taddei, *Tetrahedron Lett.*, **28**, 2751 (1987).
[781] J. I. Grayson, S. Warren, and A. T. Zaslona, *J. Chem. Soc., Perkin Trans. 1*, **1987**, 967.
[782] S. J. Danishefsky and N. Mantlo, *J. Am. Chem. Soc.*, **110**, 8129 (1988).
[783] M. H. Block and D. E. Cane, *J. Org. Chem.*, **53**, 4923 (1988).
[784] R. Hunter and C. D. Simon, *Tetrahedron Lett.*, **29**, 2257 (1988).
[785] E. Ho, Y.-S. Cheng, and P. S. Mariano, *Tetrahedron Lett.*, **29**, 4799 (1988).

[786] W. Kitching, K. G. Penman, B. Laycock, and I. Maynard, *Tetrahedron*, **44**, 3819 (1988).
[787] S.-F. Chen, E. Ho, and P. S. Mariano, *Tetrahedron*, **44**, 7013 (1988).
[788] J.-B. Verlhac, J.-P. Quintard, and M. Pereyre, *J. Chem. Soc., Chem. Commun.*, **1988**, 503.
[789] D. C. Billington, W. J. Kerr, and P. L. Pauson, *J. Organomet. Chem.*, **341**, 181 (1988).
[790] G. Maier, H. P. Reissonauer, W. Schwab, P. Carsky, B. A. Hess, and L. J. Schaad, *J. Am. Chem. Soc.*, **109**, 5183 (1987).
[791] A. B. Smith, III, Y. Yokoyama, N. K. Dunlap, A. Hadener, and C. Tamm, *Tetrahedron Lett.*, **28**, 3663 (1987).
[792] D. P. Curran and S.-C. Kuo, *Tetrahedron*, **43**, 5653 (1987).
[793] K. Suzuki, T. Matsumoto, K. Tomooka, K. Matsumoto, and G. Tsuchihashi, *Chem. Lett.*, **1987**, 113.
[794] P. Binger, E. Sternberg, and U. Wittig, *Chem. Ber.*, **120**, 1933 (1987).
[795] S. J. Danishefsky, H. G. Selnick, R. E. Zelle, and M. P. DeNinno, *J. Am. Chem. Soc.*, **110**, 4368 (1988).
[796] J. M. Dener, D. J. Hart, and S. Ramesh, *J. Org. Chem.*, **53**, 6022 (1988).
[797] J. A. Soderquist and C. L. Anderson, *Tetrahedron Lett.*, **29**, 2777 (1988).
[798] G. Angellini, Y. Keheyan, G. Laguzzi, and G. Lilla, *Tetrahedron Lett.*, **29**, 4159 (1988).
[799] K. Suzuki, M. Miyazawa, M. Shimazaki, and G. Tsuchihashi, *Tetrahedron*, **44**, 4061 (1988).
[800] M. Ochiai, M. Kunishima, K. Fuji, M. Shiro, and Y. Nagao, *J. Chem. Soc., Chem. Commun.*, **1988**, 1076.
[801] K. Ritter, *Synthesis*, **1989**, 218.
[802] H. Mayr and G. Hagen, *J. Chem. Soc., Chem. Commun.*, **1989**, 91.
[803] P. C. B. Page, S. Rosenthal, and R. V. Williams, *Synthesis*, **1988**, 621.
[804] D. A. Becker and R. L. Danheiser, *J. Am. Chem. Soc.*, **111**, 389 (1989).
[805] S. L. Schreiber, M. T. Klimas, and T. Sammakia, *J. Am. Chem. Soc.*, **109**, 5749 (1987).
[806] W. S. Johnson, S. D. Lindell, and J. Steele, *J. Am. Chem. Soc.*, **109**, 5852 (1987).
[807] N. Ono, A. Kamimura, H. Sasatani, and A. Kaji, *J. Org. Chem.*, **52**, 4133 (1987).
[808] N. Ono, T. X. Jun, T. Hashimoto, and A. Kaji, *J. Chem. Soc., Chem. Commun.*, **1987**, 947.
[809] M. Murokaim, T. Kato, and T. Mukaiyama, *Chem. Lett.*, **1987**, 1167.
[810] M. Hayashi, A. Inubushi, and T. Mukaiyama, *Chem. Lett.*, **1987**, 1975.
[811] H. Ishibashi, H. Nakatani, Y. Umei, W. Yamamoto, and M. Ikeda, *J. Chem. Soc., Perkin Trans. 1*, **1987**, 589.
[812] H. Ishibashi, T. Sato, M. Irie, M. Ito, and M. Ikeda, *J. Chem. Soc., Perkin Trans. 1*, **1987**, 1095.
[813] R. J. Giguere, S. M. Duncan, J. M. Bean, and L. Purvis, *Tetrahedron Lett.*, **29**, 6071 (1988).
[814] H. M. R. Hoffmann, U. Eggert, U. Gibbels, K. Giesel, O. Koch, R. Lies, and J. Rabe, *Tetrahedron*, **44**, 3899 (1988).
[815] C. Blankenship, G. J. Wells, and L. A. Paquette, *Tetrahedron*, **44**, 4023 (1988).
[816] S.-F. Chen, E. Ho, and P. S. Mariano, *Tetrahedron*, **44**, 7013 (1988).
[817] M. Hayashi, A. Inubushi, and T. Mukaiyama, *Bull. Chem. Soc. Jpn.*, **61**, 4037 (1988).
[818] M. Kawashima and T. Fujisawa, *Bull. Chem. Soc. Jpn.*, **61**, 4051 (1988).
[819] L. Pettersson, T. Frejd, and G. Magnusson, *Tetrahedron Lett.*, **28**, 2753 (1987).
[820] Y. Naruta and K. Maruyama, *Chem. Lett.*, **1987**, 963.
[821] X. Xiao, S.-K. Park, and G. D. Prestwich, *J. Org. Chem.*, **53**, 4869 (1988).
[822] G. Procter, A. T. Russell, P. J. Murphy, T. S. Tan, and A. N. Mather, *Tetrahedron*, **44**, 3953 (1988).
[823] G. A. Molander and D. C. Shubert, *J. Am. Chem. Soc.*, **109**, 576 (1987).
[824] S. E. Denmark, B. R. Henke, and E. Weber, *J. Am. Chem. Soc.*, **109**, 2512 (1987).
[825] G. A. Molander and D C Shubert, *J. Am. Chem. Soc.*, **109**, 6877 (1987).
[826] H. Hamana, N. Ikota, and B. Ganem, *J. Org. Chem.*, **52**, 5492 (1987).
[827] Z. Y. Wei, J. S. Li, D. Wang, and T. H. Chan, *Tetrahedron Lett.*, **28**, 3441 (1987).
[828] M. Kira, M. Kobayashi, and H. Sakurai, *Tetrahedron Lett.*, **28**, 4081 (1987).
[829] I. Fleming, N. D. Kindon, and A. K. Sarkar, *Tetrahedron Lett.*, **28**, 5921 (1987).
[830] M. J. Wanner, N. P. Willard, G.-J. Koomen, and U. K. Pandit, *Tetrahedron*, **43**, 2549 (1987).

[831] A. Hosomi, K. Hoashi, S. Kohra, Y. Tominaga, K. Otaka, and H. Sakurai, *J. Chem. Soc., Chem. Commun.*, **1987**, 570.
[832] A. Hosomi, S. Kohra, and Y. Tominaga, *J. Chem. Soc., Chem. Commun.*, **1987**, 1517.
[833] G. Cerveau, C. Chuit, R. J. P. Corriu, and C. Reye, *J. Organomet. Chem.*, **328**, C17 (1987).
[834] K. Soai, M. Ishizaki, and S. Yokoyama, *Chem. Lett.*, **1987**, 341.
[835] T. Mukaiyama, M. Ohshima, and N. Miyoshi, *Chem. Lett.*, **1987**, 1121.
[836] O. Tsuge, S. Kanemasa, T. Naritomi, and J. Tanaka, *Bull. Chem. Soc. Jpn.*, **60**, 1497 (1987).
[837] A. Hosomi, S. Kohra, and Y. Tominaga, *Chem. Pharm. Bull.*, **35**, 2155 (1987).
[838] M. Kira, K. Sato, and H. Sakurai, *J. Am. Chem. Soc.*, **110**, 4599 (1988).
[839] T. V. Lee, R. J. Boucher, and C. J. M. Rockell, *Tetrahedron Lett.*, **29**, 689 (1988).
[840] J. Pornet, A. Rayadh, and L. Miginiac, *Tetrahedron Lett.*, **29**, 4717 (1988).
[841] H. Sugimura and M. Uematsu, *Tetrahedron Lett.*, **29**, 4953 (1988).
[842] A. Mordini, G. Palio, A. Ricci, and M. Taddei, *Tetrahedron Lett.*, **29**, 4991 (1988).
[843] T. Hayashi, Y. Matsumoto, T. Kiyoi, and Y. Ito, *Tetrahedron Lett.*, **29**, 5667 (1988).
[844] T. H. Chan in *Silicon Chemistry*, E. R. Corey, J. Y. Corey, and P. P. Gaspar, Eds., Ellis Horwood, Chichester, 1988, p. 49.
[845] G. A. Molander and S. W. Andrews, *Tetrahedron*, **44**, 3869 (1988).
[846] M. T. Reetz, A. Jung, and C. Bolm, *Tetrahedron*, **44**, 3889 (1988).
[847] T. V. Lee, R. J. Boucher, and C. J. M. Rockell, *Tetrahedron*, **44**, 4233 (1988).
[848] M. Hayashi, A. Inubushi, and T. Mukaiyama, *Bull. Chem. Soc. Jpn.*, **61**, 4037 (1988).
[849] T. Kunz and H.-U. Reissig, *Angew. Chem. Int. Ed. Engl.*, **27**, 268 (1988).
[850] M. G. Ranasinghe and P. L. Fuchs, *J. Am. Chem. Soc.*, **111**, 779 (1989).
[851] M. Kira, T. Hino, and H. Sakurai, *Tetrahedron Lett.*, **30**, 1099 (1989).
[852] M. Fujita, M. Obayashi, and T. Hiyama, *Tetrahedron*, **44**, 4135 (1988).
[853] J. S. Panek and M. A. Sparks, *Tetrahedron Lett.*, **28**, 4649 (1987).
[854] J. A. Oplinger and L. A. Paquette, *Tetrahedron Lett.*, **28**, 5441 (1987).
[855] E. J. Corey and R. M. Burk, *Tetrahedron Lett.*, **28**, 6413 (1987).
[856] T. Tokoroyama, M. Tsukamoto, T. Asada, and H. Iio, *Tetrahedron Lett.*, **28**, 6645 (1987).
[857] G. Majetich and K. Hull, *Tetrahedron*, **43**, 5621 (1987).
[858] M. Hayashi and T. Mukaiyama, *Chem. Lett.*, **1987**, 289.
[859] G. Majetich, J. Defauw, and C. Ringold, *J. Org. Chem.*, **53**, 50 (1988).
[860] D. Schinzer, G. Dettmer, M. Ruppelt, S. Solyom, and J. Steggen, *J. Org. Chem.*, **53**, 3823 (1988).
[861] K. Nikisch and H. Laurent, *Tetrahedron Lett.*, **29**, 1533 (1988).
[862] G. Majetich and K. Hull, *Tetrahedron Lett.*, **29**, 2773 (1988).
[863] J. S. Panek and M. A. Sparks, *Tetrahedron Lett.*, **29**, 4517 (1988).
[864] G. Majetich and K. Hull, *Tetrahedron*, **44**, 3833 (1988).
[865] D. Schinzer, C. Allagiannis, and S. Wichmann, *Tetrahedron*, **44**, 3851 (1988).
[866] S. E. Denmark and R. C. Klix, *Tetrahedron*, **44**, 4043 (1988).
[867] D. N. Kirk and B. W. Miller, *J. Chem. Res. (S)*, **1988**, 278.
[868] R. Bambal and R. D. W. Kemmitt, *J. Chem. Soc., Chem. Commun.*, **1988**, 734.
[869] K. Mizuon, M. Ikeda, and Y. Orsuji, *Chem. Lett.*, **1988**, 1507.
[870] D. Schinzer, *Synthesis*, **1988**, 263.
[871] S. R. Angle and K. D. Turnbull, *J. Am. Chem. Soc.*, **111**, 1136 (1989).
[872] T. Tokoroyama and L.-R. Pan, *Tetrahedron Lett.*, **30**, 197 (1989).
[873] S. E. Denmark, K. L. Habermas, and G. A. Hite, *Helv. Chim. Acta*, **71**, 168 (1988).
[874] S. E. Denmark and G. A. Hite, *Helv. Chim. Acta*, **71**, 195 (1988).
[875] G. A. Artamkina, S. V. Kovalenko, I. P. Beletskaya, and O. A. Reutov, *J. Organomet. Chem.*, **329**, 139 (1987).
[876] C. Hippeli and H.-U. Reissig, *Synthesis*, **1987**, 77.
[877] I. Mori, P. A. Bartlett, and C. H. Heathcock, *J. Am. Chem. Soc.*, **109**, 7199 (1987).
[878] J. A. Bennek and G. R. Gray, *J. Org. Chem.*, **52**, 892 (1987).
[879] M. C. Pirrung and P. M. Kenney, *J. Org. Chem.*, **52**, 2335 (1987).
[880] T. D. Aicher and Y. Kishi, *Tetrahedron Lett.*, **28**, 3463 (1987).

[881] M. Asaoko and H. Takei, *Tetrahedron Lett.*, **28**, 6343 (1987).
[882] R. G. Andrew, R. E. Conrow, J. D. Elliott, W. S. Johnson, and S. Ramezani, *Tetrahedron Lett.*, **28**, 6535 (1987).
[883] Y. Yamamoto and J. Yamada, *J. Chem. Soc., Chem. Commun.*, **1987**, 1218.
[884] D. Seebach, R. Imwinkelried, and G. Stucky, *Helv. Chim. Acta*, **70**, 448 (1987).
[885] Y. Hayashi and T. Mukaiyama, *Chem. Lett.*, **1987**, 1811.
[886] T. Takeda, Y. Kaneko, H. Nakagawa, and T. Fujiwara, *Chem. Lett.*, **1987**, 1963.
[887] R. D. Walkup and N. Uobeysekere, *Synthesis*, **1987**, 607.
[888] T. Mukaiyama and M. Murakami, *Synthesis*, **1987**, 1043.
[889] S. Torii, T. Inokuchi, S. Takagishi, H. Horike, H. Kuroda, and K. Uneyama, *Bull. Chem. Soc. Jpn.*, **60**, 2173 (1987).
[890] O. R. Martin, S. P. Rao, K. G. Kurtz, and H. A. El-Shenawy, *J. Am. Chem. Soc.*, **110**, 8698 (1988).
[891] C. Brückner, H. Holzinger, and H.-U. Reissig, *J. Org. Chem.*, **53**, 2450 (1988).
[892] A. Mann, C. Nativi, and M. Taddei, *Tetrahedron Lett.*, **29**, 3247 (1988).
[893] A. Mann, A. Ricci, and M. Taddei, *Tetrahedron Lett*,, **29**, 6175 (1988).
[894] L. D. M. Lolkema, H. Hiemstra, H. M. Mooiweer, and W. N. Speckamp, *Tetrahedron Lett.*, **29**, 6365 (1988).
[895] T. Mukaiyama, K. Homma, and H. Takenoshita, *Chem. Lett.*, **1988**, 1725.
[896] L. E. Overman and A. S. Thompson, *J. Am. Chem. Soc.*, **110**, 2248 (1988).
[897] O. R. Martin, S. P. Rao, K. G. Kurz, and H. A. El-Shenawy, *J. Am. Chem. Soc.*, **110**, 8698 (1988).
[898] S. Danishefsky, H. G. Selnick, M. P. DeNinno, and R. E. Zelle, *J. Am. Chem. Soc.*, **109**, 1572 (1987).
[899] S. Danishefsky, S. DeNinno, and P. Lartey, *J. Am. Chem. Soc.*, **109**, 2082 (1987).
[900] S. Danishefsky, D. M. Armitstead, F. E. Wincott, H. G. Selnick, and R. Hungate, *J. Am. Chem. Soc.*, **109**, 8117 (1987).
[901] S. Danishefsky, H. G. Selnick, R. E. Zelle, and M. P. DeNinno, *J. Am. Chem. Soc.*, **110**, 4368 (1988).
[902] C.-L. Tu and P. S. Mariano, *J. Am. Chem. Soc.*, **109**, 5287 (1987).
[903] G. D. Hartman, B. T. Philips, and W. Halczensko, *J. Org. Chem.*, **52**, 1136 (1987).
[904] R. Ahmed-Schofield and P. S. Mariano, *J. Org. Chem.*, **52**, 1478 (1987).
[905] H. M. Mooiweer, H. Hiemstra, H. P. Fortgens, and W. N. Speckamp, *Tetrahedron Lett.*, **28**, 3285 (1987).
[906] P. A. Grieco and W. F. Fobare, *J. Chem. Soc., Chem. Commun.*, **1987**, 185.
[907] S. Torii, T. Inokuchi, S. Takagishi, F. Akahoshi, and K. Uneyama, *Chem. Lett.*, **1987**, 639.
[908] P. G. M. Wuts and Y.-W. Jung, *J. Org. Chem.*, **53**, 1957 (1988).
[909] P. G. M. Wuts and Y.-W. Jung, *J. Org. Chem.*, **53**, 5989 (1988).
[910] T. W. Bell and L.-Y. Hu, *Tetrahedron Lett.*, **29**, 4819 (1988).
[911] S. F. Martin and L. S. Geraci, *Tetrahedron Lett.*, **29**, 6725 (1988).
[912] F. P. J. T. Rutjes, H. Hiemstra, H. M. Mooiweer, and W. N. Speckamp, *Tetrahedron Lett.*, **29**, 6975 (1988).
[913] M. J. O'Donnell and W. D. Bennett, *Tetrahedron*, **44**, 5389 (1988).
[914] A. L. Castelhano, S. Horne, G. J. Taylor, R. Billedeau, and A. Krantz, *Tetrahedron*, **44**, 5451 (1988).
[915] W. J. Klavier, H. Hiemstra, and W. N. Speckamp, *Tetrahedron*, **44**, 6729 (1988).
[916] C. Flann, T. C. Malone, and L. E. Overman, *J. Am. Chem. Soc.*, **109**, 6097 (1987).
[917] S. F. McCann and L. E. Overman, *J. Am. Chem. Soc.*, **109**, 6107 (1987).
[918] C. J. Flann and L. E. Overman, *J. Am. Chem. Soc.*, **109**, 6115 (1987).
[919] R. M. Lett, L. E. Overman, and J. Zablocki, *Tetrahedron Lett.*, **29**, 6541 (1988).
[920] G. W. Daub, D. A. Heerding, and L. E. Overman, *Tetrahedron*, **44**, 3919 (1988).
[921] L. E. Overman and A. J. Robichaud, *J. Am. Chem. Soc.*, **111**, 300 (1989).
[922] H. Uno, N. Watanabe, S. Fujiki, adn H. Suzuki, *Synthesis*, **1987**, 471.
[923] T. Hayashi, Y. Matsumoto, and Y. Ito, *Chem. Lett.*, **1987**, 2037.

[924] T. Hayashi, Y. Matsumoto, and Y. Ito, *Organometallics*, **6**, 884 (1987).
[925] M. Ohno, S. Matsuoka, and S. Eguchi, *Synthesis*, **1987**, 1092.
[926] K. Hiroi, H. Sato, L.-M. Chen, and K. Kotsuji, *Chem. Pharm. Bull.*, **35**, 1413 (1987).
[927] A. Tubul and M. Santelli, *Tetrahedron*, **44**, 3975 (1988).
[928] E. Negishi, L. D. Boardman, H. Sawada, V. Bagheri, A. T. Stoll, J. M. Tour, and C. L. Rand, *J. Am. Chem. Soc.*, **110**, 5383 (1988).
[929] M. Franck-Neumann, M. Sedrati, and A. Abdali, *J. Organomet. Chem.*, **339**, C9 (1988).
[930] T. Kämpchen, G. Modelmog, D. Schulz, and G. Seitz, *Liebigs Ann. Chem.*, **1988**, 855.
[931] A. Cambanis, E. Bäuml, and H. Mayr, *Synthesis*, **1989**, 128.
[932] H. Mayr and U. von der Brüggen, *Chem. Ber.*, **121**, 339 (1988).
[933] H. Mayr, A. Cambanis, and E. Bäuml, *Synthesis*, **1988**, 962.
[934] P. Mohr and C. Tamm, *Tetrahedron Lett.*, **28**, 391 (1987).
[935] T. Matsumoto, Y. Kitano, and F. Sato, *Tetrahedron Lett.*, **29**, 5685 (1988).
[936] G. Procter, A. T. Russell, P. J. Murphy, T. S. Tan, and A. N. Mather, *Tetrahedron*, **44**, 3953 (1988).
[937] M. A. Avery, C. Jennings-White, and W. K. M. Chong, *Tetrahedron Lett.*, **28**, 4629 (1987).
[938] M. Shimizu and H. Yoshioka, *Tetrahedron Lett.*, **30**, 967 (1989).
[939] T. Kusumoto and T. Hiyama, *Tetrahedron Lett.*, **28**, 1807; ibid., 1811 (1987).
[940] T. Takanami, K. Suda, H. Ohmori, and M. Masui, *Chem. Lett.*, **1987**, 1335.
[941] J. I. Grayson, S. Warren, and A. T. Zaslona, *J. Chem. Soc., Perkin Trans. 1*, **1987**, 967.
[942] E. Vedejs, J. D. Rogers, and S. J. Wittenberger, *J. Am. Chem. Soc.*, **110**, 4822 (1988).
[943] E. Vedejs, J. D. Rogers, and S. J. Wittenberger, *Tetrahedron Lett.*, **29**, 2287 (1988).
[944] E. Vedejs and S. Ahmad, *Tetrahedron Lett.*, **29**, 2291 (1988).
[945] N. X. Hu, Y. Aso, T. Otsubo, and F. Ogura, *Tetrahedron Lett.*, **29**, 4949 (1988).
[946] M. Arimoto, H. Yamaguchi, E. Fujita, M. Ochiai, and Y. Nagao, *Tetrahedron Lett.*, **28**, 6289 (1987).
[947] J. I. Grayson, S. Warren, and A. T. Zaslona, *J. Chem. Soc., Perkin Trans. 1*, **1987**, 967.
[948] K. Lee, D. Y. Kim, and D. Y. Oh, *Tetrahedron Lett.*, **29**, 667 (1988).
[949] K. Tamao, M. Akita, K. Maeda, and M. Kumada, *J. Org. Chem.*, **52**, 1100 (1987).
[950] S. Okamoto, T. Shimazaki, Y. Kobayashi, and F. Sato, *Tetrahedron Lett.*, **28**, 2033 (1987).
[951] A. B. Smith, III, Y. Yokoyama, D. M. Huryn, and N. K. Dunlap, *Tetrahedron Lett.*, **28**, 3659 (1987).
[952] Y. Kobayashi, T. Shimazaki, and F. Sato, *Tetrahedron Lett.*, **28**, 5849 (1987).
[953] A. Oliva and A. Molinari, *Synth. Commun.*, **7**, 837 (1987).
[954] M. Ochiai, Y. Takaoka, and Y. Nagao, *J. Am. Chem. Soc.*, **110**, 6565 (1988).
[955] M. Rowley and Y. Kishi, *Tetrahedron Lett.*, **29**, 4909 (1988).
[956] K. Tamao, K. Maeda, T. Tanaka, and Y. Ito, *Tetrahedron Lett.*, **29**, 6955 (1988).
[957] E. J. Grayson and G. H. Whitham, *Tetrahedron*, **44**, 4087 (1988).
[958] M. Ochiai, K. Sumi, Y. Takaoka, M. Kunishima, Y. Nagao, M. Shiro, and E. Fujita, *Tetrahedron*, **44**, 4095 (1988).
[959] T. Shimazaki, Y. Kobayashi, and F. Sato, *Chem. Lett.*, **1988**, 1785.
[960] K. Fugami, K. Oshima, K. Utimoto, and H. Nozaki, *Bull. Chem. Soc. Jpn.*, **60**, 2509 (1987).
[961] Y. Hatanaka and T. Hiyama, *J. Org. Chem.*, **53**, 918 (1988).
[962] S. R. Wilson and C. E. Augelli-Szafran, *Tetrahedron*, **44**, 3983 (1988).
[963] K. Ikenaga, K. Kikukawa, and T. Matsuda, *J. Org. Chem.*, **52**, 1276 (1987).
[964] Y. Hatanaka and T. Hiyama, *J. Org. Chem.*, **53**, 918 (1988).
[965] K. Ikenaga, S. Matsumoto, K. Kikukawa, and T. Matsuda, *Chem. Lett.*, **1988**, 873.

AUTHOR INDEX, VOLUMES 1–37

Volume number only is designated in this index.

CHAPTER AND TOPIC INDEX, VOLUMES 1–37

Many chapters contain brief discussions of reactions and comparisons of alternative synthetic methods related to the reaction that is the subject of the chapter. These related reactions and alternative methods are not usually listed in this index. In this index, the volume number is in **BOLDFACE**, the chapter number is in ordinary type.